Elements of
Numerical Analysis

Second Edition

Radhey S. Gupta

CAMBRIDGE
UNIVERSITY PRESS

CAMBRIDGE
UNIVERSITY PRESS

University Printing House, Cambridge CB2 8BS, United Kingdom

One Liberty Plaza, 20th Floor, New York, NY 10006, USA

477 Williamstown Road, Port Melbourne, VIC 3207, Australia

4843/24, 2nd Floor, Ansari Road, Daryaganj, Delhi - 110002, India

79 Anson Road, #06-04/06, Singapore 079906

Cambridge University Press is part of the University of Cambridge.

It furthers the University's mission by disseminating knowledge in the pursuit of education, learning and research at the highest international levels of excellence.

www.cambridge.org
Information on this title: www.cambridge.org/9781107500495

Second edition first published 2015

A catalogue record for this publication is available from the British Library

Library of Congress Cataloging in Publication data
Gupta, Radhey S.,
1937- Elements of numerical analysis. – Second edition / Radhey S. Gupta.
pages cm
Includes bibliographical references and index.
Summary: "Offers detailed discussion on difference equations, Fourier series, discrete Fourier transforms and finite element methods"– Provided by publisher.
ISBN 978-1-107-50049-5 (pbk.)
1. Numerical analysis–Textbooks. 2. Mathematics–Study and teaching (Higher)–Textbooks. 3. Mathematics–Study and teaching (Graduate)–Textbooks. I. Title.
QA297.G87 2015
518–dc23
2014038362

ISBN 978-1-107-50049-5 Paperback

Additional resources for this publication available at www.cambridge.org/9781107500495

To my supervisor
(Late) Professor John Crank

and
To my family
Wife: (Late) Pramila
Children: Virna, Neelam, Aditi and Mayank

Contents

Preface

As I think of writing about the present work I am pleasantly reminded of a few names which occupy very special place in my academic and professional career. In 1966, while I was working at the Post Office Research Station in London (now most probably at Martlesham) I got introduced to computers. I cannot forget, with how much patience and perseverence, B.E. Surtees had not only helped but had almost taught me Algol programming. Later I used the computer extensively for solving scientific problems. Further, the Research Station generously granted me day-release to attend MSc (Comp.Sc.) course at the City University, London. Although I could not complete the course, I developed a strong liking for Numerical Analysis. It would be my privilege to mention the name of Professor V.E. Price who taught the subject with full devotion and dedication. I must admit that I learnt the basics of Numerical Analysis from there and much of it makes part of Chapters 1 to 7 and 10 of the book. I was greatly impressed by the book *Modern Computing Methods* by E.T. Goodwin, and still am. My intense desire for working in Numerical Analysis was fulfilled when I joined PhD in 1969 at Brunel University under the guidance of Professor J. Crank who was known internationally in the field. Luckily, a very challenging problem came my way to work upon from Hammersmith Hospital, London. The problem required knowledge for solving partial differential equations numerically. The first book on p.d.e., I read was G.D. Smith's who was coincidently teaching in the same department. Therefore no wonder, my treatment for solving p.d.e.'s in Chapter 11 may be biased towards his book. I worked with Professor Crank for five years – three years for my PhD and two years as a postdoctoral research fellow. It was only his constant inspiration that kept me going and galloping. Those five years, I may call the most precious years of my life. I came to know with deep sense of sorrow and grief that Professor Crank passed away in October 2006. This book is a humble tribute to him. Same time when I was doing PhD, Professor J.R. Whiteman joined the department. He taught splines to the students of MSc (Numerical Analysis) and gave lectures on variational principle applied to Finite Element Method. I came to know about these techniques through him which have been provided in Chapters 8 and 12. Nick Papàmichael was another good fellow in the department who taught Integral Equations to MSc (Numerical Analysis). I learnt initially about this topic from his notes which have been useful in writing Chapter 13.

After coming to India I joined I.I.T., Delhi as Pool Officer, with Professor M.P. Singh who was heading a Centre, concerned with problems in bio-mathematics and atmospheric science. I had a good interaction with the members of his team working on diffusion problems. Professor Singh was extremely helpful in providing me all facilities – academic and otherwise. He had been a source of constant encouragement and inspiration during that period and afterwards also, as I stayed there for less than $1^1/_2$ years only. I joined in 1976, the Department of Mathematics at University of Roorkee (now I.I.T., Roorkee). There I got an opportunity to hone and extend my knowledge of Numerical Analysis further through teaching and guiding research. Professor C. Prasad, Head of Department, wanted to see an all-round expansion in Numerical Analysis and Computer Science in the department. I was entrusted to carry out various activities in these areas. A postgraduate diploma course in Computer Science was started in the department in 1978. Same year, I also organised a short term course on Numerical Solution of Partial Differential Equations under Quality Improvement Program. Professor M.B. Kanchi (Civil Engineering Department) gave lectures on Finite Element Method (FEM) in this program. It inspired me to broaden my knowledge on FEM which has been included in Chapter 12. Further, I thought to provide the reader an exposure to a very important class of problems known as free and moving boundary problems. Such problems arise in almost all branches of engineering and applied sciences. A brief introduction to these problems is given in Chapter 15. I have included a list of my research papers on moving boundary problems so that the interested readers may search other papers through cross references. My stay of 21 years at University of Roorkee (I.I.T., Roorkee) had been extremely fruitful academically as well as in personal relations. I have always cherished its memories in my heart and will continue to do so all my life. As would have been clear, the present book is, in a way, direct or indirect contribution from various people — to whom I feel greatly indebted. Whatever faults are there, they are mine — criticism and suggestions would be most welcome. I do hope the book will be useful to students, to teachers and to those who want to use Numerical Analysis as a tool for solving practical problems.

Preface to Second Edition

It gave me a great sense of satisfaction and happiness to hear some good words about the book from my old colleagues and acquaintances working in the field of Numerical Analysis. In this edition I have included a new chapter dealing with Fourier Series, Fourier Transform and Fast Fourier Transform (FFT). In fact I wanted to include this topic in the first edition itself but I was not truly prepared then. Now I have also added first order hyperbolic equation as a new section in the chapter on partial differential equations. I thank Cambridge University Press for bringing out this revised edition.

1

Errors in Computation

1.1 Introduction

For solving a mathematical problem by numerical method, an input is provided in the form of some numerical data or it is generated/created as called for by the problem. The input is processed through arithmetic operations together with logical operations, which are performed in a systematic manner and the output is produced in the form of some numbers. Thus the whole exercise in Numerical Analysis is all about manipulation of numbers. Whether we are working by hand or on a computing machine, there is always a constraint in regard to physical size of the numbers, i.e., the number of digits a number can contain. Inside a computer the size of the number is dependent on its word-length (number of bits) which also puts a limit on the range of numbers that can be represented in a particular computer. Further, it may be noted that all numbers are not represented exactly inside the computer and that the input given in the decimal form is converted to binary in the computer. It should also be remembered that fractions cannot be stored in their natural form; they are converted to decimals, for example 2/5 is input as 0.4 and 1/3 as 0.333... up to a finite number of digits acceptable by a computer.

1.2 Floating Point Representation of Number

When a number x is expressed as,

$$x = p \times 10^q$$

where $0.1 \leq |p| < 1.0$ and q is an integer (positive (+ve) or negative (−ve)), it is called 'floating point' representation of number x. A floating point form consists of two parts; the fractional part p (alongwith the sign) is known as mantissa and the other part q as exponent, a power raised to a radix (in the case of decimal system, 10). At some places it is referred to as 'normalised floating point' and when $1 \le p < 10$, the form is called 'scientific notation'. A few examples of floating point representation, $fl(x)$ of number x are given as follows:

x	$fl(x)$	*Mantissa*(p)	*Exponent*(q)
2.0456	0.20456×10^1	0.20456	1
−32.7652	-0.327652×10^2	−0.327652	2
0.00234	0.234×10^{-2}	0.234	−2
0.000000034	0.34×10^{-7}	0.34	−7
34000000	0.34×10^8	0.34	8

1.3 Binary Numbers

The decimal numbers (radix 10) are converted to binary form with digits 0 and 1 (radix 2) in the computer. An integer decimal number, may be converted to binary equivalent by following procedure:

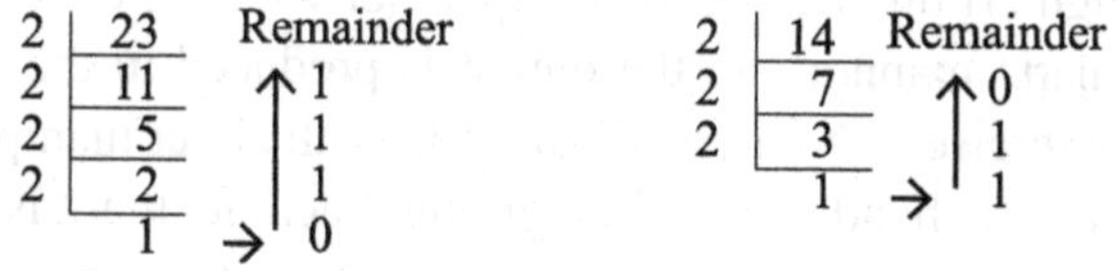

Divide repeatedly by 2 until last quotient is 1, keeping the remainder against the quotient; read the binary digits in the direction of arrow. Thus we get,

$23 = 10111$; $14 = 1110$.

The fractional decimal number is converted to binary form in the following manner:

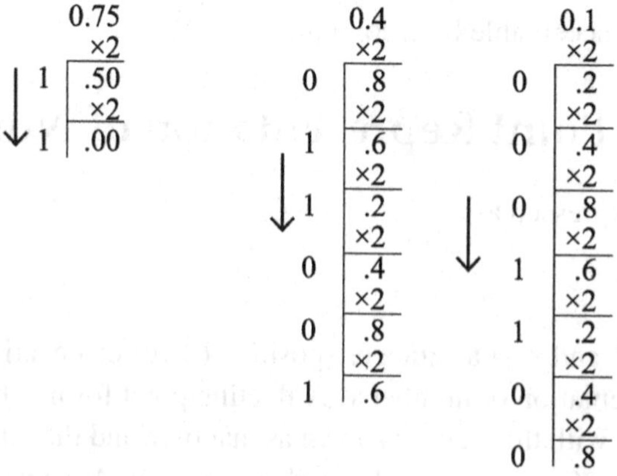

Multiply by 2 until the decimal part is zero, saving digit 0 or 1 before the decimal point. Read the digits saved in a top-down manner. Thus the converted numbers are

$$0.75 = 0.11;\ 0.4 = 0.0110011 \text{ (0011 recurring)};\ 0.1 = 0.0001100 \text{ (1100 recurring)}$$

When a decimal number consists of both parts, integral as well as fractional, then both parts are converted to binary forms separately. For example, 23.75 will convert to 10111.11. It should be clear from the above examples that the integer numbers in the decimal system can be converted exactly in the binary system but most of the non-integers may be represented approximately due to non-terminating character of the converted numbers.

For conversion from binary to decimal, we simply multiply the binary digits by their respective place-value and add. For example, 10111.11 can be converted to decimal form as,

$$\begin{array}{ccccccccl} 2^4 & 2^3 & 2^2 & 2^1 & 2^0 & 2^{-1} & 2^{-2} & & \\ 1 & 0 & 1 & 1 & 1 & 1 & 1 & = & 1\times 2^4+0\times 2^3+1\times 2^2+1\times 2^1+1\times 2^0+1\times 2^{-1}+1\times 2^{-2} \\ & & & & & & & = & 16+0+4+2+1+.5+.25 \\ & & & & & & & = & 23.75 \end{array}$$

It may also be noted that largest k-digit binary integer will have the value $2^k - 1$ in decimal. For example, the largest 2-digit binary number will be $11 = 2^2 - 1 = 3$ and a 3-digit largest binary number will be $111 = 2^3 - 1 = 7$ and so on. Obviously all the k digits will be binary 1's. A k-digit binary number can represent 2^k decimal numbers from 0 to $2^k - 1$.

1.3.1 Binary number representation in computer

As stated earlier, all the input data is converted to binary inside the computer; while the decimal integers are represented exactly in the computer memory, the non-integers are represented in floating point form. We would like to explain very briefly as how the floating point numbers are stored in the computer memory. Consider the floating point representation of binary numbers given below:

Binary number	*Floating point form*	*Mantissa*	*Exponent*
0.0111	0.1110×10^{-01}	+0.1110	−01
−1.101	$-0.1101 \times 10^{+01}$	−0.1101	+01
11.1	$0.1110 \times 10^{+10}$	+0.1110	+10

It may be noted that all numbers are in binary so that 10 is equal to 2 in decimal. The other thing to be noted is that mantissa is expressed in four digits and exponent in two digits, in each case.

Let us now consider a hypothetical case of a computer having a word length of 8 bits only. Out of eight bits, the left-most bit is used for storing the sign of mantissa. Let 0 denote positive (+ve) and 1 denote negative (−ve) sign of mantissa. The next four bits are used for storing the binary digits of mantissa. The right-most 3 bits are used for storing the exponent part; the first bit for storing its sign and last two bits for its value, digit 0 showing positive (+ve) and digit 1 showing negative (−ve) exponent (See Fig. 1.1).

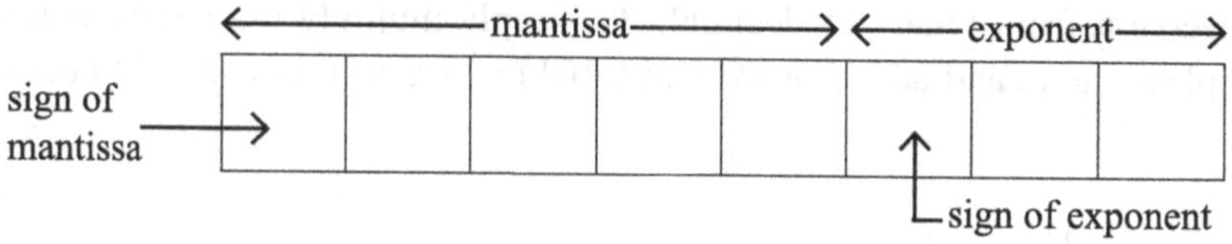

Figure 1.1 Floating point representation in 8-bit computer memory.

According to the memory configuration of Fig. 1.1 the binary numbers given above will be represented in the floating point form as follows:

	Binary number with decimal equivalent	*Representation in 8-bit memory*
(*a*)	0.0111 (0.4375)	01110101
(*b*)	−1.101 (−1.875)	11101001
(*c*)	11.1 (3.5)	01110010

It may be stated that the positive exponent varies from 000 to 011, i.e., from 0 to 3 in decimal. The negative exponent should vary from 101 to 111, i.e., from −1 to −3. But 100 may be considered as −4, since 000 is already zero, hence negative exponent varies from −1 to −4 in decimal.

It may be noted that the largest positive number that can be stored under present configuration would be, $0.1111 \times 10^{11} = 111.1$ (binary) = 7.5 (decimal). The algebraically smallest number that can be stored would be −7.5 (in decimal). Thus the range of numbers that can be represented in the computer memory would be $-7.5 \leq x \leq 7.5$. The smallest positive non-zero number represented in the above memory configuration would be, $0.1000 \times 10^{100} = 0.00001$ (binary) $= 2^{-5} = 0.03125$ (decimal). However, it may also be mentioned that even the simplest computer has a memory of 32-bit word and two or more words can be adjoined to store a number in floating point. Thus the space (number of bits) occupied by the mantissa and the exponent would be manifolds that of shown in Fig. 1.1

but the logic remains same. When a fixed number of decimal digits are kept in all numbers, it is called 'Fixed Point' representation.

1.4 Significant Digits

All the digits from 0 to 9 in a number, except the zeros which are used for fixation of decimal point, are called significant digits (or figures). For example, in the number .003456, the first two zeros are not significant since we can also express the number as $.3456 \times 10^{-2}$, while the other four digits, namely, 3, 4, 5 and 6 are significant. But in the number 20.003456, all the eight digits are significant. In order to find the number of significant digits in a number, express it in floating point; the mantissa part gives the number of significant digits.Whether the last zeros in a number are significant or not may depend on the context. For example, in measuring the heights of the students, in 168.00 cm, the last zero may not be significant and we can express the height as 168.0 cm, showing that the height is being measured nearest to the $\frac{1}{10}$th part of the centimeter, so that zero in 168.0 cm is significant.

1.5 Rounding and Chopping a Number

In scientific computing we are encountered by numbers with too many digits. More often than not, we have to shorten/reduce them to a size which may not affect the end result within a desired accuracy. There are two ways of reducing the size of or truncating the number, viz., (*i*) rounding (*ii*) chopping. Let us first discuss the procedure for rounding off a number x in decimal system.

Let the number x be expressed in floating point form with s digits in mantissa and with exponent q, as

$$x = \cdot d_1 d_2 \ldots d_n d_{n+1} \ldots . d_s \times 10^q.$$

If the number x is to be rounded to n significant digits, following procedure would be adopted:

(*i*) if $d_{n+1} < 5$, then no change in any of the digits from d_1 to d_n and the rounded number would be,

$$x \simeq .d_1 d_2 \ldots . d_n \times 10^q.$$

(*ii*) if $d_{n+1} > 5$, then digit d_n is incremented by 1, i.e. d_n becomes $d_n + 1$; as a cosequence of this other digits may get affected and even the exponent may have to be adjusted accordingly.

(*iii*) if $d_{n+1} = 5$, then there will be two ways for rounding, depending upon d_n being an even digit $(0,2,4,6,8)$ or an odd digit (1, 3, 5, 7, 9). If d_n is even, then case (i) applies and if d_n is odd then case (ii) applies. Thus probability of both cases is $\frac{1}{2}$ when $d_{n+1} = 5$.

Given below are some examples of rounding the numbers to four places of decimal (four significant digits):

	Floating point number	*Rounded to four decimals*
(*a*)	0.245684×10^2	0.2457×10^2
(*b*)	0.245629×10^{-2}	0.2456×10^{-2}
(*c*)	0.245659×10^2	0.2456×10^2
(*d*)	0.245750×10^2	0.2458×10^2
(*e*)	0.999951×10^2	0.1000×10^3
(*f*)	0.999858×10^2	0.9998×10^2

The difference between examples (e) and (f) may be noted. It may also be observed that in example (e) all zeros in the rounded number are significant.

However, a more conventional way for rounding, is straight in that if $d_{n+1} < 5$, then all the digits from d_1 to d_n remain unaltered [case (i)] while if $d_{n+1} \geq 5$, then d_n is incremented by 1 and necessary changes are made in the digits d_1 to d_n and also in the exponent, if necessary [case (ii)].

When all the digits after d_n are ignored, irrespective of whatever value d_{n+1} has, the procedure for truncating the number is known as 'chopping off' the number or simply 'chopping'. If there are sufficient number of significant digits in a number, like in a computer, the process of chopping may not affect the result in normal circumstances.

1.6 Errors due to Rounding/Chopping

Suppose a number x is rounded to x^*, then the modulus of the difference between x and x^*, i.e. $|x - x^*|$ is known as rounding error or error due to rounding in x^*.

Let x be a number which has been rounded to 4 decimals, say $x^* = 0.4387$. Then lower and upper bounds for the actual number x would be,

$$0.43865 \leq x < 0.43875$$

or $$0.43865 - 0.4387 \leq x - x^* < 0.43875 - 0.4387$$

or $$-0.00005 \le x - x^* < 0.00005$$

or $$|x - x^*| \le 0.00005 = \frac{1}{2} \times 10^{-4}$$

$$= \frac{1}{2} \times \text{ unit at } 4^{\text{th}} \text{ decimal place.}$$

The above result can be generalised for a number x represented in the floating point form as,

$$x = \cdot d_1 d_2 \ldots . d_n d_{n+1} \ldots . d_s \times 10^q.$$

If x is rounded to n decimals, then the maximum rounding error would be,

$$|x - x^*| \le \frac{1}{2} \times 10^{-n} \times 10^q$$

$$= \frac{1}{2} \times 10^{q-n}. \qquad (1.1)$$

If the number x is chopped off to n decimals, then it is easy to see that the maximum error due to chopping would be,

$$|x - x^*| \le 10^{q-n}. \qquad (1.2)$$

That is, the error in chopping a number is twice that in the rounding.

1.7 Measures of Error in Approximate Numbers

Let x^* be an approximation of exact number x, then we can measure the magnitude of error in three different forms:

(*i*) absolute error (a.e.) $= |x - x^*|$ (1.3a)

(*ii*) relative error (r.e.) $= \left|\frac{x - x^*}{x}\right|$ or $\left|\frac{x - x^*}{x^*}\right|$ (1.3b)

(*iii*) percentage error (p.e.) $=$ r.e. $\times 100$ (1.3c)

1.8 Errors in Arithmetic Operations

Let x_1 and x_2 be two numbers which are rounded to x_1^* and x_2^* respectively and let ε_1 and ε_2 be the corresponding rounding errors in them, such that $x_1 = x_1^* + \varepsilon_1$ and $x_2 = x_2^* + \varepsilon_2$. We are going to study below, the effects of rounding errors on arithmetic operations, viz., addition, subtraction, multiplication and division.

(*i*) Addition

$$x_1 + x_2 = x_1^* + \varepsilon_1 + x_2^* + \varepsilon_2$$

$$= x_1^* + x_2^* + \varepsilon_1 + \varepsilon_2$$

or $\quad |(x_1 + x_2) - (x_1^* + x_2^*)| = |(x_1 - x_1^*) + (x_2 - x_2^*)| = |\varepsilon_1 + \varepsilon_2| \leq |\varepsilon_1| + |\varepsilon_2|.$

This can be generalised to n numbers as,

$$\left|\sum_{i=1}^{n} x_i - \sum_{i=1}^{n} x_i^*\right| = \left|\sum_{i=1}^{n} (x_i - x_i^*)\right| = \left|\sum_{i=1}^{n} \varepsilon_i\right| \leq \sum_{i=1}^{n} |\varepsilon_i| \tag{1.4}$$

That is, the total absolute error in the sum of n numbers will be less than or equal to the sum of the absolute errors in each of them. Although it gives the upper bound for the absolute error in the sum, the actual error will be much smaller since some of the errors may be positive and some negative so that cumulative effect would be much reduced.

(*ii*) Subtraction

$$x_1 - x_2 = x_1^* + \varepsilon_1 - x_2^* - \varepsilon_2$$

or $\quad |(x_1 - x_2) - (x_1^* - x_2^*)| = |(x_1 - x_1^*) - (x_2 - x_2^*)| = |\varepsilon_1 - \varepsilon_2| \leq |\varepsilon_1| + |\varepsilon_2|. \qquad (1.5)$

Thus the absolute error in subtraction of two approximate numbers can be as great as the sum of their individual absolute errors; since the errors can be positive or negative and if ε_1 and ε_2 are of opposite signs, then under subtraction, they will be added up.

(*iii*) Multiplication

$$x_1 \cdot x_2 = (x_1^* + \varepsilon_1) \cdot (x_2^* + \varepsilon_2)$$

or $\quad x_1 \cdot x_2 - x_1^* \cdot x_2^* = x_1^* \varepsilon_2 + x_2^* \varepsilon_1$, neglecting $\varepsilon_1 \varepsilon_2$

$$\text{a.e.} = |x_1 x_2 - x_1^* \cdot x_2^*| \leq |x_1^* \varepsilon_2| + |x_2^* \varepsilon_1| \tag{1.6a}$$

$$\text{r.e.} = \left|\frac{x_1x_2 - x_1^*x_2^*}{x_1^*x_2^*}\right| \le \left|\frac{\varepsilon_1}{x_1^*}\right| + \left|\frac{\varepsilon_2}{x_2^*}\right|. \tag{1.6b}$$

Thus the maximum relative error in the product of two approximate numbers will be less than or equal to the sum their individual relative errors. This can be generalised to n numbers.

(*iv*) Division

$$\frac{x_1}{x_2} = \frac{x_1^* + \varepsilon_1}{x_2^* + \varepsilon_2} = \frac{x_1^*\left(1 + \dfrac{\varepsilon_1}{x_1^*}\right)}{x_2^*\left(1 + \dfrac{\varepsilon_2}{x_2^*}\right)} = \frac{x_1^*}{x_2^*}\left(1 + \frac{\varepsilon_1}{x_1^*}\right)\left(1 + \frac{\varepsilon_2}{x_2^*}\right)^{-1}$$

$$= \frac{x_1^*}{x_2^*}\left(1 + \frac{\varepsilon_1}{x_1^*}\right)\left(1 - \frac{\varepsilon_2}{x_2^*}\right), \text{ neglecting } {\varepsilon_2}^2 \text{ and higher powers}$$

$$= \frac{x_1^*}{x_2^*}\left(1 + \frac{\varepsilon_1}{x_1^*} - \frac{\varepsilon_2}{x_2^*}\right), \text{ neglecting } \varepsilon_1\varepsilon_2 \text{ term.}$$

$$\text{a.e.} = \left|\frac{x_1}{x_2} - \frac{x_1^*}{x_2^*}\right| \le \left|\frac{\varepsilon_1}{x_2^*}\right| + \left|\frac{\varepsilon_2 x_1^*}{x_2^{*2}}\right|. \tag{1.7a}$$

$$\text{r.e.} = \left|\left(\frac{x_1}{x_2} - \frac{x_1^*}{x_2^*}\right) \div \frac{x_1^*}{x_2^*}\right| \le \left|\frac{\varepsilon_1}{x_1^*}\right| + \left|\frac{\varepsilon_2}{x_2^*}\right|. \tag{1.7b}$$

Like multiplication, the relative error in the division of a number by another number cannot exceed the sum of their individual relative errors.

1.9 Computation of Errors Using Differentials

Let z be a function of two variables x and y defined as $z = f(x, y)$. If increments δx and δy are given to x and y respectively, then the corresponding increment δz in z is given by,

$$\delta z = f(x + \delta x,\ y + \delta y) - f(x, y).$$

Expanding the first term by Taylor's series (see Appendix A),

$$\delta z = \left[f(x, y) + \frac{\partial f}{\partial x}\delta x + \frac{\partial f}{\partial y}\delta y + \frac{1}{2}\left(\frac{\partial^2 f}{\partial x^2}\delta x^2 + 2\frac{\partial^2 f}{\partial x \partial y}\delta x \cdot \delta y + \frac{\partial^2 f}{\partial y^2}\delta y^2\right) + \cdots\right] -$$

$$f(x, y).$$

Neglecting higher powers of δx and δy and their products, assuming they are small, above may be written as,

$$\delta z \simeq \frac{\partial f}{\partial x}\delta x+\frac{\partial f}{\partial y}\delta y. \tag{1.8}$$

The error in arithmetic operations can be explained with the help of formula (1.8) considering δx and δy as errors in x and y respectively:

(*i*) & (*ii*) Addition/Subtraction

$$z=f(x,\, y)=x\pm y;\quad \frac{\partial f}{\partial x}=1,\ \frac{\partial f}{\partial y}=\pm 1.$$

a.e. $\quad = |\delta z| \le |\delta x|+|\delta y|$

(*iii*) Multiplication

$$z=f(x,\, y)=xy;\quad \frac{\partial f}{\partial x}=y,\ \frac{\partial f}{\partial y}=x.$$

a.e. $\quad = |\delta z| \le |y\delta x|+|x\delta y|$ and r.e. $=\left|\frac{\delta z}{z}\right| \le \left|\frac{\delta x}{x}\right|+\left|\frac{\delta y}{y}\right|.$

(*iv*) Division

$$z=f(x,\, y)=\frac{x}{y};\quad \frac{\partial f}{\partial z}=\frac{1}{y},\ \frac{\partial f}{\partial y}=-\frac{x}{y^2}.$$

$$\delta z=\frac{\delta x}{y}-\frac{x\delta y}{y^2}\ \text{ or }\ \frac{\delta z}{z}=\frac{\delta x}{x}-\frac{\delta y}{y}$$

$$\text{r.e.}=\left|\frac{\delta z}{z}\right| \le \left|\frac{\delta x}{x}\right|+\left|\frac{\delta y}{y}\right|.$$

Note: The analysis can be extended for n variables $x_1,\, x_2,\, \ldots x_n$.

If $z=f(x_1,\, x_2,\, \ldots x_n)$ then

$$\delta z=\frac{\partial f}{\partial x_1}\cdot\delta x_1+\frac{\partial f}{\partial x_2}\cdot\delta x_2+\ldots+\frac{\partial f}{\partial x_n}\cdot\delta x_n.$$

1.10 Errors in Evaluation of Some Standard Functions

Let y be a function of x defined by $y = f(x)$. Then for a small change δx in x, the corresponding change δy in y is given by,

$$\delta y = \frac{df}{dx} \cdot \delta x. \tag{1.9}$$

Let us apply formula (1.9) to compute the errors in some functions of one variable.

(*i*) Power Function $f(x) = x^a$
Differentiating $y = x^a$, we get using (1.9)

$$\text{a.e.} = |\delta y| = |ax^{a-1} \cdot \delta x|$$

$$\text{r.e.} = \left|\frac{\delta y}{y}\right| = \left|a \cdot \frac{\delta x}{x}\right| \leq |a| \left|\frac{\delta x}{x}\right|$$

$$\leq |a|. \text{ relative error in } x. \tag{1.10}$$

It means that the relative error in x^2 will be twice and in $\sqrt{x}$ it will be half that of the relative error in x. In x^{-1}, the r.e. will be same as in x and twice in x^{-2}.

(*ii*) Exponential Function $f(x) = a^x$

$$y = a^x; \quad \frac{df}{dx} = a^x \cdot \ln a.$$

$$\text{a.e.} \quad = |a^x \ln a.\ \delta x| \text{ and r.e.} \quad = |\ln a . \delta x| = |\ln a||\delta x|. \tag{1.11}$$

(*iii*) Logarithmic Function $f(x) = \ln x$

$$y = \ln x; \quad \frac{df}{dx} = \frac{1}{x}$$

$$\text{a.e.} \quad = |\delta y| = \left|\frac{\delta x}{x}\right|. \tag{1.12}$$

It shows that the absolute error in $\ln x$ will be same as the relative error in x.

(*iv*) Trigonometric Function $f(x) = \sin x (\text{or } \cos x)$

$$y = \sin x; \quad \frac{df}{dx} = \cos x$$

$$\text{a.e.} \quad = |\delta y| = |\cos x \cdot \delta x| \leq |\cos x| \cdot |\delta x| \leq |\delta x|. \tag{1.13}$$

Since the value of $\cos x$ (or $\sin x$) does not exceed 1, the absolute error in $\sin x$(or $\cos x$) does not exceed the absolute error in x.

Example 1.1

Find the sum of the following 10 approximate numbers: 0.248, 1.1524, 31.3, 9.75, 74.2, 8.14, 0.0767, 1.00621, 1.000245, 14.8 and round the sum to one place of decimal. Also add up the rounding errors in all these numbers up to 4 decimal places only.

Solution These numbers have different rounding errors. The max. error is in numbers rounded to one decimal place which is in $\frac{1}{2} \times 10^{-1} = 0.05$ in each of the numbers 31.3, 74.2 and 14.8. The min. rounded error is $\frac{1}{2} \times 10^{-6} = 0.0000005$ in 1.000245. The final sum should be computed up to one decimal only. For this it may be sufficient to retain 2 decimal digits (or at the most 3 in a number since the numbers are only 10.

Number rounded to 3 decimals	*Rounding error up to 4 decimal*
0.248	0.0005
1.152	0.0005
31.3- -	0.05- -
9.75-	0.005
74.2- -	0.05- -
8.14-	0.005-
0.077	0.0005
1.006	0.0005
1.000	0.0000
14.8- -	0.05- -
141.673	0.1620

$$\text{sum} = 141.673 \simeq 141.7$$

Max. Rounding error in the sum $= 0.162 \simeq 0.16$ or $\simeq 0.2$.
Actual error in the sum due to rounding is

$$141.7 - 141.673 = 0.027 \text{ which is much less than } 0.2 \text{ (or } 0.16)$$

$$\text{r.e.} = \frac{0.027}{141.67} \simeq 0.0002$$

Show that it will be less than the max. r.e. of any of the given number.

Example 1.2

Two numbers x_1 and x_2 are given which are correct up to their last digit: $x_1 = 12.47$, $x_2 = 10.3$. Estimate the max. absolute error, relative error and percentage error in computing $x_1 - x_2$.

Solution Max rounding error in 12.47 is $\pm$ 0.005
Max rounding error in 10.3 is $\pm$ .05

$$x_1 - x_2 = 12.47 - 10.3 = 2.17$$

$$\text{a.e.} = 0.005 + 0.05 = 0.055$$

$$\text{r.e.} = \frac{0.055}{2.17} \simeq 0.022$$

$$\text{p.e.} = 0.022 \times 100 = 2.2\%.$$

Example 1.3

Multiply two numbers $x_1 = 2.47$ and $x_2 = 1.6$ and estimate the relative error in the product. The numbers are correct to their last digit. Also compute absolute error and percentage error.

Solution Max. rounding error in $x_1 = 2.47$ is $\pm$ 0.005
Max. rounding error in $x_2 = 1.6$ is $\pm$ 0.05

$$x_1 x_2 = 2.47 \times 1.6 = 3.952 \simeq 3.95$$

$$\begin{aligned}
\text{r.e.} &= \frac{0.005}{2.47} + \frac{0.05}{1.6} \\
&\simeq 0.002 + 0.031 \\
&\simeq 0.033 \\
\text{a.e.} &= (\text{r.e.}) \times x_1 x_2 \\
&= 0.033 \times 3.95 = 0.130 \\
\text{p.e.} &= (\text{r.e.}) \times 100 \\
&= 0.033 \times 100 = 3.3\%
\end{aligned}$$

Example 1.4

Two approximate numbers x_1 and x_2 are given correct to their last digit, $x_1 = 5.16$ and $x_2 = 1.2$. Find the r.e., a.e. and p.e. in computing x_1/x_2.

Solution Max. error in 5.16 is ± 0.005
Max. error in 1.2 is ± 0.05

$$\frac{x_1}{x_2} = \frac{5.16}{12} = 4.3$$

$$\text{r.e.} = \frac{0.005}{5.16} + \frac{0.05}{1.2}$$

$$= 0.00097 + 0.0417$$

$$= 0.0427$$

$$\text{p.e.} = 0.0427 \times 100 = 4.27\%$$

$$\text{a.e.} = 0.0427 \times 4.3$$

$$\simeq 0.184.$$

Example 1.5

To what accuracy can we expect the number x to be correct if its logarithm ($\ln x$) is read from a four-place log table for $x < 100$.

Solution When $y = \ln x$

$$\delta y = \frac{\delta x}{x} \quad \text{or} \quad \delta x = x \cdot \delta y.$$

If $\ln x$ is correct up to 4 decimals the error in it will be 0.00005 $(= \delta y)$.

$$\delta x_{\max} = \delta y \cdot x = 0.00005 \times 100 = 0.005$$

For the error to be less than 0.005 the value of x may be expected to be correct up to 2 decimals.

Example 1.6

What will be the percentage error in the area of a rectangle if there is an error of 1% in the measurement of its sides?

Solution Let the accurate sides of the rectangle be x and y, then its area $A = xy$. If there is error δx in x and δy in y, the corresponding error in A in given by,

$$\delta A = x\delta y + y\delta x$$

or

$$\frac{\delta A}{A} = \frac{\delta x}{x} + \frac{\delta y}{y}$$

$$\text{Percentage error in } A = \frac{\delta A}{A} \times 100$$

Thus we get the percentage error in A as

$$\frac{\delta A}{A} \times 100 = \frac{\delta x}{x} \times 100 + \frac{\delta y}{y} \times 100$$

$$= 1 + 1 = 2\%.$$

Example 1.7

What will be the percentage error in the time period T of a pendulum where $T = 2\pi\sqrt{\frac{l}{g}}$ if there is an error of 1% in l and 2% in g.

Solution $$T = 2\pi\sqrt{\frac{l}{g}}$$

Taking log on both sides we get

$$\ln T = \ln 2\pi + \frac{1}{2}[\ln l - \ln g]$$

Its differential will give,

$$\frac{1}{T}\delta T = 0 + \frac{1}{2}\left[\frac{\delta l}{l} - \frac{\delta g}{g}\right]$$

or $$\frac{\delta T}{T} \times 100 = \frac{1}{2}\left[\frac{\delta l}{l} \times 100 - \frac{\delta g}{g} \times 100\right]$$

Max. Percentage error in T is given by

$$\frac{\delta T}{T} \times 100 = \frac{1}{2}[(\pm 1) - (\pm 2)]$$

$$= \frac{1}{2}[3] = 1.5\%.$$

1.11 Truncation Error and Taylor's Theorem

Quite often a function or a formula is expressed in the form of a series which may contain finite or infinite number of terms. Depending on the accuracy warranted by the problem or due to practical considerations, this series is truncated consisting of a first few terms of the original series, neglecting all the remaining terms. The difference between the original series and the truncated series is known as 'truncation error' or 'remainder'. Obviously, if the series is infinite, it should be convergent, otherwise truncation error will be infinite.

Let us consider the exponential function e^x which is expressed by an infinite series as,

$$e^x = 1 + x + \frac{x^2}{2!} + \frac{x^3}{3!} + \frac{x^4}{4!} + \cdots \tag{1.14}$$

Suppose we want to evaluate e^x at $x = -0.1$ from (1.14), then

$$e^{-0.1} = 1 - 0.1 + 0.005 - 0.000167 + 0.0000042\ldots. \tag{1.15}$$

Truncated up to four terms it gives,

$$e^{-0.1} = 1 - 0.1 + 0.005 - 0.000167 = 0.904843. \tag{1.16}$$

The infinite series (1.15) is a convergent series with alternating signs. Hence, if it is truncated up to four terms (1.16), the next neglected term will give the maximum error (truncation) in the truncated series. That means, the maximum error in the computation, $e^{-0.1} = 0.904843$ can be at the most 0.0000042, implying that our result is correct up to five places of decimal.

The infinite series (1.14) truncated to four terms may be written as,

$$e^x = 1 + x + \frac{x^2}{2!} + \frac{x^3}{3!}. \tag{1.17}$$

Again, let us evaluate e^x at $x = 0.1$ from the truncated series (1.17).

$$e^{0.1} = 1 + 0.1 + 0.005 + 0.000167 = 1.105167. \tag{1.18}$$

Although series (1.14) converges for all finite values of x, we can not ascertain the degree of accuracy of the result of (1.18), i.e., the magnitude of truncation error in the computation of $e^{0.1}$ from a truncated series (1.17). Secondly we may be interested to know the range of x, for which the truncated series (1.17) will provide the value of e^x which is correct up to a certain number of decimals, say for example, up to 4 places of decimal which means the truncation error is not greater than $\frac{1}{2} \times 10^{-4}$. The answer to these questions may be found in the Taylor's Theorem (formula) stated below:

Let $f(x)$ be a function of x, possessing derivatives of all orders up to $(n+1)$ in an interval I. If x_0 is a point in I, then for each x in I,

$$f(x) = f(x_0) + (x - x_0)f'(x_0) + \frac{(x - x_0)^2}{2!}f''(x_0) + \cdots + \frac{(x - x_0)^n}{n!}f^n(x_0) + R_{n+1}(x) \quad (1.19)$$

$$\text{where, } R_{n+1}(x) = \frac{(x - x_0)^{n+1}}{(n+1)!}f^{n+1}(\xi), \ \xi \text{ lies between } x_0 \text{ and } x. \quad (1.20)$$

The representation of the function $f(x)$ as power series in powers of $(x - x_0)$ as shown by (1.19) is known is Taylor's Theorem (formula) with the associated remainder term of order $(n+1)$, or error term, $R_{n+1}(x)$ as given by (1.20).

If a function is approximated by the first $(n+1)$ terms of (1.19), i.e., a polynomial of degree n, then the error in the polynomial approximation would be given by (1.20) which will be of order $(n+1)$ while the formula will be of order n.

If $R_{n+1}(x)$ tends to zero as $n \to \infty$ for all x in the interval I, then formula (1.19) may be represented by an infinite series which converges to $f(x)$, i.e.,

$$f(x) = f(x_0) + (x - x_0)f'(x_0) + \frac{(x - x_0)^2}{2!}f''(x_0) + \frac{(x - x_0)^3}{3!}f'''(x_0) + \cdots \quad (1.21)$$

The series given by (1.21) is known as Taylor's series which is also called Taylor's expansion of the function $f(x)$ about the point x_0.

Alternatively, if we put $x = x_0 + h, \ h > 0$, the Taylor's formula can be written as,

$$f(x_0 + h) = f(x_0) + hf'(x_0) + \frac{h^2}{2!}f''(x_0) + \frac{h^3}{3!}f'''(x_0) + \cdots \frac{h^n}{n!}f^n(x_0) + R_{n+1} \quad (1.22)$$

where the remainder term,

$$R_{n+1}(h) = \frac{h^{n+1}}{(n+1)!}f^{n+1}(\xi), \qquad x_0 \le \xi \le x_0 + h. \quad (1.23)$$

Similarly, expansion for $f(x_0 - h)$ can be written as,

$$f(x_0 - h) = f(x_0) - hf'(x_0) + \frac{h^2}{2!}f''(x_0) - \frac{h^3}{3!}f'''(x_0) + \cdots (-1)^n \frac{h^n}{n!}f^n(x_0) + R_{n+1}, \quad (1.24)$$

with associated remainder term,

$$R_{n+1}(h) = (-1)^{n+1}\frac{h^{n+1}}{(n+1)!}f^{n+1}(\xi), \quad x_0 - h \le \xi \le x_0. \quad (1.25)$$

If point x_0 is taken at the origin in formula (1.19), it is called Maclaurin's formula which can be written as,

$$f(x) = f_0 + xf_0' + \frac{x^2}{2!}f_0'' + \frac{x^3}{3!}f_0''' + \cdots + \frac{x^n}{n!}f_0^n + R_{n+1}(x), \tag{1.26}$$

where $$R_{n+1}(x) = \frac{x^{n+1}}{(n+1)!}f^{n+1}(\xi), \qquad 0 \le \xi \le x, \tag{1.27}$$

and f_0^k, $k = 1(1)n$, denotes the kth derivative of $f(x)$ at $x = 0$.

When a function $f(x)$ is approximated by a polynomial $P_n(x)$ of degree n by Taylor's formula, the truncation error is given by the remainder term $R_{n+1}(x)$, i.e.,

$$f(x) = P_n(x) + R_{n+1}(x)$$

or $$R_{n+1}(x) = f(x) - P_n(x). \tag{1.28}$$

But we can not compute $R_{n+1}(x)$ as its value is dependent on ξ which is a point we have no knowledge of. Therefore, we compute the upper bound of $R_{n+1}(x)$ to give the magnitude of truncation error, denoted by $R(x)$, i.e.,

$$R(x) = \max |R_{n+1}(x)|, \text{ over domain of } x. \tag{1.29}$$

Now, we revert back to our problem of computing the truncation error in the computation of e^x, at $x = 0.1$, by the truncated series (1.17).

We see that the expansion (1.16) of e^x is a Taylor's (Maclaurin's) series for $f(x) = e^x$. The truncation error in (1.17) will be given by,

$$R(x) = \left| \frac{x^4}{4!} f^{iv}(\xi) \right|, \; 0 \le \xi \le 0.1.$$

$f^{\text{iv}}(x) = e^x$, has maximum value at $x = 0.1$, for $0 \le \xi \le 0.1$ giving $f^{\text{iv}}(0.1) = e^{0.1} = 1.1052$. Thus the truncation error is given by

$$R = \frac{(0.1)^4}{24} \times 1.1052 = 0.000004605 = 0.46 \times 10^{-5}.$$

As the truncation error is less than $\frac{1}{2} \times 10^{-5}$, the value of $e^{0.1}$ computed from (1.17) is correct up to five decimal places, i.e., $e^{0.1} = 1.10517$.

Suppose we want range of x for which the truncated formula provides values of e^x correct up to 4 places of decimal, then we solve,

$$|R(x)| \le \frac{1}{2} \times 10^{-4} \text{ or } \left| \frac{x^4}{24} \times f^{iv}(x) \right| \le \frac{1}{2} \times 10^{-4} \text{ giving}$$

$$x^4 \le 0.0012 \times e^{-x} \text{ or } x \le 0.1861 \times e^{-x/4}. \qquad \because e^{\xi} \text{ will be maximum at } x.$$

To get an approximate value of x we can solve,

$$x = 0.1861\left(1 - \frac{x}{4}\right) \text{ giving } x \simeq 0.1778.$$

Thus for values approximately $0 \le x \le 0.18$, the truncation error will be less than 0.00005.

Example 1.8

Compute $(1.1)^{-1}$ from Taylor's expansion of the function $f(x) = \dfrac{1}{x}$ about $x_0 = 1$, truncated up to four terms. Also compute the truncation error and compare your result with the exact value.

Solution

$$f(x) = \frac{1}{x};\ f'(x) = -\frac{1}{x^2},\ f''(x) = +\frac{2}{x^3},\ f'''(x) = -\frac{6}{x^4},\ f^{iv}(x) = \frac{24}{x^5}.$$

$$f(1) = 1;\ f'(x) = -1,\ f''(1) = 2,\ f'''(1) = -6.$$

The truncated series about $x_0 = 1$, up to four terms is,

$$f(x) = f(1) + (x-1)f'(1) + \frac{(x-1)^2}{2}f''(1) + \frac{(x-1)^3}{6}f'''(1)$$

$$f(1.1) = 1 + 0.1 \times (-1) + \frac{0.01}{2} \times 2 + \frac{0.001}{6} \times (-6) = 0.909$$

The truncation error is given by,

$$R = \max\left|\frac{(x-1)^4}{24} \times f^{iv}(\xi)\right|,\ 1 \le \xi \le 1.1$$

$$= \frac{(1.1-1)^4}{24} \times 24 = 0.0001, \qquad \because f^{iv}(x) \text{ is maximum at } x = 1.$$

Exact value = $1/1.1 = 0.909090$ (90 recurring)
Computed value = 0.909
Actual error = $0.909090 - 0.909 = 0.00009$.
The actual error is less than the truncation error 0.0001.

Example 1.9

Using Taylor's expansion for $f(x) = \dfrac{1}{x}$ about $x_0 = 1$, truncated up to four terms, compute inverse of 2. Discuss the result.

Solution Taylor's series up to four terms is,

$$f(x) = f(x_0) + (x - x_0)f'(x_0) + \frac{(x - x_0)^2}{2!}f''(x_0) + \frac{(x - x_0)^3}{3!}f'''(x_0)$$

$$x_0 = 1;\ f(x_0) = 1,\ f'(x_0) = -1,\ f''(x_0) = 2,\ f'''(x_0) - 6.$$

$$f(2) = 1 + 1(-1) + \frac{1}{2} \times 2 - \frac{1}{6} \times 6$$

$$= 0$$

Truncation error is given by,

$$R(x) = \left| \frac{(x-1)^4}{4!} \times 24 \right|$$

For $\quad x = 2, |R| = 1.$

Exact value $\quad = \frac{1}{2} = 0.5$

Computed value $= 0$

Truncation error $= 1$

Thus we see that the truncation error is very large; the difference between the exact value and computed value could be as large as 1.0. Further, even if we take infinite number of terms, the sum will oscillate between 0 and 1 since $f(2) = 1 - 1 + 1 - 1 + \ldots$. The point x should be close to x_0, i.e. $(x - x_0)$ should not be too large.

Exercise 1

1.1 Find the range of number x, if it has been (*i*) rounded off to 3.14 (*ii*) chopped off to 3.14.

1.2 Express the number 0.007856 in floating point form; round the number to two significant digits and find the absolute error.

1.3 Let $x = 9.5$ be an approximate number which has an error of at most 5%. Find the range of the exact number.

1.4 Find the maximum value of the expression given below when all the numbers have been rounded,

$$x = \frac{1.25(4.0 - 2.25)}{10}$$

(Hint: For computing maximum value of x, take largest numerator and smallest denominator)

1.5 The diameter and height of a right circular cylinder are measured as 4 cm and 10 cm. respectively. If the possible error in each measurement is 0.1 cm, find the maximum absolute error in its volume. ($\pi = 3.14$).

1.6 Obtain a quadratic approximation for e^x near $x = 0$ by truncating the Taylor's series. Use the approximation to find the range of x so that the error does not exceed 0.005 (or approximation computes values correct up to 2 decimal places).
Compute your answer correct up to one place of decimal only.

1.7 If $y = x^{1/3}$, show that the relative error in y will be $\frac{1}{3}$rd of the relative error in x.
Hence compute $(1003)^{1/3}$.
[Hint: Take $x = 1000$, $\delta x = 3$].

References and Some Useful Related Books/Papers

1. Hartree, D.R., *Numerical Analysis*, Oxford University Press.
2. Hildebrand, F.B., *Introduction to Numerical Analysis*, Tata McGraw-Hill.
3. Scarborough, J.B., *Numerical Mathematical Analysis*, Oxford Book Company.

2

Linear Equations and Eigenvalue Problem

2.1 Introduction

Let us consider the following system of equations,

$$\left.\begin{aligned} a_{11}x_1 + a_{12}x_2 + a_{13}x_3 + \ldots. + a_{1n}x_n &= b_1 \\ a_{21}x_1 + a_{22}x_2 + a_{23}x_3 + \ldots. + a_{2n}x_n &= b_2 \\ a_{31}x_1 + a_{32}x_2 + a_{33}x_3 + \ldots. + a_{3n}x_n &= b_3 \\ \vdots \quad\quad \vdots \quad\quad \vdots \quad\quad\quad\quad \vdots \quad &\quad \vdots \\ a_{n1}x_1 + a_{n2}x_2 + a_{n3}x_3 + \ldots. + a_{nn}x_n &= b_n \end{aligned}\right\} \tag{2.1}$$

The system of equations given by (2.1) is a set of n algebraic equations which are linear in $x_1,\ x_2 \ldots.\ x_n$; while the values of a_{ij} and $b_i,\ i = 1(1)n,\ j = 1(1)n$ are prescribed, the values of the unknowns $x_1,\ x_2,\ \ldots.\ x_n$ are to be determined such that all the n equations are satisfied simultaneously.

The system of equations (2.1) can be expressed in matrix form as,

$$\mathrm{A}\boldsymbol{x} = \boldsymbol{b}, \tag{2.2}$$

$$\text{where} \quad \mathrm{A} = \begin{bmatrix} a_{11} & a_{12} & a_{13} & \cdots & a_{1n} \\ a_{21} & a_{22} & a_{23} & \cdots & a_{2n} \\ a_{31} & a_{32} & a_{33} & \cdots & a_{3n} \\ \vdots & & & & \\ a_{n1} & a_{n2} & a_{n3} & \cdots & a_{nn} \end{bmatrix} \tag{2.3}$$

$$\boldsymbol{x}^{\mathrm{T}} = (x_1 \;\; x_2 \;\; x_3 \;\; .... \;\; x_n), \tag{2.3a}$$

$$\boldsymbol{b}^{\mathrm{T}} = (b_1 \;\; b_2 \;\; b_3 \;\; .... \;\; b_n), \tag{2.3b}$$

the superscript T denotes transpose of a matrix. We assume that not all the elements of b are zero; that is, $\boldsymbol{b}$ is not a 'null' vector ($\boldsymbol{b} \neq 0$).

The matrix A is called coefficient matrix; the column vector (matrix) $\boldsymbol{x}$ which is to be determined, is the solution vector and $\boldsymbol{b}$ is the given right side.

2.2 Ill-conditioned Equations

When a minor change in the given value (s) results in a drastic change in the true solution of a system, it is called 'ill-conditioned'. Such a set of simultaneous equations is very sensitive to small errors.

2.3 Inconsistency of Equations

When a system of equations provide contradicting solutions or the equations themselves are self-contradictory we say that the equations are inconsistent.

The problem of inconsistency may arise when the system is over-determined, i.e., when there are more equations than the number of unknowns. For example let us consider following three equations in two unknowns;

$$2x_1 + x_2 = 4, \; 3x_1 - x_2 = 1, \; x_1 + x_2 = 7.$$

From the first two equations we get $x_1 = 1, \; x_2 = 2$ but they do not satisfy the third. Similarly, the second and third equations give $x_1 = 2, \; x_2 = 5$ which do not satisfy the first. Thus, there is no solution which can satisfy all the three equations simultaneously; hence the equations are 'inconsistent'. If however, there exists a solution which satisfies all the given equations, then the system would be called 'consistent'. The two equations $x_1 + x_2 = 1$ and $x_1 + x_2 = 2$ are inconsistent.

2.4 Linear Dependence

If some of the equations in a system are linearly related, the equations of such a system are said to be 'linearly dependent'; otherwise linearly independent. That means, in a linearly dependent system, at least one equation can be expressed as linear sum of some other equations. For example, consider the following equations,

$$2x_1 + 3x_2 + x_3 = 4, \; 5x_1 + 2x_2 + 2x_3 = 5, \; 3x_1 + 10x_2 + 2x_3 = 11.$$

We see that the third equation can be expressed as, four times the first minus the second. Or, first equation can be obtained by adding second and third equations and then dividing by four. As there exists a linear relation between the equations, they are not linearly independent. Effectively, there are only two linearly independent equations as third can be expressed in terms of the other two. In such cases an arbitrary value can be assigned to one of the variables, say, $x_3 = k$; then values of x_1 and x_2 can be computed in terms of k. Thus, the system will have infinite number of solutions depending on k.

It is easy to visualise that since the rows of the coefficient matrix of linearly dependent system are linearly related, its determinant will vanish. In order to know as how many of rows are linearly independent, we are lead to the notion of 'rank' of a matrix. A square matrix whose determinant vanishes is called 'singular'; otherwise 'non-singular' or 'regular'.

2.5 Rank of a Matrix

Let A be a $m \times n$ matrix. From matrix A we can form square submatrices by removing some of its rows and/or some columns including the matrix A itself (when $m = n$). The matrix A is said to have rank k if it has at least one submatrix of order k which is non-singular while all submatrices of order greater than k are singular. It is denoted as,

$$\text{rank (A) or } r(\text{A}) = \text{k}.$$

It may be noted that the number linearly independent equations in a system of equations is equal to the rank of its coefficient matrix.

2.6 Augmented Matrix

Referring (2.3) and (2.3b), when the matrix A is augmented by adjoining vector b as $(n+1)^{\text{th}}$ column, we call matrix A as augmented matrix and is denoted as,

$$\text{aug }(\text{A}/\boldsymbol{b}) = \left[\begin{array}{ccccc|c} a_{11} & a_{12} & a_{13} & \cdots & a_{1n} & b_1 \\ a_{21} & a_{22} & a_{23} & \cdots & a_{3n} & b_2 \\ a_{31} & a_{32} & a_{33} & \cdots & a_{3n} & b_3 \\ \vdots & \vdots & \vdots & & & \vdots \\ a_{n1} & a_{n2} & a_{n3} & \cdots & a_{nn} & b_n \end{array}\right]. \tag{2.4}$$

As regards existence/uniqueness of the solution of the system of equations (2.2), following three cases arise; for brevity/clarity we have used aug A for aug $(\text{A}/\boldsymbol{b})$:

(*i*) $r(\text{A}) = r\,(\text{aug A}) =$ number of unknowns
The solution is unique. The number of linearly independent equations is same as the number of unknowns.

(*ii*) $r(\text{A}) = r\,(\text{aug A}) <$ number of unknowns
There are infinite number of solutions. The number of linearly independent equations is less than the number of unknowns.

(*iii*) $r(\text{A}) < r\,(\text{aug A})$
There exists no solution. There are more linearly independent equations than there are unknowns.

Note: In the ensuing discussions we will assume that a unique solution exists of the system (2.2) which means A is regular and its inverse exists.

2.7 Methodology for Computing $\mathbf{A^{-1}}$ by Solving $\mathbf{A}\boldsymbol{x} = \boldsymbol{b}$

Let A be a regular (non-singular) matrix of order n and its inverse denoted as

$$\text{A}^{-1} = \begin{bmatrix} \alpha_{11} & \alpha_{12} & \alpha_{13} & \cdots & \alpha_{1n} \\ \alpha_{21} & \alpha_{22} & \alpha_{23} & \cdots & \alpha_{2n} \\ \alpha_{31} & \alpha_{32} & \alpha_{33} & \cdots & \alpha_{3n} \\ \vdots & \vdots & \vdots & & \vdots \\ \alpha_{n1} & \alpha_{n2} & \alpha_{n3} & \cdots & \alpha_{nn} \end{bmatrix}.$$

Then from $\text{A}\boldsymbol{x} = \boldsymbol{b}$, i.e., from (2.2) we can write,

$$\boldsymbol{x} = \text{A}^{-1}\boldsymbol{b} = \begin{bmatrix} \alpha_{11} & \alpha_{12} & \alpha_{13} & \cdots & \alpha_{1n} \\ \alpha_{21} & \alpha_{22} & \alpha_{23} & \cdots & \alpha_{2n} \\ \alpha_{31} & \alpha_{32} & \alpha_{33} & \cdots & \alpha_{3n} \\ \vdots & \vdots & \vdots & & \vdots \\ \alpha_{n1} & \alpha_{n2} & \alpha_{n3} & \cdots & \alpha_{nn} \end{bmatrix} \begin{bmatrix} b_1 \\ b_2 \\ b_3 \\ \vdots \\ b_n \end{bmatrix}.$$

We observe that if $\boldsymbol{b}$ is chosen to be such that $b_1 = 1$ and $b_2 = b_3 = \ldots. = b_n = 0$, then $\boldsymbol{x}$ is simply the first column of A^{-1}. Thus if we solve (2.2) by taking $\boldsymbol{b}$ as the first column of a unit/Identity matrix, the solution vector will give the first column of A^{-1}. Similarly, if we solve $\text{A}\boldsymbol{x} = \boldsymbol{b}$ by choosing vector $\boldsymbol{b}$ as , i.e., $b_2 = 1$ and $b_1 = b_3 = b_4 = \ldots = b_n = 0$, then, the solution vector $\boldsymbol{x}$ will provide the second column of A^{-1}; and so on. Thus solving the system of equations $A\boldsymbol{x} = \boldsymbol{b}$, n times, by taking $\boldsymbol{b}$ to be the various columns of a unit matrix will provide respective columns of A^{-1}.

2.8 Cramer's Rule

Denoting the determinant of a square matrix A by $|A|$ or $\det A$, the solution to the system of equations (2.2) by Cramer's rule, is given as,

$$x_j = \frac{|A_j|}{|A|}, \; j = 1(1)n \tag{2.5}$$

where $|A_j|$ denotes the determinant of the matrix A with its jth column replaced by the right side $\boldsymbol{b}$. Since (2.5) involves evaluation of determinants, the method is not suitable for larger systems.

2.9 Inverse of Matrix by Cofactors

The adjoint of a matrix A is defined as,

$$\text{adj } A = \begin{bmatrix} A_{11} & A_{21} & A_{31} & \cdots & A_{n1} \\ A_{12} & A_{22} & A_{32} & \cdots & A_{n2} \\ A_{13} & A_{23} & A_{33} & \cdots & A_{n3} \\ \vdots & \vdots & \vdots & & \vdots \\ A_{1n} & A_{2n} & A_{3n} & \cdots & A_{nn} \end{bmatrix} \tag{2.6}$$

where A_{ij} is the cofactor of a_{ij}. We know that if M_{ij} is the minor of a_{ij}, then $A_{ij} = (-1)^{i+j}M_{ij}$. The inverse of matrix A is given by,

$$A^{-1} = \frac{1}{|A|} \cdot \text{adj } A. \tag{2.7}$$

After computing A^{-1} the solution of $A\boldsymbol{x} = \boldsymbol{b}$ may be obtained from $\boldsymbol{x} = A^{-1}\boldsymbol{b}$. But again, the evaluation minors is a very time-consuming process. Therefore this method is also of limited use and is not suitable for larger systems.

The method described in Sec 2.7 will be taken up in detail in Sec 2.17 along with other computer-oriented methods. Now, let us present a brief review of some special square matrices and properties associated with them.

2.10 Definitions of Some Matrices

(*i*) Symmetric Matrix
A matrix A is called 'symmetric' if

$$\mathrm{A} = \mathrm{A}^{\mathrm{T}} \text{ or } [a_{ij}] = [a_{ji}]$$

Also if $\mathrm{A} = -\mathrm{A}^{\mathrm{T}}$, then it is called skew-symmetric.

(*ii*) Diagonal Matrix
A matrix A is called 'diagonal' if all its off-diagonal elements are zero, i.e.,

$$a_{ij} = 0, \ i \neq j.$$

When all the diagonal elements are unity, it is called unit or Identity matrix and when the diagonal elements are also zero it is known as Null matrix. The diagonal matrix is generally denoted by symbol D, a unit/Identity matrix by I while a Null matrix by letter O and a null vector by **0**.

(*iii*) Lower Triangular/Upper Triangular Matrices
When all the elements in a matrix above its main diagonal are zero, it is called Lower Triangular; on the other hand if all the elements below the main diagonal are zero, the matrix is called Upper Triangular. The forms of 4×4 Lower Triangular matrix (L) and Upper Triangular matrix (U) are shown below:

$$\mathrm{L} = \begin{bmatrix} l_{11} & 0 & 0 & 0 \\ l_{21} & l_{22} & 0 & 0 \\ l_{31} & l_{32} & l_{33} & 0 \\ l_{41} & l_{42} & l_{43} & l_{44} \end{bmatrix}, \ \mathrm{U} = \begin{bmatrix} u_{11} & u_{12} & u_{13} & u_{14} \\ 0 & u_{22} & u_{23} & u_{24} \\ 0 & 0 & u_{33} & u_{34} \\ 0 & 0 & 0 & u_{44} \end{bmatrix}.$$

If the diagonal elements are 1's, the matrices are called unit lower triangular and unit upper triangular. When diagonal elements are zero, they are called 'strictly' lower and upper triangular.

(*iv*) Tri-diagonal Matrix/Band matrices
A matrix a is called tri-diagonal if, for any i,

$$a_{ij} = 0, \ j < i - 1 \text{ and } j > i + 1.$$

Thus, there are at the most three non-zero terms in each row—diagonal term, one before the diagonal term and one after. The first and last rows have two terms only. A (5×5) tri-diagonal matrix will look like the following

$$A = \begin{bmatrix} a_{11} & a_{12} & 0 & 0 & 0 \\ a_{21} & a_{22} & a_{23} & 0 & 0 \\ 0 & a_{32} & a_{33} & a_{34} & 0 \\ 0 & 0 & a_{43} & a_{44} & a_{45} \\ 0 & 0 & 0 & a_{54} & a_{55} \end{bmatrix}.$$

Symbolically, it is also represented as,

A =
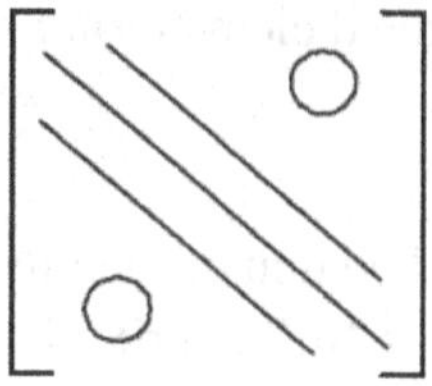

The big zeros show the zero elements and the straight lines along the main diagonal and paralled to it (sub-diagonal and super-diagonal) show the elements which may or may not be zero. We may also have matrices when there are at the most five non-zero terms in any row in the above fashion. Such matrices are called 'Penta-diagonal'. In general, when there are non-zero terms along the main diagonal, sub-diagonal and super-diagonal while all the remaining terms are zero in a matrix it is called 'Band-matrix'. These type of matrices arise in solving ordinary and partial differential equations.

(*v*) Sparse Matrix

A matrix is called 'sparse' if most of its elements are zero.

(*vi*) Orthogonal Matrix

Matrix A is said to be orthogonal, if

$$A^T = A^{-1}, \text{ implying } AA^T = A^T A = I.$$

For example, following matrix is orthogonal

$$A = \begin{bmatrix} \cos\theta & \sin\theta \\ -\sin\theta & \cos\theta \end{bmatrix}.$$

(*vii*) Hermitian Matrix

If elements of a matrix A are complex numbers and $\overline{A}$ represents complex conjugate of A, i.e., when all the elements of A are replaced by their complex conjugates, then

A is Hermitian matrix, if

$(\overline{A})^T = A^* = A$, where $*$ denotes conjugation and transposition (tranjugation) of matrix A.

It is equivalent to symmetric matrix if A is real.

(*viii*) Unitary Matrix
Matrix A is called Unitary, if

$$A^* = A^{-1}$$

It is equivalent to orthogonal if A is real.

(*ix*) Involutary Matrix

If $A = A^{-1}$, implying $A^2 = I$.

(*x*) Positive Definite Matrix
Matrix A is said to be positive definite if for any non-zero column vector $\boldsymbol{x}$,

$$\boldsymbol{x}^T A \boldsymbol{x} > 0.$$

Sometimes we use condition $\boldsymbol{x}^T A \boldsymbol{x} \geq 0$ for positive definite and $\boldsymbol{x}^T A \boldsymbol{x} > 0$ for strictly positive definite matrix. If all the leading principal minors of a matrix are positive (+ve), the matrix will be positive definite.

2.11 Properties of Matrices

We discuss below some of the properties of $(n \times n)$ square matrices. Let it be reminded that the determinant of a matrix, say, A is denoted by $|A|$.

(*i*) $|A| = |A^T|$

(*ii*) $|A \cdot B| = |A| \cdot |B| = |B \cdot A|$.
This can be generalised to any number of matrices.

(*iii*) The determinant of a Lower Triangular/Upper Triangular matrix is equal to product of its diagonal elements. That is, if L is a Lower Triangular matrix with its diagonal elements as $l_{11}, l_{22} \dots l_{nn}$, then

$$|L| = l_{11} l_{22} \dots l_{nn}.$$

Similarly if U is an Upper Triangular matrix with its diagonal elements as, $u_{11}, u_{22}, \dots u_{nn}$, then

$$|U| = u_{11} u_{22} \dots u_{nn}.$$

(*iv*) Transpose of product of matrices is equal to the product of their transposes taken in reverse order, i.e.,

$$(A \cdot B)^T = B^T \cdot A^T$$

$$(A \cdot B \cdot C)^T = C^T B^T A^T \text{ etc.}$$

(*v*) Inverse of product of matrices is equal to the product of their inverses taken in reverse order, i.e.,

$$(A \cdot B)^{-1} = B^{-1} A^{-1}$$

$$(A \cdot B \cdot C)^{-1} = C^{-1} B^{-1} A^{-1} \text{ etc.}$$

(*vi*) Inverse of a transpose of a matrix is the same as transpose of its inverse, i.e.,

$$(A^T)^{-1} = (A^{-1})^T.$$

This can be verified by multiplying both sides by A^T and using property (*iv*) which gives $I = I$.

(*vii*) If A is symmetric then A^{-1} is also symmetric, for if $A = A^T$, then

$$A^{-1} = (A^T)^{-1} = (A^{-1})^T \text{ from property } (vi).$$

(*viii*) If A is an orthogonal matrix, then

$$(A^T)^{-1} = A, \text{ since } A^T = A^{-1} \text{ and } (A^{-1})^{-1} = A.$$

(*ix*) Product of Lower Triangular matrices is also a Lower Triangular matrix. Similarly product of upper triangular matrices is an Upper Triangular matrix.

(*x*) Inverse of a lower triangular matrix is also a lower triangular and similarly the inverse of an upper triangular matrix is also an upper triangular matrix.

2.12 Elementary Transformations

The elementary operations on a matrix like, interchanging its rows (or columns) or adding a multiple of a row (or column) to another are known as elementary transformation on the matrix. The desired elementary row transformations can be performed on a matrix by premultiplying it by a unit matrix which has undergone the same transformation. Similarly elementary column operations can be performed by postmultiplying the matrix by a unit matrix with same operations.

Suppose we want to interchange the first row of a 4×4 matrix A by its third row, we can choose a 4×4 unit matrix with its first and third rows interchanged, say, I_{13}, where

$$I_{13} = \begin{bmatrix} 0 & 0 & 1 & 0 \\ 0 & 1 & 0 & 0 \\ 1 & 0 & 0 & 0 \\ 0 & 0 & 0 & 1 \end{bmatrix}.$$

The matrix obtained after pre-multiplication of A by I_{13}, i.e., $I_{13}A$ will be matrix A with its first and third rows interchanged. Similarly the matrix obtained after postmultiplication of A by I_{13}, i.e., AI_{13} will be matrix A with its first and third columns interchanged. The matrix which is used for interchanging rows (or columns) is known as permutation matrix.

Now suppose we want to add p times of the second row of A to its third row and q times of the second row to its fourth row. This can be achieved by elementary transformation by choosing a matrix I_{2R} (say) obtained by doing the same operations on a unit matrix, i.e.,

$$I_{2R} = \begin{bmatrix} 1 & 0 & 0 & 0 \\ 0 & 1 & 0 & 0 \\ 0 & p & 1 & 0 \\ 0 & q & 0 & 1 \end{bmatrix}.$$

Then we will have,

$$I_{2R}A = \begin{bmatrix} a_{11} & a_{12} & a_{13} & a_{14} \\ a_{21} & a_{22} & a_{23} & a_{24} \\ a_{31}+pa_{21} & a_{32}+pa_{22} & a_{33}+pa_{23} & a_{34}+pa_{24} \\ a_{41}+qa_{21} & a_{42}+qa_{22} & a_{43}+qa_{23} & a_{44}+qa_{24} \end{bmatrix}.$$

Similarly if we want to add p times of the second column of A to its third column and q times of its second column to the fourth column, then the desired transformation matrix will be,

$$I_{2C} = \begin{bmatrix} 1 & 0 & 0 & 0 \\ 0 & 1 & p & q \\ 0 & 0 & 1 & 0 \\ 0 & 0 & 0 & 1 \end{bmatrix}.$$

The matrix AI_{2C} will be the matrix with desired result, i.e.,

$$AI_{2C} = \begin{bmatrix} a_{11} & a_{12} & a_{13}+pa_{12} & a_{14}+qa_{12} \\ a_{21} & a_{22} & a_{23}+pa_{22} & a_{24}+qa_{22} \\ a_{31} & a_{32} & a_{33}+pa_{32} & a_{34}+qa_{32} \\ a_{41} & a_{42} & a_{43}+pa_{42} & a_{44}+qa_{42} \end{bmatrix}.$$

2.13 Methods for Solving Equations (Direct Methods)

There are two approaches for solving the system of equations $Ax = b$, known as (1) Direct Methods and (2) Iterative Methods. In the 'Direct methods' the solution is obtained in some definite number of steps while in the 'iterative methods' the process is started from an initial guess (usually taking $x = 0$) which is improved in an interative manner until the solution agrees in two successive iterations within the desired accuracy. Thus the number of steps (iterations) in an iterative method can not be predicted beforehand.

Let us first discuss the Direct Methods.

2.13.1 Gaussian elimination method (Basic)

We describe the Gaussian Elimination method, in its basic form by taking a 4×4 system of equations, i.e.,

$$\begin{bmatrix} a_{11} & a_{12} & a_{13} & a_{14} \\ a_{21} & a_{22} & a_{23} & a_{24} \\ a_{31} & a_{32} & a_{33} & a_{34} \\ a_{41} & a_{42} & a_{43} & a_{44} \end{bmatrix} \begin{bmatrix} x_1 \\ x_2 \\ x_3 \\ x_4 \end{bmatrix} = \begin{bmatrix} b_1 \\ b_2 \\ b_3 \\ b_4 \end{bmatrix}. \tag{2.8}$$

Stage 1. It consists of three steps:

(*i*) Take a multiplier $m_{21} = -\dfrac{a_{21}}{a_{11}}$.

Multiply 1st row of (2.8) by m_{21} and add to the 2nd.

i.e., $R_2 \leftarrow R_2 + m_{21}R_1$.

(*ii*) Take a multiplier $m_{31} = -\dfrac{a_{31}}{a_{11}}$.

Multiply 1st row by m_{31} and add to 3rd.

i.e., $R_3 \leftarrow R_3 + m_{31} \cdot R_1$.

(*iii*) Take a multiplier $m_{41} = -\dfrac{a_{41}}{a_{11}}$.

Multiply 1st row by m_{41} and add to 4th

i.e., $R_4 \leftarrow R_4 + m_{41} \cdot R_1$.

It may be noted that each time, first row is being multiplied by appropriate multiplier and added to different rows. The row which is being multiplied (first in this case) is known as 'pivotal' row and the element a_{11}, the divisor, is called the 'pivotal element' or simply 'pivot'. Next, we need to do operations of multiplications and additions from second column onwards since elements in the first column are made zero.

After execution of the first stage, the system will have the following form,

$$\begin{bmatrix} a_{11} & a_{12} & a_{13} & a_{14} \\ 0 & a_{22}^{(1)} & a_{23}^{(1)} & a_{24}^{(1)} \\ 0 & a_{32}^{(1)} & a_{33}^{(1)} & a_{34}^{(1)} \\ 0 & a_{42}^{(1)} & a_{43}^{(1)} & a_{44}^{(1)} \end{bmatrix} \begin{bmatrix} x_1 \\ x_2 \\ x_3 \\ x_4 \end{bmatrix} = \begin{bmatrix} b_1 \\ b_2^{(1)} \\ b_3^{(1)} \\ b_4^{(1)} \end{bmatrix}. \tag{2.9}$$

The elements with superscript 1 indicate that they have changed after stage 1.

Stage 2. We omit the first row and first column of the coefficient matrix in (2.9), thereby dealing with a 3×3 system of equations. Thus stage 2 has two steps.

(*i*) Take multiplier $m_{32} = -\dfrac{a_{32}^{(1)}}{a_{22}^{(1)}}$.

Multiply 2nd row of (2.9) by m_{32} and add to the 3rd.

i.e., $R_3 \leftarrow R_3 + m_{32} \cdot R_2$.

(*ii*) Take multiplier $m_{42} = -\dfrac{a_{42}^{(1)}}{a_{22}^{(1)}}$.

Multiply 2nd row by m_{42} and add to 4th.

i.e., $R_4 \leftarrow R_4 + m_{42} \cdot R_2$

Here second row is the pivotal row and $a_{22}^{(1)}$, the pivot.

Note that we have to do the operations of multiplications and additions from the third column onwards, since the elements in the first columns are zero which are not affected by these operations and elements of second column are reduced to zero due to choice of multipliers.

After Stage 2, the system (2.9) will assume the following form,

$$\begin{bmatrix} a_{11} & a_{12} & a_{13} & a_{14} \\ 0 & a_{22}^{(1)} & a_{23}^{(1)} & a_{24}^{(1)} \\ 0 & 0 & a_{33}^{(2)} & a_{34}^{(2)} \\ 0 & 0 & a_{43}^{(2)} & a_{44}^{(2)} \end{bmatrix} \begin{bmatrix} x_1 \\ x_2 \\ x_3 \\ x_4 \end{bmatrix} = \begin{bmatrix} b_1 \\ b_2^{(1)} \\ b_3^{(2)} \\ b_4^{(2)} \end{bmatrix}. \tag{2.10}$$

Here, the elements which have changed during Stage 2 are shown with superscript 2.

Stage 3. We ignore first two rows and first two columns in (2.10) and deal with a 2×2 system having superscript 2. This stage has only one step.

Take the multiplier $m_{43} = -\dfrac{a_{43}^{(2)}}{a_{33}^{(2)}}$.

Multiply the 3rd row of (2.10) by m_{43} and add to 4th

i.e., $\quad R_4 \leftarrow R_4 + m_{43} \cdot R_3$.

This operation has to be made only on the fourth column and on the right side, of course. Third row is pivotal row and $a_{33}^{(2)}$ is the pivot. After stage 3, the system (2.10) finally reduces to the following form,

$$\begin{bmatrix} a_{11} & a_{12} & a_{13} & a_{14} \\ 0 & a_{22}^{(1)} & a_{23}^{(1)} & a_{24}^{(1)} \\ 0 & 0 & a_{33}^{(2)} & a_{34}^{(2)} \\ 0 & 0 & 0 & a_{44}^{(3)} \end{bmatrix} \begin{bmatrix} x_1 \\ x_2 \\ x_3 \\ x_4 \end{bmatrix} = \begin{bmatrix} b_1 \\ b_2^{(1)} \\ b_3^{(2)} \\ b_4^{(3)} \end{bmatrix} \tag{2.11}$$

We see that the coefficient matrix in (2.11) has been reduced to an Upper Triangular matrix. This process of reduction of the coefficient matrix to an Upper Triangular form is known as 'Gaussian Elimination' or 'pivotal condensation'.

Having got the final reduced form (2.11), the solution is obtained by the process of 'back-substitution'. That is, we compute the value of x_4 from the last equation as,

$$x_4 = b_4^{(3)} / a_{44}^{(3)}.$$

Substituting the value of x_4 in the third equation, gives the value of x_3. After putting the values of x_3 and x_4 in the second equation we get x_2. And finally the value of x_1 is obtained from first equation, using the values of x_2, x_3 and x_4.

Caution: Some caution has to be exercised while using the basic Gaussian Elimination method in that the pivotal element should not be zero (or very small) since it will result in infinitely large multipliers. In order to avoid it, the pivotal row should be exchanged with the next row such that the pivotal element does not remain too small.

Note for Computer Algorithm: We see that the zeros once produced do not play any role in the future computations. Therefore, in order to save computer space, the multipliers may be stored in their place. For example, the multipliers m_{21}, m_{31} and m_{41} may be stored in the space previously occupied by a_{21}, a_{31} and a_{41} respectively. Similarly, m_{32} and m_{42} may be stored in the space previously occupied by $a_{32}^{(1)}$ and $a_{42}^{(1)}$ respectively. Finally, m_{43} may be stored in the space earlier occupied by $a_{43}^{(2)}$.

Thus in a compact form, we can store the coefficient matrix and the multipliers inside the computer in a 4×4 array in the following manner,

$$\begin{bmatrix} a_{11} & a_{12} & a_{13} & a_{14} \\ m_{21} & a_{22}^{(1)} & a_{23}^{(1)} & a_{24}^{(1)} \\ m_{31} & m_{32} & a_{33}^{(2)} & a_{34}^{(2)} \\ m_{41} & m_{42} & m_{43} & a_{44}^{(3)} \end{bmatrix}. \tag{2.12}$$

The method described above for (4×4) system can be easily generalised for $(n \times n)$ system. There will be $(n-1)$ stages to reduce the original coefficient matrix to upper triangular form. At the kth stage, $k = 1(1)n-1$, we compute in the following manner:

For the rows, $i = k+1(1)n$,

$$m_{ik} = -a_{ik}{}^{(k-1)}/a_{kk}{}^{(k-1)}$$

$$a_{ij}{}^{(k)} = a_{ij}{}^{(k-1)} + m_{ik}a_{kj}{}^{(k-1)}, \; j = k+1(1)n,$$

$$b_i{}^{(k)} = b_i{}^{(k-1)} + m_{ik}b_k{}^{(k-1)},$$

$$a_{ik}{}^{(k)} = m_{ik}(\text{optional}).$$

The superscript zero corresponds to original values of A and $\boldsymbol{b}$. The basic concept in the method is that the elements once made zero, remain zero throughout the subsequent operations.

2.13.2 Gaussian elimination (with row interchanges)

Gaussian Elimination with row interchanges, called 'partial pivoting' ensures that all the multipliers are less than 1 in absolute value. This is achieved as described below:

At the kth stage, all the elements in the kth column, i.e., a_{ik}, $i = k(1)n$ are scanned and the numerically largest element is selected; suppose it is a_{pk}, i.e., element in the pth row. Then kth row is interchanged with the pth row so that the pivotal element is numerically largest, rendering the multipliers less than 1 in absolute value. The objective of affecting the interchanges is to reduce the rounding errors in the subsequent computations. When a number is rounded it has certain rounding error and when this number is multiplied by a multiplier greater than 1, the rounding error also increases but if the number is multiplied by a multiplier smaller than 1, the rounding error does not increase. Moreover, the partial pivoting automatically excludes the possibility of the pivotal element being zero.

A further modification to the method is 'complete pivoting' in that all the elements $a_{i,j}^{(k-1)}$, $i = k(1)n, j = k(1)n$ are scanned and numerically largest element $a_{pq}^{(k-1)}$, is selected; then kth row is interchanged with the pth row and k^{th} column with the qth column. In this manner element $a_{pq}^{(k-1)}$ is brought at the $(k,\ k)$ position. Obviously, it requires recording of interchanges of rows and particularly of columns at each stage; however, it may not be worth except in some rare cases.

Example 2.1

Solve the following system of simultaneous equations by Gaussian elimination method,

$$3x_1 + 2x_2 + x_3 - 4x_4 = 5$$

$$x_1 - 5x_2 + 2x_3 + x_4 = 18$$

$$5x_1 + x_2 - 3x_3 + 2x_4 = -4$$

$$2x_1 + 3x_2 + x_3 + 5x_4 = 11$$

Compute up to 3 places of decimal without using fractions. Round the final answer to two decimals.

Solution Since vector x plays no part in computations we may work out with numbers only.

$$\begin{array}{cccc:c} x_1 & x_2 & x_3 & x_4 & \boldsymbol{b} \end{array}$$

$$\left[\begin{array}{cccc:c} 3 & 2 & 1 & -4 & 5 \\ 1 & -5 & 2 & 1 & 18 \\ 5 & 1 & -3 & 2 & -4 \\ 2 & 3 & 1 & 5 & 11 \end{array}\right]$$

Forward Elimination:

$$m_{21} = -\frac{1}{3} = -0.333; \qquad \mathrm{R}_2 \leftarrow \mathrm{R}_2 - 0.333\mathrm{R}_1$$

$$m_{31} = -\frac{5}{3} = -1.667; \qquad R_3 \leftarrow R_3 - 1.667R_1$$

$$m_{41} = -\frac{2}{3} = -0.667; \qquad R_4 \leftarrow R_4 - 0.667R_1$$

$$\begin{array}{c} \begin{matrix} x_1 & x_2 & x_3 & x_4 & b \end{matrix} \\ \left[\begin{array}{cccc|c} 3 & 2 & 1 & -4 & 5 \\ 0 & -5.666 & 1.667 & 2.332 & 16.335 \\ 0 & -2.334 & -4.667 & 8.668 & -12.335 \\ 0 & 1.666 & 0.333 & 7.668 & 7.665 \end{array}\right] \end{array}$$

$$m_{32} = -\frac{2.334}{5.666} = -0.412; \qquad R_3 \leftarrow R_3 - 0.412R_2$$

$$m_{42} = \frac{1.666}{5.666} = 0.294; \qquad R_4 \leftarrow R_4 + 0.294R_2$$

$$\begin{array}{c} \begin{matrix} x_1 & x_2 & x_3 & x_4 & b \end{matrix} \\ \left[\begin{array}{cccc|c} 3 & 2 & 1 & -4 & 5 \\ 0 & -5.666 & 1.667 & 2.332 & 16.335 \\ 0 & 0 & -5.354 & 7.707 & -19.065 \\ 0 & 0 & 0.823 & 8.354 & 12.467 \end{array}\right] \end{array}$$

$$m_{43} = \frac{0.823}{5.354} = 0.154; \qquad R_4 \leftarrow R_4 + 0.154R_3$$

$$\begin{array}{c} \begin{matrix} x_1 & x_2 & x_3 & x_4 & b \end{matrix} \\ \left[\begin{array}{cccc|c} 3 & 2 & 1 & -4 & 5 \\ 0 & -5.666 & 1.667 & 2.335 & 16.335 \\ 0 & 0 & -5.354 & 7.707 & -19.065 \\ 0 & 0 & 0 & 9.541 & 9.531 \end{array}\right] \end{array}$$

Back Substitution:

$$x_4 = \frac{9.531}{9.541} = 0.999$$

$$x_3 = (-19.065 - 7.707 \times 0.999)/(-5.354) = 4.999$$

$$x_2 = (16.335 - 2.332 \times 0.999 - 1.667 \times 4.999)/(-5.666) = -1.001$$

$$x_1 = (5 + 4 \times 0.999 - 1 \times 4.999 - 2 \times (-1.001)/3 = 2.000$$

Final answer rounded to two places of decimal is:

$$x_1 = 2.00,\ x_2 = -1.00,\ x_3 = 5.00 \text{ and } x_4 = 1.00.$$

(exact answer is: $x_1 = 2,\ x_2 = -1,\ x_3 = 5,\ x_4 = 1$)

Note: If we compute $a_{21} \leftarrow a_{21} + m_{21} \times a_{11}$, we get $a_{21} = 1 - 0.333 \times 3 = 0.0001$, which is not zero although we have assumed it to be so. Similar argument holds for the other zeros produced. Thus the equations are satisfied approximately. If higher accuracy is required, we may work with more decimal places and in double precision on a computer, if needed.

Example 2.2

Solve the following system of simultaneous equations by Gaussian elimination method,

$$4.3x_1 - 3.5x_1 - 1.2x_3 = 10.90$$

$$18.4x_1 + 2.1x_2 - x_3 = 7.80$$

$$7.2x_1 + 1.8x_2 + 3.4x_3 = 23.22$$

Perform computations up to 2 decimal places. Round the final answer to one decimal.

Solution

$$\begin{array}{cccc} x_1 & x_2 & x_3 & b \end{array}$$

$$\left[\begin{array}{ccc|c} 4.3 & -3.5 & -1.2 & 10.9 \\ 18.4 & 2.1 & -1.0 & 7.8 \\ 7.2 & 1.8 & 3.4 & 23.22 \end{array}\right]$$

$$m_{21} = -\frac{18.4}{4.3} = -4.28; \qquad R_2 \leftarrow R_2 - 4.28R_1.$$

$$m_{31} = -\frac{7.2}{4.3} = -1.67; \qquad R_3 \leftarrow R_3 - 1.67R_1.$$

$$\begin{array}{cccc} x_1 & x_2 & x_3 & b \end{array}$$

$$\left[\begin{array}{ccc|c} 4.3 & -3.5 & -1.2 & 10.90 \\ 0 & 17.08 & 4.14 & -38.85 \\ 0 & 7.64 & 5.40 & 5.02 \end{array}\right]$$

$$m_{32} = -\frac{7.64}{17.08} = -0.45; \qquad R_3 \leftarrow R_3 - 0.45R_2.$$

$$\begin{array}{ccc|c} x_1 & x_2 & x_3 & \boldsymbol{b} \end{array}$$

$$\left[\begin{array}{ccc|c} 4.3 & -3.5 & -1.2 & 10.90 \\ 0 & 17.08 & 4.14 & -38.85 \\ 0 & 0 & 3.54 & 22.50 \end{array}\right]$$

$x_3 = 6.36;\ x_2 = -3.82;\ x_1 = 1.20.$

Rounding to one decimal,

$$x_1 = 1.2,\ x_2 = -3.8,\ x_3 = 6.4$$

(exact answer is $x_1 = 1.2,\ x_2 = -3.8,\ x_3 = 6.3$)

Example 2.3

Solve the following system of simultaneous equations by Gaussian elimination method:

$$0.12x_2 + 0.15x_3 = 0.33$$

$$0.56x_1 + 0.40x_2 - 0.18x_3 = 2.34$$

$$0.20x_1 + 0.71x_2 + x_3 = 2.04$$

Compute up to 3 places of decimal and round the final answer to two decimal places.

Solution

$$\begin{array}{ccc|c} x_1 & x_2 & x_3 & \boldsymbol{b} \end{array}$$

$$\left[\begin{array}{ccc|c} 0 & 0.12 & 0.15 & 0.33 \\ 0.56 & 0.40 & -0.18 & 2.34 \\ 0.20 & 0.71 & 1.0 & 2.04 \end{array}\right]$$

Here first equation can not be treated as pivotal equation as $m_{21} \to \infty$. Therefore, we interchange it with the next row which has non-zero as its first element. Thus we write,

$$\begin{array}{ccc|c} x_1 & x_2 & x_3 & \boldsymbol{b} \end{array}$$

$$\left[\begin{array}{ccc|c} 0.56 & 0.40 & -0.18 & 2.34 \\ 0 & 0.12 & -0.15 & 0.33 \\ 0.20 & 0.71 & 1.0 & 2.04 \end{array}\right]$$

$m_{21} = 0,\qquad \mathrm{R}_2 \leftarrow \mathrm{R}_2$ (remains unchanged)

$$m_{31} = -\frac{0.20}{0.56} = -0.357; \qquad \mathrm{R}_3 \leftarrow \mathrm{R}_3 - 0.357\mathrm{R}_1$$

$$\begin{array}{cccc} x_1 & x_2 & x_3 & b \\ \end{array}$$

$$\left[\begin{array}{ccc|c} 0.56 & 0.40 & -0.18 & 2.34 \\ 0 & 0.12 & 0.15 & 0.33 \\ 0 & 0.567 & 1.064 & 1.205 \end{array}\right]$$

$$m_{32} = -\frac{0.567}{0.12} = -4.725; \qquad R_3 \leftarrow R_3 - 4.725R_2$$

$$\begin{array}{cccc} x_1 & x_2 & x_3 & b \\ \end{array}$$

$$\left[\begin{array}{ccc|c} 0.56 & 0.40 & -0.18 & 2.34 \\ 0 & 0.12 & -0.15 & 0.33 \\ 0 & 0 & 0.355 & -0.354 \end{array}\right]$$

$x_3 = -0.997$; $x_2 = 3.996$; $x_1 = 1.004$

After rounding to two decimals

$$x_1 = 1.00,\ x_2 = 4.00,\ x_3 = -1.00$$

(exact values are $x_1 = 1,\ x_2 = 4,\ x_3 = -1$)

Example 2.4

Solve the following system of simultaneous equations using Gaussian elimination with row interchanges:

$$0.3x_1 + 2.6x_2 + 1.3x_3 = 7.65$$

$$8.3x_1 + 8.2x_2 + 5.6x_3 = 43.17$$

$$12.7x_2 + 3.5x_2 + 7.4x_3 = 49.68$$

Compute up to two significant figures after decimal. Also solve the system without interchanging the rows.

Solution (by row interchanges)

$$\begin{array}{cccc} x_1 & x_2 & x_3 & b \\ \end{array}$$

$$\left[\begin{array}{ccc|c} 0.3 & 2.6 & 1.3 & 7.65 \\ 8.3 & 8.2 & 5.6 & 43.17 \\ 12.7 & 3.5 & 7.4 & 49.68 \end{array}\right]$$

Since $|12.7|$ is largest in absolute value in the first column, we interchange first row with third.

$$\begin{array}{ccc|c} x_1 & x_2 & x_3 & \boldsymbol{b} \end{array}$$
$$\left[\begin{array}{ccc|c} 12.7 & 3.5 & 7.4 & 49.68 \\ 8.3 & 8.2 & 5.6 & 43.17 \\ 0.3 & 2.6 & 1.3 & 7.65 \end{array}\right]$$

$$m_{21} = -\frac{8.3}{12.7} = -0.65, \qquad R_2 \leftarrow R_2 - 0.65R_1$$

$$m_{31} = -\frac{0.3}{12.7} = -0.024, \text{ (two significant figures after decimal)}$$

$R_3 \leftarrow R_3 - 0.024R_1$

$$\begin{array}{ccc|c} x_1 & x_2 & x_3 & \boldsymbol{b} \end{array}$$
$$\left[\begin{array}{ccc|c} 12.7 & 3.5 & 7.4 & 49.68 \\ 0 & 5.92 & 0.79 & 10.88 \\ 0 & 2.52 & 1.12 & 6.46 \end{array}\right]$$

Since $|5.92|$ is larger than $|2.52|$, there is no need to interchange rows.

$$m_{32} = -\frac{2.52}{5.92} = -0.42; \qquad R_3 \leftarrow R_3 - 0.42R_2$$

$$\begin{array}{ccc|c} x_1 & x_2 & x_3 & \boldsymbol{b} \end{array}$$
$$\left[\begin{array}{ccc|c} 12.7 & 3.5 & 7.4 & 49.68 \\ 0 & 5.92 & 0.79 & 10.88 \\ 0 & 0 & 0.79 & 1.89 \end{array}\right]$$

By back substitution,

$$x_3 = 2.39,\ x_2 = 1.52,\ x_1 = 2.10$$

(exact solution is $x_1 = 2.1$, $x_2 = 1.5$, $x_3 = 2.4$)

Solution (without row interchanges)

$$\begin{array}{ccc|c} x_1 & x_2 & x_3 & \boldsymbol{b} \end{array}$$
$$\left[\begin{array}{ccc|c} 0.3 & 2.6 & 1.3 & 7.65 \\ 8.3 & 8.2 & 5.6 & 43.17 \\ 12.7 & 3.5 & 7.4 & 49.68 \end{array}\right]$$

$$m_{21} = -\frac{8.3}{0.3} = -27.67; \qquad R_2 \leftarrow R_2 - 27.67R_1$$

$$m_{31} = -\frac{12.7}{0.3} = -42.33; \qquad R_3 \leftarrow R_3 - 42.33R_1$$

$$\begin{array}{cccc} x_1 & x_2 & x_3 & b \end{array}$$
$$\left[\begin{array}{ccc|c} 0.3 & 2.6 & 1.3 & 7.65 \\ 0 & -63.74 & -30.37 & -168.50 \\ 0 & -106.56 & -47.63 & -274.14 \end{array}\right]$$

$$m_{32} = -\frac{106.56}{63.74} = -1.67; \qquad R_3 \leftarrow R_3 - 1.67R_2$$

$$\begin{array}{cccc} x_1 & x_2 & x_3 & b \end{array}$$
$$\left[\begin{array}{ccc|c} 0.3 & 2.6 & 1.3 & 7.65 \\ 0 & -63.74 & -30.37 & -168.50 \\ 0 & 0 & 3.09 & 7.26 \end{array}\right]$$

$$x_3 = 2.35,\ x_2 = 1.52,\ x_1 = 2.14$$

(exact solution is, $x_1 = 2.1$, $x_2 = 1.5$, $x_3 = 2.4$)

2.14 LU Decomposition/Factorisation

A square matrix A can be decomposed/factorised (with conditions) into a product of two matrices L and U, i.e., A = LU, where L and U are lower triangular and upper triangular matrices respectively. We will discuss three methods in this regard (*i*) By Gaussian Elimination method (*ii*) Crout's method and (*iii*) Cholesky's method.

2.14.1 By Gaussian elimination method

We assume that no interchange of rows has taken place at any stage. Let us recall that in Gaussian Elimination method, matrix A is reduced to an upper triangular matrix U by a series of elementary row operations/transformations. For a 4×4 matrix, these transformations may be expressed in following way,

$$L_3L_2L_1A = U \tag{2.13}$$

where L_1, L_2 and L_3 are lower triangular matrices as,

$$L_1 = \begin{bmatrix} 1 & 0 & 0 & 0 \\ m_{21} & 1 & 0 & 0 \\ m_{31} & 0 & 1 & 0 \\ m_{41} & 0 & 0 & 1 \end{bmatrix}, \; L_2 = \begin{bmatrix} 1 & 0 & 0 & 0 \\ 0 & 1 & 0 & 0 \\ 0 & m_{32} & 1 & 0 \\ 0 & m_{42} & 0 & 1 \end{bmatrix}, \; L_3 = \begin{bmatrix} 1 & 0 & 0 & 0 \\ 0 & 1 & 0 & 0 \\ 0 & 0 & 1 & 0 \\ 0 & 0 & m_{43} & 1 \end{bmatrix}.$$

The matrix A is original matrix and U the final upper triangular, i.e.,

$$A = \begin{bmatrix} a_{11} & a_{12} & a_{13} & a_{14} \\ a_{21} & a_{22} & a_{23} & a_{24} \\ a_{31} & a_{32} & a_{33} & a_{34} \\ a_{41} & a_{42} & a_{43} & a_{44} \end{bmatrix}, U = \begin{bmatrix} a_{11} & a_{12} & a_{13} & a_{14} \\ 0 & a_{22}^{(1)} & a_{23}^{(1)} & a_{24}^{(1)} \\ 0 & 0 & a_{33}^{(2)} & a_{34}^{(2)} \\ 0 & 0 & 0 & a_{44}^{(3)} \end{bmatrix}.$$

From (2.13) we have,

$$A = (L_3L_2L_1)^{-1}U = L_1^{-1}L_2^{-1}L_3^{-1}\, U.$$

It is easy to see that the inverses of L_1, L_2 and L_3 can be obtained simply by changing the signs of multipliers. It can also be shown by the logic of elementary transformations that,

$$L = L_1^{-1}L_2^{-1}L_3^{-1} = \begin{bmatrix} 1 & 0 & 0 & 0 \\ -m_{21} & 1 & 0 & 0 \\ -m_{31} & 0 & 1 & 0 \\ -m_{41} & 0 & 0 & 1 \end{bmatrix} \begin{bmatrix} 1 & 0 & 0 & 0 \\ 0 & 1 & 0 & 0 \\ 0 & -m_{32} & 1 & 0 \\ 0 & -m_{42} & 0 & 1 \end{bmatrix} \begin{bmatrix} 1 & 0 & 0 & 0 \\ 0 & 1 & 0 & 0 \\ 0 & 0 & 1 & 0 \\ 0 & 0 & -m_{43} & 1 \end{bmatrix}$$

$$= \begin{bmatrix} 1 & 0 & 0 & 0 \\ -m_{21} & 1 & 0 & 0 \\ -m_{31} & -m_{32} & 1 & 0 \\ -m_{41} & -m_{42} & -m_{43} & 1 \end{bmatrix}. \tag{2.14}$$

Thus we have the desired decomposition $A = LU$ where L is given by (2.14).

The above may be generalised for $n \times n$ matrix straight away.

After reducing the matrix to LU form we can solve the system of equations $A\boldsymbol{x} = \boldsymbol{b}$ in the following manner:

$$A\boldsymbol{x} = LU\boldsymbol{x} = \boldsymbol{b} \tag{2.15}$$

Put $\mathrm{U}\boldsymbol{x} = \boldsymbol{y}$, in (2.15); (2.16)

then, $\mathrm{L}\boldsymbol{y} = \boldsymbol{b}$, from (2.15). (2.17)

Solve (2.17) for $\boldsymbol{y}$, then solve (2.16) to obtain the required solution $\boldsymbol{x}$. The system (2.17) will be solved in a top-down manner while (2.16) in a bottom-up manner. Evidently the pivotal element should not be zero or very small since it will give rise to infinitely large multiplier, an element of L.

2.14.2 Crout's method

Let us consider each of L, U and A as 4×4 matrices, i.e.,

$$\begin{bmatrix} 1 & 0 & 0 & 0 \\ l_{21} & 1 & 0 & 0 \\ l_{31} & l_{32} & 1 & 0 \\ l_{41} & l_{42} & l_{43} & 1 \end{bmatrix} \begin{bmatrix} u_{11} & u_{12} & u_{13} & u_{14} \\ 0 & u_{22} & u_{23} & u_{24} \\ 0 & 0 & u_{33} & u_{34} \\ 0 & 0 & 0 & u_{44} \end{bmatrix} = \begin{bmatrix} a_{11} & a_{12} & a_{13} & a_{14} \\ a_{21} & a_{22} & a_{23} & a_{24} \\ a_{31} & a_{32} & a_{33} & a_{34} \\ a_{41} & a_{42} & a_{43} & a_{44} \end{bmatrix} \tag{2.18}$$

On the L.H.S. there are 16 unknowns, 6 elements of L and 10 of U. In order to evaluate them, we perform the product LU and equate, term by term, with the 16 elements of A. This is done in a systematic manner given as under:

Compute row-wise in the following order by equating the corresponding terms of LU and A:

(*i*) $l_{11} = 1;\ u_{11},\ u_{12},\ u_{13},\ u_{14}$

(*ii*) $l_{21},\ l_{22} = 1;\ u_{22},\ u_{23},\ u_{24}$

(*iii*) $l_{31},\ l_{32},\ l_{33} = 1,\ u_{33},\ u_{34}$

(*iv*) $l_{41},\ l_{42},\ l_{43},\ l_{44} = 1,\ u_{44}.$

Note: In practical computations we put $l_{11} = l_{22} = l_{33} = l_{44} = 1$ to start with. Thus we get,

(*i*) $u_{11} = a_{11},\ u_{12} = a_{12},\ u_{13} = a_{13},\ u_{14} = a_{14}$

(ii) $l_{21} \cdot u_{11} = a_{21}$ giving $l_{21} = a_{21}/u_{11};\ u_{11}(= a_{11})$ should not be zero.

$l_{21}a_{12} + u_{22} = a_{22}$, giving $u_{22} = a_{22} - l_{21}a_{12};\ u_{22}$ should not be zero.

$l_{21}a_{13} + u_{23} = a_{23}$, giving $u_{23} = a_{23} - l_{21} \cdot a_{13}$

$l_{21}a_{14} + u_{24} = a_{24}$, giving $u_{24} = a_{24} - l_{21}a_{14}$

(*iii*) $l_{31}u_{11} = a_{31}$, giving $l_{31} = a_{31}/u_{11}$

$l_{31}u_{12} + l_{32}u_{22} = a_{32}$, giving $l_{32} = (a_{32} - l_{31}u_{12})/u_{22}$; $u_{22} \neq 0$.

$l_{31}u_{13} + l_{32}u_{23} + u_{33} = a_{33}$, giving $u_{33} = a_{33} - (l_{31}u_{13} + l_{32}u_{23})$; u_{33} should not be zero.

$l_{31}u_{14} + l_{32}u_{24} + u_{34} = a_{34}$, giving $u_{34} = a_{34} - (l_{31}u_{14} + l_{32}u_{24})$

(*iv*) $l_{41}u_{11} = a_{41}$, giving $l_{41} = a_{41}/u_{11}$

$l_{41}u_{12} + l_{42}u_{22} = a_{42}$, giving $l_{42} = (a_{42} - l_{41}u_{12})/u_{22}$

$l_{41}u_{13} + l_{42}u_{23} + l_{43}u_{33} = a_{43}$, giving $l_{43} = (a_{43} - l_{41}u_{13} - l_{42}u_{23})/u_{33}$; $u_{33} \neq 0$.

$l_{41}u_{14} + l_{42}u_{24} + l_{43}u_{34} + u_{44} = a_{44}$, giving $u_{44} = a_{44} - (l_{41}u_{14} + l_{42}u_{24} + l_{43}u_{34})$.

It may be noted that each element is computed before it is used.
The above can be generalised when A is an $(n \times n)$ matrix:

$$l_{ii} = 1, \qquad i = 1(1)n.$$

$$u_{1j} = a_{1j}, \quad j = 1(1)n.$$

For the ith row, $i = 2(1)n$,

$$l_{ij} = [a_{ij} - \{l_{i1}u_{1j} + l_{i2}u_{2j} + \ldots + l_{i,i-1}u_{i-1,j}\}]/u_{jj}$$

$$= \left[a_{ij} - \sum_{k=1}^{j-1} l_{ik}u_{kj}\right] \Big/ u_{jj}, \quad j = 1(1)i-1. \tag{2.19}$$

$$u_{ij} = a_{ij} - \sum_{k=1}^{i-1} l_{ik}u_{kj}, \quad j = i(1)n. \tag{2.20}$$

Note: (*i*) If it is required that A = LU where L is a Lower triangular and U a Unit upper triangular matrix, then we can proceed as follows:

Put $u_{ii} = 1,\ i = 1(1)n.$

Compute the elements column-wise, i.e., elements of kth column of U and elements of kth column of L in that order for $k = 1(1)n$.

(*ii*) Matrix A can also be reduced to the form

A = LDU

where L and U are unit lower and unit upper triangular matrices and D, a diagonal matrix. Let A be a matrix of order n. It is easy to see that there will be $\frac{(n-1)n}{2}$ elements in the matrices L and U each and n elements in D. Thus there are total n^2 unknowns in the product LDU which can be matched with the n^2 elements of A. In practice however, we can reduce the matrix A to LU form where L is a unit lower triangular matrix and U, an upper triangular. We choose the elements of the diagonal matrix D to be the diagonal elements of U, i.e., $d_{ii} = u_{ii}$ and divide each element of U in the ith row by u_{ii}. It may again be emphasised that value of none of the dividing element u_{ii} should be zero; otherwise the process will break down.

This method is also attributed to Doolittle.

2.14.3 Cholesky's method

The Cholesky's method deals with a special case when the given matrix is symmetric and positive definite. If A is a symmetric matrix, $A = A^T$ then it can be expressed as product of two matrices L and L^T where L is a lower triangular matrix, i.e.,

$$A = LL^T \text{ or } (U^TU). \tag{2.21}$$

For example, for a 4×4 matrix,

$$\underbrace{\begin{bmatrix} l_{11} & 0 & 0 & 0 \\ l_{21} & l_{22} & 0 & 0 \\ l_{31} & l_{32} & l_{33} & 0 \\ l_{41} & l_{42} & l_{43} & l_{44} \end{bmatrix}}_{\text{L or } U^T} \underbrace{\begin{bmatrix} l_{11} & l_{21} & l_{31} & l_{41} \\ 0 & l_{22} & l_{32} & l_{42} \\ 0 & 0 & l_{33} & l_{43} \\ 0 & 0 & 0 & l_{44} \end{bmatrix}}_{L^T \text{ or U}} = \underbrace{\begin{bmatrix} a_{11} & a_{21} & a_{31} & a_{41} \\ a_{21} & a_{22} & a_{32} & a_{42} \\ a_{31} & a_{32} & a_{33} & a_{43} \\ a_{41} & a_{42} & a_{43} & a_{44} \end{bmatrix}}_{\text{A}} \tag{2.22}$$

In (2.22), there are 10 elements in matrix L and 16 in A. But due to symmetry six elements above the diagonal are same as below the diagonal, i.e., $a_{ij} = a_{ji}$, $i \neq j$. Thus equating the corresponding terms of LL^T and A gives the elements of L. When A is an $(n \times n)$ matrix, the number of elements to be evaluated are $n(n+1)/2$.

We evaluate the elements of L by equating the corresponding elements column-wise (or row-wise since LL^T is a symmetric matrix). Let us say, we are equating column-wise.

For each value of $j = 1(1)n$, compute for $i = j(1)n$, the values of l_{ij} by the following formulae:

$$\text{for } i = j, \quad l_{jj} = \left[a_{jj} - \sum_{k=1}^{j-1} l_{jk}^2 \right]^{1/2}, \tag{2.23}$$

$$\text{for } i > j, \quad l_{ij} = \left[a_{ij} - \sum_{k=1}^{j-1} l_{ik} \cdot l_{jk} \right] \Big/ l_{jj} \tag{2.24}$$

In case of 4×4 matrix, i.e., (2.22) we compute the elements of L using formulae (2.23) and (2.24) in the following manner:

For $j = 1$:

$$i = j = 1 \Rightarrow l_{11} = \sqrt{a_{11}},$$

$$i = 2(j = 1) \Rightarrow l_{21} = a_{21}/l_{11},$$

$$i = 3(j = 1) \Rightarrow l_{31} = a_{31}/l_{11},$$

$$i = 4(j = 1) \Rightarrow l_{41} = a_{41}/l_{11}.$$

For $j = 2$:

$$i = j = 2 \Rightarrow l_{22} = [(a_{22} - l_{21})^2]^{1/2},$$

$$i = 3(j = 2) \Rightarrow l_{32} = (a_{23} - l_{31} \cdot l_{21})/l_{22},$$

$$i = 4(j = 2) \Rightarrow l_{42} = (a_{24} - l_{41} \cdot l_{21})/l_{22}.$$

For $j = 3$:

$$i = j = 3 \Rightarrow l_{33} = [a_{33} - (l_{31}^2 + l_{32}^2)]^{1/2},$$

$$i = 4(j = 3) \Rightarrow l_{43} = [a_{43} - (l_{41} l_{31} + l_{42} \cdot l_{32})]/l_{33}.$$

For $j = 4$:

$$i = j = 4 \Rightarrow l_{44} = [a_{44} - (l_{41}^2 + l_{42}^2 + l_{43}^2)]^{1/2}.$$

Example 2.5

Reduce the following matrix A to LU form by Gaussian elimination where L is a Unit Lower Triangular matrix and U is an upper triangular; also solve the system $A\boldsymbol{x} = \boldsymbol{b}$ where,

$$A = \begin{bmatrix} 3 & 2 & 1 & -4 \\ 1 & -5 & 2 & 1 \\ 5 & 1 & -3 & 2 \\ 2 & 3 & 1 & 5 \end{bmatrix},$$

$$\boldsymbol{x}^{\mathrm{T}} = (x_1 \;\; x_2 \;\; x_3 \;\; x_4) \text{ and } \boldsymbol{b}^{\mathrm{T}} = (5 \;\; 18 \;\; -4 \;\; 11).$$

Solution It is same coefficient matrix as in Example 2.1.

From (2.14) and (2.13),

$$\mathrm{L} = \begin{bmatrix} 1 & 0 & 0 & 0 \\ 0.333 & 1 & 0 & 0 \\ 1.667 & 0.412 & 1 & 0 \\ 0.667 & -0.294 & -0.0154 & 1 \end{bmatrix} \text{ and } \mathrm{U} = \begin{bmatrix} 3 & 2 & 1 & -4 \\ 0 & -5.666 & 1.667 & 2.332 \\ 0 & 0 & -5.354 & 7.707 \\ 0 & 0 & 0 & 9.541 \end{bmatrix}$$

In $\mathrm{LU}\boldsymbol{x} = \boldsymbol{b}$, putting $\mathrm{U}\boldsymbol{x} = \boldsymbol{y}$, the system becomes $\mathrm{L}\boldsymbol{y} = \boldsymbol{b}$, where $\boldsymbol{y}^{\mathrm{T}} = (y_1 \;\; y_2 \;\; y_3 \;\; y_4)$. Solving $\mathrm{L}\boldsymbol{y} = \boldsymbol{b}$ by forward substitution gives,

$$y_1 = 5,\; y_2 = 16.335,\; y_3 = -19.065,\; y_4 = 9.531$$

Finally on solving $\mathrm{U}\boldsymbol{x} = \boldsymbol{y}$, by backward substitution, we get,

$$x_4 = 0.999,\; x_3 = 4.999,\; x_2 = -1.001,\; x_1 = 2.000$$

Example 2.6

Decompose the following matrix A to LU form by Crout's method where L is a unit lower triangular and U an upper triangular matrix,

$$\mathrm{A} = \begin{bmatrix} 3 & 2 & 1 & -4 \\ 1 & -5 & 2 & 1 \\ 5 & 1 & -3 & 2 \\ 2 & 3 & 1 & 5 \end{bmatrix}.$$

Solution Let us assume,

$$\begin{bmatrix} 1 & 0 & 0 & 0 \\ l_{21} & 1 & 0 & 0 \\ l_{31} & l_{32} & 1 & 0 \\ l_{41} & l_{42} & l_{43} & 1 \end{bmatrix} \begin{bmatrix} u_{11} & u_{12} & u_{13} & u_{14} \\ 0 & u_{22} & u_{23} & u_{24} \\ 0 & 0 & u_{33} & u_{34} \\ 0 & 0 & 0 & u_{44} \end{bmatrix} = \begin{bmatrix} 3 & 2 & 1 & -4 \\ 1 & -5 & 2 & 1 \\ 5 & 1 & -3 & 2 \\ 2 & 3 & 1 & 5 \end{bmatrix}.$$

1st row:

$$u_{11} = 3,\; u_{12} = 2,\; u_{13} = 1,\; u_{14} = -4$$

2nd row:

$$l_{21} = a_{21}/u_{11} = 1/3 = 0.333$$

$$u_{22} = a_{22} - l_{21} \cdot u_{12} = -5 - 0.333 \times 2 = -5.666$$

$$u_{23} = a_{23} - l_{21} \cdot u_{13} = 2 - 0.333 \times 1 = 1.667$$

$$u_{24} = a_{24} - l_{21} \cdot u_{14} = 1 - 0.333 \times (-4) = 2.332$$

3rd row:

$$l_{31} = a_{31}/u_{11} = 5/3 = 1.667$$

$$l_{32} = (a_{32} - l_{31} \cdot u_{12})/u_{22} = (1 - 1.667 \times 2)/(-5.667) = 0.412$$

$$u_{33} = a_{33} - (l_{31}u_{13} + l_{32}u_{23}) = -3 - (1.667 \times 1 + 0.412 \times 1.667) = -5.354$$

$$u_{34} = a_{34} - (l_{31}u_{14} + l_{32}u_{24}) = 2 - (1.667 \times (-4) + 0.412 \times 2.332) = 7.707$$

4th row:

$$l_{41} = a_{41}/u_{11} = 2/3 = 0.667$$

$$l_{42} = (a_{42} - l_{41} \cdot u_{12})/u_{22} = (3 - 0.667 \times 2)/(-5.667) = -0.294$$

$$\begin{aligned} l_{43} &= [a_{43} - (l_{41} \cdot u_{13} + l_{42} \cdot u_{23})]/u_{33} \\ &= [1 - (0.667 \times 1 - 0.294 \times 1.667)]/(-5.354) = -0.154 \end{aligned}$$

$$\begin{aligned} u_{44} &= a_{44} - (l_{41} \cdot u_{14} + l_{42} \cdot u_{24} + l_{43} \cdot u_{34}) \\ &= 5 - (-0.667 \times 4 - 0.294 \times 2.332 - 0.154 \times 7.707) \\ &= 9.540 \end{aligned}$$

$$L = \begin{bmatrix} 1 & 0 & 0 & 0 \\ 0.333 & 1 & 0 & 0 \\ 1.667 & 0.412 & 1 & 0 \\ 0.667 & -0.294 & -0.154 & 1 \end{bmatrix}; U = \begin{bmatrix} 3 & 2 & 1 & -4 \\ 0 & -5.666 & 1.667 & 2.332 \\ 0 & 0 & -5.354 & 7.707 \\ 0 & 0 & 0 & 9.540 \end{bmatrix}$$

Note: This example is same as 2.5. There is a difference of .001 in u_{44} due to rounding, which may be expected.

Example 2.7

Using the lower and upper triangular matrices obtained in Example 2.6, reduce the matrix A to LDU form where L and U are unit lower and unit upper triangular matrices respectively.

Further, using LDU, express A = LU where L is a lower triangular and U a unit upper triangular matrix.

Solution In order to express A = LDU, we divide each row of U by its diagonal element, i.e., we divide the ith row by u_{ii}. The elements of D are given by $d_{ii} = u_{ii}$. Thus we have,

$$\text{LDU} = \underset{\text{L}}{\begin{bmatrix} 1 & 0 & 0 & 0 \\ 0.333 & 1 & 0 & 0 \\ 1.667 & 0.412 & 1 & 0 \\ 0.667 & -0.294 & -0.154 & 1 \end{bmatrix}} \underset{\text{D}}{\begin{bmatrix} 3 & 0 & 0 & 0 \\ 0 & -5.666 & 0 & 0 \\ 0 & 0 & -5.354 & 0 \\ 0 & 0 & 0 & 9.540 \end{bmatrix}}$$

$$\underset{\text{U}}{\begin{bmatrix} 1 & 0.667 & 0.333 & -1.333 \\ 0 & 1 & -0.294 & -0.412 \\ 0 & 0 & 1 & -1.439 \\ 0 & 0 & 0 & 1 \end{bmatrix}}$$

To further reduce it to A = LU form, we have to multiply the elements of the jth column of unit lower triangular matrix by $d_{jj},\ j = 1(1)4$. Thus we get

$$\text{LU} = \begin{bmatrix} 3 & 0 & 0 & 0 \\ 0.999 & -5.666 & 0 & 0 \\ 5.001 & -2.334 & -5.354 & 0 \\ 2.001 & 1.666 & 0.824 & 9.540 \end{bmatrix} \begin{bmatrix} 1 & 0.667 & 0.333 & -1.333 \\ 0 & 1 & -0.294 & -0.412 \\ 0 & 0 & 1 & -1.439 \\ 0 & 0 & 0 & 1 \end{bmatrix}$$

2.14.4 Reduction to PA = LU

In Sec. 2.14.1, we had reduced matrix A to LU form by Gaussian Elimination assuming there were no interchanges of rows at any stage. But if interchanges are employed, then the matrix A can be reduced to the form PA = LU where P is a permutation matrix, i.e., a matrix obtained from a unit matrix with its rows interchanged (permuted). We illustrate the reduction by considering a 4×4 matrix A.

Let I_{pq} denote a unit matrix with its pth and q^{th} rows interchanged and that matrix A has been reduced to an upper triangular matrix U under following transformations,

$$\text{L}_3\text{I}_{34}\text{L}_2\text{I}_{24}\text{L}_1\text{I}_{13}\text{A} = \text{U}, \tag{2.25}$$

where L_1, L_2 and L_3 are unit lower triangular matrices as given in (2.14).
First we should note that $\text{I}_{pq}^{-1} = \text{I}_{pq}$, i.e., I_{pq} is an involutary matrix.
From (2.25) we can write,

$$I_{34}I_{24}I_{13}A = \text{I}_{34} \cdot \text{I}_{24}\text{L}_1^{-1}\text{I}_{24}^{-1}\text{L}_2^{-1}\text{I}_{34}^{-1}\text{L}_3^{-1}\text{U}$$

$$= I_{34}I_{24}L_1^{-1}I_{24}L_2^{-1}I_{34}L_3^{-1}U$$

$$= [I_{34}\{(I_{24}L_1^{-1}I_{24})L_2^{-1}\}I_{34}]L_3^{-1}U. \tag{2.26}$$

We see that,

$$I_{24}L_1^{-1} = \begin{bmatrix} 1 & 0 & 0 & 0 \\ -m_{41} & 0 & 0 & 1 \\ -m_{31} & 0 & 1 & 0 \\ -m_{21} & 1 & 0 & 0 \end{bmatrix}, \quad I_{24}L_1^{-1}I_{24} = \begin{bmatrix} 1 & 0 & 0 & 0 \\ -m_{41} & 1 & 0 & 0 \\ -m_{31} & 0 & 1 & 0 \\ -m_{21} & 0 & 0 & 1 \end{bmatrix}$$

Now, the bracketed term,

$$(I_{24}L_1^{-1}I_{24})_{L_2}^{-1} = \begin{bmatrix} 1 & 0 & 0 & 0 \\ -m_{41} & 1 & 0 & 0 \\ -m_{31} & 0 & 1 & 0 \\ -m_{21} & 0 & 0 & 1 \end{bmatrix} \begin{bmatrix} 1 & 0 & 0 & 0 \\ 0 & 1 & 0 & 0 \\ 0 & -m_{32} & 1 & 0 \\ 0 & -m_{42} & 0 & 1 \end{bmatrix}$$

$$= \begin{bmatrix} 1 & 0 & 0 & 0 \\ -m_{41} & 1 & 0 & 0 \\ -m_{31} & -m_{32} & 1 & 0 \\ -m_{21} & -m_{42} & 0 & 1 \end{bmatrix}.$$

The term within square bracket in (2.26) becomes,

$$I_{34}\{(I_{24}L_1^{-1}I_{24})L_2{}^{-1}\}I_{34} = \begin{bmatrix} 1 & 0 & 0 & 0 \\ -m_{41} & 1 & 0 & 0 \\ -m_{21} & -m_{42} & 1 & 0 \\ -m_{31} & -m_{32} & 0 & 1 \end{bmatrix}$$

Finally the right side of (2.26) will be,

$$\left[I_{34}\left\{\left(I_{24}L_1^{-1}I_{24}\right)L_2{}^{-1}\right\}I_{34}\right]L_3^{-1}U = \begin{bmatrix} 1 & 0 & 0 & 0 \\ -m_{41} & 1 & 0 & 0 \\ -m_{21} & -m_{42} & 1 & 0 \\ -m_{31} & -m_{32} & -m_{43} & 1 \end{bmatrix} U.$$

The left side of (2.26) will be,

$$I_{34}I_{24}I_{13}A = \begin{bmatrix} 1 & 0 & 0 & 0 \\ 0 & 1 & 0 & 0 \\ 0 & 0 & 0 & 1 \\ 0 & 0 & 1 & 0 \end{bmatrix} \begin{bmatrix} 1 & 0 & 0 & 0 \\ 0 & 0 & 0 & 1 \\ 0 & 0 & 1 & 0 \\ 0 & 1 & 0 & 0 \end{bmatrix} \begin{bmatrix} 0 & 0 & 1 & 0 \\ 0 & 1 & 0 & 0 \\ 1 & 0 & 0 & 0 \\ 0 & 0 & 0 & 1 \end{bmatrix} A$$

$$= \begin{bmatrix} 0 & 0 & 1 & 0 \\ 0 & 0 & 0 & 1 \\ 0 & 1 & 0 & 0 \\ 1 & 0 & 0 & 0 \end{bmatrix} A$$

Thus we have the reduction,

$$PA = LU \tag{2.27}$$

where P is the permutation matrix and L is a unit lower triangular matrix given as

$$P = \begin{bmatrix} 0 & 0 & 1 & 0 \\ 0 & 0 & 0 & 1 \\ 0 & 1 & 0 & 0 \\ 1 & 0 & 0 & 0 \end{bmatrix}, \quad L = \begin{bmatrix} 1 & 0 & 0 & 0 \\ -m_{41} & 1 & 0 & 0 \\ -m_{21} & -m_{42} & 1 & 0 \\ -m_{31} & -m_{32} & -m_{43} & 1 \end{bmatrix}. \tag{2.28}$$

The reduction $PA = LU$ can also be expressed as $A = P^{-1}LU$ where $P^{-1} = I_{13}I_{24}I_{34}$.

In practice, to compute the permutation matrix P, we maintain the record of the interchanges of rows of the unit matrix by a one-dimensional array (vector) storing the row number in it. In the above case of 4×4 matrix we may have an array row number (1 : 4) having values 1, 2, 3, 4 initially. We change these numbers according to change of rows. In the present case I_{13} will change the order to 3, 2, 1, 4. Then I_{24} will change it to 3, 4, 1, 2 and finally I_{34} will change it to 3, 4, 2, 1. Then the four rows of matrix P can be written as row number 3, 4, 2 and 1 of the unit matrix. For complete details of computational procedure see Example 2.8.

Example 2.8

Reduce the following matrix A in the form PA = LU, by Gaussian Elimination method with partial pivoting, where L is a unit lower triangular and U an upper triangular matrix; P is a permutation matrix.

$$A = \begin{bmatrix} 2 & 5 & 1 & 8 \\ 1 & 6 & 3 & 5 \\ 7 & 2 & 6 & 3 \\ 4 & 8 & 1 & 2 \end{bmatrix}$$

Also solve the system of equations $A\boldsymbol{x} = \boldsymbol{b}$ where $\boldsymbol{b}^T = (5\ 11\ 14\ 19)$. Compute up to two decimals only.

Solution

Row number of unit matrix — Cols of multipliers — Transformed matrix — $\boldsymbol{b}$

$$\begin{bmatrix} 1 \\ 2 \\ 3 \\ 4 \end{bmatrix} \quad - \quad \begin{bmatrix} 2 & 5 & 1 & 8 \\ 1 & 6 & 3 & 5 \\ 7 & 2 & 6 & 3 \\ 4 & 8 & 1 & 2 \end{bmatrix} \quad \begin{bmatrix} 5 \\ 11 \\ 14 \\ 19 \end{bmatrix}$$

Interchanging 1st row by 3rd,

$$\begin{bmatrix} 3 \\ 2 \\ 1 \\ 4 \end{bmatrix} \quad - \quad \begin{bmatrix} 7 & 2 & 6 & 3 \\ 1 & 6 & 3 & 5 \\ 2 & 5 & 1 & 8 \\ 4 & 8 & 1 & 2 \end{bmatrix} \quad \begin{bmatrix} 14 \\ 11 \\ 5 \\ 19 \end{bmatrix}$$

Elimination of coeff. of x_1;

$$\begin{bmatrix} 3 \\ 2 \\ 1 \\ 4 \end{bmatrix} \quad \begin{matrix} 1 \\ -0.14 \\ -0.28 \\ -0.57 \end{matrix} \quad \begin{bmatrix} 7 & 2 & 6 & 3 \\ 0 & 5.72 & 2.16 & 4.58 \\ 0 & 4.44 & -0.68 & 7.16 \\ 0 & 6.86 & -2.42 & 0.29 \end{bmatrix} \begin{bmatrix} 14 \\ 9.04 \\ 1.08 \\ 11.02 \end{bmatrix}$$

Interchanging 2nd by 4th row,

$$\begin{bmatrix} 3 \\ 4 \\ 1 \\ 2 \end{bmatrix} \quad \begin{matrix} 1 \\ -0.57 \\ -0.14 \\ -0.14 \end{matrix} \quad \begin{bmatrix} 7 & 2 & 6 & 3 \\ 0 & 6.86 & -2.42 & 0.29 \\ 0 & 4.44 & -0.68 & 7.16 \\ 0 & 5.72 & 2.16 & 4.58 \end{bmatrix} \begin{bmatrix} 14 \\ 11.02 \\ 1.08 \\ 9.04 \end{bmatrix}$$

Elimination of coeff. of x_2;

$$\begin{bmatrix} 3 \\ 4 \\ 1 \\ 2 \end{bmatrix} \quad \begin{matrix} 1 & 0 \\ -0.57 & 1 \\ -0.28 & -0.65 \\ -0.14 & -0.83 \end{matrix} \quad \begin{bmatrix} 7 & 2 & 6 & 3 \\ 0 & 6.86 & -2.42 & 0.29 \\ 0 & 0 & 0.89 & 6.97 \\ 0 & 0 & 4.17 & 4.34 \end{bmatrix} \begin{bmatrix} 14 \\ 11.02 \\ -6.08 \\ -0.11 \end{bmatrix}$$

Interchanging 3rd by 4th row,

$$\begin{bmatrix} 3 \\ 4 \\ 2 \\ 1 \end{bmatrix} \quad \begin{matrix} 1 & 0 \\ -0.57 & 1 \\ -0.14 & -0.83 \\ -0.28 & -0.65 \end{matrix} \quad \begin{bmatrix} 7 & 2 & 6 & 3 \\ 0 & 6.86 & -2.42 & 0.29 \\ 0 & 0 & 4.17 & 4.34 \\ 0 & 0 & 0.89 & 6.97 \end{bmatrix} \begin{bmatrix} 14 \\ 11.02 \\ -0.11 \\ -6.08 \end{bmatrix}$$

Eliminating coeff. of x_3,

$$\begin{bmatrix} 3 \\ 4 \\ 2 \\ 1 \end{bmatrix} \quad \begin{matrix} 1 & 0 & 0 \\ -0.57 & 1 & 0 \\ -0.14 & -0.83 & 1 \\ -0.28 & -0.65 & -0.21 \end{matrix} \quad \begin{bmatrix} 7 & 2 & 6 & 3 \\ 0 & 6.86 & -2.42 & 0.29 \\ 0 & 0 & 4.17 & 4.34 \\ 0 & 0 & 0 & 6.06 \end{bmatrix} \begin{bmatrix} 14 \\ 11.02 \\ 0.11 \\ -6.06 \end{bmatrix}$$

Solving by back-substitution gives

$$x_4 = -1.00,\ x_3 = 1.01,\ x_2 = 2.00,\ x_1 = 0.99$$

(exact solution is $x_1 = 1,\ x_2 = 2,\ x_3 = 2,\ x_3 = 1,\ x_4 = -1$)

$$P = \begin{bmatrix} 0 & 0 & 1 & 0 \\ 0 & 0 & 0 & 1 \\ 0 & 1 & 0 & 0 \\ 1 & 0 & 0 & 0 \end{bmatrix} \begin{matrix} \leftarrow 3^{\text{rd}} \text{ row of I} \\ \leftarrow 4^{\text{th}} \text{ row of I} \\ \leftarrow 2^{\text{nd}} \text{ row of I} \\ \leftarrow 1^{\text{st}} \text{ row of I} \end{matrix} \ ;\ PA = \begin{bmatrix} 7 & 2 & 6 & 3 \\ 4 & 8 & 1 & 2 \\ 1 & 6 & 3 & 5 \\ 2 & 5 & 1 & 8 \end{bmatrix};$$

$$L = \begin{bmatrix} 1 & 0 & 0 & 0 \\ 0.57 & 1 & 0 & 0 \\ 0.14 & 0.83 & 1 & 0 \\ 0.28 & 0.65 & 0.21 & 1 \end{bmatrix};U = \begin{bmatrix} 7 & 2 & 6 & 3 \\ 0 & 6.86 & -2.42 & 0.29 \\ 0 & 0 & 4.17 & 4.34 \\ 0 & 0 & 0 & 6.06 \end{bmatrix}.$$

Note: Multipliers can be stored in space occupied by zeros in the appropriate column.

2.15 Gauss–Jordan (or Jordan's) Method

Let us again consider solving the system of equations (2.2), i.e., $\mathrm{A}\boldsymbol{x} = \boldsymbol{b}$. In the Gaussian elimination method we reduce the system to an upper triangular form by making elementary operations on the rows below the pivotal row and then obtain the solution by the process of back-substitution. In the Jordan's method operations are made on all the rows below the pivotal row as well as on the rows above it such that the system reduces to an Identity matrix form, $\mathrm{I}\boldsymbol{x} = \boldsymbol{b}'$ and the solution is straight away given by the transformed right side $\boldsymbol{b}'$. In order to achieve this, the pivotal row is divided by the pivotal element throughout, thus making the pivot as unity. Then operations are made, choosing suitable multipliers so that all the elements below the pivot as well as above it are reduced to zero. Of course same operations are to be made simultaneously on the right side. For solving an $n \times n$ system, there will be $(n-1)$ stages and in each stage there will be $(n-1)$ steps to be performed to make $(n-1)$ elements zero. The computations are made on the augmented matrix aug $(\mathrm{A}/\boldsymbol{b})$ as shown in the following example.

Example 2.9

Solve by Jordan's method $\mathrm{A}\boldsymbol{x} = \boldsymbol{b}$ where

$$\mathrm{A} = \begin{bmatrix} 2 & 5 & 1 & 8 \\ 1 & 6 & 3 & 5 \\ 7 & 2 & 6 & 3 \\ 4 & 8 & 1 & 2 \end{bmatrix}; \boldsymbol{b}^{\mathrm{T}} = (5\ \ 11\ \ 14\ \ 19)$$

Solution

$$\left[\begin{array}{cccc|c} 2 & 5 & 1 & 8 & 5 \\ 1 & 6 & 3 & 5 & 11 \\ 7 & 2 & 6 & 3 & 14 \\ 4 & 8 & 1 & 2 & 19 \end{array}\right] \xrightarrow{\mathrm{R}_1 \leftarrow \mathrm{R}_1/2} \left[\begin{array}{cccc|c} 1 & 2.5 & 0.5 & 4 & 2.5 \\ 1 & 6 & 3 & 5 & 11 \\ 7 & 2 & 6 & 3 & 14 \\ 4 & 8 & 1 & 2 & 19 \end{array}\right]$$

$\mathrm{R}_2 \leftarrow \mathrm{R}_2 - \mathrm{R}_1$; $\mathrm{R}_3 \leftarrow \mathrm{R}_3 - 7\mathrm{R}_1$; $\mathrm{R}_4 \leftarrow \mathrm{R}_4 - 4\mathrm{R}_1$

$$\left[\begin{array}{cccc|c} 1 & 2.5 & 0.5 & 4 & 2.5 \\ 0 & 3.5 & 2.5 & 1 & 8.5 \\ 0 & -15.5 & 2.5 & -25 & -3.5 \\ 0 & -2 & -1 & -14 & 9 \end{array}\right] \xrightarrow{\mathrm{R}_2 \leftarrow \mathrm{R}_2/3.5} \left[\begin{array}{cccc|c} 1 & 2.5 & 0.5 & 4 & 2.5 \\ 0 & 1 & 0.71 & 0.28 & 2.43 \\ 0 & -15.5 & 2.5 & -25 & -3.5 \\ 0 & -2 & -1 & -14 & 9 \end{array}\right]$$

$R_1 \leftarrow R_1 - 2.5R_2$; $R_3 \leftarrow R_3 + 15.5R_2$; $R_4 \leftarrow R_4 + 2R_2$

$$\left[\begin{array}{cccc|c} 1 & 0 & -1.28 & 3.3 & -3.58 \\ 0 & 1 & 0.71 & 0.28 & 2.43 \\ 0 & 0 & 13.50 & -20.66 & 34.16 \\ 0 & 0 & 0.42 & -13.44 & 13.86 \end{array}\right] \xrightarrow{R_3 \leftarrow R_3/13.50} \left[\begin{array}{cccc|c} 1 & 0 & -1.28 & 3.3 & -3.58 \\ 0 & 1 & 0.71 & 0.28 & 2.43 \\ 0 & 0 & 1 & -1.53 & 2.53 \\ 0 & 0 & 0.42 & -13.44 & 13.86 \end{array}\right]$$

$R_1 \leftarrow R_1 + 1.28R_1$; $R_2 \leftarrow R_2 - 0.71R_3$; $R_4 \leftarrow R_4 - 0.42R_3$

$$\left[\begin{array}{cccc|c} 1 & 0 & 0 & 1.34 & -0.34 \\ 0 & 1 & 0 & 1.37 & 0.63 \\ 0 & 0 & 1 & -1.53 & 12.53 \\ 0 & 0 & 0 & -12.80 & 12.80 \end{array}\right] \xrightarrow{R_4 \leftarrow R_4/(-12.8)} \left[\begin{array}{cccc|c} 1 & 0 & 0 & 1.34 & -0.34 \\ 0 & 1 & 0 & 1.37 & 0.63 \\ 0 & 0 & 1 & -1.53 & 2.53 \\ 0 & 0 & 0 & 1 & -1.00 \end{array}\right]$$

$R_1 \leftarrow R_1 - 1.34R_4$; $R_2 \leftarrow R_2 - 1.37R_4$; $R_3 \leftarrow R_3 - 1.53R_4$

$$\left[\begin{array}{cccc|c} 1 & 0 & 0 & 0 & 1.00 \\ 0 & 1 & 0 & 0 & 2.00 \\ 0 & 0 & 1 & 0 & 1.00 \\ 0 & 0 & 0 & 1 & -1.00 \end{array}\right].$$

$x_1 = 1,\ x_2 = 2,\ x_3 = 1,\ x_4 = -1$

(correct answer is $x_1 = 1,\ x_2 = 2,\ x_3 = 1,\ x_4 = -1$)

Note: Change of rows may be performed if required.

2.16 Tridiagonal System

The solution of second order boundary value problems by numerical method reduces to solving a system of linear equations which is tridiagonal in nature. A (4×4) tri-diagonal system of equations may be written as,

$$\begin{bmatrix} a_{11} & a_{12} & 0 & 0 \\ a_{21} & a_{22} & a_{23} & 0 \\ 0 & a_{32} & a_{33} & a_{34} \\ 0 & 0 & a_{43} & a_{44} \end{bmatrix} \begin{bmatrix} x_1 \\ x_2 \\ x_3 \\ x_4 \end{bmatrix} = \begin{bmatrix} b_1 \\ b_2 \\ b_3 \\ b_4 \end{bmatrix}. \tag{2.29}$$

From Gaussian elimination or Crout's method, it is easy to see that the coefficient matrix in (2.29) can be reduced to LU where L and U have following forms,

$$\begin{bmatrix} a_{11} & a_{12} & 0 & 0 \\ a_{21} & a_{22} & a_{23} & 0 \\ 0 & a_{32} & a_{33} & a_{34} \\ 0 & 0 & a_{43} & a_{44} \end{bmatrix} = \begin{bmatrix} 1 & 0 & 0 & 0 \\ l_{21} & 1 & 0 & 0 \\ 0 & l_{32} & 1 & 0 \\ 0 & 0 & l_{43} & 1 \end{bmatrix} \begin{bmatrix} u_{11} & u_{12} & 0 & 0 \\ 0 & u_{22} & u_{23} & 0 \\ 0 & 0 & u_{33} & u_{34} \\ 0 & 0 & 0 & u_{44} \end{bmatrix}.$$

We can easily find the elements of L and U as given below:

$$u_{11} = a_{11},\ u_{12} = a_{12};\ a_{11} \neq 0.$$

$$l_{21} = a_{21}/a_{11};\ u_{22} = a_{22} - \frac{a_{12} \cdot a_{21}}{u_{11}};\ u_{23} = a_{23}.$$

$$l_{32} = \frac{a_{32}}{a_{22}};\ u_{33} = a_{33} - \frac{a_{23} \cdot a_{32}}{u_{22}};\ u_{34} = a_{34}.$$

$$l_{43} = \frac{a_{43}}{a_{33}};\ u_{44} = a_{44} - \frac{a_{34} \cdot a_{43}}{u_{33}}.$$

We solve $A\boldsymbol{x} = \boldsymbol{b}$ or $LU\boldsymbol{x} = \boldsymbol{b}$ by putting $U\boldsymbol{x} = \boldsymbol{y}$ so that $L\boldsymbol{y} = \boldsymbol{b}$ has to be solved first for $\boldsymbol{y}$ giving,

$$y_1 = b_1,\ y_2 = b_2 - \frac{a_{21}}{u_{11}} y_1;\ y_3 = b_3 - \frac{a_{32}}{u_{22}} y_2,\ y_4 = b_4 - \frac{a_{43}}{u_{33}} y_3.$$

Further, on solving $U\boldsymbol{x} = \boldsymbol{y}$, we get,

$$x_4 = y_4/u_{44};\ x_3 = (y_3 - a_{34}x_4)/u_{33}$$

$$x_2 = (y_2 - a_{23} \cdot x_3)/u_{22};\ x_1 = (y_1 - a_{12} \cdot x_2)/u_{11}.$$

Thus we are required to compute the following in that order:

(1) $u_{11} \quad u_{22} \quad u_{33} \quad u_{44}$

(2) $y_1 \quad y_2 \quad y_3 \quad y_4$

(3) $x_4 \quad x_3 \quad x_2 \quad x_1$

However in order to save memory space in the computer, we may store the elements of A by three one-dimensional arrays for storing the diagonal, subdiagonal and super diagonal elements. Let us now consider an $n \times n$ tridiagonal system which is stored as follows:

$$\begin{bmatrix} b_1 & c_1 & 0 & \cdots & 0 \\ a_2 & b_2 & c_2 & \cdots & 0 \\ 0 & a_3 & b_3 & c_3 \cdots & 0 \\ \vdots & & & \ddots & \\ 0 & 0 & & a_n & b_n \end{bmatrix} \begin{bmatrix} x_1 \\ x_2 \\ x_3 \\ \vdots \\ x_n \end{bmatrix} \begin{bmatrix} r_1 \\ r_2 \\ r_3 \\ \vdots \\ r_n \end{bmatrix} \tag{2.30}$$

Compute the following in that order:

(1) $u_1 = b_1;\ u_i = b_i - \dfrac{a_i \cdot c_{i-1}}{u_{i-1}},\ i = 2(1)n.$ (2.30a)

(2) $y_1 = r_1;\ y_i = r_i - \dfrac{a_i y_{i-1}}{u_{i-1}},\ i = 2(1)n.$ (2.30b)

(3) $x_n = y_n/u_n;\ x_i = (y_i - c_i x_{i+1})/u_i,\ i = n-1(-1)1.$ (2.30c)

It may be noted that instead of $n \times n$, only $3n-2$ computer locations are required for storing the elements of the coefficient matrix.

Example 2.10

Solve the following tridiagonal system of equations,

$$\begin{aligned} 1.98x_1 - 1.01x_2 &= 0.985 \\ 0.98x_1 - 1.98x_2 + 1.02x_3 &= 0.01 \\ 0.97x_2 - 1.98x_3 + 1.03x_4 &= 0.015 \\ 0.96x_3 - 1.98x_4 &= -1.540 \end{aligned}$$

Compute up to four places of decimal.

Solution

$$u_1 = 1.98$$

$$u_2 = -1.98 - \frac{0.98 \times (-1.01)}{1.98} = -1.4801$$

$$u_3 = -1.98 - \frac{0.97 \times (1.02)}{-1.4801} = -1.3115$$

$$u_4 = -1.98 - \frac{0.96 \times (1.03)}{-1.3115} = -1.2260$$

$$y_1 = 0.985$$

$$y_2 = 0.01 - \frac{0.98}{1.98} \times 0.985 = -0.4775$$

$$y_3 = 0.015 - \frac{0.97 \times (-0.4775)}{-1.4801} = -0.2979$$

$$y_4 = -1.540 - \frac{0.96 \times (-0.2979)}{-1.3115} = -1.7580$$

$$x_4 = \frac{-1.7580}{-1.2260} = 1.4339$$

$$x_3 = (-0.2979 - 1.03 \times 1.4339)/(-1.3115) = 1.3533$$

$$x_2 = (-0.4775 - 1.02 \times 1.3533)/(-1.4801) = 1.2552$$

$$x_1 = \{0.985 - (-1.01) \times 1.2552\}/1.98 = 1.1378$$

Note: The above equations have arrived at in solving a differential equation described in Chapter 7.

2.17 Inversion of Matrix

We can find the solution of $Ax = b$ using A^{-1} as $x = A^{-1}b$, although it will not be wise to first compute A^{-1} and then form $A^{-1}b$. However, if we have got a software/subroutine/procedure for A^{-1}, we can use it easily on any number of different b vectors. We had also said in Sec 2.9 that to compute A^{-1} by adj $A/|A|$ is most uneconomical computation-wise. We had also mentioned in Sec 2.7 that A^{-1} can be found by solving $Ax = b$, taking b as different columns of a unit matrix. That is, solution x of $Ax = I_k$, where I_k is the kth column of the unit/identity matrix I, renders the kth column of A^{-1}. Thus we have to work with the augmented matrix aug (A/I). When matrix A is of order 4, we have,

$$\text{aug}\,(A/I) = \begin{array}{c} \begin{matrix} x_1 & x_2 & x_3 & x_4 & I_1 & I_2 & I_3 & I_4 \end{matrix} \\ \left[\begin{array}{cccc|cccc} a_{11} & a_{12} & a_{13} & a_{14} & 1 & 0 & 0 & 0 \\ a_{21} & a_{22} & a_{23} & a_{24} & 0 & 1 & 0 & 0 \\ a_{31} & a_{32} & a_{33} & a_{34} & 0 & 0 & 1 & 0 \\ a_{41} & a_{42} & a_{43} & a_{44} & 0 & 0 & 0 & 1 \end{array}\right] \end{array}.$$

All the methods discussed before can be employed to solve the 4×4 system with four right hand sides, namely (*i*) Gauss Elimination (*ii*) LU decomposition and (*iii*) Jordan's. See Examples 2.11 and 2.12.

Example 2.11

Find the inverse of the following matrix A by Gaussian elimination method, where

$$A = \begin{bmatrix} 4.3 & -3.5 & -1.2 \\ 18.4 & 2.1 & -1.0 \\ 7.2 & 1.8 & 3.4 \end{bmatrix}.$$

Using A^{-1}, compute $\boldsymbol{x}$ form $A\boldsymbol{x} = \boldsymbol{b}$, when $\boldsymbol{b}^T = (10.90 \quad 7.80 \quad 23.22)$. Compute up to 2 decimals only.

Solution We apply Gaussian elimination method taking $\boldsymbol{b}$ as columns of unit matrix I, i.e., considering augmented matrix:

$$\left[\begin{array}{ccc|ccc} 4.3 & -3.5 & -1.2 & 1 & 0 & 0 \\ 18.4 & 2.1 & -1.0 & 0 & 1 & 0 \\ 7.2 & 1.8 & 3.4 & 0 & 0 & 1 \end{array}\right]$$

$R_2 \leftarrow R_2 - 4.28R_1$; $R_3 \leftarrow R_3 - 1.67R_1$

$$\left[\begin{array}{ccc|ccc} 4.3 & -3.5 & -1.2 & 1 & 0 & 0 \\ 0 & 17.08 & 4.14 & -4.28 & 1 & 0 \\ 0 & 7.64 & 5.40 & -1.67 & 0 & 1 \end{array}\right]$$

$R_3 \leftarrow R_3 - 0.45R_2$

$$\left[\begin{array}{ccc|ccc} 4.3 & -3.5 & -1.2 & 1 & 0 & 0 \\ 0 & 17.08 & 4.14 & -4.28 & 1 & 0 \\ 0 & 0 & 3.54 & 0.26 & -0.45 & 1 \end{array}\right]$$

Solving by back-substitution, for 3 right sides, we get the respective 3 columns of A^{-1}, i.e.,

$$A^{-1} = \begin{bmatrix} 0.03 & 0.04 & 0.02 \\ -0.27 & 0.09 & -0.07 \\ 0.07 & -0.13 & 0.28 \end{bmatrix}.$$

For given $\boldsymbol{b}^T$,

$$\boldsymbol{x} = \begin{bmatrix} 0.03 & 0.04 & 0.02 \\ -0.27 & 0.09 & -0.07 \\ 0.07 & -0.13 & 0.28 \end{bmatrix} \begin{bmatrix} 10.90 \\ 7.80 \\ 23.22 \end{bmatrix} = \begin{bmatrix} 1.10 \\ -3.87 \\ 6.25 \end{bmatrix}.$$

correct answer is: $\boldsymbol{x}^T = (1.2 \quad -3.8 \quad 6.3)$

Note: This is the same problem as in Example 2.2. The values obtained here are inferior to those obtained earlier. Firstly, the computations have been done with only two places of decimal. Secondly the number of arithmetic operations performed in the present approach are far more than in Example 2.2; in computing A^{-1} only, the number of computations are three times. Then, further multiplications of A^{-1} and $\boldsymbol{b}$ adds to more rounding errors. Therefore this aprroach is useful only when there are a number of right sides to be operated upon by A^{-1}.

Example 2.12

Find the inverse of matrix A, given in Example 2.11 by Gauss–Jordan's method. Compute up to 2 decimals only.

Solution

$$\left[\begin{array}{ccc|ccc} 4.3 & -3.5 & -1.2 & 1 & 0 & 0 \\ 18.4 & 2.1 & -1.0 & 0 & 1 & 0 \\ 7.2 & 1.8 & 3.4 & 0 & 0 & 1 \end{array}\right]$$

$R_1 \leftarrow R_1/4.3$

$$\left[\begin{array}{ccc|ccc} 1 & -0.81 & -0.28 & 0.23 & 0 & 0 \\ 18.4 & 2.1 & -1.0 & 0 & 1 & 0 \\ 7.2 & 1.8 & 3.4 & 0 & 0 & 1 \end{array}\right]$$

$R_2 \leftarrow R_2 - 18.4R_1;\; R_3 \leftarrow R_3 - 7.2R_1$

$$\left[\begin{array}{ccc|ccc} 1 & -0.81 & -0.28 & 0.23 & 0 & 0 \\ 0 & 17.00 & 4.15 & -4.23 & 1 & 0 \\ 0 & 7.63 & 5.42 & -1.66 & 0 & 1 \end{array}\right]$$

$R_2 \leftarrow R_2/17.00$

$$\left[\begin{array}{ccc|ccc} 1 & -0.81 & -0.28 & 0.23 & 0 & 0 \\ 0 & 1 & 0.24 & -0.25 & 0.06 & 0 \\ 0 & 7.63 & 5.42 & -1.66 & 0 & 1 \end{array}\right]$$

$R_1 \leftarrow R_1 + 0.81R_2;\; R_3 \leftarrow R_3 - 7.63R_2$

$$\left[\begin{array}{ccc|ccc} 1 & 0 & -0.08 & 0.03 & 0.05 & 0 \\ 0 & 1 & 0.24 & -0.25 & 0.06 & 0 \\ 0 & 0 & 3.59 & 0.25 & -0.46 & 1 \end{array}\right]$$

$R_3 \leftarrow R_3/3.59$

$$\left[\begin{array}{ccc|ccc} 1 & 0 & -0.08 & 0.03 & 0.05 & 0 \\ 0 & 1 & 0.24 & -0.25 & 0.06 & 0 \\ 0 & 0 & 1 & 0.07 & -0.13 & 0.28 \end{array}\right]$$

$R_1 \leftarrow R_1 + 0.08R_3$; $R_2 \leftarrow R_2 - 0.24R_3$

$$\left[\begin{array}{ccc|ccc} 1 & 0 & 0 & 0.04 & 0.04 & 0.02 \\ 0 & 1 & 0 & -0.27 & 0.09 & -0.07 \\ 0 & 0 & 1 & 0.07 & -0.13 & 0.28 \end{array}\right]$$

The inverse is given by,

$$A^{-1} = \begin{bmatrix} 0.04 & 0.04 & 0.02 \\ -0.27 & 0.09 & -0.07 \\ 0.07 & -0.13 & 0.28 \end{bmatrix}.$$

2.18 Number of Arithmetic Operations in Gaussian Elimination

Let us count the number of arithmetic operations, i.e., multiplication/division and addition/subtraction performed in the solution of $A\boldsymbol{x} = \boldsymbol{b}$ by Gaussian elimination method when A is a $n \times n$ matrix. In first stage we compute the multipliers m_{i1}, $i = 2(1)n$, i.e., $(n-1)$ divisions. Then $(n-1)$ elements of A and one of b in the first row are multiplied by m_{21} and added to the second row of the augmented matrix. Thus n multiplications are performed with each of m_{i1}, i.e., $(n-1)n$ multiplications. Hence the number of multiplications/divisions performed in the first stage are $(n-1)+(n-1)n = n^2-1$. This process is repeated in the subsequent stages on the reduced matrix until $n = 2$. Thus in the Gaussian elimination,

Total number of multiplications/divisions $= \sum_{k=2}^{n}(k^2-1) = \dfrac{n(2n^2+3n-5)}{6}$

Similarly total number of additions/subtractions $= \sum_{k=2}^{n} k(k-1) = \dfrac{n(n^2-1)}{6}$

In the process of back-substitution:

Total number of multiplications/divisions $= \sum_{k=1}^{n} k = \frac{n(n+1)}{2}$

Total number of additions/subtractions $= \sum_{k=1}^{n-1} k = \frac{(n-1)n}{2}$

In the solution of system of equations

$$\text{Number of multiplications/divisions} = \frac{n(2n^2+3n-5)}{6} + \frac{n(n+1)}{2} = \frac{n^3+3n^2-n}{3}$$

$$\text{Number additions/subtractions} = \frac{n(n^2-1)}{6} + \frac{n(n-1)}{2} = \frac{n^3+3n^2-4n}{6}.$$

2.19 Eigenvalues and Eigenvectors

The system of equations $\mathrm{A}\boldsymbol{x} = \boldsymbol{b}$ is called homogeneous if the vector $\boldsymbol{b}$ is a null vector, i.e., $\boldsymbol{b} = 0$. Let us suppose we have a set of n homogeneous equations in n unknowns, $\mathrm{A}\boldsymbol{x} = \mathbf{0}$. If rank of the coefficient matrix A is n then we know that there exists a unique solution since

$$r(\mathrm{A}) = r\{\text{aug}\ (\mathrm{A}\ |\mathbf{0})\} = n.$$

It is also obvious that this solution is $\boldsymbol{x} = \mathbf{0}$, i.e., $x_1 = x_2 = \ldots = x_n = 0$, as it satisfies the system. This solution is called 'trivial' and is of no use in practical problems (as will be seen later). Therefore, for the system $\mathrm{A}\boldsymbol{x} = \mathbf{0}$ to have a non-trivial solution it is necessary that $|\mathrm{A}| = 0$. It is a different matter however, that there will exist more than one non-trivial solutions, since $r(\mathrm{A}) < n$.

In solving differential equations, particular type of homogeneous equations occur. As an example, a 3×3 system is given as follows:

$$\begin{aligned} a_{11}x_1 + a_{12}x_2 + a_{13}x_3 &= \lambda x_1 \\ a_{21}x_1 + a_{22}x_2 + a_{23}x_3 &= \lambda x_2 \\ a_{31}x_1 + a_{32}x_2 + a_{33}x_3 &= \lambda x_3, \end{aligned} \tag{2.31}$$

where λ is a parameter to be determined such that the system (2.31) possesses a non-trival solution.

The system (2.31) may be re-written as,

$$(a_{11} - \lambda)x_1 + a_{12}x_2 + a_{13}x_3 = 0$$

$$a_{21}x_1 + (a_{22} - \lambda)x_2 + a_{23}x_3 = 0 \tag{2.32}$$

$$a_{31}x_1 + a_{32}x_2 + (a_{33} - \lambda)x_3 = 0.$$

Now, for the system (2.32) to possess a non-trivial solution, the determinant of the coefficient matrix should vanish, i.e.,

$$\begin{vmatrix} a_{11} - \lambda & a_{12} & a_{13} \\ a_{21} & a_{22} - \lambda & a_{23} \\ a_{31} & a_{32} & a_{33} - \lambda \end{vmatrix} = 0. \tag{2.33}$$

Expanding the determinant will give a cubic in λ, say

$$P_3(\lambda) \equiv \lambda^3 + a_2\lambda^2 + a_1\lambda + a_0 = 0. \tag{2.34}$$

The equation (2.34) is called characteristic equation of matrix A where,

$$\mathrm{A} = \begin{bmatrix} a_{11} & a_{12} & a_{13} \\ a_{21} & a_{22} & a_{23} \\ a_{31} & a_{32} & a_{33} \end{bmatrix}. \tag{2.35}$$

The characteristic equation (2.34) will have three roots of λ, say λ_1, λ_2, λ_3. These roots which may be simple or multiple (repeated) are called characteristic roots, latent roots or eigenvalues of matrix A.

For each of the eigenvalues λ_1, λ_2 and λ_3 there can be found a corresponding solution vector from (2.32), say $\boldsymbol{x}_1$, $\boldsymbol{x}_2$ and $\boldsymbol{x}_3$ which may be represented as,

$$\boldsymbol{x}_1 = \begin{bmatrix} x_{11} \\ x_{21} \\ x_{31} \end{bmatrix}, \boldsymbol{x}_2 = \begin{bmatrix} x_{12} \\ x_{22} \\ x_{32} \end{bmatrix}, \boldsymbol{x}_3 = \begin{bmatrix} x_{13} \\ x_{23} \\ x_{33} \end{bmatrix}. \tag{2.36}$$

It should be remembered that the rank of the coefficient matrix in (2.33) is at the most two which means the number of linearly independent equations will be either two or one. If there are two linearly independent equations but three unknowns, then an arbitrary value may be assigned to one of the variables, say $x_3 = k$ and the other two values may be found in terms of k. Similarly if the rank is 1 then we can assign arbitrary values to two unknowns, e.g., $x_2 = 0$, $x_3 = k$ or $x_2 = k$, $x_3 = 0$ and the value of x_1 can be computed for both of these cases. This shows that the eigenvectors are not unique; they can be scaled/normalised in any manner we like.

The above analysis can be generalised for an $n \times n$ matrix A when the eigenvalue problem may be defined by,

$$\mathrm{A}\boldsymbol{x} = \lambda \boldsymbol{x} \text{ or } (\mathrm{A} - \lambda \mathrm{I})\boldsymbol{x} = 0, \tag{2.37}$$

where λ is a parameter to be determined.

The existence of a non-trial solution to (2.37) implies that,

$$|\mathrm{A} - \lambda \mathrm{I}| = 0 \tag{2.38}$$

which will give rise to a polynomial equation of degree n in λ, i.e., the characteristic equation, say,

$$P_n(\lambda) \equiv \lambda^n + a_{n-1}\lambda^{n-1} + a_{n-2}\lambda^{n-2} + \ldots + a_1\lambda + a_0 = 0. \tag{2.39}$$

The roots of the equation (2.39), $\lambda = \lambda_i$, $i = 1(1)n$ will be the eigenvalues/latent roots/ characteristic roots of matrix A. Using the property for roots of a polynomial, we have

$$\text{Sum of the roots} = \sum_{i=1}^{n} \lambda_i = \sum a_{ii} = \mathrm{Tr}\ (\mathrm{A}) = \text{Trace of A.} \tag{2.39a}$$

$$\text{Product of the roots} = \prod_{i=1}^{n} \lambda_i = |\mathrm{A}| = \text{Determinant of A.} \tag{2.39b}$$

The eigenvector (or latent vector) corresponding to the eigenvalue (or e.value for short) λ_i will be given by solving the system (2.37) for $\boldsymbol{x}$ after substituting the value of λ as λ_i and will be denoted as,

$$\boldsymbol{x}_i^{\mathrm{T}} = (x_{1i} \quad x_{2i} \quad x_{3i} \quad \ldots \quad x_{ni}),\ i = 1(1)n. \tag{2.40}$$

It may be noted from (2.37) that if $\boldsymbol{x}$ is an e.vector corresponding to the e.value λ of A, then k$\boldsymbol{x}$ is also an e.vector as it also satisfies (2.37), i.e.,

$$\mathrm{Ak}\boldsymbol{x} = \lambda \mathrm{k}\boldsymbol{x} \text{ or } \mathrm{kA}\boldsymbol{x} = \mathrm{k}\lambda \boldsymbol{x} \text{ or } \mathrm{A}\boldsymbol{x} = \lambda \boldsymbol{x}.$$

This shows that e.vector $\boldsymbol{x}$ is not unique; it can be normalised or scaled as desired.

Further, if an e.value is a simple (not repeated) root, then there will be $(n-1)$ linearly independent equations corresponding to it to be solved. As there are n unknowns, one may be chosen arbitrarily. If an e.value has a multiplicity of m (repeated m times) then there will be at least $(n-m)$ linearly independent equations, i.e., $r(\mathrm{A} - \lambda \mathrm{I}) \geq n - m$. If however, $r(\mathrm{A} - \lambda \mathrm{I}) = n - m$, then there will be m unknowns free to be assigned arbitrary values.

As a simple case, we can assign value 1 to any one of the m unknowns and 0 to the other $m-1$ unknowns and then solving (2.37) for the remaining $n-m$ unknowns. By assigning arbitrary value 1 to different unknowns, we can get n different e.vectors for an e.value of multiplicity m. In fact, the number of independent e.vectors corresponding to an e.value of multiplicity m will be k, $1 \leq k \leq m$ where $k = n - r(\mathrm{A} - \lambda \mathrm{I})$. Nevertheless there are methods to find m independent e.vectors. See [1], [3] and [7].

Example 2.13

Find the eigenvalues and eigenvectors of the following matrix,

$$\mathrm{A} = \begin{bmatrix} 5 & -2 & 0 \\ 1 & 2 & -3 \\ 1 & -2 & 4 \end{bmatrix}$$

Solution The characteristic equation is given by

$$|\mathrm{A} - \lambda \mathrm{I}| = \begin{vmatrix} 5-\lambda & -2 & 0 \\ 1 & 2-\lambda & -3 \\ 1 & -2 & 4-\lambda \end{vmatrix} = 0$$

or $(5-\lambda)[(2-\lambda)(4-\lambda)-6]+2[4-\lambda+3]=0$

or $\lambda^3 - 11\lambda^2 + 34\lambda - 24 = 0$

Factorising gives, $(\lambda-1)(\lambda-4)(\lambda-6)=0$.

The eigenvalues are,

$$\lambda_1 = 1,\ \lambda_2 = 4,\ \lambda_3 = 6.$$

The eigenvector corresponding to $\lambda_1 = 1$ is given by the solution of the system of equations,

$$(\mathrm{A} - \lambda_1 \mathrm{I})\boldsymbol{x} = 0$$

Putting $\lambda_1 = 1$, $(\mathrm{A} - \mathrm{I})\boldsymbol{x} = 0$.

$$4x_1 - 2x_2 + 0x_3 = 0 \qquad \text{....}(i)$$

$$x_1 + x_2 - 3x_3 = 0 \qquad \text{....}(ii)$$

$$x_1 - 2x_2 + 3x_3 = 0 \qquad \text{....}(iii)$$

Subtracting (*iii*) from (*ii*) gives,

$$x_2 = 2x_3;$$

Adding (*ii*) and (*iii*) gives equation (*i*); it is linearly dependent. It gives,

$$2x_1 = x_2.$$

Thus the solution is,

$$2x_1 = x_2 = 2x_3.$$

Choosing $x_3 = 1$, gives $x_2 = 2$ and $x_1 = 1$.
The eigenvector $\boldsymbol{x}_1$ corresponding to $\lambda = \lambda_1 = 1$ is,

$$\boldsymbol{x}_1^{\mathrm{T}} = (1 \;\; 2 \;\; 1) \text{ or } (0.5 \;\; 1 \;\; 0.5).$$

The eigenvector corresponding to $\lambda_2 = 4$ is given by the solution of,

$$x_1 - 2x_2 + 0x_3 = 0 \qquad ....(i)$$

$$x_1 - 2x_2 - 3x_3 = 0 \qquad ....(ii)$$

$$x_1 - 2x_2 + 0x_3 = 0. \qquad ....(iii)$$

(*ii*) and (*iii*) give $x_3 = 0$ and (*i*) give $x_1 = 2x_2$. Thus, choosing $x_2 = 1$, we get,

$$\boldsymbol{x}_2^{\mathrm{T}} = (2 \;\; 1 \;\; 0) \text{ or } (1 \;\; 0.5 \;\; 0).$$

Similarly, the equations corresponding to $\lambda = \lambda_3 = 6$ are given by,

$$-x_1 - 2x_2 + 0x_3 = 0 \qquad ....(i)$$

$$x_1 - 4x_2 - 3x_3 = 0 \qquad ....(ii)$$

$$x_1 - 2x_2 - 2x_3 = 0. \qquad ....(iii)$$

From (*ii*) & (*iii*) we get by taking $x_3 = 1,\ x_1 = 1,\ x_2 = -\frac{1}{2}$.

This also satisfies eq. (*i*). Thus, after normalising,

$$\boldsymbol{x}_3^{\mathrm{T}} = (1 \;\; -0.5 \;\; 1)$$

Note: The rank of $(\mathrm{A} - \lambda_i \mathrm{I})$, $i = 1, 2, 3$ is 2. Therefore for each eigenvalue there is only one (i.e. $3 - 2 = 1$) eigenvector.

Example 2.14

Find the eigenvalue and eigenvector of the following matrix,

$$\mathrm{A} = \begin{bmatrix} 5 & -2 & 0 \\ 1 & 2 & -3 \\ -1 & 1 & 0 \end{bmatrix}.$$

Solution The eigenvalues are given by

$$\begin{vmatrix} 5-\lambda & -2 & 0 \\ 1 & 2-\lambda & -3 \\ -1 & 1 & -\lambda \end{vmatrix} = 0$$

The characteristic equation is,

$$\lambda^3 - 7\lambda^2 + 15\lambda - 9 = 0$$

The eigenvalues are,

$$\lambda_1 = 1,\ \lambda_2 = 3,\ \lambda_3 = 3.$$

For $\lambda_1 = 1$, the associated eigenvector is,

$$\boldsymbol{x}_1^{\mathrm{T}} = (1\ \ 2\ \ 1).$$

For $\lambda_2,\ \lambda_3 = 3$, the resulting equations are,

$$2x_1 - 2x_2 + 0x_3 = 0$$

$$x_1 - x_2 - 3x_3 = 0$$

$$-x_1 + x_2 - 3x_3 = 0.$$

The above system gives,

$$x_3 = 0,\ x_1 = x_2 = 1 \text{ (say)},$$

The eigenvector is,

$$\boldsymbol{x}_2^{\mathrm{T}} = \boldsymbol{x}_3^{\mathrm{T}} = (1\ \ 1\ \ 0).$$

Note: The rank of the matrix A − 3I is 2. Therefore there is only one independent eigenvector (i.e. $3 - 2 = 1$).

Example 2.15

Find the eigen values and eigenvectors of the matrix

$$\mathrm{A} = \begin{bmatrix} 3 & 2 & 2 \\ 2 & 3 & 2 \\ 2 & 2 & 3 \end{bmatrix}.$$

Solution The characteristic equation is given by,

$$\begin{vmatrix} 3-\lambda & 2 & 2 \\ 2 & 3-\lambda & 2 \\ 2 & 2 & 3-\lambda \end{vmatrix} = 0$$

The characteristic equation is,

$$\lambda^3 - 9\lambda^2 + 15\lambda - 7 = 0$$

The roots are,

$$(\lambda - 1)^2(\lambda - 7) = 0$$

or
$$\lambda_1 = \lambda_2 = 1,\ \lambda_3 = 7.$$

For $\lambda_3 = 7$, the associated eigenvector after choosing $x_3 = 1$, is,

$$\boldsymbol{x}_1^{\mathrm{T}} = (1\ \ 1\ \ 1).$$

For $\lambda_1 = \lambda_2 = 1$, the system of homogeneous equations is,

$$2x_1 + 2x_2 + 2x_3 = 0$$

$$2x_1 + 2x_2 + 2x_3 = 0$$

$$2x_1 + 2x_2 + 2x_3 = 0$$

The rank of the above system is 1. Therefore, there will be two linearly independent e.vectors: Choosing $x_2 = 1,\ x_3 = 0$ gives $x_1 = -1$. Hence one e.vector is

$$\boldsymbol{x}_2^{\mathrm{T}} = (-1\ \ 1\ \ 0).$$

Next time choosing $x_2 = 0,\ x_3 = 1$ gives, $x_1 = -1$. The associated e.vector is,

$$\boldsymbol{x}_3^{\mathrm{T}} = (-1\ \ 0\ \ 1).$$

Note: Here rank of $(\mathrm{A} - \lambda_1\mathrm{I}) = (\mathrm{A} - \mathrm{I})$ is one. Therefore, there are two linearly independent e.vectors.

2.20 Power Method to Find Dominant Eigenvalue/Latent Root

The Power Method is a computer-oriented method that finds the numerically largest (dominant) latent root (e.value) of a given matrix and the associated latent vector (e.vector). In order to find the subdominant (second largest) root, the size of the matrix is reduced by one and a new matrix is formed which has the same e.values as the original matrix without its dominant root. The power method may be applied again on the reduced matrix to find its dominant root which will be subdominant root of the original matrix. This process may be repeated to find the other roots one by one by reducing the size of the matrix each time. Let us first explain the power method to find the dominant root and the associated e.vector. The procedure for finding the subdominant root will be explained later.

The Power Method is an iterative procedure described by the sequence of vectors as follows:

$$\left.\begin{aligned} z_{k+1} &= A\boldsymbol{y}_k \\ \boldsymbol{y}_{k+1} &= \frac{1}{\alpha_{k+1}} \cdot z_{k+1} \end{aligned}\right\} k \geq 0 \tag{2.41}$$

where α_{k+1} is the numerically largest element of z_{k+1} i.e., $A\boldsymbol{y}_k$. Thus $\boldsymbol{y}_k$ is a normalised vector whose largest element is 1. The computing process is started from an arbitrary vector $\boldsymbol{y}_0$ which is usually taken to be with all its elements unity. It will be shown that the sequence of vectors $\boldsymbol{y}_k$ converges, as $k \to \infty$, to the e.vector corresponding to the numerically largest e.value of A while the divisor α_k tends to the numerically largest e.value.

In order to prove the above let us suppose that a matrix A has n distinct e.values, say $\lambda_1, \lambda_2, \ldots \lambda_n$ with corresponding e.vectors $\boldsymbol{x}_1, \boldsymbol{x}_2, \ldots \boldsymbol{x}_n$ respectively. Further, without loss of generality, let $|\lambda_1| > |\lambda_2| \ldots > |\lambda_n|$. It is easy to see that all the e.vectors will be linearly independent, so that an arbitrary vector $\boldsymbol{y}_0$ may be expressed as linear combination of these vectors, i.e.

$$\boldsymbol{y}_0 = c_1\boldsymbol{x}_1 + c_2\boldsymbol{x}_2 + c_3\boldsymbol{x}_3 + \ldots. + c_n\boldsymbol{x}_n, \tag{2.42}$$

where $c_1, c_2 \ldots. c_n$ are constants, not all zero.

Premultiplication of both sides of (2.42) by A, gives,

$$z_1 = A\boldsymbol{y}_0 = c_1 A\boldsymbol{x}_1 + c_2 A\boldsymbol{x}_2 + c_3 A\boldsymbol{x}_3 + \ldots + c_n A\boldsymbol{x}_n$$

$$\boldsymbol{y}_1 = \frac{1}{\alpha_1}[c_1\lambda_1\boldsymbol{x}_1 + c_2\lambda_2\boldsymbol{x}_2 + c_3\lambda_3\boldsymbol{x}_3 + \ldots + c_n\lambda_n\boldsymbol{x}_n]$$

$$= \frac{\lambda_1}{\alpha_1}\left[c_1\boldsymbol{x}_1 + c_2\frac{\lambda_2}{\lambda_1}\boldsymbol{x}_2 + c_3\frac{\lambda_3}{\lambda_1}\boldsymbol{x}_3 + \cdots + c_n\frac{\lambda_n}{\lambda_1}\boldsymbol{x}_n\right]$$

where α_1 is numerically largest element of z_1.

Repeating the process, k times will yield,

$$z_k = \frac{\lambda_1^k}{\alpha_1\alpha_2\ldots\alpha_{k-1}}\left[c_1\boldsymbol{x}_1 + c_2\left(\frac{\lambda_2}{\lambda_1}\right)^k\boldsymbol{x}_2 + c_3\left(\frac{\lambda_3}{\lambda_1}\right)^k\boldsymbol{x}_3 + \cdots + c_n\left(\frac{\lambda_n}{\lambda_1}\right)^k\boldsymbol{x}_n\right]$$

$$y_k = \frac{\lambda_1{}^k}{\alpha_1 \cdots \alpha_k}\left[c_1\boldsymbol{x}_1 + c_2\left(\frac{\lambda_2}{\lambda_1}\right)^k \boldsymbol{x}_2 + c_3\left(\frac{\lambda_3}{\lambda_1}\right)^k \boldsymbol{x}_3 + \cdots + c_n\left(\frac{\lambda_n}{\lambda_1}\right)^k \boldsymbol{x}_n\right]. \tag{2.43}$$

We see from (2.43) that as $k \to \infty$, the vector $\boldsymbol{y}_k \to \lambda_1^k c_1 \boldsymbol{x}_1/\alpha$ where $\alpha = \alpha_1 . \alpha_2 \ldots \alpha_k$ since the ratio $(\lambda_i/\lambda_1) < 1,\ i = 2, 3,\ \ldots n$. The vector $\lambda_1^k c_1 \boldsymbol{x}_1/\alpha$ is parallel to $\boldsymbol{x}_1$ which is normalised such that its largest element remains unity. Thus $\boldsymbol{y}_k$ tends to the e.vector $\boldsymbol{x}_1$ corresponding to the largest e.value. Like (2.43) we can write the relation for $\boldsymbol{y}_{k+1}$ as,

$$y_{k+1} = \frac{\lambda_1^{k+1}}{\alpha_1 \cdots \alpha_{k+1}}\left[c_1\boldsymbol{x}_1 + c_2\left(\frac{\lambda_2}{\lambda_1}\right)^{k+1} \boldsymbol{x}_2 + c_3\left(\frac{\lambda_3}{\lambda_1}\right)^{k+1} \boldsymbol{x}_3 + \cdots + c_n\left(\frac{\lambda_n}{\lambda_1}\right)^{k+1} \boldsymbol{x}_n\right] \tag{2.44}$$

From (2.43) and (2.44) we conclude that,

$$\alpha_{k+1}\boldsymbol{y}_{k+1} \simeq \lambda_1 \boldsymbol{y}_k. \tag{2.45}$$

But vector $\boldsymbol{y}_k$ and $\boldsymbol{y}_{k+1}$ are the normalised vectors whose largest element is unity and they both tend to the e.vector $\boldsymbol{x}_1$, with largest element unity, corresponding to the largest e.value. Therefore on comparing the largest element of $\boldsymbol{y}_k$ and $\boldsymbol{y}_{k+1}$ in (2.45) we conclude that α_{k+1} will tend to λ_1, the numerically largest e.value.

2.20.1 To find smallest eigenvalue by power method

The power method can also be used to find the numerically smallest e.value of a matrix A. It is easy to see that if λ is an e.value of A then the corresponding e.value of A^{-1} will be $1/\lambda$ with eigenvector remaining same since if $A\boldsymbol{x} = \lambda\boldsymbol{x}$ then $A^{-1}\boldsymbol{x} = 1/\lambda\boldsymbol{x}$. Thus the largest e.value of A^{-1} will be the smallest e.value of A. Hence to find the numerically smallest e.value of A, we can use the formula,

$$y_{k+1} = \frac{1}{\alpha_{k+1}} A^{-1} y_k,\ k \geq 0.$$

where α_{k+1} is the numerically largest element of $A^{-1}y_k$, i.e., y_{k+1} is a vector whose largest element is unity. We can also proceed by solving the system $Ay_{k+1} = y_k$ for y_{k+1} and then dividing it by its numerically largest element. In practice we reduce A to LU form and solve the system of equations by forward and backward substitution.

If the e.values are well-separated, we can also find out the e.value closest to a given number β. If λ is an e.value of A then $\lambda - \beta$ is the corresponding e.value of matrix $A - \beta I$ with e.vector remaining same since if $A\boldsymbol{x} = \lambda\boldsymbol{x}$, then $(A - \beta I)\boldsymbol{x} = (\lambda - \beta)\boldsymbol{x}$. Also $\lambda - \beta$ will be the smallest (numerically) e.value of $(A - \beta I)$.

Hence, we perform $z_{k+1} = (A - \beta I)^{-1} y_k$ or $(A - \beta I) z_{k+1} = y_k$

and $$y_{k+1} = \frac{1}{\alpha_{k+1}} z_{k+1},\ k \geq 0$$

where α_{k+1} is the numerically largest element of z_{k+1}; or solve $(A - \beta I) y_{k+1} = y_k$ and divide y_{k+1} by its numerically largest element (see [3]). Again, it will be efficient computationally to factorise $(A - \beta I)$ to LU form since the formula is to be used repeatedly.

Example 2.16

Compute the numerically largest e.value and the corresponding e.vector of the following matrix A by Power method,

$$A = \begin{bmatrix} 5 & -2 & 0 \\ 1 & 2 & -3 \\ 1 & -2 & 4 \end{bmatrix}.$$

Solution Choosing the initial vector $y_0^T = (1\ 1\ 1)$,

$$Ay_0 = \begin{bmatrix} 5 & -2 & 0 \\ 1 & 2 & -3 \\ 1 & -2 & 4 \end{bmatrix} \begin{bmatrix} 1 \\ 1 \\ 1 \end{bmatrix} = \begin{bmatrix} 3 \\ 0 \\ 1 \end{bmatrix}$$

$$y_1 = \frac{1}{3} \begin{bmatrix} 3 \\ 0 \\ 1 \end{bmatrix} = \begin{bmatrix} 1 \\ 0 \\ 0.33 \end{bmatrix}$$

$$Ay_1 = \begin{bmatrix} 5 \\ 0 \\ 2.32 \end{bmatrix};\ y_2 = \frac{1}{5} \begin{bmatrix} 5 \\ 0 \\ 2.32 \end{bmatrix} = \begin{bmatrix} 1 \\ 0 \\ 0.46 \end{bmatrix}$$

$$Ay_2 = \begin{bmatrix} 5 \\ -0.41 \\ 2.88 \end{bmatrix};\ y_3 = \frac{1}{5} \begin{bmatrix} 5 \\ -0.41 \\ 2.88 \end{bmatrix} = \begin{bmatrix} 1 \\ -0.07 \\ 0.53 \end{bmatrix}.$$

Continuing in this way we get,

$$Ay_8 = \begin{bmatrix} 5.72 \\ -2.30 \\ 5.16 \end{bmatrix};\ y_9 = \frac{1}{5.72} \begin{bmatrix} 5.72 \\ -2.30 \\ 5.16 \end{bmatrix} = \begin{bmatrix} 1 \\ -0.46 \\ 0.90 \end{bmatrix}$$

$$\mathrm{A}y_{11} = \begin{bmatrix} 5.88 \\ -2.73 \\ 5.68 \end{bmatrix}; \quad y_{12} = \frac{1}{5.88}\begin{bmatrix} 5.88 \\ -2.73 \\ 5.68 \end{bmatrix} = \begin{bmatrix} 1 \\ -0.46 \\ 0.96 \end{bmatrix}.$$

We may retain more decimal figures to achieve greater accuracy; however, on the computer we may compute straight away up to more decimals.

$$\mathrm{A}y_{13} = \begin{bmatrix} 5.946 \\ -2.865 \\ 5.838 \end{bmatrix}; y_{14} = \frac{1}{5.946}\begin{bmatrix} 5.946 \\ -2.865 \\ 5.838 \end{bmatrix} = \begin{bmatrix} 1 \\ -0.482 \\ 0.982 \end{bmatrix}$$

$$\mathrm{A}y_{14} = \begin{bmatrix} 5.946 \\ -2.910 \\ 5.892 \end{bmatrix}; \quad y_{15} = \frac{1}{5.964}\begin{bmatrix} 5.964 \\ -2.910 \\ 5.892 \end{bmatrix} = \begin{bmatrix} 1 \\ -0.488 \\ 0.988 \end{bmatrix}$$

Approximately, $\lambda_1 = 6$, $\boldsymbol{x}_1^{\mathrm{T}} = (1 \quad -0.5 \quad 1)$; it is Example 2.13 and correct answer is same.

2.20.2 Determination of subdominant eigenvalues

Let us now discuss as how to find the subdominant (next largest) latent root using power method. The procedure is illustrated by way of an example as follows:

We continue with the matrix of Example 2.16. Having found the numerically largest e.value $\lambda_1 = 6$ and the associated e.vector $\boldsymbol{x}_1^{\mathrm{T}} = (1 \quad -0.5 \quad 1)$ we would like to compute the other e.values λ_2 and λ_3 and the corresponding e.vectors, $\boldsymbol{x}_2$ and $\boldsymbol{x}_3$.

First, a new matrix A_1 is formed as,

$$\mathrm{A}_1 = \mathrm{A} - \boldsymbol{x}_1 \boldsymbol{p}_i^{\mathrm{T}} \tag{2.46}$$

where $\boldsymbol{p}_i^{\mathrm{T}}$ is the ith row of A which corresponds to the largest element 1 of $\boldsymbol{x}_1$. Here we can choose anyone of the $\boldsymbol{p}_1^{\mathrm{T}}$ or $\boldsymbol{p}_3^{\mathrm{T}}$ since $\boldsymbol{x}_1$ has unit element in its 1st as well as in 3rd place.

Let us choose $\boldsymbol{p}_1^{\mathrm{T}}$ (say), then,

$$\mathrm{A}_1 = \begin{bmatrix} 5 & -2 & 0 \\ 1 & 2 & -3 \\ 1 & -2 & 4 \end{bmatrix} - \begin{bmatrix} 1 \\ -0.5 \\ 1 \end{bmatrix} [5 \ -2 \ \ 0] = \begin{bmatrix} 0 & 0 & 0 \\ 3.5 & 1 & -3 \\ -4 & 0 & 4 \end{bmatrix}. \tag{2.47}$$

The matrix A_1 is such that its e.values are same as those of A except that the e.value corresponding to λ_1 of A is reduced to zero in A_1. Although e.values of A_1 are same λ_2 and λ_3, as that of A, the e.vectors corresponding to λ_2 and λ_3 will not be same, $\boldsymbol{x}_2$ and

x_3. Instead, the e.vectors will be given by $x_1 - x_2$ and $x_1 - x_3$ respectively. However, it will be true only when the e.vectors x_2 and x_3 are normalised in such a way that they have unit element in the same position as x_1 has. To see that $x_1 - x_2$ is the e.vector of A_1 corresponding to e.value λ_2, we get from (2.46)

$$\begin{aligned} A_1(x_1 - x_2) &= (A - x_1 p_1^T)(x_1 - x_2) \\ &= A(x_1 - x_2) - x_1 p_1^T (x_1 - x_2) \\ &= Ax_1 - Ax_2 - x_1(p_1^T x_1 - p_1^T x_2) \\ &= \lambda_1 x_1 - \lambda_2 x_2 - x_1(\lambda_1 x_1 - \lambda_2 x_2) \because \text{1st element of } x_1 \text{ and } x_2 \text{ is unity} \\ &= \lambda_2 (x_1 - x_2). \end{aligned}$$

Similarly for $x_1 - x_3$. See [3].

Let us denote the e.vectors corresponding to λ_2 and λ_3 given by A_1, as z_2 and z_3 respectively where $z_2 = x_1 - x_2$ and $z_3 = x_1 - x_3$. Since x_1 has its first element 1 as also x_2 and x_3 are supposed to have their first elements unity, the first element of z_2 and z_3 will be zero. We leave out first row and first column of A_1 and work with a reduced matrix A_1, of order 2, i.e.,

$$A_1 = \begin{bmatrix} 1 & -3 \\ 0 & 4 \end{bmatrix}. \tag{2.47a}$$

This process of reducing the size of the matrix is known as 'deflation' of matrix.

Now applying power method on A_1,

$$A_1 y_0 = \begin{bmatrix} 1 & -3 \\ 0 & 4 \end{bmatrix} \begin{bmatrix} 1 \\ 1 \end{bmatrix} = \begin{bmatrix} -2 \\ 4 \end{bmatrix},$$

$$y_1 = \frac{1}{4} \begin{bmatrix} -2 \\ 4 \end{bmatrix} = \begin{bmatrix} -0.5 \\ 1 \end{bmatrix}.$$

Continuing further we get,

$$A_1 y_1 = \begin{bmatrix} -3.5 \\ 4 \end{bmatrix}, \; y_2 = \frac{1}{4} \begin{bmatrix} -3.5 \\ 4 \end{bmatrix} = \begin{bmatrix} -0.88 \\ 1 \end{bmatrix}$$

$$A_1 y_3 = \begin{bmatrix} -3.97 \\ 4 \end{bmatrix}, \; y_4 = \frac{1}{4}\begin{bmatrix} -3.97 \\ 4 \end{bmatrix} = \begin{bmatrix} -0.99 \\ 1 \end{bmatrix}$$

$$A_1 y_5 = \begin{bmatrix} -4 \\ 4 \end{bmatrix}, \; y_6 = \frac{1}{4}\begin{bmatrix} -4 \\ 4 \end{bmatrix} = \begin{bmatrix} -1 \\ 1 \end{bmatrix}.$$

Thus the largest e.value of reduced matrix A_1 is 4 with the associated e.vector, $z_2^T = (-1 \;\; 1)$. The e.value of A_1 is also 4 which is its largest e.value with the associated e.vector

$z_2^T = (0 \;\; -1 \;\; 1)$ since the first element is zero.

We have got the subdominant latent root of the original matrix, i.e. $\lambda_2 = 4$. Now to find the corresponding e.vector $\boldsymbol{x}_2$, we use the relation,

$$\boldsymbol{x}_1 - \boldsymbol{x}_2 = kz_2, \tag{2.48}$$

where k is a scaling factor to be determined.
Multiplying (2.48) by the first row of A, we get

$$\boldsymbol{p}_1^T \boldsymbol{x}_1 - \boldsymbol{p}_1^T \boldsymbol{x}_2 = k\boldsymbol{p}_1^T z_2$$

or $$\lambda_1 - \lambda_2 = k\boldsymbol{p}_1^T z_2, \tag{2.48a}$$

since the first elements of $\boldsymbol{x}_1$ and $\boldsymbol{x}_2$ are unity.
Using (2.48a) gives,

$$6 - 4 = k[5 \;\; -2 \;\; 0]\begin{bmatrix} 0 \\ -1 \\ 1 \end{bmatrix} = 2k$$

or $$k = 1.$$

Thus from (2.48),

$$\boldsymbol{x}_2 = \begin{bmatrix} 1 \\ -0.5 \\ 1 \end{bmatrix} - 1 \times \begin{bmatrix} 0 \\ -1 \\ 1 \end{bmatrix} = \begin{bmatrix} 1 \\ 0.5 \\ 0 \end{bmatrix}.$$

To find the third e.value λ_3 and the associated e.vector we consider the reduced matrix A_1 given by (2.47a). It is a 2×2 matrix with its largest e.value 4 and the associated e.vector $z_2^T = (-1 \ \ 1)$.

In this case second element is unity, therefore the reduced matrix will be formed by taking second row of A_1, so that,

$$A_2 = \begin{bmatrix} 1 & -3 \\ 0 & 4 \end{bmatrix} - \begin{bmatrix} -1 \\ 1 \end{bmatrix} [0 \quad 4] = \begin{bmatrix} 1 & 1 \\ 0 & 0 \end{bmatrix}. \tag{2.49}$$

Now the matrix A_2 is such that the e.value corresponding to λ_2 is zero and it has same e.value λ_3 as that of A_1. But again the corresponding e.vector will be $v_3 = z_2 - z_3$ provided z_3 is normalised such that its second element is unity, i.e., same element as that of z_2.

From (2.49) we get the reduced matrix A_2 after removing second row and second column of (2.49), i.e. reduced A_2 just the single element,

$$A_2 = [1].$$

This has e.value $\lambda_3 = 1$ and the associated e.vector $v_3 = (1)$. Extending v_3 by adding 0 element, the e.vector corresponding to λ_3 of (2.49) is $v_3 = (1 \quad 0)$.

As before we have,

$$z_2 - z_3 = kv_3$$

Premultiplying it by second row of A_1 gives,

$$\lambda_2 - \lambda_3 = k[0 \quad 4] \begin{bmatrix} 1 \\ 0 \end{bmatrix}$$

or $$4 - 1 = k \times 0.$$

That is, $k \to \infty$ or is very large.

From (2.50) we have,

$$z_3 = \begin{bmatrix} -1 \\ 1 \end{bmatrix} - k \begin{bmatrix} 1 \\ 0 \end{bmatrix} = \begin{bmatrix} -1-k \\ 1 \end{bmatrix}$$

Normalising z_3 by dividing by $-1-k$,

$$z_3^T = (1 \quad 0).$$

For getting $\boldsymbol{x}_3$ we proceed as before, i.e., extend z_3 by adding a 0 as the first element of z_3, i.e., $z_3^T = (0 \quad 1 \quad 0)$.

$$\lambda_1 - \lambda_3 = \text{k}[5 \quad -2 \quad 0] \begin{bmatrix} 0 \\ 1 \\ 0 \end{bmatrix}$$

or $\quad 6 - 1 = -2\text{k}$

or $\quad \text{k} = -2.5$

$$\text{Hence,} \quad \boldsymbol{x}_3 = \begin{bmatrix} 1 \\ -0.5 \\ 1 \end{bmatrix} + 2.5 \begin{bmatrix} 0 \\ 1 \\ 0 \end{bmatrix}$$

$$= \begin{bmatrix} 1 \\ 2 \\ 1 \end{bmatrix}$$

or $\quad \boldsymbol{x}_3^T = (0.5 \quad 1 \quad 0.5)$.

The procedure is known as method of 'root removal' or 'exhaustion' method.

2.21 Iterative Methods

In Sec 2.13 we had stated that there are two approaches for solving linear simultaneous equations, viz., direct and iterative. So far we have discussed direct methods in which the solution was obtained in a definite number of steps. In an iterative method we start initially from an approximate solution which is improved by repeated use of the method, called iterations. The process is terminated when two successive solutions agree within a desired accuracy. The application of the iterative methods is however restricted only to a particular class of system of equations, in that in the system $\text{A}\boldsymbol{x} = \boldsymbol{b}$, the coefficient matrix A should be diagonally dominant. That is, the modulus of the diagonal element of A in any row should be greater than the sum of the modulii of the other elements in that row, i.e., for an $n \times n$ system,

$$|a_{ii}| > \sum_{j \neq i} |a_{ij}|, \qquad i = 1(1)n. \tag{2.50}$$

The condition (2.50) is sufficient (not necessary) for the convergence of the iterative process to the true solution of the system. A slightly milder condition for the convergence is that sign $>$ in (2.50) may be replaced by $\geq$ but for at least one of the rows (2.50) should hold.

It may be stated that the equations $\mathrm{A}\boldsymbol{x} = \boldsymbol{b}$ in their original form may not satisfy condition (2.50) but after re-arranging them they may. Therefore our first attempt would be to see if the equations can be re-ordered to satisfy the condition (2.50). Secondly, as stated, condition (2.50) is not a necessary (is sufficient) condition for convergence of the iterative process which means that there may be systems which do not satisfy condition (2.50) but the process of iteration does converge. There are basically two approaches of iterative processes for solving the system $\mathrm{A}\boldsymbol{x} = \boldsymbol{b}$, called as:

(*i*) Gauss–Jacobi method.

(*ii*) Gauss–Seidel method.

Assuming that A is diagonally dominant, we start, in both cases, from an arbitrarily chosen initial solution vector, say, $x^{(0)}$ which is usually taken to be a zero vector, i.e., $\boldsymbol{x}^{(0)} = 0$. Then using interative methods, improved values for the solution vector are obtained, say, $\boldsymbol{x}^{(1)}, \boldsymbol{x}^{(2)}, \cdots \boldsymbol{x}^{(k)} \cdots$ and the process is stopped as soon as two successive vectors agree within the desired accuracy. The difference between the two methods is that while performing $(n+1)^{\text{th}}$ iteration to compute $\boldsymbol{x}^{(n+1)}$, the values of nth iteration $\boldsymbol{x}^{(n)}$ are used in the Gauss–Jacobi method while in the Gauss–Seidel method, the most recent values of $\boldsymbol{x}^{(n+1)}$ are used, i.e., the values computed at the $(n+1)^{\text{th}}$ iteration are used as soon as they are computed alongwith the values at n^{th} iteration that are not known at the $(n+1)^{th}$ iteration.

We will illustrate these methods for a 3×3 system, $\mathrm{A}\boldsymbol{x} = \boldsymbol{b}$ which is given by,

$$a_{11}x_1 + a_{12}x_2 + a_{13}x_3 = b_1$$

$$a_{21}x_1 + a_{22}x_2 + a_{23}x_3 = b_2$$

$$a_{31}x_1 + a_{32}x_2 + a_{33}x_3 = b_3.$$

It is assumed that the coefficient matrix is diagonally-dominant. We solve first equation for x_1, second for x_2 and third for x_3; thus getting,

$$\left.\begin{aligned} x_1 &= (b_1 - a_{12}x_2 - a_{13}x_3)/a_{11} \\ x_2 &= (b_2 - a_{21}x_1 - a_{23}x_3)/a_{22} \\ x_3 &= (b_3 - a_{31}x_1 - a_{32}x_2)/a_{33}. \end{aligned}\right\} \tag{2.51}$$

Suppose values at n^{th} iteration have already been computed and we are required to compute the values at the $(n+1)^{\text{th}}$ iteration. That is, $\boldsymbol{x}^{(n)}$ is known and $\boldsymbol{x}^{(n+1)}$ is to be computed. The schemes for computing $\boldsymbol{x}^{(n+1)}$ by the two methods are given below:

2.21.1 Gauss–Jacobi method

$$\left.\begin{aligned} x_1{}^{(n+1)} &= [b_1 - a_{12}x_2{}^{(n)} - a_{13}x_3{}^{(n)}]/a_{11} \\ x_2{}^{(n+1)} &= [b_2 - a_{21}x_1{}^{(n)} - a_{23}x_3{}^{(n)}]/a_{22} \\ x_3{}^{(n+1)} &= [b_3 - a_{31}x_1{}^{(n)} - a_{32}x_2{}^{(n)}]/a_{33} \end{aligned}\right\} \tag{2.52}$$

The superscript denotes the iteration number and the value of the unknown corresponds to that iteration. Thus we should note that in order to compute values of unknowns at the $(n+1)^{\text{th}}$ iteration, their values at previous iteration n^{th}, are used.

2.21.2 Gauss–Seidel method

The scheme for computing $x^{(n+1)}$ from the Gauss–Seidel method is as follows:

$$\left.\begin{aligned} x_1{}^{(n+1)} &= [b_1 - a_{12}x_2{}^{(n)} - a_{13}x_3{}^{(n)}]/a_{11} \\ x_2{}^{(n+1)} &= [b_2 - a_{21}x_1{}^{(n+1)} - a_{23}x_3{}^{(n)}]/a_{22} \\ x_3{}^{(n+1)} &= [b_3 - a_{31}x_1{}^{(n+1)} - a_{32}x_2{}^{(n+1)}]/a_{33} \end{aligned}\right\} \tag{2.53}$$

As can be noticed that in the Gauss–Jacobi iteration the value of x_1, x_2 and x_3 at the nth iteration are used while in the Gauss–Seidel scheme, the most recent values are used. That is, $x_1^{(n+1)}$ is used in computing $x_2^{(n+1)}$ and values of $x_1^{(n+1)}$ and $x_2^{(n+1)}$ are used in computing $x_3^{(n+1)}$. The Gauss–Jacobi scheme is known as method of 'simultaneous displacement' while Gauss–Seidel scheme as method of 'successive displacement'.

Example 2.17

Solve the following equations by (*i*) Gauss–Jacobi method (*ii*) Gauss–Seidel method:

$$3x_1 - 2x_2 + 8x_3 = -4$$

$$5x_1 + x_2 - x_3 = 12$$

$$x_1 + 6x_2 + 2x_3 = 6$$

Write down the two schemes clearly for $(n+1)^{\text{th}}$ iteration and compute up to two decimals only.

Solution First, we re-arrange the equations such that thecoefficient matrix is diagonally dominant.

$$5x_1 + x_2 - x_3 = 12$$

$$x_1 + 6x_2 + 2x_3 = 6$$

$$3x_1 - 2x_2 + 8x_3 = -4$$

Solving first equation for x_1, second for x_2 and third for x_3,

$$x_1 = \frac{1}{5}[12 - x_2 + x_3]$$

$$x_2 = \frac{1}{6}[6 - x_1 - 2x_3]$$

$$x_3 = \frac{1}{8}[-4 - 3x_1 + 2x_2]$$

(*i*) Gauss–Jacobi scheme

$$x_1{}^{(n+1)} = \frac{1}{5}[12 - x_1{}^{(n)} + x_3{}^{(n)}]$$

$$x_2{}^{(n+1)} = \frac{1}{6}[6 - x_1{}^{(n)} - 2x_3{}^{(n)}]$$

$$x_3{}^{(n+1)} = \frac{1}{8}[-4 - 3x_1{}^{(n)} + 2x_2{}^{(n)}].$$

Taking initial guess as, $x^{(0)} = 0$, i.e., $x_1 = 0,\ x_2 = 0,\ x_3 = 0$, the subsequent computed values are shown in the following table,

$k \rightarrow$ / x	0	1	2	3	4	5	6	7
$x_1^{(k)}$	0	2.40	2.10	2.02	1.97	1.99	2.00	2.00
$x_2^{(k)}$	0	1.00	0.77	1.03	1.03	1.00	1.00	1.00
$x_3^{(k)}$	0	−0.50	−1.15	−1.10	−1.00	−1.00	−1.00	−1.00

The iterations 6 and 7 give same values up to two decimal places.

$$\therefore \quad x_1 = 2.00,\ x_2 = 1.00,\ x_3 = -1.00$$

(correct answer is same)

Note: If the answer was required within an accuracy of one decimal then iterations 4 and 5 give the same values, so that we could stop at the 5th iteration.

(*ii*) Gauss–Seidel scheme

$$x_1{}^{(n+1)} = \frac{1}{5}\left[12 - x_2{}^{(n)} + x_3^{(n)}\right]$$

$$x_2{}^{(n+1)} = \frac{1}{6}\left[6 - x_1{}^{(n+1)} - 2x_3{}^{(n)}\right]$$

$$x_3{}^{(n+1)} = \frac{1}{8}\left[-4 - 3x_1{}^{(n+1)} + 2x_2{}^{(n+1)}\right].$$

Taking $x_1 = 0$, $x_2 = 0$ and $x_3 = 0$, the iterated values are shown in the table:

x \ $k \rightarrow$	0	1	2	3	4	5
$x_1^{(k)}$	0	2.50	1.98	1.99	2.00	2.00
$x_2^{(k)}$	0	0.58	1.10	0.99	1.00	1.00
$x_3^{(k)}$	0	−1.29	−0.97	−1.00	−1.00	−1.00

Note: In general, under convergence conditions, the Gauss–Seidel scheme converges faster than Gauss–Jacobi. See Exercise 2.14.

2.22 Condition for Convergence of Iterative Methods

In the previous section it was mentioned that diagonal dominance of the coefficient matrix in the solution of $\mathrm{A}\boldsymbol{x} = \boldsymbol{b}$ is a sufficient condition for the convergence of the iterative methods described there. That is, the iterative process may converge in some cases when the condition diagonal dominance is not fulfilled. Now we would like to derive the necessary condition for the convergence of Gauss–Jacobi and Gauss–Seidel methods.

Let us consider a $(n \times n)$ system $\mathrm{A}\boldsymbol{x} = \boldsymbol{b}$ where A is not necessarily diagonally dominant matrix. We can express A as sum three matrices and write the equations in the following form,

$$(\mathrm{D} + \mathrm{L} + \mathrm{U})\boldsymbol{x} = \boldsymbol{b}, \tag{2.54}$$

where

D is the diagonal matrix consisting of the diagonal elements of A; no diagonal element is zero.

L is the strictly lower triangular matrix consisting of the elements of A below the diagonal (diagonal terms of L are zero)

U is strictly upper triangular matrix consisting of the elements of A above the diagonal (diagonal terms of U are zero)

The Gauss–Jacobi scheme may be written for the kth iteration, as

$$\mathrm{D}\boldsymbol{x}^{(k)} = \boldsymbol{b} - (\mathrm{L}+\mathrm{U})\boldsymbol{x}^{(k-1)}$$

or
$$\boldsymbol{x}^{(k)} = \mathrm{D}^{-1}\boldsymbol{b} - \mathrm{D}^{-1}(\mathrm{L}+\mathrm{U})\boldsymbol{x}^{(k-1)} \tag{2.55}$$

Also from (2.54)

$$\boldsymbol{x} = \mathrm{D}^{-1}\boldsymbol{b} - \mathrm{D}^{-1}(\mathrm{L}+\mathrm{U})\boldsymbol{x}. \tag{2.56}$$

Subtracting (2.55) from (2.56) gives,

$$\boldsymbol{x} - \boldsymbol{x}^{(k)} = -\mathrm{D}^{-1}(\mathrm{L}+\mathrm{U})(\boldsymbol{x} - \boldsymbol{x}^{(k-1)})$$

or
$$\boldsymbol{e}^{(k)} = -\mathrm{D}^{-1}(\mathrm{L}+\mathrm{U})\boldsymbol{e}^{(k-1)}$$

$$= \mathrm{G}\boldsymbol{e}^{(k-1)}, \tag{2.57}$$

where $\boldsymbol{e}^{(k)}$ and $\boldsymbol{e}^{(k-1)}$ denote the error between the true solution and k^{th} and $(k-1)^{\text{th}}$ iterates; $\mathrm{G} = -\mathrm{D}^{-1}(\mathrm{L}+\mathrm{U})$ is called convergence matrix for Gauss–Jacobi scheme.

Applying (2.57) repeatedly, we can write,

$$\boldsymbol{e}^{(k)} = \mathrm{G}^k\boldsymbol{e}^{(0)}, \tag{2.58}$$

where $\boldsymbol{e}^{(0)} = \boldsymbol{x} - \boldsymbol{x}^{(0)}$, is the error in the initial vector $\boldsymbol{x}^{(0)}$.

In order to see that the error vector tends to a zero vector as k increases, i.e. the solution converges to the true solution as k increases. We proceed as follows:

Let us assume that the matrix G which is of order n, has n linearly independent e.vectors. Then vector $\boldsymbol{e}_0$ can be expressed as linear combination of these vectors, say $\boldsymbol{v}_i,\ i = 1(1)n$, i.e.

$$\boldsymbol{e}^{(0)} = \sum_{i=1}^{n} c_t\boldsymbol{v}_i$$

where $c_i,\ i = 1(1)n$ are constants.

Let λ_i be an e.value corresponding to e.vector $\boldsymbol{v}_i,\ i = 1(1)n$. Then

$$\mathrm{G}\boldsymbol{e}^{(0)} = \Sigma c_i \mathrm{G}\boldsymbol{v}_i = \Sigma \lambda_i c_i \boldsymbol{v}_i \text{ since } \mathrm{G}\boldsymbol{v}_i = \lambda_i \boldsymbol{v}_i$$

or
$$\boldsymbol{e}^{(1)} = \sum_{i=1}^{n} \lambda_i c_i \boldsymbol{v}_i$$

Proceeding in this manner we get,

$$\boldsymbol{e}^{(k)} = \sum_{i=1}^{n} \lambda_i^k c_i \boldsymbol{v}_i. \tag{2.59}$$

Obviously if the error vector $\boldsymbol{e}^{(k)}$ has to tend to zero as k increases, the moduli of λ_i has to be less than 1. That is, the numerically largest e.value of the convergence matrix G should not exceed unity. In other words, the 'spectral radius' $\rho(\mathrm{G})$ of G, should be less than 1 or

$$\rho(\mathrm{G}) < 1, \tag{2.60}$$

which is the necessary condition for convergence of an iterative scheme whose convergence matrix is G. In the case of Gauss–Jacobi method $\mathrm{G} = -\mathrm{D}^{-1}(\mathrm{L}+\mathrm{U})$.

Let us consider the Gauss–Seidel scheme which can be expressed, for k^{th} iteration as,

$$\mathrm{D}\boldsymbol{x}^{(k)} = \boldsymbol{b} - \mathrm{L}\boldsymbol{x}^{(k)} - \mathrm{U}\boldsymbol{x}^{(k-1)}$$

or
$$(\mathrm{D}+\mathrm{L})\boldsymbol{x}^{(k)} = \boldsymbol{b} - \mathrm{U}\boldsymbol{x}^{(k-1)}$$

or
$$\boldsymbol{x}^{(k)} = (\mathrm{D}+\mathrm{L})^{-1}\boldsymbol{b} - (\mathrm{D}+\mathrm{L})^{-1}\mathrm{U}\boldsymbol{x}^{(k-1)} \tag{2.61}$$

As before, the error vector,

$$\begin{aligned} \boldsymbol{e}^{(k)} &= -(\mathrm{D}+\mathrm{L})^{-1}\mathrm{U}\boldsymbol{e}^{(k-1)} \\ &= \mathrm{H}^k \boldsymbol{e}^{(0)}, \end{aligned}$$

where $\mathrm{H} = -(\mathrm{D}+\mathrm{L})^{-1}\mathrm{U}$ is the convergence matrix for the Gauss–Seidel scheme, and for its convergence $\rho(\mathrm{H}) < 1$.

It may be emphasised here that since the convergence matrices in both the methods are different, therefore it is quite likely that when the coefficient matrix is not diagonally dominant, one method may converge but the other may not.

2.23 Successive Over-Relaxation (S.O.R.) Method

The iterative methods for solving linear system of equations, namely Gauss–Jacobi and Gauss–Seidel can be modified to increase their rate of convergence. These improvised methods are known as Successive Over-Relaxation (S.O.R.) methods. They are also called extrapolated Liebmann's method. If $\boldsymbol{x}^{*(n+1)}$ denotes the $(n+1)^{\text{th}}$ iterated value from the above methods, then the corresponding S.O.R. method can be written as

$$\boldsymbol{x}^{(n+1)} = \boldsymbol{x}^{(n)} + \omega(\boldsymbol{x}^{*(n+1)} - \boldsymbol{x}^{(n)}) = \boldsymbol{x}^{(n)} + \omega \boldsymbol{r}_n, \text{ say} \tag{2.62}$$

or $$\boldsymbol{x}^{(n+1)} = (1-\omega)\boldsymbol{x}^{(n)} + \omega \boldsymbol{x}^{*(n+1)} \tag{2.63}$$

where $\boldsymbol{r}_n$ is the displacement/residual (relaxation) vector and ω is a parameter which is the accelerating factor of the SOR scheme. Its value has to be chosen depending on the iterative scheme which involves the 'spectral radius' (modulus of largest e.value) of the coefficient matrix of the system of equations. It may be noted that for $\omega = 1$ the SOR scheme will be same as the iterative scheme, and also that for $\omega < 1$ the residual vector will be under-relaxed.

Let us consider the solution of $\mathrm{A}\boldsymbol{x} = \boldsymbol{b}$ which can also be written as $(\mathrm{L}+\mathrm{D}+\mathrm{U})\boldsymbol{x} = \boldsymbol{b}$ where L and U are strictly lower and upper triangular and D is the diagonal matrix. The SOR scheme for Gauss–Jacobi method may be written as,

$$\begin{aligned}\boldsymbol{x}^{(n+1)} &= \boldsymbol{x}^{(n)} + \omega[\mathrm{D}^{-1}\{\boldsymbol{b} - \mathrm{L}\boldsymbol{x}^{(n)} - \mathrm{U}\boldsymbol{x}^{(n)}\} - \boldsymbol{x}^{(n)}]\\ &= (1-\omega)\boldsymbol{x}^{(n)} + \omega\mathrm{D}^{-1}(\boldsymbol{b} - \mathrm{L}\boldsymbol{x}^{(n)} - \mathrm{U}\boldsymbol{x}^{(n)})\\ &= \{(1-\omega)\mathrm{I} - \omega\mathrm{D}^{-1}(\mathrm{L}+\mathrm{U})\}\boldsymbol{x}^{(n)} + \omega\mathrm{D}^{-1}\boldsymbol{b} \qquad (2.64)\\ &= \mathrm{G}\boldsymbol{x}^{(n)} + \omega\mathrm{D}^{-1}\boldsymbol{b}, \qquad (2.64\text{a})\end{aligned}$$

where $\mathrm{G} = (1-\omega)\mathrm{I} - \omega\mathrm{D}^{-1}(\mathrm{L}+\mathrm{U})$ is the convergence matrix for the SOR (Gauss–Jacobi) method.

For Gauss–Seidel method the SOR scheme would be

$$\boldsymbol{x}^{(n+1)} = \boldsymbol{x}^{(n)} + \omega[\mathrm{D}^{-1}\{\boldsymbol{b} - \mathrm{L}\boldsymbol{x}^{(n+1)} - \mathrm{U}\boldsymbol{x}^{(n)}\} - \boldsymbol{x}^{(n)}]$$

or $$(\mathrm{D}+\omega\mathrm{L})\boldsymbol{x}^{(n+1)} = \{(1-\omega)\,\mathrm{D} - \omega\mathrm{U}\}\boldsymbol{x}^{(n)} + \omega\boldsymbol{b}$$

or $$\begin{aligned}\boldsymbol{x}^{(n+1)} &= (\mathrm{D}+\omega\mathrm{L})^{-1}\{(1-\omega)\,\mathrm{D} - \omega\mathrm{U}\}\boldsymbol{x}^{(n)} + \omega(\mathrm{D}+\omega\mathrm{L})^{-1}\boldsymbol{b} \qquad (2.65)\\ &= \mathrm{H}\boldsymbol{x}^{(n)} + \omega(\mathrm{D}+\omega\mathrm{L})^{-1}\boldsymbol{b} \qquad (2.65\text{a})\end{aligned}$$

where $\mathrm{H} = (\mathrm{D}+\omega\mathrm{L})^{-1}\{(1-\omega)\mathrm{D} - \omega\mathrm{U}\}$ is the convergence matrix for the corresponding SOR scheme.

The optimum value of ω for best convergence of the Gauss–Jacobi SOR scheme is given by

$$\omega_{\text{opt}} = \frac{2}{1+\sqrt{1-\mu^2}} \tag{2.66}$$

where μ is the 'spectral radius' of Gauss–Jacobi convergence matrix, i.e., $\mu = \rho(\mathrm{G})$. It lies between 1 and 2. Also in practical cases the relation between the spectral radii of the Gauss–Jacobi and Gauss–Seidel iteration matrices is given by

$$\rho(\mathrm{H}) = \rho^2(\mathrm{G}) \tag{2.67}$$

where G and H are their respective convergence matrices. See [7]

The application of SOR method is shown under Elliptic equations in Chapter 11.

2.24 Norms of Vectors and Matrices

Norms define magnitudes of vectors and matrices. It will be assumed that the vectors and matrices belong to n-dimensional space. The norm of a vector or a matrix will be denoted by enclosing it between pairs of parallel lines $\|\cdot\|$.

2.24.1 Vector norm

The norm of a vector is a real-valued function which satisfies the following axioms:

$$\left.\begin{aligned} &(i)\ \|\boldsymbol{x}\| \geq 0 \text{ with } \|\boldsymbol{x}\| = 0 \text{ only when } \boldsymbol{x} = 0 (\text{a null vector}) \\ &(ii)\ \|\alpha \boldsymbol{x}\| = |\alpha| \cdot \|\boldsymbol{x}\| \text{ where } \alpha \text{ is a scalar.} \\ &(iii)\ \|\boldsymbol{x}+\boldsymbol{y}\| \leq \|\boldsymbol{x}\| + \|\boldsymbol{y}\|, \text{ (known as triangular inequality).} \end{aligned}\right\} \tag{2.68}$$

There can be many ways for expressing norm of a vector but following are more important from the point of view of applications.

Let $\boldsymbol{x}$ be a vector $\boldsymbol{x}^{\mathrm{T}} = (x_1 \ \ x_2 \ \ \ldots. \ \ x_n)$, then

$$(i)\ \ \|\boldsymbol{x}\|_1 = \sum_{i=1}^{n} |x_i| \tag{2.69a}$$

$$(ii)\ \|\boldsymbol{x}\|_2 = \left\{ \sum_{i=1}^{n} |x_i|^2 \right\}^{1/2}, \text{ known as Euclidean norm and is also denoted as } \|\boldsymbol{x}\|_{\mathrm{E}}. \tag{2.69b}$$

$$(iii)\ \|\boldsymbol{x}\|_\infty = \max_{\mathrm{i}} |x_i|. \tag{2.69c}$$

In fact (2.69a, b and c) are particular cases of the Hölder's norm which is defined as

$$\|\boldsymbol{x}\|_p = \left\{ \sum_{i=1}^{n} |x_i|^p \right\}^{1/p}, \ 1 \leq p < \infty \tag{2.70}$$

for $p = 1, 2$ and ∞.

When the elements of $\boldsymbol{x}$ are real

$$\|\boldsymbol{x}\|_2 = \left\{ \sum_{i=1}^{n} x_i^2 \right\}^{1/2} = (\boldsymbol{x}^{\mathrm{T}}\boldsymbol{x})^{1/2}. \tag{2.71}$$

Thus $\|\boldsymbol{x}\|_2$ is used to define the length/magnitude of the vector $\boldsymbol{x}$. In vector analysis it may denote, for example, magnitude of a force vector. It is also used to define the distance between two vectors $\boldsymbol{x}$ and $\boldsymbol{y}$ which is given by $\|\boldsymbol{x}-\boldsymbol{y}\|_2$. In vector analysis if $\boldsymbol{x}$ is the position vector of a point P and $\boldsymbol{y}$ is the position vector of Q, then $\|\boldsymbol{x}-\boldsymbol{y}\|_2$ denotes the distance between P and Q.

We often make use of the infinity norm $\|\cdot\|_\infty$ to examine the convergence of a vector in iterative processes. For example, in the Gauss–Jacobi/Gauss–Seidel or Power Method we obtain $\boldsymbol{x}^{(k)}$ and $\boldsymbol{x}^{(k+1)}$ as k^{th} and $(k+1)^{\text{th}}$ interates for computing vector $\boldsymbol{x}$, $k = 0, 1, \ldots$. The difference between two iterates is denoted by the vector $\boldsymbol{e}^{(k+1)} = \boldsymbol{x}^{(k+1)} - \boldsymbol{x}^{(k)}$. The process of iteration is continued until the modulus of the largest element of $\boldsymbol{e}^{(k+1)}$ becomes less than or equal to a prescribed positive number ε, i.e., $\|\boldsymbol{e}^{(k+1)}\|_\infty \le \varepsilon$

A simple example for using $\|\cdot\|_1$ may be when a vector $\boldsymbol{x}$ is closely approximated by a vector $\boldsymbol{x}^*$, suppose after rounding. Let us express the error by $\varepsilon = \boldsymbol{x} - \boldsymbol{x}^*$. Then the actual error incurred in the summation $\Sigma \boldsymbol{x}_i$ and $\Sigma \boldsymbol{x}_i^*$ will be given by $\Sigma|\varepsilon_i|$, i.e., $\|\varepsilon\|_1$.

2.24.2 Matrix norm

The norm of a matrix may be defined analogously in the following manner:

That is, the norm of a matrix is a real-valued function which satisfies the following properties:

$$\left.\begin{array}{l}
(i)\ \|\mathrm{A}\| \ge 0 \text{ with} \|\mathrm{A}\| = 0 \text{ only when } \mathrm{A} = \mathrm{O} \text{ (Null matrix)}.\\
(ii)\ \|\alpha\mathrm{A}\| = |\alpha| \cdot \|\mathrm{A}\|, \text{ where } \alpha \text{ is a scalar}.\\
(iii)\ \|\mathrm{A}+\mathrm{B}\| \le \|\mathrm{A}\| + \|\mathrm{B}\|, \text{ (known as triangular inequality)}.\\
(iv)\ \|\mathrm{A}\cdot\mathrm{B}\| \le \|\mathrm{A}\| \cdot \|\mathrm{B}\|, \text{ (Schwarz inequality)}.
\end{array}\right\} \tag{2.72}$$

$\|\mathrm{A}\|$ is also denoted by $\mathrm{N(A)}$.

Example 2.18

Prove the following inequalities

(i) $\|\mathrm{A}-\mathrm{B}\| \ge \big|\|\mathrm{A}\| - \|\mathrm{B}\|\big|$

(ii) $\|A-B\| \le \|A\| + \|B\|$

Proof: We know from the above property (*iii*)

$$\|A+B\| \le \|A\| + \|B\| \quad \text{.... (1)}$$

(*i*) Putting $A = A - B$ in the above inequality (1)

$$\|A\| \le \|A-B\| + \|B\|$$

or $\|A-B\| \ge \|B\| - \|A\|$

or $\ge |\|A\| - \|B\|| \because \|A-B\| \ge 0$

(*ii*) Putting $B = -B$ in (1)

$$\|A-B\| \le \|A\| + \|-B\|$$

$$= \|A\| + \|B\| \because \|-B\| = |-1| \cdot \|B\| = \|B\|, \text{ property } (ii).$$

Example 2.19

The iterative scheme for improving the elements of an approximate inverse B_0 of matrix A is given by,

$$E_k = I - AB_k$$

and $B_{k+1} = B_k + B_kE_k = B_k(I + E_k)$.

Prove that $B_k \to A^{-1}$ if norm of the initial error matrix $\|E_0\| < 1$.

Proof: $E_{k+1} = I - AB_{k+1}$

$$= I - A\ (B_k + B_kE_k)$$

$$= I - AB_k - AB_kE_k = E_k - AB_kE_k$$

$$= (I - AB_k)E_k = E_k \cdot E_k = E_k^2$$

$$= {E_0}^{2(k+1)}$$

$$\|E_{k+1}\| = \|E_0^{2(k+1)}\|$$

$$\le \|E_0\|^{2(k+1)} \to 0 \text{ as } k \to \infty \text{ since } \|E_0\| < 1.$$

It implies $\|I - AB_k\| \to 0$

But it can be zero only if matrix $I - AB_k \to O$ from property (*i*).

or $AB_k \to I$

or $B_k \to A^{-1}$.

Example 2.20

Given $(I-A)^{-1} = I + A + A^2 + \ldots. + A^n + R_n$

Prove that $\|R_n\| \to 0$ as $n \to \infty$, if $\|A\| < 1$.

Proof: $(1-A)^{-1} = I + A + A^2 + \ldots. + A^n + R_n$

Premultiplying both sides by I−A,

$$I = (I + A + A^2 + \ldots. + A^n + R_n) - (A + A^2 + \ldots. + A^n + A^{n+1} + AR_n)$$

$$= I + R_n - A^{n+1} - AR_n$$

or $$R_n = (I-A)^{-1}A^{n+1}$$

Taking norm

$$\|R_n\| = \|(I-A)^{-1}A^{n+1}\|$$

$$\leq \|(I-A)^{-1}\|.\|A\|^{n+1}, \text{ property (iv)}$$

Let $$B = (I-A)^{-1}$$

Then $$B - I = AB$$

or $$\|B - I\| \leq \|A\|.\|B\|$$

$$|\|B\| - \|I\|| \leq \|A\|.\ \|B\|$$

or $$-\ \|A\|.\ \|B\| \leq \|B\| - \|I\| \leq \|A\|.\ \|B\|$$

or $$\frac{\|I\|}{1+\|A\|} \leq \|(I-A)^{-1}\| \leq \frac{\|I\|}{1-\|A\|}$$

Hence, $$\|R_n\| \leq \|A\|^{n+1} \cdot \frac{\|I\|}{1-\|A\|}$$

If $\|A\| < 1$, $\|R_n\| \to 0$ as $n \to \infty$, since $\|I\|$ is finite.

Note: Replacing A by −A, we have the same result.

2.24.3 Forms of matrix norm

Again, there may be numerous ways to express a matrix norm but following forms are more useful from the application point of view:

(*i*) $\|A\|_1 = \max_j \sum_{i=1}^{n} |a_{ij}|$ (i.e., maximum column sum) (2.73a)

$$(ii)\ \|A\|_\infty = \max_i \sum_{i=1}^{n} |a_{ij}| \text{ (i.e., maximum row sum)} \tag{2.73b}$$

$$(iii)\ \|A\|_E = \left\{ \sum_{i,j=1}^{n} |a_{ij}|^2 \right\}^{1/2} \text{ (known as Euclidean norm; also known as Schur or Frobenius norm)} \tag{2.73c}$$

It may be shown that all the norms defined from (2.73a) through (2.73c) satisfy the properties stated under (2.72).

The matrix norms are used to check the convergence of sequence of matrices, say $A^{(k)}$ to a particular matrix B under specific transformations. It is discussed in Chapter 10. They may also be used to make a check on the size of the elements of the matrix when it is raised to a certain power, i.e. $A^m = A \times A \times \ldots .m$ times.

Among the norms (2.73a-c), Euclidean norm is the most commonly used norm which can also be written as

$$\|A\|_E = [\mathrm{Tr}(A^*A)]^{1/2} = [\mathrm{Tr}(AA^*)]^{1/2} \tag{2.74a}$$

where A^* denotes complex conjugation and transposition of matrix A and Tr means the 'Trace' of the matrix, i.e., sum of its diagonal elements. When elements of A are real, then (2.74a) may be written as

$$\|A\|_E = [\mathrm{Tr}(A^TA)]^{1/2} = [\mathrm{Tr}(AA^T)]^{1/2}. \tag{2.74b}$$

To show that the norms mentioned from (2.73a) through (2.73c) satisfy the conditions of a norm as given in (2.72), we make use of Schwarz inequality which is given as follows:

The inner (scalar) product of two real vectors $\boldsymbol{x}$ and $\boldsymbol{y}$ in n-dimensional space is defined as

$$< \boldsymbol{x}, \boldsymbol{y} > = \boldsymbol{x}^T\boldsymbol{y} = \sum_{i=1}^{n} x_i y_i = \boldsymbol{y}^T\boldsymbol{x}.$$

Express magnitudes of vectors $\boldsymbol{x}$ and $\boldsymbol{y}$ as

$$\|\boldsymbol{x}\| = \|\boldsymbol{x}\|_2 = \sqrt{< \boldsymbol{x}, \boldsymbol{x} >}$$

$$\|\boldsymbol{y}\| = \|\boldsymbol{y}\|_2 = \sqrt{< \boldsymbol{y}, \boldsymbol{y} >}.$$

The Schwarz inequality is given as

$$| < \boldsymbol{x}, \boldsymbol{y} > | \le \|\boldsymbol{x}\| \cdot \|\boldsymbol{y}\|, \tag{2.75a}$$

or

$$|\Sigma x_i y_i| \le \sqrt{\Sigma x_i^2} \cdot \sqrt{\Sigma y_i^2} \tag{2.75b}$$

Example 2.21

Prove that Euclidean norm $||A||_E$ satisfies all the properties of a norm.

Proof: Property (*i*)

$$||A||_E = \left[\sum_{i,j} |a_{ij}|^2\right]^{1/2}$$

As the right side is sum of squares, it can not be zero unless all the elements a_{ij} of A are zero. It is ≥ 0.

Property (*ii*)

$$||\alpha A||_E = |\alpha| \cdot ||A||_E$$

$$||\alpha A||_E = \left[\sum_{i,j} |\alpha a_{ij}|^2\right]^{1/2} = |\alpha| \cdot \left[\sum_{i,j} |a_{ij}|^2\right]^{1/2}$$

Property (*iii*)

$$||A+B||_E \leq ||A||_E + ||B||_E$$

$$||A+B||_E^2 = \sum_{i,j} |a_{ij} + b_{ij}|^2$$

$$\leq \sum_{i,j} \{|a_{ij}| + |b_{ij}|\}^2 \quad \because \quad |z_1 + z_2| \leq |z_1| + |z_2|$$

$$= \Sigma|a_{ij}|^2 + \Sigma|b_{ij}|^2 + 2\Sigma|a_{ij}| \cdot |b_{ij}|$$

$$\leq \Sigma|a_{ij}|^2 + \Sigma|b_{ij}|^2 + 2\sqrt{\Sigma|a_{ij}|^2}\sqrt{\Sigma|b_{ij}|^2}, \text{ by Schwarz inequality}$$

$$= ||A||_E^2 + ||B||_E^2 + 2||A||_E \cdot ||B||_E$$

$$= \{||A||_E + ||B||_E\}^2$$

or $||A+B||_E \leq ||A||_E + ||B||_E$.

Property (*iv*)

$$||A \cdot B||_E \leq ||A||_E \cdot ||B||_E$$

$$||A \cdot B||_E^2 = \sum_{i,\,j} \left|\sum_k a_{ik} \cdot b_{kj}\right|^2$$

$$\leq \sum_{i,\,j}\left(\sum_{k}|a_{ik}\cdot b_{kj}|^2\right)$$

$$\leq \sum_{i,\,j}\left[\sum_{k}|a_{ik}|^2\cdot|b_{kj}|^2\right] \qquad \because |z_1\cdot z_2|\leq|z_1|\cdot|z_2|$$

$$\leq \sum_{i}\sum_{j}|a_{ij}|^2\cdot\sum_{j}\sum_{i}|b_{ij}|^2 \text{ by Schwarz inequality}$$

$$= \|\mathrm{A}\|_{\mathrm{E}}^2\cdot\|\mathrm{B}\|_{\mathrm{E}}^2$$

or $\|\mathrm{A}\cdot\mathrm{B}\|\leq\|\mathrm{A}\|_{\mathrm{E}}\cdot\|\mathrm{B}\|_{\mathrm{E}}$

2.24.4 Compatibility of matrix and vector norms

Since in most of the problems, matrix and vector occur together, their norms have to be dealt with simultaneously.

The norm of a matrix $\|\mathrm{A}\|$ is said to be compatible (consistent) with the norm of a vector $\|\boldsymbol{x}\|$, if

$$\|\mathrm{A}\boldsymbol{x}\|\leq\|\mathrm{A}\|\cdot\|\boldsymbol{x}\|, \tag{2.76}$$

where $\boldsymbol{x}$ is an arbitrary vector.

Following (2.76) the norm of a matrix is defined in terms of a vector norm as

$$\|\mathrm{A}\| = \max_{x}\frac{\|\mathrm{A}\boldsymbol{x}\|}{\|\boldsymbol{x}\|},\ \boldsymbol{x}\neq 0. \tag{2.77a}$$

If $\boldsymbol{x}$ is replaced by $\boldsymbol{x}/\|\boldsymbol{x}\|$, (2.77a) may also be written as

$$\|\mathrm{A}\| = \max\|\mathrm{A}\boldsymbol{x}\|,\ \|\boldsymbol{x}\| = 1. \tag{2.77b}$$

The matrix norm defined in the manner (2.77a or b) is said to be subordinate to vector norm. The matrix norms mentioned by (2.73a, b and c) satisfy property (2.76) with the corresponding vector norms.

Note: Defined in the manner (2.77a), $\|\mathrm{I}\| = 1$. Also $\|\mathrm{I}\|_{\mathrm{E}} = \sqrt{n}$ which is not subordinate to any vector. In general we will take $\|\mathrm{I}\| = 1$.

Example 2.22

Show that the matrix norm defined by $\|\mathrm{A}\| = \max(\|\mathrm{A}\boldsymbol{x}\|/\|\boldsymbol{x}\|)$, $\boldsymbol{x}\neq 0$ satisfies all the properties of a norm.

Solution (*i*) $||A|| \geq 0,\ ||A|| = 0$ only if $A = O$ (null matrix)

$||\boldsymbol{x}||$ is positive (+ve) and cannot be zero since $\boldsymbol{x} \neq 0$.

Since $\boldsymbol{x}$ is arbitrary, all the elements of vector $A\boldsymbol{x}$ can not be zero unless all the elements of matrix A are zero. Since $||A\boldsymbol{x}||$ is positive, max $(||A\boldsymbol{x}||/||\boldsymbol{x}||)$ will be positive (+ve) and will be zero only when $A = O$.

$$(ii)\ ||\alpha A|| = \max \frac{||\alpha A\boldsymbol{x}||}{||\boldsymbol{x}||} = \max \frac{|\alpha| \cdot ||A\boldsymbol{x}||}{||\boldsymbol{x}||} \qquad \because ||\alpha \boldsymbol{x}|| = |\alpha| \cdot ||\boldsymbol{x}||$$

$$= |\alpha| \cdot \max \frac{||A\boldsymbol{x}||}{||\boldsymbol{x}||}$$

$$(iii)\ ||A+B|| = \max \frac{||(A+B)\boldsymbol{x}||}{||\boldsymbol{x}||}$$

$$= \max \frac{||A\boldsymbol{x} + B\boldsymbol{x}||}{||\boldsymbol{x}||}$$

$$\leq \max \left\{ \frac{||A\boldsymbol{x}||}{||\boldsymbol{x}||} + \frac{||B\boldsymbol{x}||}{||\boldsymbol{x}||} \right\} \qquad \because ||\boldsymbol{x}+\boldsymbol{y}|| \leq ||\boldsymbol{x}|| + ||\boldsymbol{y}||$$

$$\leq \max \left(\frac{||A\boldsymbol{x}||}{||\boldsymbol{x}||} \right) + \max \left(\frac{||B\boldsymbol{y}||}{||\boldsymbol{x}||} \right)$$

$$= ||A|| + ||B||$$

$$(iv)\quad ||A \cdot B|| = \max \frac{||AB\boldsymbol{x}||}{||\boldsymbol{x}||}$$

$$\leq \max \frac{||A|| \cdot ||B\boldsymbol{x}||}{||\boldsymbol{x}||}, \quad B\boldsymbol{x} \neq 0 \text{ since } \boldsymbol{x} \neq 0 \text{ and } B \neq O$$

$$= ||A|| \cdot \max \frac{||B\boldsymbol{x}||}{||\boldsymbol{x}||}$$

$$= ||A|| \cdot ||B||.$$

Example 2.23

Show that

(*i*) Norm $||A||_1$ is compatible with norm $||\boldsymbol{x}||_1$.

(*ii*) Norm $||A||_\infty$ is compatible with norm $||\boldsymbol{x}||_\infty$.

(*i*) To prove $\|\mathrm{A}\boldsymbol{x}\|_1 \le \|\mathrm{A}\|_1 \cdot \|\boldsymbol{x}\|_1$

$$\begin{aligned}
\|\mathrm{A}\boldsymbol{x}\|_1 &= \sum_i \left|\sum_j a_{ij}x_j\right| \\
&\le \sum_i \sum_j |a_{ij}x_j| \\
&= \sum_j \sum_i |a_{ij}x_j| \\
&\le \sum_j \sum_i |a_{ij}| \cdot |x_j| \\
&= \sum_j |x_j| \sum_i |a_{ij}| \\
&\le \sum_j |x_j| \cdot \max_j \sum |a_{ij}| \\
&= \|\mathrm{A}\|_1 \cdot \|\boldsymbol{x}\|_1
\end{aligned}$$

(*ii*) To prove $\|\mathrm{A}\boldsymbol{x}\|_\infty \le \|\mathrm{A}\|_\infty \cdot \|\boldsymbol{x}\|_\infty$.

$$\begin{aligned}
\|\mathrm{A}\boldsymbol{x}\|_\infty &= \max_i \left|\sum_j a_{ij}x_j\right| \\
&\le \max_i \sum_j |a_{ij}x_j| \\
&\le \max_i \sum_j |a_{ij}| \cdot |x_j| \\
&\le \max_i \left\{\sum_j |a_{ij}| \cdot \left(\max_j |x_j|\right)\right\} \\
&= \max_i \left(\sum_j |a_{ij}|\right) \cdot \max_j |x_j| \\
&= \max_i \left(\sum_j |a_{ij}|\right) \cdot \max_i |x_i| \\
&= \|\mathrm{A}\|_\infty \cdot \|\boldsymbol{x}\|_\infty.
\end{aligned}$$

2.24.5 Spectral norm

Let us find the norm of a real matrix A which is subordinate to vector norm $\|\boldsymbol{x}\|_2$ (Euclidian norm).

$$\|A\|_2 = \max\{\|A\boldsymbol{x}\|_2\}, \ \|\boldsymbol{x}\|_2 = 1,$$

or $$\|A\|_2{}^2 = \max\{\|A\boldsymbol{x}\|_2\}^2, \ \|\boldsymbol{x}\|_2 = 1,$$

$$= \max < A\boldsymbol{x}, \ A\boldsymbol{x} >$$

$$= \max\{(A\boldsymbol{x})^T(A\boldsymbol{x})\}$$

$$= \max(\boldsymbol{x}^T A^T A\boldsymbol{x})$$

$$= \max(\boldsymbol{x}^T B\boldsymbol{x}), \text{ where } B = A^T A. \tag{2.78}$$

Note that matrix B is symmetric since $B^T = A^T A = B$.

Further, matrix B is positive definite since $\boldsymbol{x}^T B\boldsymbol{x} = \boldsymbol{x}^T A^T A\boldsymbol{x} = (A\boldsymbol{x})^T(A\boldsymbol{x}) \geq 0$; therefore all its e.values will be positive. Let μ_r be an e.value of B with corresponding e.vector $\boldsymbol{x}_r$. Then we have

$$B\boldsymbol{x}_r = \mu_r\boldsymbol{x}_r, \ r = 1(1)n. \tag{2.79}$$

or $$\boldsymbol{x}_r^T B\boldsymbol{x}_r = \mu_r\boldsymbol{x}_r^T\boldsymbol{x}_r$$

$$= \mu_r, \quad \text{since } \boldsymbol{x}_r^T\boldsymbol{x}_r = 1, \text{ i.e., } \|\boldsymbol{x}_r\|_2 = 1.$$

It may be noted that e.vector $\boldsymbol{x}_r$ and normalised e.vector $\boldsymbol{x}_r/\|\boldsymbol{x}_r\|$ both satisfy (2.79). Hence

$$\|A\|_2{}^2 = \max_r(\boldsymbol{x}_r B\boldsymbol{x}_r)$$

$$= \max_r(\mu_r), \text{ where } \mu_r \text{ is an e.value of } B = A^T A.$$

Let μ_1 be the largest e.value of B, then

$$\|A\|_2 = \sqrt{\mu_1}. \tag{2.80}$$

Further, if A is symmetric, i.e., $A^T = A$, then $B = A^T \cdot A = A^2$. In that case, if λ_r is an e.value of A, then corresponding e.value of B will be $\lambda_r{}^2$. Hence if λ_1 is the largest e.value of A, the largest e.value of B will be $\lambda_1{}^2$. Hence from (2.80)

$$\|A\|_2 = \sqrt{\lambda_1^2} = \lambda_1, \ A = A^T. \tag{2.81}$$

The norm defined by $\|A\|_2$ is called 'Spectral' norm. When matrix A is symmetric, then $\|A\|_2 = \lambda_1$, i.e. largest e.value of A which is the 'spectral radius' of A.

The above analysis is true even when elements of A are complex. In that case $B = A^*A$ (where $A^* = \overline{A}^T$).

Example 2.24

Prove the inequality for a real matrix A,

$$\|A\|_2 \leq \|A\|_E \leq \sqrt{n}\|A\|_2.$$

Proof: We know

$$\|A\|_2 = \sqrt{\mu_1}, \text{ where } \mu_1 \text{ is the largest e.value of } A^TA$$

$$\|A\|_E^2 = \text{Tr}\,(A^TA), \text{ See (2.74a)}$$

$$= \sum_{i=1}^{n} \mu_i; \ \mu_i, \ i = 1(1)n \text{ are the e.values of } A^TA, \text{ See (2.39a)}$$

$$\sqrt{\mu_1} \leq \sqrt{\Sigma\mu_i}$$

$\therefore$ $\quad \|A\|_2 \leq \|A\|_E$

Also $\quad \sqrt{n\mu_1} \geq \sqrt{\Sigma\mu_i}$

$$\sqrt{n}\|A\|_2 \geq \|A\|_E.$$

Example 2.25

Prove that

$$\|A\|_2 \leq \sqrt{\|A\|_1\|A\|_\infty}$$

Proof: Let $Ax = \lambda x$. Then for any norm

$$\|A\boldsymbol{x}\| = \|\lambda\boldsymbol{x}\| = |\lambda| \cdot \|\boldsymbol{x}\|$$

or $\quad \|A\| \cdot \|\boldsymbol{x}\| \geq |\lambda| \cdot \|\boldsymbol{x}\|$

or $\quad \|A\| \geq |\lambda|. \quad$ .... (1)

$$\|A\|_2{}^2 = \mu_1, \text{ largest e.value of } A^TA.$$

But $\quad \|A^TA\| \geq \mu_1$ from (1), for any norm.

Hence $\|A\|_2^2 \le \|A^T A\|_\infty$

$$\le \|A^T\|_\infty \cdot \|A\|_\infty$$

$$= \|A\|_1 \cdot \|A\|_\infty$$

or $\|A\|_2 \le \sqrt{\|A\|_1 \|A\|_\infty}$

Example 2.26

Prove that if $\|S\| < 1$, then matrix $(I+S)$ is non-singular. Also prove that,

(*i*) $\|(I+S)^{-1}\| \le \dfrac{1}{1-\|S\|}$,

(*ii*) $\|I-(I+S)^{-1}\| \le \dfrac{\|S\|}{1-\|S\|}$.

Proof: For any matrix S, $\|S\| \ge |\lambda|$ where λ is an e.value of S. The e.values of matrix $(I+S)$ are given by $1+\lambda$.

We know from (2.39b) that

$$|I+S| = \prod_{i=1}^{n}(1+\lambda_i) \qquad \text{.... (1)}$$

Since $\|S\| < 1$ and $\|S\| \ge |\lambda|$ we have

$$|\lambda| \le \|S\| < 1, \text{ i.e., } |\lambda| < 1.$$

$$-1 < \lambda < 1$$

As factor $1+\lambda_i$ can not be zero in (1), the determinant $|(I+S)|$ does not vanish and is non-singular, i.e., its inverse $(I+S)^{-1}$ exists.

(*i*) Let $G = (I+S)^{-1}$

Proceeding in the same manner as in Example 2.20 we can prove, taking $\|I\| = 1$,

$$\frac{1}{1+\|S\|} \le \|(1+S)^{-1}\| \le \frac{1}{1-\|S\|}$$

or $$\|(I+S)^{-1}\| \le \frac{1}{1-\|S\|}, \quad \|S\| < 1.$$

(*ii*) $$I - G = SG \quad \text{since } (I+S)G = I$$

$$\|I-G\| = \|SG\|$$

$$\leq ||S|| \cdot ||G||$$

$$\leq \frac{||S||}{1 - ||S||}$$

or $$||I - (I + S)^{-1}|| \leq \frac{||S||}{1 - ||S||}.$$

2.25 Sensitivity of Solution of Linear Equations

Let us suppose that the system of linear equations is given by

$$A\boldsymbol{x} = \boldsymbol{b}. \tag{2.82}$$

We want to compute the resulting error in the true solution $\boldsymbol{x}$ when there is perturbation (*i*) in the right side of (2.82) (*ii*) in the left side (2.82) and (*iii*) in the left and right sides of (2.82).

(*i*) Perturbation in $\boldsymbol{b}$

Let us suppose that the rightside of (2.82) is perturbed such that $\boldsymbol{b}$ becomes $\boldsymbol{b} + \boldsymbol{k}$ and let the corresponding error in $\boldsymbol{x}$ be $\boldsymbol{h}$ so that

$$A\,(\boldsymbol{x} + \boldsymbol{h}) = \boldsymbol{b} + \boldsymbol{k}. \tag{2.83}$$

From (2.83) and (2.82) we get

$$A\boldsymbol{h} = \boldsymbol{k}$$

or $$\boldsymbol{h} = A^{-1}\boldsymbol{k} \tag{2.84}$$

Taking norm

$$||\boldsymbol{h}|| = ||A^{-1}\boldsymbol{k}||$$

$$\leq ||A^{-1}|| \cdot ||\boldsymbol{k}||. \tag{2.85}$$

Also from (2.82)

$$||A\boldsymbol{x}|| = ||\boldsymbol{b}||$$

or $$||A|| \cdot ||\boldsymbol{x}|| \geq ||\boldsymbol{b}|| \tag{2.86}$$

Dividing (2.85) by (2.86)

$$\frac{||\boldsymbol{h}||}{||\boldsymbol{x}||} \leq ||\mathrm{A}^{-1}|| \cdot ||\ \mathrm{A}\ || \frac{||\boldsymbol{k}||}{||\boldsymbol{b}||} = \kappa(\mathrm{A}) \frac{||\boldsymbol{k}||}{||\boldsymbol{b}||}, \tag{2.87}$$

where the product $||\mathrm{A}^{-1}|| \cdot ||\mathrm{A}||$ is known as 'condition number' of matrix A and is denoted as

$$\kappa(\mathrm{A}) = ||\mathrm{A}^{-1}|| \cdot ||\mathrm{A}||. \tag{2.88}$$

The expression (2.87) gives the magnitude of the relative error in the solution $\boldsymbol{x}$ corresponding to the relative perturbation in $\boldsymbol{b}$. The greater the condition number, larger is the relative error. Its value is usually greater than unity, since $||\mathrm{I}|| = ||\mathrm{A}^{-1}\mathrm{A}|| \leq ||\mathrm{A}||.||\mathrm{A}^{-1}||$ or $1 \leq ||\mathrm{A}||.||\mathrm{A}^{-1}||$.

(*ii*) Perturbation in A

Let us suppose that there is perturbation E in A and the corresponding perturbation in $\boldsymbol{x}$ is $\boldsymbol{h}$ such that

$$(\mathrm{A}+\mathrm{E})(\boldsymbol{x}+\boldsymbol{h}) = \boldsymbol{b}. \tag{2.89}$$

From (2.89) after using (2.82) we get

$$(\mathrm{A}+\mathrm{E})\boldsymbol{h} = -\mathrm{E}\boldsymbol{x}$$

or

$$\begin{aligned} \boldsymbol{h} &= -(\mathrm{A}+\mathrm{E})^{-1}\mathrm{E}\boldsymbol{x} \\ &= -\{\mathrm{A}(\mathrm{I}+\mathrm{A}^{-1}\mathrm{E})\}^{-1}\mathrm{E}\boldsymbol{x} \\ &= -(\mathrm{I}+\mathrm{A}^{-1}\mathrm{E})^{-1}\mathrm{A}^{-1}\mathrm{E}\boldsymbol{x} \end{aligned}$$

Taking norm

$$\begin{aligned} ||\boldsymbol{h}|| &= ||-(\mathrm{I}+\mathrm{A}^{-1}\mathrm{E})^{-1}\mathrm{A}^{-1}\mathrm{E}\boldsymbol{x}|| \\ &\leq ||(\mathrm{I}+\mathrm{A}^{-1}\mathrm{E})^{-1}|| \cdot ||\mathrm{A}^{-1}\mathrm{E}|| \cdot ||\boldsymbol{x}||, \\ &\leq \frac{||\mathrm{A}^{-1}\mathrm{E}||}{1-||\mathrm{A}^{-1}\mathrm{E}||}||\boldsymbol{x}||, \text{ if } ||\mathrm{A}^{-1}\mathrm{E}|| < 1. \qquad \text{(See Example 2.26)} \\ &\leq \frac{||\mathrm{A}^{-1}|| \cdot ||\mathrm{E}||}{1-||\mathrm{A}^{-1}|| \cdot ||\mathrm{E}||}||\boldsymbol{x}||, \end{aligned}$$

$$\text{or}\quad \frac{\|\boldsymbol{h}\|}{\|\boldsymbol{x}\|} \le \frac{\|A^{-1}\|\cdot\|A\|\cdot\dfrac{\|E\|}{\|A\|}}{1-\|A^{-1}\|\cdot\|A\|\cdot\dfrac{\|E\|}{\|A\|}}$$

$$= \frac{\kappa(A)\dfrac{\|E\|}{\|A\|}}{1-\kappa(A)\dfrac{\|E\|}{\|A\|}}. \tag{2.90}$$

The expression (2.90) gives relative error in the solution in terms of the relative perturbation in the elements of matrix. The perturbation is expected to be small which may usually accrue due to rounding.

(iii) Perturbation in A and $\boldsymbol{b}$

Let E be a matrix showing perturbation in A and $\boldsymbol{k}$ be a vector showing perturbation in $\boldsymbol{b}$ while $\boldsymbol{h}$ denotes the corresponding error in the solution vector $\boldsymbol{x}$. The system of equations should satisfy.

$$(A+E)(\boldsymbol{x}+\boldsymbol{h}) = \boldsymbol{b}+\boldsymbol{k}$$

$$\text{or}\quad (A+E)\boldsymbol{h} = \boldsymbol{k}-E\boldsymbol{x}$$

$$\text{or}\quad \boldsymbol{h} = (A+E)^{-1}(\boldsymbol{k}-E\boldsymbol{x})$$

$$= (I+A^{-1}E)^{-1}A^{-1}(\boldsymbol{k}-E\boldsymbol{x})$$

Taking norm

$$\|\boldsymbol{h}\| \le \|(I+A^{-1}E)^{-1}\|\cdot\|A^{-1}\|\cdot\{\|\boldsymbol{k}\|+\|E\|\cdot\|\boldsymbol{x}\|\}$$

$$\le \frac{\|A^{-1}\|\{\|\boldsymbol{k}\|+\|E\|\cdot\|\boldsymbol{x}\|\}}{1-\|A^{-1}E\|}$$

$$\le \frac{\|A^{-1}\|\cdot\|A\|\left\{\dfrac{\|\boldsymbol{k}\|}{\|A\|}+\dfrac{\|E\|\cdot\|\boldsymbol{x}\|}{\|A\|}\right\}}{1-\|A^{-1}\|\cdot\|E\|},\qquad \text{if } \|A^{-1}E\|<1.$$

$$\le \frac{\|A^{-1}\|\cdot\|A\|\left\{\dfrac{\|\boldsymbol{k}\|}{\|\boldsymbol{b}\|}+\dfrac{\|E\|}{\|A\|}\right\}\|\boldsymbol{x}\|}{1-\|A^{-1}\|\|A\|\cdot\frac{\|E\|}{\|A\|}},\quad \|A\|\cdot\|\boldsymbol{x}\|\ge\|\boldsymbol{b}\| \text{ or } \|\boldsymbol{b}\|/\|\boldsymbol{x}\|\le\|A\|.$$

or
$$\frac{\|\boldsymbol{h}\|}{\|\boldsymbol{x}\|} \leq \frac{\kappa(\mathrm{A}) \cdot \left\{ \frac{\|\boldsymbol{k}\|}{\|\boldsymbol{b}\|} + \frac{\|\mathrm{E}\|}{\|\mathrm{A}\|} \right\}}{1 - \kappa(\mathrm{A}) \cdot \frac{\|\mathrm{E}\|}{\|\mathrm{A}\|}}. \tag{2.91}$$

The expression (2.91) gives the magnitude of relative error in the solution $\boldsymbol{x}$ in terms of the relative perturbation in A and $\boldsymbol{b}$.

A numerical problem is said to be 'ill-conditioned', if a minor change in the data causes a large change in the actual solution. If the problem is not too sensitive to such a change, it is called 'well-conditioned'. The condition number is a measure for determining the sensitivity of the solution. It may be stated that if A is symmetric (Hermitian), then using spectral norm, the condition number is given by

$$\kappa(\mathrm{A}) = \|\mathrm{A}^{-1}\|_2 \cdot \|\mathrm{A}\|_2 = \lambda_{\max}(\mathrm{A}^{-1}) \cdot \lambda_{\max}(\mathrm{A}) = \lambda_{\max}(\mathrm{A})/\lambda_{\min}(\mathrm{A}). \tag{2.92}$$

For more details see [6]

Exercise 2

2.1 Solve the following equations by Gaussian elimination method without row interchange,

$$\begin{aligned} 5x_1 + 2x_2 + 4x_3 &= 24 \\ 3x_1 + x_2 + 7x_3 &= 18 \\ 8x_1 + 4x_2 + 5x_3 &= 41 \end{aligned}$$

Compute up to two decimals only and give your answer to nearest integer.

2.2 Solve the following equations by Gaussian elimination method using row interchanges (partial pivoting)

$$\begin{aligned} 3x_2 + 4x_3 &= 2 \\ 4x_1 - 2x_2 + x_3 &= 18 \\ 3x_1 + 4x_2 + 5x_3 &= 11 \end{aligned}$$

Compute up to two decimals only and give your answer to the nearest integer.

2.3 Solve by Gauss–Jordan method the following system of equations,

$$\begin{aligned} 4x_1 + 5x_2 - 2x_3 &= 16.8 \\ -5x_1 + x_2 + 4x_3 &= 7.4 \\ 3x_1 - 2x_2 + x_3 &= -0.7 \end{aligned}$$

Compute up to three decimals only and give your answer rounding up to one place of decimal.

2.4 Find A^{-1} up to two decimal places (no fraction) by Gaussian elimination method when,

$$A = \begin{bmatrix} 6 & 7 & 8 \\ 3 & 5 & 2 \\ 2 & 4 & 5 \end{bmatrix}.$$

Also solve $Ax = b$ using A^{-1} where $b^T = (15 \;\; 2 \;\; 8)$.

2.5 Find A^{-1} using Gauss–Jordan's method where

$$A = \begin{bmatrix} 2 & 4 & 5 \\ 3 & 5 & 2 \\ 6 & 7 & 8 \end{bmatrix}.$$

Also solve $Ax = b$ using A^{-1} where $b^T = (8 \;\; 2 \;\; 15)$.
Compute up to 2 decimal places (no fractions).

2.6 Decompose the following matrix A into LU form, by Gaussian elimination method, where L is a unit lower triangular matrix and U an upper triangular.

$$A = \begin{bmatrix} 5 & 4 & -2 \\ 8 & 3 & 1 \\ 2 & 5 & 6 \end{bmatrix}.$$

Using decomposition solve $Ax = b$ where $b^T = (21 \;\; 41 \;\; 56)$.

2.7 Using Crout's method factorise A to LU form where L is a lower triangular matrix and U, a unit upper triangular,

$$A = \begin{bmatrix} 12 & 7 & 5 \\ 8 & 15 & 10 \\ 10 & 6 & 15 \end{bmatrix}$$

Also solve $Ax = b$ where $b^T = (26 \;\; 21 \;\; 11)$. Compute up to 2 decimals only.

2.8 Use Cholesky's method to reduce the symmetric matrix A to LL^T form where L is a lower triangular matrix and

$$A = \begin{bmatrix} 4 & 5 & 3 & 4 \\ 5 & 8 & 4 & 6 \\ 3 & 4 & 5 & 7 \\ 4 & 6 & 7 & 12 \end{bmatrix}.$$

2.9 Reduce the following matrix A to the form PA = LU by Gaussian elimination with partial pivoting; P is a permutation matrix and L is a unit lower triangular and U an upper triangular matrix,

$$A = \begin{bmatrix} 4 & 2 & 3 & 4 \\ 6 & 5 & 4 & 5 \\ 3 & 5 & 4 & 7 \\ 3 & 6 & 8 & 3 \end{bmatrix}$$

Also solve $A\boldsymbol{x} = \boldsymbol{b}$ where $\boldsymbol{b}^T = (9 \quad 10 \quad 9 \quad 8)$

2.10 Solve the following equation, correct up to two places of decimal, by
(*i*) Gauss–Jacobi method (*ii*) Gauss–Seidel method
Write the appropriate iterative schemes in both cases.

$$-4x_1 + x_2 + 10x_3 = 21$$
$$5x_1 - x_2 + x_3 = 14$$
$$2x_1 + 8x_2 - x_3 = -7$$

2.11 Using Power method find the dominant latent root and associated latent vector of the matrix A, where

$$A = \begin{bmatrix} 5 & 10 \\ 2 & 6 \end{bmatrix}.$$

Also find the subdominant root and the associated latent vector by the process of root removal. Compute up to two decimal places only.

2.12 Prove the following inequalities

(*i*) $\|\boldsymbol{x}\|_\infty \le \|\boldsymbol{x}\|_1 \le n\|\boldsymbol{x}\|_\infty$

(*ii*) $\|\boldsymbol{x}\|_\infty \le \|\boldsymbol{x}\|_2 \le \sqrt{n}\|\boldsymbol{x}\|_\infty$.

2.13 If E is perturbation in matrix A, show that the magnitude of relative error in the computation of A^{-1} is given by

$$\frac{\kappa(A)\cdot\|E\|/\|A\|}{1-\kappa(A)\|E\|/\|A\|},\quad \|A^{-1}E\|<1.$$

Hint: $\|A^{-1}-(A+E)^{-1}\| = \|\{I-(I+A^{-1}E)^{-1}\}A^{-1}\|$; (Use result of Example 2.26).

2.14 Using ∞-norm, show that the iterative schemes (a) Gauss–Jacobi and (b) Gauss–Seidel for solving $Ax = \mathbf{b}$, converge to the true solution when the coefficient matrix $A(n\times n)$ is diagonally dominant. Also prove that (b) converges faster.
Hint: Since A is diagonally dominant

$$|a_{ii}| > \sum_{j=1}^{i-1}|a_{i,j}| + \sum_{j=i+1}^{n}|a_{i,j}| or \sum_{j=1}^{i-1}\left|\frac{a_{ij}}{a_{ii}}\right| + \sum_{j=i+1}^{n}\left|\frac{a_{ij}}{a_{ii}}\right| < 1.$$

Let $p_i = \sum_{j=1}^{i-1}\left|\frac{a_{i,j}}{a_{ii}}\right|$ and $q_i = \sum_{j=i+1}^{n}\left|\frac{a_{i,j}}{a_{ii}}\right|, i = 1(1)n.$

Then $p_i + q_i < 1, i = 1, 2, \cdots n.$ (A)

Note that,

(*i*) the diagonal matrix $D^{-1} = [1/a_{ii}]$.

(*ii*) $D^{-1}L$ is a strictly lower triangular and $D^{-1}U$ a strictly upper triangular matrix i.e., diagonal elements are zero.

(a) Convergence of Gauss–Jacobi Scheme

The error vector $e^{k+1} = x^{k+1} - x$, satisfies

$$\boldsymbol{e}^{(k+1)} = G.\boldsymbol{e}^{(k)} = D^{-1}(L+U)\boldsymbol{e}^{(k)} = (D^{-1}L + D^{-1}U)\boldsymbol{e}^{(k)}$$

$$\left\|\boldsymbol{e}^{(k+1)}\right\| = \left\|(D^{-1}L + D^{-1}U)\boldsymbol{e}^{(k)}\right\| \cdots \qquad \text{(B)}$$

Let s^{th} be the largest element (in modulus) of $\boldsymbol{e}^{(k+1)}$ i.e.,

$$\left|e_s^{(k+1)}\right| = \max_i \left|e_i^{(k+1)}\right| = \left\|e^{(k+1)}\right\| \cdots \qquad \text{(C)}$$

Hence from (B) and (C)

$$\left\|e^{(k+1)}\right\| \leq \left|e_s^{(k+1)}\right| = \sum_{j=1}^{i-1}\left|\frac{a_{sj}}{a_{ss}}e_j^{(k)}\right| + \sum_{j=i+1}^{n}\left|\frac{a_{sj}}{a_{ss}}e_j^{(k)}\right|$$

$$\leq \sum_{j=1}^{i-1}\left|\frac{a_{sj}}{a_{ss}}\right|\cdot\left|e_j^{(k)}\right| + \sum_{j=i+1}^{n}\left|\frac{a_{sj}}{a_{ss}}\right|\cdot\left|e_j^{(k)}\right|$$

i.e., $\leq (p_s + q_s).\left\|e^{(k)}\right\|$.

But $p_s + q_s < 1$ from A; hence proved

Remember that (s,s) element of $\mathrm{D}^{-1}\mathrm{L}$ and $\mathrm{D}^{-1}\mathrm{U}$ is zero.

(b) Convergence of Gauss–Seidel Scheme

$$\boldsymbol{e}^{(k+1)} = \mathrm{H}\boldsymbol{e}^{(k)} = -(\mathrm{D}+\mathrm{L})^{-1}\mathrm{U}\boldsymbol{e}^{(k)} = -(\mathrm{I}+\mathrm{D}^{-1}\mathrm{L})^{-1}\mathrm{D}^{-1}\mathrm{U}\boldsymbol{e}^{(k)}$$

or $$\boldsymbol{e}^{k+1} = -\mathrm{D}^{-1}L\boldsymbol{e}^{(k+1)} - \mathrm{D}^{-1}\mathrm{U}\boldsymbol{e}^{(k)} \quad \text{(D)}$$

As in (a) let s^{th} be the numerically largest element of e^{k+1}. Proceeding in the same manner as before we can write

$$\left\|e^{(k+1)}\right\| \leq \sum_{j=1}^{i-1}\left|\frac{a_{sj}}{a_{ss}}\right|\cdot\left\|e^{(k+1)}\right\| + \sum_{j=i+1}^{n}\left|\frac{a_{sj}}{a_{ss}}\right|\cdot\left\|e^{(k)}\right\|$$

$$= p_s\left\|e^{k+1}\right\| + q_s\left\|e^{(k)}\right\|$$

or $$\left\|e^{(k+1)}\right\| \leq \frac{q_s}{1-p_s} < 1 \quad \text{from (A).}$$

To prove that (b) converges faster than (a) we have to show that the convergence factor in Gauss–Seidel scheme is smaller than in Gauss–Jacobi i.e.,

$$\frac{q_i}{1-p_i} < p_i + q_i \quad \text{or} \quad q_i < p_i + q_i - p_i(p_i + q_i)$$

or $p_i + q_i < 1$, provided $p_i \neq 0$, which is true.

Hence, for any diagonally dominant coefficient matrix Gauss–Seidel scheme will converge faster than Gauss–Jacobi's. Also remember that the condition (A) is sufficient, not necessary.

References and Some Useful Related Books/Papers

1. Datta K.B., *Matrix and Linear Algebra*, Prentice Hall of India.
2. Froberg C.E., *Introduction to Numerical Analysis*, Addison-Wesley Pub. Co. Inc.
3. Goodwin E.T., *Modern Computing Methods*, HMSO, London.
4. Householder A.S., *Principles of Numerical Analysis*, McGraw-Hill.
5. Schwaz H.R., Rutishauser *H.* and Stiefel E. (Translated by Hertelendy *P.*), *Numerical Analysis Symmetric Matrices*, Prentice-Hall Inc.
6. Varga R.S., *Matrix Iterative Analysis*, Prentice-Hall.
7. Wilkinson J.H., *The Algebraic Eigenvalue Problem*, Oxford University Press.

3

Nonlinear Equations

3.1 Introduction

In this chapter we deal with methods for solving nonlinear equations which may be algebraic or may include transcendental functions, i.e., the functions which are expressed by infinite series, e.g., $e^x, \log x$, $\sin x$, etc.

A number α is said to be the root of an equation $f(x) = 0$, if it satisfies the equation, i.e., $f(\alpha) = 0$; it is also called a zero of the function $f(x)$. There can be one or more roots of an equation in a given domain. The roots may be simple (non-repeated) or repeated and real or complex. Although we shall be mainly concerned with the real roots, some of the methods discussed may be applicable for determining the complex roots as well. It will also be assumed that the function $f(x)$ is continuous in a finite domain containing the root and that all the roots are isolated meaning thereby that there exists a finite neighbourhood of the root in which no other root lies.

All the methods to be discussed for finding the roots will be iterative. That is, we start from an initial estimate (reasonably good, not arbitrary) of the root or a small interval in which the root lies; then improve it using a method in an iterative manner. As an inherent criterion of the iterative techniques, we stop the process as soon as the two successive values agree within a pre-assigned accuracy. It may be emphasised that we get an approximate root $\bar{x}$ which is as close to the exact root α as we desire.

Thus the root $\bar{x}$ will satisfy the given equation to only a certain degree of accuracy, i.e., $f(\bar{x}) = \delta$, where δ is quite small in magnitude and is generally known as residual error or simply residual. It may also be noted that the roots obtained for the equations $f(x) = 0$ and $\mathrm{k}f(x) = 0$, will not be same since in the latter case the residual δ will get multiplied by a factor k. In general, we may not know from the residual error as how much accurate an approximate root is but we can ascertain whether one estimate is better or worse vis-a-vis another estimate by their residuals.

If $\bar{x}$ is an estimate of the exact root α and ε denotes the error in $\bar{x}$ such that $\alpha - \bar{x} = \varepsilon$, then the magnitude of the error may be determined as follows:

Assuming ε small we can write,

$$f(\bar{x}) = f(\alpha - \varepsilon)$$

$$= f(\alpha) - \varepsilon f'(\alpha), \text{ neglecting higher powers of } \varepsilon.$$

or $$|\varepsilon| = |f(\bar{x})/f'(\alpha)|, \text{ since } f(\alpha) = 0$$

or $$|\varepsilon| \simeq |f(\bar{x})/f'(\bar{x})|, \tag{3.1}$$

provided $\bar{x}$ is close to α and function $f(x)$ is smooth near the root α.

In all the methods to be discussed, our first step would be to identify an interval, say, $I = (x_1, x_2)$ in which the root, we are interested in, lies and that there is no other root in I. Looking at the general nature of the function $y = f(x)$, we search for two points x_1 and x_2 enclosing the root such that $f(x_1)$ and $f(x_2)$ have opposite signs ascertaining at the same time that there is only one root of $f(x) = 0$ in I. There are a few simple rules regarding the roots when we have a polynomial equation,

$$f(x) \equiv p_n(x) \equiv a_n x^n + a_{n-1} x^{n-1} + \ldots. + a_1 x + a_0 = 0, \; a_n > 0.$$

According to those rules we may get an idea about the number of positive and negative roots a polynomial equation might have. However, that information may not be of much consequence as in any case we shall have to tabulate the values of the polynomial to look for an interval in which the desired root lies.

3.2 Order of Convergence of Iterative Method

The order of convergence of an iterative method signifies the rate at which the nth iterate x_n approaches the exact root α as n increases. Let ε_n be the error in the nth iterate x_n so that $\varepsilon_n = \alpha - x_n$ and similarly the error in the $(n-1)^{\text{th}}$ iterate x_{n-1}, i.e., $\varepsilon_{n-1} = \alpha - x_{n-1}$. Then an iterative method is said to be of order p, if

$$\varepsilon_n = C\varepsilon_{n-1}^p \tag{3.2}$$

where C is a constant called 'convergence factor'. It means that the error at any step is C times the pth power of the error that of the previous step. It should be noted that the rate of convergence of a method is dependent on both C and p, since (3.2) can be written as,

$$\varepsilon_n = C^n \varepsilon_0^p. \tag{3.2a}$$

It will be assumed that the error ε_0 is small.

Now, we discuss a few methods to find the roots.

3.3 Method of Successive Substitution

Let us suppose, we have to find the roots of the equation,

$$f(x) = 0. \tag{3.3}$$

Express (3.3) in the form,

$$x = \phi(x), \tag{3.4}$$

and the required iterative scheme is given by,

$$x_{n+1} = \phi(x_n),\ n \geq 0 \tag{3.5}$$

where x_n denotes the nth iterated value which is known and the $(n+1)^{\text{th}}$ approximation x_{n+1} is to be computed from (3.5); x_0 is the initial estimate of the root α which we are to find. Obviously, the root is the point of intersection of the curves $y = x$ and $y = \phi(x)$. However, there can be many (infinite) possibilities for expressing (3.3) in the form (3.4) and consequently the iterative scheme (3.5). But all iterative schemes may not converge to a solution or rather they may diverge, that is, x_n may start giving absurd values as n increases.

The necessary and sufficient condition for the scheme (3.5) to converge is that the modulus of the first derivative of $\phi(x)$ at $x = \alpha$ should be less than 1, i.e.,

$$|\phi'(\alpha)| < 1.$$

To prove the above we note that if α is a root of $f(x) = 0$, then (3.4) becomes,

$$\alpha = \phi(\alpha). \tag{3.6}$$

From the iterative scheme (3.5),

$$x_n = \phi(x_{n-1}). \tag{3.7}$$

Subtracting (3.7) from (3.6),

$$\alpha - x_n = \phi\ (\alpha) - \phi\ (x_{n-1})$$

or
$$\begin{aligned} \varepsilon_n &= \phi(\alpha) - \phi(\alpha - \varepsilon_{n-1}) \\ &= \phi(\alpha) - \phi(\alpha) + \varepsilon_{n-1}\phi'(\alpha) + \text{ higher powers of } \varepsilon_{n-1} \\ &\simeq \phi'(\alpha)\varepsilon_{n-1}, \text{ assuming error is small.} \end{aligned} \tag{3.8}$$

Applying (3.8) repeatedly, we get,

$$\varepsilon_n = \{\phi'(\alpha)\}^n \varepsilon_0$$

For convergence ε_n should tend to zero as $n \to \infty$. This can happen only if,

$$|\phi'(\alpha)| < 1, \tag{3.9}$$

which is the required condition for convergence.

As α is not known, we test the condition (3.9) at a point x_0, the initial estimate which is close to α. Thus in practice to choose appropriate function $\phi(x)$, we make sure that,

$$|\phi'(x_0)| < 1. \tag{3.10}$$

As far as order of convergence of the method is concerned it is clear from (3.8) that,

$$\varepsilon_{n+1} = C\varepsilon_n, \tag{3.11}$$

where $\quad C = \phi'(\alpha).$ (3.11a)

Comparing with (3.2) we see that the method has linear convergence; its rate depends on the value of $|\phi'(\alpha)|$, i.e., smaller its value faster would be the convergence.

This method is also known as 'Fixed Point' method for the reason that the mapping $x \to \phi(x)$ maps the root α to itself since it satisfies $\alpha = \phi(\alpha)$. That is, the point (root) $x = \alpha$ remains 'fixed' (unchanged) under the mapping $x \to \phi(x)$.

Example 3.1

Find the real roots of $x^3 - 2x - 8 = 0$ by method of successive substitution, correct up to two places of decimal only.

Solution $f(x) = x^3 - 2x - 8$

To locate the roots we form the table:

x	$f(x)$
0	−8
1	−9
2	−4
3	13

x	$f(x)$
0	−8
−1	−7
−2	−12
−3	−29

There is only one real root which lies between 2 and 3; and no negative roots at all.

Now to express $f(x) = 0$ in the form $x = \phi(x)$, let us try

(*i*) $x = \frac{1}{2}(x^3 - 8)$

(*ii*) $x = x^3 - x - 8$

(*iii*) $x = (2x + 8)^{1/3}$

Choose an estimate x_0 somewhere between 2 and 3, say $x_0 = 2.5$ (although it will be better to take x_0 closer to 2 since the value of $f(x)$ is smaller at $x = 2$ as compared at $x = 3$).

Now to check which of the forms out of (*i*), (*ii*) or (*iii*) is acceptable, we check the criterion (3.9), i.e., $|\phi'(x_0)| < 1$.

case (*i*) $\quad \phi(x) = \frac{1}{2}(x^3 - 8);\ \phi'(x) = \frac{3}{2}x^2$

$|\phi'(2.5)| > 1$, therefore not acceptable

case (*ii*) $\quad \phi(x) = x^3 - x - 8;\ \phi'(x) = 3x^2 - 1$

$|\phi'(2.5)| > 1$, therefore not acceptable.

case (*iii*) $\quad \phi(x) = (2x + 8)^{1/3};\ \phi'(x) = \dfrac{1}{3(2x+8)^{2/3}}$

$|\phi'(2.5)| < 1$; this form should be acceptable.

Therefore, we select the iterative scheme,

$$x_{n+1} = (2x_n + 8)^{1/3}.$$

Starting from $x_0 = 2.5$, the iterated values are given as follows:

n	0	1	2	3
x_n	2.5	2.351	2.333	2.331

We compute up to 3 places of decimal and continue until two successive values agree within two places of decimal, after rounding. In this case iterations 2 and 3 agree and hence the approximate value of the root is,

$$\alpha \simeq 2.33.$$

3.4 Bisection Method (Method of Halving)

The Bisection method may be explained by the following steps:

(*i*) Find an interval I = $(x_1,\ x_2)$ in which the desired root of $f(x) = 0$ lies and that there is no other root in I. Obviously $f(x_1)$ and $f(x_2)$ have opposite signs.

(*ii*) Find midpoint of x_1 and x_2, say $x = \dfrac{x_1 + x_2}{2}$, and compute value of $f(x)$. If $f(x) = 0$ (or some very small pre-assigned value), then x is the desired root.

(*iii*) If $f(x)$ and $f(x_2)$ have same sign, i.e., sign $\{f(x)\} = $ sign $\{f(x_2)\}$ then change x_2 by x else change x_1 by x. In the computer either sign function may be used or condition $f(x) \times f(x_2) > 0$ may be used.

(*iv*) Check if the new interval $|x_2 - x_1|$ is within the desired accuracy. If criterion is fulfilled then stop else go to step (*ii*) and repeat the process until the condition is satisfied. For example if we want the answer correct up to two places of decimal then we should continue the process of bisecting the interval until its length is ≤ 0.005.

If the length of the initial interval is |I|, then after n iterations its length would be

$$|\mathrm{I}_n| = \frac{|\mathrm{I}|}{2^n} = \varepsilon \text{ (say)}. \tag{3.12}$$

It is easy to calculate the number of iterations, n required to obtain the value of $|\mathrm{I}_n|$ such that it is less than or equal to ε.

Example 3.2

Compute the positive root of $x^3 - 2x - 8 = 0$ by Bisection method, correct up to two decimal places.

Solution From Example 3.1 we know that the root lies between 2 and 3.

$$x_1 = 2,\ f(x_1) = \text{negative} -\text{ve}$$

$$x_2 = 3,\ f(x_2) = \text{positive} +\text{ve}$$

Computations are shown in the following table:

Iteration number	x	*Sign* $f(x)$	*Sign* $f(x) \times f(x_2)$	x_1	x_2
1	2.5	+	+	2.00	2.50
2	2.25	−	−	2.25	2.50
3	2.375	+	+	2.25	2.375
4	2.312	−	−	2.312	2.375
5	2.344	+	+	2.312	2.344
6	2.328	−	−	2.328	2.344
7	2.336	+	+	2.328	2.336
8	2.332	+	+	2.328	2.332

$|2.332 - 2.328| = 0.004 \leq 0.005;\ x_1 = x_2 = 2.33$ approx.

Note: Initial length of interval $I_0 = |3-2| = 1$

After eighth iteration, the length of the interval is, $|I_8| = \frac{1}{2^8} = 0.004 (= \varepsilon)$.

$\varepsilon = 0.004$ is the less than 0.005 which is the maximum error in x_1 and x_2.

3.5 Regula–Falsi Method (or Method of False Position)

Let the function be expressed as $y = f(x)$ and we want to find the zeros of the function. Proceed as follows:

(*i*) Find x_1 and x_2 where $y_1 = f(x_1)$ and $y_2 = f(x_2)$ have opposite signs and that there is only one value of x between x_1 and x_2 where $f(x)$ vanishes.

(*ii*) Approximate the function between x_1 and x_2 by a straight line joining points (x_1, y_1) and (x_2, y_2). The point where the straight line cuts the x-axis is an estimate of the root. The equation of the straight line is,

$$y - y_1 = \frac{y_2 - y_1}{x_2 - x_1}(x - x_1), \tag{3.13}$$

and the point where it intersects the x-axis is given by putting $y = 0$ in (3.13), as

$$x = \frac{x_1 y_2 - x_2 y_1}{y_2 - y_1}. \tag{3.14}$$

Thus the value of x given by (3.14) is the approximate root of $f(x) = 0$.

(*iii*) Compute $f(x)$ and check its sign.

If $f(x) \times f(x_2) > 0$, then replace x_2 by x and $y_2 = f(x)$ else replace x_1 by x and $y_1 = f(x)$.

(*iv*) Repeat (*ii*) and (*iii*) until two consecutive values of x agree within desired accuracy. See Fig. 3.1.

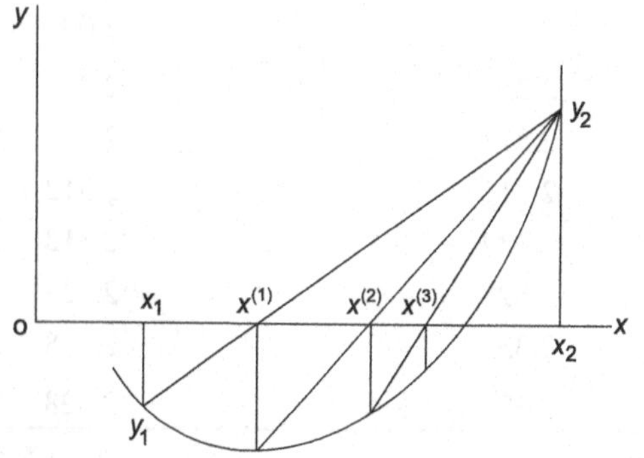

Figure 3.1 Regula–Falsi method (superscript shows iteration number).

Example 3.3

Find the positive root of the equation. $x^2 - 6e^{-x} = 0$, by Regula–Falsi method correct up to two places of decimal.

Solution We have $y = f(x)$ where $f(x) = x^2 - 6e^{-x}$. As the answer is required up to two places of decimal, we will compute up to at least 3 places of decimal and stop the process when two successive values coincide after rounded to two decimal places.

First we crudely find the values of x_1 and x_2 where $f(x_1)$ and $f(x_2)$ have opposite signs:

x	$f(x)$
0	−6
1	−1.207
2	+3.188

The root lies between 1 and 2. Computations are shown in the table.

Itera-tion	x_1	$y_1 = f(x_1)$	x_2	$y_2 = f(x_2)$	x	$f(x)$	$f(x) \times f(x_2)$	*Change*
1	1	−1.207	2	3.188	1.275	−0.051	−	x_1, y_1
2	1.275	−0.051	2	3.188	1.327	0.169	+	x_2, y_2
3	1.275	−0.051	1.327	0.169	1.287	−0.0002	−	x_1, y_1
4	1.287	−0.0002	1.327	0.169	1.287			

In the fourth iteration the value is exactly repeated. Therefore the root is 1.287.

Note: In the third iteration the value of root satisfies the equation almost exactly. Therefore we may conclude from there also, that $x = 1.287$. However, even if the equation was satisfied exactly we may give any sign to $f(x) \times f(x_2)$ and make appropriate changes. The next iteration will give the same value of x as in the previous iteration. Or we may stop at third iteration.

3.6 Secant Method

In the secant method also two values x_1 and x_2 are taken in the neighbourhood of the root but they are not ought to be on the opposite sides of the root like Regula–Falsi method. That is, $f(x_1)$ and $f(x_2)$ may have same sign, or opposite signs. Then a straight line (secant) is drawn through (x_1, y_1) and (x_2, y_2) intersecting the x-axis at a point x. One of the points, say (x_1, y_1) is discarded and again a line is drawn through (x_2, y_2) and (x, y). In actual computations (x_1, y_1) is replaced by (x_2, y_2) and the new point (x, y) replaces (x_2, y_2).

The process is repeated until two successive values of x agree within desired accuracy. To start with we may choose x_2 to be closer to the root. See Fig. 3.2.

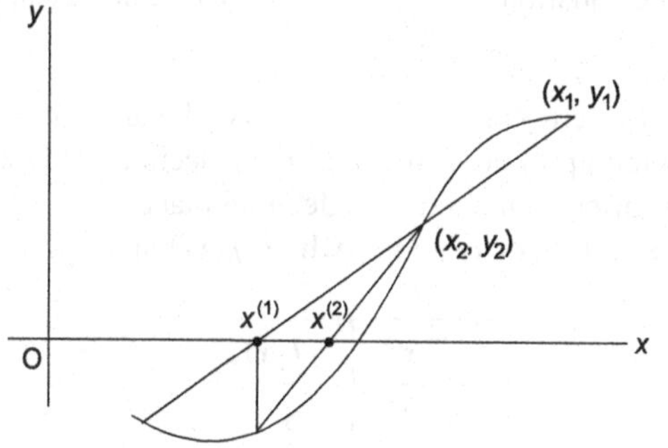

Figure 3.2 Secant Method (superscript shows iteration number).

Example 3.4

Using Secant method find the positive root of $x^2 - 6e^{-x} = 0$ correct up to two places of decimal. Take the initial values $x_1 = 2.5$ and $x_2 = 2$.

Solution It is the same problem as Example 3.3. We will compute up to 3 decimals and check the accuracy rounding to 2 decimal places.

Iteration	x_1	$y_1 = f(x_1)$	x_2	$y_2 = f(x_2)$	x	$y = f(x)$
1	2.5	5.758	2	3.188	1.380	0.395
2	2	3.188	1.380	0.395	1.292	0.021
3	1.380	0.395	1.292	0.021	1.287	−0.0002
4	1.292	0.021	1.287	−0.0002	1.287	

As discussed in Example 3.3, the answer has been obtained correct up to 3 decimal places, i.e., the root is 1.287.

3.7 Convergence of Secant/Regula–Falsi Methods

Let x_n denote the nth iterate for the root of $f(x) = 0$ or zero of the function $y = f(x)$. If α is the exact root, let $\alpha = x_n + \varepsilon_n$ where ε_n is the error in x_n.

Let us suppose that x_{n-1} and x_n have been computed, then the next iterate x_{n+1} is given by the formula (3.14), i.e.,

$$x_{n+1} = \frac{x_{n-1}y_n - x_n y_{n-1}}{y_n - y_{n-1}} = \frac{x_{n-1}f(x_n) - x_n f(x_{n-1})}{f(x_n) - f(x_{n-1})}$$

or $\alpha - \varepsilon_{n+1} = \dfrac{(\alpha-\varepsilon_{n-1})f(\alpha-\varepsilon_n) - (\alpha-\varepsilon_n)f(\alpha-\varepsilon_{n-1})}{f(\alpha-\varepsilon_n) - f(\alpha-\varepsilon_{n-1})}$

$$= \frac{(\alpha-\varepsilon_{n-1})\left\{f(\alpha)-\varepsilon_n f'(\alpha)+\frac{\varepsilon_n^2}{2}f''(\alpha)+\cdots\right\} - (\alpha-\varepsilon_n)\left\{f(\alpha)-\varepsilon_{n-1}f'(\alpha)+\frac{\varepsilon_{n-1}^2}{2}f''(\alpha)+\cdots\right\}}{\left\{f(\alpha)-\varepsilon_n f'(\alpha)+\frac{\varepsilon_n^2}{2}f''(\alpha)+\cdots\right\} - \left\{f(\alpha)-\varepsilon_{n-1}f'(\alpha)+\frac{\varepsilon_{n-1}^2}{2}f''(\alpha)+\cdots\right\}}$$

expanding by Taylor's series about α up to second powers of ε_n and ε_{n-1}.

$$= \frac{[-(\alpha-\varepsilon_{n-1})\varepsilon_n + (\alpha-\varepsilon_n)\varepsilon_{n-1}]f'(\alpha) + \frac{f''(\alpha)}{2}[(\alpha-\varepsilon_{n-1})\varepsilon_n^2 - (\alpha-\varepsilon_n)\varepsilon_{n-1}^2]}{(-\varepsilon_n+\varepsilon_{n-1})f'(\alpha) + \frac{f''(\alpha)}{2}(\varepsilon_n^2 - \varepsilon_{n-1}^2)}$$

since $f(\alpha) = 0$.

$$= \frac{-\alpha(\varepsilon_n-\varepsilon_{n-1})f' + \frac{f''}{2}[\alpha(\varepsilon_n^2-\varepsilon_{n-1}^2) - \varepsilon_{n-1}\cdot\varepsilon_n(\varepsilon_n-\varepsilon_{n-1})]}{-(\varepsilon_n-\varepsilon_{n-1})f' + \frac{f''}{2}(\varepsilon_n-\varepsilon_{n-1})(\varepsilon_n+\varepsilon_{n-1})}$$

where $f'' = f''(\alpha); f' = f'(\alpha)$

$$= \frac{-\alpha f' + \frac{f''}{2}[\alpha(\varepsilon_n+\varepsilon_{n-1}) - \varepsilon_{n-1}\cdot\varepsilon_n]}{-f' + \frac{f''}{2}(\varepsilon_n+\varepsilon_{n-1})}$$

$$= \frac{\alpha - \mathrm{k}[\alpha(\varepsilon_n+\varepsilon_{n-1}) - \varepsilon_{n-1}\cdot\varepsilon_n]}{1 - k(\varepsilon_n+\varepsilon_{n-1})}, \text{ where } \quad \mathrm{k} = \frac{f''}{2f'}$$

$$= [\alpha - \mathrm{k}\{\alpha(\varepsilon_n+\varepsilon_{n-1}) - \varepsilon_{n-1}\varepsilon_n\}][1-\mathrm{k}(\varepsilon_n+\varepsilon_{n-1})]^{-1}$$

$$= [\alpha - \mathrm{k}\{\alpha(\varepsilon_n+\varepsilon_{n-1}) - \varepsilon_{n-1}\varepsilon_n\}][1+\mathrm{k}(\varepsilon_n+\varepsilon_{n-1}) + \mathrm{k}^2(\varepsilon_n+\varepsilon_{n-1})^2 + \ldots]$$

expanding binomially and leaving out third and higher order terms in ε_{n-1} and ε_n.

$$= \alpha + \mathrm{k}\varepsilon_{n-1}\varepsilon_n$$

or $\quad \varepsilon_{n+1} = -\mathrm{k}\varepsilon_n\varepsilon_{n-1}$ where $\mathrm{k} = \dfrac{f''(\alpha)}{2f'(\alpha)}$

or $\qquad = K\varepsilon_n\varepsilon_{n-1}$ where $K = -\dfrac{f''(\alpha)}{2f'(\alpha)}$. (3.15)

To determine the order of convergence of the method, we express (3.15) in accordance of (3.2) as,

$\varepsilon_{n+1} = C\varepsilon_n^p$ where C is a constant and p is the order of the method which is to be determined.

As we also have $\varepsilon_n = C\varepsilon_{n-1}^p$ or $\varepsilon_{n-1} = \left(\dfrac{\varepsilon_n}{C}\right)^{1/p}$,

(3.15) may be written after substituting the values of ε_n and ε_{n-1}, as,

$$C\varepsilon_n^p = K\varepsilon_n\left(\frac{\varepsilon_n}{C}\right)^{1/p}$$

or $C^{1+\frac{1}{p}}\varepsilon_n^p = K\varepsilon_n^{1+\frac{1}{p}}$.

This gives $p = 1 + \dfrac{1}{p}$ and $C^{1+\frac{1}{p}} = C^p = K$.

From the above relation,

$p = \frac{1+\sqrt{5}}{2} = 1.62$, considering only positive (+ve) sign. The negative value of p means the error will increase with each iteration, hence ignored.

Also, we have,

$$C = K^{1/p} = K^{0.62} \text{ where } K = \left|\frac{f''(\alpha)}{2f'(\alpha)}\right|.$$

Thus the order of the convergence of the Secant method is 1.62 and we have,

$$\varepsilon_{n+1} = C\varepsilon_n^{1.62}, \tag{3.16}$$

$$C = K^{0.62} \text{ where } K = \left|\frac{f''(\alpha)}{2f'(\alpha)}\right|. \tag{3.17}$$

The convergence of the Regula–Falsi method may be at the most as high as of Secant method. However, in general, the Secant method should converge faster as compared to Regula–Falsi but its convergence is not guaranteed while the latter may be a little slower but will definitely converge to the true root. These methods are also called ‘method of chords’.

It may also be noted that since $f'(\alpha)$ appears in the denominator in K, it should not be too small for convergence.

3.8 Newton–Raphson (N–R) Method

In the foregoing methods, namely, Bisection, Regula–Falsi and Secant, we needed two values in the neighbourhood of the root while in the N–R method only one value is required. Let us suppose that x_0 is the initial estimate for the root α of $f(x) = 0$. We draw a tangent at the point $(x_0,\ y_0)$ of the curve $y = f(x)$,

$$y - y_0 = f'(x_0)(x - x_0) \tag{3.18}$$

where $f'(x_0) = \left(\dfrac{dy}{dx}\right)_{x = x_0}$

The point where the tangent (3.18) cuts the x-axis $(y = 0)$, say $x = x_1$, is the next estimate of the root i.e.,

$$x_1 = x_0 - \frac{f(x_0)}{f'(x_0)},\ y_0 = f(x_0). \tag{3.19}$$

This process may be repeated to get the next estimate x_2 using x_1 and so on. Let us suppose that we have computed the nth estimate x_n, the next estimate may be computed from,

$$x_{n+1} = x_n - \frac{f(x_n)}{f'(x_n)},\ n = 0, 1,\ \cdots \tag{3.20}$$

The iterative formula (3.20) is the well-known Newton–Raphson's formula. A few values $x_0,\ x_1,\ x_2,\ x_3$ are shown in Fig. 3.3.

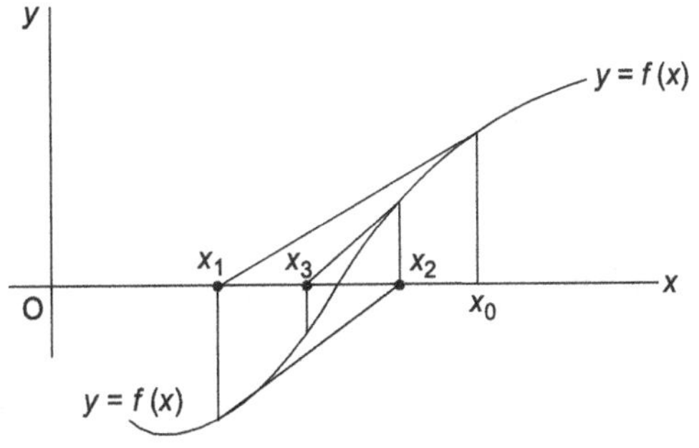

Figure 3.3 Newton–Raphson method (suffix shows iteration number).

The interpretation given above for deriving N–R method was meant to show the process geometrically like other methods. Analytically it may be explained as follows:

Let x_0 be an estimate for the root α and h be the error in it such that $x_0 + h = \alpha$. The Taylor's series expansion gives,

$$f(\alpha) = f(x_0 + h) = f(x_0) + hf'(x_0) + \frac{h^2}{2!} f''(x_0) + \cdots \tag{3.21}$$

Neglecting second and higher powers of h and remembering that $f(\alpha) = 0$, we get approximate value of h, as

$$h \simeq -\frac{f(x_0)}{f'(x_0)}, \tag{3.22}$$

and the next estimate,

$$x_1 = x_0 + h = x_0 - \frac{f(x_0)}{f'(x_0)}.$$

In general the above may be written as,

$$x_{n+1} = x_n - \frac{f(x_n)}{f'(x_n)}, \quad n = 0,\ 1,\ 2,\ \cdots \tag{3.23}$$

which is same as (3.20).

3.8.1 Evaluation of some arithmetical functions

(*i*) To evaluate q^{th} root of a given number N we wish to evaluate $\mathrm{N}^{\frac{1}{q}}$, where q is an integer greater than 1. That is, we have to find x when,

$$x = \mathrm{N}^{\frac{1}{q}} \quad \text{or} \quad x^q = \mathrm{N}.$$

In other words, we have to find the roots of the equation,

$$x^q - \mathrm{N} = 0.$$

To apply N–R method we have,

$$f(x) = x^q - \mathrm{N} \text{ and } f'(x) = qx^{q-1}.$$

The iterative scheme is given by

$$x_{n+1} = x_n - \frac{x_n^q - \mathrm{N}}{qx_n^{q-1}} \tag{3.24}$$

or

$$= \frac{(q-1)x_n^q + \mathrm{N}}{qx_n^{q-1}}. \tag{3.24a}$$

Note: (*i*) We can also evaluate $N^{p/q}$ where p and q are integers greater than 1, by putting

$$x = N^{p/q} = (N^p)^{\frac{1}{q}} = (A)^{\frac{1}{q}}, \text{ where } A = N^p.$$

(*ii*) To find reciprocal of a given number N.

Putting $x = \frac{1}{N} = N^{-1}$ or $x^{-1} = N$, we have,

$$f(x) = x^{-1} - N \text{ and } f'(x) = -\frac{1}{x^2}.$$

Hence N–R scheme becomes,

$$\begin{aligned} x_{n+1} &= x_n + (x_n - Nx_n^2) \\ &= x_n(2 - Nx_n). \end{aligned} \tag{3.25}$$

(*iii*) To find reciprocal of $N^{\frac{1}{q}}$

Let, $x = 1/N^{\frac{1}{q}}$

or $x^{-1} = N^{\frac{1}{q}}$ or $x^{-q} = N$.

Hence taking,

$f(x) = x^{-q} - N$ and $f'(x) = -qx^{-q-1}$, we can write the N–R scheme as,

$$x_{n+1} = x_n - \frac{x_n^{-q} - N}{-qx^{-q-1}}$$

In order to avoid computing negative powers, we express the formula as,

$$x_{n+1} = \frac{x_n}{q}(q + 1 - Nx_n^q) \tag{3.26}$$

Example 3.5

Find the roots of the equation $x^2 - \cos x = 0$ by Newton–Raphson's method correct up to 3 places of decimal.

Solution $y = f(x) \equiv x^2 - \cos x$

$f'(x) = 2x + \sin x.$

In order to locate the approximate position of the root it may be worth-while to plot the graphs of the two functions separately,

$$y = x^2 \qquad \text{.... (1)}$$

$$y = \cos x \qquad \text{.... (2)}$$

Obviously, subtracting one from the other, i.e., eliminating y gives the solution of $x^2 - \cos x = 0$. That is the x-coordinate of the points of intersection is the required root. The rough plots of (1) and (2) are as follows:

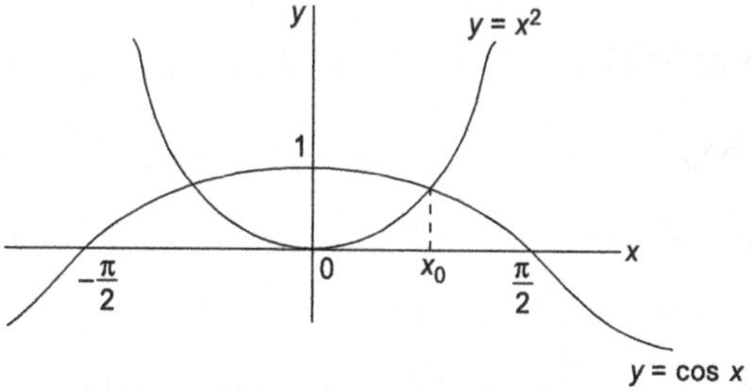

Figure 3.4 Graphs of $y = x^2$ and $y = \cos x$

We see that the function $y = f(x) \equiv x^2 - \cos x$ is symmetrical about the y-axis, i.e., $f(x) = f(-x)$. Therefore, we need only to find the positive root. We see from the figure above that the root lies between 0 and $\pi/2$. Therefore we try to compute $f(x)$ for a few values of x.

Important: The values of $\cos x$ and $\sin x$ are to be computed in radians, e.g., $\cos 1 = 0.5403$ and $\cos(0.5) = 0.8776$

x	$f(x)$
0	−1
0.5	−0.6276
1.0	0.4597

We can take any value for x_0 between 0.5 and 1.0. Let us take $x_0 = 1.0$. We get following iterated values:

Iteration	x	$f(x)$	$f'(x)$	$h = -f(x)/f'(x)$	
1	1	0.4597	2.8415	−0.1618	$\cos 0.8382 = 0.6690$
2	0.8382	0.0337	2.4197	−0.0139	$\sin 0.8382 = 0.7433$
3	0.8243	0.00037	2.3826	−0.00016	$\cos 0.8243 = 0.6791$
	0.8241	×	×	×	$\sin 0.8243 = 0.7340$

The roots are +0.824 and −0.824.

Example 3.6

Using N–R method, find the root of the equation $\log_e x + x - 3 = 0$, correct up to two decimal places only.

Solution To find the initial estimate we should remember that $\log 0 = -\infty$ and $\log 1 = 0$. Then we try the following values to locate the approximate value of the root.

x	$f(x)$
0	$-\infty$
1	-2
2	-0.3068
3	1.0986

We can start from $x_0 = 2$

$$f(x) = \log_e x + x - 3; \ f'(x) = \frac{1}{x} + 1$$

Iteration	x	$f(x)$	$f'(x)$	$h = -\dfrac{f(x)}{f'(x)}$
1	2	-0.3068	1.5	0.2045
2	2.2045	-0.0050	1.4536	0.0034
3	2.2079	0.0004	1.4529	-0.0003
	2.2076	×	×	×

The root is 2.21 (correct to two places of decimal).

3.8.2 Convergence of Newton–Raphson method

Assuming that iterations up to n have already been performed and we compute the $(n+1)^{\text{th}}$ iteration by N–R formula,

$$x_{n+1} = x_n - \frac{f(x_n)}{f'(x_n)}.$$

Like in Sec 3.7, let $\alpha = x_{n+1} + \varepsilon_{n+1}$ and $\alpha = x_n + \varepsilon_n$ where ε_{n+1} and ε_n denote errors in the $(n+1)^{\text{th}}$ and n^{th} iterates respectively. Then from the formula,

$$\alpha - \varepsilon_{n+1} = \alpha - \varepsilon_n - \frac{f(\alpha - \varepsilon_n)}{f'(\alpha - \varepsilon_n)}.$$

$$\text{or}\quad \varepsilon_{n+1} = \varepsilon_n + \frac{f(\alpha-\varepsilon_n)}{f'(\alpha-\varepsilon_n)}.$$

$$= \frac{\varepsilon_n f'(\alpha-\varepsilon_n) + f(\alpha-\varepsilon_n)}{f'(\alpha-\varepsilon_n)}.$$

Expanding $f(\alpha-\varepsilon_n)$ and $f'(\alpha-\varepsilon_n)$ by Taylor's series about α,

$$\varepsilon_{n+1} = \frac{\varepsilon_n\{f'(\alpha) - \varepsilon_n f''(\alpha) + \cdots\} + \left\{f(\alpha) - \varepsilon_n f'(\alpha) + \frac{\varepsilon_n^2}{2} f''(\alpha)\cdots\right\}}{f'(\alpha) - \varepsilon_n f''(\alpha) + \frac{\varepsilon_n{}^2}{2} f'''(\alpha) - \cdots}.$$

$$= -\frac{\varepsilon_n^2}{2}\frac{f''(\alpha)}{f'\alpha)}\left\{1 - \frac{\varepsilon_n f''(\alpha)}{f'(\alpha)} + \frac{\varepsilon_n^2}{2}\frac{f'''(\alpha)}{f'(\alpha)} - \cdots\right\}^{-1}$$

$$\simeq -\frac{\varepsilon_n^2}{2}\frac{f''(\alpha)}{f'(\alpha)}, \text{ neglecting } \varepsilon_n^3 \text{ and higher powers.}$$

$$= \mathrm{K}\varepsilon_n^2 \quad \text{where} \quad \mathrm{K} = -\frac{1}{2}\frac{f''(\alpha)}{f'(\alpha)}. \tag{3.27}$$

As may be seen from (3.27) the N–R method has quadratic convergence. That is, N–R method should converge faster as compared to Regula–Falsi or Secant methods. Further, it should be noted, as in the case of Regula–Falsi and Secant methods, that $f'(\alpha)$ should not be too small and convergence factor should not be too large.

Comparing with method of successive substitution where $f(x) = 0$ is put in the form $x = \phi(x)$, in the N–R method,

$$\phi(x) = x - \frac{f(x)}{f'(x)}.$$

Further, since for convergence $\left|\frac{\partial\phi}{\partial x}\right|_{x=\alpha} < 1$, we see that in the N–R method the condition for convergence is,

$$\left|\frac{ff''}{f'^2}\right| < 1.$$

3.8.3 Convergence when roots are repeated

Let us consider the convergence of the following formula,

$$x_{n+1} = x_n - m \cdot \frac{f(x_n)}{f'(x_n)}.$$

Proceeding as in Sec 3.8.2, we have

$$\varepsilon_{n+1} = \frac{-\varepsilon_n(m-1)f'(\alpha) + \left(\frac{m}{2}-1\right)\varepsilon_n^2 f'(\alpha) - \frac{1}{2}\left(\frac{m}{3}-1\right)\varepsilon_n^3 f'(\alpha) + \cdots}{f'(\alpha) - \varepsilon_n f'(\alpha) + \frac{\varepsilon_n^2}{2} f'''(\alpha) - \cdots}. \tag{3.28}$$

If the root α is repeated twice, i.e., α has a multiplicity of 2, then $f(\alpha) = f'(\alpha) = 0$.

In this case, unless $m = 2$, the highest power of ε_n in the numerator will be 2 and ε_n is of first power in the denominator and so the convergence with be linear. Therefore in order to have quadratic convergence the value of m must be equal to 2.

Similarly, if α is a root of multiplicity 3, then $f(\alpha) = f'(\alpha) = f''(\alpha) = 0$. In this case following the same argument as above we must have $m = 3$, and in general for α with multiplicity m, the N–R formula with second order convergence should be modified as follows:

$$x_{n+1} = x_n - m\frac{f(x_n)}{f'(x_n)}. \tag{3.29}$$

However, we may not know in advance in most of the cases about the multiplicity of a root, if there is any.

3.9 Simultaneous Equations

Let us consider a system of two equations,

$$\left.\begin{aligned} f(x,\ y) &= 0 \\ g\ (x,\ y) &= 0. \end{aligned}\right\} \tag{3.30}$$

We want to find out the values of x and y which satisfy both of these equations simultaneously. There may be more than one set of such values. As we did in case of finding the roots of $f(x) = 0$, we determine the values $(x,\ y)$, point by point, i.e., one at a time.

3.9.1 Method of successive substitution

We express equations (3.30) in the form,

$$\left.\begin{aligned} x &= \phi(x, y) \\ y &= \psi(x, y) \end{aligned}\right\} \tag{3.31}$$

As before we try to get an initial estimate (x_0, y_0) which is close to the exact solution, say, (α, β). Then iterations may be carried as follows:

$$\left.\begin{aligned} x_{n+1} &= \phi(x_n, y_n) \\ y_{n+1} &= \psi(x_n, y_n) \end{aligned}\right\} n = 0, 1, 2, \ldots. \tag{3.32}$$

Following conditions are to be satisfied for convergence of the iteration scheme (3.32) at (α, β). But as (α, β) are not known, these conditions may be made to satisfy at (x_0, y_0) which is a close approximation of (α, β).

$$\left.\begin{aligned} &\left|\frac{\partial \phi}{\partial x}\right| + \left|\frac{\partial \phi}{\partial y}\right| < 1 \\ \text{and} \quad &\left|\frac{\partial \psi}{\partial x}\right| + \left|\frac{\partial \psi}{\partial y}\right| < 1. \end{aligned}\right\} \tag{3.33}$$

The above conditions may also be put in another form as,

$$\left.\begin{aligned} &\left|\frac{\partial \phi}{\partial x}\right| + \left|\frac{\partial \psi}{\partial x}\right| < 1 \\ \text{and} \quad &\left|\frac{\partial \phi}{\partial y}\right| + \left|\frac{\partial \phi}{\partial y}\right| < 1. \end{aligned}\right\} \tag{3.34}$$

For derivation of convergence condition, See [1]. This method may not be applied in a straight forward manner on account of the ambiguity in the choice of the functions $\phi(x, y)$ and $\psi(x, y)$.

3.9.2 Newton–Raphson method

This method has natural extension in two and higher dimensions. After finding the initial estimate (x_0, y_0) by trial and error or graphically, we try to compute h, k such that

$$\left.\begin{aligned} x_0 + h &= \alpha \\ y_0 + k &= \beta. \end{aligned}\right\} \tag{3.35}$$

Being exact solution, $(\alpha,\ \beta)$ should satisfy the given system of equations (3.30), i.e.,

$$\left.\begin{aligned} f(\alpha,\ \beta) &= 0 \\ g(\alpha,\ \beta) &= 0. \end{aligned}\right\} \tag{3.36}$$

Using (3.35) and (3.36) we write the Taylor's series expansion in two variables about $(x_0,\ y_0)$ in the following manner,

$$0 = f(\alpha,\ \beta) = f(x_0 + h,\ y_0 + k)$$

$$\text{or } 0 = f(x_0,\ y_0) + \left\{h\frac{\partial f}{\partial x} + k\frac{\partial f}{\partial y}\right\}_{\substack{x=x_0\\ y=y_0}} + \frac{1}{2}\left\{h^2\frac{\partial^2 f}{\partial x^2} + 2hk\frac{\partial^2 f}{\partial x \partial y} + k^2\frac{\partial^2 f}{\partial y^2}\right\}_{\substack{x=x_0\\ y=y_0}} + \cdots \tag{3.37a}$$

$$\text{and } \quad 0 = g(\alpha,\ \beta) = g(x_0 + h,\ y_0 + k)$$

$$\text{or } 0 = g(x_0,\ y_0) + \left\{h\frac{\partial g}{\partial x} + k\frac{\partial g}{\partial y}\right\}_{\substack{x=x_0\\ y=y_0}} + \frac{1}{2}\left\{h^2\frac{\partial^2 g}{\partial x^2} + 2hk\frac{\partial^2 g}{\partial x \partial y} + k^2\frac{\partial^2 g}{\partial y^2}\right\}_{\substack{x=x_0\\ y=y_0}} + \cdots \tag{3.37b}$$

Neglecting the second and higher order terms in (3.37a and b) we compute h and k approximately by solving the following two simultaneous linear equations in $(h,\ k)$,

$$\left.\begin{aligned} \frac{\partial f}{\partial x}h + \frac{\partial f}{\partial y}k &= -f(x_0,\ y_0) \\ \frac{\partial g}{\partial x}h + \frac{\partial g}{\partial y}k &= -g(x_0,\ y_0), \end{aligned}\right\} \tag{3.38}$$

where values of the partial derivatives are to be taken at (x_0, y_0).

Since values of h and k computed from (3.38) are approximate they will not satisfy (3.35) exactly and thus we shall get a new estimate for $(\alpha,\ \beta)$ instead, denoted by, say $(x_1,\ y_1)$. This process can be repeated and another estimate $(x_2,\ y_2)$ obtained using $(x_1,\ y_1)$. The process may be continued until two successive estimates agree within desired accuracy. In general, the iterative scheme may be written as,

$$\left.\begin{aligned} x_{n+1} &= x_n + h \\ \text{and } \quad y_{n+1} &= y_n + k \end{aligned}\right\} \tag{3.39}$$

where
$$\left.\begin{aligned} h &= -\frac{fg_y - gf_y}{f_x g_y - g_x f_y} \\ k &= -\frac{gf_x - fg_x}{f_x g_y - g_x f_y} \end{aligned}\right\} \tag{3.40}$$

$$f = f(x,\ y),\ f_x = \frac{\partial f}{\partial x},\ f_y = \frac{\partial f}{\partial y}$$

$$g = g(x,\ y),\ g_x = \frac{\partial g}{\partial x},\ g_y = \frac{\partial g}{\partial y}$$

and the values of functions and their partial derivatives to be computed at $(x_n,\ y_n)$.

3.10 Complex Roots

Let us suppose we have to find the zeros of the function $F(z)$ or roots of the equation $F(z) = 0$ where z is a complex variable.

We can separate the real and imaginary parts of the function $F(z) = F(x + iy) = f(x,\ y) + ig(x,\ y)$. This function will vanish when the real and imaginary parts both vanish separately, i.e.,

$$\left.\begin{aligned} f(x,\ y) &= 0 \\ g\ (x,\ y) &= 0. \end{aligned}\right\} \tag{3.41}$$

Now, we can apply the Newton–Raphson method of Sec. 3.9 to solve the simultaneous equations (3.41).

Example 3.7

Using Newton–Raphson method find the solution of the following system of equations near $x = 2,\ y = 1$ correct up to two places of decimal only:

$$x^2 + y^2 - y = 5$$

$$y - e^{-x} = 1.$$

Solution

$$f = f(x,\ y) = x^2 + y^2 - y - 5;\ f_x = \frac{\partial f}{\partial x} = 2x;\ f_y = \frac{\partial f}{\partial y} = 2y - 1$$

$$g = g(x,\ y) = y - e^{-x} - 1;\ \ g_x = \frac{\partial g}{\partial x} = e^{-x};\ g_y = \frac{\partial g}{\partial y} = 1$$

We will use (3.39) and (3.40)

1st Iteration $x_0 = 2,\ y_0 = 1$

$$e^{-2} = 0.1353$$

$$f = -1;\ f_x = 4;\ f_y = 1$$

$$g = -0.1353;\ g_x = 0.1353;\ g_y = 1$$

$$h = -\frac{-1\times 1 - (-0.1353\times 1)}{4\times 1 - 0.1353} = \frac{0.8647}{3.8647} = 0.2237$$

$$k = -\frac{(-0.1353)\times 4 - (-1)\times 0.1353}{3.8647} = \frac{0.4059}{3.8647} = 0.1050$$

$$x_1 = 2 + 0.2237 = 2.2237;\ y_1 = 1 + 0.1050 = 1.1050$$

2nd Iteration

$$e^{-2.2237} = 0.1083$$

$$f = 0.0609;\ f_x = 4.4474;\ f_y = 1.21$$

$$g = -0.0033;\ g_x = 0.1083;\ g_y = 1$$

$$h = -\frac{0.0609\times 1 - (-0.0033)\times 1.21}{4.4474\times 1 - (1.21)\times 0.1083} = -\frac{0.0040}{4.3200} = -0.0009$$

$$k = -\frac{(-0.0033)\times 4.4474 - 0.0609\times 0.1083}{4.3200} = \frac{0.0213}{4.3200} = 0.0049$$

$$x_2 = 2.2237 - 0.0009 = 2.2228$$

$$y_2 = 1.1050 + 0.0049 = 1.1099$$

Since $(x_1,\ y_1)$ and $(x_2,\ y_2)$ agree up to two places of decimal after rounding, the solution is,

$$x = 2.22,\ y = 1.11.$$

Example 3.8

Find the zero of the complex function $F(z) = z^3 - 3z + 52$ in the neighbourhood of the point, $z = 2.2 + 2.8i$. Show only two iterations of N–R method computing up to four places of decimal.

Solution

$$F(z) = F(x+iy) = (x+iy)^3 - 3(x+iy) + 52$$

$$= (x^3 - 3xy^2 - 3x + 52) + i(3x^2y - y^3 - 3y)$$

$$f(x, y) = x^3 - 3xy^2 - 3x + 52$$

$$g(x, y) = 3x^2y - y^3 - 3y$$

$$f_x = 3x^2 - 3y^2 - 3; \; f_y = -6xy$$

$$g_x = 6xy; \; g_y = 3x^2 - 3y^2 - 3$$

(It may be noted that $\frac{\partial f}{\partial x} = \frac{\partial g}{\partial y}$ and $\frac{\partial f}{\partial y} = -\frac{\partial g}{\partial x}$. They are known as Cauchy-Riemann's conditions for a function $f(z)$ to be analytic. We assume that these will hold althrough).

1st Iteration

$$x_0 = 2.2, \; y_0 = 2.8$$

$$f = 4.304, \; f_x = -12.0, \; f_y = -36.96$$

$$g = 10.304, \; g_x = 36.96, \; g_y = -12.0$$

$$h = -\frac{4.304 \times (-12) - 10.304 \times (-36.96)}{(-12) \times (-12) - 36.96 \times (36.96)} = -\frac{329.1878}{1510.0416} = -0.2180$$

$$k = -\frac{10.304 \times (-12) - 4.304 \times 36.96}{1510.0416} = \frac{282.7238}{1510.0416} = 0.1872$$

$$x_1 = 2.2 - 0.2180 = 1.9820$$

$$y_1 = 2.8 + 0.1872 = 2.9872.$$

2nd Iteration

$$x_1 = 1.9820, \; y_1 = 2.9872$$

$$f = 0.7816, \; f_x = -17.9851, \; f_y = -35.5238$$

$$g = -0.4134, \; g_x = 35.5238, \; g_y = -17.9851$$

$$h = -\frac{0.7816 \times (-17.9851) - (-0.4134) \times (-35.5238)}{(-17.9851) \times (-17.9851) - (35.5238) \times (-35.5238)}$$

$$= \frac{28.7427}{1585.4042} = 0.0181$$

$$k = -\frac{(-0.4134) \times (-17.9851) - 0.7816 \times 35.5238}{1585.4042}$$

$$= \frac{20.3304}{1585.4042} = 0.0128$$

$$x_2 = 1.9820 + 0.0181 = 2.0001$$

$$y_2 = 2.9872 + 0.0128 = 3.0000$$

(Exact answer is $x = 2,\ y = 3$, i.e., root is $2+3i$).

3.11 Bairstow's Method

Let us consider an algebraic equation with real coefficients or a real polynomial equation of degree n

$$P_n(x) \equiv x^n + a_1x^{n-1} + a_2x^{n-2} + a_3x^{n-3} + \ldots. + a_{n-1}x + a_n = 0. \tag{3.42}$$

The equation (3.42) is written in its 'normal form', i.e., coefficient of the leading term x^n is unity (monomial). We know that when $n = 2$, the roots of the equation (3.42) can be written as

$$x_{1,\,2} = \frac{-a_1 \pm \sqrt{a_1^2 - 4a_2}}{2}. \tag{3.43}$$

When $n = 3$ or 4, the roots of (3.42) can be written like $n = 2$, in closed form (in terms of its coefficients) with some exceptions. But, when $n \geq 5$, they can not be expressed in closed form as proved by Abel. Hence numerical methods have to be adopted in these cases.

The Bairstow's method finds a quadratic factor of $P_n(x)$ which gives two roots using (3.43). Because of this property of the method, no special procedure is needed to deal with imaginary (complex) roots of the equation (3.42). For, if the coefficients $a_1, a_2, \ldots a_n$ are real then complex roots will occur in pairs, like $\alpha \pm i\beta$. The quadratic factor having roots $\alpha \pm i\beta$ will be of the form $\{x^2 - 2\alpha x + (\alpha^2 + \beta^2)\}$ where α and β are real. The Bairstow's method computes $-2\alpha = p$ and $\alpha^2 + \beta^2 = q$ instead of $\alpha + i\beta$ and $\alpha - i\beta$. That is, it finds a factor $x^2 + px + q$ of $P_n(x)$. It is an iterative procedure requiring an initial guess for p and q say $(p_0,\ q_0)$ which is improved successively in an iterative manner.

Let $x^2 + px + q$ be an arbitrary quadratic. Dividing $P_n(x)$ by this quadratic, we can express

$$P_n(x) = (x^2 + px + q)Q_{n-2}(x) + Rx + S \tag{3.44}$$

where $Q_{n-2}(x)$ is the quotient, a polynomial of degree $n-2$ and $Rx + S$ is the remainder, R and S being parameters depending on p and q.

Let $$Q_{n-2}(x) = x^{n-2} + b_1x^{n-3} + b_2x^{n-4} + \ldots. + b_{n-2} \tag{3.45}$$

Then (3.44) may be written as

$$x^n + a_1x^{n-1} + a_2x^{n-2} + \dots + a_kx^{n-k} + \dots + a_n$$
$$= (x^2 + px + q)(x^{n-2} + b_1x^{n-3} + \dots + b_{n-2}) + Rx + S. \qquad (3.46)$$

Equating the coefficients we get

$$\left.\begin{aligned} a_1 &= b_1 + p \\ a_2 &= b_2 + pb_1 + q \\ a_k &= b_k + pb_{k-1} + qb_{k-2},\ k = 3, 4, \dots n-2 \\ a_{n-1} &= R + pb_{n-2} + qb_{n-3} \\ a_n &= S + qb_{n-2} \end{aligned}\right\} \qquad (3.47)$$

We can find the values of the unknowns $b_1,\ b_2,\ \dots\ b_{n-2},\ R$ and S for some values of p and q by forward substitution, i.e., starting from the top and going downwards. However, we would like to see the recursive formula given in (3.47) for $k = 3,\ 4,\ \dots\ n-2$, becomes also valid for $k = 1,\ 2,\ n-1$ and n. In that case we put the following:

Define $b_{-1} = 0,\ b_0 = 1$ to satisfy (3.47) for $k = 1,\ 2$ (3.48a)

and $R = b_{n-1},\ S = b_n + pb_{n-1}$ to satisfy for $k = n-1,\ n$ (3.48b)

With the assumptions (3.48a) and (3.48b), the following formula may be used for determining the unknowns:

$$b_k = a_k - pb_{k-1} - qb_{k-2},\ k = 1,\ 2,\ \dots\ n. \qquad (3.49)$$

It must be remembered that b_k's are dependent on the choice of p and q in the quadratic factor or in other words b_k's are functions of p and q.

Our object is to determine the values of p and q such that $x^2 + px + q$ is a factor of $P_n(x)$. It means remainder should be zero, i.e., $R(p,\ q) = 0$ and $S(p,\ q) = 0$. We start the procedure by taking initial approximate values $(p_0,\ q_0)$ which are sufficiently close to $(p,\ q)$. To get the next improved value $p_1 = p_0 + \delta p$ and $q_1 = q_0 + \delta q$ we follow the procedure explained in Sec 3.9.2 giving

$$\left.\begin{aligned} \frac{\partial R}{\partial p}\delta p + \frac{\partial R}{\partial q}\delta q &= -R(p_0,\ q_0) \\ \frac{\partial S}{\partial p}\delta p + \frac{\partial S}{\partial q}\delta q &= -S(p_0,\ q_0) \end{aligned}\right\} \qquad (3.50)$$

Remember that partial derivatives are to be computed at (p_0, q_0).

Using (3.48b), i.e., $R = b_{n-1}$ and $S = b_n + pb_{n-1} = b_n + pR$, the equations (3.50) may be written in terms of b_{n-1} and b_n and their partial derivatives as

$$\left.\begin{aligned} \frac{\partial b_{n-1}}{\partial p}\delta p + \frac{\partial b_{n-1}}{\partial q}\delta q &= -b_{n-1} \\ \left(\frac{\partial b_n}{\partial p} + b_{n-1}\right)\delta p + \frac{\partial b_n}{\partial q}\delta q &= -b_n \end{aligned}\right\} \tag{3.51}$$

We determine partial derivatives w.r.t. p from (3.49),

$$-\frac{\partial b_k}{\partial p} = p\frac{\partial b_{k-1}}{\partial p} + q\frac{\partial b_{k-2}}{\partial q} + b_{k-1},\ k = 1,\ 2,\ \cdots n. \tag{3.52a}$$

Writing $-\dfrac{\partial b_k}{\partial p} = c_k$, etc., relation (3.52a) may be expressed as

$$c_k = b_{k-1} - pc_{k-1} - qc_{k-2},\ k = 1,\ 2,\ \ldots.\ n, \tag{3.52b}$$

where $c_{-1} = 0$ and $c_0 = 0$ since $b_{-1} = 0,\ b_0 = 1$. Hence values of $c_k,\ k = 1,\ 2,\ \ldots.\ n$ can be computed from (3.52b).

Again differentiating (3.49) w.r.t. q and putting $-\dfrac{\partial b_k}{\partial q} = d_k$, we get

$$-\frac{\partial b_k}{\partial q} = p\frac{\partial b_{k-1}}{\partial q} + q\frac{\partial b_{k-2}}{\partial q} + b_{k-2} \tag{3.53a}$$

or $$d_k = b_{k-2} - pd_{k-1} - qd_{k-2},\ k = 1,\ 2,\ \ldots.\ n. \tag{3.53b}$$

Writing eq. (3.52b) for $k-1$ and subtracting from (3.53b) we get

$$(d_k - c_{k-1}) = -p(d_{k-1} - c_{k-2}) - q(d_{k-2} - c_{k-3}),\quad k = 2,\ 3,\ \cdots n. \tag{3.54}$$

Equation (3.54) is satisfied identically if

$$d_k = c_{k-1},\ k = 2,\ 3,\ \ldots.\ n.$$

Also, $d_1 = 0 = c_0$, from (3.53b)

Thus we can find c_k from (3.52b) and straight away d_k from the relation

$$d_k = c_{k-1},\ k = 1,\ 2,\ \ldots.\ n. \tag{3.55}$$

Now, the equations (3.51) may be written as,

$$\left.\begin{aligned} c_{n-1}\delta p + c_{n-2}\delta q &= b_{n-1} \\ (c_n - b_{n-1})\delta p + c_{n-1}\delta q &= b_n \end{aligned}\right\} \tag{3.56}$$

Solving (3.56) we get

$$\left.\begin{aligned} \delta p &= \frac{b_{n-1}c_{n-1} - b_n c_{n-2}}{c_{n-1}^2 - c_{n-2}(c_n - b_{n-1})} \\ \delta q &= \frac{b_n c_{n-1} - b_{n-1}(c_n - b_{n-1})}{c_{n-1}^2 - c_{n-2}(c_n - b_{n-1})}. \end{aligned}\right\} \tag{3.57}$$

Thus, after computing b_k and c_k from (3.49) and (3.52b) respectively for $p = p_0$ and $q = q_0$, we can compute δp and δq from (3.57) giving the improved value for p and q as $p_1 = p_0 + \delta p$ and $q_1 = q_0 + \delta p$. The process is repeated until desired accuracy is achieved in the values of p and q in two successive iterations. Having got the quadratic factor the process may be applied on $Q_{n-2}(x)$ and so on until linear or quadratic polynomial is left at the end.

Example 3.9

Show only one iteration of Bairstow's method to find a quadratic factor $x^2 + px + q$ of the polynomial $x^4 - 5x^3 + 13x^2 - 19x + 10$. Take the initial values of p and q as $p_0 = -2.8$ and $q_0 = 1.5$ and compute p_1 and q_1.

Solution $n = 4$.

a_1	a_2	a_3	a_4
−5	13	−19	10

Formula:

$$b_k = a_k - pb_{k-1} - qb_{k-2},\ k = 1,\ 2,3,4$$

$$= a_k + 2.8b_{k-1} - 1.5b_{k-2}$$

$$b_{-1} = 0$$

$$b_0 = 1$$

$$b_1 = -5 + 2.8 = -2.2$$

$$b_2 = 13{+}2.8\ (-2.2) - 1.5 = 5.34$$

$b_3 = -19+2.8\ (5.34) - 1.5 \times (-2.2) = -0.748$

$b_4 = 10+2.8\ (-0.748) - 1.5 \times 5.34 = -0.1044$

Formula: $c_k = b_{k-1} - pc_{k-1} - qc_{k-2},\ k = 1,\ 2, 3, 4$

$= b_{k-1} + 2.8c_{k-1} - 1.5c_{k-2}$

$c_{-1} = 0$

$c_0 = 0$

$c_1 = 1$

$c_2 = -2.2 + 2.8 \times 1 - 1.5 \times 0 = 0.6$

$c_3 = 5.34 + 2.8 \times 0.6 - 1.5 \times 1 = 5.52$

$c_4 = -0.748 + 2.8 \times 5.52 - 1.5 \times 0.6 = 13.808$

${c_{n-1}}^2 - c_{n-2}(c_n - b_{n-1}) = 5.52^2 - .6(13.808 + 0.748) = 21.7368$

$b_{n-1}c_{n-1} - b_nc_{n-2} = (-0.748) \times 5.52 - (-0.1044)(0.6) = -4.06632$

$\delta p\ = -4.06632/21.7368 = -0.1871$

$b_nc_{n-1} - b_{n-1}(c_n - b_{n-1}) = -0.1044 \times 5.52 - (-0.748)(13.808 + 0.748) = 10.3116$

$\delta q = 10.3116/21.7368 = 0.4744$

$p_1 = -2.8 - 0.1871 = -2.9871$

$q_1 = 1.5 + 0.4744 = 1.9744$

Actual factors are : $(x^2 - 3x\ + 2)(x^2 - 2x\ + 5)$.

Exercise 3

3.1 Using Bisection method find an interval of length 0.1 or less in which the positive root of the equation $x^3 - 2x^2 - 3x - 10 = 0$, lies.

3.2 Write down the scheme (formula) to find the root of the equation $xe^x - 1 = 0$ by method of successive substitution. Show that the scheme will converge to the actual root near $x_0 = 0.5$. Hence find the root correct to three decimals.

3.3 Find the minimum value of the function $f(x) = x^2 - 2\sin x,\ x > 0$ which occurs for $0.5 < x < 1.0$, using Regula–Falsi method correct up to 3 decimals.
Hint: Find zero of $f'(x) = x - \cos x$.

3.4 Find the root of the equation $\ln x - x + 1.5 = 0$ by Secant method starting from $x_1 = 1.98$, $x_2 = 2.00$. Also solve by Regula–Falsi method taking $x_1 = 2.00$ and $x_2 = 2.50$. Compute up to 3 decimals.

3.5 Using N–R method find the zero of the function $f(x) = e^{-x} - \tan x,\ 0 < x < \frac{\pi}{2}$, correct up to 2 decimals.

3.6 Write down N–R scheme to find the cube-root of a number N. Hence, find the cube-root of 100 starting from the initial guess $x_0 = 4.5$. Compute your answer correct up to 3 decimals.

Hint: $x_{n+1} = x_n - \dfrac{x_n^3 - 100}{3x_n^2} = \dfrac{2x_n^3 + 100}{3x_n^2} = \dfrac{2(x_n^3 + 50)}{3x_n^2}$

3.7 Solve the simultaneous equations $y^2 = x^3$ and $\cos y = x$ by N–R method taking the initial values of x and y as $x_0 = 0.8$ and $y_0 = 0.7$. Show only one iteration computing, up to 3 decimals.

3.8 Solve the simultaneous equations $x^2y + y - 4 = 0$ and $x^3 + xy^2 - 4y^2 = 0$ by N–R method taking the initial estimate as $x_0 = 1.5$ and $y_0 = 1.2$. Perform only one iteration to compute the values up to 4 decimals.

3.9 Find the point of intersection of the curves $x^{\frac{1}{2}} + y^{\frac{1}{2}} = 1$ and $y^2 = x^3$ by N–R method taking initial guess as $x_0 = 0.45$ and $y_0 = 0.3$. Compute the values of x_1 and y_1 (first estimate) correct to three places of decimal.

References and Some Useful Related Books/Papers

1. Hildebrand F.B., *Introduction to Numerical Analysis*, Tata McGraw-Hill.
2. Scarborough J.B., *Numerical Mathematical Analysis*, Oxford Book Company.
3. Zurmuhl R., *Numerical Analysis for Engineers and Physicists* (Translated by Subramanian R., Achuthan P. and Venkatesan K.), Springer-Verlag.

4

Interpolation

4.1 Introduction

Let us suppose, there are given $(n+1)$ pair of values $(x_i,\ y_i),\ i = 0(1)n$. This data may be an outcome of an experiment in which for different values of a variable x, the values of y are observed; thus no relation between variables x and y is known. Alternatively, it may be that the values of a known function $y = f(x)$ are given for specific values of x. The abscissas x_i, $i = 0(1)n$ are called tabular points or nodal/pivotal points. Without loss of generality, we can assume $x_0 < x_1 < x_2 \ldots < x_n$, i.e., the values of y are prescribed at $(n+1)$ points in the interval $[x_0,\ x_n]$. Interpolation means to find the value of y for some intermediate value of x in $(x_0,\ x_n)$. If x lies outside the interval $(x_0,\ x_n)$, the process is called 'extrapolation'.

The methods for interpolation may be put into two categories according to whether the tabular points are equidistant (equally spaced or equi-spaced or evenly/uniformly spaced) or they are not necessarily at equal interval; in other words, whether the interval $x_i - x_{i-1}$, $i = 1(1)n$ is same throughout or not. In any case, it will be assumed that the behaviour of y w.r.t. x is smooth i.e. there are no sudden variations in the value of y.

The basis for an interpolation method is to approximate the data by some function $y = F(x)$, say, which may satisfy all the data points or only some of them. The function is called interpolating function and the points on which the function $F(x)$ is based are called interpolating points. Although there may be several functions interpolating the same data, we shall be confined to polynomial approximation in one form or another. Before describing various interpolation methods let us give some essential preliminaries which will be required in the development of the methods.

4.2 Some Operators and their Properties

Let us suppose that the tabular points are evenly spaced with a common interval h, i.e., $x_i - x_{i-1} = h,\ i = 1(1)n$, and let $y = f(x)$ be some function defined for $x_0 \le x \le x_n$.

Following operators may be defined at a general point x:

(*i*) Forward Difference (FD) Operator: Δ (Delta)

$$\Delta f(x) = f(x+h) - f(x)$$

(*ii*) Backward Difference (BD) Operator: ∇ (Inverted Delta or del or nabla)

$$\nabla f(x) = f(x) - f(x-h)$$

(*iii*) Central Difference (CD) Operator: δ (small Delta)

$$\delta f(x) = f\left(x + \frac{1}{2}h\right) - f\left(x - \frac{1}{2}\right)h$$

(*iv*) Averaging Operator: μ (mew)

$$\mu f(x) = \frac{1}{2}\left[f\left(x + \frac{1}{2}h\right) + f\left(x - \frac{1}{2}h\right)\right]$$

(*v*) Shift/Displacement Operator: E

$$\mathrm{E}f(x) = f(x+h)$$

(*vi*) Differential Operator: D

$$\mathrm{D}f(x) = \frac{d}{dx}f(x)$$

The operators (*i*) to (*v*) are called finite difference operators.

4.2.1 Linearity and commutativity of operators

The operators defined above satisfy the properties of Linearity and Commutativity. That is, if L, L_1, L_2 are operators and a, b scalars, then following relations hold:

(*i*) Linearity

$$(a\mathrm{L}_1 + b\mathrm{L}_2)f(x) = a\mathrm{L}_1 f(x) + b\mathrm{L}_2 f(x) \tag{4.1}$$

$$\mathrm{L}[af_1(x) + bf_2(x)] = a\mathrm{L}f_1(x) + b\mathrm{L}f_2(x) \tag{4.2}$$

(*ii*) Commutativity

$$L_1L_2f(x) = L_2L_1f(x) \text{ and } (L_1 + L_2)f(x) = (L_2 + L_1)f(x). \tag{4.3}$$

4.2.2 Repeated application and exponentiation of operators

The operators can be applied repeatedly on a given function $f(x)$, e.g., $\Delta\{\Delta f(x)\}$. It can be represented by $\Delta^2 f(x)$, and is called second forward difference of $f(x)$, thus the nth forward difference of $f(x)$ will be denoted by $\Delta^n f(x)$.

Similarly, the nth backward difference (BD) and central difference (CD) of $f(x)$ may be represented by $\nabla^n f(x)$ and $\delta^n f(x)$ respectively. It should be noted that n has to be an integer.

The operator E has got special significance in that $E^p f(x)$ will mean $f(x+ph)$; that is shifting the function by a distance p times the common interval h. The exponent p may be integer or a fraction and positive or negative. For example,

$$E^2 f(x) = f(x+2h);\ E^{-2}f(x) = f(x-2h);$$

$$E^{1/2}f(x) = f\left(x+\frac{1}{2}h\right);\ E^{-1/2}f(x) = f\left(x-\frac{1}{2}h\right).$$

Note: $E^p(1) = E^p(x^0) = (x+ph)^0 = 1$ and $E^p\alpha = \alpha E^p 1 = \alpha$ where α is constant.

4.2.3 Interrelations between operators

The operators defined in Sec 4.2 are interrelated as we can express one operator in terms of the other. Let us give a few examples:

(*i*) Δ and E

$$\Delta f(x) = f(x+h) - f(x) = Ef(x) - f(x)$$

$$= (E-1)\, f(x)$$

$$\therefore \quad \Delta = E - 1 \text{ or } E = 1 + \Delta.$$

(*ii*) ∇ and E

$$\nabla f(x) = f(x) - f(x-h) = f(x) - E^{-1}f(x)$$

$$= (1 - E^{-1})f(x)$$

$$\therefore \quad \nabla = 1 - E^{-1} \text{ or } E = (1-\nabla)^{-1}$$

(iii) Δ, δ and E

$$\delta f(x) = f\left(x+\frac{1}{2}h\right) - f\left(x-\frac{1}{2}h\right) = \mathrm{E}^{1/2}f(x) - \mathrm{E}^{-1/2}f(x)$$

$$= (\mathrm{E}^{1/2} - \mathrm{E}^{-1/2})f(x)$$

$$\delta = \mathrm{E}^{1/2} - \mathrm{E}^{-1/2}$$

To express Δ and E in terms of δ, we square both sides, giving

$$\mathrm{E}^2 - (2+\delta^2)\mathrm{E} + 1 = 0$$

and
$$\mathrm{E} = 1 + \frac{\delta^2}{2} \pm \frac{\delta}{2}\sqrt{\delta^2+4}.$$

or
$$\mathrm{E} - 1 = \Delta = \frac{\delta}{2}\{\delta \pm \sqrt{\delta^2+4}\}$$

To decide about the sign ($\pm$) in the above expression we may proceed as follows:

$$\sqrt{\delta^2+4} = \left\{(\mathrm{E}^{1/2} - \mathrm{E}^{-1/2})^2 + 4\right\}^{1/2} = \mathrm{E}^{1/2} + \mathrm{E}^{-1/2}.$$

Now, $\mathrm{E} - 1 = \frac{1}{2}(\mathrm{E}^{1/2} - \mathrm{E}^{-1/2})\left\{(\mathrm{E}^{1/2} - \mathrm{E}^{-1/2}) \pm (\mathrm{E}^{1/2} + \mathrm{E}^{-1/2})\right\}$

If we choose minus ($-$) sign in the above expression, we get

$$\mathrm{E} - 1 = \Delta = \frac{1}{2}(\mathrm{E}^{1/2} - \mathrm{E}^{-1/2})(-2\mathrm{E}^{-1/2}) = -\nabla, \text{ which is not true.}$$

The plus ($+$) sign gives $\mathrm{E} - 1 = \Delta = \frac{1}{2}(\mathrm{E}^{1/2} - \mathrm{E}^{-1/2})(2\mathrm{E}^{1/2}) = \Delta$, which is correct.

Hence $\mathrm{E} - 1 = \Delta = \frac{\delta^2}{2} + \frac{\delta}{2}\sqrt{\delta^2+4}.$

and
$$\mathrm{E} = 1 + \frac{\delta^2}{2} + \delta\sqrt{1+\frac{\delta^2}{4}}.$$

(iv) μ, E and δ

$$\mu = \frac{1}{2}\left[f\left(x+\frac{1}{2}h\right) + f\left(x-\frac{1}{2}h\right)\right] = \frac{1}{2}(\mathrm{E}^{1/2} + \mathrm{E}^{-1/2})f(x)$$

or $$\mu = \frac{E^{1/2} + E^{-1/2}}{2} \text{ or } 2\mu = E^{1/2} + E^{-1/2}.$$

also $$\delta = E^{1/2} - E^{-1/2}.$$

Squaring and subtracting one from the other, gives

$$4\mu^2 - \delta^2 = 4$$

or $$\mu = \sqrt{1 + \frac{\delta^2}{4}}.$$

(*v*) E and D.

From Taylor's series,

$$f(x+h) = f(x) + hf'(x) + \frac{h^2}{2!}f''(x) + \frac{h^3}{3!}f'''(x) + \cdots$$

or $$Ef(x) = f(x) + hDf(x) + \frac{h^2D^2}{2!}f(x) + \frac{h^3D^3}{3!}f(x) + \cdots$$

$$= \left(1 + hD + \frac{h^2D^2}{2!} + \frac{h^3D^3}{3!} + \cdots\right) f(x)$$

$$= e^{hD} f(x).$$

Hence $$E = e^{hD} \text{ or } hD = \ln E.$$

From the above relation we can also derive other relations, like,

(*a*) $$hD = \ln E = \ln(1 + \Delta) = -\ln(1 - \nabla).$$

(*b*) $$\delta = E^{1/2} - E^{-1/2} = e^{1/2hD} - e^{-1/2hD}$$

$$= 2\sinh\frac{hD}{2}$$

(*c*) $$\mu = \frac{1}{2}(E^{1/2} + E^{-1/2}) = \frac{1}{2}(e^{1/2hD} + e^{-1/2hD})$$

$$= \cosh\frac{hD}{2}.$$

Interrelations between various operators are shown in Table 4.1.

Table 4.1. Interrelations between various operators

	E	Δ	∇	δ	hD
E	E	$1+\Delta$	$(1-\nabla)^{-1}$	$1+\frac{1}{2}\delta^2+\delta\sqrt{1+\frac{\delta^2}{4}}$	e^{hD}
Δ	$E-1$	Δ	$\nabla(1-\nabla)^{-1}$	$\delta\sqrt{1+\frac{\delta^2}{4}}+\frac{1}{2}\delta^2$	$e^{hD}-1$
∇	$1-E^{-1}$	$\Delta(1+\Delta)^{-1}$	∇	$\delta\sqrt{1+\frac{\delta^2}{4}}-\frac{1}{2}\delta^2$	$1-e^{-hD}$
δ	$E^{1/2}-E^{-1/2}$	$\Delta(1+\Delta)^{-1/2}$	$\nabla(1+\nabla)^{-1/2}$	δ	$2\sinh\frac{hD}{2}$
μ	$\frac{1}{2}(E^{1/2}+E^{-1/2})$	$\left(1+\frac{\Delta}{2}\right)(1+\Delta)^{-1/2}$	$(1-\frac{\nabla}{2})(1+\nabla)^{-1}$	$\sqrt{1+\frac{\Delta^2}{4}}$	$\cosh\frac{hD}{2}$
hD	$ln E$	$ln(1+\Delta)$	$-ln(1-\nabla)$	$2\sinh^{-1}\frac{\delta}{2}$	hD

4.2.4 Application of operators on some functions

We would now see applications of operators on some functions, particularly polynomials.

(*i*) $f(x)=a$ (const)

Applying FD operator

$$\Delta f(x) = \Delta a = a - a = 0$$

It can also be explained as,

$$\Delta f(x) = \Delta a x^0$$

$$= a\Delta x^0 = a\,[(x+h)^0 - x^0] = a\,(1-1) = 0$$

(*ii*) $f(x)=ax$, a is const.

$$\Delta f(x) = \Delta ax$$

$$= a\Delta x = a[(x+h)-x] = ah \text{ (const)}$$

$$\Delta^2 f(x) = \Delta^2 ax = \Delta.\Delta ax = \Delta ah = 0 \text{ from } (i)$$

Thus first FD of ax is ah and second and higher order differences are zero.

(*iii*) $f(x) = x^2$

The first FD is given by

$$\Delta x^2 = (x+h)^2 - x^2 = 2hx + h^2$$

and second FD will be

$$\Delta^2 x^2 = \Delta \cdot \Delta x^2 = \Delta(2hx + h^2)$$

$$= 2h\Delta x + \Delta h^2 = 2h^2 \text{ (const)}$$

The third and higher differences will be zero.

(*iv*) $f(x) = x^n$.

To find the nth difference (or n^{th} forward difference) can be put in the form of a theorem.

Theorem: The nth forward difference of $f(x) = x^n$ is $\Delta^n x^n = n!h^n$.

Proof: The theorem is true for $n = 0, 1$ and 2 as shown above. Let us assume that it is true up to $(n-1)$, i.e.,

$$\Delta^{n-1} x^{n-1} = (n-1)!h^{n-1}.$$

Then,

$$\Delta^n x^n = \Delta^{n-1} \cdot \Delta x^n = \Delta^{n-1}\{(x+h)^n - x^n\}$$

$$= \Delta^{n-1}\left[\left\{x^n + \binom{n}{1} x^{n-1}h + \binom{n}{2} x^{n-2}h^2 + \cdots + \binom{n}{n-1} xh^{n-1} + h^n\right\} - x^n\right]$$

$$= \Delta^{n-1}\left[\binom{n}{1} x^{n-1}h + \binom{n}{2} x^{n-2}h^2 + \cdots + nxh^{n-1} + h^n\right]$$

$$= nh\Delta^{n-1}x^{n-1} + \binom{n}{2} h^2\Delta^{n-1}x^{n-2} + \ldots + nh^{n-1}\Delta^{n-1}x + \Delta^{n-1}h^n$$

Since the theorem is true up to $(n-1)$, i.e., for

$$x^0,\ x^1,\ x^2,\ \ldots.\ x^{n-1},\ \Delta^k x^k = (k-1)!h^{k-1},\ k = 0,1,2,\ \ldots.\ (n-1)$$

and $\Delta^{k+i}x^k = 0$, for i positive integer. All the terms from second onwards will be zero, while first term gives,

$$\Delta^n x^n = nh(n-1)!h^{n-1} = n!h^n. \tag{4.4}$$

Corollary: When $f(x)$ is a polynomial of degree n, say

$$P_n(x) = a_n x^n + a_{n-1}x^{n-1} + \ldots. + a_1 x + a_0, \text{ then}$$

$$\Delta^n P_n(x) = a_n n!h^n.$$

To show it we have

$$\begin{aligned}\Delta^n P_n(x) &= a_n \Delta^n x^n + a_{n-1}\Delta^n x^{n-1} + \ldots. + a_1 \Delta^n x + \Delta^n a_0 \\ &= a_n n!h^n,\end{aligned} \tag{4.5}$$

since all the terms from second onwards will be zero.

Note: The above results can also be proved for backward/central difference.

Representation of Differences in Terms of Function Values

Assuming that values of y are tabulated at equal intervals, we can express the differences in terms of function values, e.g.,

$$\begin{aligned}\Delta^k y_i &= (\mathrm{E}-1)^k y_i \\ &= \left[\mathrm{E}^k - \binom{k}{1}\mathrm{E}^{k-1} + \binom{k}{2}\mathrm{E}^{k-2} + \cdots + (-1)^r \binom{k}{r}\mathrm{E}^{k-r} + \cdots + (-1)^k \mathrm{E}^0\right] y_i \\ &= \sum_{r=0}^{k} (-1)^r \binom{k}{r} \mathrm{E}^{k-r} y_i = \sum_{r=0}^{k} (-1)^r \binom{k}{r} y_{k+i-r}.\end{aligned} \tag{4.6}$$

When differences are taken about y_0, i.e., $i = 0$, then above becomes,

$$\Delta^k y_0 = \sum_{r=0}^{k} (-1)^r \binom{k}{r} y_{k-r}. \tag{4.6a}$$

For $k = 3$ and $i = 2$, we can write from (4.6)

$$\Delta^3 y_2 = (\mathrm{E}-1)^3 y_2 = (\mathrm{E}^3 - 3\mathrm{E}^2 + 3\mathrm{E}^1 - \mathrm{E}^0)y_2$$

$$= \mathrm{E}^3 y_2 - 3\mathrm{E}^2 y_2 + 3\mathrm{E}^1 y_2 - \mathrm{E}^0 y_2$$

$$= y_5 - 3y_4 + 3y_3 - y_2.$$

When the differences are taken of $f(x)$ about any general point we can express it as follows:

$$\Delta^2 f(x) = (\mathrm{E}-1)^2 f(x)$$

$$= (\mathrm{E}^2 - 2\mathrm{E} + 1)f(x)$$

$$= \mathrm{E}^2 f(x) \ - 2\mathrm{E}f(x) \ + f(x)$$

$$= f(x+2h) - 2f(x+h) + f(x)$$

If $f(x) = x^2$, the above becomes,

$$\Delta^2 x^2 = \ (x+2h)^2 - 2(x+h)^2 + x^2 = 2h^2$$

This may be compared with Sec 4.2.4. (*iii*).

4.3 Finite Difference Table

The finite difference table plays a central role in the implementation of various interpolation formulae when the tabular points are uniformly/evenly spaced. Let us suppose, only five points are provided, say, $(x_i,\ y_i)$, $i = 0(1)4$, $x_{i+1} = x_i + h$. In order to construct a difference table, we proceed as follows (see Table 4.2):

1st differences are obtained as: $y_{i+1} - y_i = d_{i1},\ i = 0(1)3$

2nd differences are obtained as: $d_{i+1,1} - d_{i,1} = d_{i2},\ i = 0, 1, 2$

3rd differences are obtained as: $d_{i+1,2} - d_{i,2} = d_{i3},\ i = 0, 1$

4th differences are obtained as: $d_{i+1,3} - d_{i,3} = d_{i4},\ i = 0$

Table 4.2. Finite difference table

i	x_i	y_i	*Differences*			
			1st *Diff*	2nd *Diff*	3rd *Diff*	4th *Diff*
0	1.2	1.0627				
			0.0560			
1	1.4	1.1187		−0.0051		
			0.0509		0.0010	
2	1.6	1.1696		−0.0041		−0.0002
			0.0468		0.0008	
3	1.8	1.2164		−0.0033		
			0.0435			
4	2.0	1.2599				

It is important to note that the differences in a difference table can be expressed in terms of any of the difference operators, viz., forward, backward or central. For example, the first difference 0.0560 can be expressed as,

$$0.0560 = y_1 - y_0 = \Delta y_0 = \nabla y_1 = \delta y_{1/2}.$$

Similarly, the second difference −0.0051, can be expressed as,

$$-0.0051 = \Delta y_1 - \Delta y_0 = \Delta(y_1 - y_0) = \Delta \cdot \Delta y_0 = \Delta^2 y_0$$

or $$= \nabla y_2 - \nabla y_1 = \nabla(y_2 - y_1) = \nabla \cdot \nabla y_2 = \nabla^2 y_2$$

or $$= \delta y_{3/2} - \delta y_{1/2} = \delta(y_{3/2} - y_{1/2}) = \delta . \delta y_1 = \delta^2 y_1.$$

The other differences can be expressed in the same manner. Tables 4.3, 4.4 and 4.5 show the shape of the difference table when the differences are expressed by forward, backward and central differences operators respectively.

Table 4.3. Differences expressed by forward difference operator

i	x_i	y_i	Δ	Δ^2	Δ^3	Δ^4
0	x_0	$\underline{y_0}$				
			$\underline{\Delta y_0}$			
1	x_1	y_1		$\underline{\Delta^2 y_0}$		
			Δy_1		$\underline{\Delta^3 y_0}$	
2	x_2	y_2		$\Delta^2 y_1$		$\underline{\Delta^4 y_0}$
			Δy_2		$\Delta^3 y_1$	
3	x_3	y_3		$\Delta^2 y_2$		
			Δy_3			
4	x_4	y_4				

Table 4.4. Differences expressed by backward difference operator

i	x_i	y_i	∇	∇^2	∇^3	∇^4
0	x_0	y_0				
			∇y_1			
1	x_1	y_1		$\nabla^2 y_2$		
			∇y_2		$\nabla^3 y_3$	
2	x_2	y_2		$\nabla^2 y_3$		$\nabla^4 y_4$
			∇y_3		$\nabla^3 y_4$	
3	x_3	y_3		$\nabla^2 y_4$		
			∇y_4			
4	x_4	y_4				

Table 4.5. Differences expressed by central difference operator

i	x_i	y_i	δ	δ^2	δ^3	δ^4
0	x_0	y_0				
			$\delta y_{1/2}$			
1	x_1	y_1		$\delta^2 y_1$		
			$\delta y_{3/2}$		$\delta^3 y_{3/2}$	
2	x_2	y_2		$\delta^2 y_2$		$\delta^4 y_2$
			$\delta y_{5/2}$		$\delta^3 y_{5/2}$	
3	x_3	y_3		$\delta^2 y_3$		
			$\delta y_{7/2}$			
4	x_4	y_4				

It may be noticed that 4th difference of Table 4.2 can be expressed in three different systems as,

$$-0.0002 = \Delta^4 y_0 = \nabla^4 y_4 = \delta^4 y_2.$$

Further, it should also be observed that y_0 and its differences Δy_0, $\Delta^2 y_0$, $\Delta^3 y_0$ and $\Delta^4 y_0$ appear in a row slanting downwards (Table 4.3), while y_4 and ∇y_4, $\nabla^2 y_4$, $\nabla^3 y_4$ and $\nabla^4 y_4$ appear in a row upwards (Table 4.4). In central differences (Table 4.5) y_2 and its even differences $\delta^2 y_2$ and $\delta^4 y_2$ appear on a horizontal line. These are underlined in the respective tables. The above is true for y_i, and its differences in general.

Example 4.1

Construct a finite difference table for the polynomial function

$y = 3x^3 - 5x + 4$ for $x = -3(2)7$ and verify that $\Delta^n P_n(x) = a_0 n! h^n$ where

$P_n(x) = a_0 x^n + a_1 x^{n-1} + \ldots + a_n$.

Solution

Difference table

x	$y = P(x)$	Δ	Δ^2	Δ^3
−3	−62			
		68		
−1	6		−72	
		−4		144
1	2		72	
		68		144
3	70		216	
		284		144
5	354		360	
		644		
7	998			

Here, $a_0 = 3,\ h = 2$ and $n = 3$

$$\Delta^3 P_3(x) = 3 \times 3! \times 2^3$$

$$= 3 \times 6 \times 8 = 144$$

Example 4.2

The population of a certain city is given below for various years at equal intervals except for one year which is to be estimated.

Year	1951	1961	1971	1981	1991
Population (in lakhs)	45	43	?	52	55

Solution

First we assume $x_0 = 1951,\ x_1 = 1961,\ x_2 = 1971,\ x_3 = 1981$ and $x_4 = 1991$ and $h = 10$.

We have to find $y(x_2) = y_2$.

Four values are given, so we can fit a polynomial of degree 3 passing through four points. Then differences of order three will be constant and fourth differences will be zero.

Assuming that the value of y corresponding to x_2 also lie on this polynomial we write,

$$0 = \Delta^4 y_0 = (\mathrm{E} - 1)^4 y_0 = (\mathrm{E}^4 - 4\mathrm{E}^3 + 6\mathrm{E}^2 - 4\mathrm{E} + 1) y_0$$

or
$$y_4 - 4y_3 + 6y_2 - 4y_1 + y_0 = 0$$

Putting the values we get,

$$55 - 4 \times 52 + 6 \times y_2 - 4 \times 43 + 45 = 0$$

$$\therefore \qquad y_2 = 46.7 \text{ lakhs.}$$

Note: If two values are missing and there are say, 6 known values then two simultaneous equations may be obtained arising from,

$$\Delta^6 y_0 = 0 \text{ and } \Delta^6 y_1 = 0.$$

By solving these two equations we get the required values of two unknowns.

4.3.1 Propagation of error in a difference table

Let us suppose that values of y are tabulated against the values of x at equal interval. In general, we expect the magnitude of the differences getting smaller with the increasing order of differences showing the smoothness of the data. It may happen that the differences may start behaving erratically at some stage. This may be attributed to the error in the data (y values) due to rounding or otherwise. In Table 4.6, values (x_i, y_i) are given, $i = 0(1)8$, and an error s is introduced in y_4 to become $y_4 + \varepsilon$. The effect of the error, as how it propagates in the finite differences is shown in the Table.

Table 4.6. Propagation of error in a difference table

x_0	y_0			
		Δy_0		
x_1	y_1		$\Delta^2 y_0$	
		Δy_1		$\Delta^3 y_0$
x_2	y_2		$\Delta^2 y_1$	
		Δy_2		$\Delta^3 y_1 + \varepsilon$
x_3	y_3		$\Delta^2 y_2 + \varepsilon$	
		$\Delta y_3 + \varepsilon$		$\Delta^3 y_2 - 3\varepsilon$
x_4	$y_4 + \varepsilon$		$\Delta^2 y_3 - 2\varepsilon$	
		$\Delta y_4 - \varepsilon$		$\Delta^3 y_3 + 3\varepsilon$
x_5	y_5		$\Delta^2 y_4 + \varepsilon$	
		Δy_5		$\Delta^3 y_4 - \varepsilon$
x_6	y_6		$\Delta^2 y_5$	
		Δy_6		$\Delta^4 y_5$
x_7	y_7		$\Delta^2 y_6$	
		Δy_7		
x_8	y_8			

Table 4.6 shows the behaviour of the propagation of error in the Table. In any column, the errors are distributed in a binomial fashion. For example in the first difference they appear as $\varepsilon(1 \;\; -1)$; in the second difference as $\varepsilon(1 \;\; -2 \;\; 1)$ and in the third as $\varepsilon(1 \;\; -3 \;\; 3 \;\; -1)$. They can be written as $\varepsilon(1-1)^1$, $\varepsilon(1-1)^2$ and $\varepsilon(1-1)^3$ respectively in binomial form. Evidently the sum of errors in any column comes to zero. Secondly the errors are distributed symmetrically about the incorrect value y_4 above and below it (See Examples 4.3 and 4.4).

Example 4.3

Locate the error in the following data and correct it.

x	−1	0	1	2	3	4	5	6	7	8
y	7	5	3	7	25	57	115	203	327	493

Solution

Difference table

x	y	Δ	Δ^2	Δ^3	Δ^4
−1	7				
		−2			
0	5		0		
		−2		6	
1	3		6		2
		4		8	
2	7		14		−8
		18		0	
3	25		14		12
		32		12	
4	57		26		−8
		58		4	
5	115		30		2
		88		6	
6	203		36		0
		124		6	
7	327		42		
		166			
8	493				

The fourth differences are distributed according to binomial distribution, i.e., $2\,(1 \;\; -4 \;\; 6 \;\; -4 \;\; 1)$

$\therefore$ $\varepsilon = 2$ and it is against the middle term in the binomial distribution, i.e., in 25.

Therefore the correct value should be

$$25 - 2 = 23.$$

Example 4.4

Locate the error in the following data and correct it.

x	0	1	2	3	4	5	6
y	1	0	3	10	21	34	55

Solution

x	y	Δ	Δ^2	Δ^3
0	1			
		−1		
1	0		4	
		3		0
2	3		4	
		7		0
3	10		4	
		11		−2
4	21		2	
		13		6
5	34		8	
		21		
6	55			

Here, the third difference should have been all zero, but two terms are not zero due to error in the data.

If we compare with the binomial expansion, only two terms are available, i.e., $-2, 6$. Had we more terms in the data then we would have got all the terms of the binomial expansion. In this case we see that -2, 6 can be seen as $-2\ (1\ -3)$, i.e., first two terms of binomial expansion. Therefore, $\varepsilon = -2$ and the error should be against middle term, i.e., between -3 and 3, i.e., in 34.

$\therefore$ correct value is $34-(-2)=36$.

4.4 Error in Approximating a Function by Polynomial

Let us suppose that the values of a function $y=f(x)$ are prescribed at $(k+1)$ points $x=x_i$, $i=0(1)k$. We can approximate the function $f(x)$ by a polynomial $P(x)$ of highest degree k which passes through the points $(x_i,\ y_i)$ satisfying $y_i=f(x_i),\ i=0(1)k$. Let us express,

$$f(x)=P(x)+R(x), \tag{4.7}$$

where function $R(x)$ is error in the approximation.

Since $f(x_i)=P(x_i),\ i=0(1)k$, the error $R(x)$ should vanish at these points. Therefore, we can write,

$$R(x)=K(x)\cdot(x-x_0)(x-x_1)\ldots.(x-x_k), \tag{4.8}$$

and $\quad f(x)=P(x)+K(x)(x-x_0)(x-x_1)\ldots(x-x_k),$

$$=P(x)+K(x)\pi(x-x_i) \tag{4.9}$$

where $K(x)$ is not known and $\pi(x-x_i)=(x-x_0)(x-x_1)\ldots(x-x_k)$.

Further, let $x_0 < x_1 < x_2 \ldots < x_k$ and $\bar{x}$ be another point in $(x_0,\ x_k)$. We can choose $K(x)$ such that (4.9) is satisfied at $x = \bar{x}$, giving

$$K(\bar{x}) = \frac{f(\bar{x}) - P(\bar{x})}{\pi(\bar{x} - x_i)}. \tag{4.10}$$

Putting this value of $K(\bar{x})$ in (4.9),

$$f(x) = P(x) + K(\bar{x})\pi(x - x_i) \tag{4.11}$$

while $\quad R(x) = K(\bar{x})\pi(x - x_i).$ (4.11a)

Now, to determine the magnitude of $K(\bar{x})$, let us consider a function

$$\phi(x) = f(x) - P(x) - \frac{f(\bar{x}) - P(\bar{x})}{\pi(\bar{x} - x_i)} \cdot \pi(x - x_i). \tag{4.12}$$

The function $\phi(x)$ given by (4.12) vanishes at $(k+2)$ points, namely $x_0, x_1, \ \ldots.\ x_k$ and one more point $\bar{x}$. Let us assume that $\phi(x)$ and its derivatives are continuous and Rolle's theorem is applicable (see Appendix-A). Since $\phi(x)$ vanishes at $(k+2)$ points, $\phi'(x)$ must vanish at least at $(k+1)$ points and $\phi''(x)$ at k points and so on. Continuing in this manner $\phi^{k+1}(x)$ must vanish at least at one point, say ξ in $(x_0,\ x_k)$. Remembering that $P(x)$ is a polynomial of highest degree k, we get from (4.12) after differentiating it $(k+1)$ times and putting $\phi^{k+1}(\xi) = 0$,

$$\frac{f(\bar{x}) - P(\bar{x})}{\pi(\bar{x} - x_i)} = \frac{f^{k+1}(\xi)}{(k+1)!},\ x_0 \le \xi \le x_k. \tag{4.13}$$

From (4.10) and (4.13),

$$K(\bar{x}) = \frac{f^{k+1}(\xi)}{(k+1)!},\ x_0 \le \xi \le x_k.$$

Since $\bar{x}$ may be any point in $(x_0,\ x_k)$, the error $R(x)$ will be given from (4.11a), as,

$$R(x) = (x - x_0)(x - x_1)\cdots(x - x_k) \cdot \frac{f^{k+1}(\xi)}{(k+1)!}, \tag{4.14}$$

where ξ is a point in $[x_0,\ x_k]$.

We compute the upper bound of $R(x)$ for given x, taking the maximum value of $|f^{k+1}(\xi)|$ for ξ in $[x_0,\ x_k]$.

It may be noted that as $x_1,\ x_2,\ \ldots.\ x_k$ approach x_0, the error $R(x)$ approaches truncation error R_{n+1} in the Taylor's expansion of $f(x)$ about x_0 (See Appendix-B).

It must be realised that $R(x)$ gives an upper bound for the error while the actual error may be much smaller. Further, there are two components in $R(x)$ which determine its magnitude. First component is the positions of the tabular points, $x_0,\ x_1,\ \dots.\ x_n$. If they are located too far apart, then the value of the factor $(x-x_0)(x-x_1)\dots.\ (x-x_n)$ will be too large. Moreover, the oscillations of the approximating polynomial may attain too big amplitudes which may result in totally absurd result (See Example 4.15). The second component is the value of the derivative $f^{k+1}(x)$ in the interval of interest. Its value should remain small in the given interval otherwise, the magnitude of the error $R(x)$ may be too large.

4.4.1 Justification for approximation by polynomial

The famous Weierstrass's theorem states that any function $f(x)$ which is continuous in a closed interval $[a,\ b]$, can be approximated by a polynomial $P(x)$, uniformly over that interval, such that for a positive number ε, howsoever small,

$$|f(x)-P(x)|<\varepsilon,\ a\ \leq x \leq b.$$

Mathematically elegant, but from Numerical Analysis viewpoint, the theorem does not tell us, as for a given ε (accuracy) what degree of polynomial should be chosen. Nevertheless, the approximation of a function by a polynomial is justified.

Now, we derive some interpolation formulae.

4.5 Newton's (Newton–Gregory) Forward Difference (FD) Formula

Let us suppose we are given $(n+1)$ values $(x_i,\ y_i),\ i=0(1)n$, tabulated at an equal interval h, such that $x_i-x_{i-1}=h,\ i=1(1)n$ and that $x_0<x_1<x_2\dots.<x_n$. We want to determine the value of y at some point x in $(x_0,\ x_n)$. Let us denote the exact value of y for any general point x by $y(x)$, and the computed (estimated) value by y_x. It may be noted that $y(x_i)=y_{x_i}=y_i,\ i=0(1)n$.

$$\text{Now, let}\quad x_p=x=x_0+ph \ \text{ or } \ p=\frac{x-x_0}{h}, \tag{4.15}$$

where p may be, in general, positive or negative and integer or a fraction.

We can express,

$$\begin{aligned} y(x) &= y(x_0+ph)=\mathrm{E}^p y(x_0)=(1+\Delta)^p y_0 \\ &= y_0+p\Delta y_0+\frac{p(p-1)}{2!}\Delta^2 y_0+\frac{p(p-1)(p-2)}{3!}\Delta^3 y_0+\dots \end{aligned}$$

But since we can employ only up to a certain number of differences $k(\leq n)$, the above series is a terminating series; hence the estimated value of $y(x)$ will be,

$$y_{x_p} = y_p = y_0 + p\Delta y_0 + \frac{p(p-1)}{2!}\Delta^2 y_0 + \cdots + \binom{p}{k}\Delta^k y_0 \tag{4.16}$$

where $x = x_p$ and

$$\binom{p}{k} = \frac{p(p-1)(p-2)\cdots(p-k+1)}{1\cdot 2\cdot 3\cdots k}, \quad k \leq n. \tag{4.16a}$$

Formula (4.16) is the required Newton's FD formula. It can also be expressed in terms of x and the tabular points as follows:

Note that, $p = \dfrac{x-x_0}{h}$; $p-1 = \dfrac{x-x_0-h}{h} = \dfrac{x-(x_0+h)}{h} = \dfrac{x-x_1}{h}$;

$$p-2 = \frac{x-x_0}{h} - 2 = \frac{x-x_2}{h} \text{ etc.}$$

Putting the values of $p,\ p-1,\ p-2$, etc., in (4.16), the formula becomes,

$$y_x = y_0 + \frac{x-x_0}{h}\Delta y_0 + \frac{(x-x_0)(x-x_1)}{2!h^2}\Delta^2 y_0 + \cdots + \frac{(x-x_0)\cdots(x-x_{k-1})}{k!h^k}\Delta^k y_0 \tag{4.17}$$

Following points must be kept in mind while using Newton's FD formula:

(*i*) Formula (4.16) is used to compute value of y for a given value of x and formula (4.17) is used only when y is required to be expressed as a polynomial in x.

(*ii*) We try to retain as many differences as possible without losing accuracy. In Newton's FD formula, the number of differences for particular y, decreases as we go downwards in the FD Table 4.3. Therefore, this formula is suitable only for the values to be computed near the upper end of the table. There are other formulae for the values near the lower end or in the middle of the table.

(*iii*) If the differences start behaving erratically or increasing in magnitude at any stage, we should leave out those differences and higher order differences in the formula.

(*iv*) Most important point is that we should shift the origin so that $0 < p < 1$. This means that it is not necessary to have first tabular point as x_0. If required we can make second point as x_0 and then order of the tabular points will be $x_{-1},\ x_0,\ x_1, \cdots\cdots$ If third point is made x_0, then the order of tabular points would be $x_{-2},\ x_{-1},\ x_0,\ x_1,\ \cdots\cdots$ That way we can have $0 < p < 1$ by choosing x_0 suitably. For extrapolation before the first tabular point p will be negative.

4.5.1 Error in Newton's FD formula

Let us suppose that the tabulated values $(x_i,\ y_i),\ i = 0(1)n$ are represented by a function, known or unknown, $y = f(x)$. The function $f(x)$ is approximated by y_x given by (4.17) which is expressed as a terminating/truncated series. Let us suppose that the terms only up to k^{th} difference $\Delta^k y_0,\ k \leq n$, are retained in y_x. Then we want to find the truncation error in y_x, given that $x_i = x_0 + ih,\ i = 1(1)n$.

From the mean value theorem (See Appendix B),

$$f(x_0+h) - f(x_0) = hf'(x_0 + \theta_1 h),\ 0 \leq \theta_1 \leq 1,$$

where f' denoted the first derivative.

or $$\Delta f(x_0) = hf'(x_0 + \theta_1 h),\ 0 \leq \theta_1 \leq 1.$$

Applying this result again, we get

$$\Delta^2 f(x_0) = h^2 f''(x_0 + \theta_1 h + \theta' h),\ 0 \leq \theta' \leq 1.$$
$$= h^2 f''[x_0 + (\theta_1 + \theta')h].$$

Now since θ_1 and θ' lie between 0 and 1, we have

$$0 \leq \theta_1 + \theta' \leq 2 \text{ or } 0 \leq \frac{\theta_1 + \theta'}{2} \leq 1$$

or $$0 \leq \theta_2 \leq 1 \text{ where } \frac{\theta_1 + \theta'}{2} = \theta_2 \text{ or } \theta_1 + \theta' = 2\theta_2.$$

Thus, the above can be written as,

$$\Delta^2 f(x_0) = h^2 f''(x_0 + \theta_2 \cdot 2h),\ 0 \leq \theta_2 \leq 1$$

Proceeding in this manner, we can get

$$\Delta^k f(x_0) = h^k f^k(x_0 + \theta_k \cdot kh),\ 0 \leq \theta_k \leq 1,$$

or, $$\frac{1}{h^k} \cdot \Delta^k f(x_0) = f^k(x_0 + \theta_k \cdot kh),\ 0 \leq \theta_k \leq 1, \tag{4.18}$$

where f^k is the kth derivatives of $f(x)$.

It may be noted that when $\theta_k = 0$, then $x_0 + \theta_k \cdot kh = x_0$ and when $\theta_k = 1$, then $x_0 + \theta_k \cdot kh = x_0 + kh = x_k$. Thus, as $x_0 \leq x_0 + \theta_k \cdot kh \leq x_k$, (4.18) may be written as,

$$\frac{1}{h^k} \cdot \Delta^k f(x_0) = f^k(\xi),\ x_0 \leq \xi \leq x_k. \tag{4.19}$$

Further, it is easy to verify that the FD formula given by (4.16) or (4.17) is a polynomial of degree k, (or less) passing through the points (x_i, y_i), $i = 0(1)k$, since by putting $x = x_i (p = i)$, all the terms from containing $\Delta^{i+1} y_0$ onwards will be zero due to a factor $x - x_i \{\text{or } (p - i)\}$ and the terms up to $\Delta^i y_0$ will add up to y_i.

Hence from the result (4.14) of Sec 4.4, the error in FD formula is given by

$$R(x) = (x - x_0)(x - x_1) \ldots . (x - x_k) \frac{f^{k+1}(\xi)}{(k+1)!}, \; x_0 \leq \xi \leq x_k \tag{4.20}$$

If the function $f(x)$ is not known, then by using (4.19) in (4.20) we get, the error in terms of differences by replacing the derivative, as

$$R(x) = (x - x_0)(x - x_1) \ldots . (x - x_k) \frac{\Delta^{k+1} y_0}{(k+1)! h^{k+1}}. \tag{4.21}$$

Using the transformation $p - i = \dfrac{x - x_i}{h}$, $i = 0(1)k$, (4.21) may be expressed as,

$$R(p) = p(p-1)(p-2) \ldots . (p-k) \cdot \frac{\Delta^{k+1} y_0}{(k+1)!}. \tag{4.22}$$

It may be noted that as the tabular points $x_1, x_2, \ldots . x_k$ approach x_0 (i.e., h is small), then (4.19) may be written as,

$$\frac{1}{h^k} \cdot \Delta^k f(x_0) = f^k(x_0), \; k = 0, 1, 2, \ldots . \tag{4.23}$$

In that case, (4.17) reduces to truncated Taylor's series,

$$y_x = y_0 + (x - x_0) f'(x_0) + \frac{(x - x_0)^2}{2!} f''(x_0) + \cdots + \frac{(x - x_0)^k}{k!} f^k(x_0),$$

with truncation error, given by (4.20), as

$$R(x) = R_{k+1}(x) = \frac{(x - x_0)^{k+1}}{(k+1)!} f^{k+1}(\xi), \; x_0 \leq \xi \leq x_k. \tag{4.24}$$

Thus (4.21) or (4.22) should give a good estimate of the error (truncation error) in the Newton's FD formula.

4.6 Newton's (Newton–Gregory) Backward Difference (BD) Formula

As stated, the FD formula should not be used near the lower end of the table as there may be only a few differences available. Therefore Newton's BD formula is to be used for interpolation near the lower end of the table. The formula may be derived in a manner similar to FD formula, only expressing E in terms of BD operator ∇. Thus,

$$\begin{aligned} y_p &= \mathrm{E}^p y_0 = (1-\nabla)^{-p} y_0 \\ &= y_0 + p\nabla y_0 + \frac{p(p+1)}{2!}\nabla^2 y_0 + \cdots + \frac{p(p+1)\cdots(p+k-1)}{k!}\nabla^k y_0 \end{aligned} \tag{4.25}$$

Formula (4.25) may be expressed in terms of x, using the relations,

$$p = \frac{x-x_0}{h},\ p+1 = \frac{x-x_{-1}}{h},\ p+2 = \frac{x-x_{-2}}{h}, \text{ etc.}$$

and in general $p+i = \dfrac{x-x_{-i}}{h}$, $i = 0, 1, 2, \ \ldots . \ k-1$. Thus,

$$y_x = y_0 + \frac{x-x_0}{h}\nabla y_0 + \frac{(x-x_0)(x-x_{-1})}{2!h^2}\nabla^2 y_0 + \ldots + \frac{(x-x_0)(x-x_{-1})\ldots(x-x_{-k+1})}{k!h^k}\nabla^k y_0 \tag{4.26}$$

Obviously, in using the formula (4.25) or (4.26), the point x_0 has to be taken near the lower end of the table so that the points above it will be correspondingly referred to as x_{-1}, x_{-2}, ..., etc. Also, the point x_0 should be chosen such that $-1 < p < 0$. For extrapolation, after the last point p will be positive.

The error in the formula will be same as given in the FD formula except that FD operator Δ will be replaced by the BD operator ∇.

4.7 Central Difference (CD) Formulae

The central difference formulae are used to interpolate the value anywhere in the middle of the table. There are six formulae based on central differences which are given below along with the approximate range of p for their application:

(*i*) Gauss's Backward	$-.5 < p < 0$
(*ii*) Gauss's Forward	$0 < p < .5$
(*iii*) Stirling's	$-.25 < p < .25$
(*iv*) Bessel's	$.25 < p < .75$
(*v*) Everett's	$0 < p < 1$
(*vi*) Steffensen's	$-.5 < p < .5$

These formulae have been designed on the choice of specific differences in the formulae. Fig. 4.1 shows the structure of various formulae using the values of y and their differences. Let us now show how these formulae can be constructed.

Following properties of combination will be used in the derivation of formulas:

$$\binom{n}{r} + \binom{n}{r-1} = \binom{n+1}{r} \tag{4.27}$$

or $$\binom{n+1}{r} - \binom{n}{r-1} = \binom{n}{r} \tag{4.27a}$$

or $$\binom{n+1}{r} - \binom{n}{r} = \binom{n}{r-1} \tag{4.27b}$$

4.7.1 Gauss's Backward (GB) formula

The GB formula uses y_0 and its even differences plus odd differences of $y_{-1/2}$ (See Fig. 4. 1c). It can be expressed as,

$$y_p = a_0 y_0 + a_1 \delta y_{-1/2} + a_2 \delta^2 y_0 + a_3 \delta^3 y_{-1/2} + a_4 \delta^4 y_0 + \cdots, \tag{4.28}$$

where a_0, a_1, a_2, etc., are constants to be determined. Using operator E, formula (4.28) may be converted as

$$\mathrm{E}^p y_0 = (a_0 + a_1 \delta \mathrm{E}^{-1/2} + a_2 \delta^2 + a_3 \delta^3 \mathrm{E}^{-1/2} + a_4 \delta^4 + \cdots) y_0. \tag{4.28a}$$

From (4.28a) we get the following relation between operators,

$$(1+\Delta)^p = a_0 + a_1 \delta \mathrm{E}^{-1/2} + a_2 \delta^2 + a_3 \delta^3 \mathrm{E}^{-1/2} + a_4 \delta^4 + \cdots \tag{4.29}$$

Since left side of (4.29) can be expressed as power series in Δ, we convert the right side also as powers of Δ, so that a's may be determined by comparing the coefficients.

Note the following relations

(i) $$\delta \mathrm{E}^{1/2} = \Delta, \text{ giving } \delta^2 = \frac{\Delta^2}{1+\Delta}$$

(ii) $$\delta \mathrm{E}^{-1/2} = \frac{\delta \mathrm{E}^{1/2}}{\mathrm{E}} = \frac{\Delta}{1+\Delta}$$

(iii) $$\delta^3 \mathrm{E}^{-1/2} = \delta^2 \cdot \delta \mathrm{E}^{-1/2} = \frac{\Delta^3}{(1+\Delta)^2}$$

(iv) $$\delta^5 \mathrm{E}^{-1/2} = \delta^2 \cdot \delta^3 \mathrm{E}^{-1/2} = \frac{\Delta^5}{(1+\Delta)^3}, \text{ etc.}$$

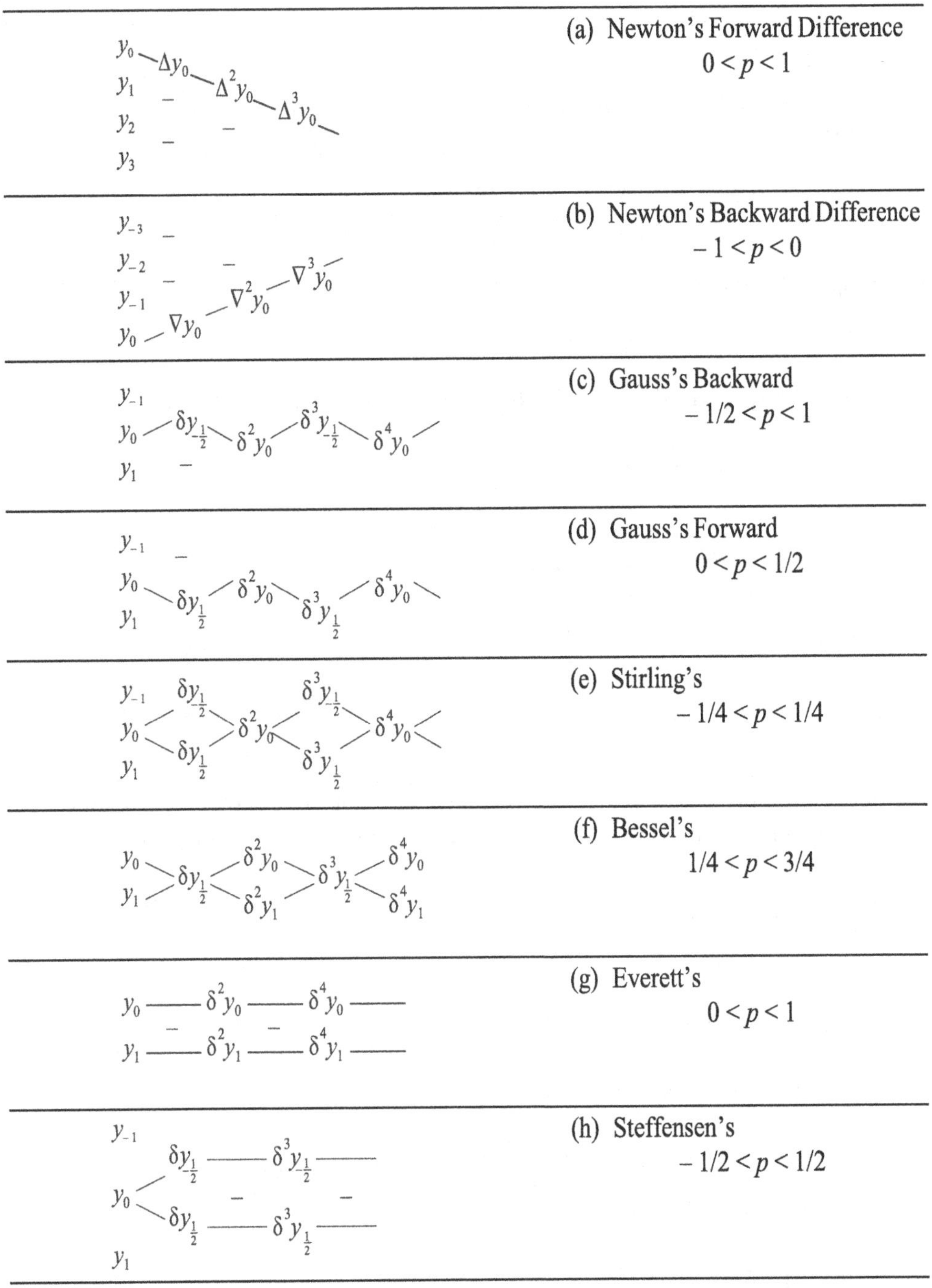

Figure 4.1 Differences used in various formulae

Expanding the left side of (4.29) and putting the values from the above relations in the right side we get,

$$1+p\Delta+\frac{p(p-1)}{2!}\Delta^2+\frac{p(p-1)(p-2)}{3!}\Delta^3+\frac{p(p-1)(p-2)(p-3)}{4!}\Delta^4+\cdots$$

$$=a_0+a_1\frac{\Delta}{1+\Delta}+a_2\frac{\Delta^2}{1+\Delta}+a_3\frac{\Delta^3}{(1+\Delta)^2}+a_4\frac{\Delta^4}{(1+\Delta)^2}+\cdots \tag{4.30}$$

On comparing the constant term,

$$a_0=1.$$

Now multiplying both sides of (4.30) by $(1+\Delta)$ and expanding left side,

$$1+(p+1)\Delta+\frac{(p+1)p}{2!}\Delta^2+\frac{(p+1)p(p-1)}{3!}\Delta^3+\cdots$$

$$=1\ (1+\Delta)+a_1\Delta^2+a_2\frac{\Delta^3}{(1+\Delta)^2}+\cdots \tag{4.30a}$$

Comparing the coefficients of Δ and Δ^2, we obtain,

$$a_1=p \text{ and } a_2=\binom{p+1}{2}.$$

Again, we multiply both sides of (4.30) by $(1+\Delta)^2$ {or (4.30a) by $(1+\Delta)$} and expand the left side by binomial series. After putting the values of a_0, a_1 and a_2 on the right side, we get

$$1+(p+2)\Delta+\frac{(p+2)(p+1)}{2}\Delta^2+\frac{(p+2)(p+1)p}{3!}\Delta^3+\frac{(p+2)(p+1)p(p-1)}{4!}\Delta^4+\cdots$$

$$=(1+\Delta)^2+p\Delta(1+\Delta)+\frac{(p+1)p}{2}\Delta^2(1+\Delta)+a_3\Delta^3+a_4\Delta^4+\cdots$$

Comparision of coefficients of Δ^3 and Δ^4 gives,

$$a_3=\binom{p+2}{3}-\binom{p+1}{2}=\binom{p+1}{3}, \text{ from (4.27a)},$$

$$a_4=\binom{p+2}{4}.$$

Proceeding in this manner and comparing the coefficients of Δ^{2m-1} and Δ^{2m}, we get

$$a_{2m-1} = \binom{p+m}{2m-1} - \binom{p+m-1}{2m-2} = \binom{p+m-1}{2m-1}, \tag{4.31a}$$

$$a_{2m} = \binom{p+m}{2m}, \quad m = 1, 2, \cdots \tag{4.31b}$$

Substituting the values of a's in (4.28), the GB Formula becomes,

$$y_p = y_0 + p\delta y_{-1/2} + \binom{p+1}{2}\delta^2 y_0 + \binom{p+1}{3}\delta^3 y_{-1/2}\binom{p+2}{4}\delta^4 y_0 \cdots \tag{4.32}$$

4.7.2 Gauss's Forward (GF) formula

Refering to Fig. 4.1d, the GF formula is written as,

$$y_p = a_0 y_0 + a_1 \delta y_{1/2} + a_2 \delta^2 y_0 + a_3 \delta^3 y_{1/2} + a_4 \delta^4 y_0 \cdots \tag{4.33}$$

where a_0, a_1, a_2 etc. are to be determined.

Using the relation $\delta \mathrm{E}^{1/2} = \Delta$, following equality may be set up,

$$(1+\Delta)^p = a_0 + a_1\Delta + a_2\frac{\Delta^2}{1+\Delta} + a_3\frac{\Delta^3}{1+\Delta} + a_4\frac{\Delta^4}{(1+\Delta)^2} + a_5\frac{\Delta^5}{(1+\Delta)^2} + \cdots \tag{4.34}$$

Expanding left side and comparing coefficients of Δ^0 and Δ gives,

$$a_0 = 1 \text{ and } a_1 = p.$$

Multiplying both sides of (4.34) by $(1+\Delta)$ and comparing coefficients, we get

$$a_2 = \binom{p}{2}; \; a_3 = \binom{p+1}{3}.$$

In general,

$$a_{2m} = \binom{p+m-1}{2m}; a_{2m+1} = \binom{p+m}{2m+1}, \quad m = 1, 2, 3, \ldots.. \tag{4.35}$$

Hence the GF formula may be written as,

$$y_p = y_0 + \binom{p}{1}\delta y_{1/2} + \binom{p}{2}\delta^2 y_0 + \binom{p+1}{3}\delta^3 y_{1/2} + \binom{p+1}{4}\delta^4 y_0$$

$$+\binom{p+2}{5}\delta^5 y_{1/2} + \ldots \qquad (4.36)$$

4.7.3 Stirling's formula

This formula uses y_0 and its even differences plus average of the odd differences of $y_{-1/2}$ and $y_{1/2}$ (See Fig. 4.1e). It is written as,

$$y_p = a_0 y_0 + a_1 \mu\delta y_0 + a_2 \delta^2 y_0 + a_3 \mu\delta^3 y_0 + a_4 \delta^4 y_0 + \ldots, \qquad (4.37)$$

where a_0, a_1, a_2 are constants to be determined and μ is averaging operator, i.e., $\mu = (\mathrm{E}^{1/2} + \mathrm{E}^{-1/2})/2$. Obviously Stirling's formula is the average of GB and GF formulae. Therefore, the coefficients in (4.37) are as given below:

$$a_0 = 1;$$

The other coefficients are given by

$$a_{2m-1} = \binom{p+m-1}{2m-1},$$

$$a_{2m} = \frac{1}{2}\left[\binom{p+m}{2m} + \binom{p+m-1}{2m}\right] \text{ from (4.31a) and (4.35)}$$

$$m = 1, 2, 3, \ldots, \text{ etc.}$$

Now, $$\binom{p+m}{2m} = \frac{(p+m)(p+m-1)\cdots(p+1)p(p-1)\cdots(p-m+1)}{(2m)!}$$

$$= \frac{(p+m)p(p^2-1^2)(p^2-2^2)\cdots(p^2-\overline{m-1}^2)}{(2m)!}$$

Similarly

$$\binom{p+m-1}{2m} = \frac{(p+m-1)(p+m-2)\cdots(p+1)p(p-1)\cdots(p-m-1)(p-m)}{(2m)!}$$

$$= \frac{(p-m)p(p^2-1^2)(p^2-2^2)\cdots(p^2-\overline{m-1}^2)}{(2m)!}$$

$$a_{2m} = \frac{p^2(p^2-1^2)(p^2-2^2)\cdots(p^2-\overline{m-1}^2)}{(2m)!}.$$

Putting the values of a_0, a_{2m-1} and $a_{2m}m = 1, 2, \ldots.$ in (4.37), gives

$$y_p = y_0 + p\mu\delta y_0 + \frac{p^2}{2!}\delta^2 y_0 + \binom{p+1}{3}\mu\delta^3 y_0 + \frac{p^2(p^2-1^2)}{4!}\delta^4 y_0 + \cdots \quad (4.38)$$

Instead of taking the average of GB and GF formula we can derive the Stirling's formula independently, as follows:

First let us note,

$$\mu\delta = \frac{(E^{1/2}+E^{-1/2})}{2}(E^{1/2}+E^{-1/2}) = \frac{E-E^{-1}}{2}$$

$$= \frac{\Delta+\Delta^2/2}{1+\Delta}.$$

$$\mu\delta^3 = \mu\delta\cdot\delta^2 = \frac{\Delta+\Delta^2/2}{1+\Delta}\cdot\frac{\Delta^2}{1+\Delta} \quad \because \delta^2 = \frac{\Delta^2}{1+\Delta}, \text{ etc.}$$

Putting these values in (4.37), we get

$$y_p = a_0 y_0 + a_1\left(\frac{\Delta+\Delta^2/2}{1+\Delta}\right)y_0 + a_2\frac{\Delta^2}{1+\Delta}y_0 + a_3\left(\frac{\Delta+\Delta^2/2}{1+\Delta}\right)\frac{\Delta^2}{1+\Delta} + a_4\frac{\Delta^4}{(1+\Delta)^2}y_0 + \cdots$$

The relation between the operators may be expressed as,

$$(1+\Delta)^p = a_0 + a_1\frac{\Delta}{1+\Delta} + \left(\frac{a_1}{2}+a_2\right)\frac{\Delta^2}{1+\Delta} + a_3\frac{\Delta^3}{(1+\Delta)^2} + \left(\frac{a_3}{2}+a_4\right)\frac{\Delta^4}{(1+\Delta)^2} + \cdots \quad (4.39)$$

Of course, $a_0 = 1$.

Multiplying both sides by $(1+\Delta)$ and comparing the coefficients of Δ and Δ^2,

$$a_1 = p \text{ and } \frac{a_1}{2} + a_2 = \frac{(p+1)p}{2} \text{ giving}$$

$$a_2 = \frac{(p+1)p}{2} - \frac{p}{2} = \frac{p^2}{2}.$$

Again multiplying both sides by $(1+\Delta)^2$ and comparing the coefficients of Δ^3 and Δ^4, gives,

$$\binom{p+1}{2} + a_3 = \binom{p+2}{3} \text{ giving } a_3 = \binom{p+2}{3} - \binom{p+1}{2} = \binom{p+1}{3}; \text{ and}$$

$$\frac{a_3}{2} + a_4 = \binom{p+2}{4} \text{ giving } a_4 = \binom{p+2}{4} - \frac{1}{2}\binom{p+1}{3} = \frac{p^2(p^2-1^2)}{4!}$$

Proceeding in the same manner, we can get, by comparing the coefficients of Δ^{2m-1} and Δ^{2m},

$$a_{2m-1} = \binom{p+m}{2m-1} - \binom{p+m-1}{2m-2} = \binom{p+m-1}{2m-1} \tag{4.40a}$$

$$a_{2m} = \binom{p+m}{2m} - \frac{1}{2}\binom{p+m-1}{2m-1} \tag{4.40b}$$

$$= \frac{(p+m)(p)(p^2-1^2)\cdots(p^2-\overline{m-1}^2)}{(2m)!} - \frac{1}{2}\frac{p(p^2-1^2)\cdots(p^2-\overline{m-1}^2)}{(2m-1)!}$$

$$= \frac{p^2(p^2-1^2)(p^2-2^2)\cdots(p^2-\overline{m-1}^2)}{(2m)!}$$

Substituting the above values in (4.37), the Stirling's formula (4.38) follows.

4.7.4 Bessel's formula

The Bessel's formula is converse of the Stirling's in the sense that in the Stirling's formula average of the odd differences was used while in the Bessel's, average of the even differences is used. Thus the Bessel's formula is written as (See Fig 4.1f),

$$y_p = a_0\mu y_{1/2} + a_1\delta y_{1/2} + a_2\mu\delta^2 y_{1/2} + a_3\delta^3 y_{1/2} + a_4\mu\delta^4 y_{1/2} + \cdots \tag{4.41}$$

where a_0, a_1, a_2, etc. are constants to be determined.

Note the relations,

$$\mu E^{1/2} = \frac{1}{2}(E^{1/2} + E^{-1/2})E^{1/2} = 1 + \frac{\Delta}{2}$$

$$\delta E^{1/2} = (E^{1/2} - E^{-1/2})E^{1/2} = \Delta$$

$$\delta^2 = \frac{\Delta^2}{1+\Delta}, \quad \delta^4 = \frac{\Delta^4}{(1+\Delta)^2} \text{ etc.}$$

Using above relations in (4.41), we get following equality,

$$(1+\Delta)^p = a_0\left(1+\frac{\Delta}{2}\right) + a_1\Delta + a_2\left(1+\frac{\Delta}{2}\right)\frac{\Delta^2}{1+\Delta} + a_3\Delta\cdot\frac{\Delta^2}{1+\Delta} + a_4\left(1+\frac{\Delta}{2}\right)\frac{\Delta^4}{(1+\Delta)^2} + \cdots$$

$$= a_0 + \left(\frac{a_0}{2} + a_1\right)\Delta + a_2\frac{\Delta^2}{1+\Delta} + \left(\frac{a_2}{2} + a_3\right)\frac{\Delta^3}{1+\Delta} + a_4\frac{\Delta^4}{(1+\Delta)^2} + \cdots \tag{4.42}$$

On comparing the constant term and coefficient of Δ, we get

$$a_0 = 1 \text{ and } \frac{a_0}{2} + a_1 = p \text{ or } a_1 = p - \frac{1}{2}.$$

Multiplying (4.42) by $(1+\Delta)$ throughout and comparing coefficients of Δ^2 and Δ^3, gives,

$$p + a_2 = \frac{(p+1)p}{2!} \text{ giving } a_2 = \binom{p+1}{2} - \binom{p}{1} = \binom{p}{2}$$

and, $$\frac{a_2}{2} + a_3 = \frac{(p+1)p(p-1)}{3!} \text{ giving } a_3 = \frac{(p+1)p(p-1)}{3.2} - \frac{p(p-1)}{2.2}$$

$$= \binom{p}{2}\cdot\frac{p-\frac{1}{2}}{3}.$$

Again, multiplying (4.42) by $(1+\Delta)^2$ and comparing the coefficients of Δ^4 and Δ^5, we get

$$\binom{p+1}{3} + a_4 = \binom{p+2}{4} \text{ or } a_4 = \binom{p+2}{4} - \binom{p+1}{3} = \binom{p+1}{4}$$

and $\frac{a_4}{2}+a_5=\begin{pmatrix} p+2 \\ 5 \end{pmatrix}$ or $a_5=\begin{pmatrix} p+2 \\ 5 \end{pmatrix}-\frac{1}{2}\begin{pmatrix} p+1 \\ 3 \end{pmatrix}$

$$=\begin{pmatrix} p+1 \\ 4 \end{pmatrix}\frac{p-\frac{1}{2}}{5}.$$

The general term can be written as,

$$a_{2m}=\begin{pmatrix} p+m-1 \\ 2m \end{pmatrix};\ a_{2m+1}=\begin{pmatrix} p+m-1 \\ 2m \end{pmatrix}\frac{p-\frac{1}{2}1}{2m+1},\quad m=1,2,\ \ldots.$$

Putting the above values in (4.41) the Bessel's formula may be written as,

$$y_p=\mu y_{1/2}+\left(p-\frac{1}{2}\right)\delta y_{1/2}+\begin{pmatrix} p \\ 2 \end{pmatrix}\mu\delta^2 y_{1/2}+\begin{pmatrix} p \\ 2 \end{pmatrix}\frac{p-\frac{1}{2}}{3}\delta^3 y_{1/2}$$

$$+\begin{pmatrix} p+1 \\ 4 \end{pmatrix}\mu\delta^4 y_{1/2}+\begin{pmatrix} p+1 \\ 4 \end{pmatrix}\frac{p-\frac{1}{2}}{5}\delta^5 y_{1/2}+\cdots \tag{4.43}$$

Bessel's formula also can be derived from the GB and GF formula.

4.7.5 Everett's formula

In this formula only even order differences are involved and no odd differences are used (See Fig. 4.1g).

It can be written as,

$$y_p=a_0y_0+a_1\delta^2 y_0+a_2\delta^4 y_0+\ldots+b_0y_1+b_1\delta^2 y_1+b_2\delta^4 y_1+\ldots. \tag{4.44}$$

We get the relation,

$$(1+\Delta)^p=(a_0+b_0)+b_0\Delta+(a_1+b_1)\frac{\Delta^2}{1+\Delta}+b_1\frac{\Delta^3}{1+\Delta}+(a_2+b_2)\frac{\Delta^4}{(1+\Delta)^2}+b_2\frac{\Delta^5}{(1+\Delta)^2}+\ldots \tag{4.44a}$$

Comparison of coefficient of Δ and the constant term gives,

$b_0=p$ and $a_0+b_0=1$ giving $a_0=1-p$.

Multiplying (4.44a) by $(1+\Delta)$ and comparing the coefficients of Δ^3 and Δ^2 respectively, gives,

$$b_1 = \binom{p+1}{3};$$

and $$b_0 + a_1 + b_1 = \binom{p+1}{2}$$

or $$a_1 = \binom{p+1}{2} - \binom{p}{1} - \binom{p+1}{3}$$

$$= \binom{p}{2} - \binom{p+1}{3} = -\binom{p}{3}$$

Similarly, after multiplying (4.44a) by $(1+\Delta)^2$ and comparing the coefficients of Δ^5 and Δ^4 respectively, we get

$$b_2 = \binom{p+2}{5};$$

$$b_1 + a_2 + b_2 = \binom{p+2}{4} \text{ or } a_2 = \binom{p+2}{4} - \binom{p+1}{3} - \binom{p+2}{5}$$

or $$a_2 = \binom{p+1}{4} - \binom{p+2}{5} = -\binom{p+1}{5}.$$

Proceeding in this manner we can get the general term by comparing the coefficients of Δ^{2m+1} and Δ^{2m} respectively, as

$$b_m = \binom{p+m}{2m+1};\ a_m = -\binom{p+m-1}{2m+1},\ m = 0, 1, 2 \cdots \tag{4.45}$$

Putting the above values in (4.44), the formula becomes,

$$y_p = -(p-1)y_0 - \binom{p}{3}\delta^2 y_0 - \binom{p+1}{5}\delta^4 y_0 - \cdots$$

$$+py_1 + \binom{p+1}{3}\delta^2 y_1 + \binom{p+2}{5}\delta^4 y_1 + \cdots$$

If we put $q = 1 - p$, the Everett's formula may be written after using,

$$\binom{p}{m} = (-1)^m \binom{-p+m-1}{m}, \text{ as}$$

$$y_p = qy_0 + \binom{q+1}{3}\delta^2 y_0 + \binom{q+2}{5}\delta^4 y_0 + \ldots + py_1 + \binom{p+1}{3}\delta^2 y_1$$

$$+ \binom{p+2}{5}\delta^4 y_1 + \ldots \tag{4.46}$$

It may be noted that Everett's formula (4.46) is symmetrical about $p = \frac{1}{2}(= q)$ in that the coefficients of y_0 and its differences become same as that of y_1 and its differences. The formula can also be written in the following form,

$$y_p = (qy_0 + py_1) + \left\{\binom{q+1}{3}\delta^2 y_0 + \binom{p+1}{3}\delta^2 y_1\right\} + \left\{\binom{q+2}{5}\delta^4 y_0\right.$$

$$\left. + \binom{p+2}{5}\delta^4 y_1\right\} + \cdots \tag{4.46a}$$

4.7.6 Steffensen's formula

As opposed to Everett's which uses even differences the Steffensen's formula uses odd differences (See Fig. 4.1h). It can be written as,

$$y_p = c_0 y_0 + a_1 \delta y_{1/2} + a_2 \delta^3 y_{1/2} + a_3 \delta^5 y_{1/2} + \cdots + b_1 \delta y_{-1/2} + b_2 \delta^3 y_{-1/2} + b_3 \delta^5 y_{-1/2} + \cdots \tag{4.47}$$

Note the following relations,

$$(i)\quad \delta E^{1/2} = \Delta;\ \delta^3 E^{1/2} = \frac{\Delta^3}{1+\Delta};\ \delta^5 E^{1/2} = \frac{\Delta^5}{(1+\Delta)^2}, \text{ etc.}$$

$$(ii)\ \delta E^{-1/2} = \frac{\Delta}{1+\Delta};\ \delta^3 E^{-1/2} = \frac{\Delta^3}{(1+\Delta)^2};\ \delta^5 E^{-1/2} = \frac{\Delta^5}{(1+\Delta)^3}, \text{ etc.}$$

Using the above relations we write from (4.47) the following equality,

$$(1+\Delta)^p = c_0 + a_1\Delta + a_2\frac{\Delta^3}{1+\Delta} + a_3\frac{\Delta^5}{(1+\Delta)^2} + \cdots$$

$$+b_1\frac{\Delta}{1+\Delta} + b_2\frac{\Delta^3}{(1+\Delta)^2} + b_3\frac{\Delta^5}{(1+\Delta)^3} + \cdots$$

$$= c_0 + a_1\Delta + (b_1 + a_2\Delta^2)\frac{\Delta}{1+\Delta} + (b_2 + a_3\Delta^2)\frac{\Delta^3}{(1+\Delta)^2} + \cdots \qquad (4.48)$$

Comparing the constant term in (4.48) gives,

$$c_0 = \binom{p}{0} = 1.$$

Now, multiplying both sides of (4.48) by $(1+\Delta)$ we get

$$(1+\Delta)^{p+1} = (1+a_1\Delta)(1+\Delta) + (b_1 + a_2\Delta^2)\Delta + (b_2 + a_3\Delta^2)\frac{\Delta^3}{(1+\Delta)} + \cdots \qquad (4.48a)$$

Comparison of the coefficients of Δ^2 and Δ gives respectively,

$$a_1 = \binom{p+1}{2};\ c_0 + a_1 + b_1 = \binom{p+1}{1}$$

$$\text{or}\quad b_1 = \binom{p+1}{1} - \binom{p}{0} - \binom{p+1}{2} = \binom{p}{1} - \binom{p+1}{2} = -\binom{p}{2}.$$

Again multiply (4.48a) by $(1+\Delta)$ or (4.48) by $(1+\Delta)^2$, we get

$$(1+\Delta)^{p+2} = (1+a_1\Delta)(1+\Delta)^2 + (b_1 + a_2\Delta^2)\Delta(1+\Delta) + (b_2 + a_3\Delta^2)\Delta^3 + \ldots.$$

Comparison of coefficients of Δ^4 and Δ^3 respectively, gives,

$$a_2 = \binom{p+2}{4};\ a_1 + a_2 + b_2 = \binom{p+2}{3}$$

or $$b_2 = \binom{p+2}{3} - \binom{p+1}{2} - \binom{p+2}{4} = \binom{p+1}{3} - \binom{p+2}{4}$$

$$= -\binom{p+1}{4}$$

Proceeding in this manner we can get the general term as,

$$a_m = \binom{p+m}{2m};\ b_m = -\binom{p+m-1}{2m},\quad m = 1, 2, \cdots$$

Putting the above values in (4.47), the formula may be written as,

$$y_p = y_0 + \binom{p+1}{2}\delta y_{1/2} + \binom{p+2}{4}\delta^3 y_{1/2} + \ldots - \binom{p}{2}\delta y_{-1/2} + \binom{p+1}{4}\delta^3 y_{-1/2} - \ldots$$

Using the equality, as in the Everett's formula,

$$\binom{p}{m} = (-1)^m \binom{-p+m-1}{m}$$

we have,

$$y_p = y_0 + \binom{1+p}{2}\delta y_{1/2} + \binom{2+p}{4}\delta^3 y_{1/2} + \cdots$$
$$- \binom{1-p}{2}\delta y_{-1/2} - \binom{2-p}{4}\delta^3 y_{-1/2} - \cdots \quad (4.49)$$

This is the required Steffensen's formula, which may also be written as,

$$y_p = y_0 + \left\{ \binom{1+p}{2} \delta y_{1/2} - \binom{1-p}{2} \delta y_{-1/2} \right\} + \left\{ \binom{2+p}{4} \delta^3 y_{1/2} - \binom{2-p}{4} \delta^3 y_{-1/2} \right\} + \cdots \tag{4.49a}$$

4.7.7 Comments on central difference formulae

(*a*) Like forward/backward difference formulae, the error in a central difference formula is given by the mganitude of the first neglected term in the formula. For example, if terms up to second differences are taken into account, then the error in various formulae will be as follows:

(*i*) Stirling's $\binom{p+1}{3} \mu \delta^3 y_0$ see (4.33)

(*ii*) Bessel's $\binom{p}{2} \dfrac{p-\frac{1}{2}}{3} \delta^3 y_{1/2}$ see (4.43)

(*iii*) Everett's $\binom{q+2}{5} \delta^4 y_0 + \binom{p+2}{5} \delta^4 y_1$ see(4.46a)

(*iv*) Steffensen's $\binom{2+p}{4} \delta^3 y_{1/2} - \binom{2-p}{4} \delta^3 y_{-1/2}$ see(4.49a)

It may be noted that formulae (*i*), (*ii*) and (*iv*) will be exact for a polynomial of degree 2 while the Everett's formula (*iii*) will be exact for a polynomial of degree up to three.

(*b*) We see that the domain of application of the Stirling's formula is $-\frac{1}{2} < p < \frac{1}{2}$ and that of Bessel's formula is $0 < p < 1$. In order to achieve better accuracy we reduce their domain of application as $-\frac{1}{4} < p < \frac{1}{4}$ for Stirling's and $\frac{1}{4} < p < \frac{3}{4}$ for Bessel's. We can adopt similar approach for other formulae wherever necessary.

4.8 General Comments on Interpolation

In interpolation, the error may enter from two sources; one, due to truncation of the interpolation formula by neglecting the higher order differences and two, due to rounding errors introduced in computation of various terms in the formula. As regards truncation error we can neglect the terms/differences whose contribution does not affect the interpolated value within the desired accuracy. In order to minimise the rounding error, the coefficients of differences in the formula should remain small in magnitude. Therefore choice of appropriate formula and that of x_0 is necessary.

It should be clear that all the formulae are equivalent as they are derived by comparing the powers of Δ in the expansion of $(1+\Delta)^p$. But the coefficients of the differences in all the formulae are different. A formula is considered better if it provides higher accuracy with less number of terms/differences, and if the same number of terms/differences are used then a formula with smaller truncation error is better.

Example 4.5

Following table gives the value of ln x for $x = 2.00\ (0.05)\ 2.25$.

x	2.00	2.05	2.10	2.15	2.20	2.25
ln x	0.69315	0.71784	0.74194	0.76547	0.78846	0.81093

Compute the value of (*i*) ln (2.07) and (*ii*) ln (2.21) using appropriate formulae.

Solution

x	$y = \ln x$	1st *Diff.*	2nd *Diff.*	3rd *Diff.*	4th *Diff.*	5th *Diff.*
2.00	0.69315					
		0.02469				
2.05	0.71784		−0.00059			
		0.02410		0.00002		
2.10	0.74194		−0.00057		0.00001	
		0.02353		0.00003		−0.00002
2.15	0.76547		−0.00054		−0.00001	
		0.02299		0.00002		
2.20	0.78846		−0.00052			
		0.02247				
2.25	0.81093					

Note:

(*i*) Since the values are given up to 5 decimals we should compute up to 6 or 7 decimals to account for rounding errors.

(*ii*) Since the 5th difference is greater in absolute value from the 4th difference we should leave it out. We should consider up to 4th differences only.

To compute $y(2.07)$ we will use Newton's Forward Difference Formula since it is near the upper end of the table. Further in order to keep $0 < p < 1$ we take

$$x_0 = 2.05$$

$$p = \frac{2.07 - 2.05}{0.05} = \frac{0.02}{0.05} = 0.4$$

Now substituting this value in the formula

$$y_p = y_0 + p\Delta y_0 + \frac{p(p-1)}{2!}\Delta^2 y_0 + \frac{p(p-1)(p-2)}{3!}\Delta^3 y_0 + \frac{p(p-1)(p-2)(p-3)}{4!}\Delta^4 y_0$$

we get

$$y(2.07) = 0.71784 + 0.4 \times 0.02410 + \frac{0.4(0.4-1)}{2}(-0.00057)$$

$$+ \frac{0.4(0.4-1)(0.4-2)}{3!}(0.00003) + \frac{0.4(0.4-1)(0.4-2)(0.4-3)}{4!} \times (-0.00001)$$

$$= 0.71784 + 0.009640 + 0.000068 + 0.000002 + 0.0000004$$

$$= 0.72755$$

The actual value obtained from the table is also 0.72755.

(*ii*) For computing value of ln (2.21) we use Newton's Backward Difference formula since the value is near the lower end of the table.

$$y_p = y_0 + p\nabla y_0 + \frac{p(p+1)}{2!}\nabla^2 y_0 + \frac{p(p+1)(p+2)}{3!}\nabla^3 y_0 +$$

$$\frac{p(p+1)(p+2)(p+3)}{4!}\nabla^4 y_0, \ -1 < p < 0.$$

We take $x_0 = 2.25$ which gives

$$p = \frac{2.21 - 2.25}{0.05} = \frac{-0.04}{0.05} = -0.8$$

$$y_p = 0.81093 + (-0.8)(0.02247) + \frac{(-0.8)(0.2)}{2}(-0.00052)$$

$$+ \frac{(-0.8)(0.2)(1.2)}{6}(0.00002) + \frac{(-0.8)(1.2)(1.2)(2.2)}{24}(-0.00001)$$

$$= 0.81093 - 0.017976 + 0.0000416 + 0.00000064 + 0.000000176$$

We could neglect the last term and also can round the term to 6 decimals (since the answer is required up to 5 decimals)

$$y(2.21) = 0.81093 - 0.017976 + 0.000042 + 0.000001$$

$$= 0.792997 \simeq 0.79300 \text{ (after rounding to 5 places)}$$

The actual value read from the table is 0.79299.

Example 4.6

From the data of Example 4.5, compute the value of ln x for $x = 2.07$ by linear interpolation, i.e., using first two terms only in the respective formulae. Also estimate the magnitude of error using derivative as well as differences.

Solution $x = 2.07;\ x_0 = 2.05, p = 0.4$

$$y = 0.71784 + 0.00964 = 0.72748$$

The truncation error given by formulae (4.14) is

$$R(x) = (x - x_0)(x - x_1)\frac{f''(\xi)}{2!}, \quad 2.05 \le \xi \le 2.10.$$

$$f(x) = \ln x;\ f'(x) = \frac{1}{x};\ f''(x) = -\frac{1}{x^2}$$

$$|f''(x)|_{\max} = \frac{1}{(2.05)^2}$$

$$R(2.07) = \left|\frac{(2.07 - 2.05)(2.07 - 2.10)}{2} \times \frac{1}{(2.05)^2}\right|$$

$$= 0.00007$$

The truncation error given by formula (4.22) is,

$$R(p) = \frac{p(p-1)}{2} \cdot \Delta^2 y_0$$

$$= \left|\frac{(0.4)(0.4-1)}{2} \times (-0.00057)\right|$$

$$= 0.000068$$

$$\simeq 0.00007$$

Example 4.7

Given below are the values of e^x tabulated for $x = 1.00\ (.20)\ 2.00$.

x	1.00	1.20	1.40	1.60	1.80	2.00
e^x	2.7183	3.3201	4.0552	4.9530	6.0496	7.3891

Find the values of e^x for $x = 1.12$ and $x = 1.68$.

Solution

Difference table

x	$y = e^x$	1st *Diff.*	2nd *Diff.*	3rd *Diff.*	4th *Diff.*	5th *Diff.*
1.00	2.7183					
		0.6018				
1.20	3.3201		0.1333			
		0.7351		0.0294		
1.40	4.0552		0.1627		0.0067	
		0.8978		0.0361		0.0013
1.60	4.9530		0.1988		0.0080	
		1.0966		0.0441		
1.80	6.0496		0.2429			
		1.3395				
2.00	7.3891					

(*i*) Newton's Forward Difference formula to be used at $x = 1.12$

$$p = \frac{1.12 - 1.00}{0.20} = \frac{0.12}{0.20} = 0.6$$

$$y(1.12) = 2.7183 + 0.6 \times 0.6018 + \frac{0.6(0.6-1)}{2} \times 0.1333 + \frac{0.6(-0.4)(-1.4)}{6} \times 0.0294$$

$$+ \frac{0.6(-0.4)(-1.4)(-2.4)}{24} \times 0.0067 + \frac{0.6(-0.4)(-2.4)(-3.4)}{5 \times 24} \times 0.0013$$

$$= 2.7183 + 0.36108 - 0.015996 + 0.016464 - 0.000225 + 0.000030$$

$= 3.0648$ (correct value from the table is 3.0649).

Note: Since the result is required correct up to 4 places of decimal, we ought to compute each term up to 5 or 6 decimal places. Therefore, none of the terms above can be disregarded.

(*ii*) For computing $y(1.68)$ Newton's Backward Difference Formula will be used.

$$p = \frac{1.68 - 1.80}{0.20} = -\frac{0.12}{0.20} = -0.6$$

$$y(1.68) = 6.0496 + (-0.6)1.0966 + \frac{(0.6)(-0.6+1)}{2} \times 0.1988$$

$$+ \frac{(-0.6)(-0.6+1)(-0.6+2)}{6} \times 0.0361 + \frac{(-0.6)(0.4)(1.4)(2.4)}{24} \times 0.0067$$

$$= 6.0496 - 0.65796 - 0.023856 - 0.002022 - 0.000225$$

$$= 5.36554 \simeq 5.3655 \text{ (correct answer is 5.3656).}$$

Example 4.8

From the table of Example 4.7, evaluate $e^{1.44}$ and $e^{1.52}$.

Solution Since these points are in the middle of the table we will use central difference formula. For Stirling's formula $-0.25 < p < 0.25$ and for Bessel's formula $0.25 < p < 0.75$.

(*i*) When $x = 1.44$

Taking $x_0 = 1.40$ so that

$$p = \frac{1.44 - 1.40}{0.20} = \frac{0.04}{0.20} = 0.2$$

Stirling's formula (4.38)

$$y_p = y_0 + p\mu\delta y_0 + \frac{p^2}{2}\delta^2 y_0 + \binom{p+1}{3}\mu\delta^3 y_0 + \frac{p^2(p^2-1^2)}{24}\delta^4 y_0 + \cdots$$

$$= y_0 + p\frac{\delta y_{1/2} + \delta y_{-1/2}}{2} + \frac{p^2}{2}\delta^2 y_0 + \frac{(p+1)p(p-1)}{6} \cdot \frac{\delta^3 y_{1/2} + \delta^3 y_{-1/2}}{2}$$

$$+ \frac{p^2(p^2-1^2)}{24}\delta^4 y_0$$

$$= 4.0552 + 0.2 \times \frac{0.7351 + 0.8978}{2} + \frac{(0.2)^2}{2} \times 0.1627$$

$$+ \frac{(1.2)(0.2)(-0.8)}{6} \cdot \frac{0.0294 + 0.0361}{2} + \frac{(0.2)^2(0.2^2-1)}{24} \times 0.0067$$

$$= 4.0552 + 0.16329 + 0.003254 - 0.001048 - 0.000011$$

$$= 4.22068 \simeq 4.2207 \text{ (correct value is 4.2207)}$$

(*ii*) When $x = 1.52$

Taking $x_0 = 1.40$,

$$p = \frac{1.52 - 1.40}{0.20} = \frac{0.12}{0.20} = 0.6$$

Since p lies between 0.25 and 0.75 we will use Bessel's formula. It is given by (4.43) as,

$$y_p = \mu y_{1/2} + \left(p - \frac{1}{2}\right)\delta y_{1/2} + \binom{p}{2}\mu\delta^2 y_{1/2} + \binom{p}{2}\frac{p - \frac{1}{2}}{3}\delta^3 y_{1/2}$$

$$+ \binom{p+1}{4}\mu\delta^4 y_{1/2} + \binom{p+1}{4}\frac{p - \frac{1}{2}}{5}\delta^5 y_{1/2} + \cdots$$

$$= \frac{y_0 + y_1}{2} + \left(p - \frac{1}{2}\right)\delta y_{1/2} + \frac{p(p-1)}{2} \times \frac{\delta^2 y_0 + \delta^2 y_1}{2} + \frac{p(p-1)}{2}\frac{p - \frac{1}{2}}{2}\delta^3 y_{1/2}$$

$$+ \frac{(p+1)p(p-1)(p-2)}{24} \times \frac{\delta^4 y_0 + \delta^4 y_1}{2} + \frac{(p+1)p(p-1)(p-2)}{24}\frac{p - \frac{1}{2}}{5}\delta^5 y_{1/2} + \cdots$$

$$= \frac{4.0552 + 4.9530}{2} + (0.6 - 0.5)(0.8978) + \frac{0.6(0.6-1)}{2}\frac{(0.1627 + 0.1988)}{2}$$

$$+ \frac{0.6(0.6-1)}{2}\frac{(0.6 - 0.5)}{2}(0.0361) + \frac{(0.6+1)(0.6)(0.6-1)(0.6-2)}{24}\left(\frac{0.0067 + 0.0080}{2}\right)$$

$$+ \frac{(0.6+1)(0.6)(0.6-1)(0.6-2)}{24}\frac{(0.6-0.5)}{5} \times 0.0013$$

$$= 4.5041 + 0.08978 - 0.021690 - 0.0002166 + 0.000165 + 0.00000062$$

$$= 4.57214 \simeq 4.5721 \text{ (correct value is 4.5722)}$$

Note: Last term can be neglected.

Example 4.9

Values of the reciprocal function $f(x) = \dfrac{1}{x}$ are given below for $x = 1\ (0.10)\ 1.50$.

x	1.00	1.10	1.20	1.30	1.40	1.50
$\frac{1}{3}$	1.000	0.9091	0.8333	0.7692	0.7143	0.6667

Compute reciprocal of 1.26 using only even order differences in an interpolation formula. Use only second order difference to compute the value. Also find the error.

Solution

Difference table

x	$y=\frac{1}{x}$	1st Diff.	2nd Diff.	3rd Diff.	4th Diff.	5th Diff.
1.00	1.000					
		−0.0909				
1.10	0.9091		0.0151			
		−0.0758		−0.0034		
1.20	0.8333		0.0117		0.0009	
		−0.0641		−0.0025		−0.0003
1.30	0.7692		0.0092		0.0006	
		−0.0549		−0.0019		
1.40	0.743		0.0073			
		−0.0476				
1.50	0.6667					

When $x = 1.26,\ p = \dfrac{1.26-1.20}{0.10} = \dfrac{0.06}{0.10} = 0.6.$

We use Everett's formula since it contains even differences only

$$q = 1 - p = 0.4$$

Everett's formula is given by (4.46a)

$$y_p = (qy_0 + py_1) + \left\{\binom{q+1}{3}\delta^2 y_0 + \binom{p+1}{3}\delta^2 y_1\right\}$$

$$+\left\{\binom{q+2}{5}\delta^4 y_0 + \binom{p+2}{5}\delta^4 y_1\right\} + \cdots$$

$$= 0.4 \times (0.8333) + 0.6 \times 0.7692 + \frac{(1.4)(0.4)(-0.6)}{6} \times 0.0117$$

$$+\frac{(1.6)(0.6)(-0.4)}{6} \times 0.0092$$

$$= 0.33332 + 0.46152 - 0.0006552 - 0.0005888$$

$$= 0.79360 \simeq 0.7936 \text{ (correct value is 0.7936).}$$

Estimation of error:

The truncation error can be obtained by the next neglected term in the formula (taking maximum value of the difference)

$$R(p) = \binom{q+2}{5}\delta^4 y_0 + \binom{p+2}{5}\delta^4 y_1,\ q = 1-p,$$

$$R(0.6) = \frac{(0.4+2)(0.4+1)(0.4)(-0.6)(-1.6)}{5\times 24}\times 0.0009$$

$$+\frac{(0.6+2)(0.6+1)(0.6)(-0.4)(-1.4)}{5\times 24}\times 0.0006$$

$$= 0.9684\times 10^{-5} + 0.6990\times 10^{-5} = 0.1667\times 10^{-4}$$

which is less than $\frac{1}{2}\times 10^{-4}$ so the answer should be correct up to 4 decimal places.

Example 4.10

Taking the values of the reciprocal function $f(x) = \frac{1}{x}$ as in Example 4.9, compute the reciprocal of $x = 1.22$, using only odd order differences (up to 1st difference) of Steffensen's formula. Also estimate the error.

Solution Steffensen's formula is given by (4.49a),

$$y_p = y_0 + \left\{\binom{1+p}{2}\delta y_{1/2} - \binom{1-p}{2}\delta y_{-1/2}\right\}$$

$$+\left\{\binom{2+p}{4}\delta^3 y_{1/2} - \binom{2-p}{4}\delta^3 y_{1/2}\right\}+\cdots$$

$$p = \frac{1.22-1.20}{0.10} = \frac{0.02}{0.10} = 0.2$$

Substituting in the first two terms of the formula,

$$y(1.22) = 0.8333 + \frac{(1+0.2)(0.2)}{2}\times(-0.0641) - \frac{(1-0.2)(-0.2)}{2}\times(-0.0758)$$

$$= 0.8333 - 0.007692 - 0.006064$$

$$= 0.819544 \simeq 0.8195 \text{ (correct value is 0.8195)}$$

From the next term neglected we will get an estimate of the error, i.e.,

$$R(p) = \binom{2+p}{4}\delta^3 y_{1/2} - \binom{2-p}{4}\delta^3 y_{-1/2}$$

$$= \frac{(2+p)(1+p)(p)(p-1)}{24} \times (-0.0025) - \frac{(2-p)(1-p)(-p)(-p-1)}{24} \times (-0.0034)$$

$$R(0.2) = \frac{(2.2)(1.2)(0.2)(-0.8)}{24} \times (-0.0025) - \frac{(1.8)(0.8)(-0.2)(-1.2)}{24} \times (-0.0034)$$

$$= 0.00009296 + 0.00004896$$

$$= 0.0009296 = 0.93 \times 10^{-3}$$

The actual error is much less than the upper bound of the error.

Example 4.11

In a class of 100, the students are placed into following categories according to the marks they have obtained in a test out of 60.

Marks obtained	0 – 9	10 – 19	20 – 29	30 – 39	40 – 49	50 – 59
Number of Students	3	12	15	35	25	10

Find the number of students who have secured 75% and above marks. Use backward difference formula.

Solution 75% of 60 $= \frac{60 \times 75}{100} = 45$

We have to find the number of students who have secured 45 marks and above. We reconstruct the above table as follows and will compute the number of students who have got less than 45 marks.

Marks obtained x	*Number of students* y	1st *Diff.*	2nd *Diff.*	3rd *Diff.*	4th *Diff.*
Below 10	3				
		12			
Below 20	15		3		
		15		17	
Below 30	30		20		−47
		35		−30	
Below 40	65		−10		25
		25		−5	
Below 50	90		−15		
		10			
Below 60	100				

Taking $x_0 = 50;\ p = \dfrac{45-50}{10} = -0.5$

$$y(45) = y_0 + p\nabla y_0 + \frac{p(p+1)}{2!}\nabla^2 y_0$$

$$= 90 + (-0.5)25 + \frac{(-0.5)(0.5)}{2} \times (-10)$$

$$= 90 - 12.5 + 1.25$$

$$= 78.75 = 79$$

79 students have got less than 45 marks.
Therefore, $100 - 79 = 21$ students have got 75% and above.

Note: We should leave out 3rd and 4th differences as they start increasing.

Example 4.12

The population of a certain city according to census is given below

Year	1961	1971	1981	1991	2001
Population (in Lacs)	46	66	81	93	101

Extrapolate the population for the year 2007. Also fit a cubic taking first four values and years as $0, 1, 2$ and 3.

Solution We take 1961 as year zero as shown in table below:

Year	*x*	*y*	1st *Diff.*	2nd *Diff.*	3rd *Diff.*	4th *Diff.*
1961	0	46				
			20			
1971	1	66		−5		
			15		2	
1981	2	81		−3		−3
			12		−1	
1991	3	93		−4		
			8			
2001	4	101				

Let us take $x_0 = 2001$

$$p = \frac{2007-2001}{10} = \frac{6}{10} = 0.6$$

Using backward difference formula we get,

$$y(2007) = 101 + 0.6 \times 8 + \frac{0.6(0.6+1)}{2!} \times (-4) + \frac{0.6(0.6+1)(0.6+2)}{6} \times (-1)$$

$$= 101 + 4.8 - 1.92 - 0.416 = 103.5 \text{ lacs.}$$

Note: Last difference may be neglected as its value is greater than the lower order differences.

Cubic: $y(x) = 46 + \frac{x}{1} \times 20 + \frac{x(x-1)}{2}(-5) + \frac{x(x-1)(x-2)}{6} \times 2$

$$= \frac{1}{6}(2x^3 - 21x^2 + 139x + 276)$$

4.9 Lagrange's Method

So far we have discussed methods suitable for interpolation when the data is prescribed at equal interval. The Lagrange's method can be applied when the tabular points are not necessarily uniformly spaced. The method simply suggests to represent the given data $(x_i,\ y_i)$, $i = 0(1)n$ by a polynomial function $y = P(x)$ such that $y_i = P(x_i),\ i = 0(1)n$. Obviously the degree of the polynomial $P(x)$ can not be higher than n. Before we describe the Lagrangian representation of the polynomial, a naive approach to fit a polynomial passing through the $(n+1)$ points, $(x_i,\ y_i),\ i = 0(1)n$ would be to assume,

$$P(x) = a_0 + a_1 x + a_2 x^2 + \ldots + a_n x^n \tag{4.50}$$

The constants $a_0, a_1, \ \ldots.\ a_n$ can be determined from the system of $(n+1)$ linear equations (4.50) after substituting $(x_i,\ y_i),\ i = 0(1)n$,

$$y_i = a_0 + a_1 x_i + a_2 x_i^2 + \ldots. + a_n x_i^n. \tag{4.51}$$

Putting the values of the constants back in (4.50) gives the polynomial function $y = P(x)$ which can be used for interpolating the value of y for a given x.

In the Lagrangian form the polynomial is not represented by a power series like (4.50). Instead, a Lagrange's polynomial satisfying the points $(x_i,\ y_i),\ i = 0(1)n$, is expressed as,

$$P(x) = L_0(x)y_0 + L_1(x)y_1 + \ldots. + L_n(x)y_n \tag{4.52}$$

$$= \sum_{i=0}^{n} L_i(x) y_i$$

where $L_i(x) = \dfrac{(x-x_0)(x-x_1)\cdots(x-x_{i-1})(x-x_{i+1})\cdots(x-x_n)}{(x_i-x_0)(x_i-x_1)\cdots(x_i-x_{i-1})(x_i-x_{i+1})\cdots(x_i-x_n)},\ i = 0(1)n. \quad (4.53)$

The functions, $L_i(x),\ i = 0(1)n$ are called Lagrange's coefficients.

It is important to note that each of the $L_i(x)$'s are polynomial of degree n and satisfy the property of Kronecker's delta, i.e.,

$$\left.\begin{aligned} L_i(x_j) &= 1,\ i = j \\ &= 0,\ i \neq j \end{aligned}\right\} \tag{4.54}$$

Thus, it is easy to verify that the polynomial (4.52) passes through the points $(x_i,\ y_i), i = 0(1)n$. Further, in order to express the polynomial as power series each of the $L_i(x)$'s has to be expanded and then terms of the same powers collected. However, as far as interpolation is concerned we do not expand $L_i(x)$'s. Instead, we put the value of x in each of $L_i(x)$ and compute their values and finally obtain the corresponding value of y by simple multiplications and additions.

As per discussion in Sec. 4.4, the error in Lagrange's method would be,

$$R(x) = (x - x_0)(x - x_1)\cdots(x - x_n)\frac{f^{n+1}(\xi)}{(n+1)!},\ x_0 \leq \xi \leq x_n. \tag{4.55}$$

The drawback of the method is that even for a moderately large value of n, it may be a tedious job to represent the polynomial as power series in x due to multiplication of n factors, $(n+1)$ number of times.

The Lagrange's method may be used for inverse interpolation, i.e., to find the value of x for a given value of y. In that case we have to fit another polynomial $Q(y)$ such that, function is defined as $x = Q(y)$, i.e., taking y as independent variable and x as dependent variable. Or, it will be still better if we interchange the names of variables and fit $y = Q(x)$.

4.10 Divided Differences (DD)

In Sections 4.5, 4.6 and 4.7 we discussed various interpolation methods suitable for the data tabulated at equal interval. These methods approximate the function in terms of finite differences and it was illustrated in Sec 4.3 as how to construct a finite difference table when the function values were prescribed at equal interval. However in many practical problems we may not get the values of the function prescribed at uniform interval. In such cases we can form a table, called 'Divided Difference Table' which is going to be described as follows:

Let us suppose that the values of a function $f(x)$ are tabulated for $x = x_i,\ i = 0(1)n$, which are not necessarily equidistant. Then we define first (order) divided difference as,

$$f(x_i,\ x_{i+1}) = \frac{f(x_{i+1}) - f(x_i)}{x_{i+1} - x_i},\ i = 0(1)n-1. \tag{4.56}$$

Similarly, the second divided differences are defined as,

$$f(x_i,\ x_{i+1},\ x_{i+2}) = \frac{f(x_{i+1},\ x_{i+2}) - f(x_i,\ x_{i+1})}{x_{i+2} - x_i},\ i = 0(1)n-2 \tag{4.56a}$$

The third (order) divided difference will be,

$$f(x_i,\ x_{i+1},\ x_{i+2},\ x_{i+3}) = \frac{f(x_{i+1},\ x_{i+2},\ x_{i+3}) - f(x_i,\ x_{i+1},\ x_{i+2})}{x_{i+3} - x_i},\ i = 0(1)n-3 \tag{4.56b}$$

and so on.

In general the kth divided difference may be written as,

$$f(x_i,\ x_{i+1},\ \ldots.\ x_{i+k}) = \frac{f(x_{i+1},\ x_{i+2} \cdots x_{i+k}) - f(x_i, x_{i+1} \cdots x_{i+k-1})}{x_{i+k} - x_i}, \tag{4.57}$$

$$i = 0(1)n-k \text{ and } k = 1(1)n-1$$

Thus following the criteria (4.57) for divided difference, we can construct the Divided Difference table as given in Table 4.7.

Table 4.7. Divided Difference table

i	x_i	$f(x_i)$	1st DD	2nd DD	3rd DD	4th DD
0	x_0	$f(x_0)$				
			$f(x_0, x_1)$			
1	x_1	$f(x_1)$		$f(x_0, x_1, x_2)$		
			$f(x_1, x_2)$		$f(x_0, x_1, x_2, x_3)$	
2	x_2	$f(x_2)$		$f(x_1, x_2, x_3)$		$f(x_0, x_1, x_2, x_3, x_4)$
			$f(x_2, x_3)$		$f(x_1, x_2, x_3, x_4)$	
3	x_3	$f(x_3)$		$f(x_2, x_3, x_4)$		
			$f(x_3, x_4)$			
4	x_4	$f(x_4)$				

Note: DD stands for Divided Difference.

Let us consider the second divided differences,

$$f(x_1,\ x_2,\ x_3) = \frac{f(x_2,\ x_3) - f(x_1,\ x_2)}{(x_3,\ x_1)} \tag{4.58}$$

In forming the difference (4.58) we have taken arguments x_2 and x_3 in $f(x_2, x_3)$ and subtracted $f(x_1, x_2)$ which has arguments x_1, x_2. In the denominator we take the argument x_3 which is missing in $f(x_1, x_2)$ and subtract x_1 which is missing in $f(x_2, x_3)$.

In fact, we can take any arguments and define the divided differences (4.58) as follows:

$$f(x_1, x_2, x_3) = \frac{f(x_1, x_3) - f(x_2, x_3)}{(x_1, x_2)} \tag{4.58a}$$

The criterion is that take any two arguments in the numerator but in the denominator the first term x_1 is the one which is missing in $f(x_2, x_3)$ and second term x_2 is missing in the arguments of $f(x_1, x_3)$. The difference formulae (4.58) and (4.58a) are equivalent. This will be clear from the following.

4.10.1 Divided differences are independent of order of arguments

For example, in case of second divided difference

$$f(x_0, x_1, x_2) = f(x_0, x_2, x_1) = f(x_1, x_0, x_2) = f(x_2, x_0, x_1), \text{ etc.}$$

We will prove the above statement by induction.

The first divided difference is given by,

$$\begin{aligned} f(x_0, x_1) &= \frac{f(x_1) - f(x_0)}{x_1 - x_0} \\ &= \frac{f(x_0)}{x_0 - x_1} + \frac{f(x_1)}{x_1 - x_0} \end{aligned}$$

We see that if the arguments x_0 and x_1 are interchanged, the expressions on the right side remain same.

Further, consider second divided difference

$$\begin{aligned} f(x_0, x_1, x_2) &= \frac{f(x_1, x_2) - f(x_0, x_1)}{x_2 - x_0} \\ &= \frac{1}{(x_2 - x_0)}\left[\frac{f(x_2) - f(x_1)}{x_2 - x_1} - \frac{f(x_1) - f(x_0)}{x_1 - x_0}\right] \\ &= \frac{f(x_0)}{(x_0 - x_1)(x_0 - x_2)} + \frac{f(x_1)}{(x_1 - x_0)(x_1 - x_2)} + \frac{f(x_2)}{(x_2 - x_0)(x_2 - x_1)}. \end{aligned}$$

Again, interchanging of arguments does not affect the expressions on the right; only their positions will interchange.

Let us suppose that the divided difference up to order $(n-1)$ are independent of the order of their aurguments. We will show that the nth order divided difference will also be independent of the order of its arguments.

Thus $(n-1)^{\text{th}}$ order divided differences may be written as,

$$f(x_1, x_2, \cdots x_n) = \sum_{i=1}^{n} \frac{f(x_i)}{(x_i - x_1)\cdots(x_i - x_{i-1})(x_i - x_{i+1})\cdots(x_i - x_n)}$$

$$= \sum_{r=1}^{n-1} \frac{f(x_i)}{\underset{r\neq i}{\pi'}(x_i - x_r)} + \frac{f(n)}{\underset{r\neq n}{\pi'}(x_n - x_r)}, \tag{4.59}$$

where $\underset{r\neq i}{\pi'}(x_i - x_r) = (x_i - x_1)\ldots(x_i - x_{i-1})(x_i - x_{i+1})\ldots(x_i - x_n)$

and $\underset{r\neq n}{\pi'}(x_n - x_r) = (x_n - x_1)(x_n - x_2)\ldots.(x_n - x_{n-1})$

$$f(x_0, x_1, \ldots. x_{n-1}) = \sum_{i=0}^{n-1} \frac{f(x_i)}{(x_i - x_0)\cdots(x_i - x_{i-1})(x_i - x_{i+1})\cdots(x_i - x_{n-1})}$$

$$= \sum_{i=1}^{n-1} \frac{f(x_i)}{\underset{r\neq i}{\pi''}(x_i - x_r)} + \frac{f(x_0)}{\underset{r\neq 0}{\pi''}(x_0 - x_r)}, \tag{4.60}$$

where $\underset{r\neq i}{\pi''}(x_i - x_r) = (x_i - x_0)\ldots.(x_i - x_{i-1})(x_i - x_{i+1})\ldots..(x - x_{n-1})$

and $\underset{r\neq 0}{\pi''}(x_0 - x_r) = (x_0 - x_1)(x_0 - x_2)\ldots(x_0 - x_{n-1})$

Now the nth order divided difference may be written as,

$$f(x_0, x_1, \ldots. x_n) = \frac{f(x_1, x_2, \cdots x_n) - f(x_0, x_1, \cdots x_{n-1})}{x_n - x_0}. \tag{4.61}$$

Substituting the values from (4.59) and (4.60) in (4.61), we get,

$$f(x_0, x_1, \ldots. x_n) = \frac{1}{(x_n - x_0)}\left[\sum_{i=1}^{n-1} \frac{f(x_i)}{\underset{r\neq i}{\pi'}(x_i - x_r)} - \sum_{i=1}^{n-1} \frac{f(x_i)}{\underset{r\neq i}{\pi''}(x_i - x_r)}\right]$$

$$+ \frac{1}{(x_n - x_0)}\left[\frac{f(x_n)}{\underset{r\neq n}{\pi'}(x_n - x_r)} - \frac{f(x_0)}{\underset{r\neq 0}{\pi''}(x_0 - x_r)}\right]. \tag{4.62}$$

Let us consider second bracketed term in (4.62)

$$\frac{1}{x_n - x_0}\left[\frac{f(x_n)}{(x_n - x_1)(x_n - x_2)\cdots(x_n - x_{n-1})} - \frac{f(x_0)}{(x_0 - x_1)(x_0 - x_2)\cdots(x_0 - x_{n-1})}\right]$$

$$= \frac{f(x_0)}{\underset{i=1}{\overset{n-1}{\pi}}(x_0 - x_i)} + \frac{f(x_n)}{\underset{i=0}{\overset{n-1}{\pi}}(x_n - x_i)}. \tag{4.63}$$

Let us obtain the ith term from the first bracketed term of (4.62) for $i = 1(1)n-1$,

$$\frac{f(x_i)}{x_n - x_0}\left[\frac{1}{(x_i - x_1)(x_i - x_2)\cdots(x_i - x_{i-1})(x_i - x_{i+1})\cdots(x_i - x_n)}\right.$$
$$\left. - \frac{1}{(x_i - x_0)(x_i - x_1)\cdots(x_i - x_{i-1})(x_i - x_{i+1})\cdots(x_i - x_{n-1})}\right]$$

$$= \frac{f(x_i)}{(x_n - x_0)}\left[\frac{x_i - x_0 - x_i + x_n}{(x_i - x_0)(x_i - x_1)\cdots(x_{i-1} - x_i)(x_i - x_{i+1})\cdots(x_i - x_{n-1})}\right]$$

$$= \frac{f(x_i)}{\underset{r\neq i}{\pi'}(x_i - x_r)},\ i = 1(1)n-1, \tag{4.64}$$

where $\pi'(x_i - x_r) = (x_i - x_0)\ldots.(x_i - x_{i-1})(x_i - x_{i+1})\ldots.(x_i - x_n)$.

After combining (4.63) and (4.64) we can write (4.62), as

$$f(x_0,\ x_1,\ \ldots.x_n) = \sum_{i=0}^{n} \frac{f(x_i)}{\underset{r\neq i}{\pi'}(x_i - x_r)}, \tag{4.65}$$

where $\underset{r\neq i}{\pi'}(x_i - x_r) = (x_i - x_0)\ldots.(x_i - x_{i-1})(x_i - x_{i+1})\ldots.(x_i - x_n)$

We see that if we change the order of the arguments, the expression on the right side of (4.65) remains unaffected.

Thus we can now realise that following are equivalent,

$$f(x_0,\ x_1,\ x_2) = f(x_0,\ x_2,\ x_1) = f(x_1,\ x_0,\ x_2) = f(x_1,\ x_2,\ x_0)$$
$$= f(x_2,\ x_0,\ x_1) = f(x_2,\ x_1,\ x_0).$$

4.10.2 Newton's Divided Difference (DD) formula

Let $f(x)$ be a function whose values are tabulated for the abscissas $x = x_i$, $i = 0(1)n$ which are not necessarily equidistant, nor in any particular order. We want to approximate the function $f(x)$ by a polynomial $P(x)$ of highest degree n which satisfies the function at $(n+1)$ points (x_i, f_i), $i = 0(1)n$, i.e., $P(x_i) = f(x_i) = f_i$. Let us suppose, without loss of generality, that x_0 and x_n are the algebrically smallest and largest values respectively and $\bar{x}$ is a point in $(x_0,\ x_n)$. Then we can write $(k+1)^{\text{th}}$ divided difference as,

$$f(\bar{x},\ x_0,\ x_1,\ \ldots.\ x_k) = \frac{f(\bar{x},\ x_0,\ x_1, \cdots x_{k-1}) - f(x_0,\ x_1,\ \cdots x_k)}{\bar{x} - x_k}, \qquad (4.66)$$

$$k = 0, 1, 2,\ \ldots.\ n.$$

After transposing the terms, (4.66) can be written in the following manner,

$$f(\bar{x},\ x_0,\ \ldots.\ x_{k-1}) = f(x_0,\ x_1,\ \ldots.\ x_k) + (\bar{x} - x_k) f(\bar{x},\ x_0,\ x_1,\ \ldots.\ x_k). \qquad (4.66a)$$

Writing (4.66a) separately, corresponding to $k = 0(1)n$, we get the following system:

$$\begin{aligned}
k = 0: &\quad f(\bar{x}) = f(x_0) + (\bar{x} - x_0) f(\bar{x},\ x_0)\\
k = 1: &\quad f(\bar{x},\ x_0) = f(x_0,\ x_1) + (\bar{x} - x_1) f(\bar{x},\ x_0,\ x_1)\\
k = 2: &\quad f(\bar{x},\ x_0,\ x_1) = f(x_0,\ x_1,\ x_2) + (\bar{x} - x_2) f(\bar{x},\ x_0,\ x_1,\ x_2)\\
&\ldots\ldots\ldots\ldots\ldots\ldots\ldots\ldots\ldots\ldots\ldots\ldots\\
k = n-1: &\quad f(\bar{x},\ x_0,\ x_1,\ \ldots.\ x_{n-2}) = f(x_0,\ x_1,\ \ldots.\ x_{n-1}) + (\bar{x} - x_{n-1}) f(\bar{x},\ x_0,\ x_1,\ \ldots.\ x_{n-1})\\
k = n: &\quad f(\bar{x},\ x_0,\ x_1,\ \ldots.\ x_{n-1}) = f(x_0,\ x_1,\ \ldots.\ x_n) + (\bar{x} - x_n) f(\bar{x},\ x_0,\ x_1,\ \ldots.\ x_n)
\end{aligned}$$

We multiply second equation ($k = 1$) by $(\bar{x} - x_0)$; third equation ($k = 2$) by $(\bar{x} - x_0)(\bar{x} - x_1)$ and so on and the last equation ($k = n$) by n factors $(\bar{x} - x_0)(\bar{x} - x_1) \cdots (\bar{x} - x_{n-1})$, and add all equations. All the terms, except $f(\bar{x})$, on the left cancel out with the second terms on the right and finally we get,

$$\begin{aligned}
f(\bar{x}) = f(x_0) &+ (\bar{x} - x_0) f(x_0,\ x_1) + (\bar{x} - x_0)(\bar{x} - x_1) f(x_0,\ x_1,\ x_2) + \ldots.\\
&+ (\bar{x} - x_0)(\bar{x} - x_1) \ldots.\ (\bar{x} - x_{n-1}) f(x_0,\ x_1,\ \ldots.\ x_n)\\
&+ (\bar{x} - x_0)(\bar{x} - x_1) \ldots.\ (\bar{x} - x_n) f(\bar{x},\ x_0,\ x_1,\ \ldots.\ x_n).
\end{aligned} \qquad (4.67)$$

Now, since $\bar{x}$ is any general point, we can replace it by x. Further, the last term in (4.67) can not be computed as $f(\bar{x})$ is not known. Therefore (4.67) may be written as,

$$
\begin{aligned}
f(x) &= f(x_0) + (x-x_0)f(x_0,\ x_1) + (x-x_0)(x-x_1)f(x_0,\ x_1,\ x_2) + \ldots. \\
&\quad + (x-x_0)(x-x_1)\ldots.\ (x-x_{n-1})f(x_0,\ x_1,\ \ldots.\ x_n) + R(x)
\end{aligned}
\tag{4.68}
$$

where

$$R(x) = (x-x_0)(x-x_1)\ldots.\ (x-x_n)f(x,\ x_0,\ x_1,\ \ldots.\ x_n). \tag{4.68a}$$

The expression given by (4.68) is called the Newton's Divided Difference (DD) formula with remainder/error term $R(x)$ given by (4.68a).

As stated, the remainder term in the form (4.68a) cannot be computed. But we should note that $f(x)$ in (4.68), without the remainder term, represents a polynomial passing through the $(n+1)$ points $(x_i,\ f_i),\ i = 0(1)n$. Thus recalling the result of Sec 4.4, the error in the approximation is given by,

$$R(x) = (x-x_0)(x-x_1)\ldots.\ (x-x_n)\cdot\frac{f^{n+1}(\xi)}{(n+1)!}, \tag{4.69}$$

$$x_0 \le \xi \le x_n.$$

From (4.68a) and (4.69) we infer that,

$$f(x,\ x_0,\ x_1,\ \ldots.\ x_n) = \frac{f^{n+1}(\xi)}{(n+1)!},\ x_0 \le \xi \le x_n. \tag{4.70}$$

If the tabular points $x_0,\ x_1,\ \ldots.\ x_n$ approach x, then following equality is obtained,

$$\underbrace{f(x,x,\cdots x)}_{(n+2)\text{ times}} = \frac{f^{n+1}(x)}{(n+1)!}.$$

Or, in general

$$\underbrace{f(x,x,x,\cdots x)}_{(k+1)\text{ times}} = \frac{f^{k}(x)}{k!},\ k = 1(1)n. \tag{4.71}$$

Formula (4.71) is equivalent to (4.23). Further, it is easy to verify that Newton's DD Formula converts to Newton's FD formula for equidistant tabular points.

4.11 Lagrange's Formula Versus Newton's DD Formula

(*i*) In Newton's DD formula, the degree of the polynomial increases by one each time a new term is included in the formula. That is, if the function $f(x)$ is approximated by the first term only, it is a constant $f(x_0)$. If two terms are taken then $f(x)$ is approximated by a straight line $f(x) = f(x_0) + (x - x_0)f(x_0, x_1)$. Similarly, if we take three terms $f(x)$ will be approximated by a polynomial of degree two. That is, for each one additional tabular point, one extra term is added in the formula without affecting the previous terms. This is a great advantage, in that, if a new point is added in the table of given values, the formula can be modified without affecting its previous structure, while on the other hand in the Lagrange's formula the whole exercise will have to be repeated all over again.

(*ii*) If the polynomial has to be expressed as a power series, then in the Lagrange's method we have to multiply n linear factors $(n+1)$ times, while in the case of Newton's DD formula we have to multiply 2, 3, 4, ... n factors only once which means less than half times of the former.

Example 4.13

Using Lagrange's formula find the value of $y(2)$ from the following data:

x	0	1	4	5
y	8	11	68	123

Solution The Lagrange's formula is

$$y = \mathrm{P}(x) = \frac{(x-1)(x-4)(x-5)}{(0-1)(0-4)(0-5)} \times 8 + \frac{(x-0)(x-4)(x-5)}{(1-0)(1-4)(1-5)} \times 11$$

$$+ \frac{(x-0)(x-1)(x-5)}{(4-0)(4-1)(4-5)} \times 68 + \frac{(x-0)(x-1)(x-4)}{(5-0)(5-1)(5-4)} \times 123$$

$$y(2) = \frac{(2-1)(2-4)(2-5)}{(-1)(-4)(-1)} \times 8 + \frac{(2-0)(2-4)(2-5)}{1(-3)(-4)} \times 11$$

$$+ \frac{(2-0)(2-1)(2-5)}{4(3)(-1)} \times 68 + \frac{(2-0)(2-1)(2-4)}{5 \times 4 \times 1} \times 123$$

$$= -\frac{1(-2)(-3) \times 8}{1 \times 4 \times 5} + \frac{2(-2)(-3) \times 11}{3 \times 4} + \frac{2 \times (3) \times 68}{4 \times 3} + \frac{2 \times (-2) \times 123}{5 \times 4}$$

$$= -\frac{12}{5} + 11 + 34 - \frac{123}{5}$$

$$= 45 - \frac{135}{5} = 45 - 27 = 18.$$

Note: The determination of polynomial should not be tried unless specifically required.

Example 4.14

Following values of the function $y = x^3$ are provided,

x	1	2	3
y	1	8	27

Compute the cube-root of 21 from the above data using Lagrange's method. Also discuss the error in the result.

Solution Let us interchange the variables for convenience

x	1	8	27
y	1	2	3

We have to find y for $x = 21$.

$$y = \frac{(x-8)(x-27)}{(1-8)(1-27)} \times 1 + \frac{(x-1)(x-27)}{(8-1)(8-27)} \times 2 + \frac{(x-1)(x-8)}{(27-1)(27-8)} \times 3$$

$$y(21) = \frac{(21-8)(21-27)}{(-7)(-26)} + \frac{(21-1)(21-27)}{7 \times (-19)} \times 2 + \frac{(21-1)(21-8)}{26 \times 19} \times 3$$

$$= -\frac{13 \times 6}{7 \times 26} + \frac{20 \times 6 \times 2}{7 \times 19} + \frac{20 \times 13 \times 3}{26 \times 19}$$

$$= -0.4286 + 1.8045 + 1.5789$$

$$= 2.9548 \text{ (exact value is } (21)^{1/3} = 2.7589)$$

The maximum error is given by,

$$R(x) = (x-1)(x-8)(x-27)\frac{f'''(\xi)}{3!}, \quad 1 \le \xi \le 27.$$

Here $\quad y = x^{1/3}; f''' = \frac{10}{27}x^{-8/3}$

$$R(21) = (21-1)(21-8)(21-27) \cdot \frac{1}{6} \cdot \frac{10}{27}, \text{ taking max value of } f'''(x) \text{ at } x = 1.$$

$$\simeq 9.63.$$

The value of the maximum error is very high 9.6 while the actual error is 0.2 approx.

Example 4.15

Following values of the function $y = x^3$ are provided,

x	0	1	2	3
y	0	1	8	27

Compute cube-root of 21 from the above data using Lagrange's method. Discuss the error in the result.

Solution Let us interchange the variables as,

x	0	1	8	27
y	0	1	2	3

$$y = \frac{(x-1)(x-8)(x-27)}{(0-1)(0-8)(0-27)} \times 0 + \frac{(x-0)(x-8)(x-27)}{(1-0)(1-8)(1-27)} \times 1$$

$$+ \frac{(x-0)(x-1)(x-27)}{(8-0)(8-1)(8-27)} \times 2 + \frac{(x-0)(x-1)(x-8)}{(27-0)(27-1)(27-8)} \times 3$$

$$y(21) = 21\left[-\frac{13 \times 6}{7 \times 26} + \frac{20 \times 6 \times 2}{8 \times 7 \times 19} + \frac{20 \times 13 \times 3}{27 \times 26 \times 19}\right]$$

$$= 21\left[-\frac{3}{7} + \frac{39}{7 \times 19} + \frac{10}{9 \times 19}\right]$$

$$= -3.0345 \text{ (exact value is 2.7589)}$$

$$R(x) = (x-0)(x-1)(x-8)(x-27) \times \frac{1}{24} \times \frac{80}{81} \times \frac{1}{x^{11/3}}$$

At $x = 0$, $R(x) \to \infty$.

Since the tabular points are far apart, the amplitudes of the oscillations of the approximating cubic polynomial are very large. So much so that the curve assumes negative values between $x = 12$ and 24 (approx.).

Example 4.16

Find the polynomial satisfying the following data

x	0	1	4	5	7
y	8	11	68	123	323

Also compute $y(2)$.

Solution

x	y	1st DD	2nd DD	3rd DD
0	8			
		3		
1	11		4	
		19		1
4	68		9	
		55		1
5	123		15	
		100		
7	323			

$$\begin{aligned} y = f(x) &= 8 + (x-0)\times 3 + (x-0)(x-1)4 + (x-0)(x-1)(x-4)\times 1 \\ &= 8 + 3x + 4x^2 - 4x + x^3 - 5x^2 + 4x \\ &= x^3 - x^2 + 3x + 8 \end{aligned}$$

$$\begin{aligned} y(2) &= 8 - 4 + 6 + 8 \\ &= 18 \end{aligned}$$

4.12 Hermite's Interpolation

Let us suppose that $f(x)$ is a known function of x and its values are tabulated for $x = x_i,\ i = 0\ (1)\ n$ not necessarily equidistant. A polynomial of degree n can be found by Lagrange's method that interpolates $f(x)$ at $x = x_i,\ i = 0(1)n$, i.e., agrees with $f(x)$ for $x = x_i$. In Hermite's interpolation the degree of the polynomial is increased without increasing the number of tabular points x_i. It uses the function values and its derivatives (say, first), i.e., $f(x_i)$ and $f'(x_i)$ at $x = x_i,\ i = 0(1)n$. Now, since there are $(2n+2)$ conditions to be satisfied, a polynomial of degree $(2n+1)$ would approximate the function $f(x)$. Let $y(x)$ be a polynomial of degree $(2n+1)$ which approximates the function $f(x)$ satisfying the conditions

$$\left.\begin{aligned} y(x_i) &= f(x_i) \ \text{ or } \ y_i = f_i \\ y'(x_i) &= f'(x_i) \ \text{ or } \ y_i' = f_i' \end{aligned}\right\} i = 0(1)n. \tag{4.72}$$

The polynomial is expressed as

$$y(x) = \sum_{i=0}^{n} u_i(x) y_i + \sum_{i=0}^{n} v_i(x) y_i'. \tag{4.73}$$

where $u_i(x)$ and $v_i(x)$, in general, are polynomials of degree $(2n+1)$. In order that first of (4.72) is satisfied we should have in (4.73) for $i = 0, 1, 2, \ldots. n$,

$$u_i(x_i) = 1;\ u_i(x_j) = 0,\ i \neq j \tag{4.74a}$$

and $$v_i(x_j) = 0;\ j = 0(1)n. \tag{4.74b}$$

Similarly second of (4.72) will be satisfied if

$$u_i'(x_j) = 0, j = 0(1)n. \tag{4.75a}$$

$$v_i'(x_i) = 1; v_i'(x_j) = 0,\ i \neq j. \tag{4.75b}$$

Let the polynomials $u_i(x)$ and $v_i(x)$ be expressed as

$$u_i(x) = (a_i x + b_i)L_i^2(x) \tag{4.76a}$$

$$v_i(x) = (c_i x + d_i)L_i^2(x) \tag{4.76b}$$

where a_i, b_i and c_i, d_i are constants to be determined and $\mathrm{L}_i(x)$ is Lagrange's coefficient (polynomial of degree n) given as

$$L_i(x) = \frac{\pi'(x-x_i)}{\pi'_{j\neq i}(x_i-x_j)} = \frac{(x-x_0)(x-x_1)\cdots(x-x_{i-1})(x-x_{i+1})\cdots(x-x_n)}{(x_i-x_0)(x_i-x_1).(x_i-x_{i-1})(x-x_{i+1})\cdots(x_i-x_{i+1})}$$

Differentiating (4.76a) and (4.76b) we get respectively

$$u_i'(x) = \{2x\,L_i(x)\,L_i'(x)\ + L_i^2(x)\}a_i + 2L_i(x)L_i'(x)b_i \tag{4.77a}$$

$$v_i'(x) = \{2x\,L_i(x)\,L_i'(x)\ + L_i^2(x)\}c_i + 2L_i(x)L_i'(x)d_i. \tag{4.77b}$$

Using conditions (4.74a) and (4.74b) in (4.76a and b)

$$a_i x_i + b_i = 1 \tag{4.78a}$$

$$c_i x_i + d_i = 0. \tag{4.78b}$$

Similarly, using (4.75a) and (4.75b) in (4.77a) and (4.77b)

$$\{2x_i L_i'(x_i) + 1\}a_i + 2L_i'(x_i)b_i = 0 \tag{4.79a}$$

$$\{2x_i L_i'(x_i) + 1\}c_i + 2L_i'(x_i)d_i = 1 \tag{4.79b}$$

From (4.78a) and (4.79a) we get the values of a_i and b_i; and from (4.78b) and (4.79b), the values of c_i and d_i.

$$a_i = -2L_i'(x_i),\ b_i = 1 + 2x_i L_i'(x_i); \tag{4.80a}$$

$$c_i = 1,\ d_i = -x_i. \tag{4.80b}$$

Putting values of a_i, b_i and c_i, d_i in (4.76a) and (4.76b) the polynomial $y(x)$ may be written from (4.73) as

$$y(x) = \sum_{i=0}^{n} \{1 - 2L_i'(x_i)(x - x_i)L_i^2(x)\}y_i + \sum_{i=0}^{n}(x - x_i)L_i^2(x)y_i'. \tag{4.81}$$

It is known as Hermite's interpolation formula.

The error in the formula is given by

$$R(x) = \frac{f^{2n+2}(\xi)}{(2n+2)!}[\pi(x - x_i)]^2, \because R(x_i) = R'(x_i) = 0,\ i = 0(1)n \tag{4.82}$$

where ξ lies in the interval for which $f(x)$ is defined.

Example 4.17

Using Hermite's interpolation formula, find the value of $\sin(0.5)$ from the following data:

x	$\sin x$	$\cos x$
-1	-0.8415	0.5403
0	0	1
1	0.8415	0.5403

Solution $f(x) = \sin(x), f'(x) = \cos x$

$$L_0(x) = \frac{x(x-1)}{2},\ L_0(0.5) = -0.125;$$

$$L_1(x) = -(x+1)(x-1),\ L_1(0.5) = 0.75;$$

$$L_2(x) = \frac{(x+1)(x)}{2},\ L_2(0.5) = 0.375.$$

$$L_0'(x) = \frac{2x-1}{2},\ L_0'(-1) = -\frac{3}{2};$$

$$L_1'(x) = -2x,\ L_1'(0) = 0;$$

$$L_2'(x) = \frac{2x+1}{2},\ L_2'(1) = \frac{3}{2}.$$

$$y(x) = \sum_{i=0}^{2}\{1 - 2L_i'(x_i)(x - x_i)L_i^2(x)\}y_i + \sum_{i=0}^{2}(x - x_i)L_i^2(x)y_i'$$

$$y(0.5) = \left\{1 - 2\left(-\frac{3}{2}\right)(0.5+1)(-0.125)^2\right\} \times (-0.8415) + 0.0$$

$$+\{1 - 2(\frac{3}{2})(0.5+1)(0.375)^2\} \times (0.8415) + (0.5+1)(-0.125)^2 \times 0.5403$$

$$+(0.5)(0.75)^2 \times 1 + (0.5-1)(0.375)^2 \times 0.5403$$

$$= 1.07031 \times (-0.8415) + 1.21094 \times 0.8415 + 0.01266 + 0.28125 - 0.03799$$

$$= 0.3743$$

Exact value is $\sin(0.5) = 0.4794$

Note: If Lagrange's formula is used we get, $\sin(0.5) = 0.4207$. The higher degree polynomial does not necessarily give better result.

Exercise 4

4.1 If $f_k = f(x_0 + kh)$ and $g_k = g(x_0 + kh)$, etc., show that

(*i*) $\Delta^r f_k = \nabla^r f_{k+r} = \delta^r f_{k+r/2}$

(*ii*) $\Delta(f_k g_k) = f_k \Delta g_k + g_{k+1}\Delta f_k$

(*iii*) $\Delta\left(\frac{f_k}{g_k}\right) = \frac{g_k \Delta f_k - f_k \Delta g_k}{g_k \cdot g_{k+1}}$

(*iv*) $\Delta\nabla = \nabla\Delta = \Delta - \nabla = \delta^2$

(*v*) $\delta(f_k g_k) = \mu f_k \delta g_k + \mu g_k \delta f_k$

(*vi*) $2\mu\delta = \Delta + \nabla = \frac{\Delta}{\nabla} - \frac{\nabla}{\Delta}$

(*vii*) $\delta^{2r} y_0 = \Delta^{2r} y_{-r}$ and $\delta^{2r+1} y_{1/2} = \Delta^{2r+1} y_{-r}$.

4.2 Show that

$$\Delta^r\left(\frac{1}{x}\right)=\frac{(-1)^r r! h^r}{x(x+h)\cdots(x+rh)}.$$

4.3 Evaluate the following:

(i) $\left(\frac{\Delta^2}{\mathrm{E}}\right)x^3$

(ii) $\Delta^6(a_0+a_1x)(b_0+b_1x+b_2x^2)(c_0+c_1x+c_2x^2+c_3x^3)$

4.4 There is error in a function value tabulated below:

x	0	1	2	3	4	5	6
$f(x)$	6	10	16	32	58	106	180

Locate the error and correct the same.

4.5 Five consecutive terms in a sequence are given as 5, 9, 23, 53, 105. Form a difference table and get terms one before and one after the given sequence.

4.6 The following table gives the production of steel in different years while the data for two years is not available:

Year	1990	1991	1992	1993	1994	1995	1996
Production (in '000 tons)	45	48	?	63	?	72	80

Estimate the production in 1992 and 1994.

4.7 Following table gives the value of cube root of 10x for $x = 2.00(0.50)4.00$.

x	2.00	2.50	3.00	3.50	4.00
$(10x)^{1/3}$	2.714	2.924	3.107	3.271	3.420

Estimate the value of cube root of 24 and 38. Apply appropriate formula with first four terms only.

4.8 Values of $\log_e x$ are tabulated below for $x = 1.25(0.25)2.50$:

x	1.25	1.50	1.75	2.00	2.25	2.50
$\log_e x$	0.2234	0.4055	0.5596	0.6932	0.8109	0.9163

Compute the values of $\log_e x$ at $x = 1.70$ and 2.15 using Bessel's and Stirling's formulae. Use up to 4th term in the formula.

4.9 Use Everett's and Steffensen's formulae in Exercise 4.8, for $x = 2.15$ and $x = 1.70$ respectively using up to 3rd differences only.

4.10 Compute $y(3)$ by Lagrange's method from the data given below:

x	0	2	4	5
y	5	9	85	180

4.11 Prepare a divided difference table for the data given below:

x	−3	0	1	3	4
$f(x)$	−28	2	4	32	70

Change the order of these values according to abscissas $x = 0, -3, 3, 4, 1$. Again prepare a divided difference table and show that $f(1,3,4) = f(3,4,1)$ and $f(-3,0,1,3,4) = f(0,-3,3,4,1)$. Also find $f(2)$ from both.

4.12 Using Divided Difference formula estimate the reciprocal of 6 from the following table of values:

x	1	2	4	8
$1/x$	1	0.5	0.25	0.125

How does it compare with the exact value? Compute the maximum error from the formula.

4.13 Show that for $f(x) = \dfrac{1}{1-x}$,

$$f(x_0, x_1, \ldots. x_n) = \frac{1}{(1-x_0)(1-x_1)\cdots(1-x_n)}$$

(Hint: Prove by induction)

Also show that if $f(x) = \dfrac{1}{1+x}$, then

$$f(x_0, x_1, \ldots. x_n) = \frac{(-1)^n}{(1+x_0)(1+x_1)\cdots(1+x_n)}.$$

4.14 Show that $f(x_0, x_0, x_1) = \dfrac{1}{(x_1 - x_0)^2}[f(x_1) - f(x_0) - (x_1 - x_0)f'(x_0)]$

4.15 Prove that $\delta^n x^n = n!h^n$.

Hint: Prove by induction assuming the relation to be true upto $(n-1)$. To prove for n consider

$$\delta^n x^n = \delta^{n-1}.\delta x^n = \delta^{n-1}\left\{\left(x+\frac{1}{2}h\right)^n - \left(x-\frac{1}{2}h\right)^n\right\}$$

$$(x+\frac{1}{2}h)^n = x^n + nx^{n-1}.\frac{h}{2} + \text{ terms of powers } (n-2) \text{ and lower.}$$

$$(x-\frac{1}{2}h)^n = x^n - nx^{n-1}.\frac{h}{2} \pm \text{ terms of powers } (n-2) \text{ and lower.}$$

$$(x+\tfrac{1}{2}h)^n - (x-\tfrac{1}{2}h)^n = nhx^{n-1} \pm \text{ terms of powers } (n-2) \text{ and lower etc.}$$

References and Some Useful Related Books/Papers

1. Froberg C.E., *Introduction to Numerical Analysis*, Addison-Wesley Pub. Co. Inc.
2. Goodwin E.T., *Modern Computing Methods*, HMSO, London.

5

Numerical Differentiation

5.1 Introduction

As in Chapter 4, let there be prescribed $(n+1)$ pair of values $(x_i,\ y_i),\ i = 0(1)n$ and that $x_0 < x_1 < x_2 \ldots . < x_n$. We assume that there exists some functional relationship between x and y, say, $y = f(x)$ and that $f(x)$ is differentiable in $(x_0,\ x_n)$. As the function $f(x)$ is not known in general, we resort to numerical methods to approximate the first and higher order derivatives of y with respect to x, viz., $\dfrac{dy}{dx}, \dfrac{d^2y}{dx^2}$ etc. at some point x in $[x_0,\ x_n]$ where x may be a tabular or non-tabular point, including x_0 and x_n.

5.2 Methodology for Numerical Differentiation

The methodology for computing the derivatives at a given point is rather simple. In Chapter 4, we discussed various methods for approximating the function $f(x)$ by a polynomial in one way or another. In order to find derivative, we can differentiate this polynomial, put the value of x in it and get the required value of the derivative. But the process of numerical differentiation is very likely to incur greater error as compared to interpolation. Suppose a function $f(x)$ is approximated by a polynomial $P(x)$, then the graph of $y = P(x)$, generally oscillates about the curve of $y = f(x)$. See Fig. 5.1. Although the polynomial $P(x)$ approximates $f(x)$ closely, the directions of their tangents may differ significantly- even at the tabular points where $P(x)$ and $f(x)$ both agree. For example, at point A, the derivative dy/dx, i.e., the gradients of the tangents, AF and AP at point A, to the function $f(x)$ and $P(x)$ are far apart.

5.3 Differentiation by Newton's FD Formula

The Newton's FD interpolation formula is given as,

$$y_p = y_0 + p\Delta y_0 + \frac{p(p-1)}{2!}\Delta^2 y_0 + \frac{p(p-1)(p-2)}{3!}\Delta^3 y_0 + \cdots \tag{5.1}$$

$$\text{where } p = \frac{x - x_0}{h} \text{ or } x = x_0 + ph \text{ and } \frac{dp}{dx} = \frac{1}{h}. \tag{5.1a}$$

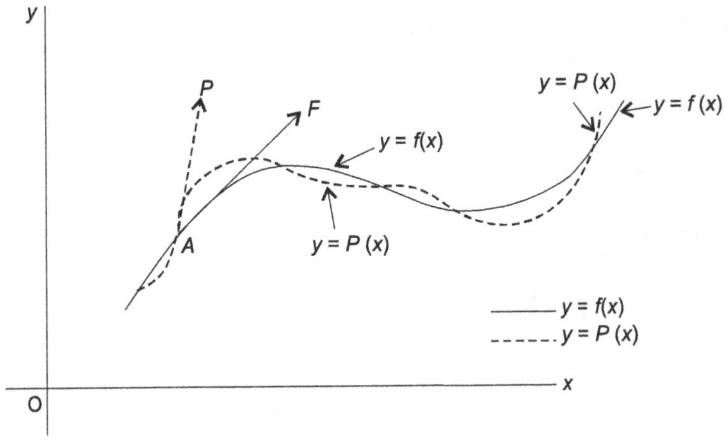

Figure 5.1 Tangents at point A on f(x) and P(x).

Using (5.la) we get,

$$\frac{dy}{dx} = \frac{dy}{dp}.\frac{dp}{dx} = \frac{1}{h}\cdot\frac{dy}{dp}, \tag{5.2}$$

$$\frac{d^2y}{dx^2} = \frac{1}{h}\frac{d^2y}{dp^2}\cdot\frac{dp}{dx} = \frac{1}{h^2}\cdot\frac{d^2y}{dp^2}. \tag{5.3}$$

Thus, differentiating (5.1) we get from (5.2) and (5.3) respectively,

$$\frac{dy}{dx} = \frac{1}{h}\left(\Delta y_0 + \frac{2p-1}{2}\Delta^2 y_0 + \frac{3p^2-6p+2}{6}\Delta^3 y_0 + \frac{2p^3-9p^2+11p-3}{12}\Delta^4 y_0 + \cdots\right), \tag{5.4}$$

$$\frac{d^2y}{dx^2} = \frac{1}{h^2}\left(\Delta^2 y_0 + (p-1)\Delta^3 y_0 + \frac{6p^2-18p+11}{12}\Delta^4 y_0 + \cdots\right). \tag{5.5}$$

Formulae (5.4) and (5.5) may be used to compute first and second derivatives respectively near the upper end of the table. At the tabular point $x = x_0$ where $p = 0$, these formulae reduce to the following:

$$\frac{dy}{dx} = \frac{1}{h}\left(\Delta y_0 - \frac{1}{2}\Delta^2 y_0 + \frac{1}{3}\Delta^3 y_0 - \frac{1}{4}\Delta^4 y_0 + \cdots\right) \tag{5.6}$$

$$\frac{d^2y}{dx^2} = \frac{1}{h^2}\left(\Delta^2 y_0 - \Delta^3 y_0 + \frac{11}{12}\Delta^4 y_0 - \cdots\right). \tag{5.6a}$$

Formulae (5.6) and (5.6a), giving derivatives at the tabular point can also be obtained through operators in the following manner.

Note the relation,

$$e^{h\mathrm{D}} = \mathrm{E} = 1 + \Delta.$$

Taking log on both sides we get,

$$h\mathrm{D} = \ln(1+\Delta) = \Delta - \frac{\Delta^2}{2} + \frac{\Delta^3}{3} - \frac{\Delta^4}{4} + \cdots$$

$$\text{or} \quad \mathrm{D} = \frac{1}{h}\left(\Delta - \frac{\Delta^2}{2} + \frac{\Delta^3}{3} - \frac{\Delta^4}{4} + \cdots\right). \tag{5.7}$$

Squaring (5.7) gives,

$$\mathrm{D}^2 = \frac{1}{h^2}\left(\Delta^2 - \Delta^3 + \frac{11}{12}\Delta^4 - \cdots\right). \tag{5.7a}$$

Operating on y_0, relations (5.7) and (5.7a) give (5.6) and (5.6a), respectively.

5.3.1 Error in differentiation

The error in approximating a function $f(x)$ by a polynomial which interpolates it at $(k+1)$ points x_i, $i = 0(1)k$ is given by (4.14), i.e.,

$$R(x) = (x - x_0)(x - x_1)\ldots.(x - x_k)\frac{f^{k+1}(\xi)}{(k+1)!}, \quad x_0 \le \xi \le x_k. \tag{5.8}$$

The error in approximating a derivative of $f(x)$ may be obtained by differentiating $R(x)$; once, if error in first derivative is required and twice if in the second derivative.

In case of FD formula, the error can also be written as given by (4.21) and (4.22), i.e.,

$$R(x) = (x-x_0)(x-x_1)\ldots.(x-x_k)\cdot\frac{\Delta^{k+1}y_0}{(k+1)!h^{k+1}}, \tag{5.9}$$

$$R(p) = p(p-1)(p-2)\ldots.(p-k)\frac{\Delta^{k+1}y_0}{(k+1)!}, \tag{5.9a}$$

$$\text{where} \quad p = \frac{x-x_0}{h} \text{ or } hp = x-x_0. \tag{5.9b}$$

Thus in order to compute first and second derivatives, at any general point, we can use formulae (5.4) and (5.5) respectively for forward difference approximation; and at the tabular point, put $p=0$ in the concerned formula. The associated error may be computed by formula (5.9) or (5.9a). A formula approximating k^{th} order derivative is called of order p, if it is exact for all polynomials of degree up to p, i.e., $y(x) = x^0,\ x^1,\ x^2,\ \ldots.\ x^p$.

In particular, the one-term approximations for first and second derivatives, at the tabular point $x = x_0$, are given below alongwith associated error in them:

$$\frac{dy}{dx} = \frac{1}{h}\Delta y_0 = \frac{y_1-y_0}{h}, \tag{5.10}$$

with error, $R = -\dfrac{h}{2}f''(\xi),\ x_0 \le \xi \le x_1$

or $\quad = -\dfrac{\Delta^2 y_0}{2h}.$

It is a first order formula since it is exact for $y(x) = x^0,\ x^1$.

$$\frac{d^2y}{dx^2} = \frac{1}{h^2}\Delta^2 y_0 = \frac{y_0-2y_1+y_2}{h^2}, \tag{5.10a}$$

with error, $R = -hf'''(\xi),\ x_0 \le \xi \le x_2,$

or $\quad = -\dfrac{\Delta^3 y_0}{h^2}.$

It is a formula of order two since it is exact for $y(x) = x^0,\ x^1,\ x^2$.

Similarly, the two-terms approximations can also be written from formulae (5.6) and (5.6a) respectively alongwith the order of the error term as,

$$\frac{dy}{dx} = \frac{4y_1-3y_0-y_2}{2h} + O(h^2) \tag{5.11}$$

$$\frac{d^2y}{dx^2} = \frac{2y_0 - 5y_1 + 4y_2 - y_3}{h^2} + O(h^2), \tag{5.11a}$$

where $O(h^2)$ denotes terms of h^2 and higher powers.

Formulae (5.11) and (5.11a) are higher order formulae for first and second derivatives at the tabular point $x = x_0$, using forward difference. Formula (5.11) is of order two as it is exact for $y(x) = x^{-0}, x^1, x^2$ and formula (5.11a) is order three since it gives exact differential for $y(x) = x^0, x^1, x^2, x^3$.

5.4 Differentiation by Newton's BD Formula

Differentiating the Newton's BD formula (4.25), we get

$$\frac{dy}{dx} = \frac{1}{h}\left[\nabla y_0 + \frac{2p+1}{2}\nabla^2 y_0 + \frac{3p^2+6p+2}{6}\nabla^3 y_0 + \frac{2p^3+9p^2+11p+3}{12}\nabla^4 y_0 + \cdots\right] \tag{5.12}$$

$$\frac{d^2y}{dx^2} = \frac{1}{h^2}\left[\nabla^2 y_0 + (p+1)\nabla^3 y_0 + \frac{6p^2+18p+11}{12}\nabla^4 y_0 + \cdots\right] \tag{5.13}$$

Like FD formula the estimate of error will be given by the first neglected term in the formula.

Thus the one-term approximation at the tabular point $x = x_0$ may be written for the first and second derivatives, as

$$\left.\frac{dy}{dx}\right|_{x=0} = \frac{1}{h}\nabla y_0 = \frac{y_0 - y_{-1}}{h} \tag{5.14}$$

$$\left.\begin{aligned} \text{with an error} \quad &= \frac{h}{2}f''(\xi),\ x_0 \le \xi < x_1 \\ \text{or} \quad &= \frac{1}{2h}\nabla^2 y_0 \end{aligned}\right\} \tag{5.14a}$$

$$\left.\frac{d^2y}{dx^2}\right|_{x=x_0} = \frac{1}{h^2}\nabla^2 y_0 = \frac{y_0 - 2y_{-1} + y_{-2}}{h^2} \tag{5.15}$$

$$\left.\begin{aligned} \text{with an error} \quad &= \frac{1}{h^2}\nabla^3 y_0, \\ \text{or} \quad &= hf'''(\xi),\ x_0 \le \xi \le x_2. \end{aligned}\right\} \tag{5.15a}$$

The formulae similar to (5.11) and (5.11a) can be easily written for BD formulae. In general, the relations between the differential operator D and ∇ can be set up as,

$$e^{hD} = E = (1-\nabla)^{-1} \text{ or } hD = -\log(1-\nabla)$$

or $$D = \frac{1}{h}\left(\nabla + \frac{\nabla^2}{2} + \frac{\nabla^3}{3} + \cdots\right) \tag{5.16}$$

and $$D^2 = \frac{1}{h^2}\left(\nabla^2 + \nabla^3 + \frac{11}{12}\nabla^4 + \cdots\right) \tag{5.17}$$

Example 5.1

From the following data find y' and y'' at $x = 2.00$ using up to third differences only.

x	2.00	2.20	2.40	2.60	2.80	3.00
y	0.6932	0.7885	0.8755	0.9555	1.0296	1.0986

Also compute the truncation error.

Solution

Difference Table

x	y	1st *Diff.*	2nd *Diff.*	3rd *Diff.*	4th *Diff.*
2.00	0.6932				
		0.0953			
2.20	0.7885		−0.0083		
		0.0870		0.0013	
2.40	0.8755		−0.0070		−0.0002
		0.0800		0.0011	
2.60	0.9555		−0.0059		−0.0003
		0.0741		0.0008	
2.80	1.0296		−0.0051		
		0.0690			
3.00	1.0986				

At the tabular point x_0, up to 3rd differences,

$$y'(x_0) = \frac{1}{h}\left(\Delta y_0 - \frac{1}{2}\Delta^2 y_0 + \frac{1}{3}\Delta^3 y_0\right)$$

Substituting the values,

$$y'(2.00) = \frac{1}{20}\left[0.0953 - \frac{1}{2}(-0.0083) + \frac{1}{3}(0.0013)\right]$$

$$= 0.4994 \text{ (correct value is 0.5000)}$$

The truncation error is given by the next neglected term,

$$R'(x_0) = \left|-\frac{1}{h}\frac{\Delta^4 y_0}{4}\right|$$

$$R'(2.00) = \left| -\frac{1}{0.20} \times \frac{0.0002}{4} \right|$$

$$= 0.00025$$

The formula for second derivative, up to 3rd differences, is

$$y''(x_0) = \frac{1}{h^2}(\Delta^2 y_0 - \Delta^3 y_0)$$

$$= \frac{1}{(0.20)^2}(-0.0083 - 0.0013)$$

$$= -0.24 \text{ (correct value is } -0.25)$$

The error given by the next neglected term is,

$$R''(x_0) = \left| \frac{1}{h^2} \cdot \frac{11}{12} \Delta^4 y_0 \right|$$

$$= \left| \frac{1}{(0.2)^2} \cdot \frac{11}{12} \times (-0.0002) \right|$$

$$= 0.004583.$$

Example 5.2

From the data of Example 5.1, compute y' and y'' at $x = 2.10$, using difference up to third order only. Also compute trucation error; $p = 0.5$.

Solution Let $x_0 = 2.00$ (since $x = 2.10$ is near upper end of table)

$$p = \frac{2.10 - 2.00}{0.20} = 0.5$$

The formula up to third differences for first derivative is,

$$y'(p) = \frac{1}{h}\left(\Delta y_0 + \frac{2p-1}{2}\Delta^2 y_0 + \frac{3p^2 - 6p + 2}{6}\Delta^3 y_0 \right)$$

Putting $p = 0.5$

$$y'(0.5) = \frac{1}{0.2}\left[0.0953 + \frac{2 \times 0.5 - 1}{2} \times (-0.0083) + \frac{3 \times 0.5^2 - 6 \times 0.5 + 2}{6} \times (0.0013) \right]$$

$$= 0.4762 \text{ (correct value is } 0.4762)$$

The truncation error is given by the first neglected term

$$R'(p) = \frac{1}{h}\left[\frac{2p^3 - 9p^2 + 11p - 3}{12} \cdot \Delta^4 y_0\right]$$

$$= \left|\frac{1}{0.2 \times 12}[0.250 - 2.25 + 5.5 - 3] \times (-0.0002)\right| = 0.000042$$

Formula for second derivative is,

$$y''(p) = \frac{1}{h^2}[\Delta^2 y_0 + (p-1)\Delta^3 y_0]$$

$$y''(.5) = \frac{1}{(0.2)^2}[-0.0083 - 0.5 \times 0.0013]$$

$$= 0.2238 \text{ (correct value is 0.2267)}$$

The truncation error,

$$R''(p) = \frac{1}{h^2}\left[\frac{6p^2 - 18p + 11}{12}\right]\Delta^4 y_0$$

$$= \left|\frac{1}{(0.2)^2 \times 12}[6 \times 0.25 - 18 \times 0.5 + 1] \times (-0.0002)\right|$$

$$= 0.001458.$$

Example 5.3

The values of the function $y = \ln x$ are tabulated in Example 5.1. The first and second derivatives are computed at $x = 2.10$ using up to third differences as in Example 5.2. Compute the error using the function $y = \ln x$; $p = 0.5$.

Solution. The error in $y(x)$ is given as,

$$R(x) = (x - x_0)(x - x_1)(x - x_2)(x - x_3)\frac{f^{iv}(\xi)}{4!}, \; x_0 \le \xi \le x_3.$$

$$R(p) = p(p-1)(p-2)(p-3)\frac{h^4 f^{iv}(\xi)}{4!}$$

$$R'(p) = \frac{1}{h}\mathrm{R}'(p) = (2p^3 - 9p^2 + 11p - 3) \times \frac{h^3 f^{iv}(\xi)}{12},$$

$$f(x) = \ln x;\; f'(x) = \frac{1}{x};\; f''(x) = -\frac{1}{x^2};\; f'''(x) = \frac{2}{x^3};\; f^{iv}(x) = -\frac{6}{x^4}$$

$|f^{iv}(x)|_{\max}$ in [2.00, 2.60] is given at $\xi = 2.00$

$$f^{iv}(\xi) = -\frac{6}{16} = -0.375;\ p = 0.5$$

$$R'(x) = |(2\times 0.5^3 - 9\times 0.5^2 + 11\times 0.5 - 3)\times (0.2)^3 \times (-0.375)/12|$$
$$= 0.000125$$

Similarly error in the second derivative,

$$R''\ (p) = (6p^2 - 18p + 11)\times h^2 \times \frac{f^{iv}(\xi)}{12}$$
$$= |(6\times 0.5^2 - 18\times 0.5 + 11)\times (0.2)^2 \times (-0.375)/12|$$
$$= 0.004375$$

Example 5.4

Compute y' and y'' at $x = 2.25$ using appropriate formula up to third difference only, from the following data:

x	1.00	1.25	1.50	1.75	2.00	2.25
y	2.7183	3.4903	4.4817	5.7546	7.3891	9.4877

Solution

x	y	1st *Diff*	2nd *Diff.*	3rd *Diff.*	4th *Diff.*
1.00	2.7183				
		0.7720			
1.25	3.4903		0.2194		
		0.9914		0.0621	
1.50	4.4817		0.2815		0.0180
		1.2729		0.0801	
1.75	5.7546		0.3616		0.0204
		1.6345		0.1025	
2.00	7.3891		0.4641		
		2.0986			
2.25	9.4877				

We have to use Newton's BD formula near the lower end of the table.

$$x_0 = 2.25;\ p = 0;\ h = 0.25$$

$$y'(x) = \frac{1}{h}\left(\nabla y_0 + \frac{1}{2}\nabla^2 y_0 + \frac{1}{3}\nabla^3 y_0\right)$$

$$y'(2.25) = \frac{1}{0.25}\left[2.0986 + \frac{1}{2}(0.4641) + \frac{1}{3}(0.1025)\right]$$

$$= 9.4593 \text{ (correct value is 9.4877)}$$

$$y''(p) = \frac{1}{h^2}[\nabla^2 y_0 + (p+1)\nabla^3 y_0]$$

$$y''(2.25) = 16\ [0.4641 + 0.1025]$$

$$= 9.0656$$

(Function values are taken from $y = e^x$).

Example 5.5

From data of Example 5.4, compute $y'(x)$ and $y''(x)$ at $x = 2.15$, using appropriate formula up to 3rd differences. Also estimate the truncation error.

Solution We use Newton's BD formula near lower end of the table.

$$x_0 = 2.25;\ h = 0.25;\ p = \frac{2.15 - 2.25}{0.25} = -0.4$$

The formula for first derivative is,

$$y'(p) = \frac{1}{h}\left[\nabla y_0 + \frac{2p+1}{2}\nabla^2 y_0 + \frac{3p^2+6p+2}{6}\nabla^3 y_0\right]$$

Putting $p = -0.4$

$$y'(p) = \frac{1}{0.25}\left[2.0986 + \frac{-0.8+1}{2}(0.4641) + \frac{3(0.4)^2 - 0.4 \times 6 + 2}{6} \times 0.1025\right]$$

$$= 8.5855 \text{ (correct value is 8.5849)}$$

Formula for second derivative is,

$$y''(p) = \frac{1}{h^2}[\nabla^2 y_0 + (p+1)\nabla^3 y_0]$$

$$= \frac{1}{(0.25)^2}[0.4641 + 0.6 \times 0.1025]$$

$$= 8.4096 \text{ (correct value is 8.5849)}$$

The error in first derivative is given by,

$$R'(p) = \frac{1}{12h}[2p^3 + 9p^2 + 11p + 3]\nabla^4 y_0$$

$$= \left|\frac{1}{12 \times 0.25}[-2 \times (0.4)^3 + 9 \times 0.4^2 - 11 \times 0.4 + 3] \times 0.0204\right|$$

$$= 0.0005984$$

The error in second derivative,

$$R''(p) = \frac{1}{12h^2}(6p^2 + 18p + 11)\nabla^4 y_0$$

$$= \left| \frac{1}{12 \times (0.25)^2}[6 \times 0.4^2 - 18 \times 0.4 + 11) \times 0.0204 \right|$$

$$= 0.12947$$

Note: Errors can also be computed if $f(x) = e^x$ is given.

5.5 Differentiation by Central Difference Formulae

In Chapter 4, several central difference formulae were given for interpolating the function values. By differentiating we can use them for computing the derivatives of the tabulated function, at the tabular as well as at non-tabular points. Further, like interpolation these formulae are used at the points in the middle, away from the ends of the table.

5.5.1 At tabular points

Let us set up a relation between the differential operator D and the central difference operator δ, which may be employed for approximating the derivatives at the tabular points. We know that,

$$\delta = \mathrm{E}^{1/2} - \mathrm{E}^{-1/2} = e^{1/2h\mathrm{D}} - e^{-1/2h\mathrm{D}} = 2\sinh\left(\frac{1}{2}h\mathrm{D}\right)$$

or $$\mathrm{D} = \frac{2}{h}\sinh^{-1}\frac{\delta}{2}$$

$$= \frac{2}{h}\left[\frac{\delta}{2} - \frac{1^2}{3!}\left(\frac{\delta}{2}\right)^3 + \frac{1^2 \cdot 3^2}{5!}\left(\frac{\delta}{2}\right)^5 - \frac{1^2 \cdot 3^2 \cdot 5^2}{7!}\left(\frac{\delta}{2}\right)^7 + \ldots\right]$$

$$= \frac{1}{h}\left[\delta - \frac{\delta^3}{24} + \frac{3\delta^5}{640} - \frac{5\delta^7}{7168} + \cdots\right] \tag{5.18}$$

Operating on y_0, formula gives the first derivative,

$$y'(x_0) = \frac{1}{h}\left[\delta y_0 - \frac{1}{24}\delta^3 y_0 + \frac{3}{640}\delta^5 y_0 - \cdots\right] \tag{5.18a}$$

However, formula (5.18a) as it stands, can not be used to compute the first derivative, as the odd differences at the tabular points are not available, i.e., there is no difference available, like $\delta y_0 = y_{1/2} - y_{-1/2}$, in the difference table. Instead we have $\delta y_{1/2} = y_1 - y_0$, etc., tabulated.

Nevertheless we can modify the formula in the following manner. We multiply the right side of relation (5.18) by $\mu\mu^{-1}$. Using the relation $\mu = \left(1+\frac{\delta^2}{4}\right)^{1/2}$, the relation (5.18) is equivalent to,

$$\begin{aligned}
\mathrm{D} &= \frac{\mu}{h}\left(\delta - \frac{\delta^3}{24} + \frac{3}{640}\delta^5 - \cdots\right)\left(1+\frac{\delta^2}{4}\right)^{-1/2} \\
&= \frac{\mu}{h}\left(\delta - \frac{\delta^3}{24} + \frac{3}{640}\delta^5 - \cdots\right)\left(1 - \frac{\delta^2}{8} + \frac{3\delta^4}{128} - \cdots\right) \\
&= \frac{\mu}{h}\left(\delta - \frac{\delta^3}{6} + \frac{\delta^5}{30} - \frac{\delta^7}{140} + \cdots\right).
\end{aligned} \tag{5.19}$$

Operating on y_0, we get the first derivative from (5.19) as

$$y'(x_0) = \mathrm{D}y_0 = \frac{1}{h}\left[\mu\delta y_0 - \frac{1}{6}\mu\delta^3 y_0 + \frac{1}{30}\mu\delta^5 y_0 - \cdots\right]. \tag{5.19a}$$

Now, formula (5.19a) can be used to determine $y'(x_0)$ since the values of $\mu\delta y_0$, $\mu\delta^3 y_0$, etc., are available in the difference table, since $\mu\delta y_0 = \frac{1}{2}(\delta y_{1/2} + \delta y_{-1/2})$ and $\mu\delta^3 y_0 = \frac{1}{2}\left(\delta^3 y_{1/2} + \delta^3 y_{-1/2}\right)$, etc.

A one-term approximation for the first derivative may be written from (5.19a) as,

$$y'(x_0) = \frac{1}{2h}\left(\delta y_{1/2} + \delta y_{-1/2}\right) = \frac{y_1 - y_{-1}}{2h}. \tag{5.20}$$

The error in the formula (5.20) is given by,

$$R'(x_0) = -\frac{1}{6h}\mu\delta^3 y_0 = -\frac{1}{12h}\left(\delta^3 y_{1/2} + \delta^3 y_{-1/2}\right) \tag{5.20a}$$

Note that,

$$\delta^3 y_{1/2} = y_2 - 3y_1 + 3y_0 - y_{-1} = \Delta^3 y_{-1} \text{ and}$$

$$\delta^3 y_{-1/2} = y_1 - 3y_0 + 3y_{-1} - y_{-2} = \Delta^3 y_{-2}$$

$$\frac{\delta^3 y_{1/2}}{h^3} = \frac{\Delta^3 y_{-1}}{h^3} \simeq f'''(\xi_1),\ x_{-1} \le \xi_1 \le x_2.$$

$$\frac{\delta^3 y_{-1/2}}{h^3} = \frac{\Delta^3 y_{-2}}{h^3} \simeq f'''(\xi_2),\ x_{-2} \le \xi_2 \le x_1.$$

In general, we can convert the CD to FD as follows:

$$\delta^3 y_{1/2} = \left(\mathrm{E}^{1/2} - \mathrm{E}^{-1/2}\right)^3 \mathrm{E}^{1/2} y_0 = (\mathrm{E}-1)^3 \mathrm{E}^{-1} y_0 = \Delta^3 y_{-1}$$

$$\delta^4 y_0 = \left(\mathrm{E}^{1/2} - \mathrm{E}^{-1/2}\right)^4 y_0 = (\mathrm{E}-1)^4 \mathrm{E}^{-2} y_0 = \Delta^4 y_{-2}. \quad \text{[See Exercise 4.1 } (vii)\text{].}$$

Therefore, if the function $f(x)$ is known, the error may be written from (5.20a) as,

$$R'(x_0) = \frac{h^2}{6} f'''(\xi),\ x_{-2} \le \xi \le x_2. \tag{5.20b}$$

Formula (5.20) is a second order formula since it is exact for $f(x) = x^0,\ x^1,\ x^2$.
In order to obtain second derivative we square both sides of (5.18), giving

$$\mathrm{D}^2 = \frac{1}{h^2}\left[\delta^2 - \frac{\delta^4}{12} + \frac{\delta^6}{90} - \cdots\right] \tag{5.21}$$

After operating (5.21) on y_0, we get

$$y''(x_0) = \mathrm{D}^2 y_0 = \frac{1}{h^2}\left[\delta^2 y_0 - \frac{1}{12}\delta^4 y_0 + \frac{1}{90}\delta^6 y_0 - \cdots\right]. \tag{5.21a}$$

It may be noted that even order differences at the tabular points are available in the difference table. A one-term approximation for the second derivative may be written from (5.21a) as,

$$y''(x_0) = \frac{1}{h^2}\delta^2 y_0 = \frac{y_{-1} - 2y_0 + y_1}{h^2}. \tag{5.22}$$

The error in the second derivative approximated by (5.22) is given by,

$$R''(x_0) = -\frac{1}{12h^2}\delta^4 y_0 = -\frac{1}{12h^2}\cdot \Delta^4 y_{-2}. \tag{5.22a}$$

If the function $f(x)$ is known, then the error will be

$$R''(x_0) = -\frac{h^2}{12} f^{iv}(\xi),\ x_{-2} \le \xi \le x_2. \tag{5.22b}$$

For getting third order derivative we can multiply (5.19) and (5.21); for fourth order derivative, square (5.21). And so on.

5.5.2 At non-tabular points

The approximations for derivatives at the non-tabular points may be obtained by differentiating various central difference interpolation formulae discussed in Chapter 4. The approximations for first and second derivatives are given as follows:

(*i*) By Stirling's Formula

$$y'(x) = \frac{1}{h}\left[\mu\delta y_0 + p\delta^2 y_0 + \frac{3p^2-1}{6}\mu\delta^3 y_0 + \frac{2p^3-p}{12}\delta^4 y_0 + \cdots\right]. \tag{5.23}$$

$$y''(x) = \frac{1}{h^2}\left[\delta^2 y_0 + p\mu\delta^3 y_0 + \frac{6p^2-1}{12}\delta^4 y_0 + \frac{2p^3-3p}{12}\mu\delta^5 y_0 + \cdots\right]. \tag{5.24}$$

(*ii*) By Bessel's Formula

$$y'(x) = \frac{1}{h}\left[\delta y_{1/2} + \frac{2p-1}{2}\mu\delta^2 y_{1/2} + \frac{6p^2-6p+1}{12}\delta^3 y_{1/2} + \frac{2p^3-3p^2-p+1}{12}\mu\delta^4 y_{1/2} + \ldots\right]. \tag{5.25}$$

$$y''(x) = \frac{1}{h^2}\left[\mu\delta^2 y_{1/2} + \frac{2p-1}{2}\delta^3 y_{1/2} + \frac{6p^2-6p-1}{12}\mu\delta^4 y_{1/2} + \cdots\right] \tag{5.26}$$

(*iii*) By Everett's Formula

$$\begin{aligned} y'(x) &= \frac{1}{h}\Big[(y_1 - y_0) + \frac{1}{6}\{(3p^2-1)\delta^2 y_1 - (3p^2-6p+2)\delta^2 y_0\} \\ &+ \frac{1}{120}\{(5p^4-15p^2+4)\delta^4 y_1 - (5p^4-20p^3+15p^2+10p-6)\delta^4 y_0\} + \ldots.\Big] \end{aligned} \tag{5.27}$$

$$\begin{aligned} y''(x) &= \frac{1}{h^2}\Big[p\delta^2 y_1 - (p-1)\delta^2 y_0 + \frac{1}{12}\{(2p^3-3p^2)\delta^4 y_1 \\ &\quad -(2p^3-6p^2+3p+10)\delta^4 y_0\} + \ldots.\Big] \end{aligned} \tag{5.28}$$

(*iv*) By Steffensen's Formula

$$\begin{aligned} y'(x) &= \frac{1}{h}\Big[p(\delta y_{1/2} + \delta y_{-1/2}) + \frac{1}{12}\{(2p^3+3p^2-p-1)\delta^3 y_{1/2} \\ &\quad +(2p^3-3p^2-p+1)\delta^3 y_{-1/2}\} + \cdots\Big] \end{aligned} \tag{5.29}$$

$$y''(x) = \frac{1}{h^2}\left[\delta y_{1/2} + \delta y_{-1/2} + \frac{1}{12}\{(6p^2+6p-1)\delta^3 y_{1/2} + (6p^2-6p-1)\delta^3 y_{-1/2}\} + \ldots\right] \tag{5.30}$$

Example 5.6

Given below are the values of $y = \sin x$ for $x = 10°(10°)60°$.

x	10°	20°	30°	40°	50°	60°
y	0.1736	0.3420	0.5000	0.6428	0.7660	0.8660

Compute the first and second derivatives of $\sin x$ from the above table at $x = 40°$, using only two terms in the finite differences formula. ($\pi = 3.1416$)

Solution

Difference Table

x	$y = \sin x$	*I Diff.*	*II Diff.*	*III Diff.*	*IV Diff.*
10°	0.1736				
		0.1684			
20°	0.3420		−0.0104		
		0.1580		−0.0048	
30°	0.5000		−0.0152		0.0004
		0.1428		−0.0044	
40°	0.6428		−0.0196		0.0008
		0.1232		−0.0036	
50°	0.7660		−0.0232		0.0005
		0.1000		−0.0031	
60°	0.8660		−0.0263		
		0.0737			
70°	0.9397				

First of all we should change the interval $h = 10^{\text{o}}$ into radians so that,

$$h = 10° = \frac{\pi}{180} \times 10 = \frac{\pi}{18} \text{ radians}$$

Using only first two terms in (5.19a),

$$y'(x_0) = \frac{1}{h}\left(\mu\delta y_0 - \frac{1}{6}\mu\delta^3 y_0\right)$$

$$= \frac{18}{\pi}\left(\frac{0.1428+0.1232}{2} - \frac{1}{6}\cdot\frac{-0.0044-0.0036}{2}\right)$$

$$= 18 \times 0.3183\ (0.1330+0.00067)$$

$$= 0.7658 \text{ (correct value is 0.7660).}$$

The second derivative is given by (5.21a) as,

$$y''(x_0) = \frac{1}{h^2}\left[\delta^2 y_0 - \frac{1}{12}\delta^4 y_0\right]$$

$$= (18 \times 0.3183)^2 \left[-0.0196 - \frac{1}{12} \times 0.0008\right]$$

$$= (32.8260)\ (-0.01967)$$

$$= -0.6459 \text{ (correct value is } -0.6428).$$

Note: The magnitude of error in the computation of $y'(x_0)$ will be given by the next neglected term in (5.19a), i.e., $\frac{1}{h} \times \frac{1}{30} \times \mu\delta^5 y_0$ and in that of $y''(x_0)$ by the next neglected term in (5.21a), i.e., $\frac{1}{h^2} \times \frac{1}{90} \times \delta^6 y_0$. If function $f(x) = \sin x$ is to be used then $\delta^5 y_0$ will be replaced by $h^5 f^v(\xi)$ and $\delta^6 y_0$ by $h^6 f^{vi}(\xi)$.

Example 5.7

Following table gives the values of a function $y(x)$ for $x = 1.00(.50)3.50$

x	1.00	1.50	2.00	2.50	3.00	3.50
$y(x)$	1.000	1.1447	1.2599	1.3572	1.4422	1.5183

Compute $y'(x)$ and $y''(x)$ at $x = 2.40$ and 2.80, using appropriate CD formula.

Solution

x	y	*I Diff.*	*II Diff.*	*III Diff.*	*IV Diff.*	*V Diff.*
1.00	1.000					
		0.1447				
1.50	1.1447		−0.0295			
		0.1152		0.0116		
2.00	1.2599		−0.0179		−0.0060	
		0.0973		0.0056		0.0038
2.50	1.3572		−0.0123		−0.0022	
		0.0850		0.0034		
3.00	1.4422		−0.0089			
		0.0761				
3.50	1.5183					

Note: Since fifth difference is larger than the fourth we neglect it.

At $x = 2.40$, we choose the initial point $x_0 = 2.50$.

$$\text{Then } p = \frac{2.40 - 2.50}{0.50} = -\frac{0.10}{0.50} = -0.2$$

Since for $-0.25 < p < 0.25$, Stirling's formula should be used, we get from (5.23),

$$y'(x) = \frac{1}{h}\left[\mu\delta y_0 + p\delta^2 y_0 + \frac{3p^2 - 1}{6}\mu\delta^3 y_0\right]$$

$$= \frac{1}{0.5}\left[\frac{0.0973+0.0850}{2} + (-0.2)(-0.0123) + \frac{3(0.2)^2-1}{6}\left(\frac{0.0056+0.0034}{2}\right)\right]$$

$$= 2[0.09115 + 0.00246 - 0.00029]$$

$$= 0.1866 \text{ (correct value is 0.1860)}$$

From (5.24) we can write the second derivative,

$$y''(x) = \frac{1}{h^2}\left[\delta^2 y_0 + p\mu\delta^3 y_0 + \frac{p^2}{2}\delta^4 y_0\right]$$

$$= \frac{1}{(0.5)^2}\left[-0.0123 + (-0.2)\frac{0.0056+0.0034}{2} + \frac{6\times(0.2)^2-1}{12}(-0.0022)\right]$$

$$= 4[-0.0123 - 0.0008 + 0.00014]$$

$$= -0.0518 \text{ (correct value is } -0.0516).$$

At $x = 2.80$, choose $x_0 = 2.50$ which gives,

$$p = \frac{2.80-2.50}{0.50} = \frac{0.30}{0.50} = 0.6.$$

Since p lies in the range $0.25 < p < 0.75$ we use Bessel's formula, (5.25) and (5.26) respectively for first and second derivative.

$$y'(x) = \frac{1}{h}\left[\delta y_{1/2} + \frac{2p-1}{2}\mu\delta^2 y_{1/2} + \frac{6p^2-6p+1}{12}\delta^3 y_{1/2}\right]$$

$$= \frac{1}{0.5}\left[(0.0850) + \frac{2\times 0.6-1}{2} \times \frac{-0.0089-0.0123}{2} + \frac{6\times(0.6)^2-6\times 0.6+1}{12} \times (0.0034)\right]$$

$$= 2[0.0850 - 0.00106 - 0.00012]$$

$$= 0.1676 \text{ (correct value is 0.1860)}$$

$$y''(x) = \frac{1}{h^2}\left[\mu\delta^2 y_{1/2} + \frac{2p-1}{2}\delta^3 y_{1/2}\right]$$

$$= \frac{1}{(0.5)^2}\left[\frac{-0.0089-0.0123}{2} + \frac{2\times 0.6-1}{2} \times 0.0034\right]$$

$$= 4[-0.0106 + 0.00034]$$

$$= -0.0410 \text{ (correct value is } -0.0516).$$

Example 5.8

The speed of a car km/ min is recorded for $1\frac{1}{2}$ minutes as given below:

time(t)(min)	0.00	0.25	0.50	0.75	1.00	1.25	1.50
Speed(u) (km/ min)	1.000	0.784	0.649	0.617	0.718	0.990	1.482

Compute after how much time from the start the accelaration of the car will be zero?

Solution We are required to compute the point where $\frac{du}{dt} = 0.$

t	u	1st *Diff.*	2nd *Diff.*	3rd *Diff.*
0.00	1.000			
		−0.216		
0.25	0.784		0.081	
		−0.135		0.022
0.50	0.649		0.103	
		−0.032		0.030
0.75	0.617		0.133	
		0.101		0.038
1.00	0.718		0.171	
		0.272		0.039
1.25	0.990		0.220	
		0.492		
1.50	1.482			

We can see from the first difference that the point where $\frac{du}{dt} = 0$, should be near $t = 0.75$. From Newton's FD formula we have, taking $t_0 = 0.75$,

$$\frac{du}{dt} = \frac{1}{h}\left[\Delta u_0 + \frac{2p-1}{2}\Delta^2 u_0 + \frac{3p^2-6p+2}{6}\Delta^3 u_0\right]$$

$$= \frac{1}{0.25}\left[0.101 + \frac{2p-1}{2}(0.171) + \frac{3p^2-6p+2}{6}(0.039)\right]$$

$$0 = \frac{4}{6}[0.606{+}0.513\ (2p-1) + (3p^2-6p+2)\ (0.039)\ ]$$

or $\quad 0 = \frac{4}{6}[0.171 + 0.792p + 0.117p^2]$

or $\quad 0.117p^2 + 0.792p + 0.171 = 0$

On solving the above quadratic equation, we get

$$p = \frac{-0.792 \pm \sqrt{0.792^2 - 4 \times 0.171 \times 0.117}}{2 \times 0.117}$$

$$= \frac{-0.792 \pm 0.7397}{0.234} = -0.2235 \text{ or } -7.4004$$

We take $p = -0.2235$ since the other value is too large (in modulus).

The required time is,

$$t = t_0 + ph = 0.75 - 0.2235 \times 0.25$$

$$= 0.6941 \text{ (correct value is 0.6932)}$$

Alternatively, applying Stirling's formula at $t_0 = 0.75$,

$$\frac{du}{dt} = \frac{1}{0.25}\left[\frac{0.101 - 0.032}{2} + p \times (0.133) + \frac{(3p^2 - 1)}{6}\left(\frac{0.030 + 0.038}{2}\right)\right]$$

$$= \frac{4}{6}[0.034p^2 + 0.798p + 0.173]$$

Putting $\frac{du}{dt} = 0$ and solving the quadratic, we get

$$p = -0.2191, \text{ giving } t = 0.6952.$$

Again, suppose we choose $t_0 = 0.50$, then using FD formula, gives,

$$0 = \frac{du}{dt} = \frac{1}{0.25}\left[-0.032 + \frac{2p-1}{2}(0.133) + \frac{3p^2 - 6p + 2}{6}(0.038)\right]$$

or $0.114p^2 + 0.630p - 0.515 = 0$

On solving the above,

$$p = 0.7228 \text{ and } -6.2491$$

The acceptable value is $p = 0.7228$, which gives,

$$t = 0.50 + 0.7228 \times 0.25$$

$$= 0.6807 \text{ (which is not so accurate)}$$

Note: The data has been taken from $u = e^t - 2t$.

5.6 Method of Undetermined Coefficients

Sometimes we may be interested to devise a differentiation formula consisting of some specific ordinates. Suppose we want a formula of the type,

$$f'(x_0) = af(x_0) + bf(x_0 + h) + cf(x_0 + 3h).$$

where $a,\ b$ and c are constants.

In order to determine $a,\ b$ and c we expand $f(x_0+h)$ and $f(x_0+3h)$ by Taylor's series about the point $x=x_0$,

$$f(x_0+h)=f(x_0)+hf'(x_0)+\frac{h^2}{2}f''(x_0)+\frac{h^3}{6}f'''(x_0)+\cdots$$

$$f(x_0+3h)=f(x_0)+3hf'(x_0)+\frac{9h^2}{2}f''(x_0)+\frac{27h^3}{6}f'''(x_0)+\cdots.$$

Eliminate the term containing f''_0 from these equations by multiplying first equation by 9 and subtracting the second from it, giving,

$$f'(x_0)=\frac{9f_1-8f_0-f_3}{6h}+\frac{h^2}{2}f'''(x_0)+\cdots$$

$$=\frac{9f_1-8f_0-f_3}{6h}+0(h^2)$$

This is the required formula and the truncation error is given by the first neglected non-zero term, i.e.,

$$R'(x_0)=\frac{h^2}{2}f'''(\xi),\ x_0\le\xi\le x_3.$$

Alternatively, $R(x)=(x-x_0)(x-x_1)(x-x_3)\dfrac{f'''(\xi)}{3!}$ giving $R'(x_0)=\dfrac{h^2}{2}f'''(\xi)$.

The above formula can also be obtained by equating the coefficients after expanding the functions by Taylor's series,

$$f_0'=af_0+b\left[f_0+hf_0'+\frac{h^2}{2}f_0''+\frac{h^3}{6}f_0'''+\cdots\right]+c\left[f_0+3hf_0'+\frac{9h^2}{2}f_0''+\frac{27h^3}{6}f_0'''+\cdots\right]$$

$$=(a+b+c)f_0+h(b+3c)f_0'+\frac{h^2}{2}(b+9c)f_0''+\frac{h^3}{6}(b+27c)f_0'''+\cdots$$

After comparing coefficients on both sides the resulting equations in $a,\ b$ and c are,

$$a+b+c=0;\ b+3c=1 \text{ and } (b+9c)=0.$$

From these three equations, values of $a,\ b$ and c may be determined.

5.7 Comments on Differentiation

For h small, the one term approximation in different schemes provides a fairly good estimate for the first and second derivatives at the tabular points, i.e., $y'(x_0)$ and $y''(x_0)$. For the first derivative, FD formula (5.10), BD formula (5.14) and CD formula (5.20) may be used. However, on comparison it may be noted that the error in the FD and BD formula is of $O(h)$ while in the CD formula it is $O(h^2)$. Therefore wherever possible CD formula should be preferred. Obviously at the top and bottom ends of the table only FD and BD formulae may be applied. As $(dy/dx)_{x=x_0}$ denotes slope of the tangent to the curve $y = f(x)$ at $x = x_0$, it can be seen from Fig. 5.2, as how it is approximated in three different schemes, viz., FD, BD and CD. AB, AC and CB approximate the slope AP by FD, BD and CD formulae, respectively. Further, for approximation of the second derivatives also, same is true as for the first derivatives.

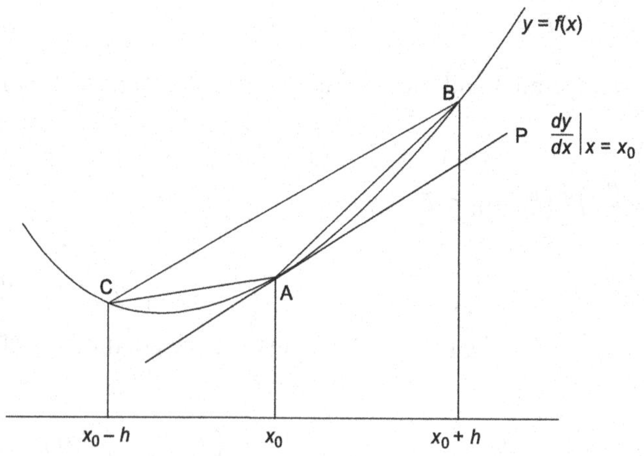

Figure 5.2 Approximation of first derivative by FD, BD and CD.

5.8 Derivatives with Unequal Intervals

So far we have considered the approximation of derivatives when the function values were prescribed at equal intervals. In many practical problems the function values are given at the nodal (tabular) points which are not necessarity evenly spaced. We will derive the formulae for first and second derivatives using Taylor's series and formulae for higher derivatives can be easily deduced if required.

5.8.1 Forward Difference formulae

Let us suppose the values of the function $f(x)$ are given for the abscissas $x = x_0,\ x_1$ and x_2 where $x_1 = x_0 + h_1$ and $x_2 = x_1 + h_2 = x_0 + h_1 + h_2$ (see Fig. 5.3a).

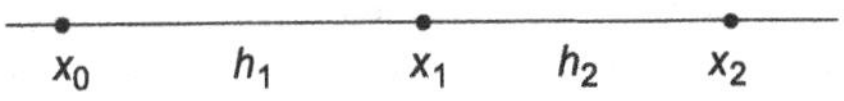

Figure 5.3a Forward Difference approximation with unequal intervals.

Writing Taylor's expansion of $f(x_1)$ and $f(x_2)$ about x_0,

$$f(x_1) = f(x_0 + h_1) = f(x_0) + h_1 f'(x_0) + \frac{h_1^2}{2} f''(x_0) + \frac{h_1^3}{6} f'''(x_0) + \cdots \tag{5.31}$$

$$f(x_2) = f(x_0 + h_1 + h_2)$$

$$= f(x_0) + (h_1 + h_2) f'(x_0) + \frac{(h_1 + h_2)^2}{2} f''(x_0) + \frac{(h_1 + h_2)^3}{6} f'''(x_0) + \cdots \tag{5.32}$$

(*i*) First Derivative

The simplest forward difference formula may be obtained from (5.31) as,

$$f'(x_0) = \frac{f(x_1) - f(x_0)}{h_1} - \frac{h_1}{2} f''(\xi),\ x_0 \le \xi \le x_1. \tag{5.33}$$

A higher order formula may be obtained by eliminating $f''(x_0)$ term from (5.31) and (5.32) giving,

$$f'(x_0) = \frac{(h_1 + h_2)^2 f_1 - h_1^2 f_2 - \{(h_1 + h_2)^2 - h_1^2\} f_0}{h_1 h_2 (h_1 + h_2)} + \frac{h_1 (h_1 + h_2)}{6} f'''(\xi),\ x_0 \le \xi \le x_2. \tag{5.33a}$$

where $f(x_0) = f_0, f(x_1) = f_1$ and $f(x_2) = f_2$.

Formulae (5.33) and (5.33a) may be compared with (5.10) and (5.10a) when $h_1 = h_2 = h$.

(*ii*) Second Derivative

Eliminate $f'(x_0)$ by multiplying (5.31) by $(h_1 + h_2)$ and (5.32) by h_1 and subtracting so that,

$$f''(x_0) = \frac{2}{h_1 h_2 (h_1 + h_2)} \{h_2 f_0 - (h_1 + h_2) f_1 + h_1 f_2\} - \frac{1}{3h_2} \{(h_1 + h_2)^2 - h_1^2\} f'''(\xi),\ x_0 \le \xi \le x_2. \tag{5.34}$$

This may be compared with (5.11) when $h_1 = h_2 = h$.

5.8.2 Backward Difference formulae

Refer Fig. 5.3b where function values are given for the abscissas $x = x_0,\ x_{-1} = x_0 - h_1$ and $x_{-2} = x_{-1} - h_2 = x_0 - (h_1 + h_2)$.

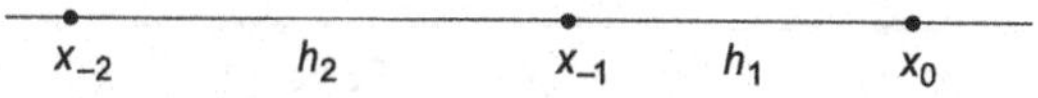

Figure 5.3b Backward Difference approximation with unequal intervals

Again, writing Taylor's expansion for $f(x_0 - h_1)$ and $f\{x_0 - (h_1 + h_2)\}$ about x_0,

$$f(x_0 - h_1) = f(x_0) - h_1 f'(x_0) + \frac{h_1^2}{2} f''(x_0) - \frac{h_1^3}{6} f'''(x_0) + \cdots \tag{5.35}$$

$$f\{x_0 - (h_1 + h_2)\} = f(x_0) - (h_1 + h_2) f'(x_0) + \frac{(h_1 + h_2)^2}{2} f''(x_0) - \frac{(h_1 + h_2)^3}{6} f'''(x_0) + \cdots \tag{5.36}$$

(*i*) First Derivative

The simplest backward difference first derivative formula would be,

$$f'(x_0) = \frac{f(x_0) - f(x_{-1})}{h_1} + \frac{h_1}{2} f''(\xi),\ x_{-1} \le \xi \le x_0. \tag{5.37}$$

A higher order formula similar to (5.33a) may be written by replacing h_1 and h_2 by $-h_1$ and $-h_2$ respectively, as

$$f'(x_0) = \frac{h_1^2 f_{-2} - (h_1 + h_2)^2 f_{-1} + h_2^2 f_0}{h_1 h_2 (h_1 + h_0)} + \frac{h_1 (h_1 + h_2)}{6} f'''(\xi), \tag{5.37a}$$

$f(x_0 - h_1) = f_{-1}$ and $f(x_0 - h_1 - h_2) = f_{-2}$.

(*ii*) Second Derivative

The second derivative in backward difference form may be written similar to (5.34) as,

$$f''(x_0) = \frac{2}{h_1 h_2 (h_1 + h_2)} \{h_2 f_0 - (h_1 + h_2) f_{-1} + h_1 f_{-2}\}$$

$$+ \frac{1}{3h_2} \{(h_1 + h_2)^2 - h_1^2\} f'''(\xi),\ x_{-2} \le \xi \le x_0. \tag{5.38}$$

where $f(x_0 - h_1) = f_{-1}$ and $f(x_0 - h_1 - h_2) = f_{-2}$.

5.8.3 Central Difference formulae

When the function values are given for abscissas, one to the right of x_0 and other to the left of it, then we can use central difference formulae for the approximation of first and second derivatives in the following manner. Let us suppose that the values are given for abscissas $x_1 = x_0 + h_1$ and $x_{-1} = x_0 - h_2$ as $f(x_1) = f_1$ and $f(x_{-1}) = f_{-1}$ (see Fig. 5.3c).

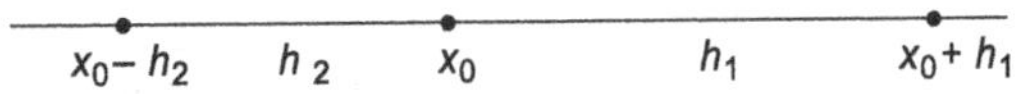

Figure 5.3c Central Difference approximation with unequal intervals

We write Taylor's expansion about x_0, as

$$f(x_0 + h_1) = f(x_0) + h_1 f'(x_0) + \frac{h_1^2}{2} f''(x_0) + \frac{h_1^3}{6} f'''(x_0) + \cdots \tag{5.39}$$

$$f(x_0 - h_2) = f(x_0) - h_2 f'(x_0) + \frac{h_2^2}{2} f''(x_0) - \frac{h_2^3}{6} f'''(x_0) + \cdots \tag{5.40}$$

(*i*) First Derivative
Formula for first derivative may be obtained by eliminating $f''(x_0)$ from (5.39) and (5.40) as,

$$f'(x_0) = \frac{h_2^2 f_1 - h_1^2 f_{-1} - (h_2^2 - h_1^2) f_0}{h_1 h_2 (h_1 + h_2)} - \frac{h_1 h_2}{6} f'''(\xi),\ x_{-1}\xi \le x_1. \tag{5.41}$$

Compare with formula (5.20) when $h_1 = h_2 = h$.

(*ii*) Second Derivative
From (5.39) and (5.40) after eliminating $f'(x_0)$ we may obtain,

$$f''(x_0) = \frac{2}{h_1 h_2 (h_1 + h_2)} \{h_2 f_1 - (h_1 + h_2) f_0 + h_1 f_{-1}\} - \frac{1}{6}(h_1 - h_2) f'''(x_0) - \frac{1}{6} \frac{h_1^3 + h_2^3}{h_1 + h_2} f^{iv}(\xi),$$
$$x_{-1} \le \xi \le x_1. \tag{5.42}$$

Compare with (5.22) when $h_1 = h_2 = h$; it may be noted that f''' term will vanish.

Exercise 5

5.1 Given below are the values of a function $y(x)$ for $x = 1.5(0.5)3.5$. Compute $y'(x)$ and $y''(x)$ at $x = 1.5$, and $x = 1.7$. Use up to 3rd differences only.
Estimate the truncation error in each case.

x	1.5	2.0	2.5	3.0	3.5
y	1.2247	1.4142	1.5811	1.7320	1.8708

(The above data is taken from $f(x) = \sqrt{x}$)

5.2 Compute $y'(x)$ and $y''(x)$ at $x = 3.0$ and $x = 3.25$ from the data of Exercise 5.1.

5.3 Values of $\tan x$ are tabulated below for $x = 10°(10°)50°$.

x	10°	20°	30°	40°	50°
$\tan x$	0.1763	0.3640	0.5774	0.8391	1.1918

Find the value of first and second derivatives for $x = 32°$ and $36°$ using appropriate formulae, up to 3rd differences only.

5.4 Following data is recorded for the speed u of a car, at various times t:

t	0	1	2	3	4	5	6
u	32	26	12	2	32	162	476

Find the time when its speed is minimum. Use up to third difference. Also find the speed at that time. Use FD formula.
(The data is taken from $u = t^4 - 4t^3 + t^2 - 4t + 32$).

5.5 The height of a ball at different times is given as follows:

t	1	2	3	4	5
h	7	15	21	19	3

Find the time when the height of the ball is maximum. Also find the height at that time. Use Stirling's formula taking $t_0 = 3$.
(The data is taken from $h = 3 + 5t^2 - t^3$).

5.6 Derive a formula and the associated truncation error in the formula

$$f'(x_0) = af(x_0) + bf(x_0 + 2h) + cf(x_0 + 3h)$$

when a, b and c are constants.

References and Some Useful Related Books/Papers

1. Jain M.K., Iyengar S.R.K. and Jain R.K., *Numerical Methods for Scientific and Engineering Computation*, New Age International (P) Limited.
2. Kopchenova N.V. and Maron I.A., *Computational Mathematics*, MIR Publishers.

6

Numerical Integration

6.1 Introduction

In this chapter we shall be concerned with the numerical evaluation of the definite integral $I = \int_a^b f(x)dx$, where the function $f(x)$ is defined at each point between $x = a$ and $x = b$. In case of a finite discontinuity, we can break the integral about the point of discontinuity and evaluate the integrals separately. We also assume that $f(x)$ possesses same sign (positive(+ve) or negative(−ve)) throughout in (a, b). Otherwise, we can divide the interval (a, b) into sub-intervals where the function remains positive (above x-axis) or negative (below x-axis). In Numerical Analysis, the problem of definite integral may arise in two ways. Either, the numerical values of $f(x)$ are provided at a finite number of points in (a, b), including at a and b or the function $f(x)$ is prescribed so that we are free to choose the points in (a, b) and get the desired values of $f(x)$. In both the cases we make use of the function values at the specified points, for evaluation of the definite integral.

Geometrically, the integral $I = \int_a^b f(x)dx$ represents the area under the curve $y = f(x)$ enclosed by the vertical lines $x = a$, $x = b$ and the x-axis. The process of determining the area in a plane is called 'Quadrature' in engineering parlance and process of determining the volume as 'Cubature'. However, we shall be confined to quadrature only.

6.2 Methodology for Numerical Integration

Let us suppose that the function values are known at $x = a$, $x = b$ and $(n-1)$ internal points in (a, b), namely $x = x_i$, $i = 1(1)n-1$. Further, denoting $x_0 = a$ and $x_n = b$, let us assume that $a = x_0 < x_1 < x_2 \ldots. < x_n = b$. These points on the x-axis are called pivotal or nodal points. Thus, there are $(n+1)$ nodal points, viz., $x_i = i = 0(1)n$ and n sub-intervals $[x_i, x_{i+1}]$, $i = 0(1)n-1$.

The common approach for evaluation of the integral would be to approximate the function $f(x)$ by a polynomial and integrate it. But instead of approximating the function by a single polynomial globally over the entire domain $a \le x \le b$, it is approximated in piecewise manner. That is, we fit a polynomial $P(x)$ over k sub-intervals passing through the points $(x_i,\ y_i)$, $i = 0(1)k$ and evaluate the integral $\mathrm{I}_1 = \int_{x_0}^{x_k} f(x)dx = \int_{x_0}^{x_k} P(x)dx$. Obviously it covers k intervals, $(x_i,\ x_{i+1})$, $i = 0(1)k-1$. The process is repeated for next k intervals and so on until the entire domain $a = x_0 \le x \le x_n = b$ is covered. It is essential that n should be a multiple of k, say, $m \times k = n$, so that the integration process may be carried out exactly m times to cover the whole domain.

We will illustrate the methods by representing the approximating polynomial in two forms: one, in the Lagrange's form and the other in the form of Newton's FD formula. In the Lagrange's form, $P(x) = \sum_{i=0}^{k} L_i(x) y_i$, so that

$$\int_{x_0}^{x_k} f(x)dx = \int_{x_0}^{x_k} \sum_{i=0}^{k} L_i(x) y_i dx$$

$$= \sum_{i=0}^{k} y_i \int_{x_0}^{x_k} L_i(x)dx, \tag{6.1}$$

where $$L_i(x) = \frac{(x-x_0)(x-x_1)\cdots(x-x_{i-1})(x-x_{i+1})\cdots(x-x_k)}{(x_i-x_0)(x_i-x_1)\cdots(x_i-x_{i-1})(x_i-x_{i+1})\cdots(x_i-x_k)} \text{ and } y_i = f(x_i).$$

When the values of the function $y = f(x)$ are provided at equally-spaced abscissas, say $h = x_i - x_{i-1}$, $i = 1(1)n$, then the approximating polynomial may be represented by a finite difference formula. In terms of Newton's FD formula it can be represented as,

$$P(x) = y_0 + \frac{(x-x_0)}{h}\Delta y_0 + \frac{(x-x_0)(x-x_1)}{2!h^2}\Delta^2 y_0 + \cdots$$

$$+ \frac{(x-x_0)(x-x_1)\cdots(x-x_{k-1})}{k!h^k}\Delta^k y_0 \tag{6.2}$$

or $$P(p) = y_0 + p\Delta y_0 + \frac{p(p-1)}{2!}\Delta^2 y_0 + \cdots + \frac{p(p-1)\cdots(p-k+1)}{k!}\Delta^k y_0, \tag{6.2a}$$

where $x = x_0 + ph$.

Using $P(x)$ or $P(p)$ the integral may be approximated as,

$$\int_{x_0}^{x_k} f(x)dx = \int_{x_0}^{x_k} P(x)dx \tag{6.3}$$

or $$= h\int_{0}^{k} P(p)dp, \quad \because dx = hdp. \tag{6.3a}$$

The error in the above approximation may be obtained by integrating $R(x)$, w.r.t. x from x_0 to x_k, i.e.,

$$E = \int_{x_0}^{x_k} R(x)dx = \int_{x_0}^{x_k} (x-x_0)(x-x_1)\cdots(x-x_k)\frac{f^{k+1}(\xi)}{(k+1)!}dx, \; x_0 \le \xi \le x_k,$$

$$= \frac{f^{k+1}(\xi)}{(k+1)!}\int_{x_0}^{x_k} (x-x_0)(x-x_1)\cdots(x-x_k)dx \tag{6.4}$$

or $$= \frac{h^{k+2} f^{k+1}(\xi)}{(k+1)}\int_{0}^{k} p(p-1)\cdots(p-k+1)dp. \tag{6.4a}$$

The total error (TE) in computing integral $I = \int_{x_0}^{x_n} f(x)dx$ would be sum of the moduli of the errors, the number of times an integration formula is used to cover the interval (x_0, x_n).

We shall now discuss some methods, that are based on equally-spaced abscissas. Let us suppose that the interval (a, b) is subdivided into n sub-intervals, each of width h, such that,

$$h = (b-a)/n; \; x_i - x_{i-1} = h, \; i = 1(1)n,$$

and also, $a = x_0 < x_1 < x_2 \ldots . < x_n = b.$

6.3 Rectangular Rule

In the rectangular rule (or method), the function $f(x)$ is approximated by a constant value of $f(x)$ at $x = x_0$, i.e., $y_0 = f(x_0)$ in the first interval (x_0, x_1). In the second interval it is approximated by $y_1 = f(x_1)$ and so on until in the last n^{th} interval (x_{n-1}, x_n) it is approximated by $y_{n-1} = f(x_{n-1})$. See Fig. 6.1a and 6.1b.

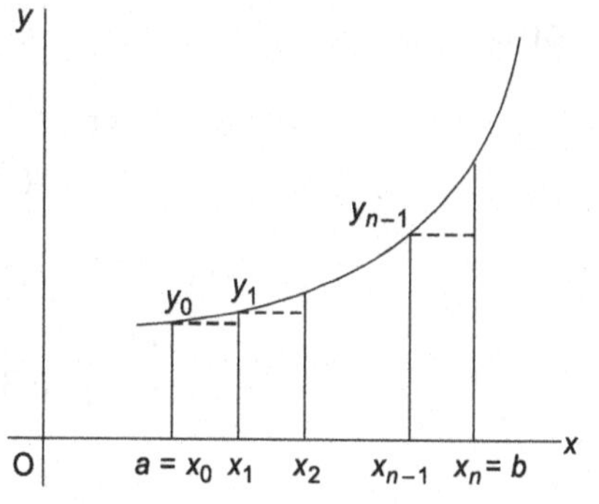

Figure 6.1a. Approximation by rectangles (monotonically increasing function).

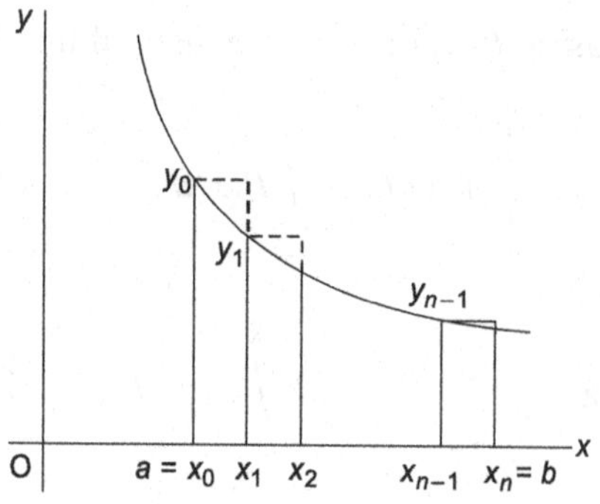

Figure 6.1b. Approximation by rectangles (monotonically decreasing function).

Thus the area in an interval is approximated by the area of a rectangle. For example in the first interval,

$$\int_{x_0}^{x_1} f(x)dx \simeq y_0(x_1 - x_0) = hy_0. \tag{6.5}$$

It is known as 'rectangular rule'.

Adding up such areas over the n intervals, we get

$$\int_{x_0}^{x_n} f(x)dx = \int_{x_0}^{x_1} + \int_{x_1}^{x_2} + \cdots + \int_{x_{n-1}}^{x_n}$$

$$= h(y_0 + y_1 + y_2 + \ldots . y_{n-1}). \tag{6.5a}$$

This is known as 'composite formula' for n intervals.

It may be noted that if the function $f(x)$ is monotonically increasing (Fig. 6.1a), then

$$\int_a^b f(x)dx > h\,(y_0 + y_1 + \ldots . + y_{n-1}).$$

Similarly if $f(x)$ is monotonically decreasing, (Fig. 6.1b), then

$$\int_a^b f(x)dx < h\,(y_0 + y_1 + \ldots .. + y_{n-1}).$$

Let us now consider the rectangular rule from Numerical Analysis viewpoint. We will discuss two approaches, viz., (*i*) Based on Lagrange's Formula (*ii*) Based on Newton's FD Formula.

(*i*) Based on Lagrange's Formula

In the first interval $(x_0,\ x_1)$, we fit a polynomial passing through the points $(x_0,\ y_0)$ and $(x_1,\ y_0)$, giving

$$P(x) = \frac{x - x_1}{x_0 - x_1} y_0 + \frac{(x - x_0)}{x_1 - x_0} \cdot y_0 = y_0 \tag{6.6}$$

The approximation (6.6) has an error,

$$R(x) = (x - x_0) f'(\xi),\ x_0 \le \xi \le x_1. \tag{6.7}$$

Approximating $f(x)$ by $P(x)$ over the interval $(x_0,\ x_1)$ gives,

$$\int_{x_0}^{x_1} f(x)dx = \int_{x_0}^{x} P(x)dx = \int_{x_0}^{x_1} y_0 dx$$

$$= hy_0 \tag{6.8}$$

Formula (6.8) is the rectangular rule. The error in this formula is given by integration of (6.7) between x_0 and x_1, i.e.,

$$E = \int_{x_0}^{x_1} R(x)dx = \int_{x_0}^{x_0+h} (x - x_0) f'(\xi) dx$$

$$= \frac{h^2}{2} f'(\xi),\ x_0 \le \xi \le x_1. \tag{6.9}$$

The composite formula may be written as before

$$\mathrm{I} = \int_{a}^{b} f(x)dx = \int_{x_0}^{x_n} f(x)dx = h\ (y_0 + y_1 + y_2 \ldots.\ + y_{n-1}). \tag{6.9a}$$

In evaluating the integral I, from (6.9a), the formula (6.8) is used n times, once for each of the n intervals. Therefore, the Total Error (TE) in formula (6.9a) would be sum of the moduli of errors for the n intervals. That is, the total error would be,

$$TE = \frac{nh^2}{2} f'(\xi),\ x_0 \le \xi \le x_n \tag{6.10}$$

or $$= \frac{(b-a)h}{2} f'(\xi). \quad \because nh = b-a. \tag{6.10a}$$

(*ii*) Based on Newton's FD Formula

Approximating $f(x)$ in $(x_0,\ x_1)$ by the first term of (6.2) or (6.2a),

$$\int_{x_0}^{x_1} f(x)dx = \int_{x_0}^{x_1} P(x)dx = h\int_0^1 P(p)dp$$

$$= h\int_0^1 y_0\, dp = hy_0. \tag{6.11}$$

The error in the formula is obtained by integrating the next term, i.e.,

$$E = h\int_0^1 p\Delta y_0 dp = \frac{h}{2}\Delta y_0 \tag{6.12}$$

$$= \frac{h^2}{2} f'(\xi), \qquad \because \frac{\Delta y_0}{h} = f'(\xi). \tag{6.13}$$

Rest of the analysis will be same as in (*i*).

6.4 Trapezoidal Rule

In the trapezoidal rule, the function $f(x)$ is approximated over the interval $(x_0,\ x_1)$ by a straight line (polynomial of degree one) passing through the points $(x_0,\ y_0)$ and $(x_1,\ y_1)$, See Fig. 6.2.

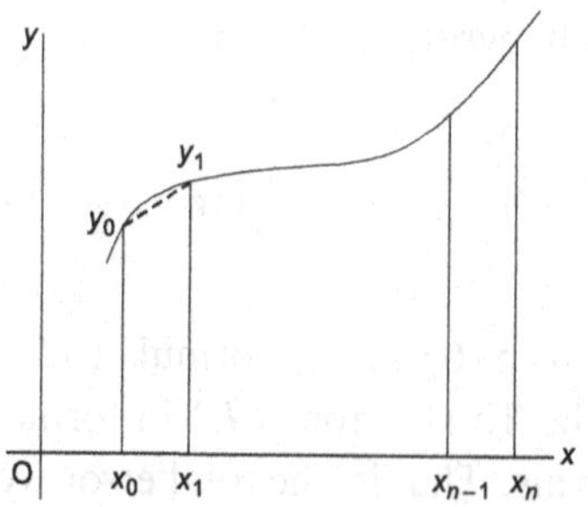

Figure 6.2. Approximation by trapezium.

Thus the area under the curve $y = f(x)$ is approximated by the area of the trapezium, i.e.,

$$\int_{x_0}^{x_1} f(x)dx = h\left(\frac{y_0 + y_1}{2}\right). \tag{6.14}$$

The value of the integral I over $(a,\ b)$ may be obtained repeating the above process for all the n intervals. Now let us discuss it analytically.

(*i*) Based on Lagrange's Method

By Lagrange's method, the equation of the straight line passing through the points $(x_0,\ y_0)$ and $(x_1,\ y_1)$ is given by,

$$y = P(x) = \frac{x - x_1}{x_0 - x_1} y_0 + \frac{x - x_0}{x_1 - x_0} y_1$$

$$= \frac{1}{h}[(x - x_0)y_1 - (x - x_1)y_0], \tag{6.15}$$

with associated error

$$R(x) = (x - x_0)(x - x_1)\frac{f''(\xi)}{2!},\ x_0 \leq \xi \leq x_1. \tag{6.16}$$

The area over the first interval is thus approximated as,

$$\int_{x_0}^{x_1} f(x)dx = \frac{1}{h}\int_{x_0}^{x_1} [(x - x_0)y_1 - (x - x_1)y_0]\,dx$$

$$= \frac{1}{2h}\left[(x - x_0)^2 y_1 - (x - x_1)^2 y_0\right]_{x_0}^{x_1} = \frac{h}{2}(y_0 + y_1). \tag{6.17}$$

This is 'trapezoidal rule'.

The error in trapezoidal rule will be given by integrating (6.16) from x_0 to x_1, i.e.,

$$E = \int_{x_0}^{x_1} (x - x_0)(x - x_1)\frac{f''(\xi)}{2}dx$$

$$= \frac{f''(\xi)}{2} \cdot h^3 \int_0^1 p(p-1)dp \quad \because dx = hdp$$

$$= -\frac{h^3}{12} f''(\xi), \; x_0 \le \xi \le x_1. \tag{6.18}$$

The composite formula for integral I may be obtained by applying the trapezoidal rule over all the n intervals and adding, i.e.,

$$\int_{a=x_0}^{b=x_n} f(x)dx = \frac{h}{2}[y_0 + y_n + 2\,(y_1 + y_2 + \ldots. + y_{n-1})]. \tag{6.19}$$

The upper bound for the error in the composite formula (6.19) may be taken as n times the largest error in any interval, i.e.,

$$TE = -\frac{nh^3}{12} f''(\xi), \; x_0 \le \xi \le x_n \tag{6.20}$$

$$= -\frac{(b-a)h^2}{12} f''(\xi). \qquad \because nh = b - a. \tag{6.21}$$

(*ii*) Based on Newton's FD Formula

$$\int_{x_0}^{x_1} f(x)dx = h \int_0^1 (y_0 + p\Delta y_0)dp$$

$$= h\left[y_0 + \frac{1}{2}(y_1 - y_0)\right]$$

$$= \frac{h}{2}(y_0 + y_1) \tag{6.22}$$

The error is given by,

$$E = h \int_0^1 \frac{p(p-1)}{2} \Delta^2 y_0 \, dp$$

$$= -\frac{h}{12} \Delta^2 y_0, \tag{6.23}$$

or
$$= -\frac{h^3}{12} f''(\xi), \quad \because \frac{\Delta^2 y_0}{h^2} = f''(\xi). \tag{6.24}$$

The composite formula and the associated error in it may be written as in (*i*).

6.5 Simpson's $\frac{1}{3}^{\text{rd}}$ Rule

Simpson's $\frac{1}{3}^{\text{rd}}$ rule (or formula) is a widely used formula. Generally when Simpson's formula is mentioned it implies $\frac{1}{3}^{\text{rd}}$ rule. In this formula, the function $f(x)$ is approximated by a quadratic, $P(x)$ passing through three points $(x_0,\ y_0)$, $(x_1,\ y_1)$ and $(x_2,\ y_2)$, covering two sub-intervals, namely $(x_0,\ x_1)$ and $(x_1,\ x_2)$. Integration is performed of the polynomial function $y = P(x)$ from x_0 to x_2. Obviously, as the formula operates on two sub-intervals, the interval $[a,\ b]$ should be subdivided into even number of sub-intervals, i.e., n should be even.

(*i*) Based on Lagrange's Formula

The quadratic passing through $(x_0,\ y_0)$, $(x_1,\ y_1)$, $(x_2,\ y_2)$ may be written as,

$$P(x) = \frac{(x-x_1)(x-x_2)}{(x_0-x_1)(x_0-x_2)}y_0 + \frac{(x-x_0)(x-x_2)}{(x_1-x_0)(x_1-x_2)}y_1 + \frac{(x-x_0)(x-x_1)}{(x_2-x_0)(x_2-x_1)}y_2 \tag{6.25}$$

The error in the quadratic approximation (6.25) is given by,

$$R(x) = (x-x_0)(x-x_1)(x-x_2)\frac{f'''(\xi)}{3!},\ x_0 \le \xi \le x_2. \tag{6.26}$$

As the function $f(x)$ is approximated by $P(x)$ we have,

$$\int_{x_0}^{x_2} f(x)dx = \int_{x_0}^{x_2} P(x)dx = \int_{x_0}^{x_2} \sum_{i=0}^{2} L_i(x)y_i\, dx, \tag{6.27}$$

where $L_i(x)$, $i = 0(1)2$ are Lagrange's coefficients.

We find the integration of $L_0(x)$, $L_1(x)$ and $L_2(x)$, as follows:

$$\int_{x_0}^{x_2} \frac{(x-x_1)(x-x_2)}{(x_0-x_1)(x_0-x_2)}dx = \frac{h}{2}\int_0^2 (p-1)(p-2)dp = \frac{h}{3}.$$

$$\int_{x_0}^{x_2} \frac{(x-x_0)(x-x_2)}{(x_1-x_0)(x_1-x_2)}dx = h\int_0^2 p(p-2)dp = \frac{4h}{3}.$$

$$\int_{x_0}^{x_2} \frac{(x-x_0)(x-x_1)}{(x_2-x_0)(x_2-x_1)}dx = \frac{h}{2}\int_0^2 p(p-1)dp = \frac{h}{3}.$$

Substituting the above values in (6.27) we get the required Simpson's $\frac{1}{3}^{\text{rd}}$ formula,

$$\int_{x_0}^{x_2} f(x)dx = \frac{h}{3}(y_0 + 4y_1 + y_2). \tag{6.28}$$

The composite formula for the entire interval may be written as,

$$\int_{x_0}^{x_n} f(x)dx = \frac{h}{3}[y_0 + y_n + 4(y_1 + y_3 + \ldots + y_{n-1}) + 2(y_2 + y_4 + \ldots + y_{n-2})], \tag{6.29}$$

n is even.

Let us now consider the error in the formula (6.28), which should be obtained by integrating (6.26), as

$$E = \frac{1}{6} f'''(\xi) \int_{x_0}^{x_2} (x - x_0)(x - x_1)(x - x_2)dx, \quad x_0 \le \xi \le x_2.$$

$$= \frac{h^4}{6} f'''(\xi) \int_0^2 p(p-1)(p-2)dp$$

Putting $p - 1 = t$ in the above integral gives,

$$E = \frac{h^4}{6} f'''(\xi) \int_{-1}^{1} (t+1)\, t(t-1)dt$$

$= 0$, since integrand is an odd function.

It shows that the formula (6.28) is exact not only for the quadratic passing through the points (x_0, y_0), (x_1, y_1) and (x_2, y_2) but also is exact for a cubic satisfying these points. Hence the error will be given by

$$R(x) = (x - x_0)(x - x_1)(x - x_2)(x - x_r)K(\xi)$$

where x_r may be any of x_0, x_1 or x_2 so that $R(x)$ remains zero for these points.

The best way, however, would be to square the factor in the middle, i.e., $(x - x_1)$ as the error becomes symmetric about $x = x_1$ (See note below). Hence we take,

$$R(x) = (x - x_0)(x - x_1)^2(x - x_2)\frac{f^{iv}(\xi)}{4!}, \; x_0 \le \xi \le x_2. \tag{6.30}$$

Now, integrating (6.30),

$$E = \int_{x_0}^{x_2} R(x)dx = \frac{h^5}{24} f^{iv}(\xi) \int_{-1}^{1} (t+1)\, t^2 (t-1) dt$$

$$= \frac{h^5}{12} f^{iv}(\xi) \int_{0}^{1} t^2(t^2 - 1)dt, \text{ since integrand is an even function}$$

$$= -\frac{h^5}{90} f^{iv}(\xi), \; x_0 \le \xi \le x_2. \tag{6.31}$$

As the Simpson's formula will be invoked $n/2$ times on the composite formula (6.29), the over-all error will be given by,

$$TE = -\frac{nh^5}{180} f^{iv}(\xi), \; x_0 \le \xi \le x_n. \tag{6.32}$$

Expressed in terms of limits of integration it can be written as,

$$TE = -\frac{(b-a)h^4}{180} f^{iv}(\xi), \; a \le \xi \le b. \quad \because nh = b - a. \tag{6.33}$$

Note: We can square any of the three factors in $R(x)$, and arriving at the same result. For instance, if we square $(x - x_2)$, then

$$\frac{1}{h^5} \int_{0}^{2} (x - x_0)(x - x_1)(x - x_2)^2 dx = \int_{-1}^{1} (t+1)\, t(t-1)^2 dt$$

$$= \int_{-1}^{1} (t+1)t^2(t-1)dt - \int_{-1}^{1} (t+1)\, t(t-1)dt.$$

The second integral is zero as the integrand is an odd function again. This can be shown for any point in $[x_0, \, x_2]$.

(*ii*) Based on Newton's FD formula

Using first three terms of the FD formula.

$$\int_{x_0}^{x_2} f(x)dx = h\int_0^2 \left[y_0 + p\Delta y_0 + \frac{p(p-1)}{2}\Delta^2 y_0\right] dp$$

$$= h\left[py_0 + \frac{p^2}{2}(y_1 - y_0) + \frac{1}{2}\left(\frac{p^3}{3} - \frac{p^2}{2}\right)(y_0 - 2y_1 + y_2)\right]_{p=0}^{p=2}$$

$$= \frac{h}{3}(y_0 + 4y_1 + y_2).$$

As the integral of the next neglected term will be zero, the error will be given by,

$$E = \frac{h}{24}\int_0^2 p(p-1)(p-2)(p-3)\Delta^4 y_0\, dp$$

$$= -\frac{h}{90}\Delta^4 y_0$$

In the composite formula, the error will be,

$$TE = -\frac{nh}{180}\Delta^4 y_0 \tag{6.34}$$

or

$$= -\frac{(b-a)}{180}\Delta^4 y_0 \tag{6.35}$$

Expressed in terms of function (6.35) becomes

$$TE = -\frac{(b-a)}{180}h^4 f^{iv}(\xi), \quad \because \quad \frac{\Delta^4 y_0}{h^4} = f^{iv}(\xi). \tag{6.35a}$$

6.5.1 Comments on Simpson's $\frac{1}{3}^{\text{rd}}$ rule

Although the function $f(x)$ is approximated by a quadratic in Simpson's formula, it is exact even when the function is a cubic. This is indicated by error reducing to zero as explained above. For example consider $f(x) = x^3$. The points on curve $y = x^3$ may be (1, 1), (2, 8), (3, 27). Then the exact value of the integral is,

$$\mathrm{I} = \int_1^3 x^3 dx = \left.\frac{x^4}{4}\right|_1^3 = 20.$$

From Simpson's formula,

$$\mathrm{I} = \frac{1}{3}(1+4\times 8+27) = 20.$$

6.6 Simpson's $\frac{3}{8}^{\text{th}}$ Rule

In this method, the area is computed over three sub-intervals at a time, approximating the function $f(x)$ by a cubic passing through four points $(x_0,\, y_0)$, $(x_1,\, y_1)$, $(x_2,\, y_2)$ and $(x_3,\, y_3)$. After integrating the relevant polynomial over $(x_0,\, x_3)$, following formula is obtained:

$$\int_{x_0}^{x_3} f(x)dx = \frac{3h}{8}(y_0+3y_1+3y_2+y_3), \tag{6.36}$$

with associated error,

$$E = -\frac{3h^5}{80}f^{iv}(\xi),\; x_0 \le \xi \le x_3. \tag{6.36a}$$

Naturally, total number of sub-invervals should be in multiples of three. There does not seem any merit in this formula over the $\frac{1}{3}^{\text{rd}}$ rule, accuracy-wise.

6.7 Weddle's Rule

In this method six intervals are covered by fitting a polynomial of degree six passing through seven points, $(x_0,\, y_0)$ through $(x_6,\, y_6)$ and integration carried between x_0 and x_6. Similarly, if Newton's FD formula is used, terms up to $\Delta^6 y_0$ are taken into account. We get

$$\int_{x_0}^{x_6} f(x)dx = \frac{h}{140}(41y_0+216y_1+27y_2+272y_3+27y_4+216y_5+41y_6). \tag{6.37}$$

and the associated error is,

$$E = -\frac{9h^9}{1400}f^{viii}(\xi),\; x_0 \le \xi \le x_6. \tag{6.37a}$$

Since here also the function $f(x)$ is approximated by an even degree polynomial like Simpson's $\frac{1}{3}^{\text{rd}}$ rule (See Sec. 6.5), the error is computed by squaring the middle term, i.e.,

$$E = \frac{h^9}{8!} f^{viii}(\xi) \int_{-3}^{3} (t+3)(t+2)(t+1)\, t^2 (t-1)(t-2)(t-3) dt.$$

Example 6.1

Evaluate the integral $I = \int_0^1 \sqrt{1-x^2}dx$ taking $h = 0.25$, by (*i*) Trapezoidal (*ii*) Simpson's Rule.
Compute up to 4 decimals and round the answer to 3 decimals.

Solution

x	0	0.25	0.50	0.75	1.00
$y = \sqrt{1-x^2}$	1.0	0.9682	0.8660	0.6614	0.0

(*i*) Trapezoidal Rule

$$I = \frac{0.25}{2}[1 + 0 + 2\,(0.9682 + 0.8660 + 0.6614)\,]$$

$$= 0.7489 \simeq 0.749$$

(*ii*) Simpson's Rule

$$I = \frac{0.25}{3}\,[1.0 + 0 + 4(0.9682 + 0.6614) + 2 \times 0.8660]$$

$$= \frac{1}{3 \times 4}(1 + 4 \times 1.6296 + 1.7320)$$

$$= 0.7709 \simeq 0.771$$

Note: Correct value is $\frac{\pi}{4} = 0.785$

Example 6.2

Evaluate the integral $I = \int_0^{1.2} e^{-x^2} dx$ by (i) Trapezoidal Rule (*ii*) Simpson's Rule. Take $h = 0.2$.
Also estimate the error in the Trapezoidal Rule.

Solution

x	0.0	0.2	0.4	0.6	0.8	1.0	1.2
$y = e^{-x^2}$	1.000	0.9608	0.8521	0.6977	0.5273	0.3679	0.2369

(*i*) Trapezoidal Rule

$$I = \frac{0.2}{2}\ [1+0.2369+2\ (0.9608+0.8521+0.6977+0.5273+0.3679)]$$

$$= 0.1 \times (1.2369 + 2 \times 3.4058)$$

$$= 0.8048$$

(*ii*) Simpson's Rule

$$I = \frac{0.2}{3}[1 + 0.2369 + 4(0.9608 + 0.6977 + 0.3679) + 2\ (0.8521 + 0.5273)]$$

$$= \frac{0.2}{3}[1.2369 + 4 \times 2.0264 + 2 \times 1.3794]$$

$$= \frac{0.2 \times 12.1013}{3}$$

$$= 0.80675 \simeq 0.8068$$

Note: Correct value is 0.8067.

Error in Trapezoidal Rule is given by,

$$TE = -\frac{nh^3}{12} f''(\xi),\ 0 \le \xi \le 1.2$$

or $$= -\frac{(b-a)h^2}{12} f''(\xi),$$

Here, $f(x) = e^{-x^2}$; $f'(x) = -2xe^{-x^2}$; $f''(x) = -2e^{-x^2}(1-2x^2)$
Since $f''(x)$ is maximum at $x = 0$,

$$f''(\xi) = -2e^0(1-0) = -2$$

$\therefore$ $$TE = -\frac{(1.2-0) \times 0.04 \times (-2)}{12}$$

$$= 0.008$$

Note: To check maximum of $f''(x)$ the stationary points of $f''(x)$ also be obtained by differentiating $f''(x)$ once more and putting $f'''(x) = 0$. But generally, when the function is monotonic, the global maximum lies at one of the end points.

Example 6.3

Find the minimum number of intervals that will be required to evaluate the integral $I = \int_0^1 \frac{dx}{1+x}$ by (i) Trapezoidal Rule (*ii*) Simpson's Rule, so that the error does not exceed 10^{-4}. Also find the value of integral by any method and compare with the analytical solution.

Solution $f(x) = \frac{1}{1+x}$; $f'(x) = \frac{-1}{(1+x)^2}$; $f''(x) = \frac{2}{(1+x)^3}$; $f'''(x) = -\frac{6}{(1+x)^4}$;

$$f^{iv}(x) = \frac{24}{(1+x)^5}$$

(*i*) Trapezoidal Rule

Maximum value of $f''(x)$ will be at $x = 0$, so that $f''(x) = 2$.

We have,

$$\left| -\frac{(b-a)h^2}{12} \times f''(\xi) \right| \le 0.0001$$

or
$$\frac{(1-0)h^2}{12} \times 2 \le 0.0001$$

or
$$h^2 \le 0.0006$$

or
$$h = 0.025 \text{ (approx.)}$$

Number of intervals required $n = \frac{b-a}{h} = \frac{1}{0.025} = 40$

(*ii*) Simpson's Rule

$f^{iv}(x)$ will be max. at $x = 0$, so that $f^{iv}(\xi) = 24$.

We have

$$\left| -\frac{(1-0)h^4}{180} \times 24 \right| \le 0.0001$$

or
$$h^4 \le 0.00075 \text{ or } h^2 \le 0.0274$$

or
$$h \le 0.16$$

Number of intervals required $n = \dfrac{1-0}{0.16} = 7$ (approx.)

As the number of intervals should be even we can take $n > 7$, i.e., $n = 8$ (say). However for ease in computation we will take $n = 10$ so that $h = 0.1$.

Note: We will evaluate the integral by Simpson's method because we have to evaluate only 11 function values while in the Trapezoidal rule we will have to use 41 function values.

x	0	0.1	0.2	0.3	0.4	0.5	0.6
$f(x) = \dfrac{1}{1+x}$	1.0000	0.9091	0.8333	0.7692	0.7143	0.6667	0.6250

x	0.7	0.8	0.9	1.0
$f(x)$	0.5882	0.5556	0.5263	0.5000

Using Simpson's rule

$$\begin{aligned} I &= \frac{0.1}{3}\,[1.0+0.5000+4\,(0.9091+0.7092+0.6667+0.5882 \\ &\quad +0.5263) + 2\,(0.8333+0.7143+0.6250+0.5556] \\ &= \frac{0.1}{3}[1.5000 + 4 \times 3.4595 + 2 \times 2.7282] \\ &= \frac{0.1}{3}\,(20.7944) \\ &= 0.69315 \end{aligned}$$

Analytically $\displaystyle\int_0^1 \frac{dx}{1+x} = [\ln(1+x)]_0^1 = \ln 2 = 0.69315$

This is exactly the same as computed by Simpson's Rule.

Example 6.4

Integrate the function $Si\,(x) = \dfrac{\sin x}{x}$ between $x = 0$ and $x = 0.8$ by (*i*) Trapezoidal Rule (*ii*) Simpson's Rule, by subdividing the interval into four equal parts.

Solution $$I = \int_0^{0.8} \frac{\sin x}{x}\,dx$$

x (radian)	$\sin(x)$	$Si(x) = \dfrac{\sin x}{x}$
0	0	1000
0.2	0.1988	0.994
0.4	0.3894	0.973
0.6	0.5645	0.941
0.8	0.7173	0.897

(*i*) Trapezoidal Rule

$$\mathrm{I} = \frac{0.2}{2}\,[1.0{+}0.897{+}2\,(0.994{+}0.973{+}0.941)]$$

$$= 0.1[1.897 + 2 \times 2.908]$$

$$= 0.7713$$

(*ii*) Simpson's Rule

$$\mathrm{I} = \frac{0.2}{3}\,[1.0 + 0.897 + 4(0.994 + 0.941) + 2 \times 0.973]$$

$$= \frac{0.2}{3}\,(1.897{+}7.740{+}1.946)$$

$$= 0.7722$$

Note: Correct answer is 0.7721.

6.8 Open-Type Formulae

The formulae discussed in Secs (6.3) through (6.7) are 'closed-type' formuale in that we had approximated the function $f(x)$ by a polynomial $P(x)$, passing through $(k+1)$ points $(x_i,\ y_i)$, $i = 0(1)k$; $k > 0$. The polynomial was then integrated between x_0 to x_k. But in open-type formulae, the approximating polynomial passes through the points $(x_i,\ y_i)$, $i = 1(1)k-1$, $k > 1$ i.e., excluding end points; and is integrated from x_0 to x_k to approximate the integral $\int_{x_0}^{x_k} f(x)dx$. The approximating polynomial is of highest degree $k-2$ and may or may not pass through the end points $(x_0,\ y_0)$ and $(x_k,\ y_k)$.

The error in the polynomial approximation is given by,

$$R(x) = (x - x_1)(x - x_2)\ldots(x - x_{k-1})\frac{f^{k-1}(\xi)}{(k-1)!},\ x_0 < \xi < x_k. \tag{6.38}$$

The error in the integration formula would be

$$E = \frac{f^{k-1}(\xi)}{(k-1)!}\int_{x_0}^{x_k}(x-x_1)(x-x_2)\cdots(x-x_{k-1})dx \tag{6.39}$$

or

$$= \frac{f^{k-1}(\xi)h^k}{(k-1)!}\int_{0}^{k}(p-1)(p-2)\cdots(p-k+1)dp. \tag{6.39a}$$

As explained earlier (Sec. 6.5) when k is even, the integral (6.39) or (6.39a) vanishes and we have to modify the error expression by squaring the middle term. In case of k even, the error will then be computed by the following integral,

$$E = \frac{f^{k}(\xi)}{k!}\int_{x_0}^{x_k}(x-x_1)(x-x_2)\cdots(x-x_{k/2})^2\cdots(x-x_{k-1})dx \tag{6.40}$$

or

$$= \frac{f^{k}(\xi)h^{k+1}}{k!}\int_{x_0}^{x_k}(p-1)(p-2)\cdots\left(p-\frac{k}{2}\right)^2\cdots(p-k+1)\,dp. \tag{6.40a}$$

Some of the popular open-type formuale alongwith associated error in them are given below:

(*i*) $$\int_{x_0}^{x_2} f(x)dx = 2hy_1 + \frac{h^3}{3}f''(\xi),\ x_0 < \xi < x_2. \tag{6.41}$$

(*ii*) $$\int_{x_0}^{x_3} f(x)dx = \frac{3h}{2}(y_1+y_2) + \frac{3h^3}{4}f''(\xi),\ x_0 < \xi < x_3. \tag{6.42}$$

(*iii*) $$\int_{x_0}^{x_4} f(x)dx = \frac{4h}{3}(2y_1-y_2+2y_3) + \frac{14}{45}h^5 f^{iv}(\xi),\ x_0 < \xi < x_4. \tag{6.43}$$

The formula (6.41) is known as 'mid-point rule', since the function $f(x)$ is approximated in the interval $(x_0,\ x_2)$ by its value y_1 at the point x_1 which is the mid-point of x_0 and x_2. Thus the integral is approximated as,

$$\int_{x_0}^{x_2} f(x)dx = \int_{x_0}^{x_2} y_1 dx = 2hy_1.$$

For computing the error we use formula (6.40a) as formula (6.39a) will give zero value. Therefore, we get,

$$E = \frac{f''(\xi)}{2} \cdot h^2 \int_0^2 (p-1)^2 dp$$

$$= \frac{h^3}{3} f''(\xi), \ x_0 < \xi < x_2.$$

It may be noted that although the function $f(x)$ has been approximated by a constant value (like rectangular rule), the error in the midpoint rule is comparable to trapezoidal rule. In fact, the formula is equivalent to approximating the function by a straight line joining the points (x_0, y_0) and (x_2, y_2). When the function $f(x)$ is given, the midpoint formula may be conveniently applied over one sub-interval of lenghth h, (x_0, x_1) in the following form,

$$\int_{x_0}^{x_1} f(x)dx = hf\left(\frac{x_0 + x_1}{2}\right) + \frac{h^3}{24} f''(\xi), \ x_0 < \xi < x_1. \tag{6.44}$$

Formula (6.42) may be derived by fitting a straight line passing through (x_1, y_1) and (x_2, y_2) and integrating it between x_0 and x_3. Similarly formula (6.43) may be derived approximating $f(x)$ by a quadratic passing through the points (x_1, y_1), (x_2, y_2) and (x_3, y_3) and then integrating from x_0 to x_4. The error term will be obtained using formula (6.40a).

The open-type formulae may be used when the values of the function are not defined at the end points.

6.9 Newton–Cotes (or Cotes) Formulae

The integration formulae discussed so far are, in general, called Newton–Cotes (or Cotes) formulae. A $(k+1)-$ point Cotes formula can be written as,

$$\int_{x_0}^{x_k} f(x)dx = kh\ (a_0 y_0 + a_1 y_1 + \ldots. + a_i y_i + \ldots. \ + a_k y_k), \tag{6.45}$$

where kh is the length of the interval $[x_0, x_k]$, i.e., $x_k - x_0 = kh$.

The weights $a_0, a_1, \ldots. a_k$ are called Cotes numbers and are denoted as $a_i = C_i^k$, $i = 0(1)k$.

There are two main properties of Cotes numbers, *viz.* $(i)\ \sum_{i=0}^{k} a_i = \sum_{i=0}^{k} C_i^k = 1$ $(ii) C_i^k = C_{k-i}^k$.

For example in Simpson's $\frac{1}{3}^{\text{rd}}$ rule,

$$k = 2;\ C_0^2 = \frac{1}{6},\ C_1^2 = \frac{4}{6},\ C_2^2 = \frac{1}{6}.\ \text{Both properties are satisfied.}$$

Let us verify the above properties in general. We have, $f(x) = \Sigma L_i(x) y_i$, so that

$$\int f(x)dx = \Sigma \int L_i(x)dx\, y_i = kh\, \Sigma a_i y_i$$

or $$kh \cdot a_i = \int_{x_0}^{xk} L_i(x)dx, \text{ where } L_i(x) \text{ is Lagrange's coefficient}$$

or $$a_i = \frac{1}{kh} \int_{x_0}^{k} L_i(x)dx,\ i = 0(1)k;$$

$$L_i(x) = \frac{(x - x_0) \cdots\cdot (x - x_{i-1})(x - x_{i+1}) \cdots (x - x_k)}{(x_i - x_0) \cdots (x_i - x_{i-1})(x_i - x_{i+1}) \cdots (x_i - x_k)}.$$

(*i*) To prove $\Sigma a_i = 1$
Formula (6.45) should be true when $f(x) = x^0 = 1$. That is,

$$\int_{x_0}^{x_k} 1 \cdot dx = kh\ (a_0 + a_1 + \ldots. + a_i + \ldots. + a_k)$$

or $$x_k - x_0 = kh = kh\, \Sigma a_i$$

or $$\Sigma_{a_i} = 1.$$

(*ii*) To prove $C_i^k = C_{k-i}^k$.
Putting $x = x_0 + ph$ and $dx = hdp$

$$C_i^k = a_i = \frac{1}{k} \int_0^k \frac{p(p-1) \cdots (p-i+1)(p-i-1) \cdots (p-k)dp}{i(i-1) \cdots (1)(-1)(-2) \cdots (i-k)}$$

$$= \frac{(-1)^{k-i}}{k} \int_0^k \frac{p(p-1) \cdots (p-i+1)(p-i-1) \cdots (p-k)dp}{(i)!(k-i)!}$$

$$C_{k-i}^{k} = a_{k-i} = \frac{1}{k}\int_{0}^{k} \frac{p(p-1)\cdots(p-k+i+1).(p-k+i-1)\cdots(p-k)dp}{(k-i)(k-i-1)\cdots 1(-1)(-2)\cdots(-i)}$$

$$= \frac{1}{k}\int_{0}^{k} \frac{p(p-1)\cdots(p-k+i-1)(p-k+i-1)\cdots(p-k)dp}{(-1)^{i}(k-i)!(i)!}$$

Now putting $p = k - q$ and $dp = -dq$

$$a_{k-i} = \frac{1}{k}\int_{0}^{k} \frac{(k-q)(k-q-1)\cdots(-q+i+1)(-q+i-1)\cdots(-q)(-dq)}{(-1)^{i}(i)!(k-i)!}$$

$$= \frac{(-1)^{k+2}}{k}\int_{0}^{k} \frac{q(q-1)\cdots(q-i+1)(q-i-1)\cdots(q-k)dq}{(-1)^{i}(i)!(k-i)!}$$

$$= \frac{(-1)^{k-i}}{k}\int_{0}^{k} \frac{q(q-1)\cdots(q-i+1)(q-i-1)\cdots(q-k)dq}{(i)!(k-i)!}.$$

This is same as a_i.

Example 6.5

Evaluate the integral $I = \int_0^1 \ln x dx$ dividing the interval into four equal parts. Compare with exact solution.

Solution As the integrand tends to infinity as $x \to 0$, a closed-type formula can not be used. Therefore, we will use open-type formula.

x	$\ln x$
0	$-\infty$
0.25	−1.3863
0.50	−0.6932
0.75	−0.2877
1.00	0.0

(*i*) Applying mid-point formula

$$\text{I} = 2 \times 0.25\{-1.3863 + (-0.2877)\}$$

$$= -0.5 \times 1.6740 = -0.8870$$

(*ii*) Applying three-point formula

$$\text{I} = \frac{4h}{3}(2y_1 - y_2 + 2y_3)$$

$$= \frac{4 \times 0.25}{3}[2 \times (-1.3863) - (-0.6932) + 2 \times (-0.2877)]$$

$$= -\frac{1}{3} \times 2.6548 = -0.8849$$

Exact solution

$$\int_0^1 \ln x dx = \left[x \cdot \ln x \; - \int \frac{1}{x} \cdot x \, dx\right]_0^1$$

$$= [x \ln x]_0^1 - [x]_0^1$$

$$= 1 \times 0 - 0 \times (-\infty) - 1$$

Since $x \ln x$ is indeterminate at $x = 0$, we find limit,

$$\lim_{x \to 0} x \ln x = \lim_{x \to 0} \frac{\ln x}{1/x} = 0$$

$$\therefore \quad \int_0^1 \ln x dx = -1.$$

6.10 Method of Undetermined Coefficients

Some times we are required to design a quadrature formula according to the prescribed specifications. The basic idea would be the same, in that we approximate the function by various powers of x. For example, if there are $k+1$ parameters to be determined, then we satisfy the formula for $f(x) = x^r$, $r = 0(1)k$. In doing so $k+1$ equations are obtained and their solution provides the values of various parameters. It may be remembered that the formula so obtained will be exact for all polynomials of degree up to k. The method has been illustrated by Examples 6.6 and 6.7.

Example 6.6

Determine the coefficients $a,\ b,\ c$ and the error in the formula

$$\int_{-h}^{h} f(x)dx = af(-h) + bf\,(0) + cf(h).$$

Solution As there are three unknowns we need three equations to determine them. Approximate the function $f(x)$ by $f(x) = x^r, r = 0(1)2$. Substituting in the formula following equations are obtained:

$$f(x) = x^0 = 1 \Rightarrow \qquad a + b + c = 2h$$

$$f(x) = x^1 \qquad \Rightarrow -ha + 0 + hc = 0$$

$$f(x) = x^2 \qquad \Rightarrow h^2a + 0 + h^2c = \frac{2}{3}h^3$$

The solution of the above equations gives,

$$a = c = \frac{h}{3};\ b = \frac{4h}{3}.$$

The required formula becomes,

$$\int_{-h}^{h} f(x)dx = \frac{h}{3}[f(-h) + 4f(0) + f(h)]$$

This of course is Simpson's $\frac{1}{3}$**rd** rule.
In order to compute the error, we evaluate

$$E = \int_{-h}^{h} (x+h)x(x-h)dx \cdot \frac{f'''(\xi)}{3!}$$

But this gives zero; therefore as explained before,

$$E = \int_{-h}^{h} (x+h)x^2(x-h)dx \cdot \frac{f^{iv}(\xi)}{4!}$$

$$= h^5 \int_{-1}^{1} (p+1)p^2(p-1)dp \cdot \frac{f^{iv}(\xi)}{4!}, \text{ putting } \frac{x}{h} = p$$

$$= -\frac{h^5}{90} f^{iv}(\xi), \; -h \le \xi \le h.$$

Alternatively, the error can also be computed in the following manner.

$$E = C\frac{f^{iv}(\xi)}{4!},$$

where, $$C = \int_{-h}^{h} f(x)dx - \frac{h}{3}[f(-h) + 4f(0) + f(h)]$$

Since C will be zero for polynomials of degree up to 3, we satisfy the formula for $f(x) = x^4$, giving

$$C = \int_{-h}^{h} x^4 dx - \frac{h}{3}[h^4 + 0 + h^4] = -\frac{4}{15}h^5, \text{ and}$$

$$E = -\frac{4}{15}h^5 \cdot \frac{f^{iv}(\xi)}{4!}$$

$$= -\frac{h^5}{90} f^{iv}(\xi).$$

Example 6.7

Derive the generalised trapezoidal formula,

$$\int_{x_0}^{x_1} f(x)dx = af_0 + bf_1 + c(f_1' - f_0').$$

Also find the error.

Solution Since there are three unknowns, we satisfy the formula for $f(x) = x^r$, $r = 0, 1, 2$.

Following equations are obtained,

$$f(x) = x^0 \Rightarrow \quad a + b + 0 \cdot c = x_1 - x_0 = h$$

$$f(x) = x \Rightarrow x_0 a + x_1 b + 0 \cdot c = \frac{h}{2}(x_1 + x_0) \quad \because \; x_1^2 - x_0^2 = h(x_1 + x_0)$$

$$f(x) = x^2 \Rightarrow x_0^2 a + x_1^2 b + 2hc = \frac{1}{3}(x_1^3 - x_0^3) = \frac{h}{3}(x_1^2 + x_0 x_1 + x_0^2)$$

The first two equations give,

$$a = b = \frac{h}{2}.$$

Putting the values of a and b in the third equation,

$$\frac{x_0^2}{2} + \frac{x_1^2}{2} + 2c = \frac{1}{3}(x_1^2 + x_0x_1 + x_0^2)$$

or $$12c = -(x_1 - x_0)^2$$

or $$c = -\frac{h^2}{12}.$$

Therefore, formula becomes,

$$\int_{x_0}^{x_1} f(x)dx = \frac{h}{2}(f_0 + f_1) - \frac{h^2}{12}(f_1' - f_0').$$

For computing error, let $E = \dfrac{Cf'''(\xi)}{3!}$

where $$C = \int_{x_0}^{x_1} f(x)dx - \frac{h}{2}(f_0 + f_1) - \frac{h^2}{12}(f_1' - f_0')$$

Putting $f(x) = x^3$

$$C = \int_{x_0}^{x_1} x^3 dx - \frac{h}{2}(x_0^3 + x_1^3) - \frac{h^2}{12}(x_1^2 - x_0^2) \times 3$$

$$= \frac{1}{2}(x_1^4 - x_0^4) - \frac{h}{2}(x_0^3 + x_1^3) - \frac{h^2}{4}(x_1^2 - x_0^2)$$

$$C = \frac{1}{4}(x_1^2 - x_0^2)(x_1^2 + x_0^2 + h^2) - \frac{h}{2}(x_1 + x_0)(x_1^2 + x_0^2 - x_0x_1)$$

$$= \frac{1}{2}(x_1^2 - x_0^2)(x_1^2 + x_0^2 - x_0x_1) - \frac{1}{2}(x_1^2 - x_0^2)(x_1^2 + x_0^2 - x_0x_1)$$

$$= 0$$

Since error becomes zero for $f(x) = x^3$, it shoud be determined by approximating $f(x) = x^4$, so that

$$E = \frac{Cf^{iv}(\xi)}{4!},\ x_0 \leq \xi \leq x_1.$$

Putting $f(x) = x^4$,

$$C = \frac{1}{5}(x_1^5 - x_0^5) - \frac{h}{2}(x_1^4 + x_0^4) + \frac{h^2}{3}(x_1^3 - x_0^3)$$

$$= \frac{1}{5}(x_1 - x_0)(x_1^4 + x_1^3 x_0 + x_1^2 x_0^2 + x_1 x_0^3 + x_0^4) - \frac{h}{2}(x_1^4 + x_0^4) + \frac{h^2}{3}(x_1^3 - x_0^3)$$

$$= \frac{h}{30}[6(x_1^4 + x_1^3 x_0 + x_1^2 x_0^2 + x_1 x_0^3 + x_0^4) - 15(x_1^4 + x_0^4) + 10(x_1 - x_0) \times (x_1^3 - x_0^3)$$

$$= \frac{h}{30}[x_1^4 + x_0^4 - 4x_1^3 x_0 - 4x_0 x_1^3 + 6x_0^2 x_1^2]$$

$$= \frac{h^3}{30}\left[(x_1^2 - x_0^2)^2 - 4x_0 x_1 (x_1^2 + x_0^2 - 2x_0 x_1)\right] = \frac{h}{30}(x_1 - x_0)^4 = \frac{h^5}{30}$$

or

$$= \frac{h^3}{30}[x_1^2 + x_0^2 - 2x_0 x_1] = \frac{h^3}{30}(x_1 - x_0)^2 = \frac{h^5}{30}$$

$$E = \frac{h^5}{30} \cdot \frac{f^{iv}(\xi)}{4!} = \frac{h^5}{720} f^{iv}(\xi),\ x_0 \le \xi \le x_1.$$

Note: Alternatively, the error can also be computed as follows:

The error in the function approximation will be given $(x - x_0)^2 (x - x_1)^2 \frac{f^{iv}(\xi)}{4!}$, since it will be zero at $x = x_0$ and $x = x_1$ as well as its derivative will be zero at these points. That is $(x - x_0)^2$ $(x - x_1)^2$ interpolates $f(x)$ and $f'(x)$ at $x = x_0$ and $x = x_1$. Thus the error in the above formula is given by,

$$E = \int_{x_0}^{x} (x - x_0)^2 (x - x_1)^2 dx \cdot \frac{f^{iv}(\xi)}{4!} = h^5 \int_0^1 p^2 (p-1)^2 dp \cdot \frac{f^{iv}(\xi)}{4!},\ (\because x = x_0 + ph)$$

$$= \frac{h^5}{720} f^{iv}(\xi),\ x_0 \le \xi \le x_1.$$ See (4.82).

Note: Above computations will be greatly simplified if we shift the origin to x_0, i.e., take $x_0 = 0$ and $x_1 = h$. Derive $\int_0^h f(x)dx = af(0) + b(h) + c\{f'(h) - f'(0)\}$.

6.11 Euler–Maclaurin Formula

Let $F(x)$ be the antiderivative of $f(x)$, then

$$\int f(x)dx = F(x) + c, \tag{6.46}$$

where c is an arbitrary constant.

In operator form we can write,

$$\frac{1}{\mathrm{D}}f(x) = \mathrm{F}(x) + \mathrm{c} \tag{6.47}$$

and $$\mathrm{DF}\,(x) = f(x). \tag{6.48}$$

Also, the definite integral

$$\int_{x_0}^{x_k} f(x)dx = \mathrm{F}(x_k) - \mathrm{F}(x_0). \tag{6.49}$$

If $x_0,\ x_1,\ x_2 \ldots.\ x_k$ are equi-spaced abscissas such that $x_i = x_0 + ih,\ i = 1(1)k$, i.e., h is the common interval and $f(x)$ is a function defined in $[x_0,\ x_k]$, then we should also note that,

$$\Delta \sum_{i=0}^{k-1} f(x_i) = \sum_{i=0}^{k-1} \Delta f(x_i) = \sum_{i=0}^{k-1} [f(x_{i+1}) - f(x_i)] = f(x_k) - f(x_0). \tag{6.50}$$

Now, let us express function $f(x)$ as,

$$f(x) = \Delta\Delta^{-1} f(x) = \Delta(E-1)^{-1} f(x) = \Delta(e^{h\mathrm{D}} - 1)^{-1} f(x)$$

$$= \Delta(e^u - 1)^{-1} f(x), \text{ where } u = h\mathrm{D}$$

$$= \Delta\left(u + \frac{u^2}{2} + \frac{u^3}{6} + \frac{u^4}{24} + \cdots\right)^{-1} f(x)$$

$$= \frac{\Delta}{u}\left(1 + \frac{u}{2} + \frac{u^2}{6} + \frac{u^3}{24} + \cdots\right)^{-1} f(x)$$

$$= \frac{\Delta}{u}\left[1 - \left(\frac{u}{2} + \frac{u^2}{6} + \frac{u^3}{24} + \cdots\right) + \left(\frac{u}{2} + \frac{u^2}{6} + \frac{u^3}{24} + \cdots\right)^2 - \cdots\right] f(x)$$

$$= \frac{\Delta}{u}\left(1 - \frac{u}{2} + \frac{u^2}{12} - \frac{u^4}{720} + \cdots\right) f(x)$$

Replacing u by hD

$$hf(x) = \frac{\Delta}{D}f(x) - \frac{h}{2}\Delta f(x) + \frac{h^2}{12}D\Delta f(x) - \frac{h^4}{720}D^3\Delta f(x) + \cdots$$

$$= \Delta[F(x) + c] - \frac{h}{2}\Delta f(x) + \frac{h^2}{12}\Delta Df(x) - \frac{h^4}{720}\Delta D^3 f(x) + \cdots$$

$$= \Delta F(x) - \frac{h}{2}\Delta f(x) + \frac{h^2}{12}\Delta f'(x) - \frac{h^4}{720}\Delta f'''(x) + \cdots \tag{6.51}$$

Now putting $x = x_0, x_1, \ldots. x_{k-1}$ in (6.51) and adding,

$$h\Sigma f(x_i) = \Sigma\Delta F(x_i) - \frac{h}{2}\Sigma\Delta f(x_i) + \frac{h^2}{12}\Sigma\Delta f'(x_i) - \frac{h^4}{720}\Sigma\Delta f'''(x_i) + \cdots,\ i = 0(1)k-1.$$

$$= F(x_k) - F(x_0) - \frac{h}{2}(f_k - f_0) + \frac{h^2}{12}(f_k' - f_0') - \frac{h^4}{720}(f_k''' - f_0''') + \cdots$$

where $f(x_i) = f_i$ and $f'(x_i) = f_i'$, etc., $i = 0(1)k$.

or $$F(x_k) - F(x_0) = h\left[\frac{f_0 + f_k}{2} + \sum_{i=1}^{k-1} f_i\right] - \frac{h^2}{12}(f_k' - f_0') + \frac{h^4}{720}(f_k''' - f_0''') + \cdots$$

or $$\int_{x_0}^{x_k} f(x)dx = h\left[\frac{f_0 + f_k}{2} + \sum_{i=1}^{k-1} f_i\right] - \frac{h^2}{12}(f_k' - f_0') + \frac{h^4}{720}(f_k''' - f_0''') + \cdots \tag{6.52}$$

This is the required Euler–Maclaurin's Formula. The above formula can also be expressed as

$$\sum_{i=0}^{k} f_i = \frac{f_0 + f_k}{2} + \frac{1}{h}\int_{x_0}^{x_k} f(x)dx + \frac{h}{12}(f_k' - f_0') - \frac{h^3}{720}(f_k''' - f_0''') + \cdots \tag{6.53}$$

Note: Alternatively we can use operator

$$\frac{E^k - 1}{E - 1}f(x_0) = (E^{k-1} + E^{k-2} + \ldots.. + E + 1)f(x_0) = \sum_{i=0}^{k-1} f(x_i)$$

and $$\frac{E^k - 1}{E - 1}f(x_0) = \frac{E^k - 1}{\Delta}f(x_0) = \frac{f(x_k) - f(x_0)}{\Delta},\ \text{etc.}$$

Example 6.8

Using Euler–Maclaurin formula, compute the value of the integral $I = \int_0^{1.0} (\sin x - \ln(1+x) + e^x)dx$ dividing the interval into five equal parts.

Solution $h = \frac{1-0}{5} = 0.2$

x	$\sin x$	e^x	$\ln(1+x)$	$\sin x + e^x - \ln(1+x)$
0.0	0	1	0	1.0000
0.2	0.1987	1.2214	0.1823	1.2378
0.4	0.3894	1.4918	0.3365	1.5447
0.6	0.5646	1.8221	0.4700	1.9167
0.8	0.7174	2.2255	0.5878	2.3551
1.0	0.8415	2.7183	0.6932	2.8666

$$f(x) = \sin x + e^x - \ln(1+x)$$

$$f'(x) = \cos x + e^x - \frac{1}{1+x};\ f'(0) = 1,\ f'(1) = 2.7586$$

$$f''(x) = -\sin x + e^x - \frac{1}{(1+x)^2}$$

$$f'''(x) = -\cos x + e^x - \frac{2}{(1+x)^3};\ f'''(0) = -2,\ f'''(1) = 1.9280$$

Using formula (6.52)

$$\int_{x_0}^{x_6} f(x)dx = 0.2\left[\frac{1+2.8666}{2} + 7.0543\right] - \frac{(0.2)^2}{12}(2.7586-1) + \frac{(0.2)^4}{720}(1.9280+2)$$

$$= 0.2 \times 8.9876 - 0.005862 + 0.87 \times 10^{-5}$$

$$= 1.7975 - 0.005862 + 0.87 \times 10^{-5} = 1.7916$$

Analytical solution.

$$\int_0^1 (\sin x - \ln(1+x) + e^x)dx = [-\cos x + e^x]_0^1 - \int_0^1 \ln(1+x)dx$$

$$= (-0.5403 + 1 + 2.7183 - 1) - [(x+1)\ln(1+x) - x]_0^1$$

$$= 2.1780 - (2\ln 2 - 1) = 2.1780 - 2 \times 0.6932 + 1$$

$$= 2.1780 - 0.3864 = 1.7916$$

Example 6.9

Show that the sum of the cubes of first n natural number, i.e., $1^3 + 2^3 + 3^3 + \dots + n^3 = \left(\frac{n(n+1)}{2}\right)^2$. Use Euler–Maclaurin formula.

Solution Consider $f(x) = x^3$.

We have to find the sum $\sum_{i=1}^{n} x_i^3$ where $x_1 = 1,\ x_n = n,\ x_i = i$ and $h = 1$. Also $f'(x) = 3x^2,\ f''(x) = 6x,\ f'''(x) = 6,\ f^{iv}(x) = 0$.

From formula (6.53)

$$\sum_{i=1}^{n} f_i = \frac{f_1 + f_n}{2} + \frac{1}{h}\int_{x_1}^{x_n} f(x)dx + \frac{h}{12}(f_n' - f_1') - \frac{h^3}{720}(f_n''' - f_1''')$$

$$\sum_{i=1}^{n} i^3 = \frac{1+n^3}{2} + \int_1^n x^3 dx + \frac{1}{12}(3n^2 - 3) - \frac{1}{720}(6-6)$$

$$= \frac{1+n^3}{2} + \frac{1}{4}(n^4 - 1) + \frac{1}{4}(n^2 - 1)$$

$$= \frac{n^4 + 2n^3 + n^2}{4} = \left(\frac{n(n+1)}{2}\right)^2.$$

Example 6.10

Find the sum of the series, using Euler–Maclaurin formula up to third derivative only,

$$\frac{1}{1.00} + \frac{1}{1.01} + \frac{1}{1.02} + \dots + \frac{1}{2.00}.$$

Solution Let us consider the function,

$$f(x) = \frac{1}{1+x}$$

$$f'(x) = -\frac{1}{(1+x)^2};\ f''(x) = \frac{2}{(1+x)^3};\ f'''(x) = \frac{-6}{(1+x)^4}.$$

We have to find the sum $\sum_{i=0}^{100} f(x_i)$, where $h = .01$ and $x_i = ih;\ x_0 = 0,\ x_{100} = 1.$

$$h = 0.01;\ f'(0) = -1;\ f'(1) = -\frac{1}{4};\ f'''(0) = -6;\ f'''(1) = -\frac{6}{16}$$

$$\Sigma f_i = \frac{1+0.5}{2} + \frac{1}{0.01}\int_0^1 \frac{dx}{1+x} + \frac{0.01}{12}\left(-\frac{1}{4}-(-1)\right) - \frac{(0.01)^3}{720}\left(-\frac{6}{16}+6\right)$$

$$= 0.75 + 100 \times \ln 2 + \frac{0.01}{12} \times \frac{3}{4} - \frac{(0.01)^3}{120} \times \left(1 - \frac{1}{16}\right)$$

$$= 0.75 + 100 \times 0.69315 + 0.000625$$

$$= 70.0644$$

6.12 Richardson's Extrapolation

Suppose we want to use a Newton–Cotes formula to evaluate the integral $\mathrm{I} = \int_a^b f(x)dx$ when the interval $(a,\ b)$ has been subdivided into sub-intervals each of width h. A formula is called of order k if the truncation error in the formula is $\mathrm{O}(h^{k+1})$. Let P be a formula of order k and the values of the integral are computed through this formula with common interval h and $h/2$. If the respective computed values are P_0 and P_1, then we can write,

$$\mathrm{I} = \mathrm{P}_0 + c_1 h^{k+1} + c_2 h^{k+2} + \ldots.. \tag{6.54a}$$

$$\mathrm{I} = \mathrm{P}_1 + c_1\left(\frac{h}{2}\right)^{k+1} + c_2\left(\frac{h}{2}\right)^{k+2} + \cdots \tag{6.54b}$$

Eliminating the lowest power of h from (6.54a) and (6.54b),

$$\mathrm{I} = \frac{2^{k+1}\mathrm{P}_1 - \mathrm{P}_0}{2^{k+1}-1} + d_2 h^{k+2} + d_3 h^{k+3} + \cdots$$

$$= \mathrm{P}^{(1)} + d_2 h^{k+2} + d_3 h^{k+3} + \ldots.$$

where $\quad \mathrm{P}^{(1)} = \dfrac{2^{k+1}\mathrm{P}_1 - \mathrm{P}_0}{2^{k+1}-1}.$

We see that the formula $P^{(1)}$ is of the order at least $(k+1)$. This process of obtaining higher order formula from two different spacings h and $h/2$ is called Richardson's extrapolation or 'deferred approach to the limit'. It may be borne in mind that this concept may be applied not only in numerical integration but in other numerical processes also.

We discuss now the Richardson's extrapolation with regard to a specific formula, say, Trapezoidal rule. It may be noted from the Euler–Maclaurin's formula (6.52) that the first term on the right is nothing but Trapezoidal formula and truncation error consists only even powers of h. If the Trapezoidal formula is denoted by T and T_0 and T_1 are the values of the integral computed with common intervals h and $h/2$ respectively then, we can write,

$$I = T_0 + c_1 h^2 + c_2 h^4 + c_3 h^6 + \ldots. \tag{6.55}$$

and

$$I = T_1 + c_1\left(\frac{h}{2}\right)^2 + c_2\left(\frac{h}{2}\right)^4 + c_3\left(\frac{h}{2}\right)^6 + \cdots \tag{6.56}$$

where c_1, c_2, c_3, etc., are known constants, independent of h.
Eliminating h^2 term from (6.55) and (6.56), we get

$$I = \frac{2^2 T_1 - T_0}{2^2 - 1} + d_2 h^4 + d_3 h^6 + \cdots \tag{6.57a}$$

$$= T^{(1)} + d_2 h^4 + d_3 h^6 + \ldots. \tag{6.57b}$$

where

$$T^{(1)} = \frac{2^2 T_1 - T_0}{2^2 - 1}. \tag{6.58}$$

Formula $T^{(1)}$ given by (6.58) is a higher order formula obtained from Trapezoidal formula with common intervals h and $h/2$. The error is of the same order as of Simpson's formula. Let us express $T^{(1)}$ in terms of function values:

$$\int_{x_0}^{x_1} f(x)dx = \frac{h}{2}(f_0 + f_1) = T_0 \text{ (Interval is } h)$$

$$= \frac{h}{4}(f_0 + 2f_{1/2} + f_1) = T_1. \text{ (Interval is } \frac{h}{2} \text{ and } T_0 \text{ is applied twice)}$$

Putting the above values in (6.58) we get formula T_1 improved over two intervals of width $\frac{h}{2}$ each, as

$$\mathrm{T}^{(1)} = \frac{h}{6}(f_0 + 4f_{1/2} + f_1).$$

or $$\int_{x_0}^{x_2} f(x)dx = \frac{h}{3}(f_0 + 4f_1 + f_2), \text{ for interval of width } h \text{ each.}$$

This is Simpson's formula.

6.13 Romberg Integration

Romberg's integration method is an iterative technique which is applied successively to obtain the improved value of the integral each time by halving the common interval between the abscissas. It is the natural follow-up of Richardson's extrapolation as it uses the same concept for computing the improved value. We discuss here the method in regard to Trapezoidal rule T for evaluating the integral $\mathrm{I} = \int_a^b f(x)dx$.

(*i*) Take $h = b - a = \frac{b-a}{2^0}$; one interval (two ordinates). Apply Trapezoidal rule and get initial estimate with one interval $\mathrm{T}_0^{(0)}$; the superscript denotes the improved value and subscript corresponds to zero in the number of intervals $1 = 2^0$.

The error in $\mathrm{T}_0^{(0)}$ is $\mathrm{O}(h^2)$.

(*ii*) Interval $(b - \mathrm{a})$ is divided into two intervals each of width $h/2$, i.e.,

$$\frac{h}{2} = \frac{b-a}{2^1}; \text{ two intervals (3 ordinates).}$$

Apply Trapezoidal rule with two intervals when common interval between abscissas is $h/2^1$; compute $\mathrm{T}_1^{(0)}$.

(*iii*) Using Richardson's extrapolation formula (6.58), compute

$$\mathrm{T}_1^{(1)} = \frac{2^2\mathrm{T}_1^{(0)} - \mathrm{T}_0^{(0)}}{2^2 - 1} = \frac{4\mathrm{T}_1^{(0)} - \mathrm{T}_0^{(0)}}{4 - 1}. \tag{6.59}$$

The error in the formula is $\mathrm{O}(h^4)$, and all powers are even. It is the first improved formula with number of intervals 2^1, each of width $h/2$.

(*iv*) To improve formula (6.59), the number of intervals are doubled to $2^2 = 4$, each of width $h/4$; use formula on four intervals and obtain,

$$T_2^{(1)} = \frac{4T_2^{(0)} - T_1^{(0)}}{4-1}.$$

Then formula $T_2^{(1)}$ is improved as,

$$T_2^{(2)} = \frac{2^4T_2^{(1)} - T_1^{(1)}}{2^4-1} = \frac{4^2T_2^{(1)} - T_1^{(1)}}{4^2-1}. \tag{6.60}$$

Formula (6.60) has truncation error of $O(h^6)$ and operates on four intervals.

(*v*) To further improve the value of the integral, the number of intervals are again doubled to $2^3 = 8$, each of width $h/8$. Compute $T_3^{(0)}$, i.e., apply Trapezoidal rule on eight intervals. Formula (6.60) is improved in following three steps:

$$T_3^{(1)} = \frac{4T_3^{(0)} - T_2^{(0)}}{4-1}, \text{ has an error of } O(h^4)$$

$$T_3^{(2)} = \frac{4^2T_3^{(1)} - T_2^{(1)}}{4^2-1}, \text{ has an error of } O(h^6)$$

$$T_3^{(3)} = \frac{4^3T_3^{(2)} - T_2^{(2)}}{4^3-1}, \text{ has an error of } O(h^8). \tag{6.61}$$

(*vi*) In general the above scheme may be written for k^{th} improved formula as follows:
Number of intervals are 2^k, each of width $(b-a) \div 2^k$ or $h/2^k$.
Apply Trapezoidal rule to get $T_k^{(0)}$ which is improved in k steps using formulae,

$$T_k^{(r)} = \frac{4^rT_k^{(r-1)} - T_{k-1}^{(r-1)}}{4^r-1}, \quad r = 1,\ 2,\ \ldots.\ k. \tag{6.62}$$

Formula (6.62) will have an error of $O(h^{2r+2})$.

The process of carrying Romberg integration is shown in Table 6.1, The computations are performed row-wise and the process is stopped as soon as two successive values agree within desired accuracy.

Table 6.1. Process for Romberg integration

Interval	*Order of truncation error*			
length	$O(h^2)$	$O(h^4)$	$O(h^6)$	$O(h^8)$
h	$T_0^{(0)}$			
$h/2$	$T_1^{(0)}$	$T_1^{(1)}$		
$h/4$	$T_2^{(0)}$	$T_2^{(1)}$	$T_2^{(2)}$	
$h/8$	$T_3^{(0)}$	$T_3^{(1)}$	$T_3^{(2)}$	$T_3^{(3)}$
etc.				

It may be noted that while computing $T_k^{(0)}$ the value of $T_{k-1}^{(0)}$ which has already been computed, can be utilised. Suppose the number of intervals is denoted by $m = 2^k$ and the width of each interval by $h_k = \dfrac{b-a}{2^k}$ then

$$T_k^{(0)} = \frac{1}{2}T_{k-1}^{(0)} + h_k(f_1 + f_3 + \cdots + f_{m-1}), \tag{6.63}$$

We can also use Simpson's $\frac{1}{3}$**rd** rule instead of the Trapezoidal rule.

Example 6.11

Compute $I = \int_0^1 e^{2x}dx$ by Romberg integration method correct up to 3 decimal places, using Trapezoidal rule.

Solution $h_0 = \dfrac{1-0}{2^0} = 1.$

(*i*) $T_0^{(0)} = \dfrac{1}{2}(e^0 + e^2) = 4.19455$

(*ii*) $h_1 = \dfrac{1-0}{2^1} = 0.5$

$$T_1^{(0)} = \frac{4.19455}{2} + 0.5 \times e^{2\times 0.5} \text{ or } \frac{0.5}{2}[e^0 + e^2 + 2e^1]$$

$$= 2.09727 + 1.35915 = 3.4564$$

$$T_1^{(1)} = \frac{4T_1^{(0)} - T_0^{(0)}}{3} = \frac{4 \times 3.4564 - 4.19455}{3} = 3.21037$$

(*iii*) $h_2 = \frac{1-0}{2^2} = 0.25$

$$\mathrm{T}_2^{(0)} = \frac{\mathrm{T}_1^{(0)}}{2} + 0.25(e^{2\times 0.25} + e^{2\times 0.75}) \text{ or } \frac{0.25}{2}[e^0 + e^2 + 2(e^{0.5} + e^{1.0} + e^{1.5})]$$

$$= \frac{3.4564}{2} + 0.25(1.6487 + 4.4817) = 3.2606$$

$$\mathrm{T}_2^{(1)} = \frac{4\mathrm{T}_2^{(0)} - \mathrm{T}_1^{(0)}}{3} = \frac{4 \times 3.2606 - 3.4564}{3} = 3.1956$$

$$\mathrm{T}_2^{(2)} = \frac{4^2\mathrm{T}_2^{(1)} - \mathrm{T}_1^{(1)}}{4^2 - 1} = \frac{16 \times 3.1956 - 3.2104}{15} = 3.1954$$

$\mathrm{T}_2^{(2)}$ and $\mathrm{T}_2^{(1)}$ are approximately equal and we can stop here.

$\mathrm{I} = 3.1954$ (correct answer is 3.1946).

6.14 Comments on Numerical Integration

(*i*) As stated earlier, the polynomial $P(x)$ approximating the function $f(x)$ generally oscillates about the function. It means that if $y = P(x)$ overestimates the function $y = f(x)$ in one interval, it would underestimate it in the next interval. As a result, while the area is overestimated in one interval, it may be underestimated in the next interval so that the overall effect of the errors in the two intervals will not be equal to the sum of their moduli, instead the effect of the error in one interval will be neutralised to some extent by the error in the next interval. Therefore, the computed estimated error in an integration formula may be unrealistically too high.

(*ii*) Increasing the degree of the approximating polynomial does not guarantee better accuracy. In a higher degree polynomial, the coefficients also get bigger which may magnify the errors. Similarly reducing the size of the sub-interval by increasing their number may also lead to accumulation of rounding errors. Therefore, a balance should be kept between the two, i.e., degree of polynomial and total number of intervals.

6.15 Gaussian Quadrature

The use of orthogonal polynomials is fundamental to Gaussian Quadrature. The orthogonal polynomials are defined as follows:

Let Φ be a set of polynomials, i.e.,

$$\Phi = \{\phi_0(x), \phi_1(x), \ldots, \phi_n(x), \ldots\},$$

where $\phi_n(x)$ is a polynomial of degree n. These polynomials are said to belong to a family of orthogonal polynomials if each of its members, is orthogonal to the other, i.e., they satisfy the condition

$$\int_a^b w(x)\phi_m(x)\phi_n(x)dx = 0, \quad m \neq n. \tag{6.64}$$

where $w(x)$ is a known function (called weight function) which has constant sign over the interval (a, b). The polynomials are said to be orthogonal with weight function $w(x)$ and with respect to integral between the limits a and b.

The Gaussian quadrature is different from the methods discussed earlier in the following ways:

(*i*) A Gaussian quadrature formula can be used only when the function is given; can not be used when the data values are prescribed at discrete points.

(*ii*) The abscissas are not equally-spaced and are given by the zeros of the orthogonal polynomials.

(*iii*) Gauss quadrature formulae are open-type in general.

There exist various quadrature formulae. It was Gauss who introduced such a formula using Legendre's orthogonal polynomials. We shall discuss this formula in detail.

6.15.1 Gauss–Legendre quadrature formula

The Gauss–Legendre quadrature formula is commonly known as Gaussian quadrature formula. The formula uses Legendre orthogonal polynomials which satisfy (6.64) for $w(x) = 1$ with limits of integration from -1 to 1. The Legendre polynomial of degree n may be expressed in the following form,

$$P_n(x) = \sum_{m=0}^{M} (-1)^m \frac{(2n-2m)!x^{n-2m}}{2^n m!(n-m)!(n-2m)!}, \tag{6.65}$$

where $M = n/2$ or $(n-1)/2$ whichever is an integer.

The Legendre polynomial $p_n(x)$ of degree n defined as $p_n(x) = \frac{1}{2^n n!}\frac{d^n}{dx^n}(x^2-1)^n$ readily provides a recurrence relation,

$$P_n(x) = \left(\frac{2n-1}{n}\right)xP_{n-1}(x) - \left(\frac{n-1}{n}\right)P_{n-2}(x), \; n > 1 \tag{6.66}$$

with $P_0(x) = 1$ and $P_1(x) = x$.

From (6.66) we can easily get

$$P_2(x) = \frac{1}{2}(3x^2 - 1); \;\; P_3(x) = \frac{1}{2}x(5x^2 - 3) \text{ etc.}$$

The Legendre polynomials possess following properties:

(*i*) $P_n(1) = 1$ and $P_n(-1) = (-1)^n$.

(*ii*) $\int\limits_{-1}^{1} x^m P_n(x)dx = 0, \; m < n$

Property (*ii*) implies that if $Q_m(x)$ is any polynomial of degree $m(m < n)$, then

$$\int\limits_{-1}^{1} Q_m(x)\, P_n(x)dx = 0$$

(*iii*) $\int\limits_{-1}^{1} P_m(x) \cdot P_n(x)dx = \begin{cases} 0 & , \; m \neq n \\ \dfrac{2}{2n+1} & , \; m = n \end{cases}$

(*iv*) $P_n(x) = 0$ has n distinct and real roots which lie in the interval $-1 < x < 1$.

The Gauss–Legendre quadrature formula is used to evaluate the integral of the form $\int\limits_{-1}^{1} f(x)dx$ where function $f(x)$ is known. If different limits are given, say $\int\limits_{a}^{b} F(t)dt$ then we convert the limits to $-1 \leq x \leq 1$ by change of variable from t to x by putting

$$x = \frac{2}{b-a}\left(t - \frac{a+b}{2}\right);$$

$$\text{or} \quad t = \frac{b-a}{2}\left(x + \frac{b+a}{b-a}\right) \text{ and } dt = \frac{b-a}{2}\, dx. \tag{6.67}$$

The n-point Gauss–Legendre quadrature formula is expressed as

$$\mathrm{I} = \int_{-1}^{1} f(x)dx = \mathrm{A}_1 f(x_1) + \mathrm{A}_2 f(x_2) + \ldots. + \mathrm{A}_n f(x_n), \tag{6.68}$$

where the abscissas $x_1,\ x_2,\ \ldots.\ x_n$ are not known 'a priori' and are to be determined alongwith the coefficients $A_1,\ A_2,\ \ldots.\ A_n$ commonly called 'weights' corresponding to function values $f(x_1),\ f(x_2),\ \ldots.\ f(x_n)$ respectively.

There are total $2n$ unknowns in the formula (6.68) n number of A's and n number of x's. In order to determine them, we make the formula satisfy for $f(x) = x^r,\ r = 0(1)2n-1$. This would provide us $2n$ equations, the solution of which should provide the values of the weights and the abscissas. It is easy to see that the formula will be exact for any polynomial of degree up to $(2n-1)$. For let us suppose that $P(x)$ is an arbitrary polynomial of degree $(2n-1)$ represented as

$$P(x) = a_0 + a_1x + \ldots.a_ix^i + \ldots. + a_{2n-1}x^{2n-1} = \sum_{i=0}^{2n-1} a_ix^i.$$

If the formula (6.68) is satisfied for $f(x) = x^r,\ r = 0(1)2n-1$, then

$$\int_{-1}^{1} x^r dx = A_1x_1^r + A_2x_2{}^r + \ldots. + A_jx_j{}^r + \ldots. + A_nx_n{}^r, r = 0(1)2n-1. \tag{6.69}$$

Multiplying first equation of (6.69) by a_0, second by a_1 and so on the last equation by a_{2n-1} and adding them vertically we get

$$\int_{-1}^{1} \sum_{r=1}^{2n-1} a_rx^r dx = A_1 \sum_{r=0}^{2n-1} a_rx_1^r + A_2 \sum_{r=0}^{2n-1} a_rx_2^r + \cdots + A_n \sum_{r=0}^{2n-1} a_rx_n^r$$

or
$$\int_{-1}^{1} P(x)dx = A_1P(x_1) + A_2P(x_2) + \ldots. + A_nP(x_n). \tag{6.70}$$

Equation (6.70) proves that if $f(x) = x^r,\ r = 0(1)2n-1$ satisfy equation (6.68) then any arbitrary polynomial of degree up to $2n-1$ will also satisfy it.

It may be noted that if the abscissas were not free, i.e., their positions were already fixed, say by dividing the interval $[-1, 1]$ into $n-1$ number of subintervals, then the formula can be made exact for polynomials of highest degree $(n-1)$ only.

Let us now write down as follows, the $2n$ equations obtained by putting $f(x) = x^r,\ r = 0$ (1) $2n-1$ in (6.68); the function $f(x) = x^r$ which the equation corresponds to, is also shown against the equation:

$$\left.\begin{array}{lll}
x^0 \rightarrow & A_1 + A_2 + A_3 + ... + A_n & = \int_{-1}^{1} x^0\, dx = 2 \\
x^1 \rightarrow & A_1x_1 + A_2x_2 + A_3x_3 + ... + A_nx_n & = \int_{-1}^{1} x\, dx = 0 \\
x^2 \rightarrow & A_1x_1^2 + A_2x_2^2 + A_3x_3^2 + ... + A_nx_n^2 & = \int_{-1}^{1} x^2\, dx = \frac{2}{3} \\
\ldots\ldots & \ldots\ldots & \ldots\ldots \\
x^n \rightarrow & A_1x_1^n + A_2x_2^n + A_3x_3^n + ... + A_nx_n^n & = \int_{-1}^{1} x^n\, dx = \begin{cases} 0, & n \text{ is odd} \\ \frac{2}{n+1}, & n \text{ is even} \end{cases} \\
\ldots\ldots & \ldots\ldots & \ldots\ldots \\
x^{2n-1} \rightarrow & A_1x_1^{2n-1} + A_2x_2^{2n-1} + A_3x_3^{2n-1} + ... + A_nx_n^{2n-1} & = \int_{-1}^{1} x^{2n-1}\, dx = 0
\end{array}\right\} \quad (6.71)$$

The system of equations given by (6.71) is not easy to solve. Therefore, special technique is required. Let $P_n(x)$ be a polynomial of degree n given by,

$$P_n(x) = c_0 + c_1x + c_2x^2 + \ldots.. + c_nx^n, \quad (6.72)$$

where c_0, c_1, c_2, etc., are constants.

Consider first $(n+1)$ equations of (6.71); multiply 1st equation by c_0, 2nd by c_1 and so on and $(n+1)^{\text{th}}$ equation by c_n; by adding these equations vertically we get the first equation of (6.73). Now leaving out first, consider next $(n+1)$ equations from 2nd to $(n+2)^{\text{th}}$; multiply 2nd eqn. by c_0, 3rd by c_1 and so on and $(n+2)^{\text{th}}$; eqn. by c_n; by adding all these equations vertically we get second equation of (6.73). Continuing in this manner, consider the last $(n+1)$ equations of (6.71), i.e., from n^{th} to $(2n)^{\text{th}}$ equations. Again multiply n^{th} equation by c_0, $(n+1)^{\text{th}}$ eqn. by c_1 and so on and $(2n)^{\text{th}}$ equation by c_n; after adding all these equations we get the last (n^{th}) equation of the set (6.73).

$$\left.\begin{array}{ll}
A_1P_n(x_1) + A_2P_n(x_2) + \ldots + A_nP_n(x_n) & = \int_{-1}^{1} P_n(x)dx \\
A_1x_1P_n(x_1) + A_2x_2P_n(x_2) + \ldots + A_nx_nP_n(x_n) & = \int_{-1}^{1} xP_n(x)dx \\
A_1x_1^2P_n(x_1) + A_2x_2^2P_n(x_2) + \ldots + A_nx_n^2P_n(x_n) & = \int_{-1}^{1} x^2P_n(x)dx \\
\ldots\ldots & \ldots\ldots \\
A_1x_1^{n-1}P_n(x_1) + A_2x_2^{n-1}P_n(x_2) + \ldots + A_nx_n^{n-1}P_n(x_n) & = \int_{-1}^{1} x^{n-1}P_n(x)dx
\end{array}\right\} \quad (6.73)$$

If abscissas $x_1, x_2, \ldots. x_n$ are chosen to be roots of $P_n(x) = 0$, then $P(x_1) = P(x_2) = \cdots P(x_n) = 0$ and hence the left sides of all the equations of (6.73) reduce to zero, no matter what the values of A's may be. Further, if $P_n(x)$ is chosen to be a Legendre polynomial then right side of (6.73) will also be zero due to property (*ii*). Thus by choosing the abscissas as zeros of the Legendre polynomial $P_n(x)$ the system of equations (6.73) is satisfied identically. Having got the abscissas fixed, we can substitute these values in the first n equations of (6.71) and solve them for the weights $A_1, A_2, \ldots. A_n$ which will be a simple task as the equations will be linear.

It may be noted that in accordance with property (*iv*) all the n roots of $P_n(x) = 0$ are distinct, real and lie between -1 and 1. That is, the abscissas, called quadrature points, are within the domain of integration, $-1 < x_i < 1,\ i = 1(1)n$. Further, since the integration is performed beyond the quadrature points, the Gauss quadrature formula is open-type.

There is also a direct formula for computing the weights once the quadrature points $x_i,\ i = 1(1)n$ are obtained from $P_n(x) = 0$, which are given by,

$$A_i = \frac{2}{(1-x_i)^2[P_n'(x_i)]^2},\ i = 1(1)n, \tag{6.74}$$

where $P_n'(x_i)$ is the value of the first derivative of $P_n(x)$ at $x = x_i$.

The associated error in the n-point Gauss quadrature formula is,

$$E = \frac{2^{n+1}(n!)^4 f^{2n}(\xi)}{(2n+1)[(2n)!]^3},\ -1 < \xi < 1. \tag{6.75}$$

Let us now consider a few particular cases of Gauss quadrature formula for different values of n.

(*i*) $n = 2$.

The requisite formula is,

$$\int_{-1}^{1} f(x)dx = A_1 f(x_1) + A_2 f(x_2) \tag{6.76}$$

The abscissas (quadrature points) are given by $P_2(x) = 0$ or $\frac{1}{2}(3x^2 - 1) = 0$ giving

$$x_1 = -\frac{1}{\sqrt{3}},\ x_2 = \frac{1}{\sqrt{3}}. \tag{6.77}$$

The equations for computing A_1 and A_2 are given by putting $f(x) = x^k$, $k = 0, 1$ in (6.76),

$$A_1 + A_2 = 2$$

$$-\frac{1}{\sqrt{3}}A_1 + \frac{1}{\sqrt{3}}A_2 = 0.$$

We get $A_1 = A_2 = 1$. Hence the 2-point Gauss quadrature formula (6.76) becomes,

$$\int_{-1}^{1} f(x)dx = f\left(-\frac{1}{\sqrt{3}}\right) + f\left(\frac{1}{\sqrt{3}}\right). \tag{6.78}$$

(*ii*) $n = 3$

The formula may be written as,

$$\int_{-1}^{1} f(x)dx = A_1 f(x_1) + A_2 f(x_2) + A_3 f(x_3). \tag{6.79}$$

The quadrature points are given by $P_3(x) = 0$ as

$$x_1 = -\sqrt{\frac{3}{5}},\ x_2 = 0,\ x_3 = \sqrt{\frac{3}{5}}. \tag{6.80}$$

The equations corresponding to $f(x) = x^k$, $k = 0, 1, 2$ are obtained from (6.79) as

$$A_1 + A_2 + A_3 = 2$$

$$-\sqrt{\frac{3}{5}}A_1 + 0 + \sqrt{\frac{3}{5}}A_3 = 0$$

$$\frac{3}{5}A_1 + 0 + \frac{3}{5}A_3 = \frac{2}{3}$$

The solution of the above gives,

$$A_1 = A_3 = \frac{5}{9};\ A_2 = \frac{8}{9}.$$

Hence, the 3-point Gauss quadrature formula (6.79) becomes,

$$\int_{-1}^{1} f(x)dx = \frac{1}{9}\left[5f\left(-\sqrt{\frac{3}{5}}\right) + 8f(0) + 5f\left(\sqrt{\frac{3}{5}}\right)\right]. \tag{6.81}$$

In Table 6.2 are given the quadrature points and values of weights for n-point quadrature formula, $n = 2(1)6$.

Table 6.2. Abscissas and weights in Gauss–Legendre quadrature formulae(along with associated error)

n	x_i $i = 1(1)n$	A_i $i = 1(1)n$	Approx. Error
2	$\pm$ 0.577350	1.00000	$0.74 \times 10^{-2} f^{iv}(\xi)$
3	$\pm$ 0.774597	0.555556	$0.64 \times 10^{-4} f^{vi}(\xi)$
	0.000000	0.888889	
4	$\pm$ 0.861136	0.347855	$0.3 \times 10^{-6} f^{viii}(\xi)$
	$\pm$ 0.339981	0.652145	
5	$\pm$ 0.906180	0.236927	$0.8 \times 10^{-8} f^{x}(\xi)$
	$\pm$ 0.538469	0.478629	
	0.000000	0.568889	
6	$\pm$ 0.932470	0.171324	$0.8 \times 10^{-10} f^{xii}(\xi)$
	$\pm$ 0.661209	0.360762	
	$\pm$ 0.238619	0.467914	

A n-point Gauss quadrature formula would evaluate the integral $\int_{-1}^{1} f(x)dx$ exactly for $f(x)$ being a polynomial of degree up to $2n-1$. On the other hand $2n$ points (or $2n-1$ points) will be required in the Cotes formula to get the same accuracy. It means that a 2-point Gauss quadrature should give comparable accuracy to Simpson's $\frac{3}{8}^{\text{th}}$ or $\frac{1}{3}^{\text{rd}}$ rule. The drawback of the Gaussian quadrature is that the quadrature points are irrational numbers, hence more digits should be retained. Further, it can not be applied if function $f(x)$ is not known.

Example 6.12

Evaluate the integral $I = \int_0^2 \sqrt{1+4x}$, by

(i) Gaussian 2-point formula (ii) Gaussian 3-point formula.
(iii) Simpson's Rule with two intervals (iv) Simpson's rule with four intervals.

Compare the result with exact value.

Solution For Gaussian formula we should change the limits from $\int_0^2$ to $\int_{-1}^1$.

Let the given integral be,

$$I = \int_0^2 \sqrt{1+4t}\, dt$$

Substituting $t = \frac{b-a}{2}x + \frac{b+a}{2}$
$= x+1$ or $x = t-1$.

So that when $t = 0,\ x = -1$ and when $t = 2,\ x = 1$ and also $dt = dx$.
Thus,

$$I = \int_0^2 \sqrt{1+4t}\, dt = \int_{-1}^1 \sqrt{5+4x}\, dx.$$

Now we apply the Gaussian formulae on this integral.

(i) Gauss 2-point formula

$$I = \int_{-1}^1 \sqrt{5+4x}\, dx = A_1 f(x_1) + A_2 f(x_2)$$

Absissas: $x_1 = -0.57735,\ x_2 = 0.57735$

Weights: $A_1 = 1,\ A_2 = 1$

$$I = \int_{-1}^1 \sqrt{5+4x}\, dx = \sqrt{2.6904} + \sqrt{7.3094}$$

$$= 1.6402 + 2.7035$$

$$= 4.3437$$

(*ii*) Gauss 3-point formula

$$I = \int_{-1}^{1} \sqrt{5+4x}\, dx = A_1 f(x_1) + A_2 f(x_2) + A_3 f(x_3)$$

Absissas: $x_1 = -0.7746,\ x_2 = 0,\ x_3 = 0.7746$

Weights: $A_1 = 0.5556 = \frac{5}{9},\ A_2 = 0.8889 = \frac{8}{9},\ A_3 = 0.5556 = \frac{5}{9}$

$$I = \frac{1}{9}\left[5\sqrt{5-4\times 0.7746} + 8\sqrt{5+0} + 5\sqrt{5+4\times 0.7746}\right]$$

$$= \frac{1}{9}[5\sqrt{1.9016} + 8\sqrt{5} + 5\sqrt{8.0984}]$$

$$= \frac{1}{9}[5\times 1.3790 + 8\times 2.2361 + 5\times 2.8456]$$

$$= \frac{39.0218}{9} = 4.3358.$$

(*iii*) Simpson's rule with 2 intervals

	x	-1	0	1
$\sqrt{5+4x} \rightarrow$	f	1	$\sqrt{5}$	3

$$I = \frac{1}{3}\left[1 + 4\sqrt{5} + 3\right]$$

$$= \frac{4}{3}[3.23607] = 4.3148$$

(*iv*) Simpson's rule with 4 intervals

	x	-1.0	-0.5	0.0	0.5	1.0
$\sqrt{5+4x} \rightarrow$	f	1	$\sqrt{3}$	$\sqrt{5}$	$\sqrt{7}$	3

$$I = \frac{0.5}{3}[1 + 3 + 4(1.73205 + 2.64575) + 2\times 2.23607]$$

$$= 4.3306$$

Exact Solution.

$$I = \int_{-1}^{1} \sqrt{5+4x}\, dx = \frac{2}{3\times 4}\left[(5+4x)^{3/2}\right]_{-1}^{1}$$

$$= \frac{1}{6}(27-1) = \frac{13}{3} = 4.3333.$$

Note: Gauss's 2-point and 3-point formulae have given better results than the 3-point Simpson's. But when 5-points are taken in the Simpson's the result gets lot better.

6.15.2 Gauss–Chebyshev quadrature formulae

The Chebyshev (in Russian Tschebycheff) polynomial of degree n, denoted by $T_n(x)$, is defined as

$$T_n(x) = \cos(n\cos^{-1}x), \quad -1 \le x \le 1. \tag{6.82}$$

By putting $x = \cos\theta$ or $\cos^{-1}x = \theta$, it can be shown straightaway that they follow the recurrence relation

$$T_{n+1}(x) + T_{n-1}(x) = 2x\,T_n(x)$$

or

$$T_{n+1}(x) = 2xT_n(x) - T_{n-1}(x), \; n \ge 1. \tag{6.83}$$

The first few polynomials are

$$\left.\begin{aligned} &T_0(x) = 1; && T_3(x) = 4x^3 - 3x \\ &T_1(x) = x; && T_4(x) = 8x^4 - 8x^2 + 1 \\ &T_2(x) = 2x^2 - 1; && \text{etc.} \end{aligned}\right\} \tag{6.84}$$

Like Legendre polynomials, all the n zeros of $T_n(x)$ are distinct, real and lie in the interval $-1 < x < 1$ and that $T_n(x)$ oscillates between -1 and 1 for $-1 \le x \le 1$.

The zeros of $T_n(x)$ may be obtained by putting

$$n\cos^{-1}x = (2r-1)\frac{\pi}{2}, \text{ giving}$$

$$x_r = \cos\frac{2r-1}{2n}\pi, \; r = 1,\, 2,\, \ldots.\, n. \tag{6.85}$$

The Chebyshev polynomials are orthogonal with weight function $w(x) = \dfrac{1}{\sqrt{1-x^2}}$ over the interval $-1 \le x \le 1$ satisfying

$$\int_{-1}^{1} T_m(x)\cdot T_n(x)\cdot\frac{dx}{\sqrt{1-x^2}} = \begin{cases} 0 & , \; m \ne n \\ \pi/2 & , \; m = n \ne 0 \\ \pi & , \; m = n = 0 \end{cases} \tag{6.86}$$

If abscissas are chosen as zeros of $T_n(x)$, then weights in n-point quadrature formula, are given by

$$A_i = \frac{\pi}{T_n'(x_i)T_{n-1}(x_i)} = \frac{\pi}{n},\ i = 1(1)n, \text{ i.e.,}$$

$$A_1 = A_2 = \ldots. = A_n = \frac{\pi}{n}. \tag{6.87}$$

Hence the n-point Gauss-Chebychev quadrature formula can be written as

$$\int_{-1}^{1} \frac{f(x)dx}{\sqrt{1-x^2}} = \frac{\pi}{n}\sum_{r=1}^{n} f(x_r) = \frac{\pi}{n}\sum_{r=1}^{n} f\left\{\frac{2r-1}{2n}\pi\right\}. \tag{6.88}$$

The formula (6.88) is exact when $f(x)$ is a polynomial of degree up to $2n-1$. The error in the formula is given by

$$E = \frac{\pi f^{2n}(\xi)}{2^{2n-1}(2n)!},\quad -1 \le \xi \le 1. \tag{6.89}$$

Note: A Chebychev quadrature formula is available for weight function $w(x) = 1$, given as

$$\int_{-1}^{1} f(x)dx = \frac{2}{n}\sum_{r=1}^{n} f(x_r), \tag{6.90}$$

where x_r are zeros of the nth degree polynomial obtained from the power series $F(x)$ given as

$$F(x) = x^n - \frac{n}{3!}x^{n-2} + \frac{n}{5!}\left(\frac{5n}{3} - 6\right)x^{n-4} - \frac{n}{7!}\left(\frac{35}{9}n^2 - 42n + 120\right)x^{n-6} + \cdots \tag{6.91}$$

The formula (6.90) is exact for polynomials of degree up to n only. Further, the formula holds for $n \le 7$ only since for $n = 8$, not all the roots are real; some roots are complex. See [2]. More details about Chebyshev polynomials are given in Chapter 9.

6.15.3 Gauss–Laguerre formula

The Laguerre polynomial of degree n is defined by

$$L_n(x) = e^x \frac{d^n}{dx^n}(e^{-x}x^n). \tag{6.92}$$

These polynomials are orthogonal with weight function $w(x) = e^{-x}$ over the interval $0 \leq x < \infty$. The n-point Gauss–Laguerre formula may be expressed as

$$\int_0^{\infty} e^{-x} f(x)dx = \sum_{i=1}^{n} A_i f(x_i) \tag{6.93}$$

where x_i's are the zeros of $L_n(x)$ and weights A_i's are given by

$$A_i = \frac{1}{x_i}\left\{\frac{n!}{L_n'(x_i)}\right\}^2, \quad i = 1(1)n. \tag{6.94}$$

6.15.4 Gauss–Hermite formula

The Hermite polynomial of degree n is given by

$$H_n(x) = (-1)^n e^{x^2} \frac{d^n}{dx^n}(e^{-x^2}). \tag{6.95}$$

These polynomials are orthogonal with weight function $w(x) = e^{-x^2}$ over the interval $-\infty < x < \infty$. The n-point quadrature formula is,

$$\int_{-\infty}^{\infty} e^{-x^2} f(x)dx = \sum_{i=1}^{n} A_i f(x_i), \tag{6.96}$$

where x_i's are the roots of $H_n(x) = 0$ and weights can be obtained from

$$A_i = \frac{2^{n+1} n! \sqrt{\pi}}{[H_n'(x_i)]^2}, \quad i = 1(1)n. \tag{6.97}$$

Besides the above, there are various other formulas of Gaussian type. For example, Radau formula integrates in the interval $-1 \leq x \leq 1$ where abscissa $x = -1$ is kept fixed while the remaining are free. In Labatto formula which is also defined for $-1 \leq x \leq 1$, both the end points are fixed and the remaining quadrature points are left free. The details of these and other methods may be found in [2] and [3].

For the quadrature points (abscissas) and weights (coefficients) reference may be made to any of the standard book on mathematical Tables.

Exercise 6

6.1 Evaluate integral $I = \int_{0.2}^{1.0} \frac{\cos x}{\sqrt{x}} dx$, dividing the domain of integration into four equal parts, by (*i*) Trapezoidal rule (*ii*) Simpson's rule.
Express the integrand in power series and find the value of the integral considering first two terms only.

6.2 Find the value of the integral $I = \int_{0}^{1} \frac{\sin x}{x} dx$, by Trapezoidal rule so that the error is not more than 0.005. Take the value of the integrand to be unity at $x = 0$.
Also expand the integrand in power series and obtain the value of the integral from the first three terms only.
For computing error, you may use the value of the second derivative from the power series.

6.3 Evaluate the integral $I = \int_{0}^{0.8} \frac{dx}{\sqrt{1+x}}$, by Simpson's $\frac{1}{3}^{\text{rd}}$ rule so that the error is less than 10^{-4}. Compare the answer with the correct value.

6.4 Evaluate the integral $I = \int_{0}^{1} \frac{dx}{\sqrt{1+x^2}}$, by Trapezoidal rule so that the error is less than 0.005. Compare the result with the analytical solution, $\ln\left(x + \sqrt{1+x^2}\right)$.

6.5 Find the value of the integral $I = \int_{0}^{1} \frac{dx}{1+x^2}$, by Romberg integration when the rounding error is approximately 0.0001 or less. Compare the result with the analytical solution.

6.6 Find the value of the integral $I = \int_{0}^{\pi/2} \frac{\cos x}{1+x} dx$ by Romberg integration correct up to 3 places of decimal.

6.7 Find the value of the integral $I = \int_{1}^{2} e^{2x} dx$ by Gaussian quadrature, using (*i*) 2-point (*ii*) 3-point formulae. Compare the result with true value.

6.8 Find the value of the integral $I = \int_{0}^{1} \sqrt{1+2x}\, dx$ by Gaussian quadrature, using (*i*) 2-point (*ii*) 3-point formulae. Compare the result with the true solution.

6.9 Evaluate the integral $I = \int_{1}^{2} \frac{dx}{\sqrt{1-x}}$, by Newton–Cotes open-type formulae (*i*) mid-point and (*ii*) 3-point, after dividing the interval into four equal parts. Compare with exact solution.

6.10 Determine the value of the integral $I = \int_0^1 x\ln(1+x)dx$ using Euler–Maclaurin's formula after dividing the interval into five equal parts. Compute up to five decimal places and use formula up to first derivative only.

6.11 Find the sum of the following series using Euler–Maclaurin formula,
$\sin 1^o + \sin 2^o + \sin 3^o + \ldots. + \sin 90^o$.
(Hint: Change degrees to radians, $1^o = \pi/180 = 0.01745$; change to $\sin 0 + \sin(0 + h) + \sin(0+2h) + \ldots. + \sin(0+90h)$, $h = 0.01745$; Consider. $\int_0^{\pi/2} \sin x dx$, taking $h = 0.01745$)

6.12 Determine the values of the constants a, b and c in the following formula so that it is exact for polynomials of degree as high as possible,

$$\int_0^{3h} f(x)dx = 3h\,[af(0) + bf(h) + cf(3h)].$$

Also compute the error in the formula.

6.13 Find the values of the constants a, b and c in the improved Simpson's formula given below:

$$\int_{-h}^{h} y\,dx = h(ay_{-1} + by_0 + ay_1) + ch^2(y_1' - y_{-1}').$$

Also determine the error in the formula.

Note: Find error by method discussed as well as by integrating $(x-h)^2x^2(x+h)^2\dfrac{h^6 f^{vi}(\xi)}{6!}$.

6.14 The speed of a train km/min. on a straight track is observed for 1.2 minutes at an interval of 0.2 minute as given below:

$t = 0.0$	0.2	0.4	0.6	0.8	1.0	1.2
$u = 0.0$	0.4472	0.6324	0.7746	0.8944	1.0000	1.0954

Find the distance covered by the train during 1.2 minutes using (i) Trapezoidal rule (ii) Simpson's rule. Compare with exact value when $u = \sqrt{t}$.

6.15 Find the integral $\int_{0.125}^{0.875} f(x)dx$ when the function values are supplied at the discrete points given below:

x	0.125	0.250	0.375	0.500	0.625	0.750	0.875
$f(x)$	2.8060	1.9379	1.5194	1.2411	1.0258	0.8450	0.6854

Use (*i*) Simpson's $\frac{3}{8}$ rule (*ii*) Weddle's rule.

(**Note:** Data is taken from $f(x) = \dfrac{\cos x}{\sqrt{x}}$).

References and Some Useful Related Books/Papers

1. Demidovich, B.P., and Maron, I.A., (translated by Yankovsky G.), *Computational Mathematics*, MIR Publishers.
2. Froberg, C.E., *Introduction to Numerical Analysis*, Addison-Wesley Pub. Co. Inc.
3. Hildebrand, F.B., *Introduction to Numerical Analysis*, Tata McGraw-Hill.
4. Jain, M.K., Iyengar, S.R.K., and Jain, R.K., *Numerical Methods for Scientific and Engineering Computations*, New Age International (P) Ltd, Publishers.

7

Ordinary Differential Equations

7.1 Introduction

An ordinary differential equation (ode) may be expressed as $\Phi(x,\ y,\ y',\ y',\ \ldots y^n) = 0$ where Φ is a function of independent variable x, dependent variable y and its derivatives; y^k denotes the kth order derivative (w.r.t. x), i.e., $\frac{d^k y}{dx^k}$, $k = 1,\ 2,\ \ldots n$. A differential equation is called of order n, if the order of the highest derivative is n. Its degree is equal to the degree of the highest order derivative provided neither y nor its derivatives appear under radical sign. That means, the equation should be made free of radical sign containing y or its derivative before determining its degree. For example, the differential equation,

$$\frac{d^2y}{dx^2} + \sqrt{\frac{dy}{dx}} + y = 0,$$

is of order two and also of degree two since after removing the square root, it can be written as,

$$\left(\frac{d^2y}{dx^2}\right)^2 + 2y\frac{d^2y}{dx^2} - \frac{dy}{dx} + y^2 = 0.$$

The ode's can be broadly classified as (*i*) Linear and (*ii*) Nonlinear. In a linear differential equation, y and its derivatives appear in linear form, i.e., y and its derivatives can occur in single power only and their product can not occur. Remember, functions like $\sin y$, $\log y$, e^y, etc., are nonlinear. If this condition is violated the equation will be nonlinear.

For example, the differential equations,

$$y\frac{d^2y}{dx^2} + x^2\frac{dy}{dx} + y = 0$$

and $$\frac{d^2y}{dx^2} + x^2\frac{dy}{dx} + \ln y = 0$$

are nonlinear due to $y\frac{d^2y}{dx^2}$ and $\ln y$ terms.

The equations given below are linear,

$$x^2\frac{d^2y}{dx^2} + \frac{dy}{dx} + y = e^x$$

and $$x\frac{d^2y}{dx^2} + e^x\frac{dy}{dx} + y = \sin x.$$

There are standard methods for finding analytical (closed form) solution to linear ode's. But no such techniques are available for the nonlinear equations except for some special cases. Sometimes the analytical solutions are presented in the form of an integral or in the form of complicated expression which is of no practical use. Moreover, the differential equations arising in the practical situations are generally quite complex and nonlinear in nature. Therefore a numerical method is the only recourse. Besides, the numerical methods are easy to apply on all kinds of differential equations and provide solutions to desired accuracy.

7.2 Initial Value and Boundary Value Problems (IVP and BVP): Solution of IVP

Let us write an nth order differential equation as $\frac{d^ny}{dx^n} = f(x, y, y', \ldots y^{n-1})$. We know its general solution will contain n arbitrary constants, and that n conditions must be prescribed to determine these constants uniquely. That is, there should be provided n values of y and its derivatives at one or more points. The conditions may be prescribed in two ways:

(*i*) All the conditions are prescribed at a single point, i.e., values of y, y', y'', $\ldots y^{n-1}$ are prescribed at some point $x = a$ (say). In this case solution is required in the domain $x > a$, which means solution domain is open. Such problems are called Initial Value Problems (IVPs) and the associated conditions are called initial conditions.

(*ii*) The conditions are prescribed at two or more (usually two) points, say $x = a$ and $x = b$ $(b > a)$ and the solution is required in the domain $a \leq x \leq b$, which is bounded. These problems are called Boundary Value Problems (BVPs) and the associated conditions as boundary conditions.

A problem is said to be well-posed if there are prescribed as many conditions as the order of the equation-no less and no more. We shall be concerned only with the well-posed problems.

First, let us discuss the methods for solving Initial Value Problems (IVPs).

7.3 Reduction of Higher-Order IVP to System of First Order Equations

Let us consider n^{th} order initial value problem

$$\frac{d^n y}{dx^n} = f(x,\, y,\, y',\, y'',\, \ldots.\, y^{n-1}) \tag{7.1}$$

with initial conditions prescribed at $x = x_0$,

$$y(x_0) = y_0,\; y'(x_0) = y_0',\; y''(x_0) = y_0'' \ldots . y^{n-1}(x_0) = y_0^{n-1}. \tag{7.2}$$

Introducing new variables, $z_1,\, z_2,\, \ldots.\, z_{n-1}$ where $z_1 = \dfrac{dy}{dx},\; z_2 = \dfrac{d^2y}{dx^2},\; \cdots z_{n-1} = \dfrac{d^{n-1}y}{dx^{n-1}}$, the differential equation (7.1) may be reduced to a system of n simultaneous equations of the first order, given as,

$$\left.\begin{aligned} \frac{dy}{dx} &= z_1, \\ \frac{dz_1}{dx} &= z_2, \\ &\vdots \\ \frac{dz_{n-1}}{dx} &= f(x,\, y,\, z_1,\, z_2, \ldots .\, z_{n-1}), \end{aligned}\right\} \tag{7.3}$$

with initial conditions prescribed at $x = x_0$, as known values of $y(x_0),\; z_1(x_0),\; z_2(x_0),\; \ldots.\; z_{n-1}(x_0)$. For generalisation $y(x_0)$ may be replaced by z_0.

Thus we see that a higher order IVP can be reduced to a set of first order equations. Therefore, our study will be confined to solving IVPs of first order only, e.g.

$$\frac{dy}{dx} = f(x,\, y), \quad x > x_0, \tag{7.4}$$

$$y = y_0, \text{ at } x = x_0. \tag{7.5}$$

7.4 Picard's Method (Method of Successive Approximations)

Integrating the differential equation (7.4) from x_0 to a general point x and using (7.5), the problem is transformed to an integral equation,

$$y(x) - y(x_0) = \int_{x_0}^{x} f(x, y)dx. \tag{7.6}$$

Solution to (7.6) is obtained in an iterative manner according to the scheme,

$$y^{(n+1)}(x) = y_0 + \int_{x_0}^{x} f[x, y^{(n)}(x)]dx, \quad n = 0, 1, 2, \cdots \tag{7.7}$$

where $y^{(n)}(x)$ denotes n^{th} iteration and $y_0 = y(x_0)$. To start the process, the initial approximation $y^{(0)}(x)$ may be taken as y_0.

Since the solution is obtained as a function of x, the difference between two successive approximations will also be a function of x. Hence the accuracy of the solution will be dependent on the value of x, and the solution obtained will be valid for certain range of x, for the prescribed accuracy. It may be mentioned that if the solution is obtained in the form of a series with alternating signs, the first neglected term gives the magnitude of the maximum error when the series converges uniformly in a certain interval.

The iterative process converges to the true solution under certain conditions. The major drawback of the method is that integration has to be performed at each stage which may not be possible when the integrand is complicated. The method is not a numerical method; it may be called semi-analytical or approximate analytical method.

Example 7.1

Find the approximate solution by Picard's method for the differential equation,

$$\frac{dy}{dx} = x^2 - y, \ y(0) = 1$$

which is correct within an accuracy of 10^{-3} for $0 \leq x \leq 0.2$.

Solution The iterative scheme for the above problem is,

$$y^{(n+1)} - 1 = \int_0^x (x^2 - y^{(n)})\, dx.$$

Taking initial estimate $y^{(0)}$ as $y(0) = 1$,

$$y^{(1)} = 1 + \int_0^x (x^2 - 1)dx = 1 - x + \frac{x^3}{3}.$$

$$y^{(2)} = 1 + \int_0^x \left\{x^2 - \left(1 - x + \frac{x^3}{3}\right)\right\} dx$$

$$= 1 - x + \frac{x^2}{2} + \frac{x^3}{3} - \frac{x^4}{12}$$

$$\text{error} = y^{(2)} - y^{(1)} = \frac{x^2}{2} - \frac{x^4}{12}$$

max error is given at $x = 0.2$, i.e.,

$$\frac{(0.2)^4}{2} - \frac{(0.2)^4}{12} > 10^{-3}$$

Since the error is more than the prescribed accuracy, we go for the next iteration,

$$y^{(3)} = 1 + \int_0^x \left\{x^2 - \left(1 - x + \frac{x^2}{2} + \frac{x^3}{3} - \frac{x^4}{12}\right)\right\} dx$$

$$= 1 - x + \frac{x^2}{2} + \frac{x^3}{6} - \frac{x^4}{12} + \frac{x^5}{60}$$

$$\text{error} = y^{(3)} - y^{(2)} = -\frac{x^3}{6} - \frac{x^5}{60}$$

$$\text{at} \quad x = 0.2, \text{ max error} = -\frac{(0.2)^3}{6} + \frac{(0.2)^5}{60} > 10^{-3}$$

Again, we go for the next iteration, giving,

$$y^{(4)} = 1 - x + \frac{x^2}{2} + \frac{x^3}{6} - \frac{x^4}{24} + \frac{x^5}{60} - \frac{x^6}{360}$$

$$\text{max error} = y^{(4)} - y^{(3)} = \frac{x^4}{24} - \frac{x^6}{360} < 10^{-3}$$

Therefore $y^{(4)}$ is the desired solution which will give an error less than 10^{-3} for values of x in the interval (0, 0.2).

(The analytical solution is $y(x) = x^2 - 2x + 3 - 2e^{-x}$).

7.5 Taylor's Series Method

Referring to equation (7.4) we are required to find the solution of $\frac{dy}{dx} = f(x, y)$, $x > x_0$ subject to condition (7.5), $y(x_0) = y_0$. The value of y at some point x, close to x_0 can be found by using a Taylor's series expansion,

$$y(x) = y_0 + (x - x_0)y'_0 + \frac{(x-x_0)^2}{2!}y''_0 + \cdots + \frac{(x-x_0)^n}{n!}y_0^n$$

with a remainder term $R = \frac{(x-x_0)^{n+1}}{(n+1)!}f^{n+1}(\xi),\ x_0 \le \xi \le x,$

$$y_0{}^k = \frac{d^k y}{dx^k},\ \text{at } x = x_0;\ \ k = 1,\ 2,\ 3,\ \ldots.\ n.$$

To find the values of $y_0,\ y'_0,\ y''_0\ \ldots$, etc., we proceed as follows:

(*i*) $y_0 = y(x_0)$, given as initial condition

(*ii*) $y'_0 = f(x_0,\ y_0)$, to be found from the differential equation.

(*iii*) y''_0 to be found by differentiating the d.e. once

$$y'' = \frac{d^2y}{dx^2} = \frac{d}{dx}f(x,\ y) = \frac{\partial f}{\partial x} + \frac{\partial f}{\partial y}\cdot\frac{dy}{dx}$$

$$= f_x + f_y \cdot y' = f_x + f_y \cdot f;\quad f = f(x, y).$$

Putting $x = x_0$ and $y = y_0$ we get,

$$y''_0 = f_x + f_y y'_0,\ \text{at } x = x_0,\ \ y = y_0.$$

(*iv*) y'''_0 is computed by differenting y'' as follows:

$$y''' = \frac{\partial}{\partial x}(f_x) + \frac{\partial}{\partial y}(f_x)\cdot\frac{dy}{dx} + \left[\frac{\partial}{\partial x}(f_y) + \frac{\partial}{\partial y}(f_y)\cdot\frac{dy}{dx}\right]y' + f_y \cdot y''$$

$$= f_{xx} + (y')^2 f_{yy} + 2y' f_{xy} + f_y \cdot y'' \qquad \because\ \frac{\partial^2 f}{\partial x \partial y} = \frac{\partial^2 f}{\partial y \partial x},\ \text{ i.e., } f_{yx} = f_{xy}$$

y'''_0 can be found by putting $x = x_0$ and $y = y_0$.

By putting the values of $y_0,\ y'_0,\ y''_0$, etc. we get the solution $y(x)$ in the form of a power series, which will be valid in the neighbourhood of x_0. This method is an approximate analytical method. It can be used to solve higher order IVP's directly without having to solve the simultaneous equations.

Example 7.2

Obtain the first five terms in the Taylor's series as solution of the equation,

$$\frac{dy}{dx} = \frac{1}{2}(x^2 + y^2),\ y(0) = 1.$$

Also discuss its truncation error in interval [0, 0.1].

Solution.

$$f(x,\ y) = \frac{1}{2}(x^2 + y^2);$$

$$f_x = x,\ f_y = y,\ f_{xx} = 1,\ f_{yy} = 1,\ f_{xy} = 0.$$

$$x_0 = 0,\ y_0 = 1$$

$$y' = \frac{1}{2}(x^2 + y^2); \qquad y'_0 = \frac{1}{2}$$

$$y'' = x + yy'; \qquad y''_0 = \frac{1}{2}$$

$$y''' = 1 + yy'' + (y')^2; \qquad y'''_0 = \frac{7}{4}$$

$$y^{iv} = yy''' + 3y'y''; \qquad y^{iv}_0 = \frac{5}{2}$$

The Taylor's series solution is given by,

$$y(x) = 1 + \frac{x}{2} + \frac{x^2}{4} + \frac{7}{24}x^3 + \frac{5}{48}x^4.$$

In general it will not be straight-forward to compute the truncation error. However, assuming that the derivatives do not increase faster than the factorial term in the Taylor's expansion, it may be stated that the truncation error will be of the order $O(x^5)$, i.e., $O(10^{-5})$ for $x = 0.1$.

Example 7.3

Find the solution of the following differential equation by Taylor's series method so that the truncation error is not greater than $\frac{1}{2} \times 10^{-4}$ for $x \leq 0.2$,

$$\frac{d^2y}{dx^2} + x\frac{dy}{dx} + y = 0;\ y(0) = 0,\ y'(0) = 1.$$

Solution The given d.e. is

$$y'' = -xy' - y.$$

$$\left.\begin{array}{ll} y(0) = 0, & y_0 = 0 \\ y'(0) = 1, & y'_0 = 1 \end{array}\right\}\text{ initial conditions}$$

$$y''(x) = -xy' - y, \qquad y_0'' = 0$$

$$y'''(x) = -xy'' - 2y', \qquad y_0''' = -2$$

$$y^{iv}(x) = -xy''' - 3y'', \qquad y_0^{iv} = 0$$

$$y^{v}(x) = -xy^{iv} - 4y''', \qquad y_0^{v} = 8$$

The Taylor's series solution is,

$$y(x) = x - \frac{x^3}{3} + \frac{x^5}{15}$$

This is an alternating series, therefore the magnitude of truncation error will not be greater than the first non-zero term neglected. In this case, the maximum value of the last term for $x \leq 0.2$, is

$$\frac{1}{15} \times (.2)^5 = \frac{1}{15} \times 0.00032 \simeq 0.00002.$$

Since it is less than the required accuracy, the solution after neglecting this term is,

$$y(x) = x - \frac{x^3}{3}.$$

7.6 Numerical Method, its Order and Stability

There are several numerical methods for solving an initial value problem $\frac{dy}{dx} = f(x, y)$ with a prescribed condition $y = y_0$ at $x = x_0$ while we are required to find the value of y for $x > x_0$. The common approach in all the methods would be to choose a suitable step size, h (say) and compute the value of y successively at an interval of h. That is, values y_i of y are computed at $x = x_i$, $x_i = x_0 + ih$ in a step-by-step manner, i.e., for $i = 1, 2, 3$ ...., etc. in that order. It should be clear that before computing y_{n+1} at x_{n+1}, the value of y would be known at all the points previous to x_{n+1}, i.e., $x_0, x_1, \ldots. x_n$.

There are two important phenomena associated with a numerical method for solving an IVP, namely, (*i*) Order and (*ii*) Stability.

(i) Order of Method

If the error in a numerical method is of the order $O(h^{k+1})$, the method is said to be of order k. It implies that if the actual solution of the differential equation is a polynomial function of degree k, then the method will provide an exact solution.

(ii) Stability of Method

As stated above an IVP is solved in a step-by-step manner. If an error, introduced at some stage grows exponentially with increasing number of steps, then the method is said to be 'Unstable'; otherwise it will be called 'Stable' or 'absolutely stable'.

Let us suppose there has been an error ε_0 in the value of y at $x = x_0$ and that it becomes $A\varepsilon_0$ in the next step where A is constant. Then at the nth step, this error will be $A^n\varepsilon_0$ and hence for the method to be stable $|A| < 1$.

7.7 Euler's Method

The Euler's method is simplest of all the numerical methods. Refer to Fig. 7.1 for better understanding of the method.

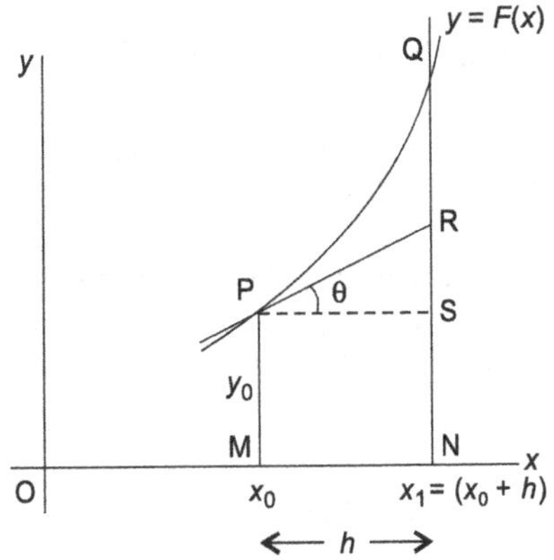

Figure 7.1 Euler's method.

Given differential equation is $\dfrac{dy}{dx} = f(x,\ y)$

Initial condition is $y(x_0) = y_0 = \text{PM}$

Let the true solution be $y = F(x)$ and the exact value of y at $x = x_1$ is given by $F(x_1) = \text{QN}$. By Euler's method we approximate the value of y at $x_1 = x_0 + h$ as follows:

The value of $\frac{dy}{dx}$ at $\text{P}(x_0,\ y_0)$ is $f(x_0,\ y_0)$ which gives the slope of the tangent PR at P, i.e., $f(x_0,\ y_0) = \tan\theta$.

Then, $\qquad \text{RS} = \text{PS}\tan\theta = h\ \tan\theta.$

Hence, $\qquad \text{RN} = \text{SN}\ (= \text{PM}) + h\ \tan\theta.$

or, $y_1 = y_0 + h\cdot f(x_0,\ y_0)$, where y_0 is known from the initial condition. The error between the exact value $y(x_1) = \text{QN}$ and the computed value $y_1 = \text{RN}$ is given by $y(x_1) - y_1 = \text{QR}$.

Mathematically, the Euler's method may be obtained expanding $y(x_0 + h)$ by Taylor's series about x_0, as

$$y(x_0 + h) = y(x_0) + hy'(x_0-) + R$$

where R is the remainder term given by,

$$R = \frac{h^2}{2} y''(\xi), \quad x_0 \le \xi \le x_1.$$

Neglecting R in the above expansion, gives the Euler's formula,

$$y_1 \simeq y_0 + hy'(x_0) = y_0 + hf(x_0, y_0) \tag{7.8}$$

with an error,

$$R = \frac{h^2}{2} y''(\xi) = \frac{h^2}{2} f'(\xi), \quad x_0 \le \xi \le x_1 \tag{7.9}$$

where $$f' = \frac{d^2 y}{dx^2} = f_x + f_y f. \tag{7.9a}$$

The general form of the Euler's formula may be written as,

$$y_{n+1} = y_n + hy'_n = y_n + hf(x_n, y_n), \quad n = 0, 1, 2, \ldots \tag{7.10}$$

with the error term,

$$R = \frac{h^2}{2} y''(\xi_n) = \frac{h^2}{2} f'(\xi_n, y), \quad x_n \le \xi_n \le x_{n+1}. \tag{7.11}$$

As the error in Euler's formula is of order $O(h^2)$, the method is of order one.

We would now like to see how the error propagates from one step to the next and in the subsequent steps.

Let us suppose that $y(x_n)$ is the exact value of y at $x_n = x_0 + nh$, while y_n is the value at x_n computed from Euler's formula (7.10) and that the error in y_n is ε_n such that,

$$y(x_n) = y_n + \varepsilon_n, \quad n = 1, 2, 3, \ldots \tag{7.12}$$

Assuming there is no error in the given value y_0, the actual value of y at x_1 may be expressed as,

$$\begin{aligned} y(x_1) &= y_0 + hf(x_0, y_0) + \frac{h^2}{2} f'(\xi_0, y), \quad x_0 \le \xi_0 \le x_1 \\ &= y_1 + \varepsilon_1, \end{aligned}$$

where $\varepsilon_1 = \frac{h^2}{2} f'(\xi_0, y)$.

At the next step the computed value y_2 at $x = x_2$ and the actual value $y(x_2)$ will be related as follows:

$$\begin{aligned}
y(x_2) &= y(x_1 + h) \\
&= y(x_1) + hy'(x_1) + \frac{h^2}{2} f'(\xi_1,\, y), \quad x_1 \le \xi_1 \le x_2. \\
&= y_1 + \varepsilon_1 + hf(x_1,\, y_1 + \varepsilon_1) + \frac{h^2}{2} f'(\xi_1,\, y) \\
&= y_1 + hf(x_1,\, y_1) + \varepsilon_1 + \varepsilon_1 h f_y(x_1,\, y_1) + \frac{h^2}{2} f'(\xi_1,\, y)
\end{aligned}$$

neglecting higher powers of ε_1

$$= y_2 + \varepsilon_2,$$

where

$$\begin{aligned}
\varepsilon_2 &= \varepsilon_1 + \varepsilon_1 h f_y(x_1,\, y_1) + \frac{h^2}{2} f'(\xi_1,\, y) \\
&= \varepsilon_1 [1 + h f_y(x_1,\, y_1)] + \frac{h^2}{2} f'(\xi_1,\, y). \qquad (7.13)
\end{aligned}$$

In (7.13), the error ε_2 in the computation of y_2 is the combination of local truncation error $\frac{h^2}{2} f'(\xi_1,\, y)$ and the error propagated from the first step, i.e., $\varepsilon_1 [1 + h f_y(x_1,\, y_1)]$. Thus the total error will get compounded with each step in the subsequent computations. It is not the simple sum of the local truncation errors. In generalised form the formula (7.13) may be written as,

$$\varepsilon_{n+1} = \varepsilon_n \{1 + h f_y(x_n,\, y_n)\} + \frac{h^2}{2} f'(\xi_n,\, y), \quad x_n \le \xi_n \le x_{n+1}, \qquad (7.14)$$

which gives the error in the computation of y_{n+1}.

Example 7.4

Using Euler's method, compute y_1 and y_2 taking $h = 0.1$ from the following differential equation,

$$\frac{dy}{dx} = 1 + xy^2, \quad y(0) = 1.$$

Also compute the error in both.

Solution $y'(x) = f(x, y) = 1 + xy^2$,

$$y''(x) = f'(x, y) = \frac{df}{dx} = \frac{\partial f}{\partial x} + \frac{\partial f}{\partial y} \cdot \frac{dy}{dx} = \frac{\partial f}{\partial x} + \frac{\partial f}{\partial y} \cdot f$$

$$= y^2 + 2xy\,(1 + xy^2)$$

For $x = 0.1$

Let, $k = hy_0' = 0.1(1 + 0 \times 1) = 0.1$

$$y_1 = y_0 + k = 1 + 0.1 = 1.1$$

Max. truncation error $\varepsilon_1 = \frac{1}{2}h^2 y''(\xi) = \frac{1}{2}h^2 f'(\xi, y),\ 0 \le \xi \le 0.1$

$$= \frac{1}{2} \times 0.01[y^2 + 2xy(1 + xy^2)]$$

max value at $x = 0.1$ is given by

$$\varepsilon_1 = \frac{1}{2} \times 0.01[1.1^2 + 2 \times 0.1 \times 1.1(1 + 0.1 \times 1.1^2)]$$

$$= 0.005 \times 1.45662 = 0.00728$$

For $x = 0.2$

$$k = 0.1(1 + 0.1 \times 1.1^2) = 0.1121.$$

$$y_2 = y_1 + k = 1.1 + 0.1121 = 1.2121.$$

Truncation error,

$$\varepsilon_2 = \varepsilon_1\{1 + hf_y(x_1, y_1)\} + \frac{h^2}{2} \cdot f'(\xi, y)\ \ x_1 \le \xi \le x_2.$$

$$= \varepsilon_1\{1 + 2hx_1y_1)\} + \frac{h^2}{2}\{y^2 + 2xy\,(1 + xy^2)\}$$

$$= 0.00728(1 + 2 \times 0.1 \times 0.1 \times 1.1) + \frac{(0.1)^2}{2}$$

$$\{1.2121^2 + 2 \times 0.2 \times 1.2121(1 + 0.2 \times 1.2121^2)\}$$

$$= 0.00728 \times 1.022 + 0.005\{1.4692 + 0.4 \times 1.2121\ (1.2938)\}$$

$$= 0.00744 + 0.005\ \{1.4692 + 0.6272\}$$

$$= 0.00744 + 0.01048$$

$$= 0.0179$$

7.8 Modified (Improved) Euler's Method

The Euler's method described in Sec. 7.7 provides a rather crude estimate for $y(x_1)$. Further, as the truncation error in formula is of order $O(h^2)$, a very small step size h has to be chosen. The modified Euler's method is an improvement over the earlier method, having an error of $O(h^3)$. In this method, first a crude estimate for y_1 is obtained from the Euler's method, i.e.,

$$y_1{}^* = y_0 + hf(x_0,\ y_0). \tag{7.15}$$

Then the value of the gradient $\dfrac{dy}{dx}$ is computed at the point $(x_1,\ y_1{}^*)$ as $f(x_1,\ y_1{}^*)$. It may be stated that point $(x_1,\ y_1{}^*)$ does not lie on the actual curve unless it happens to be straight line between x_0 and x_1. The value of y_1 is improved by taking average value of the gradients $f(x_0,\ y_0)$ and $f(x_1,\ y_1{}^*)$, i.e.,

$$y_1 = y_0 + \frac{h}{2}[f(x_0,\ y_0) + f(x_1,\ y_1{}^*)]. \tag{7.16}$$

The methods can be explained in another way also. Suppose the value of y is known to be y_0 at $x = x_0$. It becomes $y_1 = y_0 + k$ at $x = x_0 + h$ satisfying $\dfrac{dy}{dx} = f(x,\ y),\ y(x_0) = y_0$. We wish to estimate k. In the Euler's method $k = hf(x_0,\ y_0)$ while in the modified Euler's method $k = \dfrac{1}{2}(k_1 + k_2)$, where $k_1 = hf(x_0,\ y_0)$ and $k_2 = hf(x_0 + h,\ y_0 + k_1)$.

It will now be shown that error in the modified Euler's method is $O(h^3)$. First we expand the right side of the formula (7.16) by Taylor's series:

$$\begin{aligned}
y_1 &= y_0 + \frac{h}{2}[f(x_0,\ y_0) + f(x_1,\ y_1{}^*)] \\
&= y_0 + \frac{h}{2}f(x_0,\ y_0) + \frac{h}{2}f(x_0 + h,\ y_0 + k)
\end{aligned}$$

where $k = hf(x_0,\ y_0) = hf$ (say).

$$\begin{aligned}
&= y_0 + \frac{hf}{2} + \frac{h}{2}\left[f(x_0,\ y_0) + \left(h\frac{\partial f}{\partial x} + k\frac{\partial f}{\partial y}\right) + \frac{1}{2}\left(h^2\frac{\partial^2 f}{\partial x^2} + 2hk\frac{\partial^2 f}{\partial x \partial y} + k^2\frac{\partial^2 f}{\partial y^2}\right) + \cdots\right] \\
&= y_0 + hf + \frac{h^2}{2}(f_x + f_y \cdot f) + \frac{h^3}{4}(f_{xx} + 2f_{xy}f + f_{yy}f^2) + \cdots, \qquad (7.17)
\end{aligned}$$

after using $k = hf$.

Now, the exact solution $y(x_0+h) = y(x_1)$ can also be expressed by Taylor's series expansion about $x = x_0$ as follows:

$$y(x_1) = y(x_0+h) = y_0 + hy_0' + \frac{h^2}{2}y_0'' + \frac{h^3}{6}y_0''' + \cdots$$

where ${y_0}^k$ denotes the value of the kth derivative of $y(x)$ at x_0.

$$= y_0 + hf + \frac{h^2}{2}(f_x + f_y f) + \frac{h^3}{6}(f_{xx} + 2f_{xy}f + f_{yy}f^2 + f_y \cdot y'') + \cdots, \quad (7.18)$$

Subtracting (5.17) from (5.18) gives the error between the exact value and the computed value,

$$y(x_1) - y_1 = -\frac{h^3}{12}(f_{xx} + 2_{xy}f + f_{yy}f^2 - 2f_y \cdot y'')$$

$$= -\frac{h^3}{12}\left[f_{xx} + 2f_{xy} \cdot f + f_{yy} \cdot f^2 - 2f_y(f_x + f_y \cdot f)\right]$$

$$= O(h^3).$$

Example 7.5

Compute y, from the differential equation,

$$\frac{dy}{dx} = x - y, \quad y(0) = 1$$

(*i*) for $x = 0.1(0.1)0.4$, using Euler's method.

(*ii*) for $x = 0.2(0.2)0.4$, using modified Euler's method.

Solution (*i*) Euler's method ($h = 0.1$)

$$y_{n+1} = y_n + hy_n' = y_n + hf(x_n, y_n).$$

$x = 0.1$: $\quad y_1 = 1 + 0.1(0-1) = 0.9$

$x = 0.2$: $\quad y_2 = 0.9 + 0.1\ (0.1 - 0.9) = 0.82$

$x = 0.3$: $\quad y_3 = 0.82 + 0.1\ (0.2 - 0.82) = 0.758$

$x = 0.4$: $\quad y_4 = 0.758 + 0.1\ (0.3 - 0.758) = 0.7122$

(*ii*) Modified Euler's method ($h = 0.2$)

$$y_{n+1}^* = y_n + hy_n' = y_n + hf(x_n, y_n)$$

$$y_{n+1} = y_n + \frac{h}{2}[f(x_n, y_n) + f(x_{n+1}, y^*_{n+1})]$$

$x = 0.2$; $\quad y_1{}^* = 1 + 0.2(0 - 1) = 0.8$

$$y_1 = 1 + \frac{0.2}{2}\{(0-1) + (0.2 - 0.8)\}$$

$$= 1 - 0.16 = 0.84$$

$x = 0.4$ $\quad y_2{}^* = 0.84{+}0.2\ (0.2 - 0.84) = 0.712$

$$y_2 = 0.84 + \frac{0.2}{2}\{(0.2 - 0.8) + (0.4 - 0.712)\}$$

$$= 0.84 - 0.0912 = 0.7488$$

Note: The analytical solution is $y = x - 1 + 2e^{-x}$.
correct values at $x = 0.1(0.1)0.4$ are,

x	0.1	0.2	0.3	0.4
y	0.9096	0.8375	0.7816	0.7406

7.9 Runge–Kutta (R–K) Methods

In Sec. 7.8, we see that the modified Euler's method can be expressed in the form,

$y_1 = y_0 + k$, where $k = \frac{1}{2}(k_1 + k_2)$ and values of k_1, k_2 are computed as,

$k_1 = hf(x_0, y_0)$; $k_2 = hf(x_0 + h, y_0 + k_1)$.

This may be considered as R–K method of order two.
In general, a R–K method of order m can be written as follows:

$$\left.\begin{aligned}
k_1 &= hf(x_0, y_0) \\
k_2 &= hf(x_0 + \alpha_1 h, y_0 + \beta_{11}k_1) \\
k_3 &= hf(x_0 + \alpha_2 h, y_0 + \beta_{21}k_1 + \beta_{22}k_2) \\
k_4 &= hf(x_0 + \alpha_3 h, y_0 + \beta_{31}k_1 + \beta_{32}k_2 + \beta_{33}k_3) \\
&\cdots\cdots\cdots\cdots\cdots\cdots\cdots\cdots\cdots \\
k_m &= hf(x_0 + \alpha_{m-1}h, y_0 + \beta_{m-1,1}k_1 + \beta_{m-1,2}k_2 + \ldots + \beta_{m-1,m-1}k_{m-1}); \\
k &= \omega_1 k_1 + \omega_2 k_2 + \ldots + \omega_m k_m, \\
y_1 &= y_0 + k.
\end{aligned}\right\} \quad (7.19)$$

The parameters $\alpha,\ \beta$ and weights ω are chosen to satisfy certain conditions. They are determined by expanding various functions in $y_1 = y_0 + k$, about $(x_0,\ y_0)$ and comparing powers of h in the expansion of $y(x_0+h)$. We illustrate the method for $m = 2$.

(*i*) Second Order R–K Method

Let $k_1 = hf(x_0, y_0)$

$$k_2 = hf(x_0+\alpha h, y_0+\beta k_1)$$

$$k = \omega_1 k_1 + \omega_2 k_2$$

where the parameters $\alpha,\ \beta$ and weights $\omega_1,\ \omega_2$ (positive) are to be determined. Expanding the function y_1, about $(x_0,\ y_0)$

$$y_1 = y_0 + k = y_0 + \omega_1 k_1 + \omega_2 k_2$$

$$= y_0 + \omega_1 hf(x_0,\ y_0) + \omega_2 hf(x_0+\alpha h,\ y_0+\beta k_1)$$

$$= y_0 + \omega_1 hf(x_0,\ y_0) + \omega_2 h\left[f(x_0,\ y_0) + \left(\alpha h\frac{\partial f}{\partial x} + \beta k_1\frac{\partial f}{\partial y}\right) + \frac{1}{2}\left(\alpha^2 h^2\frac{\partial^2 f}{\partial x^2}\right.\right.$$

$$\left.\left. + 2\alpha\beta hk_1\frac{\partial^2 f}{\partial x\partial y} + \beta^2 k_1^2\frac{\partial^2 f}{\partial y^2}\right) + \cdots\right]$$

$$= y_0 + h(\omega_1+\omega_2)f + h\omega_2(\alpha hf_x + \beta hf\cdot f_y)$$

$$+\frac{\omega_2 h}{2}[\alpha^2 h^2 f_{xx} + 2\alpha\beta h^2 ff_{xy} + \beta^2\cdot h^2\cdot f^2\cdot f_{yy}] + \ldots;$$

the arguments $(x_0,\ y_0)$ in f and all its partial derivatives are dropped.

$$y_1 = y_0 + (\omega_1+\omega_2)hf + \omega_2 h^2(\alpha f_x + \beta f_y f) + \frac{\omega_2 h^3}{2}[\alpha^2 f_{xx} + 2\alpha\beta f_{xy} f + \beta^2 f_{yy} f^2] + \cdots \quad (7.20)$$

Now expanding $y(x_0+h)$ about x_0,

$$y(x_1) = y(x_0+h) = y_0 + hy_0' + \frac{h^2}{2}y_0'' + \frac{h^3}{6}y_0''' + \cdots$$

$$= y_0 + hf + \frac{h^2}{2}(f_x + f_y f) + \frac{h^3}{6}(f_{xx} + 2f_{xy}f + f_{yy}f^2 + f_x\cdot f_y + f_y^2\cdot f) + \ldots. \quad (7.21)$$

Comparing powers of h in (7.20) and (7.21) up to h^2, we get

$$\omega_1 + \omega_2 = 1 \text{ and } \omega_2\alpha = \frac{1}{2}, \omega_2\beta = \frac{1}{2} \text{ implying } \alpha = \beta.$$

We can assign an arbitrary value to any one of the parameters, say, $\alpha = a$; then we get $\beta = a,\ \omega_2 = \dfrac{1}{2a}$ and $\omega_1 = 1 - \dfrac{1}{2a}$. There are more unknowns than equations.

By giving different values to a $\left(a \geq \dfrac{1}{2}\right.$ since $0 \leq \omega_1,\ \omega_2 \leq 1$ such that $\omega_1 + \omega_2 = 1)$ various second order R–K formulae may be constructed. For example, when $a = 1$, we get $\alpha = \beta = 1,\ \omega_1 = \omega_2 = \dfrac{1}{2}$ and the required formula becomes,

$$\left.\begin{aligned} &k_1 = hf(x_0,\ y_0);\ \ k_2 = hf(x_0 + h,\ y_0 + k_1) \\ &k\ = \frac{1}{2}(k_1 + k_2) \\ &y_1 = y_0 + k. \end{aligned}\right\} \tag{7.22}$$

It may be noted that it is the same as modified Euler's formula. The truncation error in (7.22) will be obtained by subtracting (7.20) from (7.21) and putting the value of constants; it gives,

$$R = y(x_1) - y_1 = -\frac{h^3}{12}[f_{xx} + 2f_{xy}f + f_{yy}f^2 - 2(f_x + f_yf)f_y] + \ldots \tag{7.23}$$

$$= \mathrm{O}(h^3) \text{ i.e., terms of } h^3 \text{ and higher powers.}$$

This method is also known as Euler-Cauchy method. Third Order R–K methods can also be constructed by considering $k_1,\ k_2$ and k_3 in (7.19) and pursuing the same analysis as given above.

(*ii*) Fourth Order R–K Method (Classical)

In a fourth order R–K method, the value of k is computed in four steps, i.e., computing the values of $k_1,\ k_2,\ k_3,\ k_4$ and finally by taking their weighted average. In its simple form a fourth order method may be expressed as follows:

$$\left.\begin{aligned}
k_1 &= hf(x_0,\, y_0),\\
k_2 &= hf(x_0+\alpha_1 h,\, y_0+\beta_1 k_1),\\
k_3 &= hf(x_0+\alpha_2 h,\, y_0+\beta_2 k_2),\\
k_4 &= hf(x_0+\alpha_3 h,\, y_0+\beta_3 k_3),\\
k &= \omega_1 k_1+\omega_2 k_2+\omega_3 k_3+\omega_4 k_4\\
y_1 &= y_0+k.
\end{aligned}\right\} \tag{7.24}$$

There are 10 unknowns and procedure for computing their values is same as discussed earlier. There will be less number of equations than the unknowns. Therefore, we have freedom to assign arbitrary values to some of the unknowns. It will give rise to various fourth order R–K formulae. The most popular among them is the 'classical' method which is given below:

$$\left.\begin{aligned}
k_1 &= hf(x_0,\, y_0)\\
k_2 &= hf\left(x_0+\frac{1}{2}h,\, y_0+\frac{1}{2}k_1\right)\\
k_3 &= hf\left(x_0+\frac{1}{2}h,\, y_0+\frac{1}{2}k_2\right)\\
k_4 &= hf(x_0+h,\, y_0+k_3)\\
k &= \frac{1}{6}(k_1+2k_2+2k_3+k_4)\\
y_1 &= y_0+k.
\end{aligned}\right\} \tag{7.25}$$

The error in the above formula is $O(h^5)$.

7.9.1 Application to first order simultaneous equations

The fourth order R–K method is most popular computer method for solving first order differential equations. It can also be easily applied to solve first order simultaneous equations. Further as a higher order equation can be reduced to a system of first order equations (See Sec.7.3), the method can be applied on higher order equations also. Its application on two simultaneous equations is illustrated below which can be naturally extended to more equations.

Let the given differential equations be,

$$\frac{dy}{dx} = f(x,\ y,\ z);\ \ \frac{dz}{dx} = g(x,\ y,\ z), \tag{7.26}$$

with initial conditions,

$$y = y_0 \text{ and } z = z_0 \text{ at } x = x_0.$$

We compute the value of y and z at $x_1 = x_0 + h$ in the following manner:

$$\left.\begin{aligned}
k_1 &= hf(x_0,\ y_0,\ z_0) &&;\ l_1 = hg(x_0,\ y_0,\ z_0);\\
k_2 &= hf\left(x_0+\frac{1}{2}h,\ y_0+\frac{1}{2}k_1,\ z_0+\frac{1}{2}l_1\right) &&;\ l_2 = hg\left(x_0+\frac{1}{2}h,\ y_0+\frac{1}{2}k_1,\ z_0+\frac{1}{2}l_1\right);\\
k_3 &= hf\left(x_0+\frac{1}{2}h,\ y_0+\frac{1}{2}k_2,\ z_0+\frac{1}{2}l_2\right) &&;\ l_3 = hg\left(x_0+\frac{1}{2}h,\ y_0+\frac{1}{2}k_2,\ z_0+\frac{1}{2}l_2\right);\\
k_4 &= hf(x_0+h,\ y_0+k_3,\ z_0+l_3) &&;\ l_4 = hg(x_0+h,\ y_0+k_3,\ z_0+l_3);\\
k &= \frac{1}{6}(k_1+2k_2+2k_3+k_4) &&;\ l = \frac{1}{6}(l_1+2l_2+2l_3+l_4);\\
y_1 &= y_0+k &&;\ z_1 = z_0+l.
\end{aligned}\right\} \tag{7.27}$$

The method may be repeated to compute the values of $y_2,\ z_2;\ y_3,\ z_3$, etc. in a step-by-step manner.

Example 7.6

Compute y at $x = 0.2\,(0.2)\,0.4$ by fourth order Runge–Kutta method from the differential equation,

$$\frac{dy}{dx} = y - x,\ \ y(0) = 1.5.$$

Give your answer up to four places of decimal.

Solution $h = 0.2,\ x_0 = 0,\ y_0 = 1.5$

$$k_1 = 0.2(1.5-0) = 0.3;$$

$$k_2 = 0.2\left\{1.5+\frac{1}{2}\times 0.3-(0+0.1)\right\} = 0.2\times 1.55 = 0.310$$

$$k_3 = 0.2(1.5+0.155-0.1) = 0.2\times 1.555 = 0.3110$$

$$k_4 = 0.2(1.5+0.3110-0.2) = 0.2\times 1.6110 = 0.3222$$

$$k = \frac{1}{6}(0.3 + 2 \times 0.31 + 2 \times 0.3110 + 0.3222) = \frac{1}{6} \times 1.8642 = 0.3107$$

$$y_1 = y_0 + k = 1.5 + 0.3107 = 1.8107$$

For $x_2 = 0.4$, take initial values as $x_1 = 0.2$, $y_1 = 1.8107$.

$$k_1 = 0.2(1.8107 - 0.2) = 0.2 \times 1.6107 = 0.32214$$

$$k_2 = 0.2\{1.8107 + 0.16107 - (0.2 + 0.1)\} = 0.2 \times 1.67177 = 0.33425$$

$$k_3 = 0.2(1.8107 + 0.16712 - 0.3) = 0.2 \times 1.6778 = 0.33556$$

$$k_4 = 0.2(1.8107 + 0.33556 - 0.4) = 0.2 \times 1.7463 = 0.34926$$

$$k = \frac{1}{6}(0.32214 + 2 \times 0.33425 + 2 \times 0.33556 + 0.34926) = 0.33517$$

$$y_2 = y_1 + k = 1.8107 + 0.33517$$

$$= 2.14587 \simeq 2.1459$$

Note: The analytical solution to this problem is $y = x + 1 + \frac{1}{2}e^x$, and correct values are $y(0.2) = 1.8107$, $y(0.4) = 2.1459$, surprisingly exact.

Example 7.7

Solve the differential equation $y'' = xy$, for $x = 0.5$ in a single step, using Runge–Kutta fourth order method when the initial conditions are given to be $y(0) = 0$ and $y'(0) = 1$.

Solution We reduce the second order equation to two simultaneous equations of first order, putting $\frac{dy}{dx} = z$, so that the given d.e. becomes, $\frac{dz}{dx} = xy$. Now we have to solve simultaneous equations,

$$\frac{dy}{dx} = z \text{ and } \frac{dz}{dx} = xy$$

subject to conditions $y(0) = 0$ and $z(0) = 1$.

Using (7.27), we take $\frac{dy}{dx} = f(x, y, z) = z$ and $\frac{dz}{dx} = g(x, y, z) = xy$.

$$k_1 = 0.5(1) = 0.5 \qquad ; l_1 = 0.5(0) = 0$$

$$k_2 = 0.5(1 + \frac{1}{2} \times 0) = 0.5 \qquad ; l_2 = 0.5\left\{(0 + 0.25)(0 + \frac{1}{2} \times 0.5)\right\} = 0.03125$$

$k_3 = 0.5\left(1+\frac{1}{2}\times 0.03125\right) = 0.50781$; $l_3 = 0.5\left\{(0+0.25)(0+\frac{1}{2}\times 0.5)\right\} = 0.03125$

$k_4 = 0.5(1+0.03125) = 0.51562$; $l_4 = 0.5\{(0+0.5)(0+0.50781)\} = 0.12695$

$k = \frac{1}{6}(0.5+2\times 0.5+2\times 0.50781+0.51562)$; $l = \frac{1}{6}(0+2\times 0.03125+2\times 0.03125+0.12695)$

$= \frac{1}{6}3.03124 = 0.50521$ $= 0.04199$

$y_1 = y_0 + k = 0.50521$ $z_1 = z_0 + l = 1.04199$

Note: Value of z_1 will be required if we want to compute y_2, otherwise it is not required; so is l_4 and l.

Taylor's series soln. is $y = x + \frac{x^4}{12} + \frac{x^7}{504} + \cdots$ which gives

$$y(0.5) = 0.5052$$

7.10 Predictor–Corrector (P–C) Methods

In a Predictor–Corrector (P–C) method, the value of y is estimated (predicted) first by a coarse formula, i.e., a lower order (less accurate) formula; then the value is improved (corrected) using a higher order (more accurate) formula. Let us suppose that the values of y have already been computed up to y_n, i.e., the values of y_i are known at the points $x_i = x_0 + ih$, $i = 1(1)n$ and we are required to find y_{n+1} at $x_{n+1} = x_n + h$. The predicted and corrected values of y_{n+1} will be denoted by y_{n+1}^p and y_{n+1}^c, respectively. If required, the corrected formula may be iterated to further improve the value. In that case the scheme would be P-C-C-C-... and the process may be stopped when two successive values agree within desired accuracy.

Again, the modified Euler's formula may be considered, in a way, as a P–C formula since it can be expressed as,

$$\text{P: } y_{n+1}^p = y_n + hf(x_n,\, y_n), \tag{7.28}$$

$$\text{C}: y_{n+1}^c = y_n + \frac{h}{2}\{f(x_n,\, y_n) + f(x_{n+1},\, y_{n+1}^p)\}. \tag{7.29}$$

In the above scheme, formula P is a first order formula having an error $O(h^2)$ while formula C is a second order formula which has an error of $O(h^3)$.

Formula (7.29) may be iterated, if required, as

$$y_{n+1}^{c^{(k+1)}} = y_n + \frac{h}{2}\left\{f(x_n, y_n) + f(x_{n+1}, y_{n+1}^{c^{(k)}})\right\},$$

$k = 1, 2, 3, \ldots$.

We shall now discuss other well-known P–C formulae.

7.10.1 Milne's method

Refer to open-type quadrature formula (6.43) as given below,

$$\int_{x_0}^{x_4} y(x)dx = \frac{4h}{3}(2y_1 - y_2 + 2y_3) + \frac{14}{45}h^5 y^{iv}(\xi).$$

From the above we can write,

$$\int_{x_{n-3}}^{x_{n+1}} y(x)dx = \frac{4h}{3}(2y_{n-2} - y_{n-1} + 2y_n) + \frac{14}{45}h^5 y^{iv}(\xi). \tag{7.30}$$

Applying (7.30) on $y' = \frac{dy}{dx}$, we get

$$y_{n+1} = y_{n-3} + \frac{4h}{3}(2y'_{n-2} - y'_{n-1} + 2y'_n) + \frac{14}{45}h^5 y^{v}(\xi).$$

When the above formula is used in context of the differential equation $y' = f(x, y)$ it gives,

$$y_{n+1} = y_{n-3} + \frac{4h}{3}(2f_{n-2} - f_{n-1} + 2f_n) + \frac{14}{45}h^5 y^{v}(\xi). \tag{7.31}$$

Similarly, from the Simpson's quadrature formula (closed type), we get,

$$y_{n+1} = y_{n-1} + \frac{h}{3}(f_{n-1} + 4f_n + f_{n+1}) - \frac{h^5}{90}y^{v}(\xi). \tag{7.32}$$

It may be noted that the error in (7.31) is 28 times that of formula (7.32). Therefore, using (7.31) as predictor and (7.32) as corrector formula, the Milne's method may be written as,

$$y_{n+1}^{p} = y_{n-3} + \frac{4h}{3}(2f_{n-2} - f_{n-1} + 2f_n), \tag{7.33}$$

$$y_{n+1}^{c} = y_{n-1} + \frac{h}{3}(f_{n-1} + 4f_n + f_{n+1}). \tag{7.34}$$

The value of f_{n+1} is computed using (7.33), i.e.,

$$f_{n+1} = f(x_{n+1}, y_{n+1}^{p}). \tag{7.35}$$

The P–C method given by (7.33), (7.34) is also known as Milne-Simpson method.

It may be noted from (7.33) that in order to compute y_{n+1}^{p}, the values of y at four previous points, *viz.*, y_{n-3}, y_{n-2}, y_{n-1} and y_n are ought to be known while in all previous methods the value of y was required at one previous point, $x = x_n$ only. Thus only two values were involved in the earlier methods, one at the current level and the other at the previous level. The methods which require information at more than one previous points in order to compute the value at the current level are known as 'multi-step' methods. Milne-Simpson's is a multi-step method.

On account of the reason stated above, the Milne's method is not 'self-starting'. That is, the first three values y_1, y_2 and y_3 can not be computed by this method. These values have to be computed by any one of the methods discussed earlier. After that this method may be applied for $n = 3$ onwards.

We now discuss the stability of the corrector formula (7.34). As before, let y_n denote the computed value and $y(x_n)$, the exact value of y at $x = x_n$, and let ε_n be the error in y_n such that $y(x_n) = y_n + \varepsilon_n$. Then,

$$f\{x_n, y(x_n)\} = f\{x_n, y_n + \varepsilon_n\}$$

$$= f(x_n, y_n) + \varepsilon_n f_y(x_n, y_n) + \text{higher powers of } \varepsilon_n^2.$$

or $$f\{x_n, y(x_n)\} - f(x_n, y_n) = \varepsilon_n \theta_n, \text{ where } \theta_n = f_y(x_n, y_n)$$

Similar relations can be written for x_{n-1} and x_{n+1}, giving

$$f\{x_{n-1}, y(x_{n-1})\} - f(x_{n-1}, y_{n-1}) = \varepsilon_{n-1}\theta_{n-1}, \text{ where } \theta_{n-1} = f_y(x_{n-1}, y_{n-1})$$

and $$f\{x_{n+1}, y(x_{n+1})\} - f(x_{n+1}, y_{n+1}) = \varepsilon_{n+1}\theta_{n+1}, \text{ where } \theta_{n+1} = f_y(x_{n+1}, y_{n+1})$$

From the corrector formula (7.34), the approximate value of y at $x = x_{n+1} = x_n + h$ is given by,

$$y_{n+1} = y_{n-1} + \frac{h}{3}(f_{n-1} + 4f_n + f_{n+1}),$$

while the exact value would be,

$$y(x_{n+1}) = y(x_{n-1}) + \frac{h}{3}[f(x_{n-1}, y_{n-1} + \varepsilon_{n-1}) + 4f(x_n, y_n + \varepsilon_n) + f(x_{n+1}, y_{n+1} + \varepsilon_{n+1})].$$

Subtracting the approximate value from the exact and using above relations we get,

$$\varepsilon_{n+1} = \varepsilon_{n-1} + \frac{h}{3}(\theta_{n-1}\varepsilon_{n-1} + 4\theta_n\varepsilon_n + \theta_{n+1}\varepsilon_{n+1})$$

or $$\left(1 - \frac{h}{3}\theta_{n+1}\right)\varepsilon_{n+1} - \frac{4h}{3}\theta_n\varepsilon_n - \left(1 + \frac{h}{3}\theta_{n-1}\right)\varepsilon_{n-1} = 0.$$

Assuming h to be small such that value of $\theta = f_y$ does not vary much in $x_{n-1} \leq x \leq x_{n+1}$, the above can be written as,

$$\left(1 - \frac{h}{3}\theta\right)\varepsilon_{n+1} - \frac{4h}{3}\theta\varepsilon_n - \left(1 + \frac{h}{3}\theta\right)\varepsilon_{n-1} = 0.$$

In order to solve the above difference equation, let us assume the solution as $\varepsilon_n = \alpha^n$ (see Chapter 14). Substituting this value, we get the auxiliary/characteristic equation,

$$\left(1 - \frac{h\theta}{3}\right)\alpha^2 - \frac{4h\theta}{3}\alpha - \left(1 + \frac{h\theta}{3}\right) = 0.$$

If α_1 and α_2 are the roots of this equation, then

$$\alpha_1 + \alpha_2 = \frac{4h\theta}{3 - h\theta}; \quad \alpha_1 \cdot \alpha_2 = -\frac{3 + h\theta}{3 - h\theta},$$

and the solution will be given by,

$$\varepsilon_n = c_1(\alpha_1)^n + c_2(\alpha_2)^n,$$

where c_1 and c_2 are arbitrary constants.

For $h\theta$ small, the product of the roots $\alpha_1\alpha_2$ is nearly equal to unity in absolute value which means the roots α_1 and α_2 are almost reciprocal to each other numerically. Thus the solution can be expressed as,

$$\varepsilon_n \simeq c_1(\alpha_1)^n + c_2(\frac{1}{\alpha_1})^n, \quad \because \alpha_2 = -\frac{1}{\alpha_1}.$$

The above solution has two components which behave reciprocally. That is, if one component decreases with increasing n, the other increases. Thus, in any case, the error grows exponentially with number of steps. Hence the formula is unstable. That means, the formula should not be used for large number of steps and nor for iteration in the scheme P-C-C-C .... It may also be stated that such a situation is very likely to occur whenever a lower order differential equation is approximated by a higher order difference formula.

7.10.2 Adams–Bashforth method

Let us suppose that $y_1, y_2, \ldots, y_n$ are known and y_{n+1} is to be computed. Adams–Bashforth method is based on Newton's Backward Difference interpolation formula which can be written with x_n origin, as follows:

$$\begin{aligned} y(x) &= y(x_n + ph) = \mathrm{E}^p y_n = (1-\nabla)^{-p} y_n \\ &= y_n + p\nabla y_n + \frac{p(p+1)}{2}\nabla^2 y_n + \frac{p(p+1)(p+2)}{6}\nabla^3 y_n + \cdots \end{aligned} \tag{7.36}$$

where $\qquad x = x_n + ph.$

Applying formula (7.36) on the derivative $y'(x)$,

$$y'(x) = y'_n + p\nabla y'_n + \frac{p(p+1)}{2}\nabla^2 y'_n + \frac{p(p+1)(p+2)}{6}\nabla^3 y'_n + \cdots \tag{7.37}$$

Integrating (7.37) w.r.t. x from x_n to x_{n+1}, remembering that $dx = hdp$ and limits of integration for p, corresponding to x_n and x_{n+1} are 0 to 1 $\left(\because p = \frac{x - x_n}{h}\right)$,

$$\int_{x_n}^{x_{n+1}} y'(x)dx = h\int_0^1 \left(1 + p\nabla + \frac{p(p+1)}{2}\nabla^2 + \frac{p(p+1)(p+2)}{6}\nabla^3\right)dp \cdot y'_n$$

$$y(x_{n+1}) - y(x_n) = h\left[y'_n + \frac{1}{2}\nabla y'_n + \frac{5}{12}\nabla^2 y'_n + \frac{3}{8}\nabla^3 y'_n + \cdots\right]$$

or

$$y^p_{n+1} = y_n + h\left[y'_n + \frac{1}{2}\nabla y'_n + \frac{5}{12}\nabla^2 y'_n + \frac{3}{8}\nabla^3 y'_n + \cdots\right] \tag{7.38}$$

This is used as predictor formula.

For deriving the corrector formula, we write the backward difference formula with origin at x_{n+1}, so that (7.36) may be written as,

$$\begin{aligned} y(x) &= y(x_{n+1}+ph) = \mathrm{E}^p y_{n+1} = (1-\nabla)^{-p} y_{n+1} \\ &= y_{n+1} + p\nabla y_{n+1} + \frac{p(p+1)}{2}\nabla^2 y_{n+1} + \frac{p(p+1)(p+2)}{6}\nabla^3 y_{n+1} + \cdots \end{aligned}$$

where $x = x_{n+1} + ph$, or $p = \dfrac{x - x_{n+1}}{h}$.

Applying the formula on the derivative $y'(x)$,

$$y'(x) = y'_{n+1} + p\nabla y'_{n+1} + \frac{p(p+1)}{2}\nabla^2 y'_{n+1} + \frac{p(p+1)(p+2)}{6}\nabla^3 y'_{n+1} + \cdots \tag{7.39}$$

Integrating (7.39) w.r.t. x from x_n to x_{n+1} when the corresponding limits of p will be -1 to 0 due to $p = \dfrac{x - x_{n+1}}{h}$, we get the corrector formula,

$$y^c_{n+1} \simeq y_n + \left[y'_{n+1} - \frac{1}{2}\nabla y'_{n+1} - \frac{1}{12}\nabla^2 y'_{n+1} - \frac{1}{24}\nabla^3 y'_{n+1} - \cdots \right] \tag{7.40}$$

If terms only up to second differences are taken, then from (7.38) and (7.40) we get respectively,

$$\mathrm{P}: y^p_{n+1} = y_n + \frac{h}{12}(f_{n-2} - 16 f_{n-1} + 23 f_n). \tag{7.41}$$

$$\mathrm{C}: y^c_{n+1} = y_n - \frac{h}{12}(f_{n-1} - 8 f_n - 5 f_{n+1}). \tag{7.42}$$

Let us now examine the magnitude of errors in (7.41) and (7.42).

The error in the predictor formula (7.41) is given by,

$$\begin{aligned} E^p &= h^4 \int_0^1 \frac{p(p+1)(p+2)}{6}\,dp\, y^{iv}(\xi) \\ &= \frac{3}{8} h^4 y^{iv}(\xi),\ x_n \le \xi \le x_{n+1}. \end{aligned}$$

The error in the corrector formula is,

$$E^c = h^4 \int_{-1}^{0} \frac{p(p+1)(p+2)}{6} dp\, y^{iv}(\xi)$$

$$= -\frac{1}{24} h^4 y^{iv}(\xi),\ x_n \le \xi \le x_{n+1}.$$

As can be seen, the error in predictor formula is nine times of the error in the corrector formula. The P–C formula given by (7.41), (7.42) is the classical Adams–Bashforth formula. This formula is also multi-step; therefore, not self-starting. Three values of y are required to apply formula (7.41); therefore two additional values have to be computed by another method. Further, other higher order formulae may be obtained by incorporating higher order differences in (7.38) and (7.39). The method may be unstable. See [4]

Example 7.8

Solve the differential equation $\frac{dy}{dx} = x^2 + y^2 - 2$, for $x = 0.3$, by Milne's predictor–corrector method. Compute the starting values at $x = -0.1,\ 0,\ 0.1,\ 0.2$ by Taylor's expansion about $x = 0$ where $y(0) = 1$, taking first four non-zero terms. Show your calculations up to four decimals only.

Solution

$$y(0) = 1,$$

$$y'(x) = x^2 + y^2 - 2, \qquad y'(0) = 0 + 1 - 2 = -1$$

$$y'(x) = 2x + 2yy', \qquad y''(0) = 0 + 2 \times 1 \times (-1) = -2$$

$$y'''(x) = 2 + 2(yy'' + y'^2), \qquad y'''(0) = 2 + 2(1 \times (-2) + 1) = 0$$

$$y^{iv}(x) = 2(yy''' + 3y'y''), \qquad y^{iv}(0) = 2(1 \times 0 + 3 \times (-1) \times (-2) = 12$$

The Taylor's series is given by,

$$y(x) \simeq 1 - x - x^2 + \frac{x^4}{2}$$

The values of y computed from Taylor's series are,

$$x = -0.1, \qquad y = 1.09$$

$$x = 0, \qquad y = 1$$

$x = 0.1, \qquad y = 0.89$

$x = 0.2, \qquad y = 0.7608$

$$y_3^p = 1.09 + \frac{4 \times 0.1}{3}\{2 \times (-1) - (0.01 + 0.7921 - 2) + 2(0.04 + 0.5975 - 2)\}$$

$$= 1.09 + \frac{4 \times 0.1}{3}(-4 - 0.8021 + 1.2390)$$

$$= 0.6149$$

$$y_3^c = 0.89 + \frac{0.1}{3}\{(0.01 + 0.89^2 - 2) + 4(0.04 + 0.7608 - 2) + (0.09 + 0.6149^2 - 2)\}$$

$$= 0.89 + \frac{0.1}{3}\{-12 + 0.8021 + 4 \times 0.6195 + 0.4682\}$$

$$= 0.89 - \frac{0.1}{3} \times 8.2518$$

$$= 0.6149$$

There is no change in the corrected value to an accuracy of 4 decimals. This is expected since the error in the formula is $O(h^5)$. From Taylor's series $y(0.3) = 0.61405$.

7.11 Boundary Value Problem (BVP)

We now discuss the finite difference method for solving second order linear differential equation,

$$\frac{d^2y}{dx^2} + p(x)\frac{dy}{dx} + q(x)y = r(x), \quad a \le x \le b, \tag{7.43}$$

with appropriate boundary conditions prescribed at the end points $x = a$ and $x = b$; the functions $p(x)$, $q(x)$ and $r(x)$ which are functions of x alone, are continuous in (a, b).

The general form of the boundary condition (b.c.) may be like, $\alpha y + \beta\frac{dy}{dx} = \gamma$, where α, β and γ are constants. If $\beta = 0$, the condition is said to be of Dirichlet type, i.e., function value is prescribed. If $\alpha = 0$, then the condition is called Neumann type, i.e., derivative is prescribed. And when none of them is zero, the condition is said to be a mixed boundary condition. When $\beta = \gamma = 0$, the condition is said to be homogeneous. Let us suppose mixed conditions are prescribed at both ends, i.e.,

$$\alpha_0 y_0 + \beta_0 y_0' = \gamma_0, \; x = a \tag{7.44}$$

$$\alpha_1 y_1 + \beta_1 y_1' = \gamma_1, \; x = b. \tag{7.45}$$

First of all we subdivide the solution domain $[a,\ b]$ into suitable number of intervals, n say, each of width h, i.e., $h = (b-a)/n$, and identify (nodal/pivotal) points $x_i,\ i = 0(1)n$ such that $a = x_0 < x_1 < x_2 \ldots. < x_n = b$. The solution $y_i,\ i = 0(1)n$ will be obtained at these points. The differential equations (7.43) is then discretized at the internal points $x_i,\ i = 1(1)n-1$, i.e., derivatives are replaced by their finite difference approximations. Similarly, boundary conditions (7.44) and (7.45) are also replaced by finite differences. By doing so, we arrive at a set linear simultaneous algebraic equations which after solving, give the values of $y_0, y_1, \ldots. y_n$ which is the required solution.

Let us recall the derivatives of first and second order derived in Chapter 5 in terms of Forward, Backward and Central differences; we reproduce them below:

A. First Derivative

(*i*) By Forward Difference

$$y'(x_i) = \frac{y_{i+1} - y_i}{h} + O(h) \tag{7.46}$$

(*ii*) By Backward Difference

$$y'(x_i) = \frac{y_i - y_{i-1}}{h} + O(h) \tag{7.47}$$

(*iii*) By Central Difference

$$y'(x_i) = \frac{y_{i+1} - y_{i-1}}{2h} + O(h^2) \tag{7.48}$$

B. Second Derivative

(*i*) By Forward Difference

$$y''(x_i) = \frac{y_i - 2y_{i+1} + y_{i+2}}{h^2} + O(h) \tag{7.49}$$

(*ii*) By Backward Difference

$$y''(x_i) = \frac{y_i - 2y_{i-1} + y_{i-2}}{h^2} + O(h) \tag{7.50}$$

(*iii*) By Central Difference

$$y''(x_i) = \frac{y_{i-1} - 2y_i + y_{i+1}}{h^2} + O(h^2) \tag{7.51}$$

It may be noted that for small h, the central difference approximation for the first derivative as well as for the second would be better as compared to the forward and backward difference approximation. Therefore, central difference formula will be used where-ever possible.

Thus at the internal points x_i, $i = 1(1)n-1$, the differential equation is discretized, replacing the derivatives by central difference formulae. Consequently we get,

$$\frac{y_{i-1} - 2y_i + y_{i+1}}{h^2} + p(x_i)\frac{y_{i+1} - y_{i-1}}{2h} + q(x_i)y_i = r(x_i),\ i = 1(1)n-1. \tag{7.52}$$

After rearranging the terms, and putting $p(x_i) = p_i$, $q(x_i) = q_i$ and $r(x_i) = r_i$, equation (7.52) may be written as,

$$(2 - hp_i)y_{i-1} - 2(2 - h^2q_i)y_i + (2 + hp_i)y_{i+1} = 2h^2r_i,\ i = 1(1)n-1. \tag{7.53}$$

The boundary conditions (7.44) and (7.45) are discretized in the following manner:

(*i*) At the end-point $x = a = x_0$, as there is no pivotal point to the left of x_0, the derivative in (7.44) is replaced by forward difference formula, giving

$$\alpha_0 y_0 + \beta_0 \frac{(y_1 - y_0)}{h} = \gamma_0$$

or $$(\alpha_0 h - \beta_0)y_0 + \beta_0 y_1 = \gamma_0 h. \tag{7.54}$$

Similarly, since there is no point to the right of $x_n = b$ the derivative is replaced there by backward difference, giving,

$$\alpha_1 y_n + \beta_1 \frac{(y_n - y_{n-1})}{h} = \gamma_1$$

or $$-\beta_1 y_{n-1} + (h\alpha_1 + \beta_1)y_n = \gamma_1 h. \tag{7.55}$$

There are $(n+1)$ equations given by (7.53), (7.54) and (7.55) in $(n+1)$ unknowns y_0, y_1, y_2, ... y_n, which may be determined by solving these equations. The number of equations may reduce to n or $n-1$ if $\beta_0 = \beta_1 = 0$ or β_0 or $\beta_1 = 0$. The resulting system of equations would be tri-diagonal which can be solved as discussed in Sec 2.16.

(*ii*) There is also a technique to replace the derivative at the end point by central difference formula. Suppose we want to discretize the b.c. (7.44) at $x = x_0$, then we extend the domain to the left of x_0 and assume a fictitious point $x_{-1} = x_0 - h$. The boundary condition can now be discretized as,

$$\alpha_0 y_0 + \beta_0 \frac{(y_1 - y_{-1})}{2h} = \gamma_0$$

$$\text{or} \quad -\beta_0 y_{-1} + 2h\alpha_0 y_0 + \beta_0 y_1 = 2h\gamma_0. \tag{7.56}$$

The differential equation too is discretized at $x = x_0$, giving

$$(2 - hp_0)y_{-1} - 2(2 - h^2 q_0)y_0 + (2 + hp_0)y_1 = 2h^2 r_0, \tag{7.57}$$

where p_0, q_0, r_0 are the values of p, q and r respectively at $x = x_0 = a$.
Substituting the value of y_{-1} from (7.56) in (7.57), we get

$$\{h\alpha_0(2 - hp_0) - \beta_0(2 - h^2 q_0)\}y_0 + 2\beta_0 y_1 = \gamma_0 h(2 - hp_0) + \beta_0 h^2 r_0. \tag{7.58}$$

Similarly for discretizing the b.c. (7.45), we assume a fictitious point x_{n+1} to the right of $x_n = b$ and replace the b.c. as,

$$\alpha_1 y_n + \beta_1 \frac{y_{n+1} - y_{n-1}}{2h} = \gamma_1$$

$$\text{or} \quad -\beta_1 y_{n-1} + 2h\alpha_1 y_n + \beta_1 y_{n+1} = 2h\gamma_1. \tag{7.59}$$

Discretization of d.e. at $x = x_n = b$ gives

$$(2 - hp_n)y_{n-1} - 2(2 - h^2 q_n)y_n + (2 + hp_n)y_{n+1} = 2h^2 r_n. \tag{7.60}$$

Eliminating y_{n+1} from (7.60) using (7.59), we get

$$-2\beta_1 y_{n-1} + \{h\alpha_1(2 + hp_n) + \beta_1(2 - h^2 q_n)\}y_n = \gamma_1 h(2 + hp_n) - h^2 r_n \beta_1. \tag{7.61}$$

The $(n-1)$ equations (7.53) together with (7.58) and (7.61) form a system of $(n+1)$ linear equations in y_0, y_1, ... y_n which can be evaluated by solving the equations. The system of equations would be tri-diagonal and can be solved as discussed in Sec 2.16.

Example 7.9

Solve the differential equation,

$$y'' + 2xy' + 2y = 5x, \quad 0 \le x \le 0.5,$$

satisfying the boundary conditions,

$$y(0) = 1, \quad y(0.5) = 1.5.$$

Compute up to 4 decimals taking $h = 0.1$.

Solution Discretizing the differential equation at the internal points $x_i = x_0 + ih = 0.1 \times i,\ i = 1,\ 2,\ 3,\ 4$, i.e., $x_1 = 0.1,\ x_2 = 0.2,\ x_3 = 0.3$ and $x_4 = 0.4$

$$\frac{y_{i-1} - 2y_i + y_{i+1}}{h^2} + 2x_i \frac{y_{i+1} - y_{i-1}}{2h} + 2y_i = 5x_i$$

or $\quad (1 - hx_i)y_{i-1} - 2(1 - h^2)y_i + (1 + hx_i)y_{i+1} = 5x_i h^2$

or $\quad (1 - 0.1x_i)y_{i-1} - 1.98y_i + (1 + 0.1x_i)y_{i+1} = 0.05x_i$

For various values of x_i, we get,

$x_1 = 0.1 \Rightarrow\ 0.99y_0 - 1.98y_1 + 1.01y_2 = 0.005$

or $\quad -1.98y_1 + 1.01y_2 = -0.985$

$x_2 = 0.2 \Rightarrow\ 0.98y_1 - 1.98y_2 + 1.02y_3 = 0.01$

$x_3 = 0.3 \Rightarrow\ 0.97y_2 - 1.98y_3 + 1.03y_4 = 0.015$

$x_4 = 0.4 \Rightarrow\ 0.96y_3 - 1.98y_4 + 1.04y_5 = 0.020$

or $\quad 0.96y_3 - 1.98y_4 = -1.540$

We are required to solve following system of equations:

y_1	y_2	y_3	y_4	
1.98	−1.01	0	0	0.985
0.98	−1.98	1.02	0	0.01
0	0.97	−1.98	1.03	0.015
0	0	0.96	−1.98	−1.540

This will be reduced to the following upper triangular form

1.98	−1.01	0	0	0.985
0	−1.480	1.02	0	−0.4775
0	0	−1.3114	1.03	0.2980
0	0	0	−1.2260	−1.7581

After back substitution we get,

$$y_4 = 1.4340,\ y_3 = 1.3535,\ y_2 = 1.2554\ y_1 = 1.1378$$

$y(0.1) \simeq 1.1378;\ y(0.2) \simeq 1.2554$

$y(0.3) \simeq 1.3535;\ y(0.4) \simeq 1.4340$

Example 7.10

Solve the differential equation,

$$xy'' + (x-1)y' - y = 0, \quad 0 \le x \le 0.75,$$

subject to conditions,

$$y'(0) = 1, \ y(0.75) = 1.3125.$$

Replace the derivative boundary condition by second order formula, taking $h = 0.25$.

Solution. The pivotal points are $x_0 = 0,\ x_1 = 0.25,\ x_2 = 0.50,\ x_3 = 0.75$. Discretization at $x = x_i$ gives,

$$x_i \frac{y_{i-1} - 2y_i + y_{i+1}}{h^2} + (x_i - 1)\frac{y_{i+1} - y_{i-1}}{2h} - y_i = 0$$

or $\quad 2x_i(y_{i-1} - 2y_i + y_{i+1}) + h(x_i - 1)(y_{i+1} - y_{i-1}) - 2h^2 y_i = 0$

or $\quad (2x_i - hx_i + h)y_{i-1} - (4x_i + 2h^2)y_i + (2x_i + hx_i - h)y_{i+1} = 0$

After substituting the value of h and rationalising,

$$(7x_i + 1)y_{i-1} - (116x_i + 0.5)y_i + (9x_i - 1)y_{i+1} = 0 \qquad \text{.... (1)}$$

At $x = 0$, the discretization of b.c. gives

$$\frac{y_1 - y_{-1}}{2 \times 0.25} = 1 \text{ or } y_{-1} = y_1 - 0.5 \qquad \text{.... (2)}$$

From (1) we get at $x = 0$,

$$y_{-1} - 0.5y_0 - y_1 = 0$$

Using (2), gives $\qquad y_0 = -1 \qquad$ .... (3)

Applying (1) at $x_1 = 0.25$,

$$2.75y_0 - 4.5y_1 + 1.25y_2 = 0$$

Using (3),

$$0.9y_1 - 0.25y_2 = -0.55 \qquad \text{.... (4)}$$

Discretizing at $x_2 = 0.50$, we get from (1),

$$4.5y_1 - 8.5y_2 + 3.5y_3 = 0$$

or $0.9y_1 - 1.7y_2 + 0.7y_3 = 0$

Using boundary condition $y_3 = 0.31125$,

$$0.9y_1 + 1.7y_2 = -0.21875 \qquad \text{.... (5)}$$

Solving (4) and (5) we get,

$$y_2 = -0.2284, \; y_1 - 0.6746$$

$$y_0 = -1, \; y_1 = -0.6746, \; y_2 = -0.2284.$$

Note: The analytical solution is $y = x - 1 + x^2$.

7.12 BVP as an Eigenvalue Problem

Consider a linear second order equation

$$\frac{d^2y}{dx^2} + p(x)\frac{dy}{dx} + q(x)y = \lambda y, \;\; a \le x \le b \tag{7.62}$$

where $p(x)$, $q(x)$ are known functions of x and λ is a parameter to be determined. Let the associated boundary conditions be given as

$$\left.\begin{aligned} \alpha_0 y_a + \beta_0 y'_a = 0, \;\; x = a \\ \alpha_1 y_b + \beta_1 y'_b = 0, \;\; x = b \end{aligned}\right\} \tag{7.63}$$

where α_0, β_0 and α_1, β_1 are constants.

Such problems frequently arise in science and engineering, like vibration problems. As we have seen, the finite difference equations reduce to linear tri-diagonal system. Refer case (i) of Sec 7.11, for discretization of (7.62) and (7.63); the corresponding equations (7.54), (7.53) and (7.55) modified as follows:

Putting in these equations respectively,

$$\left.\begin{aligned} & b_0 = (\alpha_0 h - \beta_0), \; c_0 = \beta_0; \\ & a_i = 2 - hp_i, \; b_i = -2(2 - h^2 q_i), \; c_i = (2 + hp_i) \\ & \quad i = 1, 2, \cdots n-1 \\ & a_n = -\beta_1, \;\; b_n = h\alpha_1 + \beta_1. \end{aligned}\right\} \tag{7.64}$$

The relevant equations may be written in matrix from as

$$Ay = \lambda y \text{ or } (A - \lambda I)y = 0 \tag{7.65}$$

where coefficient matrix A and vector y are given as

$$A = \begin{bmatrix} b_0 & c_0 & 0 & \cdots & \cdots & \cdots & 0 \\ a_1 & b_1 & c_1 & \cdots & \cdots & \cdots & \vdots \\ 0 & a_2 & b_2 & c_2 & \cdots & \cdots & \vdots \\ \vdots & \vdots & \vdots & \ddots & & & \vdots \\ \vdots & \vdots & \vdots & & \ddots & & 0 \\ 0 & 0 & 0 & \cdots & 0 \; a_{n-1} & b_{n-1} & c_{n-1} \\ 0 & 0 & 0 & \cdots & 0 \; 0 & a_n & b_n \end{bmatrix} \tag{7.66a}$$

and $y^{\mathrm{T}} = (y_0 \;\; y_1 \;\; y_2 \;\ldots y_n); \;\; y_0 = y_a, \;\; y_n = y_b.$ (7.66b)

Obviously (7.65) is an e.value problem which requires determination of e.values of matrix A. Normally largest/smallest e.values are of interest in practical problems which can be computed by Power Method discussed in Chapter 2.

Exercise 7

7.1 Find the solution of the differential equation,

$$\frac{dy}{dx} = -2xy, \; y(0) = 1$$

by Picard's method so that the error is not more than 0.0005 for $0 \le x \le 0.4$, i.e., computed value of y is correct up to 3 decimals.

7.2 Find the solution of the differential equation,

$$\frac{dy}{dx} = 1 + y^2, \;\; y(0) = 0$$

by Picard's method so that the error is not more than $\frac{1}{2} \times 10^{-4}$ for $0 \le x \le 0.2$, i.e., approximately correct up to four decimals.

7.3 Find first four terms by Picard's method in the solution of $y' = xy^3$, $y(0) = 1$.

7.4 Using Taylor's series, find the solution up to four non-zero terms, of the differential equation,

$$y' = x + y, \;\; y(1) = 0.$$

Compute y at $x = 1.5$, up to 4 decimals. What maximum error do you expect in your result.

7.5 Solve the differential equation,

$$y' = x - y^2, \; y(0) = 1$$

by Taylor's series and compute y for $x = 0.2$ so that the error is not more than 0.0005, i.e., value is correct up to 3 places of decimal.

7.6 Compute the value of y, y' and y'' at $x = 0.1$, by Taylor series of order four, (i.e., up to x^4 terms), from the differential equation,

$$y''' + 2y'' + y' - y = \cos x, \; 0 \leq x \leq 1,$$

$$y(0) = 0, \; y'(0) = 1 \text{ and } y''(0) = 2.$$

Compute error in each of them.

7.7 Estimate the value of $y(0.1)$ and $y(0.2)$ using Euler's method, with $h = 0.1$, from the differential equation

$$\frac{dy}{dx} = x - y, \;\; y(0) = 1.$$

What is the magnitude of errors in both cases.

7.8 Compute the value of y for $x = 1.2$ with (i) $h = 0.2$ $(ii) h = 0.1$, by Euler's method, from the differential equation,

$$\frac{dy}{dx} = \frac{y - x}{y + x}, \; y(1) = 2$$

7.9 Using Euler's modified formula, find the approximate value of y (0.2), from the differential equation,

$$y' = x + |\sqrt{y}|, \; y(0) = 1$$

Take $h = 0.2$ and compute up to four decimals.

7.10 Using fourth order classical R–K method compute the values of y for $x = 0.2$ and 0.4, with $h = 0.2$, from the differential equation,

$$y' = x\,(1 + y), \; y(0) = 1.$$

7.11 Solve the system of differential equations,

$$\frac{dy}{dx} = z + 1; \quad \frac{dz}{dx} = y - x,$$

$$y(0) = 1, \quad z(0) = 1$$

for $x = 0.2$, using fourth order R–K method.

7.12 Given $\frac{dy}{dx} = -\frac{y(x+y)}{x^2}$, $y(1) = 2$, compute the approximate values of y for $x = 1.1(0.1)1.3$ by Predictor–Corrector formula. First two values should be computed by Euler's P–C formula (modified Euler's method) and then the third value by Adams–Bashforth formula.

7.13 Solve $\frac{dy}{dx} = \frac{x}{y}$, $y(0) = 1$ by Taylor's series method, taking only first three non-zero terms in the power series. Obtain values of y for $x = 0.1(0.1)0.3$ from the Taylor's series. Then use Milne's P–C method to compute the approximate value of y at $x = 0.4$.
(Hint: Differentiate $yy' = x$)

7.14 Find the approximate solution of the boundary value problem,

$$xy'' + (x-1)y' - y = 0, \quad 0 \le x \le 1.0,$$

with boundary conditions $y(0) = 1$, $y(1) = 3$, subdividing the interval into four equal parts.

7.15 Find the approximate solution of the boundary value problem,

$$y'' - xy' + 2y = x + 4, \; 0 \le x \le 0.6$$

subject to boundary conditions, $y'(0) = 1$, $y(0.6) = 1.96$, subdividing the interval into three equal sub-intervals. Discretize the derivative boundary condition by central difference.
Also solve replacing the derivative b.c. at $x = 0$ by forward difference.

References and Some Useful Related Books/Papers

1. Collatz, L., *Numerical Treatment of Differential Equations*, Springer-Verlag.
2. Fox, L., *Numerical Solution of Ordinary and Partial Differential Equations*, Pergamon.

3. Gerald, C.F., and Wheatley, P.O., *Applied Numerical Analysis*, Pearson Education Asia.
4. Jain, M.K., *Numerical Solution of Differential Equations*, Wiley Eastern Ltd.

8

Splines and their Applications

8.1 Introduction

Spline is a draftsman tool which is used to draw a smooth curve passing through the specified points in a plane. It is a flexible metal strip attached with weights which can be adjusted to keep the strip in the required shape. As an analogy, in Numerical Analysis, a function (piece-wise polynomial) that describes a smooth curve through pre-assigned points is called 'spline'. The terms 'piece-wise polynomial' and 'smooth' will be explained later in the chapter.

Let us suppose that a function $f(x)$ is defined for $(n+1)$ values of x, say $x = x_i$, $i = 0(1)n$ so that the positions of $(n+1)$ points in the plane are given by $(x_i,\ f_i)$, $i = 0(1)n$ where $f(x_i) = f_i$, while the function $f(x)$ may or may not be known. We can fit a polynomial through these $(n+1)$ points by Lagrange's formula as,

$$P(x) = \sum_{i=0}^{n} L_i(x) f_i,$$

where $$L_i(x) = \frac{(x-x_0)(x-x_1)\cdots(x-x_{i-1})(x-x_{i+1})\cdots(x-x_n)}{(x_i-x_0)(x_i-x_1)\cdots(x_i-x_{i-1})(x_i-x_{i+1})\cdots(x_i-x_n)}.$$

As discussed in Secs 4.7 and 4.10 there are two drawbacks in Lagrange's form of polynomial. First, even for moderately large value of n, it will be quite cumbersome to express the polynomial in powers of x. Therefore, it is not suitable for performing mathematical operations other than interpolation. Second, even in interpolation it is not guaranteed that a global polynomial based on all the given points will provide better estimate as compared to a lower degree polynomial based on fewer points. Therefore, instead of a single polynomial, we shall be interested in fitting a set of lower degree polynomials for approximating the function over the whole domain. This idea was originally given by Schoenberg.

8.2 A Piece-Wise Polynomial

Let us assume, without loss of generality that $x_0 < x_1 < x_2 \ldots . < x_n$. Instead of a single polynomial for the entire domain $(x_0,\ x_n)$ we can approximate the function by several polynomials defined over subdomains of $(x_0,\ x_n)$.

For example, we can fit straight lines in each of the n intervals $(x_i,\ x_{i+1}),\ i = 0(1)n-1$; or, we can fit $\frac{n}{2}$ quadratics in the intervals $(x_i,\ x_{i+2}),\ i = 0(2)n-2$, n is even; or we can fit $\frac{n}{3}$ cubics in the intervals $(x_i,\ x_{i+3}),\ i = 0(3)n-3$ where n is a multiple of 3. It may be recalled that we have already done so in cases of Trapezoidal, Simpson's and Weddle's rules in Chapter 4. When a polynomial is represented over a certain domain by means of several polynomials defined over its subdomains, it is called a piece-wise representation of the polynomial or simply a piece-wise polynomial; the different polynomials defined over the subdomains are usually of the same degree. These polynomials are, no doubt, easier to handle as compared to a single polynomial over the whole domain.

In deriving piece-wise polynomials (linear, quadratic or cubic) in the examples stated above we do not give any consideration to the smoothness of the approximating curve at the intersection of the two intervals. That is, although continuity of the function (piece-wise polynomial) is maintained throughout the interval $(x_0,\ x_n)$ the continuity of derivatives is completely overlooked. The absence of continuity of derivatives at the intersection of the two adjacent subdomains will reflect adversely upon estimation of mathematical entities, specially near these points.

Figure 8.1a shows discontinuity of first derivative and Fig. 8.1b that of second derivative at point P while Fig. 8.1c shows continuity of first and second derivatives. All the curves are continuous though.

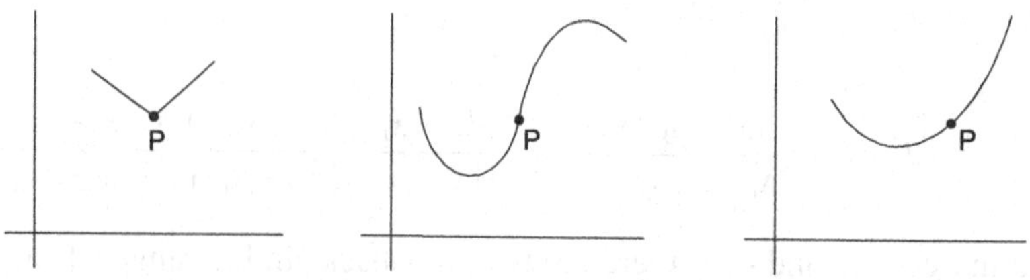

Fig.8.1a. c^0 continuity. **Fig.8.1b.** c^1continuity. **Fig.8.1c.** c^2 continuity.

8.3 Spline Approximation

A spline polynomial (or simply spline) is a piece-wise polynomial which may be defined as follows:

Given a strictly increasing sequence of values of x, say $x_0 < x_2 < x_3 \ldots. < x_n$, not necessarily equidistant, a spline $s(x)$ of degree k is a piece-wise polynomial such that it is of degree k in each of the intervals $(x_i,\ x_{i+1})$, $i = 0(1)n-1$ and that $s(x)$ and its first $(k-1)$ derivatives are continuous everywhere in $(x_0,\ x_n)$. The abscissas $x = x_i$, $i = 0(1)n$ are called 'knots'.

According to above definition, a straight line fit in each interval is a spline $s(x)$ of degree one. Although $s(x)$ will be continuous everywhere, its first derivative may not be continuous. The cubic spline $s(x)$ is most popular which provides continuity of $s(x)$, $s'(x)$ and $s''(x)$ everywhere in $(x_0,\ x_n)$. It is obtained by matching the value of the function and first and second derivatives of the cubics in the intervals $(x_{i-1},\ x_i)$ and $(x_i,\ x_{i+1})$ at the internal knots $x = x_i$, $i = 1(1)n-1$. As such the cubic spline provides a smooth curve in the sense that not only function value and gradient is same on both sides of $x = x_i$, the curvature also remains same. Thus we will be discussing cubic splines only which should be sufficient to tackle most of the practical problems where continuity of derivatives upto second order is required.

8.4 Uniqueness of Cubic Spline

Suppose we want to approximate $f(x)$ by a cubic spline $s(x)$ with knots $x_0,\ x_1,\ \ldots.\ x_n$. Let the values of the function $f(x)$ be known for $x = x_i$, $i = 0(1)n$ as $f(x_i)$ or f_i. Thus we have $(n+1)$ points $(x_i,\ f_i)$, $i = 0(1)n$, and want to fit n cubic polynomials, one each in the interval $(x_i,\ x_{i+1})$, $i = 0(1)n-1$. As there are four unknowns in a cubic, there will be total $4n$ unknown to be determined to define the cubic spline $s(x)$ in $(x_0,\ x_n)$. We can use following conditions to determine the unknowns:

(*i*) At the internal knots $x = x_i$, $i = 1(1)n-1$, following four conditions will be satisfied at each of them,

- (*a*) Cubic in the interval $(x_{i-1},\ x_i)$ will be satisfied at the point (x_i, f_i).
- (*b*) Cubic in the interval $(x_i,\ x_{i+1})$ will also be satisfied at (x_i, f_i).
- (*c*) Two conditions will be obtained by equating the first and second derivatives of the cubics in the intervals $(x_{i-1},\ x_i)$ and $(x_i,\ x_{i+1})$ at $x = x_i$.

Thus 4 $(n-1) = 4n-4$ equations can be obtained from the $(n-1)$ internal points.

(*ii*) Two conditions will be obtained from the end points, i.e., $s(x_0) = f_0$ and $s(x_n) = f_n$. In all, we have $4n-2$ equations, i.e., two short of $4n$ unknowns.
There are following possibilities for the two additional conditions which may be available at the end points x_0 and x_n:

(*a*) The value of the second derivative at the end points is zero, i.e., $s''(x_0) = 0$ and $s''(x_n) = 0$. It means, there is no change in the gradient beyond x_n and prior to x_0 and function in these domains may be approximated by straight lines. This is what precisely is the case with draftsman spline. Such a spline is called 'Natural' spline.

It may be noted that for a quintic spline, four extra conditions will be needed, two at each end. For example, we may have, $s'''(x_0) = s^{iv}(x_0) = 0$ and $s'''(x_n) = s^{iv}(x_n) = 0$. In that case it will be a quadratic in $(-\infty,\ x_0)$ and $(x_n,\ \infty)$. In general, if we are fitting a natural spline of degree $(2m-1)$ in $(x_0,\ x_n)$, it will be of degree $2m-1$ everywhere in $(x_0,\ x_n)$ but of degree $(m-1)$ in the domains $(-\infty,\ x_0)$ and $(x_n,\ \infty)$.

(*b*) The values of the second derivatives at the end points, i.e., $s''(x_0) = f_0''$ and $s''(x_n) = f_n''$ are prescribed.

(*c*) The values of the first derivatives at the end points are prescribed, i.e., $s'(x_0) = f_0'$ and $s'(x_n) = f_n'$ are known.

(*d*) First or second derivatives may be approximated at the end points by finite difference formulae.

(*e*) If $f(x)$ is periodic with period $(x_n - x_0)$ then we shall have $s(x_0) = s(x_n)$; and $s(x_1) = s(x_{n+1})$ where $x_{n+1} - x_n = x_1 - x_0$ and $s(x_{-1}) = s(x_{n-1})$ where $x_0 - x_{-1} = x_n - x_{n-1}$ (see Example 8.4). The knots are assumed at equal interval.

Thus, with the extra two conditions in any of the above forms we will have $4n$ equations in $4n$ unknowns which can be determined uniquely and hence the cubic spline in $(x_0,\ x_n)$.

8.5 Construction of Cubic Spline (Second Derivative Form)

A cubic spline $s(x)$ is a polynomial of degree three in each of the intervals $(x_i,\ x_{i+1})$, $i = 0(1)n-1$. Its first derivative $s'(x)$ will be of degree two and $s''(x)$ of degree one, i.e., linear in each interval. For sake of compactness and easy to follow, let us consider the first interval $(x_0,\ x_1)$. Let the second derivative at $x = x_0$ and $x = x_1$ be $s''(x_0)$ and $s''(x_1)$ respectively. Since $s''(x)$ is continuous and linear in $(x_0,\ x_1)$, we can write,

$$s''(x) = s''(x_0) + \frac{s''(x_1) - s''(x_0)}{x_1 - x_0}(x - x_0),\ x_0 \le x \le x_1. \tag{8.1}$$

Integrating (8.1) from x_0 to x, we get

$$s'(x) = s'(x_0) + (x - x_0)s''(x_0) + \frac{(x-x_0)^2}{2} \cdot \frac{s''(x_1) - s''(x_0)}{x_1 - x_0}, \; x_0 \le x \le x_1. \tag{8.2}$$

Further integration of (8.2) from x_0 to x gives,

$$s(x) = s(x_0) + (x - x_0)s'(x_0) + \frac{(x-x_0)^2}{2}s''(x_0) + \frac{(x-x_0)^3}{6} \cdot \frac{s''(x_1) - s''(x_0)}{x_1 - x_0}, \quad x_0 \le x \le x_1. \tag{8.3}$$

The cubic as given by (8.3) uses three values at x_0, viz., $s(x_0)$, $s'(x_0)$ and $s''(x_0)$ and only one value $s''(x_1)$ at x_1. We express the cubic (8.3) in a form containing the values of $s(x)$ and $s''(x)$ at $x = x_0$ and $x = x_1$ by eliminating $s'(x_0)$ from (8.3). In order to do this we put $x = x_1$ in (8.3) getting,

$$s(x_1) = s(x_0) + (x_1 - x_0)s'(x_0) + \frac{(x_1-x_0)^2}{2}s''(x_0) + \frac{(x_1-x_0)^2}{6} \cdot \{s''(x_1) - s''(x_0)\}.$$

$$= s(x_0) + (x_1 - x_0)s'(x_0) + \frac{(x_1-x_0)^2}{6}\{s''(x_1) + 2s''(x_0)\}.$$

or $$s'(x_0) = \frac{s(x_1) - s(x_0)}{x_1 - x_0} - \frac{(x_1 - x_0)}{6}\{s''(x_1) + 2s''(x_0)\}. \tag{8.4}$$

Using the value of $s'(x_0)$ from (8.4) in (8.2) we get the value of $s'(x)$ at $x = x_1$ as,

$$s'(x_1) = \frac{s(x_1) - s(x_0)}{x_1 - x_0} - \frac{x_1 - x_0}{6}\{s''(x_1) - 2s''(x_0)\}$$

$$+(x_1 - x_0)s''(x_0) + \frac{x_1 - x_0}{2}\{s''(x_1) - s''(x_0)\}$$

$$= \frac{s(x_1) - s(x_0)}{x_1 - x_0} + \frac{x_1 - x_0}{6}\{2s''(x_1) + s''(x_0)\}. \tag{8.5}$$

It may be noted that (8.4) gives the value of the first derivative to the right of x_0, i.e., $s'(x_0+)$ and (8.5) is the value of the first derivative to the left of x_1, i.e., $s'(x_1-)$ in the interval (x_0, x_1). We will use condition (8.5) afterwards.

Now, putting the value of $s'(x_0)$ from (8.4) in (8.3) we get,

$$s(x) = s(x_0) + (x - x_0)\left[\frac{s(x_1) - s(x_0)}{x_1 - x_0} - \frac{x_1 - x_0}{6}\{s''(x_1) + 2s''(x_0)\}\right]$$

$$+\frac{(x-x_0)^2}{2}s''(x_0) + \frac{(x-x_0)^3}{6}\left\{\frac{s''(x_1) - s''(x_0)}{x_1 - x_0}\right\}, \; x_0 \le x \le x_1. \tag{8.6}$$

To define $s(x)$ in the interval $(x_i,\ x_{i+1})$ we can replace x_0 by x_i and x_1 by x_{i+1} in (8.6) giving,

$$s(x) = s(x_i) + (x - x_i)\left[\frac{s(x_{i+1}) - s(x_i)}{x_{i+1} - x_i} - \frac{x_{i+1} - x_i}{6}\{s''(x_{i+1}) + 2s''(x_i)\}\right]$$

$$+\frac{(x-x_i)^2}{2}s''(x_i) + \frac{(x-x_i)^3}{6}\left\{\frac{s''(x_{i+1}) - s''(x_i)}{x_{i+1} - x_i}\right\},\ x_i \le x \le x_{i+1} \tag{8.7}$$

where $s(x_i) = f(x_i) = f_i$ and $s(x_{i+1}) = f(x_{i+1}) = f_{i+1}$. The required cubic spline $s(x)$ is given by (8.7) in the interval $(x_i,\ x_{i+1}),\ i = 0(1)n-1$.

However the values of the second derivatives are not given; they are unknown and are still to be determined. In order to compute the second derivatives, we equate at the internal knots $x_i,\ i = 1(1)n-1$, the first derivatives of the cubic polynomial defined in the intervals $(x_{i-1},\ x_i)$ and $(x_i,\ x_{i+1})$, i.e.,

$$s'(x_i+) = s'(x_i-),\ i = 1(1)n-1 \tag{8.8}$$

where $s'(x_i+)$ is the value of the first derivative to the right of x_i to be obtained from the interval (x_i, x_{i+1}) and $s'(x_i-)$ is the value of the first derivative to the left of x_i to be obtained from $(x_{i-1},\ x_i)$.

Thus $s'(x_i+)$ is obtained from (8.4) replacing x_0 by x_i and x_1 by x_{i+1}, giving,

$$s'(x_i+) = \frac{s(x_{i+1}) - s(x_i)}{x_{i+1} - x_i} - \frac{x_{i+1} - x_i}{6}\{s'(x_{i+1}) + 2s''(x_i)\} \tag{8.9}$$

Similarly, value of $s'(x_i-)$ may be obtained from (8.5) replacing x_0 by x_{i-1} and x_1 by x_i, giving

$$s'(x_i-) = \frac{s(x_i) - s(x_{i-1})}{x_i - x_{i-1}} + \frac{x_i - x_{i-1}}{6}\{2s''(x_i) + s''(x_{i-1})\}. \tag{8.10}$$

Equating (8.9) and (8.10) at the internal knots, we get

$$(x_i - x_{i-1})s''(x_{i-1}) + 2(x_{i+1} - x_{i-1})s''(x_i) + (x_{i+1} - x_i)s''(x_{i+1})$$

$$= 6\left[\frac{s(x_{i+1}) - s(x_i)}{x_{i+1} - x_i} - \frac{s(x_i) - s(x_{i-1})}{x_i - x_{i-1}}\right],\ i = 1(1)n-1, \tag{8.11}$$

Writing $s''(x_i) = \mathrm{M}_i,\ x_i - x_{i-1} = h_i$ and $s(x_i) = f_i$, etc., (8.11) becomes,

$$h_i M_{i-1} + 2(h_i + h_{i+1})M_i + h_{i+1}M_{i+1} = 6\left[\frac{f_{i+1} - f_i}{h_{i+1}} - \frac{f_i - f_{i-1}}{h_i}\right],\ i = 1(1)n-1. \quad (8.12)$$

If all the intervals are of the same length h, i.e., $h = x_{i+1} - x_i$, $i = 0(1)n-1$, then (8.12) may be written as,

$$M_{i-1} + 4M_i + M_{i+1} = \frac{6}{h^2}(f_{i-1} - 2f_i + f_{i+1}), \quad (8.13)$$

$$i = 1(1)n-1.$$

From (8.11), (8.12) or (8.13) we get $(n-1)$ simultaneous linear equations in $(n+1)$ unknowns M_0, M_1, ... M_n. If $M_0 = s''(x_0)$ and $M_n = s''(x_n)$ are prescribed, we get a tridiagonal system of equations which can be easily solved. Putting the values of the second derivative we can determine $s(x)$ from (8.7) in any interval. For a natural cubic spline $s''(x_0) = s''(x_n) = 0$. We can find the equation of the straight lines in the intervals $(-\infty, x_0)$ and (x_n, ∞) after computing $s'(x_0)$ from (8.9) and $s'(x_n)$ from (8.10), respectively.

8.6 Construction of Cubic Spline (First Derivative Form)

In Sec 8.5, we have constructed the spline $s(x)$ in terms of function values and second derivatives. Alternatively, we can express $s(x)$ in terms of function values and the first derivatives also. The procedure is based on Hermite's interpolation formula.

Let us consider the first interval (x_0, x_1) instead of the general interval (x_i, x_{i+1}), $i = 0\ (1)n-1$. We can express the cubic spline in (x_0, x_1) in terms of four values, viz., $s(x_0), s'(x_0)$ and $s(x_1)$ and $s'(x_1)$ as follows:

Let $L_0(x)$ and $L_1(x)$ be the Lagrange's coefficients with abcissas x_0, x_1 and let $\phi_0(x)$, $\phi_1(x)$, $\psi_0(x)$, $\psi_1(x)$ be linear functions of x. Then, we represent the cubic spline as,

$$s(x) = \phi_0(x)L_0{}^2(x)s(x_0) + \phi_1(x)L_1{}^2(x)s(x_1) + \psi_0(x)L_0{}^2(x)s'(x_0) + \psi_1(x)L_1{}^2(x)s'(x_1), \quad (8.14)$$

$$= [\phi_0(x)\ s(x_0) + \psi_0(x)\ s'(x_0)]L_0{}^2(x) + [\phi_1(x)\ s(x_1) + \psi_1(x)\ s'(x_1)]L_1{}^2(x). \quad (8.14a)$$

The Lagrange's coefficients $L_0(x)$ and $L_1(x)$ are given by,

$$L_0(x) = \frac{x - x_1}{x_0 - x_1},\ L_1(x) = \frac{x - x_0}{x_1 - x_0};$$

and the linear functions $\phi(x)$ and $\psi(x)$ are

$$\phi_0(x) = a_0x + b_0; \quad \phi_1(x) = a_1x + b_1 \text{ and}$$

$$\psi_0(x) = c_0x + d_0; \quad \psi_1(x) = c_1x + d_1.$$

The functions ϕ_0, ψ_0 and ϕ_1, ψ_1 are to be determined. It may be noted that $L_0{}^2(x)$ and its first derivative, $2L_0(x)L_0'(x)$ vanish for $x = x_1$ and $L_1{}^2(x)$ and its first derivative, $2L_1(x)L_1'(x)$ vanish at $x = x_0$, since we know that $L_0(x_1) = 0$ and $L_1(x_0) = 0$. Also, $L_0(x_0) = 1$ and $L_1(x_1) = 1$ and in general, $L_i(x_j) = 0,\ i \neq j$ and $L_i(x_i) = 1$.

Putting $x = x_0$ in (8.14) and comparing the coefficients,

$$s(x_0) = \phi_0(x_0)s(x_0) + \psi_0(x_0)s'(x_0), \text{ rendering } \phi_0(x_0) = 1 \text{ and } \psi_0(x_0) = 0.$$

$$\text{or} \quad a_0x_0 + b_0 = 1 \quad \text{and} \quad c_0x_0 + d_0 = 0. \tag{8.15}$$

Differentiating (8.14), we get,

$$s'(x) = 2[\phi_0(x)\, s(x_0) + \psi_0(x)\, s'(x_0)]L_0(x)L_0'(x) + [\phi_0'(x)\, s(x_0) + \psi_0'(x)\, s'(x_0)]L_0{}^2(x)$$
$$+2[\phi_1(x)\, s(x_1) + \psi_1(x)\, s'(x_1)]L_1(x)L_1'(x) + [\phi_1'(x)\, s(x_1) + \psi_1'(x)\, s'(x_1)]L_1{}^2(x).$$

Putting $x = x_0$ in the above equation gives

$$s'(x_0) = 2[\phi_0(x_0)s(x_0) + \psi_0(x_0)s'(x_0)]\frac{1}{x_0 - x_1} + [\phi_0'(x_0)s(x_0) + \psi_0'(x_0)s'(x_0)]$$

$$= \left[-\frac{2}{h_1}\phi_0(x_0) + \phi_0'(x_0)\right]s(x_0) + \left[-\frac{2}{h_1}\psi_0(x_0) + \psi_0'(x_0)\right]s'(x_0)$$

where $\quad h_1 = x_1 - x_0$.

On comparing the coefficients,

$$\left.\begin{aligned} &\phi_0'(x_0) - \frac{2}{h_1}\phi_0(x_0) = 0 \text{ or } a_0 = \frac{2}{h_1}(a_0x_0 + b_0) \\ \text{and}\quad &\psi_0'(x_0) - \frac{2}{h_1}\psi_0(x_0) = 1 \text{ or } c_0 - 1 = \frac{2}{h_1}(c_0x_0 + d_0) \end{aligned}\right\} \tag{8.16}$$

From (8.15) and (8.16) we get

$$\left.\begin{aligned} &a_0 = \frac{2}{h_1} \text{ and } b_0 = 1 - \frac{2x_0}{h_1} \\ &c_0 = 1 \text{ and } d_0 = -x_0. \end{aligned}\right\}. \tag{8.17}$$

Hence, we have

$$\left.\begin{aligned}\phi_0(x) &= a_0x+b_0 = 1+\frac{2(x-x_0)}{h_1},\\ \psi_0(x) &= c_0x+d_0 = x-x_0.\end{aligned}\right\}. \tag{8.18}$$

Proceeding in the same manner for $x = x_1$ we can get,

$$\left.\begin{aligned}\phi_1(x) &= a_1x+b_1 = 1-\frac{2(x-x_1)}{h_1}\\ \psi_1(x) &= c_1x+d_1 = x-x_1.\end{aligned}\right\} \tag{8.19}$$

Putting the values of $\phi(x)$ and $\psi(x)$, etc., in (8.14), cubic spline may be written in the interval $(x_i,\ x_{i+1})$ by replacing x_0 by x_i and x_1 by x_{i+1}, as

$$\begin{aligned}s(x) &= \{2(x-x_i)+h_{i+1}\}\frac{(x-x_{i+1})^2}{h_{i+1}^3}s(x_i)-\{2(x-x_{i+1})-h_{i+1}\}\frac{(x-x_i)^2}{h_{i+1}^3}s(x_{i+1})\\ &+\frac{(x-x_i)(x-x_{i+1})^2}{h_{i+1}^2}s'(x_i)+\frac{(x-x_{i+1})(x-x_i)^2}{h_{i+1}^2}s'(x_{i+1}),\ x_i\le x\le x_{i+1},\ i=0(1)n-1,\end{aligned} \tag{8.20}$$

where $h_{i+1} = x_{i+1}-x_i$; $s(x_0) = f(x_0) = f_0$, $f(x_i) = f_i$ and $s(x_{i+1}) = f(x_{i+1}) = f_{i+1}$.

The cubic spline $s(x)$ given by (8.20) is the required spline in terms of the function values and first derivatives. But again $s'(x_i)$, $i = 0(1)n$ are not known which are still to be determined. To find them we use the condition of continuity of the second derivative in $(x_0,\ x_n)$, i.e., $s''(x_i+) = s''(x_i-)$ at the internal knots x_i, $i = 1(1)n-1$.
Now, differentiating (8.20) twice w.r.t. x,

$$\begin{aligned}s''(x) &= \frac{2}{h_{i+1}^3}[2(x-x_i)+4(x-x_{i+1})+h_{i+1}]s(x_i)-\frac{2}{h_{i+1}^3}[4(x-x_i)+2(x-x_{i+1})-h_{i+1}]\\ &s(x_{i+1})+\frac{2}{h_{i+1}^2}[(x-x_i)+2(x-x_{i+1})]s'(x_i)+\frac{2}{h_{i+1}^2}[2(x-x_i)+(x-x_{i+1})]s'(x_{i+1}),\\ &\qquad x_i\le x\le x_{i+1}.\end{aligned} \tag{8.21}$$

Putting $x = x_i$ in (8.21) we get,

$$s''(x_i+) = \frac{6}{h_{i+1}^2}[s(x_{i+1})-s(x_i)]-\frac{2}{h_{i+1}}[s'(x_{i+1})+2s'(x_i)]. \tag{8.22}$$

Again, writing (8.21) for the interval (x_{i-1}, x_i) and putting $x = x_i$, we get

$$s''(x_i-) = \frac{6}{h_i^2}[s(x_{i-1}) - s(x_i)] - \frac{2}{h_i}[s'(x_{i-1}) + 2s'(x_i)],\ h_i = x_i - x_{i-1}. \tag{8.23}$$

Equating (8.22) and (8.23) we get,

$$\frac{1}{h_i}s'(x_{i-1}) + 2\left(\frac{1}{h_i} + \frac{1}{h_{i+1}}\right)s'(x_i) + \frac{1}{h_{i+1}}s'(x_{i+1}) = 3\left[\frac{s(x_{i+1}) - s(x_i)}{h_{i+1}^2} + \frac{s(x_i) - s(x_{i-1})}{h_i^2}\right],$$
$$i = 1(1)n-1, \tag{8.24}$$

where $s(x_i) = f(x_i) = f_i$ and $s(x_{i+1}) = f(x_{i+1}) = f_{i+1},\ h_i = x_i - x_{i-1}$.

Putting $s'(x_i) = m_i$ in (8.24) for brevity, it can be written as,

$$\frac{1}{h_i}m_{i-1} + 2\left(\frac{1}{h_i} + \frac{1}{h_{i+1}}\right)m_i + \frac{1}{h_{i+1}}m_{i+1} = 3\left[\frac{f_{i+1} - f_i}{h_{i+1}^2} + \frac{f_i - f_{i-1}}{h_i^2}\right]. \tag{8.25}$$

If all the intervals are of the same length, say h, i.e., $x_i - x_{i-1} = h,\ i = 1(1)n$, then (8.25) may be written as,

$$m_{i-1} + 4m_i + m_{i+1} = \frac{3}{h}(f_{i+1} - f_{i-1}),\ i = 1(1)n-1. \tag{8.26}$$

We see that there are $(n-1)$ equations in (8.25) but number of unknowns is $(n+1)$, viz., $s'(x_i),\ i = 0(1)n$. If the first derivatives are prescribed at the end points, i.e., $s'(x_0) = f_0'$ and $s'(x_n) = f_n'$ are given, then (8.25) may be solved giving a unique solution. Having got the values of the first derivatives, they can be substituted in (8.20) to get the required spline.

8.7 Minimal Property of a Cubic Spline

Let $s(x)$ be a cubic spline with knots $a = x_0 < x_1 < x_2 \ldots < x_n = b$ approximating the function $f(x)$ so that $s(x_i) = f(x_i),\ i = 0(1)n$. If $g(x)$ is any other twice differentiable function in (a, b), i.e., $g(x)\varepsilon c^2(a, b)$ which also approximates $f(x)$ satisfying $g(x_i) = f(x_i)$ implying $s(x_i) = g(x_i),\ i = 0(1)n$, then we will show that,

$$\int_a^b [s''(x)]^2 dx \le \int_a^b [g''(x)]^2 dx. \tag{8.27}$$

That is, mean square value of the second derivative of $s(x)$ is minimum as compared to any other function having continuity of order c^2. Further, the curvature of $y(x)$ is given by $(y'')^2/[1+(y')^2]^{3/2}$, i.e., curvature is dependent on the value of the second derivative, assuming that the first derivative $f'(x)$ does not change rapidly. Thus property (8.27) is known as minimum curvature property of a cubic spline.

In order to prove the inequality (8.27), we express

$$g''^2(x) = \{g''(x) - s''(x) + s''(x)\}^2$$

or

$$g''^2(x) - s''^2(x) = [g''(x) - s''(x)]^2 + 2s''(x)[g''(x) - s''(x)]. \tag{8.28}$$

Integrating (8.28) between a and b we get,

$$\int_a^b g''^2(x)dx - \int_a^b s''^2(x)dx = \int_a^b [g''(x) - s''(x)]^2 dx + 2\int_a^b s''(x)[g''(x) - s''(x)]dx. \tag{8.29}$$

The second integral in (8.29) can be evaluated as,

$$\int_a^b s''(x)[g''(x) - s''(x)]dx = [s''(x)\ g'(x) - s'(x)]_a^b - \int_a^b s'''(x)[g'(x) - s'(x)]dx$$

$$= s''(b)\ g'(b) - s'(b)\} - s''(a)\ \{g'(a)\ - s'(a)$$

$$- \sum_{i=0}^{n-1} \int_{x_i}^{x_{i+1}} s'''(x)[g'(x) - s'(x)]dx. \tag{8.30}$$

The integral on the right side (8.30) has to be evaluated as sum of integrals over each interval since $s'''(x)$ is constant in each interval and is most likely different in each interval and hence discontinuous. Let its value in the interval $(x_i,\ x_{i+1})$ be a_{i+1}, then

$$\sum_{i=0}^{n-1} \int_{x_i}^{x_{i+1}} s'''(x)[g'(x) - s'(x)]dx = \sum_{i=0}^{n-1} a_{i+1}[g(x) - s(x)]_{x_i}^{x_{i+1}} = 0, \tag{8.31}$$

since $\qquad g(x_i) = s(x_i),\ i = 0(1)n.$

Therefore, after using (8.30) and (8.31) integral (8.29) becomes,

$$\int_a^b g''^2(x)dx - \int_a^b s''^2(x)dx = \int_a^b [g'(x) - s'(x)]^2 dx + 2[s''(b)\{g'(b) - s'(b)\}$$

$$-s''(a)\{g'(a) - s'(a)\}]. \tag{8.32}$$

In case $s(x)$ is a natural spline, then $s''(a) = s''(b) = 0$; bracketed term in (8.32) become zero.

Hence, as the integral on the right side is ≥ 0, we get,

$$\int_a^b g''^2(x)dx \geq \int_a^b s''^2(x)dx,$$

the equality occurs when $g(x) = s(x)$.

Alternatively, when first derivatives is prescribed at the end points and $g(x)$ is such that $s'(a) = g'(a)$ and $s'(b) = g'(b)$, then also inequality remains true. See [1], [6].

Example 8.1

Using cubic spline find the value of the function $f(x) = \ln x$ for $x = 1.8$ from the following data:

x	1.0	1.2	1.6	2.0
$f(x)$	0.0	0.18232	0.47000	0.69315

(*i*) When values of second derivatives $f''(x)$ are given as end conditions.

(*ii*) When values of first derivatives $f'(x)$ are given as end conditions.

Solution $f'(x) = \dfrac{1}{x}$; $f'' = -\dfrac{1}{x^2}$

$$f'(1) = 1 = m_0,\ f'(2) = 0.5 = m_3;\ f''(1) = -1 = M_0,\ f''(2) = -0.25 = M_3.$$

(*i*) The governing equations for second derivatives are,

$$0.2M_0 + 2\ (0.2+0.4)\ M_1 + 0.4M_2 = 6\left[\frac{0.28768}{0.4} - \frac{0.18232}{0.2}\right]$$

or $$1.2M_1 + 0.4M_2 = -0.95440 \quad \ldots(1)$$

$$0.4M_1 + 2(0.4+0.4)M_2 + 0.4M_3 = 6\left[\frac{0.22315}{0.4} - \frac{0.28768}{0.4}\right]$$

or $$0.4M_1 + 1.6M_2 = -0.86795 \quad \ldots(2)$$

Solving (1) and (2), we get

$$M_1 = -0.67037; \;\; M_2 = -0.37488$$

Using spline in the interval (1.6, 2.0),

$$s(1.8) = 0.47 + (1.8-1.6)\left[\frac{0.22315}{0.4} - \frac{0.4}{6}\{(-0.25)+2(-0.37488)\}\right.$$

$$\left. + \frac{(1.8-1.6)^2}{2}(-0.37488) + \frac{(1.8+1.6)^3}{6}\left[\frac{-0.25-(-0.37488)}{0.4}\right]\right]$$

$$= 0.47 + 0.2(0.55788+0.06665) - 0.00750 + 0.00042$$

$$= 0.47 + 0.12491 - 0.00750 + 0.00042$$

$$= 0.58783 \text{ (exact value is 0.58779).}$$

(*ii*) Values of first derivative given as end conditions

$$f'(1) = s'(1) = 1 = m_0; \;\; f'(2) = s'(2) = 0.5 = m_3.$$

The governing equations are,

$$\frac{1}{0.2}m_0 + 2\left(\frac{1}{0.2}+\frac{1}{0.4}\right)m_1 + \frac{1}{0.4}m_2 = 3\left[\frac{0.28768}{0.4^2}+\frac{0.18232}{0.2^2}\right]$$

or

$$6m_1 + m_2 = 5.6272 \qquad \ldots (1)$$

$$\frac{1}{0.4}m_1 + 2\left(\frac{1}{0.4}+\frac{1}{0.4}\right)m_2 + \frac{1}{0.4}m_3 = \left[\frac{0.22315}{0.4^2}+\frac{0.28768}{0.4^2}\right]$$

or

$$m_1 + 4m_2 = 3.3312 \qquad \ldots (2)$$

Solving (1) and (2),

$$m_1 = 0.83382, \; m_2 = 0.62435$$

The spline in the interval $1.6 \le x \le 2.0$,

$$s(x) = \{2(x-1.6)+0.4\}\frac{(x-2)^2}{(0.4)^3}\times 0.47 - \{2(x-2)-0.4\}\frac{(x-1.6)^2}{(0.4)^3}\times 0.69315$$

$$+\frac{(x-1.6)(x-2)^2}{(0.4)^2}\times 0.62435 + \frac{(x-2)(x-1.6)^2}{(0.4)^2}\times 0.5$$

Putting $x = 1.8$

$$s(1.8) = \{2\times 0.2+0.4\}\frac{(-0.2)^2}{0.4\times 0.16}\times 0.47-\{2(-0.2)-0.4\}\frac{(0.2)^2}{0.4\times 0.16}\times 0.69315$$

$$+\frac{(0.2)(-0.2)^2}{0.16}\times 0.62435+\frac{(-0.2)(0.2)^2}{0.16}\times 0.5$$

$$=\frac{1}{0.16}\left[\frac{0.8\times 0.04}{0.4}\times 0.47+\frac{0.8\times 0.04}{0.4}\times 0.69315+0.2\times 0.04\times 0.62435-0.2\times 0.04\times 0.5\right]$$

$$=\frac{0.04}{0.16}[0.94+1.3863+0.12487-0.1]$$

$= 0.58779$ (exact value is 0.58779)

The minor difference between the values computed in two different ways is due to rounding; not that one is better or worse than the other.

Example 8.2

Find the cube-root of 21 by fitting a natural cubic spline to the following table of values of cube-root.

x	0	1	8	27
$f(x) = x^{1/3}$	0	1	2	3

Solution In the natural cubic spline, $M_0 = 0,\ M_3 = 0$
The governing simultaneous equations are:

$$1\times M_0+2(1+7)M_1+7M_2 = 6\left[\frac{2-1}{8-1}-\frac{1-0}{1-0}\right]$$

or $$16M_1+7M_2 = -5.94286 \qquad \ldots(1)$$

$$7M_1+2(7+19)M_2+19M_3 = 6\left[\frac{3-2}{19}-\frac{2-1}{8-1}\right]$$

or $$7M_1+52M_2 = -0.54135 \qquad \ldots(2)$$

Solving (1) and (2) gives $M_2 = 0.0349$
The spline in the interval $8 \le x \le 27$ is given by,

$$s(x) = 2+(x-8)\left[\frac{3-2}{27-8}-\frac{27-8}{6}\{0+2\times 0.0349\}\right.$$

$$\left.+\frac{(x-8)^2}{2}\times 0.0349+\frac{(x-8)^3}{6}\left[\frac{0-0.0349}{27-8}\right]\right.$$

Putting $x = 21$,

$$s(21) = 2 + 13\left[\frac{1}{19} - \frac{19}{3} \times 0.0349\right] + \frac{169}{2} \times 0.0349 + \frac{2197}{6}\left(-\frac{0.0349}{19}\right)$$

$$= 2 + 13(0.0526 - 0.2210) + 2.9490 - 0.6726$$

$$= 2 - 2.1892 + 2.9490 - 0.6726$$

$$= 2.0872 \text{ (exact value is 2.7589)}$$

Note: We have approximated the function $f(x) = x^{1/3}$ by a cubic. Further, we should note that the value of $f'(x)$ and $f''(x)$ at $x = 0$ is ∞. At the other end also the second derivative $f''(27)$ is not zero. But we take them to be zero for fitting a natural spline. Besides, the interval (8, 27) is too large as compared to function values. Therefore the approximation is expected to be not too good.

Compare with Lagrange's formula used in Example 4.15.

Example 8.3

From the data of Example 8.2, compute cube root of 21 using cubic spline when the derivatives at the end-points are approximated by FD/BD as follows:

(*i*) Second derivatives approximated by 3-point formula.

(*ii*) First derivative approximated by 2-point formula.

Solution (*i*) We approximate the second derivative at $x = 0$ by forward difference formula discussed in Chapter 5.

$$M_0 = f''(0) = \frac{2}{1 \times 7(1+7)}\{7 \times 0 - (1+7) \times 1 + 1 \times 2\}$$

$$= -0.2143$$

Similarly, we approximate the second derivative at $x = 27$ by backward difference formula,

$$M_3 = f''(27) = \frac{2}{7 \times 19(7+19)}(7 \times 3 - 26 \times 2 + 19 \times 1)$$

$$= -0.00694$$

The governing equations are,

$$1 \times (-0.2143) + 16M_1 + 7M_2 = 6\left[\frac{2-1}{8-1} - \frac{1-0}{1-0}\right]$$

or $$16M_1 + 7M_2 = -4.92856 \qquad \ldots (1)$$

$$7M_1 + 2(7+19)M_2 + 19 \times (-0.00694) = 6\left[\frac{3-2}{27-8} - \frac{2-1}{8-1}\right]$$

or $$7M_1 + 52M_2 = -0.40949 \qquad \ldots(2)$$

From (1) and (2) we get, $M_2 = 0.03569$

The spline in the interval (8, 27) is given by,

$$s(x) = 2 + (x-8)\left[\frac{3-2}{27-8} - \frac{27-8}{6}\{-0.00694 + 2 \times 0.03569\}\right]$$

$$+ \frac{(x-8)^2}{2} \times 0.03569 + \frac{(x-8)^3}{6}\left[\frac{-0.00694 - 0.03569}{27-8}\right]$$

$$s(21) = 2 + 13\left[\frac{1}{19} - \frac{19}{6} \times 0.06444\right] + \frac{169}{2} \times 0.03569 - \frac{2197}{6} \times \frac{0.04263}{19}$$

$$= 2 + 13[0.05263 - 0.20406] + 3.0158 - 0.82156$$

$$= 5.0158 - 1.9686 - 0.82156$$

$$= 2.2256 \text{ (exact value is 2.7589)}$$

(*ii*) $$m_0 = f'(x_0) = f'(0) = \frac{1-0}{1-0} = 1.$$

$$m_3 = f'(x_3) = f'(3) = \frac{3-2}{27-8} = \frac{1}{19} = 0.05263$$

The governing equations are,

$$\frac{1}{1}m_0 + 2\left(\frac{1}{1} + \frac{1}{7}\right)m_1 + \frac{1}{7}m_2 = 3\left[\frac{2-1}{(8-1)^2} + \frac{1-0}{(1-0)^2}\right]$$

$$\frac{16}{7}m_1 + \frac{1}{7}m_2 = \frac{3 \times 50}{49} - 1$$

or $$16m_1 + m_2 = 14.4286 \qquad \ldots(1)$$

$$\frac{1}{7}m_1 + 2\left(\frac{1}{7} + \frac{1}{19}\right)m_2 + \frac{1}{19}m_3 = 3\left[\frac{3-2}{19^2} + \frac{2-1}{7^2}\right]$$

$$19m_1 + 52m_2 = \frac{3 \times 19 \times 7}{19^2 \times 7^2}(410) - \frac{7}{19}$$

or $$19m_1 + 52m_2 = 8.8797 \qquad \ldots(2)$$

Solving (1) and (2) we get

$$m_2 = -0.16244$$

The spline in the interval $8 \le x \le 27$ is given by,

$$s(x) = \{2(x-8)+19\}\frac{(x-27)^2}{19^3}\times 2 - \{2(x-27)-19\}\frac{(x-8)^2}{19^3}\times 3$$

$$+\frac{(x-8)(x-27)^2}{19^2}(-0.16244)+\frac{(x-27)(x-8)^2}{19^2}\times(0.05263)$$

Putting $x = 21$,

$$19^3 \times s(21) = 45\times 36\times 2+31\times 169\times 3-(13\times 36\times 0.16244+6\times 169\times 0.05263)\times 19$$

$$= 3240+15717-2458.3861$$

$$= 16498.6139$$

$$s(21) = 2.4054 \text{ (exact value is 2.7589).}$$

The values of cube-root of 21 obtained in (*i*) and (*ii*) above are better than those obtained from the natural spline. The value obtained from (*ii*) is better since the amount of truncation error in computation of second derivative is more than the error in that of the first derivative. Nevertheless these values are reasonably good in comparison to the one obtained from a global polynomial fitted by Lagrange's formula, which was absurd. (Example 4.15). Another reason is that the abcissas are too far apart to provide good estimates of derivatives.

Example 8.4

Using a cubic spline, interpolate the value of the periodic function $f(x) = \cos x$ at $x = \frac{2\pi}{3}$ when the function values are tabulated as given below:

x	0	$\pi/2$	π	$3\pi/2$	2π
$f(x)$	1	0	−1	0	1

Do not convert π to decimal anywhere. Use second derivative form of spline.

Solution

i	0	1	2	3	4	5	...
x_i	0	$\pi/2$	π	$3\pi/2$	2π	$5\pi/2$	...
f_i	1	0	−1	0	1	0	...

$$h = \frac{\pi}{2},\ M_4 = M_0;\ M_5 = M_1;\ \text{since function is periodic.}$$

The governing equations for second derivatives are:

$$M_0 + 4M_1 + M_2 = \frac{24}{\pi^2} \times 0 = 0$$

$$M_1 + 4M_2 + M_3 = \frac{24}{\pi^2} \times 2 = \frac{48}{\pi^2}$$

$$M_2 + 4M_3 + M_4 = \frac{24}{\pi^2} \times 0 = 0$$

$$M_3 + 4M_4 + M_5 = \frac{24}{\pi^2}(-2) = -\frac{48}{\pi^2}$$

Using $M_4 = M_0$ and $M_5 = M_1$, above equations become

$$M_0 + 4M_1 + M_2 + 0 = 0 \quad \text{.... (1)}$$

$$0 + M_1 + 4M_2 + M_3 = \frac{48}{\pi^2} \quad \text{.... (2)}$$

$$M_0 + 0 + M_2 + 4M_3 = 0 \quad \text{.... (3)}$$

$$4M_0 + M_1 + 0 + M_3 = -\frac{48}{\pi^2} \quad \text{.... (4)}$$

From (1) and (3)

$$M_3 = M_1 \quad \text{.... (5)}$$

From (2) and (4)

$$M_2 = M_0 + \frac{24}{\pi^2} \quad \text{.... (6)}$$

Substituting from (5) and (6) equations (1) and (2) become

$$M_0 + 2M_1 = -\frac{12}{\pi^2} \quad \text{.... (7)}$$

$$2M_0 + M_1 = -\frac{24}{\pi^2} \quad \text{.... (8)}$$

From (5), (6), (7) and (8).

$$M_0 = -\frac{12}{\pi^2},\ M_1 = 0 = M_3,\ M_2 = \frac{12}{\pi^2},\ M_4 = -\frac{12}{\pi^2} = M_0$$

We have to find the value of $f(x)$ for $x = \frac{2\pi}{3}$ which lies between x_1 and x_2, i.e., $\frac{\pi}{2} \le x \le \pi$. Hence, writing the spline for $x_1 \le x \le x_2$ from (8.7)

$$s(x) = s(x_1) + (x - x_1)\left[\frac{s(x_2) - s(x_1)}{\pi} \times 2 - \frac{\pi}{2} \times \frac{1}{6}\{M_2 + 2M_1\}\right]$$

$$+ \frac{(x - x_1)^2}{2} \times M_1 + \frac{(x - x_1)^3}{6} \times \frac{2}{\pi}\{M_2 - M_1\}$$

Here, $x_1 = \frac{\pi}{2}, \; x_2 = \pi; \; x = \frac{2\pi}{3}; \; x - x_1 = \frac{\pi}{6}$

$$s(x_1) = 0, \; s(x_2) = -1; \; M_1 = 0, \; M_2 = \frac{12}{\pi^2}$$

Substituting these values

$$s\left(\frac{2\pi}{3}\right) = 0 + \frac{\pi}{6}\left[\frac{2}{\pi}(-1 - 0) - \frac{\pi}{12}\left(\frac{12}{\pi^2} - 0\right) + 0 + \frac{\pi^3}{216 \times 6} \times \frac{2}{\pi} \times \frac{12}{\pi^2}\right]$$

$$= -\frac{1}{2} + \frac{1}{54} = -\frac{13}{27}$$

$$= -0.4815 \text{ (exact value is } -0.5).$$

8.8 Application to Differential Equations

Besides their applications in interpolation, differentiation and integration the splines can be used for solving ordinary and partial differential equations. The splines have been used extensively for solving partial differential equations of all types. One reference due to the author has been cited in Chapter 16. Here we will illustrate their use in solving ordinary differential equations of boundary value type. Two cases will be considered (*i*) when first derivative is absent and (*ii*) when first derivative is present.

Case 1: First Derivative Absent

Let us consider a two-point boundary value problem (BVP) which has no first derivative term,

$$\frac{d^2y}{dx^2} + q(x)y = r(x), \; a \le x \le b \tag{8.33}$$

with associated boundary conditions,

$$y(a) = \alpha \text{ and } y(b) = \beta, \tag{8.34}$$

where $q(x)$ and $r(x)$ are the known functions of x and α, β are constants.

Subdivide the region $a \le x \le b$ into (say) n equal intervals each of width $h = \frac{b-a}{n}$ and denote the points of subdivision by $a = x_0 < x_1 < x_2 \; \ldots \; < x_n = b$. We have to compute

$y(x_i)$, $i = 1(1)n-1$ while $y(x_0)$ and $y(x_n)$ are known by virtue of (8.34). Approximate $y(x)$ by a cubic spline $s(x)$ which interpolates $y(x)$ at knots x_i, $i = 0(1)n$.

Express (8.33) at $x = x_i$, $i = 0(1)n$ as

$$y''(x_i) = r(x_i) - q(x_i)y(x_i). \tag{8.35}$$

Replacing $y(x)$ by $s(x)$ and $y''(x)$ by $s''(x)$ in (8.35),

$$s''(x_i) = r(x_i) - q(x_i)s(x_i),\ i = 0(1)n. \tag{8.36}$$

Substituting $s''(x_i)$ or M_i, etc., in (8.13) and dropping parentheses by writing $s(x_i) = s_i$, $r(x_i) = r_i$ and $g(x_i) = g_i$, etc., we get,

$$(r_{i-1} - q_{i-1}s_{i-1}) + 4(r_i - q_i s_i) + (r_{i+1} - q_{i+1}s_{i+1}) = \frac{6}{h^2}\{y_{i-1} - 2y_i + y_{i+1}\}$$

Since s_i approximates y_i at $x = x_i$, the above can be written as,

$$\left(1 + \frac{h^2}{6}q_{i-1}\right)y_{i-1} - 2\left(1 - \frac{h^2}{3}q_i\right)y_i + \left(1 + \frac{h^2}{6}q_{i+1}\right)y_{i+1} = \frac{h^2}{6}\{r_{i-1} + 4r_i + r_{i+1}\},$$
$$i = 1(1)n-1 \tag{8.37}$$

The system of equations given by (8.37) can be solved for $(n-1)$ unknowns y_1, y_2, ... y_{n-1} whence values of y_0 and y_n are known as α and β respectively in accordance with (8.34).

Case 2: First derivative also present

Let us consider the differential equation,

$$\frac{d^2y}{dx^2} + p(x)\frac{dy}{dx} + q(x)y = r(x),\ a \le x \le b \tag{8.38}$$

with associated boundary conditions

$$y(a) = \alpha,\ \ y(b) = \beta, \tag{8.39}$$

where $p(x)$, $q(x)$ and $r(x)$ are known functions of x and α, β are constants.

Again we subdivide the interval $(a,\ b)$ into n subintervals each of width $h = \frac{b-a}{n}$ and the points of subdivision are denoted by $a = x_0 < x_1 < x_2 \ldots < x_n = b$. We approximate the solution $y(x)$ by the spline $s(x)$ in $(a,\ b)$. Now, from (8.38) we can write for $x = x_i$,

$$y'(x_i) = \frac{1}{p(x_i)}\{r(x_i) - q(x_i)y(x_i) - y''(x_i)\}$$

After removing the parentheses, the above can be written as,

$$y_i' = \frac{1}{p_i}(r_i - q_i y_i - y_i''). \tag{8.40}$$

Since $y(x)$ and its derivatives satisfy the spline, we substitute y_i' from (8.40) into the expressions for $s'(x_i+)$, $i = 0(1)n-1$ and $s'(x_i-)$, $i = 1(1)n$. Thus, we get $2n$ equations and there are also $2n$ unknowns: $(n+1)$ second derivatives $s''(x_i)$, $i = 0(1)n$ and $(n-1)$ function values $s(x_i)$ or y_i, $i = 1(1)n-1$. The resulting equations are given below:

Substitution of y' for $s'(x_i+)$ in (8.9) gives,

$$\frac{1}{p_i}(r_i - q_i s_i - s_i'') = \frac{s_{i+1} - s_i}{h} - \frac{h}{6}(s_{i+1}'' + 2s_i'')$$

or $$\left(q_i - \frac{p_i}{h}\right) s_i + \frac{p_i}{h} s_{i+1} + \left(1 - \frac{p_i h}{3}\right) s_i'' - \frac{p_i h}{6} s_{i+1}'' = r_i, \; i = 0(1)n-1. \tag{8.41}$$

Similarly substituting in (8.10) gives,

$$-\frac{p_i}{h} s_{i-1} + \left(q_i + \frac{p_i}{h}\right) s_i + \frac{p_i h}{6} s_{i-1}'' + \left(1 + \frac{p_i h}{3}\right) s_i'' = r_i, \; i = 1(1)n \tag{8.42}$$

Solution of (8.41) and (8.42) will give the required solution $s(x_i) = y_i$, $i = 1(1)n-1$ plus the values of $s''(x_i)$, $i = 0(1)n$. The values of $s''(x_i)$ may be used to determine the spline in any interval which can give the value of $y(x)$ not only at the nodal points but anywhere in the interval.

Alternatively, we can use the other form of cubic spline consisting of first derivative (Sec. 8.6). In that case we can express the differential equation at $x = x_i$, as

$$y_i'' = r_i - q_i y_i - p_i y_i'.$$

Substituting the above in (8.22) gives,

$$r_i - q_i s_i - p_i s_i' = \frac{6}{h^2}(s_{i+1} - s_i) - \frac{2}{h}(s_{i+1}' + 2s_i')$$

or $$\left(q_i - \frac{6}{h^2}\right) s_i + \frac{6}{h^2} s_{i+1} + \left(p_i - \frac{4}{h}\right) s_i' - \frac{2}{h} s_{i+1}' = r_i, \; i = 0(1)n-1. \tag{8.43}$$

Similarly putting y_i'' in (8.23) gives,

$$\frac{6}{h^2} s_{i-1} + \left(q_i - \frac{6}{h^2}\right) s_i + \frac{2}{h} s_{i-1}' + \left(p_i + \frac{4}{h}\right) s_i' = r_i, \; i = 1(1)n. \tag{8.44}$$

As before we have $2n$ equations given by (8.43) and (8.44) and there are $2n$ unknowns, namely, $(n-1)$ values of $s(x_i)$, $i = 1(1)n-1$ and $(n+1)$ values of $s'(x_i)$, $i = 0(1)n$; the values of $s(x_0)$ and $s(x_n)$ are given as boundary conditions (8.39). These can be solved by Gaussian Elimination.

Other types of boundary conditions can also be handled easily. For example let us suppose a general boundary condition (mixed type) is given at $x = x_0$, as $\alpha y + \beta \frac{dy}{dx} = \gamma$. As it satisfies the spline we have $\alpha s_0 + \beta s'_0 = \gamma_0$ or $s'_0 = (\gamma_0 - \alpha s_0)/\beta$. Similarly, if the boundary condition at $x = x_n$ is given as, $\alpha' y + \beta' \frac{dy}{dx} = \gamma'$ we will have $s'_n = (\gamma_n - \alpha' s_n)/\beta'$. Then, s'_0 and s'_n are substituted in (8.43) and (8.44) and we can solve for $s_0, s_1, \ldots s_n$ and $s'_1, s'_2, \ldots s'_{n-1}$.

Note: For computations, the terms in the equations should be written in the order $s'_0, s'_1, \cdots s'_n, s_1, s_2, \ldots s_{n-1}$, so that s_i terms are computed first and s'_i terms may not be computed if not required. (See Example 8.5).

Example 8.5

Given the differential equation,

$$\frac{d^2y}{dx^2} + 4x\frac{dy}{dx} + 4x^2 y = 0, \ y(0) = 3 \text{ and } y(0.4) = 2.4684,$$

find $y(0.2)$ using a cubic spline by dividing the interval (0, 0.4) into two equal subintervals.

Solution Let us denote the nodal points as,

$$x_0 = 0, \ x_1 = 0.2 \text{ and } x_2 = 0.4.$$

The length of each interval, $h = \dfrac{0.4 - 0}{2} = 0.2.$

We have $p(x) = 4x; \ \ q(x) = 4x^2; \ \ r(x) = 0.$

Compute the following,

i	x_i	p_i	q_i
0	0	0	0
1	0.2	0.8	0.16
2	0.4	0.16	0.64

For computations we need, $\dfrac{6}{h^2} = 150.$

Using (8.43) we get following two equations,

$$(0-150)s_0+150s_1+(0-20)s'_0-10s'_1=0$$

$$(0.16\text{–}150)\ s_1+150s_2+(0.8-20)s'_1-10s'_2=0$$

Using (8.44) we get another two equations,

$$150s_0+(0.16-150)s_1+10s'_0+(0.8+20)s'_1=0$$

$$150s_1+(0.64-150)s_2+10s'_1+(0.16+20)s'_2=0$$

Now, we are given $s_0=3,\ s_2=2.4684$ and have to compute s_1. Rearranging above equations, we can write them as follows,

s'_0	s'_1	s'_2	s_1	
-20	-10	0	150	450
0	-19.2	-10	-149.84	-370.26
10	20.8	0	-149.84	-450
0	10	20.16	150	368.68

....(1)

By process of Gaussian elimination, system (1) reduces to

s'_0	s'_1	s'_2	s_1	
-20	-10	0	150	450
0	-19.2	-10	-149.84	-370.2
0	0	-8.23	-198.14	-529.69
0	0	0	-287.98	-786.33

.... (2)

By back-substitution,

$$s_1=2.73\ (\text{exact value is } 2.72)=y(0.2)$$

Analytical soln. is $y=e^{-x^2}\left(e^{\sqrt{2}x}+2e^{-\sqrt{2}x}\right)$.

Note: If value of y is required at some point also for $0\le x\le 0.4$, the system of equations given by (2) may be further solved for s'_0, s'_1 and s'_2 by back-substitution. Then cubic spline (8.20) may be used.

8.9 Cubic Spline: Parametric Form

Suppose we are given $n+1)$ points in the x-y plane $P_0\,(x_0,\ y_0)$, $P_1\,(x_1,\ y_1)$, $\ldots.\ P_n\,(x_n,\ y_n)$ and want to join these points by fitting cubic splines in that order. More often than not, the condition $x_0 < x_1 < \ldots. < x_n$ is not satisfied in practical problems (See Fig. 8.2a and Fig. 8.2b).

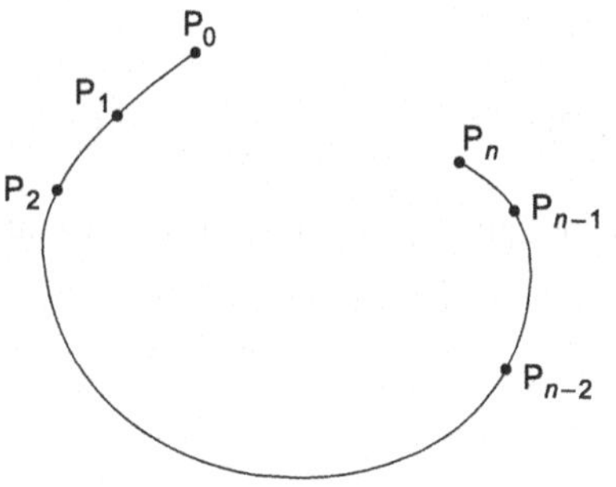

Figure 8.2(a) Open Curve.

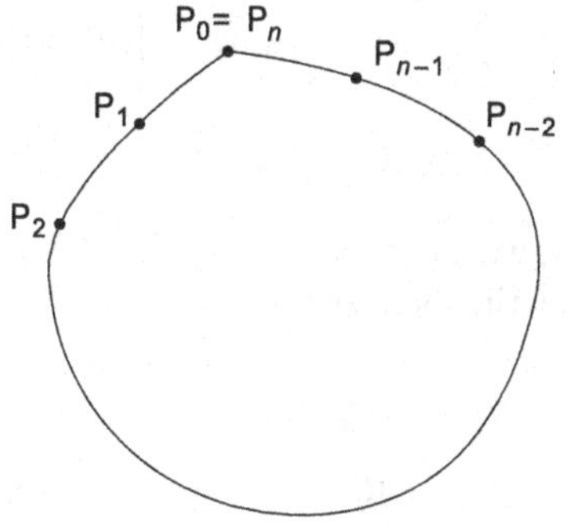

Figure 8.2(b) Closed curve.

Figure 8.2a shows an open curve while Fig. 8.2b is a closed curve where point P_n coincides with point P_0. In these cases we choose a parameter, say t which satisfies the condition $t_0 < t_1 < t_2 \ldots < t_n$ and the coordinates $x(t)$ and $y(t)$ are expressed as functions of parameter t. We fit cubic splines by computing $x(t),\ y(t)$ in the interval $t_i \le t \le t_{i+1},\ i = 0(1)n-1$. A curve may be plotted between t_i and t_{i+1} computing the coordinates $\{x(t),\ y(t)\}$ giving various values to t in $t_i \le t \le t_{i+1}$. The parameter t is generally taken to be the cumulative distances of the point starting from P_0 where $s = 0$. (We can also use angle θ as parameter, if convenient).

Denoting distance of a point $P_i\,(x_i,\ y_i)$ as s_i, we have

$$s_i = s_{i-1} + h_i \text{ where } h_i = \sqrt{(x_i - x_{i-1})^2 + (y_i - y_{i-1})^2}, \qquad i = 1,\ 2,\ \ldots n \tag{8.45}$$

with s_0 corresponding to point P_0 and s_n to last point P_n. In this way we can define $x_i = x(s_i)$, $y_i = y(s_i)$ for $s_0 < s_1 < s_2 \ldots.$ For actual computations we can proceed as follows:

In case of open curve (Fig. 8.2a) we can fit a natural spline where $\frac{d^2x}{ds^2} = 0$ and $\frac{d^2y}{ds^2} = 0$ at $s = s_0$ and $s = s_n$. The equations at the points $P_1,\ P_2, \ldots P_{n-1}$ will give $n-1$ equations (8.12) for determining $\frac{d^{2x}}{ds^2}$ and $\frac{d^2y}{ds^2}$ at these points.

In case of closed curve (Fig. 8.2b) we can proceed like a periodic function. For example for computing $M = \frac{d^{2x}}{ds^2}$, we can write $(n-1)$ equations at points $P_1, P_2, \ldots P_{n-1}$. Further two extra equations are obtained; one as $M_0 = M_n$ and the other by writing an equation at the point P_n and using $x(s_{n+1}) = x(s_1)$, $M_{n+1} = M_1$; similarly for $y(s)$.

Besides the method described above, we now derive an alternative parametric form for a cubic spline (in terms of first derivative) when the parameter has been normalised to vary between 0 and 1 in each interval. Let u be a normalised variable which takes value 0 at point P_i and 1 at P_{i+1}, i.e., $0 \le u \le 1$ between P_i and P_{i+1} and $P(x, y)$ be a point on the spline between points P_i and P_{i+1}.

$$\text{Let } x(u) = au^3 + bu^2 + cu + d. \tag{8.46}$$

After differentiating w.r.t. u we get

$$x'(u) = 3au^2 + 2bu + c. \tag{8.47}$$

Taking $u_i = 0$ and $u_{i+1} = 1$ in the interval $u_i \le u \le u_{i+1}$, we get from (8.46) and (8.47)

$$\left.\begin{aligned} x_i &= x(u_i) = d, \\ x_{i+1} &= x(u_{i+1}) = a + b + c + d, \\ x'_i &= x'(u_i) = c, \\ x'_{i+1} &= x'(u_{i+1}) = 3a + 2b + c. \end{aligned}\right\} \tag{8.48}$$

On solving (8.48) we readily get

$$\left.\begin{aligned} a &= 2x_i - 2x_{i+1} + x'_i + x'_{i+1} \\ b &= -3x_i + 3x_{i+1} - 2x'_i - x'_{i+1} \\ c &= x'_i \\ d &= x_i. \end{aligned}\right\} \tag{8.49}$$

Substituting the values from (8.49) in (8.46),

$$x(u) = (2x_i - 2x_{i+1} + x'_i + x'_{i+1})u^3 + (-3x_i + 3x_{i+1} - 2x'_i - x'_{i+1})u^2 + x'_i u + x_i \tag{8.50}$$

After re-arranging (8.50) we get the x-coordinate as

$$x(u) = (2u^3 - 3u^2 + 1)x_i + (-2u^3 + 3u^2)x_{i+1} + (u^3 - 2u^2 + u)x'_i + (u^3 - u^2)x'_{i+1},$$
$$0 \le u \le 1. \tag{8.51}$$

In computer graphics terminology the coefficients of x_i, x_{i+1}, x'_i and x'_{i+1} are called 'blending functions' or their respective 'weight functions'.

In a similar manner we can get y-coordinate as

$$y(u) = (2u^3 - 3u^2 + 1)y_i + (-2u^3 + 3u^2)y_{i+1} + (u^3 - 2u^2 + u)y'_i + (u^3 - u^2)y'_{i+1},$$
$$0 \le u \le 1. \tag{8.52}$$

Combining (8.51) and (8.52) we define in matrix form the parametric spline $P_i(u)$ by coordinates $x(u)$, $y(u)$ between points P_i and P_{i+1}, as

$$P_i(u) = \begin{bmatrix} x(u) \\ y(u) \end{bmatrix} = \begin{bmatrix} x_i & x_{i+1} & x'_i & x'_{i+1} \\ y_i & y_{i+1} & y'_i & y'_{i+1} \end{bmatrix} \begin{bmatrix} 2u^3 - 3u^2 + 1 \\ -2u^3 + 3u^2 \\ u^3 - 2u^2 + u \\ u^3 - u^2 \end{bmatrix} \tag{8.53}$$

We will discuss two cases (i) when points P'_is are equidistant and (ii) when points P'_is are not equidistant.

(*i*) Let us suppose that the points P_0, P_1, … . P_n are equidistant, i.e., $d(P_{i-1}, P_i) = d(P_i, P_{i+1})$, $i = 1(1)n-1$ where d denotes the distance between the two points.

In order to determine the derivative x'_i, x'_{i+1} and y'_i, y'_{i+1} we write down the spline between P_{i-1} and P_i,

$$P_{i-1}(u) = \begin{bmatrix} x(u) \\ y(u) \end{bmatrix} = \begin{bmatrix} x_{i-1} & x_i & x'_{i-1} & x'_i \\ y_{i-1} & y_i & y'_{i-1} & y'_i \end{bmatrix} \begin{bmatrix} 2u^3 - 3u^2 + 1 \\ -2u^3 + 3u^2 \\ u^3 - 2u^2 + u \\ u^3 - u^2 \end{bmatrix}, 0 \le u \le 1. \tag{8.54}$$

Further, we match the condition of continuity of the second derivatives at the common point P_i of the two intervals (P_{i-1}, P_i) and (P_i, P_{i+1}). Let us differentiate (8.54) twice with respect to u,

$$P''_{i-1}(u) = \begin{bmatrix} x''(u) \\ y''(u) \end{bmatrix} = \begin{bmatrix} x_{i-1} & x_i & x'_{i-1} & x'_i \\ y_{i-1} & y_i & y'_{i-1} & y'_i \end{bmatrix} \begin{bmatrix} 12u-6 \\ -12u+6 \\ 6u-4 \\ 6u-2 \end{bmatrix} \tag{8.55}$$

Similarly, we can write from (8.53)

$$P''_{i}(u) = \begin{bmatrix} x''(u) \\ y''(u) \end{bmatrix} = \begin{bmatrix} x_{i} & x_{i+1} & x'_{i} & x'_{i+1} \\ y_{i} & y_{i+1} & y'_{i} & y'_{i+1} \end{bmatrix} \begin{bmatrix} 12u-6 \\ -12u+6 \\ 6u-4 \\ 6u-2 \end{bmatrix} \tag{8.56}$$

We get $x''(\mathrm{P}_i-)$ by putting $u=1$ in (8.55) and $x''(\mathrm{P}_i+)$ by putting $u=0$ in (8.56). After equating $x''(\mathrm{P}_i-) = x''(\mathrm{P}_i+)$ we get

$$6x_{i-1} - 6x_i + 2x'_{i-1} + 4x'_i = -6x_i + 6x_{i+1} - 4x'_i - 2x'_{i+1}$$

or $$x'_{i-1} + 4x'_i + x'_{i+1} = 3(x_{i+1} - x_{i-1}),\ i = 1(1)n-1 \tag{8.57}$$

Similarly, $$y'_{i-1} + 4y'_i + y'_{i+1} = 3(y_{i+1} - y_{i-1}),\ i = 1(1)n-1. \tag{8.58}$$

If x'_0, x'_n and y'_0, y'_n are known, equations (8.57) and (8.58) can be solved for x'_i and y'_i respectively for $i = 1(1)n-1$. The corresponding values can be substituted in (8.53) to give $x(u)$ and $y(u)$ between the points P_i and P_{i+1}. The spline curve may be plotted using the coordinates $\{x(u),\ y(u)\}$ giving different values to u for $0 \le u \le 1.0$, e.g., $u = 0(0.1)1.0$.

(*ii*) If points P_0, P_1, $\ldots\mathrm{P}_n$ are not equidistant then continuity condition of the second derivative may be derived as follows:

Let us suppose that t is the parameter such that $P_i(x_i,\ y_i) = P_i(t_i)$, $i = 0(1)n$ and $t_0 < t_1 < t_2 \ldots < t_n$. Then in the interval $t_i \le t \le t_{i+1}$, i.e., between the points P_i and P_{i+1},

$$u = \frac{t - t_i}{t_{i+1} - t_i},\ \frac{du}{dt} = \frac{1}{t_{i+1} - t_i} = \frac{1}{h_{i+1}} \text{ (say)}$$

we get, $$\frac{dx}{dt} = \frac{dx}{du} \cdot \frac{du}{dt} = \frac{1}{h_{i+1}} \frac{dx}{du}.$$

and $$\frac{d^2x}{dt^2} = \frac{1}{h_{i+1}^2} \cdot \frac{d^2x}{du^2}, \quad t_i \le t \le t_{i+1}. \tag{8.59}$$

Similarly in the interval $t_{i-1} \le t \le t_i$

$$\frac{d^2x}{dt^2} = \frac{1}{h_i^2} \cdot \frac{d^2x}{du^2}, \quad t_{i-1} \le t \le t_i \tag{8.60}$$

while $$u = \frac{t - t_{i-1}}{t_i - t_{i-1}} = \frac{t - t_{i-1}}{h_i}$$

Now equating $\frac{d^2x}{dt^2}$ at $t = t_i-$ from (8.59) at $t = t_i+$ from (8.60), i.e., putting $u = 0$ for $t_i \le t \le t_{i+1}$ and $u = 1$ for $t_{i-1} \le t \le t_i$.

Consequently, we get the condition for x-derivatives,

$$\frac{1}{h_i^2}(6x_{i-1} - 6x_i + 2x'_{i-1} + 4x'_i) = \frac{1}{h_{i+1}^2}(-6x_i + 6x_{i+1} - 4x'_i - 2x'_{i+1})$$

or $$\frac{1}{h_i^2}x'_{i-1} + 2\left(\frac{1}{h_i^2} + \frac{1}{h_{i+1}^2}\right)x'_i + \frac{1}{h_{i+1}^2}x'_{i+1} = 3\left[\frac{1}{h_{i+1}^2}x_{i+1} + \left(\frac{1}{h_i^2} - \frac{1}{h_{i+1}^2}\right)x_i - \frac{1}{h_i^2}x_{i-1}\right] \tag{8.61}$$

$$i = 1(1)n - 1.$$

Similarly, we can get for y-derivatives

$$\frac{1}{h_i^2}y'_{i-1} + 2\left(\frac{1}{h_i^2} + \frac{1}{h_{i+1}^2}\right)y'_i + \frac{1}{h_{i+1}^2}y'_{i+1} = 3\left[\frac{1}{h_{i+1}^2}y_{i+1} + \left(\frac{1}{h_i^2} - \frac{1}{h_{i+1}^2}\right)y_i - \frac{1}{h_i^2}y_{i-1}\right] \tag{8.62}$$

$$i = 1(1)n - 1.$$

When points are equidistant, i.e., $h_i = h_{i+1}$, then equations (8.61) and (8.62) reduce to equations (8.57) and (8.58) respectively.

Example 8.6

The coordinates of four points on a curve are given as A (1, 0), B(0, 1), C(−1, 0) and D(0, −1). The closed curve ABCDA is to be approximated by fitting a cubic spline taking θ as a parameter where θ denotes an angle with the positive x-axis. Obtain $\frac{dx}{d\theta}$ and $\frac{dy}{d\theta}$ at these points and compute the values of x and y for $\theta = \frac{2\pi}{3}$.

Do not change π to decimal/fraction.

Solution

i	0	1	2	3	4	5
θ_i	0	$\pi/2$	π	$3\pi/2$	2π	$\pi/2$
x_i	1	0	-1	0	1	0
y_i	0	1	0	-1	0	1

Computation for x-coordinate :

$$\text{Let } m_i = \left(\frac{dx}{d\theta}\right)_{\theta=\theta_i}; \quad h = \frac{\pi}{2}$$

The relevant equations, for $i = 1, 2, 3$ and 4 are respectively

$$m_0 + 4m_1 + m_2 = \frac{6}{\pi}(-1-1) = -\frac{12}{\pi}$$

$$m_1 + 4m_2 + m_3 = \frac{6}{\pi}(0-0) = 0$$

$$m_2 + 4m_3 + m_4 = \frac{6}{\pi}(1+1) = \frac{12}{\pi}$$

$$m_3 + 4m_4 + m_5 = \frac{6}{\pi}(0-0) = 0$$

Since $m_5 = m_1$ and $m_4 = m_0$, the above equations become

$$m_0 + 4m_1 + m_2 = -\frac{12}{\pi} \quad \text{.... (1)}$$

$$m_1 + 4m_2 + m_3 = 0 \quad \text{.... (2)}$$

$$m_0 + 0 + m_2 + 4m_3 = \frac{12}{\pi} \quad \text{.... (3)}$$

$$4m_0 + m_1 + 0 + m_3 = 0 \quad \text{.... (4)}$$

(2) and (4) give $\qquad m_2 = m_0$

(1) and (3) give $\qquad m_3 = m_1 + \dfrac{6}{\pi}$

Putting these values in eqns (1) and (2)

$$m_0 + 2m_1 = -\frac{6}{\pi} \quad \text{.... (5)}$$

$$4m_0 + 2m_1 = -\frac{6}{\pi} \quad \text{.... (6)}$$

Equations (5), (6) and earlier results give

$$m_0 = 0 = m_2; \;\; m_1 = -\frac{3}{\pi}; \;\; m_3 = \frac{3}{\pi}.$$

To compute x for $\theta = \dfrac{2\pi}{3}$ we write the spline for $\dfrac{\pi}{2} \leq \theta \leq \pi$, i.e., $\theta_1 \leq \theta \leq \theta_2$.

We have $$h = \frac{\pi}{2}; \;\; \theta - \theta_1 = \frac{2\pi}{3} - \frac{\pi}{2} = \frac{\pi}{6}; \;\; \theta - \theta_2 = \frac{2\pi}{3} - \pi = -\frac{\pi}{3}.$$

Also $$x(\theta_1) = 0, \; x(\theta_2) = -1; \;\; m_1 = -\frac{3}{\pi}, \; m_2 = 0.$$

$$x(\theta) = \{2(\theta - \theta_1) + h\}\frac{(\theta - \theta_2)^2}{h^3}x(\theta_1) - \{2(\theta - \theta_2) - h\}\frac{(\theta - \theta_1)^2}{h^3}x(\theta_2)$$

$$+\frac{(\theta - \theta_1)(\theta - \theta_2)^2}{h^2}m_1 + \frac{(\theta - \theta_2)(\theta - \theta_1)^2}{h^2}m_2$$

$$x\left(\frac{2\pi}{3}\right) = 0 - \left\{2\left(-\frac{\pi}{3}\right) - \frac{\pi}{2}\right\}\left(\frac{\pi}{6}\right)^2\left(\frac{2}{\pi}\right)^3(-1) + \frac{\pi}{6} \times \left(-\frac{\pi}{3}\right)^2\left(\frac{2}{\pi}\right)^2\left(-\frac{3}{\pi}\right) + 0$$

$$= -\frac{7}{6} \times \frac{2}{9} - \frac{2}{9} = -\frac{13}{27} \text{ (same as obtained in Example 8.4)}$$

$$= -0.4815 \text{ (exact value is } -0.5).$$

Computation for y-coordinate:

Let $$m_i = \frac{dy}{d\theta} \text{ at } \theta = \theta_i.$$

Writing equations for $i = 1, 2, 3$ and 4 and using $m_4 = m_0$ and $m_5 = m_1$, etc., we get

$$m_0 + 4m_1 + m_2 = 0 \quad \text{... (7)}$$

$$m_1 + 4m_2 + m_3 = -\frac{12}{\pi} \quad \text{... (8)}$$

$$m_0 + 0 + m_2 + 4m_3 = 0 \quad \text{... (9)}$$

$$4m_0 + m_1 + 0 + m_3 = \frac{12}{\pi} \quad \text{... (10)}$$

As before, we get

$$m_0 = \frac{3}{\pi}, \; m_1 = 0 = m_3; \; m_2 = -\frac{3}{\pi}$$

The spline for y in the interval $\theta_1 \le \theta \le \theta_2$ is given by

$$y(\theta) = \{2(\theta-\theta_1)+h\}\frac{(\theta-\theta_2)^2}{h^3}y(\theta_1) - \{2(\theta-\theta_2)-h\}\frac{(\theta-\theta_1)^2}{h^3}y(\theta_2)$$

$$\frac{(\theta-\theta_1)(\theta-\theta_2)^2}{h^2}m_1 + \frac{(\theta-\theta_2)(\theta-\theta_1)^2}{h^2}m_2$$

$$y(\theta_1) = 1,\ y(\theta_2) = 0;\ m_1 = 0,\ m_2 = -\frac{3}{\pi}$$

$$\theta = \frac{2\pi}{3},\ \theta_1 = \frac{\pi}{2},\ \theta_2 = \pi;\ \theta-\theta_1 = \frac{\pi}{6},\ \theta-\theta_2 = -\frac{\pi}{3};\ h = \frac{\pi}{2}$$

Putting the above values we get

$$y\left(\frac{2\pi}{3}\right) = \left\{2\times\frac{\pi}{6}+\frac{\pi}{2}\right\}\left(\frac{\pi}{3}\right)^2 \times \left(\frac{2}{\pi}\right)^3 \times 1 + \left(\frac{2}{\pi}\right)^2\left(-\frac{\pi}{3}\right)\left(\frac{\pi}{6}\right)^2\left(-\frac{3}{\pi}\right)$$

$$= \frac{5}{6}\times\frac{8}{9}+\frac{1}{9} = \frac{23}{27} = 0.8518 \text{ (exact value is 0.8660).}$$

Example 8.7

Four points A (1, 0), B (0, 1), C (−1, 0) and D (0, −1) lie on a closed curve ABCDA in x-y plane. Find the cubic splines for x and y between the points B and C taking the parameter s where s denotes the cumulative length on the curve measured in the anticlockwise direction starting with $s = 0$ on A. Find x and y for $s = \frac{4\sqrt{2}}{3}$.

Solution

i	0	1	2	3	4	5
s_i	0	$\sqrt{2}$	$2\sqrt{2}$	$3\sqrt{2}$	$4\sqrt{2}$	$\sqrt{2}$
x_i	1	0	−1	0	1	0
y_i	0	1	0	−1	0	1

Proceeding as in Example 8.6 with $h = \sqrt{2}$ (instead of $\pi/2$) we get

$$m_2 = m_0 = 0;\ m_1 = -\frac{3\sqrt{2}}{4};\ m_3 = \frac{3\sqrt{2}}{4}$$

$$s_1 = \sqrt{2},\ s_2 = 2\sqrt{2};\ s-s_1 = \frac{4\sqrt{2}}{3} - \sqrt{2} = \frac{\sqrt{2}}{3};\ s-s_2 = -\frac{2\sqrt{2}}{3}$$

$$x\,(s) = \{2(s-s_1)+h\}\frac{(s-s_2)^2}{h^3}x(s_1) - \{2(s-s_2)-h\}\frac{(s-s_1)^2}{h^3}x(s_2)$$

$$+\frac{(s-s_1)(s-s_2)^2}{h^2}m_1 + \frac{(s-s_2)(s-s_1)^2}{h^2}m_2$$

$$x\left(\frac{4\sqrt{2}}{3}\right) = \left\{2\left(\frac{4\sqrt{2}}{3}-\sqrt{2}\right)+\sqrt{2}\right\}\left(\frac{2\sqrt{2}}{3}\right)^2\frac{1}{(\sqrt{2})^3}\times 0-$$

$$\left\{2\left(-\frac{2\sqrt{2}}{3}\right)-\sqrt{2}\right\}\left(\frac{\sqrt{2}}{3}\right)^2\left(\frac{1}{\sqrt{2}}\right)^3\times(-1)$$

$$= +\frac{\sqrt{2}}{3}\times\left(\frac{2\sqrt{2}}{3}\right)^2\cdot\left(\frac{1}{\sqrt{2}}\right)^2-\left(-\frac{3\sqrt{2}}{4}\right)+0$$

$$= -\frac{13}{27}\text{ (same as in Ex. 8.6)}$$

Similarly we can compute $y\left(\frac{4\sqrt{2}}{3}\right)$ when $m_1 = 0 = m_3$ and $m_2 = -\frac{3\sqrt{2}}{4}$; $y(s_1) = 1$, $y(s_2) = 0$.

Example 8.8

Four points A (1, 0), B (0, 1), C (−1, 0) and D (0, −1) are given on a closed curve ABCDA in the x-y plane. Using a parameter u, $0 \le u \le 1$ in each interval find the cubic spline P(u) between the points B and C. Hence find the values of x and y for $u = \frac{1}{3}$.

Solution

i	0	1	2	3	4	5
P_i	P_0	P_1	P_2	P_3	P_0	P_1
x_i	1	0	−1	0	1	0
y_i	0	1	0	−1	0	1

Using (8.57) for x-coordinate

$$m_0 + 4m_1 + m_2 = 3(-1-1) = -6$$

$$m_1 + 4m_2 + m_3 = 3(0-0) = 0$$

$$m_2 + 4m_3 + m_4 = 3(1+1) = 6$$

$$m_3 + 4m_4 + m_5 = 3(0-0) = 0$$

Using $m_4 = m_0$, $m_5 = m_1$ we get,

$$m_0 = m_2 = 0,\ m_1 = -\frac{3}{2},\ m_4 = \frac{3}{2}$$

In the interval between P_1 and P_2.

$$x_1 = 0,\ x_2 = -1,\ x_1' = -\frac{3}{2},\ x_2' = 0. \qquad \text{.... (1)}$$

Using (8.58) for y-spline

$$\begin{aligned}
m_0 + 4m_1 + m_2 &= 3(0-0) = 0\\
m_1 + 4m_2 + m_3 &= 3(-1-1) = -6\\
m_2 + 4m_3 + m_4 &= 3(0-0) = 0\\
m_3 + 4m_4 + m_5 &= 3(1+1) = 6
\end{aligned}$$

By solving the above, using $m_4 = m_0$ and $m_5 = m_1$ we get

$$m_1 = m_3 = 0,\ m_0 = \frac{3}{2},\ m_2 = -\frac{3}{2}.$$

In the interval between P_1 and P_2

$$y_1 = 1,\ y_2 = 0,\ y_1' = 0,\ y_2' = -\frac{3}{2} \qquad \text{.... (2)}$$

Putting the values from (1) and (2) in the expression (8.53), we get the spline as

$$P(u) = \begin{bmatrix} x(u) \\ y(u) \end{bmatrix} = \begin{bmatrix} 0 & -1 & -\frac{3}{2} & 0 \\ 1 & 0 & 0 & -\frac{3}{2} \end{bmatrix} \begin{bmatrix} 2u^3 - 3u^2 + 1 \\ -2u^3 + 3u^2 \\ u^3 - 2u^2 + u \\ u^3 - u^2 \end{bmatrix} \qquad \text{.... (3)}$$

Relation (3) gives

$$x(u) = \frac{1}{2}(u^3 - 3u)$$

$$y(u) = \frac{1}{2}(u^3 - 3u^2 + 2)$$

Putting $u = \frac{1}{3}$ in the above

$$x\left(\frac{1}{3}\right) = -\frac{13}{27},\ y\left(\frac{1}{3}\right) = \frac{23}{27}.$$

8.10 Introduction to B-Splines

Since their inception in 1947, the splines have been used extensively to find numerical solution of wide variety of problems arising in science and engineering. Lately, the domain of definition of splines has extended to incorporate those piece-wise polynomials also which may pass through none or only some of the given data points; however the conditions of continuity of derivatives of the polynomials defined over the adjacent intervals are satisfied at the point of their intersection. Thus the smoothness of the spline curve over the whole domain is maintained irrespective of the fact that the curve may not pass through any of the data points. There are various versions of such splines which have vast applications in the field of Computer Graphics. One of the types of such splines are B-Splines; let us discuss a cubic B-Spline in brief. See [2], [4].

Suppose we are given $(n+1)$ data points $\mathrm{P}_i\,(x_i,\ y_i),\ i=0(1)n$ in the x-y plane. These points are also called 'control points' in the Computer Graphics' terminology. In order to determine a cubic spline between points P_i and P_{i+1}, four points are used, namely $\mathrm{P}_{i-1}, \mathrm{P}_i$, P_{i+1} and P_{i+2}. Obviously, some special treatment will be needed to determine the splines in the first and the last portions as there is no extra point before P_0 and no extra point after P_n. However, there are methods to circumvent this problem. In case of a closed curve there should be no problem, anyway, as the curve repeats itself.

In order to construct a cubic B-Spline between points P_i and P_{i+1}, the continuity conditions of the function as well as their first and second derivatives corresponding to each of these points are satisfied.

Without going into mathematical details, the cubic B-Spline between points P_i and P_{i+1} may be written as

$$\mathrm{B}_i(u) = \begin{bmatrix} x(u) \\ y(u) \end{bmatrix} = \begin{bmatrix} x_{i-1} & x_i & x_{i+1} & x_{i+2} \\ y_{i-1} & y_i & y_{i+1} & y_{i+2} \end{bmatrix} \begin{bmatrix} \frac{1}{6}(1-u)^3 \\ \frac{1}{6}(3u^3-6u^2+4) \\ \frac{1}{6}(-3u^3+3u^2+3u+1) \\ \frac{1}{6}u^3 \end{bmatrix}, \quad (8.63)$$

where u varies from 0 to 1 between P_i and P_{i+1}.

As may be observed that the spline does not pass through the points P_i and P_{i+1} since by putting $u=0$ in (8.63) we do not get $x\ (0)$ equal to x_i and nor do we get $x\ (1)=x_{i+1}$. Similarly $y(0)$ and $y(1)$ are also not equal to y_i and y_{i+1} respectively. However conditions of continuity of $x(u)$, $y(u)$ and their first derivatives $x'(u)$, $y'(u)$ as well as their second derivatives $x''(u)$ and $y''(u)$ are satisfied at the intersection of the two adjacent spline curves.

To verify this let us write down the B-Spline between points P_{i-1} and P_i.

$$B_{i-1}(u) = \begin{bmatrix} x(u) \\ y(u) \end{bmatrix} = \begin{bmatrix} x_{i-2} & x_{i-1} & x_i & x_{i+1} \\ y_{i-2} & y_{i-1} & y_i & y_{i+1} \end{bmatrix} \begin{bmatrix} \frac{1}{6}(1-u)^3 \\ \frac{1}{6}(3u^3 - 6u^2 + 4) \\ \frac{1}{6}(-3u^3 + 3u^2 + 3u + 1) \\ \frac{1}{6}u^3 \end{bmatrix} \tag{8.64}$$

By putting $u = 1$ in (8.64) and $u = 0$ in (8.63) and denoting them by $B_{i-1}(1)$ and $B_i(0)$ respectively, we see that at point P_i

$$B_{i-1}(1) = B_i(0) = \begin{bmatrix} \frac{1}{6}(x_{i-1} + 4x_i + x_{i+1}) \\ \frac{1}{6}(y_{i-1} + 4y_i + y_{i+1}) \end{bmatrix} \tag{8.65}$$

Similarly, continuity of first and second derivatives

$$B'_{i-1}(1) = B'_i(0) = \begin{bmatrix} \frac{1}{2}(x_{i+1} - x_{i-1}) \\ \frac{1}{2}(y_{i+1} - y_{i-1}) \end{bmatrix} \tag{8.66}$$

$$B''_{i-1}(1) = B''_i(0) = \begin{bmatrix} x_{i-1} - 2x_i + x_{i+1} \\ y_{i-1} - 2y_i + y_{i+1} \end{bmatrix} \tag{8.67}$$

Cubic B-Spline between point P_i and P_{i+1} can also be constructed using four points P_{i-2}, P_{i-1}, P_i and P_{i+1}, i.e., two points before P_i.

8.11 Bezier Spline Curves

This is yet another technique of approximation by parametric spline curves which was developed by French engineer P. Bezier. The Bezier curves (or surfaces) are very commonly used in Computer Graphics. Considering $(n+1)$ points $P_i(x_i, y_i)$, $i = 0(1)n$ in the x-y plane, there can be a spline of any degree up to n. The Bezier curves may be drawn by taking the points in any order; however, the shape of the curves will be different for different permutations. They may be even self-intersecting. Let us suppose that an approximation is to be made taking points, P_0, P_1, $P_2 \ldots$, P_n in that order so that P_0 is the first point and P_n the last. Then the curve passes through P_0 and P_n but not necessarily through any of the other points.

Let u be a parameter so that coordinates x and y can be expressed as functions of u, i.e., $x(u)$ and $y(u)$ while point $P(x, y)$ can be represented as $P(x(u), y(u))$ or $P(u)$. Let us suppose that Bezier parametric spline of degree n is required. Then n^{th} degree spline approximation for x-coordinate by Bezier formula is expressed as

$$x(u) = \sum_{k=0}^{n} x_k B_{k,\,n}(u) \tag{8.68}$$

where $B_{k,\,n}(u)$ are Bezier blending functions (Bernstein polynomials) given by

$$B_{k,\,n}(u) = \binom{n}{k} u^k (1-u)^{n-k}; \tag{8.69}$$

parameter u varies from 0 to 1 having value 0 for x_0 and 1 for x_n, i.e., $x\,(0) = x_0$ and $x\,(1) = x_n$.

It may be noted that $B_{k,n}(u)$, $k = 0(1)n$ given by (8.69) are various terms in the binomial expansion

$$(\overline{1-u}+u)^n = \binom{n}{0}(1-u)^n + \binom{n}{1} u(1-u)^{n-1} + \ldots + \binom{n}{k} u^k (1-u)^{n-k} + \ldots + \binom{n}{n} u^n. \tag{8.70}$$

We may also state here that a function $f(u)$ defined in the interval $0 \le u \le 1$ is approximated by a Bernstein polynomial of degree n as

$$B(f,\,n) = \sum_{k=0}^{n} f\left(\frac{k}{n}\right) \cdot \binom{n}{k} u^k (1-u)^{n-k}, \tag{8.71}$$

In Bezier formula (8.68) x_k is same as $f\left(\frac{k}{n}\right)$ in (8.71).

It may be shown by differentiating $x(u)$ w.r.t. u in (8.68) that parametric first derivative at $x = x_0$, i.e., for $u = 0$ is given by

$$x'(0) = n(x_1 - x_0). \tag{8.72}$$

Similarly, the parametric first derivative at $x_n(u=1)$ is

$$x'(1) = n(x_n - x_{n-1}). \tag{8.73}$$

The expressions (8.72) and (8.73) are equivalent to replacing the derivative by forward and backward differences respectively taking $h = \frac{1}{n}$. We can also find the second derivative at the end points by differentiating (8.68) twice and putting $u = 0$ and $u = 1$.

The nth degree Bezier spline approximation can also be written for y-coordinate in parametric form as

$$y(u) = \sum_{k=0}^{n} y_k \mathrm{B}_{k,n}(u),\ 0 \le u \le 1 \tag{8.74}$$

where $\mathrm{B}_{k,n}(u)$ is same as defined by (8.69).

Giving different values to u between 0 and 1, we can compute $x(u)$ and $y(u)$ from (8.68) and (8.74) respectively and trace the spline curve by joining the points $(x(u), y(u))$ for u increasing. If n is not too large, usually a single n^{th} degree spline is used which has c^n continuity. But if lower degree spline is used it will have only c^0 continuity unless derivatives are matched at the common points where the two adjacent splines meet. See [5].

The most important property of B-Splines/Bezier curves is that they remain within the 'convex hull' of the given data points. See Sec. 8.12.

8.12 Convex Polygon and Convex Hull

Let S be a set of points in the x-y plane $\mathrm{S} = \{\mathrm{P}_0,\ \mathrm{P}_1,\ \mathrm{P}_2,\ \ldots.\mathrm{P}_n\}$. A very important property of B-Splines and Bezier curves is that they lie in the 'convex hull' of S. The 'convex hull' is a convex polygon which is discussed as follows:

For simplicity consider two different sets consisting of, say, six points each denoted by $\mathrm{P}_0,\ \mathrm{P}_1,\ \mathrm{P}_2,\ \mathrm{P}_3,\ \mathrm{P}_4,\ \mathrm{P}_5$ such that polygons $\mathrm{P}_0\mathrm{P}_1\mathrm{P}_2\mathrm{P}_3\mathrm{P}_4\mathrm{P}_5\mathrm{P}_0$ may be formed (see Figs 8.3a and 8.3b).

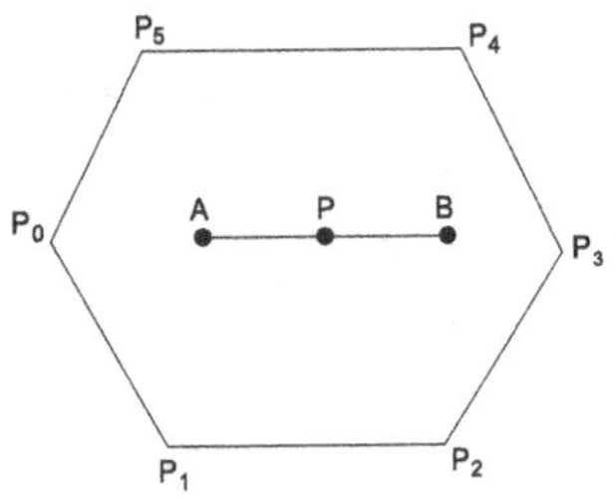

Figure 8.3a Convex polygon.

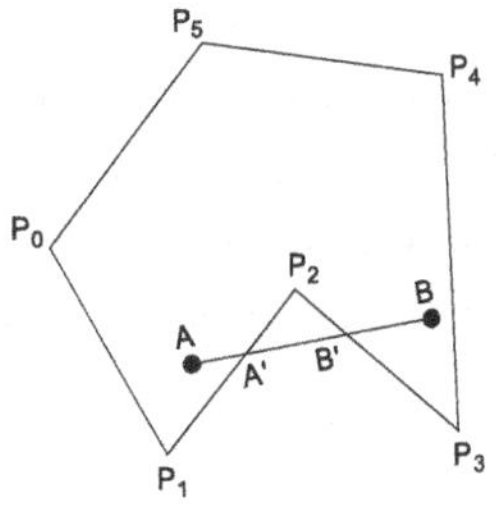

Figure 8.3b Not convex polygon.

Take any two points A and B in the interior of the polygon (or on its boundary); then if all points between A and B on the line joining them, lie in the interior of the polygon (or on its boundary) then it is said to be a 'convex' polygon. Obviously, the polygon in Fig. 8.3a is convex while polygon in Fig. 8.3b is not convex as all the points between A′ and B′ are outside the polygon. Mathematically, if the coordinates of A and B are (x_1, y_1) and (x_2, y_2) respectively, then coordinates of an arbitrary point P between A and B are given by

$$\left.\begin{aligned} x &= (1-\theta)x_1 + \theta x_2 \\ y &= (1-\theta)y_1 + \theta y_2 \end{aligned}\right\}, \quad 0 \le \theta \le 1. \tag{8.75}$$

In a convex polygon point $P(x, y)$ must lie inside the polygon (or on its boundary).

The 'convex hull' of a given set of points is the smallest convex polygon (minimum vertices/sides) such that all points are either on the boundary of the polygon or in its interior. In case of the set of points shown in Fig. 8.3a, the 'convex hull' is the same as the convex polygon while in the case of Fig. 8.3b, the 'convex hull' will be the convex polygon $P_0 P_1 P_3 P_4 P_5 P_0$ with point P_2 inside the polygon (see Fig. 8.4).

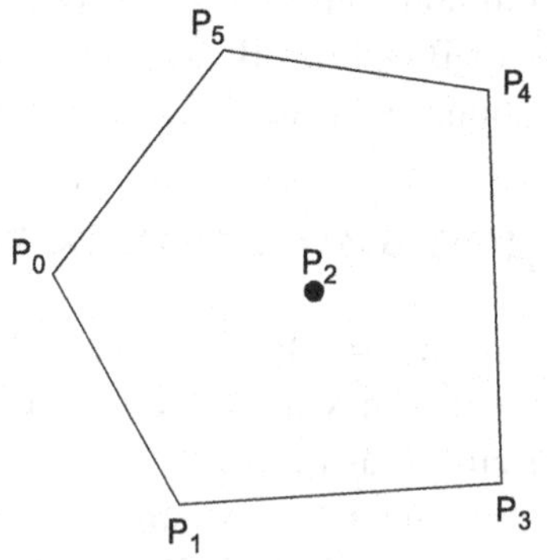

Figure 8.4 Convex hull for Fig. 8.3b.

Conceptually, consider pins sticking on a drawing board at points P_0, $P_1 \ldots$ of the set. Then 'convex hull' for these points will be the polygon formed by a rubber band stretching it over minimum number of pins such that all the pins are either touching the rubber band (on the boundary of the polygon) or are enclosed by it (in the interior of the polygon).

Exercise 8

8.1 Derive the following form of cubic spline in the interval $(x_{i-1},\ x_i)$,

$$s(x) = \frac{1}{6h_i}[(x-x_{i-1})^3 s''(x_i) - (x-x_i)^3 s''(x_{i-1})]$$

$$-\frac{h_i}{6}[(x-x_{i-1})s''(x_i) - (x-x_i)s''(x_{i-1})]$$

$$+\frac{1}{h_i}[(x-x_{i-1})s(x_i) - (x-x_i)s(x_{i-1})]$$

Hint: Take straight line equation,

$$s''(x) = \frac{x-x_i}{-h_i}s''(x_{i-1}) + \frac{x-x_{i-1}}{h_i}s''(x_i) \text{ where } h_i = x_i - x_{i-1}$$

Integrate it twice between x_{i-1} to x,

$$s(x) = s(x_{i-1}) + (x-x_{i-1})s'(x_{i-1}) - \frac{(x-x_i)^3}{6h_i}s''(x_{i-1})$$

$$+\frac{(x-x_{i-1})^3}{6h_i}s''(x_i) + \frac{h_i}{2}(x-x_i)s''(x_{i-1}) - \frac{h_i^2}{6}s''(x_{i-1})$$

Obtain $s'(x_{i-1}) = \dfrac{s(x_i) - s(x_{i-1})}{h_i} - \dfrac{h_i}{6}[s''(x_i) + 2s''(x_{i-1})]$

Substitute in the spline and also use $h_i = x_i - x_{i-1}$.

8.2 Using natural cubic spline, estimate the value of the function $f(x)$ at $x = 0.8$ from the following data,

x	0	0.2	0.6	1.0
$f(x)$	0	0.1823	0.4700	0.6932

8.3 Given below are the values of the ordinates of the normal distribution $y = \frac{1}{\sqrt{2\pi}}\exp\left(-\frac{1}{2}x^2\right)$ for $x = 0,\ 0.5,\ 1.0$:

x	0	0.5	1.0
y	0.399	0.352	0.242

Approximating the distribution by a cubic spline, compute the value of the ordinate for $x = 0.7$ using values of the first derivative at the end-points.

8.4 Solve the following boundary value problem using cubic spline,

$$\frac{d^2y}{dx^2} + 2(1-2x^2)y = 2(1-2x^2), \; 0 \le x \le 0.75$$

$$y(0) = 2, \;\; y(0.75) = 1.5698.$$

Divide the interval (0, 0.75) into three equal sub-intervals and find the value of y at the nodal points, $x = 0.25$ and 0.50.

8.5 Derive a parametric cubic spline between the points $P_i(x_i,\; y_i)$ and $P_{i+1}(x_{i+1},\; y_{i+1})$, $i = 0(1)n-1$, as

$$P(u) = \begin{bmatrix} x(u) \\ y(u) \end{bmatrix} \begin{bmatrix} x_i & x_{i+1} & x_i'' & y_{i+1}'' \\ y_i & y_{i+1} & y_i'' & y_{i+1}'' \end{bmatrix} \begin{bmatrix} -u+1 \\ u \\ -\frac{1}{6}u^3 + \frac{1}{2}u^2 - \frac{1}{3}u \\ \frac{1}{6}u^3 - \frac{1}{6}u \end{bmatrix}$$

Also derive relevant equations for determining the second derivatives for uniform distance between control points.
(Hint: Consider $x(u) = au^3 + bu^2 + cu + d$)

8.6 Using parametric spline of Exercise 8.5 solve the problem of Example 8.8.

References and Some Useful Related Books/Papers

1. Ahlberg, J.H., Nilson, E.N., and Walsh, J.L., *Theory of Splines and their Applications*, Academic Press.

2. Bartels R., Beaty J., Barsky B., *An Introduction to Splines for Use in Computer Graphics and Geometric Modeling*, Morgan Kaufmann.

3. Blum, E.K., *Numerical Analysis and Computation: Theory and Practice*, Addison-Wesley.

4. de Boor Conte, *A Practical Guide to Splines, Applied Mathematical Sciences* (vol. 27) Springer-Verlag.

5. Foley, J.D., van Dam A., Feiner S.K., Hughes, J.F., *Computer Graphics – Principles and Practice*, Pearson Education.

6. Greville, T.N.E. (Editor), *Theory and Applications of Spline Functions*, Academic Press.

7. Schultz, M.H., *Spline Analysis*, Prentice Hall, N. J.

9

Method of Least Squares and Chebyshev Approximation

9.1 Introduction

We have studied in the preceeding chapters various methods for fitting a polynomial to a given set of points, viz., Finite Difference methods, Lagrange's method, Divided Difference method and cubic splines. In all these methods (except Bezier/B-Splines) the polynomial passes through specified points. We say that the polynomial interpolates the given function (known or unknown) at the tabular points. In the method of Least Squares we fit a polynomial or some other function which may or may not pass through any of the data points.

Let us suppose we are given n pairs of values $(x_i,\ y_i)$ where for some value x_i of the variable x, the value of the variable y is given as $y_i,\ i = 1(1)n$. In the Least Squares method, sequencing/ordering of x_i's is not essential. Further, a pair of values may also repeat or there may be two or more values of y corresponding to the same value of x. If the data is derived from a single-valued function, then of course each pair will be unique. However, if an experiment is conducted several times, then the values may repeat. In the context of statistical survey, for the same value of variate x, there may be different outcomes of y; e.g., in the height versus weight study, there may be many individuals with different weights having same height and vice-versa.

9.2 Least Squares Method

Without loss of generality we shall assume in the following discussion that the values of x are not repeated and that they are exact while the corresponding values of y only, are subject to error. In the Least Squares method, we can approximate the given function (known or unknown) by a polynomial (or some other standard functions). If n data points (x_i, y_i), $i = 1(1)n$ are given, then by least squares method, we can fit a polynomial of degree m, given by $y = a_0 + a_1x + a_2x^2 + \ldots + a_mx^m$, $m \leq n-1$. When $m = n-1$, the polynomial will coincide with the Lagrange's polynomial. For $m = 0$, the approximating polynomial is a constant, for $m = 1$, a straight line and for $m = 2$, it will be a quadratic (parabola) and so on. However, we generally do not go beyond $m = 3$ or 4 since the physical interpretation may not be of much consequence and more importantly the resulting equations become ill-conditioned for higher values of m, as will be seen later. Most commonly used approximations are straight lines or parabolas. In practical applications of Least Squares method, first the points are plotted (scatter diagram) and from the behaviour of the scattering of points one guesses about the likely curve to be fitted.

We illustrate the Least Squares method below for $m = 2$, i.e., the approximating polynomial is a quadratic. Let it be represented as,

$$y^* = a_0 + a_1x + a_2x^2, \tag{9.1}$$

where a_0, a_1 and a_2 are parameters to be determined.

For $x = x_i$, the value of y from (9.1) is given as

$$y_i^* = a_0 + a_1x_i + a_2x_i^2, \; i = 1(1)n. \tag{9.2}$$

Let us denote the difference between the given value y_i and the approximate value y_i^* by e_i called 'deviation' or 'residual error' or 'error' (see Fig. 9.1), i.e.,

$$e_i = y_i - y_i^* \tag{9.3}$$

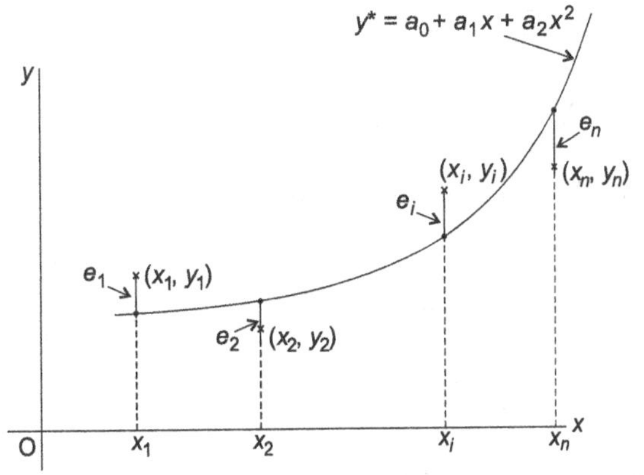

Figure 9.1 Deviations (errors) in the actual and approximated values of y.

The objective is to choose the constants in (9.1) in such a way that the deviations/errors between the given values and the approximated values is minimum. We can not simply add up the deviations as some of them will be positive and some negative and hence resultant deviation may be extremely misleading. The second option may be to take the sum of their moduli but this approach is not amenable for mathematical treatment. In order to circumvent these problems the sum of squares of deviations/errors is minimised in the method of least squares. It is known as Gaussian criterion (principle) for minimisation of error.

Let
$$s=\sum_{i=1}^{n} e_i^2=\sum_{i=1}^{n}(y_i-y_i^*)^2$$
$$=\sum_{i=1}^{n}(y_i^*-y_i)^2$$
$$=\sum_{i=1}^{n}(a_0+a_1x_i+a_2x_i{}^2-y_i)^2. \tag{9.4}$$

In (9.4) we have to determine a_0, a_1 and a_2 such that s is minimum implying that s is dependent on the parameters a_0, a_1 and a_2. For s to have minimum, its partial derivatives w.r.t. a_0, a_1 and a_2 should vanish simultaneously, i.e.,

$$\frac{\partial s}{\partial a_0}=\frac{\partial s}{\partial a_1}=\frac{\partial s}{\partial a_2}=0. \tag{9.5}$$

Differentiation of (9.4) w.r.t. a_0 and using (9.5) gives,

$$\frac{\partial s}{\partial a_0}=\sum_{i=1}^{n}2(a_0+a_1x_i+a_2x_i^2-y_i)=0.$$

or
$$na_0+\Sigma x_ia_1+\Sigma x_i^2a_2=\Sigma y_i,\ \Sigma 1=n. \tag{9.6}$$

The differentiation of (9.4) w.r.t. a_1, after using (9.5) gives,

$$\frac{\partial s}{\partial a_1}=\sum_{i=1}^{n}2(a_0+a_1x_i+a_2x_i^2-y_i)x_i=0$$

or
$$\Sigma x_ia_0+\Sigma x_i^2\cdot a_1+\Sigma x_i^3a_2=\Sigma x_iy_i. \tag{9.7}$$

Similarly, differentiation w.r.t. a_2 gives,

$$\Sigma x_i^2a_0+\Sigma x_i^3a_1+\Sigma x_i^4\cdot a_2=\Sigma x_i^2y_i. \tag{9.8}$$

From the three equations (9.6), (9.7) and (9.8) we can get the values of a_0, a_1 and a_2. These equations are called 'normal equations'. The limits of Σ are ignored as it is understood that the summation is taken over all the given points (in this case n).

It may be noted that in the quadratic approximation the power of x_i goes up to 4th. In the cubic approximation it will go up to sixth and in the quartic up to eighth. As for large value of m, higher and higher powers of x_i are involved, the normal equations become very sensitive to small errors (ill-conditioned).

For linear approximation (straight line), $y^* = a_0 + a_1 x$, the resulting normal equations would be,

$$\left.\begin{aligned} na_0 + \Sigma x_i a_1 &= \Sigma y_i \\ \Sigma x_i a_0 + \Sigma x_i^2 a_1 &= \Sigma x_i y_i. \end{aligned}\right\} \tag{9.9}$$

The goodness of fit of approximation is tested by the value of the root mean square (rms),

$$\text{where} \quad \text{rms} = \left[\frac{\Sigma e_i^2}{n}\right]^{1/2}.$$

9.3 Normal Equations in Matrix Form

Suppose there are given only four points (x_i, y_i), $i = 1(1)4$ and we are fitting a quadratic (parabola), $y^* = a_0 + a_1 x + a_2 x^2$. Then the normal equations (9.6) – (9.8) may be written in the matrix form as,

$$\begin{bmatrix} \Sigma 1 & \Sigma x_i & \Sigma x_i^2 \\ \Sigma x_i & \Sigma x_i^2 & \Sigma x_i^3 \\ \Sigma x_i^2 & \Sigma x_i^3 & \Sigma x_i^4 \end{bmatrix} \begin{bmatrix} a_0 \\ a_1 \\ a_2 \end{bmatrix} = \begin{bmatrix} 1 & 1 & 1 & 1 \\ x_1 & x_2 & x_3 & x_4 \\ x_1^2 & x_2^2 & x_3^2 & x_4^2 \end{bmatrix} \begin{bmatrix} y_1 \\ y_2 \\ y_3 \\ y_4 \end{bmatrix}, \tag{9.10}$$

$$\text{or} \quad \mathrm{A}\boldsymbol{a} = \mathrm{X}^{\mathrm{T}}\boldsymbol{y}, \tag{9.11}$$

where,

$$\mathrm{A} = \begin{bmatrix} \Sigma 1 & \Sigma x_i & \Sigma x_i^2 \\ \Sigma x_i & \Sigma x_i^2 & \Sigma x_i^2 \\ \Sigma x_i^2 & \Sigma x_i^3 & \Sigma x_i^4 \end{bmatrix}; \ \mathrm{X}^{\mathrm{T}} = \begin{bmatrix} 1 & 1 & 1 & 1 \\ x_1 & x_2 & x_3 & x_4 \\ x_1^2 & x_2^2 & x_3^2 & x_4^2 \end{bmatrix}$$

$$\boldsymbol{a}^{\mathrm{T}} = (a_0 \quad a_1 \quad a_2) \quad \text{and} \quad \boldsymbol{y}^{\mathrm{T}} = (y_1 \quad y_2 \quad y_3 \quad y_4).$$

It may be noted that A is a symmetric matrix which can be written as,

$$\mathrm{A} = \mathrm{X}^{\mathrm{T}}\mathrm{X}, \tag{9.12}$$

$$\text{i.e.,} \quad \mathrm{A} = \begin{bmatrix} \Sigma 1 & \Sigma x_i & \Sigma x_i^2 \\ \Sigma x_i & \Sigma x_i^2 & \Sigma x_i^3 \\ \Sigma x_i^2 & \Sigma x_i^3 & \Sigma x_i^4 \end{bmatrix} = \begin{bmatrix} 1 & 1 & 1 & 1 \\ x_1 & x_2 & x_3 & x_4 \\ x_1^2 & x_2^2 & x_3^2 & x_4^2 \end{bmatrix} \begin{bmatrix} 1 & x_1 & x_1^2 \\ 1 & x_2 & x_2^2 \\ 1 & x_3 & x_3^2 \\ 1 & x_4 & x_4^2 \end{bmatrix}$$

Using (9.12) in (9.11), the normal equations may be expressed as,

$$\mathrm{X}^{\mathrm{T}}\mathrm{X}\boldsymbol{a} = \mathrm{X}^{\mathrm{T}}\boldsymbol{y}, \tag{9.13}$$

which may be written as,

$$\mathrm{X}^{\mathrm{T}}(\boldsymbol{y} - \mathrm{X}\boldsymbol{a}) = 0. \tag{9.13a}$$

But X$\boldsymbol{a}$ are the approximate values $\boldsymbol{y}^*$, represented by a vector,

$$(\boldsymbol{y}^*)^{\mathrm{T}} = (y_1{}^* \quad y_2{}^* \quad y_3{}^* \quad y_4{}^*).$$

Thus, (9.13a) becomes

$$\mathrm{X}^{\mathrm{T}}(\boldsymbol{y} - \boldsymbol{y}^*) = 0 = \mathrm{X}^{\mathrm{T}}\boldsymbol{e}, \tag{9.14}$$

where $\boldsymbol{e}$ is the error (or residual) vector, given as

$$\boldsymbol{e}^{\mathrm{T}} = (e_1 \quad e_2 \quad e_3 \quad e_4).$$

Let $\boldsymbol{x}_i^{\mathrm{T}}$ denote the ith row of matrix X^{T}, $i = 1(1)3$. From eq. (9.14), the error vector $\boldsymbol{e}$ is orthogonal (normal) to each row of X^{T}. Due to this property, the equations given by (9.9) or (9.10) are called 'normal equations'.

It may be easily extended for a polynomial of degree $m(\leq n-1)$ when matrix $\mathrm{X}^{\mathrm{T}}(m \times n)$ and vectors $\boldsymbol{y}^{\mathrm{T}}$ and $\boldsymbol{a}^{\mathrm{T}}$ will be as follows:

$$X^T = \begin{bmatrix} 1 & 1 & 1 & \cdots & 1 \\ x_1 & x_2 & x_3 & \cdots & x_n \\ x_1^2 & x_2^2 & x_3^2 & \cdots & x_n^2 \\ \vdots & \vdots & \vdots & & \vdots \\ x_1^m & x_2^m & x_3^m & \cdots & x_n^m \end{bmatrix}, \tag{9.15}$$

$$\boldsymbol{y}^T = (y_1 \quad y_2 \quad y_3 \quad .... \quad y_n) \text{ and } \boldsymbol{a}^T = (a_0 \quad a_1 \quad a_2 \quad .... \quad a_m). \tag{9.15a}$$

9.4 Approximation by Standard Functions

As stated earlier, in the practical applications of the Least Squares method, the given points are plotted first and then by looking at the general pattern of the location of points as well as the nature of the experimental problem, the form of the function is guessed which is then fitted to the given data. These functions are usually standard functions which by simple transformation reduce to fitting a straight line or a parabola. Some such common functions are given below:

(*i*) $y = ae^{bx}$.

Taking log on both sides,

$$\ln y = \ln a + bx$$

Putting $\ln y = Y$; $\ln a = a_0$; $b = a_1$ and $x = x$, the above reduces to,

$$Y = a_0 + a_1 x$$

Thus we work with $\ln y$ instead of y and compute a_0 and a_1 using (9.9); then compute $a = \exp(a_0)$ and substitute the values of a and $a_1 = b$ in the desired function.

(*ii*) $y = ax^b$.

After taking log this reduces to

$$\ln y = \ln a + b \ln x.$$

Put $\ln y = Y$, $\ln x = X$, $\ln a = a_0$ and $b = a_1$. This also reduces to linear form.

In this case we work with $\ln x$ and $\ln y$ and compute a_0 and a_1; then $a = e^{a_0}$ and $b = a_1$. Instead of natural log (base e) common log (base 10) may also be used. In that case $a =$ antilog (a_0).

(*iii*) $y = ab^x$.

Taking log on both sides,

$$\ln y = \ln a + x \ln b.$$

We put $\ln y = Y$, $\ln a = a_0$, $\ln b = a_1$ and $x = X$ which reduces the above to the linear form,

$$Y = a_0 + a_1 X.$$

Again we work with $\ln y$ like in case (b). Then

$$a = \exp(a_0) \text{ and } b = \exp(a_1)$$

(*iv*) $y = a_0 + a_1 x^2 + a_2 x^4$.

We can put $x^2 = X$, which reduces it to quadratic

$$y = a_0 + a_1 X + a_2 X^2.$$

Thus we work with x^2 instead of x and get equations (9.6) – (9.8) to compute a_0, a_1 and a_2.

Example 9.1

Fit a quadratic by Least Squares method to the following data,

x	0	0.5	1.0	1.5
y	1.0	1.65	2.72	4.48

Compute up to five decimal places. Find the residual error vector $\boldsymbol{e}$ and verify that $X^T\boldsymbol{e} = \mathbf{0}$. Also compute rms.

Solution

x	y	x^2	x^3	x^4	xy	x^2y
0	1.0	0	0	0	0	0
0.5	1.65	0.25	0.125	0.0625	0.825	0.4125
1.0	2.72	1.0	1.0	1.0	2.72	2.72
1.5	4.48	2.25	3.375	5.0625	6.72	10.08
3.0	9.85	3.5	4.5	6.125	10.265	13.2125

Let the quadratic be $y^* = a_0 + a_1 x + a_2 x^2$.
The normal equations are

$$4a_0 + 3a_1 + 3.5a_2 = 9.85 \qquad \text{.... (1)}$$

$$3a_0 + 3.5a_1 + 4.5a_2 = 10.265 \qquad \text{.... (2)}$$

$$3.5a_0 + 4.5a_1 + 6.125a_2 = 13.2125 \qquad \text{.... (3)}$$

Solving the system of equations by Gaussian Elimination:

Multiply eq. (1) by $-\frac{3}{4} = -0.75$ and add to (2)

Multiply eq. (1) by $-\frac{3.5}{4} = -0.875$ and add to (3)

$$\begin{array}{ccc|c} a_0 & a_1 & a_2 & \\ 4 & 3 & 3.5 & 9.85 \\ 0 & 1.25 & 1.875 & 2.8775 \\ 0 & 1.875 & 3.0625 & 4.59375 \end{array}$$

Multiplying second row by $-\frac{1.875}{1.25} = -1.5$ and adding to third,

$$\begin{array}{ccc|c} a_0 & a_1 & a_2 & \\ 4 & 3 & 3.5 & 9.85 \\ 0 & 1.25 & 1.875 & 2.8775 \\ 0 & 0 & 0.25 & 0.2775 \end{array}$$

By back substitution,

$$a_2 = 1.11,\ a_1 = 0.637,\ a_0 = 1.0135$$

The required quadratic approximation is

$$y = 1.0135 + 0.637x + 1.11x^2$$

Computation for error vector (compute y^* from quadratic)

i	x_i	y_i	y_i^*	$e_i = y_i - y_i^*$
1	0	1.0	1.0135	−0.0135
2	0.5	1.65	1.6095	0.0405
3	1.0	2.72	2.7605	−0.0405
4	1.5	4.48	4.4665	0.0135

To show $X^T e = \mathbf{0}$.

$$\begin{bmatrix} 1 & 1 & 1 & 1 \\ 0 & 0.5 & 1.0 & 1.5 \\ 0 & 0.25 & 1.0 & 2.5 \end{bmatrix} \begin{bmatrix} -0.0135 \\ 0.0405 \\ -0.0405 \\ 0.0135 \end{bmatrix} = \begin{bmatrix} 0 \\ 0 \\ 0 \end{bmatrix}$$

$$(\text{rms})^2 = \{(-0.0135)^2 + (0.0405)^2 + (-0.0405)^2 + (0.0135)^2\}/4$$

$$= \frac{0.003645}{4} = 0.00091125$$

$$\text{rms} = 0.0302$$

Note: The values of y are taken from $y = e^x$.

Example 9.2

Fit a curve of the form $y = ax^b$ to the following data by the method of Least Squares,

x	1	2	3	4	5
y	1	3	5	8	11

Compute up to four places of decimal and round the values of a and b to two decimal places. Also compute the residual error up to two decimals.

Solution Curve to be fitted $y = ax^b$.
Taking log on both sides (base e)

$$\ln y = \ln a + b \ln x$$

Putting $\ln y = Y$, $\ln a = a_0$, $b = a_1$ and $\ln x = X$, above becomes

$$Y = a_0 + a_1 X, \text{ where } a = \exp(a_0).$$

i	x_i	y_i	$X_i = \ln x_i$	$Y_i = \ln y_i$	X_i^2	$X_i Y_i$	y_i^*	$e_i = y_i - y_i^*$
1	1	1	0	0	0	0	1.02	−0.02
2	2	3	0.6932	1.0986	0.4805	0.7615	2.84	0.16
3	3	5	1.0986	1.6094	1.2069	1.7681	5.18	−0.18
4	4	8	1.3863	2.0794	1.9218	2.8827	7.94	0.06
5	5	11	1.6094	2.3979	2.5902	3.8592	10.83	0.17
			4.7875	7.1853	6.1994	9.2715		

The normal equations are

$$5a_0 + 4.7875a_1 = 7.1853$$

$$4.7875a_0 + 6.1994a_1 = 9.2715$$

On solving the equations $a_1 = 1.4804,\ a_0 = 0.0196$

$$\therefore \quad a = \exp(0.0196) = 1.02;\ b = a_1 = 1.48$$

Curve is, $y = 1.02x^{1.48}$

Computed approximate values of y^* and errors are given in the table.

9.5 Over-Determined System of Linear Equations

Sometimes there arise in practical applications, a system of linear simultaneous equations which is over-determined, i.e., there are more equations than the number of unknowns. As they are usually inconsistent no unique solution exists which satisfies all the equations. However, from Least Squares method a plausible solution can be found. We illustrate the method by taking an example of four equations in three unknowns,

$$\left.\begin{aligned} a_{11}x_1 + a_{12}x_2 + a_{13}x_3 &= b_1 \\ a_{21}x_1 + a_{22}x_2 + a_{23}x_3 &= b_2 \\ a_{31}x_1 + a_{32}x_2 + a_{33}x_3 &= b_3 \\ a_{41}x_1 + a_{42}x_2 + a_{43}x_3 &= b_4. \end{aligned}\right\} \tag{9.16}$$

The system of equations (9.16) can be written in matrix form as,

$$\mathrm{A}\boldsymbol{x} = \boldsymbol{b}, \tag{9.17}$$

where

$$\mathrm{A} = \begin{bmatrix} a_{11} & a_{12} & a_{13} \\ a_{21} & a_{22} & a_{23} \\ a_{31} & a_{32} & a_{33} \\ a_{41} & a_{42} & a_{43} \end{bmatrix},$$

$\boldsymbol{x}^{\mathrm{T}} = (x_1 \quad x_2 \quad x_3)$ and $\boldsymbol{b}^{\mathrm{T}} = (b_1 \quad b_2 \quad b_3 \quad b_4)$.

Let $\boldsymbol{x}^*$ be a solution which satisfies (9.16) approximately where

$$\boldsymbol{x}^{*\mathrm{T}} = (x_1{}^* \ \ x_2{}^* \ \ x_3{}^*). \tag{9.18}$$

When $\boldsymbol{x}^*$ is substituted in (9.17) we get,

$$\mathrm{A}\boldsymbol{x}^* - \boldsymbol{b} = \boldsymbol{e}, \tag{9.19}$$

where $\boldsymbol{e}$ is a (residual) error vector,

$$\boldsymbol{e}^{\mathrm{T}} = (e_1 \ \ e_2 \ \ e_3 \ \ e_4).$$

In the Least Squares method we minimise the sum of squares of the residual errors, i.e.,

$$s = \sum_{i=1}^{4} e_i^2 = \sum_{i=1}^{4} (a_{i1}x_i^* + a_{i2}x_2^* + a_{i3}x_3^* - b_i)^2 \tag{9.20}$$

with respect to x_1^*, x_2^*, x_3^* which means,

$$\frac{\partial s}{\partial x_1{}^*} = \frac{\partial s}{\partial x_2{}^*} = \frac{\partial s}{\partial x_3{}^*} = 0. \tag{9.21}$$

$$\left.\begin{aligned}
\frac{\partial s}{\partial x_1{}^*} &= 0 \text{ gives, } \sum_{i=1}^{4} [(a_{i1}x_1^* + a_{i2}x_2^* + a_{i3}x_3^* - b_i)a_{i1}] = 0 \\
\frac{\partial s}{\partial x_2^*} &= 0 \text{ gives, } \sum_{i=1}^{4} [(a_{i1}x_1^* + a_{i2}x_2^* + a_{i3}x_3^* - b_i)a_{i2}] = 0 \\
\frac{\partial s}{\partial x_3^*} &= 0 \text{ gives, } \sum_{i=1}^{4} [(a_{i1}x_1^* + a_{i2}x_2^* + a_{i3}x_3^* - b_i)a_{i3}] = 0
\end{aligned}\right\}. \tag{9.22}$$

The system of equations (9.22) can be written as,

$$\left.\begin{aligned}
&\Sigma a_{i1}^2 x_1^* + \Sigma a_{i1}a_{i2}x_2^* + \Sigma a_{i1}a_{i3}x_3^* = \Sigma a_{i1}b_i \\
&\Sigma a_{i2}a_{i1}x_1^* + \Sigma a_{i2}^2 x_2^* + \Sigma a_{i2}a_{i3}x_3^* = \Sigma a_{i2}b_i \\
&\Sigma a_{i3}a_{i1}x_1^* + \Sigma a_{i3}a_{i2}x_2^* + \Sigma a_{i3}^2 x_3^* = \Sigma a_{i3}b_i.
\end{aligned}\right\} \tag{9.23}$$

In the matrix form (9.23) can be written as,

$$\mathrm{A}^{\mathrm{T}}\mathrm{A}\boldsymbol{x}^* = \mathrm{A}^{\mathrm{T}}\boldsymbol{b}. \tag{9.24}$$

We can solve the system of equations (9.23) or (9.24) to obtain the values of x_1^*, x_2^* and x_3^*.

The system of equations (9.24) can be written for m equations in n unknowns $(m > n)$ when,

$$A = \begin{bmatrix} a_{11} & a_{12} & a_{13} & \cdots & a_{1n} \\ a_{21} & a_{22} & a_{23} & \cdots & a_{2n} \\ a_{31} & a_{32} & a_{33} & \cdots & a_{3n} \\ \vdots & \vdots & \vdots & & \\ a_{m1} & a_{m2} & a_{m3} & \cdots & a_{mn} \end{bmatrix},$$

$$\boldsymbol{x}^{*\mathrm{T}} = (x_1{}^* \;\; x_2{}^* \;\; x_3{}^* \;\; .... \; x_n{}^*); \;\; \boldsymbol{b}^{\mathrm{T}} = (b_1 \;\; b_2 \;\; b_3 \;\; .... \;\; b_m).$$

For $m = n$ we will get a unique solution if A is a regular matrix and for $m < n$ there will be more than one(many) solutions.

Example 9.3

Find an approximate solution of the following system of equations by Least Squares method,

$$x_1 + 3x_2 = 5; \quad 3x_1 - x_2 = 2; \quad 2x_1 + x_2 = 3.$$

Also compute the residual error up to 3 decimals.

Solution The system of equations can be written as

$A\boldsymbol{x} = \boldsymbol{b}$, where

$$A = \begin{bmatrix} 1 & 3 \\ 3 & -1 \\ 2 & 1 \end{bmatrix}; \;\; \boldsymbol{b}^{\mathrm{T}} = (5 \;\; 2 \;\; 3); \;\; \boldsymbol{x}^{\mathrm{T}} = (x_1 \;\; x_2).$$

Let the approximate solution be $\boldsymbol{x}^{*\mathrm{T}} = (x_1{}^* \;\; x_2{}^*)$.
To find $\boldsymbol{x}^*$, we solve,

$$A^{\mathrm{T}}A\boldsymbol{x}^* = A^{\mathrm{T}}\boldsymbol{b}.$$

$$\begin{bmatrix} 1 & 3 & 2 \\ 3 & -1 & 1 \end{bmatrix} \begin{bmatrix} 1 & 3 \\ 3 & -1 \\ 2 & 1 \end{bmatrix} \begin{bmatrix} x_1^* \\ x_2^* \end{bmatrix} = \begin{bmatrix} 1 & 3 & 2 \\ 3 & -1 & 1 \end{bmatrix} \begin{bmatrix} 5 \\ 2 \\ 3 \end{bmatrix}$$

Equations to solve are,

$$14x_1^* + 2x_2^* = 17$$
$$2x_1^* + 11x_2^* = 16$$

After solving: $x_1^* = 1.032$; $x_2{}^* = 1.267$.
Residuals are: $e_1 = -0.167,\ e_2 = -0.171,\ e_3 = 0.331$.

9.6 Approximation by Linear Combination of Functions

Let us suppose n data points are given $(x_i,\ y_i),\ i = 1(1)n$. Without loss of generality we assume $x_1 < x_2 < x_3\ \ldots . < x_n$ so that function $y(x)$ is defined at n distinct points $x = x_i$, $i = 1(1)n$ in the interval $x_1 \leq x \leq x_n$. Let $\phi_0(x),\ \phi_1(x),\ \ldots .\ \phi_m(x)$ be linearly independent functions defined over $x_1 \leq x \leq x_n$. Then we can approximate $y(x)$ by the linear combination of these functions as

$$y(x) = a_0\phi_0(x) + a_1\phi_1(x) + a_2\phi_2(x) + \ldots . + a_m\phi_m(x), \quad m \leq n-1. \tag{9.25}$$

where $a_0,\ a_1,\ a_2,\ \ldots .\ a_m$ are parameters to be determined. Applying method of least squares we define

$$s = \sum_{i=1}^{n}\{a_0\phi_0(x_i)+a_1\phi_1(x_i)+a_2\phi_2(x_i)+\ldots .+a_m\phi_m(x_i)-y_i\}^2$$

For s to be minimum

$$\frac{\partial s}{\partial a_k} = 0,\ k = 0(1)m.$$

$$\frac{\partial s}{\partial a_k} = \sum_{i=1}^{n}[2\{a_0\phi_0(x_i)+a_1\phi_1(x_i)+\ldots .+a_m\phi_m(x_i)-y_i\}\phi_k(x_i)] = 0,\ k = 0(1)m.$$

The summation is taken over the n number of points $(x_i,\ y_i),\ \ i = 1(1)n$.
Writing the above in expanded form for $k = 0(1)m$

$$\left.\begin{array}{l}
a_0\Sigma\phi_0^2 + a_1\Sigma\phi_0\phi_1 + a_2\Sigma\phi_0\phi_2 + \ \ldots . \ + a_m\Sigma\phi_0\phi_m = \Sigma\phi_0 y_i \\
a_0\Sigma\phi_1\phi_0 + a_1\Sigma\phi_1^2 + a_2\Sigma\phi_1\phi_2 + \ \ldots . \ + a_m\Sigma\phi_1\phi_m = \Sigma\phi_1 y_i \\
a_0\Sigma\phi_2\phi_0 + a_1\Sigma\phi_2\phi_1 + a_2\Sigma\phi_2^2 + \ \ldots . \ + a_m\Sigma\phi_2\phi_m = \Sigma\phi_2 y_i \\
\quad\vdots \qquad\qquad \vdots \qquad\qquad \vdots \qquad\qquad\qquad \vdots \qquad\qquad \vdots \\
a_0\Sigma\phi_m\phi_0 + a_1\Sigma\phi_m\phi_1 + a_2\Sigma\phi_m\phi_2 + \ \ldots . \ + a_m\Sigma\phi_m^2 = \Sigma\phi_m y_i
\end{array}\right\} \tag{9.26}$$

The summation in the above is taken over $i = 0(1)n$.

The system of equations (9.26) can be written in matrix form as

$$\Phi^{\mathrm{T}}\Phi\boldsymbol{a} = \Phi^{\mathrm{T}}\boldsymbol{y} \tag{9.27}$$

where $\boldsymbol{a}^{\mathrm{T}} = (a_0 \;\; a_1 \;\; a_2 \;\; .... \;\; a_m)$; $\boldsymbol{y}^{\mathrm{T}} = (y_1 \;\; y_2 \;\; .... \; y_n)$ and

$$\Phi = \begin{bmatrix} \phi_0(x_1) & \phi_1(x_1) & \phi_2(x_1) & \cdots & \phi_m(x_1) \\ \phi_0(x_2) & \phi_1(x_2) & \phi_2(x_2) & \cdots & \phi_m(x_2) \\ \phi_0(x_3) & \phi_1(x_3) & \phi_2(x_3) & \cdots & \phi_m(x_3) \\ \vdots & \vdots & & & \vdots \\ \phi_0(x_n) & \phi_1(x_n) & \phi_2(x_n) & \cdots & \phi_m(x_n) \end{bmatrix} \tag{9.28}$$

We solve the system (9.27) for $\boldsymbol{a}$ and substitute the values of $a_0,\ a_1,\ \ldots.\ a_m$ in (9.25) to get the solution. If there exist more than one solution, then choose the one which makes s minimum. When we are fitting a polynomial $y = a_0 + a_1x + a_2x^2 + \ldots + a_mx^m$, then

$$\phi_0(x) = 1,\ \phi_1(x) = x,\ \phi_2(x) = x^2 \ldots,\ \phi_m(x) = x^m.$$

9.7 Approximation by Orthogonal Polynomials

Let $\mathrm{S}_m(x)$ be a polynomial of degree m; if

$$\sum_{i=1}^{n} \omega(x_i)x_i^p\mathrm{S}_m(x_i) = 0 \text{ for } p = 0,\ 1,\ 2,\ \ldots.\ m-1, \tag{9.29}$$

$x_1 < x_2 < \ldots.\ x_n$ and $\omega(x_i)$ has constant sign in $x_1 \leq x \leq x_n$, then $\mathrm{S}_n(x)$ is called orthogonal with weight function $\omega(x)$ with respect to summation for $x = x_i,\ i = 1(1)n$.

The definition (9.29) implies that

$$\sum_{i=1}^{n} \omega(x_i)\mathrm{S}_p(x_i)\mathrm{S}_m(x_i) = 0, \quad p < m. \tag{9.30}$$

Taking $\omega(x) = 1$, (9.30) may be written as

$$\sum_{i=1}^{n} \mathrm{S}_p(x_i)\mathrm{S}_m(x_i) = 0, \quad p < m. \tag{9.31}$$

Let us express the orthogonal polynomial of degree m as

$$S_m(x) = (x-\alpha_m)S_{m-1}(x) - \beta_m S_{m-2}(x) + \sum_{r=0}^{m-3} c_r S_r(x) \tag{9.32}$$

where α_m, β_m and $c_0, c_1, \ldots. c_{m-3}$ are parameters to be determined.

It can be shown that $c_0, c_1, \ldots. c_{m-3}$ are zero, for if we put $x = x_i$ in (9.32); multiply by $S_k(x_i)$, $k = 0(1)m-3$, and add over i, then we get

$$\Sigma S_k(x_i)S_m(x_i) = \Sigma x_i S_{m-1}(x_i)\cdot S_k(x_i) - \alpha_m \Sigma S_k(x_i)S_{m-1}(x_i) + c_k \Sigma S_k^2(x_i)$$
$$+\Sigma_{r\neq k} c_r \Sigma_i S_r(x_i)S_k(x_i).$$

We see that all the terms on the left side and the right side are zero except the term containing c_k, by virtue of (9.31). Hence we get

$$c_k \Sigma S_k^2(x_i) = 0,\ k = 0,\ 1,\ 2,\ \ldots.\ m-3.$$

But $\Sigma S_k^2(x_i)$ can not be zero as it is sum of squares. Hence $c_0 = c_1 = \ldots. = c_{m-3} = 0$. Therefore, recurrence formula (9.32) may be written for $m \geq 3$ as

$$S_m(x) = (x-\alpha_m)S_{m-1}(x) - \beta_m S_{m-2}(x), \tag{9.33}$$

For $m = 1$, (9.33) becomes

$$S_1(x) = (x-\alpha_1)\ S_0(x)\ - \beta_1 S_{-1}(x)$$

Let us assume $\beta_1 = 0$ since $S_{-1}(x)$ is not defined.

Now, $S_1(x) = (x-\alpha_1)S_0(x)$.

For $x = x_i$ it becomes

$$S_1(x_i) = (x-\alpha_1)\ S_0(x_i)$$

Multiplying the above by $S_0(x_i)$ and adding over i gives

$$\Sigma S_1(x_i)S_0(x_i) = \Sigma x_i {S_0}^2(x_0) - \alpha_1 \Sigma {S_0}^2(x_i)$$

or $$\alpha_1 = \Sigma x_i {S_0}^2(x_i)/\Sigma {S_0}^2(x_i) \qquad \because \Sigma S_1(x_i)\cdot S_0(x_i) = 0.$$

Let us choose $S_0(x) = 1$; then

$$\alpha_1 = \frac{1}{n}\Sigma x_i,\text{ giving}$$

$$S_1(x) = x - \bar{x},\text{ where } \bar{x} = \Sigma x_i/n. \tag{9.34}$$

Now formula (9.33) is defined for $m \geq 2$ while $S_0(x) = 1$ and $S_1(x)$ is given by (9.34). For $m \geq 2$, we can compute α_m from (9.33) by putting in it $x = x_i$, multiplying throughout by $S_{m-1}(x_i)$ and summing over i;

$$\alpha_m = \sum_i x_i S_{m-1}^2(x_i) \Big/ \sum S_{m-1}^2(x_i). \tag{9.35}$$

Similarly we can get by multiplying throughout by $S_{m-2}(x_i)$ and adding

$$\beta_m = \Sigma x_i\, S_{m-1}(x_i) S_{m-2}^2(x_i) \Big/ \Sigma S_{m-2}^2(x_i). \tag{9.36}$$

The main advantage of fitting a polynomial through orthogonal polynomials is that we do not have to decide the degree of the polynomial in advance. We can add on next higher degree polynomial starting from $y = a_0 S_0 + a_1 S_1$ to $y = a_0 S_0 + a_1 S_1 + a_2 S_2$ and so on. In computing $a_2 S_2$ we do not have to recalculate a_0 and a_1—they remain same. Secondly the problem of ill-conditioning of equations is circumvented. The values of a_j are obtained from (9.26) as

$$a_j = \sum_{i=1}^{n} S_j(x_i) \cdot y_i \Big/ \sum S_j^2(x_i), \; j = 0,\, 1,\, 2,\, \ldots. \tag{9.37}$$

remembering that $\Sigma S_p(x_i) \cdot S_q(x_i) = 0, \;\; p \neq q; \;\; \phi_j = S_j.$

Example 9.4

Fit polynomials of degree 1 and 2, using orthogonal polynomials to the data of Example 9.1.

Solution

x_i	y_i	$S_0(x_i)$	$S_1(x_i)$	$S_1^2(x_i)$	$x_i S_1^2(x_i)$	$x_i S_1(x_i)$	$S_0^2(x_i)$	$x_i S\ (x_i)\ S_0(x_i)$
0	1.0	1	−0.75	0.5625	0	0	1	0
0.5	1.65	1	−0.25	0.0625	0.03125	−0.125	1	−0.125
1.0	2.72	1	0.25	0.0625	0.06250	0.25	1	0.25
1.5	4.48	1	0.75	0.5625	0.84375	1.125	1	1.125
3.0	9.85			1.25	0.9375	1.25	4	1.25

$S_0(x) = 1$

$S_1(x) = x - \Sigma x_i/4$

$= x - 0.75$

$\alpha_2 = \Sigma x_i S_1^2(x_i)/\Sigma S_1^2(x_i) = 0.9375/1.25 = 0.75$

$\beta_2 = \Sigma x_i S_1(x_i) S_0(x_i)/\Sigma S_0^2(x_i) = 1.25/4 = 0.3125$

$S_2(x) = (x - 0.75)S_1(x) - 0.3125\, S_0(x)$

$= (x - 0.75)(x - 0.75) - 0.3125 \times 1$

$= x^2 - 1.5x + 0.25$

$S_1(x_i)y_i$	$S_2(x_i)$	$S_2(x_i)y_i$
−0.75	0.25	0.25
−0.4125	−0.25	−0.4125
0.68	−0.25	−0.68
3.36	0.25	1.12
2.8775		0.2775

$a_0 = \Sigma S_0(x_i)y_i/\Sigma S_0^2(x_i) = 9.85/4 = 2.4625$

$a_1 = \Sigma S_1(x_i)y_i/\Sigma S_1^2(x_i) = 2.8775/1.25 = 2.302$

$a_2 = \Sigma S_2(x_i)y_i/\Sigma S_2^2(x_i) = 0.2775/.25 = 1.11$

(*i*) Polynomial of degree 1

$y = a_0 S_0 + a_1 S_1 = 2.4625 + 2.302(x - 0.75)$

$= 2.302x + 0.736$

(*ii*) Polynomial of degree 2

$y = a_0 S_0 + a_1 S_1 + a_2 S_2$

$= 2.302x + 0.736 + 1.11\,(x^2 - 1.5x + 0.25)$

$= 1.11x^2 + 0.637x + 1.0135.$

9.8 Chebyshev Approximation

While in the approximation by Least Squares method, sum of squares of the errors is minimised, in the Chebyshev approximation the modulus of the maximum error over the entire domain is minimised. It may be called Chebyshev's criterion (principle) for minimisation of error. Before explaining this property let us rewrite, for sake of completeness, the definition of a Chebyshev polynomial and its simple properties, discussed already in Chapter 6.

A Chebyshev polynomial, $T_n(x)$ of degree n is defined in the interval $-1 \leq x \leq 1$, as

$$T_n(x) = \cos(n\cos^{-1}x),\ n = 0,\ 1,\ 2\ \ldots. \tag{9.38}$$

The Chebyshev polynomials are orthogonal with weight function $\omega(x) = 1/\sqrt{1-x^2}$ w.r.t. integral over the interval $-1 \le x \le 1$ and satisfy the relation

$$\int_{-1}^{1} \frac{T_m(x) \cdot T_n(x)}{\sqrt{1-x^2}} dx = \begin{cases} 0 & , \ m \neq n \\ \dfrac{\pi}{2} & , \ m = n \neq 0 \\ \pi & , \ m = n = 0. \end{cases} \tag{9.39}$$

Following recurrence relation is satisfied by these polynomials

$$T_{n+1}(x) = 2xT_n(x) - T_{n-1}(x),\ n = 1,\ 2,\ 3\ldots. \tag{9.40}$$

The first few Chebyshev polynomials may be written as

$$\left.\begin{array}{ll} T_0(x) = 1; & T_4(x) = 8x^4 - 8x^2 + 1 \\ T_1(x) = x; & T_5(x) = 16x^5 - 20x^3 + 5x \\ T_2(x) = 2x^2 - 1; & T_6(x) = 32x^6 - 48x^4 + 18x^2 - 1 \\ T_3(x) = 4x^3 - 3x; & \text{etc.} \end{array}\right\} \tag{9.41}$$

It may be noted that an even degree polynomial contains all even degree terms and an odd degree polynomial all odd degree terms. Also that the coefficients of the leading term x^n in $T_n(x)$ is 2^{n-1}. According to 'minimax' property of Chebyshev polynomial, if $p_n(x)$ is any other polynomial of degree n with leading coefficient unity (monomial), then in the interval $-1 \le x \le 1$ following is satisfied,

$$\max_{-1 \le x \le 1} \left| \frac{T_n(x)}{2^{n-1}} \right| \le \max_{-1 \le x \le 1} |p_n(x)|.$$

It may be recalled that the value of $T_n(x)$ oscillates between -1 and 1 in the interval $-1 \le x \le 1$, being $T_n(-1) = (-1)^n$ at $x = -1$ and $T_n(1) = 1$ at $x = 1$. (See Fig. 9.2). Further, the zeros of $T_n(x)$ occur where $n\cos^{-1}x = \dfrac{2r+1}{2}\pi$ or

$$x_r = \cos\left(\frac{2r+1}{2n}\pi\right),\ r = 0,\ 1,\ 2,\ \ldots(n-1) \tag{9.42}$$

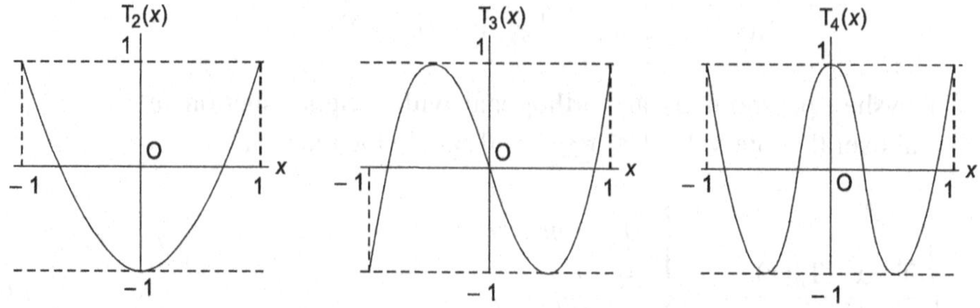

Figure 9.2 Graphs of $T_2(x)$, $T_3(x)$ and $T_4(x)$; not to scale.

and these zeros lie in the interval $-1 < x < 1$ and are real.

Suppose we want to approximate a function $f(x)$ by Chebyshev polynomials defined in an interval $-1 \leq x \leq 1$. Then the function is represented by a truncated series as

$$f(x) = a_0 \mathrm{T}_0(x) + a_1 \mathrm{T}_1(x) + a_2 \mathrm{T}_2(x) + \ldots + a_n \mathrm{T}_n(x), \tag{9.43}$$

where a_0, a_1, a_2, ... a_n are constants to be determined.

These constants can be determined using orthogonal property (9.39) of Chebyshev polynomials, giving

$$\left. \begin{aligned} a_0 &= \frac{1}{\pi} \int_{-1}^{1} \frac{f(x)}{\sqrt{1-x^2}} dx \\ a_k &= \frac{2}{\pi} \int_{-1}^{1} \frac{f(x) \mathrm{T}_k(x)}{\sqrt{1-x^2}} dx, \ k = 1, 2, \ldots n. \end{aligned} \right\} \tag{9.44}$$

To explain Chebyshev criterion for minimising the error, let us suppose that a function $f(x)$ is to be approximated by a polynomial $p_n^*(x)$ of degree n, in the interval $-1 \leq x \leq 1$, satisfying the condition

$$\max_{-1 \leq x \leq 1} |f(x) - p_n^*(x)| = \min. \tag{9.45}$$

This is known as 'minimax (or uniform) approximation' criterion. It can be shown that the Chebyshev approximation (9.43), say, $p_n(x) = \sum_{i=0}^{n} a_i T_i(x)$ will be the required polynomial of degree n that satisfies condition (9.45). See [4].

In general, evaluation of the integrals (9.44) in closed form is difficult. We can adopt numerical approach which is explained below:

Choose the abscissas as zeros of $\mathrm{T}_n(x)$, given by (9.42). Then (9.43) can be written for $x = x_r$ as

$$f(x_r) = a_0 + a_1 \mathrm{T}_1(x_r) + a_2 \mathrm{T}_2(x_r) + \ldots + a_{n-1} \mathrm{T}_{n-1}(x_r), \tag{9.46}$$

$$r = 0(1)\, n-1, \text{ since } \mathrm{T}_n(x_r) = 0.$$

To find a_k, multiply both sides of equation (9.46) by $\mathrm{T}_k(x_r)$ and add over $r = 0(1)\, n-1$. It gives

$$\sum_j a_j \sum_r \mathrm{T}_j(x_r)\mathrm{T}_k(x_r) = \sum_r f(x_r)\mathrm{T}_k(x_r), \tag{9.47}$$

$$j = 0,\ 1,\ 2, \cdots n-1.$$

Let us now evaluate the summation on the left side of (9.47)

$$\sum_{r=0}^{n-1} \mathrm{T}_j(x_r)\mathrm{T}_k(x_r) = \sum \cos\left(\frac{2r+1}{2n}\pi j\right) \cdot \cos\left(\frac{2r+1}{2n}\pi k\right)$$

$$= \frac{1}{2}\left[\sum_{r=0}^{n-1} \cos\frac{2r+1}{2n}(j+k)\pi + \sum_{r=0}^{n-1} \cos\frac{2r+1}{2n}(j-k)\pi\right] \tag{9.48}$$

Consider the first summation in (9.48)

$$\sum \cos\frac{2r+1}{2n}(j+k)\pi = \mathrm{Re}\left[\sum \exp\frac{2r+1}{2n}(j+k)\pi i\right]$$

where Re denotes real part of a complex function and $i = \sqrt{-1}$.

$$\sum_{r=0}^{n-1} \exp\frac{2r+1}{2n}(j+k)\pi i = e^{\frac{1}{2n}(j+k)\pi i} + e^{\frac{3}{2n}(j+k)\pi i} + \cdots + e^{\frac{2n-1}{2n}(j+k)\pi i}$$

which is a geometric series with common ratio $e^{\frac{j+k}{n}\pi i}$

$$= \frac{e^{\frac{1}{2n}(j+k)\pi i}\left[1 - e^{(j+k)\pi i}\right]}{1 - e^{\frac{j+k}{n}\pi i}}$$

$$= \frac{1 - e^{(j+k)\pi i}}{e^{-\frac{1}{2n}(j+k)\pi i} - e^{\frac{1}{2n}(j+k)\pi i}}$$

$$= \frac{i[1-(-1)^{j+k}]}{2\sin\frac{j+k}{2n}\pi}. \tag{9.49}$$

The real part in the above expression is zero, i.e.,

$$\text{Re}\left\{\sum_{r=0}^{n-1} e^{\frac{2r+1}{2n}(j+k)\pi i}\right\} = 0$$

Similarly, if $j \neq k$

$$\text{Re}\left\{\sum_{r=0}^{n-1} e^{\frac{2r+1}{2n}(j-k)\pi i}\right\} = 0.$$

When $j = k \neq 0$, we get from (9.48), the value of the summation as $n/2$ and when $j = k = 0$, its value is n. Thus we have the relation

$$\sum_{r=0}^{n-1} \mathrm{T}_j(x_r)\mathrm{T}_k(x_r) = \begin{cases} 0\,, & j \neq k \\ \frac{n}{2}\,, & j = k \neq 0 \\ n\,, & j = k = 0 \end{cases} \tag{9.50}$$

Now using (9.50) we can write from (9.47)

$$\left.\begin{aligned} a_0 &= \frac{1}{n}\sum_{r=0}^{n-1} f(x_r) = \frac{1}{n}\sum_{r=0}^{n-1} f\left\{\cos\frac{2r+1}{2n}\pi\right\} \\ a_k &= \frac{2}{n}\sum_{r=0}^{n-1} f(x_r)\mathrm{T}_k(x_r) = \frac{2}{n}\sum_{r=0}^{n-1} f\left\{\cos\frac{2r+1}{2n}\pi\right\}\cdot\cos\frac{2r+1}{2n}\pi k, \\ & k = 1,\,2,\,\ldots.\,(n-1). \end{aligned}\right\} \tag{9.51}$$

An estimate of the error in the approximation (9.51) may be obtained by the first neglected term $a_n\mathrm{T}_n(x)$, if it is assumed the function $f(x)$ is approximated by an infinite Chebyshev series. But the maximum value of $\mathrm{T}_n(x)$, in modulus, is 1 in the interval $-1 \leq x \leq 1$. Hence, the error estimate in the approximation (9.51) may be taken as $|a_n|$ approx. See [1].

We should note that in the process of approximating $f(x)$ in the interval $-1 \leq x \leq 1$ we make the transformation $\cos^{-1} x = \theta$ or $x = \cos\theta$, thus changing variable x to θ where

$0 \leq \theta \leq \pi$. In other words we approximate $f(\cos\theta)$ in the interval $0 \leq \theta \leq \pi$. Also that $f(\cos\theta)$ is not only an even function of θ, it is a function of an even function $\cos\theta$, hence can be approximated by a series of even functions. In the above we have chosen n nodal values of θ, say $\theta_r = \frac{2r+1}{2n}\pi$, $r = 0, 1, \ldots. n-1$ which are zeros of $T_n(x)$. Alternatively, we can also choose, if we like, $x_s = \cos\frac{s\pi}{n}$ or $\theta_s = \frac{s\pi}{n}$, $s = 0,\ 1,\ 2,\ \ldots. n$. But in this case we have to compute a_n as well since $T_n(n\theta_k)$ will not be zero; it will be $(-1)^k$. In that case the coefficients can be calculated from the following formulas. See [3]:

$$\left.\begin{aligned} a_0 &= \frac{1}{n}\sum_{s=0}^{n} f(x_s) = \frac{1}{n}\sum_{s=0}^{n} f\left\{\cos\frac{s\pi}{n}\right\} \\ a_k &= \frac{2}{n}\sum_{s=0}^{n} f(x_s)T_k(x_s) = \frac{2}{n}\sum_{s=0}^{n} f\left\{\cos\frac{s\pi}{n}\right\}\cos\left(\frac{s\pi}{n}k\right), \\ &\quad k = 1,\ 2,\ \ldots. (n-1) \\ a_n &= \frac{1}{n}\sum_{s=0}^{n}(-1)^s f(x_s) = \frac{1}{n}\sum_{s=0}^{n}(-1)^s f\left\{\cos\frac{s\pi}{n}\right\}. \end{aligned}\right\} \tag{9.52}$$

When $f(x)$ is represented by a power series in x, its various powers can be expressed by Chebyshev polynomials. After rearranging the terms, the polynomial can be represented in Chebyshev series. From (9.41) we can write the following:

$$\left.\begin{aligned} 1 &= T_0; & x^4 &= \frac{1}{2^3}\{3T_0 + 4T_2 + T_4\}; \\ x &= T_1; & x^5 &= \frac{1}{2^4}\{10T_1 + 5T_3 + T_5\}; \\ x^2 &= \frac{1}{2}\{T_0 + T_2\}; & x^6 &= \frac{1}{2^5}\{10T_0 + 15T_2 + 6T_4 + T_6\} \\ x^3 &= \frac{1}{2^2}\{3T_1 + T_3\}; & &\text{etc.} \end{aligned}\right\} \tag{9.53}$$

It may be noted that even powers are expressed by even Chebyshev polynomials and odd powers by odd polynomials. In (9.41) we saw that even Chebyshev polynomials contain even powers and odd polynomials odd powers. We shall discuss as how, in general, a power of x can be represented by Chebyshev series without using formulae (9.53)

Let us find a Chebyshev series for a general power of x, say x^m. Express,

$$x^m = a_0T_0 + a_1T_1 + \ldots. + a_kT_k + \ldots + a_mT_m \tag{9.54}$$

To determine a_k we use orthogonal property (9.39) and obtain

$$a_k = \frac{2}{\pi}\int_{-1}^{1} \frac{x^m T_k(x)dx}{\sqrt{1-x^2}}, \; k = 1,\, 2,\, \ldots.\, m.$$

$$= \frac{2}{\pi}\int_{0}^{\pi} \cos^m \theta \cos k\theta d\theta, \text{ putting } x = \cos\theta$$

$$= \frac{1}{2^m \cdot \pi}\int_{0}^{\pi} (e^{i\theta} + e^{-i\theta})^m \cdot (e^{ik\theta} + e^{-ik\theta}) d\theta. \tag{9.55}$$

Evaluate $(e^{i\theta} + e^{-i\theta})^m \, (e^{ik\theta} + e^{-ik\theta})$

$$= \sum_{r=0}^{m} \binom{m}{r} e^{i(m-2r)\theta} (e^{ik\theta} + e^{-ik\theta})$$

$$= \sum_{r=0}^{m} \binom{m}{r} e^{i(m+k-2r)\theta} + \sum_{r=0}^{m} \binom{m}{r} e^{i(m-k-2r)\theta}$$

$$= \sum_{r=0}^{m} \binom{m}{r} e^{i(m+k-2r)\theta} + \sum_{r=m}^{0} \binom{m}{m-r} e^{-i(m+k-2r)\theta}$$

after writing second term in reverse order

$$= 2\sum_{r=0}^{m} \binom{m}{r} \cos(m+k-2r)\theta, \quad \because \binom{m}{r} = \binom{m}{m-r}$$

Substituting the above result in (9.55)

$$a_k = \frac{1}{2^{m-1}\pi}\int_{0}^{\pi} \sum_{r=0}^{m} \binom{m}{r} \cos(m+k-2r)\theta \, d\theta$$

$$= \frac{1}{2^{m-1}\pi} \sum_{r=0}^{m} \left[\binom{m}{r} \frac{\sin(m+k-2r)\theta}{m+k-2r} \right]_0^{\pi}$$

The bracketed term vanishes for all values of r except when $m+k-2r=0$ or $r=\dfrac{m+k}{2}$; in which case, its value is π. Hence

$$\left.\begin{aligned} a_k &= \frac{1}{2^{m-1}}\begin{pmatrix} m \\ \frac{m+k}{2} \end{pmatrix}, \quad k=1,\ 2,\ \ldots.\ m \\ \text{and} \quad a_0 &= \frac{1}{2^m}\begin{pmatrix} m \\ m/2 \end{pmatrix}. \end{aligned}\right\} \tag{9.56}$$

From (9.56) it should be clear that both m and k should be either even or odd. Secondly we must have $m \geq \dfrac{m+k}{2}$ or $m \geq k$. It means that if $f(x)$ is a polynomial then we should include the terms $m=k,\ k+2,\ k+4$, etc. for computing a_k. For example, consider a polynomial approximated by Chebyshev series

$$\sum_{m=0}^{p} \mathrm{C}_m x^m = \sum_{r=0}^{p} a_r \mathrm{T}_r(x), \tag{9.57}$$

The coefficients are given by

$$\left.\begin{aligned} a_0 &= \sum_{m=0} \frac{\mathrm{C}_m}{2^m}\begin{pmatrix} m \\ m/2 \end{pmatrix}, \quad m=0,\ 2,\ 4,\ \ldots \\ a_k &= \sum_{m=k} \frac{\mathrm{C}_m}{2^{m-1}}\begin{pmatrix} m \\ \frac{m+k}{2} \end{pmatrix}, \quad m=k,\ k+2,\ k+4,\ \ldots. \end{aligned}\right\} \tag{9.58}$$

Thus if a function is expanded by Maclaurin's series in the interval $-1 \leq x \leq 1$, it can be represented by a Chebyshev series. It will be seen that fewer terms or lower order terms will be required in the Chebyshev series as compared to the Maclaurin's series to provide almost the same accuracy. See Example 9.6.

Example 9.5

Express x^4 and x^5 by Chebyshev series without using recurrence relation. Use appropriate formula.

Solution Let $x^4 = a_0\mathrm{T}_0 + a_2\mathrm{T}_2 + a_4\mathrm{T}_4$

$$a_0 = \frac{1}{2^4}\binom{4}{2} = \frac{3}{8}$$

$$a_2 = \frac{1}{2^3}\binom{4}{\frac{4+2}{2}} = \frac{1}{2^3}\cdot\binom{4}{3} = \frac{1}{2}$$

$$a_4 = \frac{1}{2^3}\binom{4}{\frac{4+4}{2}} = \frac{1}{2^3}\binom{4}{4} = \frac{1}{8}$$

Hence, $x^4 = \frac{1}{8}(3T_0 + 4T_2 + T_4)$

Let $x^5 = a_1T_1 + a_3T_3 + a_5T_5$

$$a_1 = \frac{1}{2^4}\binom{5}{\frac{5+1}{2}} = \frac{10}{2^4}$$

$$a_3 = \frac{1}{2^4}\binom{5}{\frac{5+3}{2}} = \frac{5}{2^4}$$

$$a_5 = \frac{1}{2^4}\binom{5}{\frac{5+5}{2}} = \frac{1}{2^4}$$

Hence, $x^5 = \frac{1}{2^4}(10T_1 + 5T_3 + T_5)$.

Note: x^4 is approximated by even and x^5 by odd Chebyshev polynomials.

Example 9.6

Express the function $f(x) = x - \frac{x^2}{2} + \frac{x^3}{3} - \frac{x^4}{4}$ by a Chebyshev series in the interval $-1 \leq x \leq 1$. Use appropriate formula and not the recurrence relation.

Solution Let

$$x - \frac{x^2}{2} + \frac{x^3}{3} - \frac{x^4}{4} = a_0T_0 + a_1T_1 + a_2T_2 + a_3T_3 + a_4T_4$$

Using (9.58) we have,

$$a_0 = -\frac{1}{2}\cdot\frac{1}{2^2}\begin{pmatrix}2\\1\end{pmatrix} - \frac{1}{4}\cdot\frac{1}{2^4}\begin{pmatrix}4\\2\end{pmatrix} = -\frac{1}{4}-\frac{3}{32} = -\frac{11}{32}$$

$$a_1 = 1\cdot\frac{1}{2^0}\begin{pmatrix}1\\1\end{pmatrix} + \frac{1}{3}\cdot\frac{1}{2^2}\begin{pmatrix}3\\2\end{pmatrix} = 1+\frac{1}{4} = \frac{5}{4}$$

$$a_2 = -\frac{1}{2}\cdot\frac{1}{2}\begin{pmatrix}2\\2\end{pmatrix} - \frac{1}{4}\cdot\frac{1}{2^3}\begin{pmatrix}4\\3\end{pmatrix} = -\frac{1}{4}-\frac{3}{8} = -\frac{3}{8}$$

$$a_3 = \frac{1}{3}\cdot\frac{1}{2^2}\begin{pmatrix}3\\3\end{pmatrix} = \frac{1}{12}$$

$$a_4 = -\frac{1}{4}\cdot\frac{1}{2^3}\begin{pmatrix}4\\4\end{pmatrix} = -\frac{1}{32}$$

Hence, $x-\frac{x^2}{2}+\frac{x^3}{3}-\frac{x^4}{4} = -\frac{11}{32}+\frac{5}{4}\mathrm{T}_1(x)-\frac{3}{8}\mathrm{T}_2(x)+\frac{1}{12}\mathrm{T}_3(x)-\frac{1}{32}\mathrm{T}_4(x)$

Discussion: Suppose we approximate the given function by truncated Chebyshev series neglecting the last term. Then

$$f(x) = x-\frac{x^2}{2}+\frac{x^3}{3}-\frac{x^4}{4} \simeq -\frac{11}{32}+\frac{5}{4}\mathrm{T}_1(x)-\frac{3}{8}\mathrm{T}_2(x)+\frac{1}{12}\mathrm{T}_3(x)$$

The error in this approximation is given by $\left|\frac{1}{32}\mathrm{T}_4(x)\right|$. But $|\mathrm{T}_4(x)| = 1$ in $-1 \le x \le 1$.

Hence, the estimated error is about 1/32.

The given function represents first four terms in the Maclaurin's expansion

$$\log(1+x) = x-\frac{x^2}{2}+\frac{x^3}{3}-\frac{x^4}{4}+\frac{x^5}{5}-\cdots$$

Suppose we want to use $f(x)$ for evaluating $\log(1+x)$ for $0 \le x \le 1$. Then for $x = 1$, the maximum error will be given by the first neglected term, i.e., 1/5 (since it is an alternating series). Compared to 1/5, the error due to truncation, i.e., 1/32 is much smaller. Extending this argument to more number of terms, it may be shown that the truncated Chebyshev series will provide almost as accurate results as obtained by a much larger Maclaurin's series.

Example 9.7

Express $\cos^{-1} x$ by Chebyshev series. Hence prove that

$$\frac{\pi^2}{8} = 1 + \frac{1}{3^2} + \frac{1}{5^2} + \frac{1}{7^2} + \cdots$$

Solution Let $\cos^{-1} x = a_0 T_0 + a_1 T_1 + a_2 T_2 + a_3 T_3 + \ldots..$

$$a_0 = \frac{1}{\pi} \int_{-1}^{1} \frac{\cos^{-1} x dx}{\sqrt{1-x^2}}$$

$$= \frac{1}{\pi} \int_{0}^{\pi} \theta \, d\theta = \frac{\pi}{2}$$

$$a_k = \frac{2}{\pi} \int_{-1}^{1} \frac{\cos^{-1} x T_k(x) dx}{\sqrt{1-x^2}}$$

$$= \frac{2}{\pi} \int_{0}^{\pi} \theta \cos k\theta \, d\theta$$

$$= \frac{2}{\pi} \left[\frac{\theta \sin k\theta}{k} - \frac{\cos k\theta}{k^2} \right]_{\theta=0}^{\pi}$$

$$= -\frac{2}{\pi} \left\{ \frac{1-(-1)^k}{k^2} \right\}, \; k = 1, 2, \ldots.$$

$$\cos^{-1} x = \frac{\pi}{2} - \frac{4}{\pi} \left\{ T_1 + \frac{1}{3^2} T_3(x) + \frac{1}{5^2} T_5(x) + \cdots \right\}$$

Putting $x = 1$, $T_1(x) = T_3(x) = \ldots. = 1$ from (9.38) and $\cos^{-1} 1 = 0$

Hence, $$\frac{\pi^2}{8} = 1 + \frac{1}{3^2} + \frac{1}{5^2} + \cdots$$

Note: From the formula $\cos^{-1} x = \frac{\pi}{2} - \sin^{-1} x$, the above expansion for $\cos^{-1} x$ should be correct since $\sin^{-1} x$ is an odd function and should contain only odd degree polynomials and also $a_0 = \frac{\pi}{2}$.

Example 9.8

Express the function $f(x) = \cos^{-1}x$ by truncated Chebyshev series $\cos^{-1}x = a_0 + a_1T_1(x) + a_3T_3(x)$. Compute a_0, a_1, a_3 numerically taking nodal points as zeros of $T_5(x)$.

Solution

$$n = 5;\ f(x) = \cos^{-1}x;\ f\{\cos x\} = x;\ x_r = \cos\frac{2r+1}{10}\pi,\ r = 0, 1, 2, 3, 4.$$

$$T_k(x_r) = \cos(k\cos^{-1}x_r) = \cos\left(\frac{2r+1}{10}\pi k\right),\ r = 0, 1, 2, 3, 4.$$

$$a_0 = \frac{1}{5}\sum_{r=0}^{4} f(x_r) = \frac{1}{5}\sum f\left\{\cos\frac{2r+1}{10}\pi\right\} = \frac{1}{5}\sum_{r=0}^{4}\left(\frac{2r+1}{10}\pi\right)$$

$$a_1 = \frac{2}{5}\sum f(x_r)\cdot T_1(x_r) = \frac{2}{5}\sum_{r=0}^{4}\left(\frac{2r+1}{10}\pi\right)\cos\left(\frac{2r+1}{10}\pi\right)$$

$$a_3 = \frac{2}{5}\sum f(x_r)\cdot T_3(x_r) = \frac{2}{5}\sum_{r=0}^{4}\left(\frac{2r+1}{10}\pi\right)\cos\left(\frac{2r+1}{10}\pi\times 3\right)$$

Note: $\pi/10 = 18^0$; $\pi = 3.1416$.

r	$\theta_r = \frac{2r+1}{10}\pi$	$\cos\theta_r$	$f(x_r) = \frac{2r+1}{10}\pi$	$T_1(x_r) = \cos\frac{2r+1}{10}\pi$	$f(x_r)T_1(x_r)$	$3\theta_r = \frac{2r+1}{10}\pi\times 3$	$T_3(x_r) = \cos\frac{2r+1}{10}\pi\times 3$	$f(x_r)\times T_3(x_r)$
0	$\pi/10$	0.95106	$\pi/10$	0.95106	$0.95106\times\frac{\pi}{10}$	$3\pi/10$	0.58779	$0.58779\times\frac{\pi}{10}$
1	$3\pi/10$	0.58779	$3\pi/10$	0.58779	$1.76337\times\frac{\pi}{10}$	$\frac{9\pi}{10} = \pi - \frac{\pi}{10}$	−0.95106	$-2.85318\times\frac{\pi}{10}$
2	$5\pi/10$	0	$5\pi/10$	0	0	$\frac{15}{10}\pi = 2\pi - \frac{\pi}{10}$	0	0
3	$7\pi/10$	−0.58779	$7\pi/10$	−0.58779	$-4.11453\times\frac{\pi}{10}$	$\frac{21\pi}{10} = 2\pi + \frac{\pi}{10}$	0.95106	$6.65742\times\frac{\pi}{10}$
4	$9\pi/10$	−0.95106	$9\pi/10$	−0.95106	$-8.55954\times\frac{\pi}{10}$	$\frac{27\pi}{10} = 2\pi + \frac{7\pi}{10}$	−0.58779	$-5.29011\times\frac{\pi}{10}$

$$a_0 = \frac{1}{5}\times\frac{\pi}{10}(1+3+5+7+9) = \frac{\pi}{2}$$

$$a_1 = \frac{2}{5}\times\frac{\pi}{10}(0.95106 + 1.76337 - 4.11453 - 8.55954) = -1.2512$$

$$a_3 = \frac{2}{5} \times \frac{\pi}{10}(0.58779 - 2.85318 + 6.65742 - 5.29011) = -0.1128$$

$$\cos^{-1} x = \frac{\pi}{2} - 1.2512T_1(x) - 0.1128T_3(x).$$

The Chebyshev series can be used to solve ordinary differential equations. For details see [3]. The application of Least Squares method in solving ordinary and partial differential equations can be found in Chapter 12.

Exercise 9

9.1 Derive normal equations for fitting a curve, $y = ax + \frac{b}{x}$ by Least Squares method. Hence fit this curve to the following data:

x	1	2	3	4	5
y	7	6	8	9	11

Also compute the approximate values of y and the residual errors.

9.2 Fit a straight line to the following data,

x	0	1	2	3	4
y	2	5	8	17	38

Compute the residual errors and find their sum. Also compute rms.

9.3 Fit a quadratic to the data of Exercise 9.2; Compute the residual error vector $\boldsymbol{e}$ and rms. Compare which one is a better fit.

9.4 Given below are the values of $y = 5e^{-x^2}$ rounded to one place of decimal. Fit a quadratic to the given data and compute the residual errors,

x	−1.0	−0.5	0.5	1.0
y	1.8	3.9	3.9	1.8

9.5 Find the quadratic approximation for $\ln x$, $1 \le x \le 3$ from the values of $\ln x$ given below,

x	1.0	1.5	2.0	2.5	3.0
$y = \ln x$	0	0.4055	0.6932	0.9163	1.0986

Also find the residual errors.

9.6 Fit an exponential curve $y = ae^{bx}$ to the following data,

x	0	0.5	1.0	1.5	2.0
y	2	2.57	3.38	4.23	5.44

9.7 Find an approximate solution of the following system of linear equations:

$x_1 + 2x_2 - x_3 = 4;\ 2x_1 - x_2 + x_3 = 2;$

$x_1 + x_2 - 2x_3 = 5;\ x_1 + 3x_2 + x_3 = 4.$

Also compute the residual errors.

References and Some Useful Related Books/Papers

1. Froberg, C.E., *Introduction to Numerical Analysis*, Addison-Wesley Publishing Co.
2. Gerald, C.F., and Wheatley, P.O., *Applied Numerical Analysis*, Pearson Education Asia.
3. Goodwin, E.T., *Modern Computing Methods*, HMSO, London.
4. Hildebrand, F.B., *Introduction to Numerical Analysis*, Tata McGraw Hill.
5. Yakowitz, S., and Szidarovszky, F., *An Introduction to Numerical Computations*, Macmillan Publishing Co. New York.

10

Eigenvalues of Symmetric Matrices

10.1 Introduction

In Chapter 2 we had given a brief introduction of eigenvalues and eigenvectors arising from solving a system of $(n \times n)$ homogeneous equations, given as

$$\left.\begin{array}{ccccccccc} a_{11}x_1 & + & a_{12}x_2 & + & \ldots. & + & a_{1n}x_n & = & \lambda x_1 \\ a_{21}x_1 & + & a_{22}x_2 & + & \ldots. & + & a_{2n}x_n & = & \lambda x_2 \\ \vdots & & \vdots & & & & \vdots & & \vdots \\ a_{n1}x_1 & + & a_{n2}x_2 & + & \ldots. & + & a_{nn}x_n & = & \lambda x_n \end{array}\right\} \tag{10.1}$$

where λ is a parameter depending upon the physical system from which these equation arise.

In matrix notation the above set of equations can be written as

$$\mathrm{A}\boldsymbol{x} = \lambda \boldsymbol{x} \tag{10.2}$$

or

$$(\mathrm{A} - \lambda \mathrm{I})\boldsymbol{x} = \mathbf{0}, \tag{10.3}$$

where A is a $(n \times n)$ matrix and $\boldsymbol{x}^{\mathrm{T}} = (x_1 \quad x_2 \quad \ldots. \quad x_n)$.

One solution of the above set is $x_1 = x_2 = \ldots. = x_n = 0$. This solution does not serve any useful purpose and is called 'trivial' solution. Physically, the trivial solution corresponds to steady state in a vibration problem. Mathematically it shows that this solution is independent of the value of λ which means for one solution there can be infinite number of values of λ. Therefore we ignore the 'trivial' and seek for a non-trivial solution. For the system (10.3) to have a non-trivial solution, the necessary condition is that the determinant of the matrix $\mathrm{A} - \lambda \mathrm{I}$ should vanish, i.e.,

$$|A - \lambda I| = 0. \tag{10.4}$$

The expansion of the left side of (10.4) gives a polynomial of degree n in λ, say $P_n(\lambda)$ which can be written as

$$P_n(\lambda) \equiv (-1)^n(\lambda^n + c_1\lambda^{n-1} + c_2\lambda^{n-2} + \ldots. + c_{n-1}\lambda + c_n). \tag{10.5}$$

Hence (10.4) can be expressed as

$$P_n(\lambda) \equiv (-1)^n(\lambda^n + c_1\lambda^{n-1} + c_2\lambda^{n-2} + \ldots. + c_{n-1}\lambda + c_n) = 0 \tag{10.6}$$

or $\quad \lambda^n + c_1\lambda^{n-1} + c_2\lambda^{n-2} + \ldots. + c_{n-1}\lambda + c_n = 0.$ (10.7)

The polynomial (10.5) is called characteristic polynomial for matrix A and the polynomial equation (10.6) or (10.7), characteristic equation. When solved, the characteristic equation would give n values of λ which should provide non-trivial solution of (10.3). As stated in Chapter 2, these values of λ are known as eigenvalues (or latent roots or characteristic roots) of matrix A. The solution vector $\boldsymbol{x}$ corresponding to a particular value of λ is called the eigenvector (or latent vector) corresponding to that eigenvalue. Thus there will be n eigenvectors corresponding to n eigenvalues of matrix A. It should however be clear that an eigenvector is not unique since if $\boldsymbol{x}$ is an eigenvector corresponding to an eigenvalue λ satisfying equation (10.2) so is k$\boldsymbol{x}$ where k is a scalar. Thus normalisation of an eigenvector is common (usually making numerically largest element to be unity). But for any given eigenvector there is a unique eigenvalue. We shall now discuss a few properties of eigenvalues and eigenvectors which are pre-requisite for further analysis.

10.2 Compact Form of Eigenvalues and Eigenvectors

Let $\boldsymbol{x}_r$ be an eigenvector corresponding to an eigenvalue λ_r of a matrix A of order n. Then from (10.2)

$$A\boldsymbol{x}_r = \lambda_r\boldsymbol{x}_r, \; r = 1, 2, \ldots. n. \tag{10.8}$$

If $\boldsymbol{x}_r^T = (x_{1r} \quad x_{2r} \quad \ldots. \quad x_{nr})$ then (10.8) can be written in the form

$$\begin{bmatrix} \boldsymbol{a}_1^T\boldsymbol{x}_r \\ \boldsymbol{a}_2^T\boldsymbol{x}_r \\ \vdots \\ \boldsymbol{a}_n^T\boldsymbol{x}_r \end{bmatrix} = \begin{bmatrix} \lambda_r x_{1r} \\ \lambda_r x_{2r} \\ \vdots \\ \lambda_n x_{nr} \end{bmatrix}, \; r = 1, 2, \ldots. n, \tag{10.9}$$

where $\boldsymbol{a}_i^{\mathrm{T}}$ is the row vector consisting of the elements of ith row of A i.e.,

$$\boldsymbol{a}_i^{\mathrm{T}} = (a_{i1} \quad a_{i2} \quad .... \quad a_{in}),\ i = 1,\ 2,\ ....\ n.$$

For various values of r, the system of equations (10.9) can be written in compact form as

$$\mathrm{AX} = \mathrm{XD} \tag{10.10}$$

where X is a matrix whose r^{th} column is the eigenvector $\boldsymbol{x}_r$ and D is a diagonal matrix consisting of λ_r in its r^{th}row/column, shown as follows,

$$\mathrm{X} = \begin{bmatrix} x_{11} & \dots & x_{1r} & \dots & x_{1n} \\ x_{21} & \dots & x_{2r} & \dots & x_{2n} \\ \vdots & & \vdots & & \vdots \\ x_{n1} & \dots & x_{nr} & \dots & x_{nn} \end{bmatrix},\ \mathrm{D} = \begin{bmatrix} \lambda_1 & 0 & \dots & 0 \\ 0 & \lambda_2 & \dots & 0 \\ \vdots & \vdots & & \vdots \\ 0 & 0 & \dots & \lambda_n \end{bmatrix}.$$

The matrix X is called 'modal' matrix and D a 'spectral' matrix. From (10.10) we can write

$$\mathrm{X}^{-1}\,\mathrm{AX} = \mathrm{D} \tag{10.11}$$

such that the rth column of X, $\boldsymbol{x}_r^{\mathrm{T}} = (x_{1r} \quad x_{2r} \quad \dots \quad x_{nr})$ is the e.vector of A corresponding to its e.value $\lambda_r = d_{rr}$, where d_{rr} is the $(r,\ r)$ element of D.

10.3 Eigenvalues of Powers of a Matrix

Premultiplying (10.2) by A we get

$$\mathrm{A}^2\boldsymbol{x} = \lambda\mathrm{A}\boldsymbol{x} = \lambda^2\boldsymbol{x},$$

which shows that e.value of A^2 is λ^2. It may be noted that e.vectors of A^2 remain same as those of A. Extending it further we can see that the e.value of A^m will be λ^m with the same e.vector $\boldsymbol{x}$.

If (10.2) is premultiplied by A^{-1}, then we get

$$\mathrm{A}^{-1}\boldsymbol{x} = \frac{1}{\lambda}\boldsymbol{x}$$

showing that e.value of A^{-1} is $\frac{1}{\lambda}(=\lambda^{-1})$ while e.vector remains same. This can also be extended to show that e.value of A^{-m} is λ^{-m} with the same e.vector as that of A.

The above analysis can also be used to find the e.values of exponential or transcendental functions of A, e.g. e^{A} and $(I-A)^{-1}$, etc., keeping in mind the conditions of convergence.

10.4 Eigenvalues of Transpose of a Matrix

The e.values of A and A^T are same for, if λ is an e.value of A, then

$$|A-\lambda I|=0$$

or $$|(A-\lambda I)^T|=0$$

or $$|A^T-\lambda I|=0,\text{ since determinant of a matrix and its transpose is same.}$$

We see that λ is the e.value of A^T also. It should be borne in mind that e.vectors of A and A^T are different in general. There is however an important relation between eigenvectors of A and A^T which is given by the following theorem.

10.5 Theorem: Eigenvectors of A and A^T are Biorthogonal

Proof: Let us suppose that the eigenvalues of A are given by $\lambda_1,\ \lambda_2,\ \ldots.\ \lambda_n$ which are distinct with corresponding e.vectors as $\boldsymbol{x}_1, \boldsymbol{x}_2, \ldots. \boldsymbol{x}_n$. Further, as $\lambda_1,\ \lambda_2,\ \ldots.\ \lambda_n$ are also the eigenvalues of A^T let us suppose that the corresponding e.vectors of A^T are $\boldsymbol{y}_1,\ \boldsymbol{y}_2,\ \ldots.$ $\boldsymbol{y}_n$ respectively (which are different from those of A). We write

$$A\boldsymbol{x}_r=\lambda_r\boldsymbol{x}_r \tag{10.12}$$

and $$A^T\boldsymbol{y}_s=\lambda_s\boldsymbol{y}_s. \tag{10.13}$$

Transposing both sides of (10.12) gives

$$\boldsymbol{x}_r^T A^T=\lambda_r\boldsymbol{x}_r^T. \tag{10.14}$$

Premultiplying (10.13) by $\boldsymbol{x}_r^T$ and postmultiplying (10.14) by $\boldsymbol{y}_s$, and subtracting we get

$$0=(\lambda_s-\lambda_r)\boldsymbol{x}_r^T\boldsymbol{y}_s. \tag{10.15}$$

Assuming that $\lambda_r \neq \lambda_s$, (i.e., e.values are distinct), it follows that

$$x_r^T y_s = 0 \tag{10.16}$$

which proves that the e.vectors of A and A^T are biorthogonal.

10.6 Corollary: Eigenvectors of Symmetric Matrix form Orthogonal Set

It is simple to see that if A is symmetric then e.vectors of A and A^T will be identical since $A = A^T$, i.e., $y_s = x_s$ and $y_r = x_r$, etc., hence $x_r^T x_s = 0$, proving that $x_1, x_2, \ldots . x_n$ will form an orthogonal set of eigenvectors.

Further, if the eigenvectors are normalised such that their length is unity, i.e., $x_r^T x_r = 1$, then it should be noted that

$$X^T X = I \quad \text{or} \quad X^T = X^{-1}. \tag{10.17}$$

Eqn. (10.17) shows that matrix X is orthogonal.

Hence if A is a symmetric matrix then we can express from (10.10)

$$X^{-1} A X = D \tag{10.18}$$

or

$$X^T A X = D, \text{ due to (10.17).} \tag{10.19}$$

Property (10.19) is an important property of a symmetric matrix which will be the basis for methods to be discussed later.

10.7 Theorem: Eigenvalues of Hermitian Matrix are Real

Proof: Let A be a Hermitian matrix ($A^* = A$, $*$ denotes conjugation and transposition) and let us assume that at least one of its e.values λ_r is complex. If x_r is the e.vector corresponding to λ_r, then

$$A x_r = \lambda_r x_r. \tag{10.20}$$

Conjugating both sides of (10.20) and transposing (conjugation denoted by a bar–)

$$(\bar{x}_r)^T (\bar{A})^T = \bar{\lambda}_r (\bar{x}_r)^T$$

or $$(\bar{x}_r)^T A^* = \bar{\lambda}_r (\bar{x}_r)^T$$

or $$(\bar{x}_r)^T\, A = \bar{\lambda}_r (\bar{x}_r)^T, \quad \because \text{A is Hermitian.} \tag{10.21}$$

Premultiplying (10.20) by $(\bar{x}_r)^T$ and postmultiplying (10.21) by x_r and subtracting, we get

$$(\lambda_r - \bar{\lambda}_r)(\bar{x}_r)^T \cdot x_r = 0.$$

The product $\bar{x}_r^T \cdot \bar{x}_r$ is sum of squares of real numbers since

$$\bar{x}_r^T \cdot x_r = \sum_m (a_m - ib_m)(a_m + ib_m) = \sum_m (a_m^2 + b_m^2),$$

where $$x_r^T = (a_1 + ib_1, \ \ldots, \ a_m + ib_m, \ \ldots).$$

As $\bar{x}_r^T \cdot x_r$ is sum of squares of real numbers it can not be zero. Therefore, we should have

$$\lambda_r - \bar{\lambda}_r = 0$$

or $$\lambda_r = \bar{\lambda}_r.$$

A complex number can be equal to its conjugate only if its imaginary part is zero. Hence our hypothesis that there exists a complex e.value is false and that all the e.values of a Hermitian matrix are real. It follows automatically that e.values of a real symmetric matrix are real as $\bar{A} = A$.

10.8 Product of Orthogonal Matrices is an Orthogonal Matrix

Let S_1 and S_2 be two orthogonal matrices. Then consider their product $S_1S_2 = S$, say.

$$S \cdot S^T = (S_1S_2)(S_1S_2)^T = S_1S_2S_2^TS_1^T$$

$$= S_1S_1^T \qquad \because \quad S_2^T = S_2^{-1} \text{ (} S_2 \text{ is orthogonal)}$$

$$= I \qquad \because \quad S_1 \text{ is orthogonal.}$$

It shows that the matrix $S = S_1 \cdot S_2$ is orthogonal.

The above result can be proved for any number of matrices. If $S_1, \ S_2, \ \ldots . \ S_n$ are n orthogonal matrices then the matrix S of their product will also be an orthogonal matrix, i.e.,

$$S = S_1S_2 \ldots . \ S_n, \text{ is orthogonal.} \tag{10.22}$$

10.9 Eigenvalues of S^TAS when S is Orthogonal

When S is an orthogonal matrix, the e.values of the transformed matrix S^TAS are the same as of A. For if μ denotes the e.value of S^TAS, the characteristic equation is given by

$$\begin{aligned} 0 &= |S^TAS - \mu I| \\ &= |S^TAS - \mu S^TS| \\ &= |S^T(A - \mu I)\, S| \\ &= |S^T| \cdot |A - \mu I| \cdot |S| \\ &= |A - \mu I|\, |S^TS| \\ &= |A - \mu I| \end{aligned}$$

Thus e.values of A and the transformed matrix S^TAS are same. The transformation of a matrix in which its e.values are preserved is called 'Similarity' transformation. We should note that if B is any non-singular matrix, then $C = B^{-1}AB$ is a similarity transformation on matrix A. The original matrix A and the transformed matrix C are said to be 'similar'.

10.10 Eigenvectors of S^TAS when S is Orthogonal

Premultiplying (10.2) by S^T, we get

$$S^TA\boldsymbol{x} = \lambda S^T\boldsymbol{x}. \tag{10.23}$$

Putting $\boldsymbol{x} = S\boldsymbol{y}$ in (10.23) gives

$$S^TAS\boldsymbol{y} = \lambda S^TS\boldsymbol{y} = \lambda \boldsymbol{y}. \tag{10.24}$$

Equation (10.24) shows that the e.value of S^TAS is λ which is same as of A, while the e.vector is $\boldsymbol{y}$ that is related with the e.vector $\boldsymbol{x}$ of A through the relation

$$S\boldsymbol{y} = \boldsymbol{x}. \tag{10.25}$$

10.11 Methods to find Eigenvalues of Symmetric Matrix

We will discuss mainly three methods to find e.values/e.vectors of a symmetric matrix, which are due to:

(*i*) Jacobi C.G.J.

(*ii*) Givens W.

(*iii*) Householder A.S.

The basic approach in all the above three methods is to perform successively a series of similarity transformations on the symmetric matrix A by choosing suitable orthogonal matrix. For example, if S_1 is an orthogonal matrix then we obtain

$$A_1 = S_1^T A S_1. \tag{10.26}$$

We should note that A_1 is symmetric since $A_1^T = S_1^T A S_1 = A_1$. Also since right side of (10.26) is a similarity transformation, as explained in Sec 10.9, the e.values of A_1 are same as of A. Again, by choosing a suitable orthogonal matrix S_2 (say), similarity transformation is made on A_1 giving

$$A_2 = S_2^T A_1 S_2$$

$$= S_2^T (S_1^T A S_1) S_2.$$

The matrix A_2 is symmetric and has the same e.value as A_1 which are same as of A. In this manner a series of transformations are made and finally matrix A is reduced to a diagonal matrix in the Jacobi method while it is reduced to tri-diagonal form in the Givens and Householder's methods. We should also remember from properties of e.values that

$$\Sigma\lambda_i = \text{Tr}\,(A) = \text{Tr}\,(A_1) = \text{Tr}\,(A_2), \text{ etc. (sum of eigenvalues)} \tag{10.27}$$

$$\overset{n}{\underset{i=1}{\pi}} \lambda_i = |A| = |A_1| = |A_2|, \text{ etc. (product of eigenvalues)} \tag{10.28}$$

10.12 Jacobi's Method (Classical)

Suggested in 1846, the classical Jacobi method transforms a symmetric matrix A to a diagonal matrix D through similarity transformations. As e.values of A and the transformed matrix D are same, they are given by the diagonal elements of D. Remember that $|D - \lambda I| = \pi(d_{ii} - \lambda) = 0$ is the characteristic equation for D. For carrying out similarity transformations, orthogonal matrix S_1, S_2, ... are used such that

$$....(S_2^T\ (S_1^T\ AS_1)S_2).... = D. \tag{10.29}$$

If A is of order n, the general form of the transformation matrix S is a unit/identity matrix of order n with four of its elements changed and other remaining same. If s_{pq} denotes (p, q) element of S, then elements of transformation matrix are,

$$s_{pp} = s_{qq} = \cos\theta;\ s_{pq} = \sin\theta; s_{qp} = -\sin\theta;$$

$$s_{ii} = 1,\ i \neq p \text{ or } q; \text{ other elements being zero.}$$

Taking $p < q$, the form of matrix S is shown below:

$$\begin{array}{ccc} & p^{\text{th}} \text{ col.} & q^{\text{th}} \text{ col.} & \\ & \downarrow & \downarrow & \end{array}$$

$$\text{S} = \begin{bmatrix} 1 \ddots & & & \bigcirc \\ & \cos\theta & \sin\theta & \\ & & \ddots & \\ & -\sin\theta & \cos\theta & \\ \bigcirc & & & \ddots 1 \end{bmatrix} \begin{array}{l} \\ \leftarrow p^{\text{th}} \text{ row} \\ \\ \leftarrow q^{\text{th}} \text{ row} \\ \\ \end{array} \tag{10.30}$$

It is easy to verify that $\text{S}^{\text{T}}\text{S} = \text{I}$, i.e., S is an orthogonal matrix.

Values of p, q and θ are chosen subject to satisfying certain criteria in different transformations. Let us consider the similarity transformation, say

$$\text{S}^{\text{T}}\text{AS} = \text{U}. \tag{10.31}$$

The method proceeds as follows:

Look for the numerically largest off-diagonal element; since A is symmetric, search is made above the principal diagonal only, i.e., for $|a_{ij}|,\ j > i$. Let us suppose $|a_{pq}|$ is the largest off-diagonal element $(p < q)$. In this manner values of $(p,\ q)$ can be fixed in (10.30).

To perform the product (10.31) let us first find, say $\text{B} = \text{S}^{\text{T}}\text{A}$ and then BS to give $\text{S}^{\text{T}}\text{AS} = \text{U}$. We should note that all the rows of A except p^{th} and q^{th} will remain unaltered in the multiplication $\text{S}^{\text{T}}\text{A}$. Thus we get

$$\left.\begin{aligned} b_{i,\,j} &= (\text{S}^{\text{T}}\text{A})_{i,\,j} = a_{i,\,j},\ i \neq p \text{ or } q,\ j = 1(1)n. \\ b_{pj} &= (\text{S}^{\text{T}}\text{A})_{pj} = a_{pj}\cos\theta - a_{qj}\sin\theta,\ \ j = 1(1)n. \\ b_{qj} &= (\text{S}^{\text{T}}\text{A})_{qj} = a_{pj}\sin\theta + a_{qj}\cos\theta,\ \ j = 1(1)n. \end{aligned}\right\} \tag{10.32}$$

Now in forming $\text{S}^{\text{T}}\text{AS} = \text{BS} = \text{U}$, we should note that only elements of B that will change belong to p^{th} and q^{th} columns. That is,

$$\left.\begin{aligned} u_{i,\,j} &= b_{i,\,j},\;\; j \neq p \text{ or } q,\;\; i = 1(1)n, \\ u_{ip} &= b_{ip}\cos\theta - b_{i,\,q}\sin\theta,\;\; i = 1(1)n, \\ u_{i,\,q} &= b_{ip}\sin\theta\; + b_{i,\,q}\cos\theta,\; i = 1(1)n. \end{aligned}\right\} \tag{10.33}$$

Substituting from (10.32) in (10.33) we get,

$$u_{i,\,j} = a_{i,\,j};\;\; i \neq p \text{ or } q,\; j \neq p \text{ or } q. \tag{10.34}$$

$$u_{ip} = a_{ip}\cos\theta - a_{iq}\sin\theta,\; i \neq p \text{ or } q. \tag{10.35}$$

$$u_{i,\,q} = a_{ip}\sin\theta + a_{iq}\cos\theta,\; i \neq p \text{ or } q. \tag{10.36}$$

$$u_{pj} = a_{pj}\cos\theta - a_{qj}\sin\theta;\; u_{qj} = a_{pj}\sin\theta + a_{qj}\cos\theta, j \neq p \text{ or } q \tag{10.37}$$

$$\begin{aligned} u_{pp} &= b_{pp}\cos\theta - b_{pq}\sin\theta \\ &= (a_{pp}\cos\theta - a_{qp}\sin\theta)\cos\theta - (a_{qp}\cos\theta - a_{qq}\sin\theta)\sin\theta \\ &= a_{pp}\cos^2\theta + a_{qq}\sin^2\theta - 2a_{pq}\sin\theta\;\cos\theta \qquad \because a_{pq} = a_{qp} \\ &= a_{pp}\cos^2\theta + a_{qq}\sin^2\theta - a_{pq}\sin 2\theta \end{aligned} \tag{10.38}$$

$$\begin{aligned} u_{qq} &= b_{qp}\sin\theta + b_{qq}\cos\theta \\ &= (a_{pp}\sin\theta + a_{qp}\cos\theta)\sin\theta + (a_{pq}\sin\theta + a_{qq}\cos\theta)\cos\theta \\ &= a_{pp}\sin^2\theta + a_{qq}\cos^2\theta + a_{pq}\sin 2\theta \end{aligned} \tag{10.39}$$

$$\begin{aligned} u_{pq} &= b_{pp}\sin\theta + b_{pq}\cos\theta \\ &= (a_{pp}\cos\theta - a_{qp}\sin\theta)\sin\theta + (a_{pq}\cos\theta - a_{qq}\sin\theta)\cos\theta \\ &= (a_{pp} - a_{qq})\sin\theta\cos\theta + a_{pq}(\cos^2\theta - \sin^2\theta) \\ &= \frac{1}{2}(a_{pp} - a_{qq})\sin 2\theta + a_{pq}\cos 2\theta. \end{aligned} \tag{10.40}$$

The elements of the transformed matrix U can be computed from formulae (10.34) through (10.40) once the value of θ is known. We should remember that since U is symmetric the elements below the principal diagonal can be written by symmetry. This similarity transformation is also called as 'Jacobi rotation' with the pair of values (p, q) known as 'plane of rotation' and θ as 'angle of rotation'.

The angle θ is chosen such that (p, q) element of U which corresponds to the numerically largest element of A is made equal to zero. This will ensure fastest convergence of (10.29). Thus by equating $u_{pq} = 0$, we get from (10.40)

$$\tan 2\theta = \frac{2a_{pq}}{a_{qq} - a_{pp}} = c \text{ (say)}. \tag{10.41}$$

The value of 2θ is chosen to lie between $-\frac{\pi}{2} \le 2\theta \le \frac{\pi}{2}$ or of θ such that $-\frac{\pi}{4} \le \theta \le \frac{\pi}{4}$ depending on the sign of right side expression of (10.41). That is, if c is negative then θ has to be negative rendering $\sin\theta$ to be negative. But $\cos\theta$ will be positive whether θ is positive (+ve) or negative (−ve).

However, instead of finding the angle θ, we can compute the values of $\sin\theta$ and $\cos\theta$ from (10.40) in the following manner. Let us take the numerical value of the right side of (10.41), ignoring its sign, i.e.,

$$\tan 2\theta = \frac{2\tan\theta}{1 - \tan^2\theta} = c$$

or $\quad c\tan^2\theta + 2\tan\theta - c = 0$, giving

$$\tan\theta = \frac{1}{c}(\pm\sqrt{1+c^2} - 1), \ c \text{ is taken positive (+ve)}. \tag{10.42}$$

Take positive sign in (10.42) so that the value of $\tan\theta$ remains between 0 and 1 or $\left(0 \le \theta \le \frac{\pi}{4}\right)$ and put

$$\tan\theta = \left(\sqrt{1+c^2} - 1\right)/c = k \tag{10.43}$$

where $0 \le k \le 1$ and c is taken positive (+ve).

From (10.43) the values of $\sin\theta$ and $\cos\theta$ may be easily computed and after assigning proper sign, they may be written as

$$\sin\theta = \frac{(\text{sign of } c)k}{\sqrt{1+k^2}}, \ \cos\theta = \frac{1}{\sqrt{1+k^2}} \tag{10.44}$$

$$\sin 2\theta = \frac{(\text{sign of } c)2k}{1+k^2}, \ \cos 2\theta = \frac{1-k^2}{1+k^2}. \tag{10.45}$$

We can use relations $\sin 2\theta = 2\sin\theta\cos\theta$ and $\cos 2\theta = \cos^2\theta - \sin^2\theta = 1 - 2\sin^2\theta =$

$2\cos^2\theta \ -1.$

The procedure for transforming a matrix A to U may be summarised as follows:

(*i*) Search for the numerically largest off-diagonal element above the principal diagonal; let it be $(p,\ q)$ element a_{pq}.

(*ii*) Find $c = 2a_{pq}/(a_{qq} - a_{pp})$; sign of c determines sign of the angle of rotation θ.

(*iii*) Compute $k = (\sqrt{1+c^2} - 1)/c$ taking c positive.

(*iv*) Compute $\sin\theta,\ \cos\theta$ and $\sin 2\theta,\ \cos 2\theta$ from (10.44) and (10.45) respectively.

(*v*) Copy unaltered elements of A into U.

$$u_{ii} = a_{ii},\ i \neq p \text{ or } q$$
$$u_{ij} = a_{ij}, j = (i+1)(1)n;\ j \neq p \text{ or } q$$

(*vi*) Compute elements in p^{th} and q^{th} columns

$$u_{ip} = a_{ip}\cos\theta - a_{iq}\sin\theta,\ 1 \leq i\ \leq p-1$$
$$u_{iq} = a_{ip}\sin\theta + a_{iq}\cos\theta,\ 1 \leq i\ \leq q-1$$

(*vii*) Compute elements along p^{th} and q^{th} rows

$$u_{pj} = a_{pj}\cos\theta - a_{qj}\sin\theta, j = (p+1)(1)n$$
$$u_{qj} = a_{pj}\sin\theta + a_{qj}\cos\theta, j = (q+1)(1)n$$

(*viii*) Compute diagonal elements

$$u_{pp} = a_{pp}\cos^2\theta + a_{qq}\sin^2\theta - a_{pq}\sin 2\theta$$
$$u_{qq} = a_{pp}\sin^2\theta + a_{qq}\cos^2\theta + a_{pq}\sin 2\theta$$

(*ix*) Zeroise $(p,\ q)$ element (although it is wrongly computed in (*vii*) but programming-wise it will be more efficient)

$$u_{pq} = 0 = \left(\frac{1}{2}(a_{pp} - a_{qq})\sin 2\theta + a_{pq}\cos 2\theta\right)$$

(Its computation will be required in Givens method).

(*x*) Copy elements below diagonal since U is symmetric

$$u_{ij} = u_{ji},\ i = 2(1)n, j = 1(1)i-1.$$

Check: $Tr(\mathrm{A}) = Tr(\mathrm{U})$.

10.12.1 Convergence of Jacobi method

We see that under the transformation $S^TAS = U$, only p^{th}, q^{th} rows and p^{th}, q^{th} columns of A change. That is, all the elements of U are same as of A except in the pth, q^{th} rows and p^{th}, q^{th} columns (see Fig. 10.1).

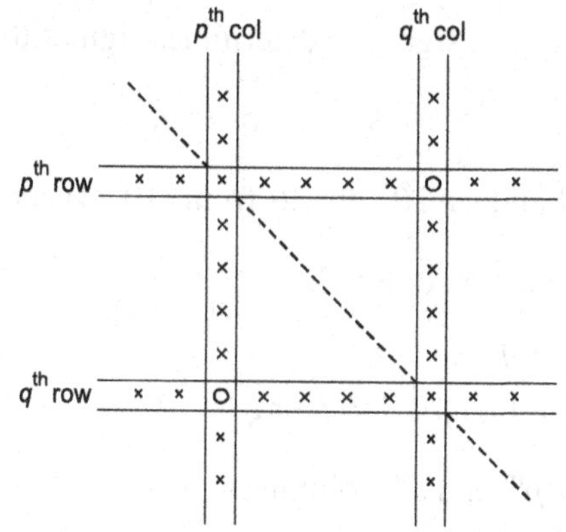

Figure 10.1 Elements of matrix U.

Note: Dotted line shows the principal diagonal

Let us consider the sum of squares of the off-diagonal elements of U vis-a-vis A. As all the elements of U are same as of A except along the pth and q^{th} rows and columns, the sum of squares of the terms in the pth and q^{th} columns of the ith row $i = 1(1)n$, $i \neq p,\ q$ is given by

$$
\begin{aligned}
u_{ip}^2 + u_{iq}^2 &= (a_{ip}\cos\theta - a_{iq}\sin\theta)^2 + (a_{ip}\sin\theta + a_{iq}\cos\theta)^2 \\
&= a_{ip}^2 + a_{iq}^2. \qquad (10.46)
\end{aligned}
$$

Similarly, the sum of squares of the terms in the pth and q^{th} row j^{th} column, $j = 1(1)n$, $j \neq p$ or q is given by

$$u_{pj}^2 + u_{qj}^2 = a_{pj}^2 + a_{qj}^2. \qquad (10.47)$$

Remembering that θ has been chosen to make $u_{pq} = 0$, we conclude in the light of (10.46) and (10.47) that sum of squares of the off-diagonal elements of A has been reduced by an amount $2a_{pq}^2$ under similarity transformation S^TAS, i.e.,

$$\sum_{i=1}^{n}\sum_{\substack{j=1\\ j\neq i}}^{n} u_{i,j}^2 = \sum_{i=1}^{n}\sum_{\substack{j=1\\ j\neq i}}^{n} a_{i,j}^2 - 2a_{pq}^2. \qquad (10.48)$$

Next time round when U undergoes similarity transformation, the sum of squares of its off-diagonal elements will be reduced by twice the square of its numerically largest off-diagonal element. Proceeding in this manner the off-diagonal elements become smaller and smaller with the number of successive rotations/transformations and the process is continued until they become negligibly small within the accuracy of working.

The chain of similarity transformations can be described by the recursive procedure

$$\mathrm{A}_r = \mathrm{S}_r^{\mathrm{T}}\mathrm{A}_{r-1}\mathrm{S}_r = \mathrm{S}_r^{\mathrm{T}}\mathrm{S}_{r-1}^{\mathrm{T}}\cdots\mathrm{S}_1^{\mathrm{T}}\mathrm{A}\mathrm{S}_1\cdots\mathrm{S}_{r-1}\mathrm{S}_r,\ r = 1,2\ldots.. \tag{10.49}$$

where $\mathrm{A}_0 = \mathrm{A}$.

If we denote by σ_k the sum of squares of the off-diagonal elements of A_k, then from (10.48)

$$\sigma_k = \sigma_{k-1} - 2(a_{pq}^{k-1})^2, \tag{10.50}$$

where a_{pq}^{k-1} denotes the largest off-diagonal element of A_{k-1}. It should be borne in mind that $(p,\ q)$ varies with k.

Obviously, if all the off-diagonal elements of A_{k-1} are replaced by a_{pq}^{k-1}, then

$$\sigma_{k-1} \le n(n-1)(a_{pq}^{k-1})^2$$

Putting this value in (10.50), we get

$$\begin{aligned}\sigma_k &\le \sigma_{k-1} - \frac{2}{n(n-1)}\sigma_{k-1}\\ &= \left\{1 - \frac{2}{n(n-1)}\right\}\sigma_{k-1}\\ &= \left\{1 - \frac{2}{n(n-1)}\right\}^k \sigma_0,\ \sigma_0 \text{ corresponds to } \mathrm{A}_0 = \mathrm{A}.\end{aligned} \tag{10.51}$$

For $n \ge 2$ (which is always so since n is the order of matrix A) the bracketed term in (10.51) is less than 1 and hence tends to zero as k tends to ∞. Hence Jacobi method converges.

10.12.2 Cyclic Jacobi method

The method described above is known as 'Classical' Jacobi method. When it comes to computer implementation the classical method becomes time-consuming on account of the procedure required for searching largest off-diagonal element at each stage of similarity transformation. Therefore, to avoid 'search', the elements are zeroised in a systematic manner. Moving along the row, in the first transformation (1, 2) element is made equal to zero; then (1, 3), (1, 4) and (1, n) elements are made zero in the successive transformations. Next we move on to second row and make (2, 3), (2, 4) ...(2, n) elements zero at each transformation. In this manner we go on up to the last element ($n-1$, n) to make it zero. It may be remembered that the element which is made zero generally changes at the subsequent transformations. Hence the process has to be repeated all over again starting from (1, 2) element and going up to ($n-1$, n). The process is continued until off-diagonal elements become small enough within accuracy of working. This is called 'Cyclic' Jacobi method. Cyclic method in general should be faster computation-wise in comparison to the classical method, although its rate of convergence may be slower. That is, number of transformations in the cyclic method may be more as compared to the classical method but its overall computing time should be less due to getting away with the 'search' procedure.

Further, it may also be more efficient time-wise to copy all the elements in U and then compute the elements in the pth and q^{th} rows and columns of U overwriting the elements earlier copied.

Let us suppose that convergence has been achieved in k transformations, i.e.,

$$\mathrm{S}_k^{\mathrm{T}}\mathrm{S}_{k-1}^{\mathrm{T}}\ldots\ldots\mathrm{S}_2^{\mathrm{T}}\mathrm{S}_1^{\mathrm{T}}\mathrm{A}\mathrm{S}_1\mathrm{S}_2\ldots\ldots\mathrm{S}_{k-1}\mathrm{S}_k=\mathrm{D}$$

or $$\mathrm{S}^{\mathrm{T}}\mathrm{AS}=\mathrm{D},\text{ where }\mathrm{S}^{\mathrm{T}}=\mathrm{S}_k^{\mathrm{T}}\mathrm{S}_{k-1}^{\mathrm{T}}\ldots\ldots\mathrm{S}_1^{\mathrm{T}}. \tag{10.52}$$

Then e.vectors are given by the rows of the matrix S^{T} or columns of S corresponding to the e.value appearing in the diagonal matrix D. See (10.19).

Check: $Tr(\mathrm{A})=Tr(\mathrm{D})$.

Example 10.1

Find by Jacobi method the e.values and e.vectors of the matrix

$$\mathrm{A}=\begin{bmatrix}1 & 0 & 0\\ 0 & 4 & \sqrt{10}\\ 0 & \sqrt{10} & 1\end{bmatrix}$$

Give your answer up to two places of decimal for e.values and three places for e.vectors.

Solution $p=2,\ q=3,\ \sqrt{10}=3.162$

$$\tan 2\theta=\frac{2\times\sqrt{10}}{1-4}=-2.108$$

hence take θ is negative.

$$c = 2.108; \quad \tan\theta = k = \frac{1}{c}(\sqrt{1+c^2} - c) = \frac{1}{2.108}(\sqrt{1+4.444} - 1) = 0.6324$$

$$\sin\theta = -\frac{1}{\sqrt{1+k^2}} = -\frac{0.6324}{1.1832} = -0.5345$$

$$\cos\theta = \frac{1}{\sqrt{1+k^2}} = \frac{1}{1.1832} = 0.8452$$

$$\sin 2\theta = -\frac{2k}{1+k^2} = -\frac{2\times 0.6324}{1+0.3999} = -0.9034$$

$$\cos 2\theta = \frac{1-k^2}{1+k^2} = \frac{1-0.3999}{1+0.3999} = 0.4287$$

$$u_{12} = a_{12}\cos\theta - a_{13}\sin\theta \ = 0; \quad u_{13} = a_{12}\sin\theta + a_{13}\cos\theta \ = 0$$

$$u_{22} = a_{22}\cos^2\theta + a_{33}\sin^2\theta - a_{23}\sin 2\theta$$

$$= 4\times 0.7144 + 1\times 0.2857 - 3.162\times(-0.9034)$$

$$= 5.9998$$

$$u_{33} = a_{22}\sin^2\theta + a_{33}\cos^2\theta + a_{23}\sin 2\theta$$

$$= 4\times 0.2857 + 1\times 0.7144 + 3.162(-0.9034)$$

$$= -0.9994$$

$$D = \begin{bmatrix} 1 & 0 & 0 \\ 0 & 5.9998 & 0 \\ 0 & 0 & -0.99994 \end{bmatrix}$$

$\lambda_1 = 1.00$, $\lambda_2 = 6.00$, $\lambda_3 = -1.00$; Tr (A) = Tr (D) = 6.

$$S = \begin{bmatrix} 1 & 0 & 0 \\ 0 & \cos\theta & \sin\theta \\ 0 & -\sin\theta & \cos\theta \end{bmatrix}$$

$$= \begin{bmatrix} 1 & 0 & 0 \\ 0 & 0.8452 & -0.5345 \\ 0 & 0.5345 & 0.8452 \end{bmatrix}$$

$$\boldsymbol{x}_1^{\mathrm{T}} = (1 \quad 0 \quad 0), \ \boldsymbol{x}_2^{\mathrm{T}} = (0 \quad 0.845 \quad 0.534), \ \boldsymbol{x}_3^{\mathrm{T}} = (0 \quad -0.534 \quad 0.845)$$

10.13 Givens Method

We have seen that in Jacobi method the symmetric matrix is transformed to a diagonal matrix by performing a series of similarity transformations. The main drawback of the method is that the zeros produced do not persist. They may change in the subsequent transformations. As a result we can not know beforehand as how many transformations are going to be made in the process of diagonalisation. In the Givens' (1954) and Householder's (1953) methods the zeros once produced do not alter in the subsequent transformations. The final matrix obtained is tridiagonal/triple diagonal/co-diagonal (not diagonal). The eigenvalues of a symmetric tridiagonal matrix can be found by a special procedure which will be described later. The advantage of these two methods is that the process of reducing matrix A of order n to tridiagonal form requires definite number of transformations.

Givens uses same orthogonal matrix as suggested in Jacobi's method but the angle θ is chosen in a manner such that $(p-1,\ q)$ element becomes zero, not $(p,\ q)$. Thus in order to reduce (1, 3) element to zero we take $p=2,\ q=3$ and to reduce (2, 4) element to zero p and q are chosen to be $p=3, q=4$, etc., and the last element $(n-2,\ n)$ will be reduced to zero by chosing $p=n-1, q=n$. Thus the total number of transformations required in reduction to tridiagonal form would be $(n-2)(n-1)/2$.

From (10.36) we get

$$u_{p-1,q} = a_{p-1,p}\sin\theta + q_{p-1,q}\cos\theta = 0, \text{ giving}$$

$$\tan\theta = -a_{p-1,q}/a_{p-1,p}. \tag{10.53}$$

Let $c=1$ if $\tan\theta \geq 0$ and -1 if $\tan\theta < 0$ and $k=(a_{p-1,p}^2 + a_{p-1,q}^2)^{1/2}$,

$$\left.\begin{aligned} \sin\theta &= c|a_{p-1},\ q|/k \\ \cos\theta &= |a_{p-1}, p|/k. \end{aligned}\right\} \tag{10.54}$$

After selecting $(p,\ q), p<q$ and evaluating $\sin\theta, \cos\theta$ by equating $u_{p-1,q}=0$, the other elements of $\mathrm{U} = \mathrm{S}^{\mathrm{T}}\mathrm{AS}$ can be computed as described in Sec. 10.12. It should however be borne in mind that u_{pq} is not zero and has to be evaluated. The elements are made zero through successive transformations in a systematic manner row-wise from left to right, i.e., in the order $(p-1,\ q), q=p+1(1)n$ for each $p=2,\ 3, \ldots (n-1)$. Table 10.1 shows the elements to be annihilated (zeroised) in the successive transformations along with the corresponding values of $(p,\ q)$.

Table 10.1

Elements to be Annihilated	*Corresponding Values of (p, q)*
(1, 3), (1, 4), (1, 5) ... (1, n)	(2, 3), (2, 4), (2, 5) ... (2, n)
(2, 4), (2, 5) ... (2, n)	(3, 4), (3, 5) ... (3, n)
(3, 5) ... (3, n)	(4, 5) ... (4, n)
⋮	⋮
($n-2$, n)	($n-1$, n)

It may be easily verified that the zeros produced are not destroyed in the subsequent transformations. In the (p, q) rotation the elements which are affected belong to p^{th} and q^{th} rows and columns. In the $(p-1)^{\text{th}}$ row, the zeros produced from $(p+1)^{\text{th}}$ to q^{th} columns remain zero and the zeros from first to $(p-2)^{\text{th}}$ row persist on accout of the formula (10.35) and (10.36). Element $(p-1,\ p)$ changes without altering tridiagonal shape. After $(n-2)$ $(n-1)/2$ transformations we finally get a symmetric tridiagonal matrix

$$T = \begin{bmatrix} \alpha_1 & \beta_1 & 0 & 0 & \cdots & 0 \\ \beta_1 & \alpha_2 & \beta_2 & 0 & & 0 \\ 0 & \beta_2 & \alpha_3 & \beta_3 & & 0 \\ & & \ddots & \ddots & \ddots & \\ & & & \beta_{n-2} & \alpha_{n-1} & \beta_{n-1} \\ & & & & \beta_{n-1} & \alpha_n \end{bmatrix} \tag{10.55}$$

Matrix T is similar to the original matrix A.

Check: Tr $(A) = $ Tr $(T) = \Sigma\alpha_i$.

For computer implementation one may compute with nested loops for $k = 1(1)n-2$, $i = k(1)n$, $j = i(1)n$.

Example 10.2

Reduce the following matrix A to tridiagonal form by Givens method:

$$A = \begin{bmatrix} 3 & 0 & 1 & 1 \\ 0 & 2 & 1 & 0 \\ 1 & 1 & 4 & 1 \\ 1 & 0 & 1 & 5 \end{bmatrix}$$

Solution (1, 3) element to be made zero.

$$p=2,\ q=3$$

$$\tan\theta = -\frac{a_{p-1,q}}{a_{p-1,p}} = -\frac{a_{13}}{a_{12}} = -\frac{1}{0} = -\infty$$

θ is negative $(-\pi/2)$; $c=-1$

$$k=\sqrt{0^2+1^2}=1$$

$$\sin\theta = \frac{c|a_{13}|}{k} = -\frac{1}{1} = -1;\ \cos\theta = \frac{|a_{12}|}{k} = \frac{0}{1} = 0$$

Let $\mathrm{A_1 = S_1^T A S_1}$

Elements of 2nd, 3rd row and 2nd, 3rd columns change.

$$a_{11}^{(1)} = a_{11} = 3 \text{ (does not change)}$$

$$a_{12}^{(1)} = a_{12}\cos\theta - a_{13}\sin\theta = 0\times 0 - 1\times(-1) = 1$$

$$a_{13}^{(1)} = 0 \text{ (is made zero)}$$

$$a_{14}^{(1)} = a_{14} = 1 \text{ (does not change)}$$

$$a_{22}^{(1)} = a_{22}\cos^2\theta + a_{33}\sin^2\theta - 2a_{23}\sin\theta\cos\theta$$

$$= 2\times 0 + 4\times 1 - 2\times 1\times(-1)\times 0 = 4$$

$$a_{23}^{(1)} = (a_{pp}-a_{qq})\sin\theta\cos\theta + a_{pq}(\cos^2\theta - \sin^2\theta) \qquad \because\ \cos 2\theta = \cos^2\theta - \sin^2\theta$$

$$= (2-4)(-1)\times 0 + 1(-1) = -1 \qquad = 1 - 2\sin^2\theta$$

$$= 2\cos^2\theta - 1$$

$$a_{24}^{(1)} = a_{24}\cos\theta - a_{34}\sin\theta = 0 - 1(-1) = 1$$

$$a_{33}^{(1)} = a_{22}\sin^2\theta + a_{33}\cos^2\theta + 2a_{23}\sin\theta\cos\theta$$

$$= 2\times 1 + 0 + 0 = 2$$

$$a_{34}^{(1)} = a_{24}\sin\theta + a_{34}\cos\theta = 0$$

$$a_{44}^{(1)} = a_{44} = 5 \text{ (does not change)}$$

$$A_1 = \begin{bmatrix} 3 & 1 & 0 & 1 \\ 1 & 4 & -1 & 1 \\ 0 & -1 & 2 & 0 \\ 1 & 1 & 0 & 5 \end{bmatrix}$$

A_1 is symmetric and is similar to A.

Check: Tr (A) = Tr $(A_1) = 14$

(1, 4) element of A_1 to be made zero; $A_2 = S_2^T A_1 S_2$.

$$p = 2,\ q = 4;$$

elements of 2, 4 rows and 2, 4 columns will change.

$$\tan\theta = \frac{a_{14}^{(1)}}{a_{12}^{(1)}} = -\frac{1}{1} = -1$$

Angle θ is negative (−ve), hence $c = -1$.

$$k = \sqrt{a_{12}^{(1)2} + a_{14}^{(1)2}} = \sqrt{1+1} = \sqrt{2} = 1.4142; \quad \sin\theta = -\frac{1}{\sqrt{2}},\ \cos\theta = \frac{1}{\sqrt{2}}.$$

$$a_{11}^{(2)} = a_{11}^{(1)} = 3 \text{ (does not change)}$$

$$a_{12}^{(2)} = a_{12}^{(1)}\cos\theta - a_{14}^{(1)}\sin\theta = 1 \times \frac{1}{\sqrt{2}} - (1)\left(-\frac{1}{\sqrt{2}}\right) = \sqrt{2} = 1.4142$$

$$a_{13}^{(2)} = a_{13}^{(1)} = 0 \text{ (does not change)}$$

$$a_{14}^{(2)} = 0 \text{ (is made zero)}$$

$$a_{22}^{(2)} = a_{22}^{(1)}\cos^2\theta - a_{44}^{(1)}\sin^2\theta - 2a_{24}^{(1)}\sin\theta\cos\theta$$

$$= 4 \times \frac{1}{2} + 5 \times \frac{1}{2} - 2 \times (1)\left(-\frac{1}{\sqrt{2}}\right)\left(\frac{1}{\sqrt{2}}\right) = \frac{11}{2} = 5.5$$

$$a_{23}^{(2)} = a_{23}^{(1)}\cos\theta - a_{43}^{(1)}\sin\theta = a_{23}^{(1)}\cos\theta - a_{34}^{(1)}\sin\theta \quad \because A_1 = A_1^T$$

$$= (-1)\left(\frac{1}{\sqrt{2}}\right) - 0 = -\frac{1}{\sqrt{2}} = -0.7071$$

$$a_{24}^{(2)} = (a_{22}^{(1)} - a_{44}^{(1)})\sin\theta\cos\theta + a_{24}^{(1)}(\cos^2\theta - \sin^2\theta)$$

$$= (4-5)\left(-\frac{1}{2}\right) + 0 = \frac{1}{2} = 0.5$$

$$a_{33}^{(2)} = a_{33}^{(1)} = 2 \text{ (does not change)}$$

$$a_{34}^{(2)} = a_{32}^{(1)} \sin\theta + a_{34}^{(1)} \cos\theta = a_{23}^{(1)} \sin\theta + a_{34}^{(1)} \cos\theta$$

$$= (-1)\left(-\frac{1}{\sqrt{2}}\right) + 0 = \frac{1}{\sqrt{2}} = 0.7071$$

$$a_{44}^{(2)} = a_{22}^{(1)} \sin^2\theta + a_{44}^{(1)} \cos^2\theta + a_{24}^{(1)} \sin\theta\cos\theta$$

$$= 4 \times \frac{1}{2} + 5 \times \frac{1}{2} + 2 \times 1 \times \left(-\frac{1}{\sqrt{2}}\right)\left(\frac{1}{\sqrt{2}}\right) = \frac{7}{2} = 3.5$$

$$A_2 = \begin{bmatrix} 3 & 1.4142 & 0 & 0 \\ 1.4142 & 5.5 & -0.7071 & 0.5 \\ 0 & -0.7071 & 2 & 0.7071 \\ 0 & 0.5 & 0.7071 & 3.5 \end{bmatrix}$$

Check: $\text{Tr}(A_2) = \text{Tr}(A_1) = \text{Tr}(A) = 14$

(2, 4) element to be made zero; $A_3 = S_3^T A_2 S_3$.

$$p = 3,\ q = 4;\ \tan\theta = -a_{24}^{(2)}/a_{23}^{(2)},\ -\text{ve};\ c = -1.$$

$$k = \sqrt{(-0.7071)^2 + (0.5)^2} = 0.866;\ \sin\theta = -\frac{0.5}{0.8660} = -0.5774,\ \cos\theta = \frac{0.7071}{0.8660} = 0.8165.$$

$$a_{23}^{(3)} = a_{23}^{(2)} \cos\theta - a_{24}^{(2)} \sin\theta = (-0.7071)(0.8165) - 0.5(-0.5774) = -0.2886$$

$$a_{24}^{(3)} = 0 \text{ (annihilated)}$$

$$a_{33}^{(3)} = a_{33}^{(2)} \cos^2\theta + a_{44}^{(2)} \sin^2\theta - 2a_{34}^{(2)} \sin\theta\cos\theta$$

$$= 2 \times 0.6667 + 3.5 \times 0.3334 - 2 \times 0.7071 \times (-0.5774)\ (0.8165)$$

$$= 3.1670$$

$$a_{34}^{(3)} = (a_{33}^{(2)} - a_{44}^{(2)}) \sin\theta\cos\theta + a_{34}^{(2)}(\cos^2\theta - \sin^2\theta)$$

$$= (2-3.5)(-0.5774)(0.8165) + (0.7071)(0.3333) = 0.9429$$

$$a_{44}^{(3)} = a_{33}^{(2)} \sin^2\theta + a_{44}^{(2)} \cos^2\theta + 2a_{34} \sin\theta\cos\theta$$

$= 2 \times 0.3334 + 3.5 \times 0.6667 + 2 \times 0.7071 \times (-0.5774)(0.8165)$

$= 2.3335$

$$A_3 = \begin{bmatrix} 3 & 1.4142 & 0 & 0 \\ 1.4142 & 5.5 & -0.2886 & 0 \\ 0 & -0.2886 & 3.1670 & 0.9429 \\ 0 & 0 & 0.9429 & 2.3335 \end{bmatrix}$$

It is final tridiagonal matrix (symmetric).

Check: $\text{Tr}(A_3) = 14.0005 \simeq 14$

10.14 Householder's Method

This method also reduces the symmetric matrix A to a symmetric tridiagonal matrix T as shown by (10.55), through a series of similarity transformations. The transformation matrix is totally different from that of Jacobi and is such that all the desired zeros in a row are produced in a single transformation. Thus only $(n-2)$ transformations are needed to reduce a matrix A of order n to tridiagonal form T while $(n-2)(n-1)/2$ transformations were required in the Givens method.

The transformation matrix S used by Householder has the form

$$S = I - 2\omega\omega^T, \tag{10.56}$$

where ω is a particular normalised vector such that the scalar product

$$\omega^T\omega = 1. \tag{10.57}$$

It may be noted that if ω has n components, the 'dyadic product' $\omega\omega^T$ is a matrix of order n. Thus with the unit matrix I being of order n, the transformation matrix S is of order n as well.

10.14.1 Matrix S is symmetric

First we see that $\omega\, \omega^T$ is symmetric since

$$(\omega\, \omega^T)^T = (\omega^T)^T(\omega)^T = \omega\, \omega^T.$$

Further, as I is symmetric and sum (difference) of two symmetric matrices is also a symmetric matrix, the matrix S should be symmetric.

10.14.2 Matrix S is orthogonal

We have to show that $S^TS = I$ or $S^T = S^{-1}$. Since S is symmetric,

$$\begin{aligned} S^TS = SS &= (I - 2\omega\omega^T)(I - 2\omega\omega^T) \\ &= I - 4\omega\omega^T + 4\omega(\omega^T\omega)\omega^T \\ &= I - 4\omega\,\omega^T + 4\omega\,\omega^T, \because \omega^T\omega = 1 \\ &= I \end{aligned}$$

Therefore S is an orthogonal matrix, i.e., $S^T = S^{-1}$. We should also note that S is an involutary matrix, i.e., $S = S^{-1} \because S^T = S$.

10.14.3 Similarity transformation

Remembering that S is a symmetric orthogonal matrix we perform similarity transformation on A as follows:

$$S^TAS = SAS \tag{10.58}$$

$$\begin{aligned} &= (I - 2\omega\omega^T)A(I - 2\omega\omega^T) \\ &= A - 2A\omega\omega^T - 2\omega\omega^TA + 4\omega\omega^TA\omega\omega^T \\ &= A - 2(A\omega)\omega^T - 2\omega(A\omega)^T + 4\omega(\omega^TA\omega)\omega^T \end{aligned}$$

$$= A - 2\boldsymbol{p}\omega^T - 2\omega\boldsymbol{p}^T + 4c\omega\omega^T \tag{10.59}$$

where $\boldsymbol{p} = A\omega$ is a column vector (10.60)

and $c = \omega^TA\omega = \omega^T\boldsymbol{p}$ is a scalar. (10.61)

$$\begin{aligned} &= A - 2\,(\boldsymbol{p} - c\,\omega\,\omega^T) - 2\omega(\boldsymbol{p}^T - c\omega^T) \\ &= A - 2\,(\boldsymbol{p} - c\omega)\,\omega^T - 2\omega(\boldsymbol{p} - c\omega)^T \end{aligned}$$

$$= A - 2\boldsymbol{q}\omega^T - 2\omega\boldsymbol{q}^T, \tag{10.62}$$

where $\boldsymbol{q} = \boldsymbol{p} - c\omega$. (10.63)

Thus the elements of the transformed matrix can be determined from (10.59) or (10.62) provided vector ω is known. It should be kept in mind that A is symmetric and the transformed matrix S^TAS is also symmetric.

10.14.4 First transformation

Now let us consider the first transformation on A that makes the transformed matrix conforming to the first row of tridiagonal matrix T. See (10.55). It should be noted that since $S^T AS$ is symmetric its first column will also reduce to the desired tridiagonal form.

We choose the vector ω such that its first one component is zero, i.e.,

$$\omega^T = (0,\ \omega_2,\ \omega_3,\ \ldots.\ \omega_n), \tag{10.64}$$

and make the transformation

$$S^T AS = U, \tag{10.65}$$

where $S = I - 2\omega\omega^T$ with ω^T given by (10.64).

Before attempting to determine components of ω, we should make the following two observations:

(i) Element a_{11} does not change under transformation $S^T AS = U$.

That is, we have to show that $u_{11} = a_{11}$. Note that the matrix $\omega\ \omega^T$ with ω given by (10.64) has all zero elements in first row and first column, so that (1, 1) element of $S = I - 2\omega\omega^T$ is 1 and all other elements in the first row (and column since S is symmetric) will be zero. In the multiplication $S^T A$, the elements of first row of A do not change and in the postmultiplication $S^T AS$, the elements of the first column of $S^T A$ do not change. That is, in the transformation $S^T AS = U$, the element in the first row and first column of A does not change. Hence we have,

$$\boldsymbol{u}_{11} = \boldsymbol{a}_{11}. \tag{10.66}$$

(ii) Sum of squares of the elements in the first row of A remains invariant under transformation $S^T AS = U$.

We have to prove that

$$\sum_{j=1}^{n} u_{1j}^2 = \sum_{j=1}^{n} a_{1j}^2. \tag{10.67}$$

Let us denote the first row of matrix A by $\boldsymbol{a}_1^T$ and that of U by $\boldsymbol{u}_1^T$. Then (10.67) can be written as

$$\boldsymbol{u}_1^T \boldsymbol{u}_1 = \boldsymbol{a}_1^T \cdot \boldsymbol{a}_1. \tag{10.68}$$

The elements of first row of U can be written from (10.62) as

$$u_{1j} = a_{1j} - 2q_1\omega_j - 2\omega_1 q_j, \;\; j = 1(1)n$$

$$= a_{1j} - 2q_1\omega_j, \;\; \because \omega_1 = 0. \tag{10.69}$$

From (10.69) we can write

$$\boldsymbol{u}_1 = \boldsymbol{a}_1 - 2q_1\boldsymbol{\omega} \tag{10.70a}$$

and
$$\boldsymbol{u}_1^{\mathrm{T}} = \boldsymbol{a}_1^{\mathrm{T}} - 2q_1\boldsymbol{\omega}^{\mathrm{T}} \tag{10.70b}$$

From (10.70a and b) we get

$$\boldsymbol{u}_1^{\mathrm{T}}\boldsymbol{u}_1 = (\boldsymbol{a}_1^{\mathrm{T}} - 2q_1\boldsymbol{\omega}^{\mathrm{T}})(\boldsymbol{a}_1 - 2q_1\boldsymbol{\omega})$$

$$= \boldsymbol{a}_1^{\mathrm{T}}\boldsymbol{a}_1 - 2q_1\boldsymbol{a}_1^{\mathrm{T}}\boldsymbol{\omega} - 2q_1\boldsymbol{\omega}^{\mathrm{T}}\boldsymbol{a}_1 + 4q_1^2\boldsymbol{\omega}^{\mathrm{T}}\boldsymbol{\omega}$$

$$= \boldsymbol{a}_1^{\mathrm{T}}\boldsymbol{a}_1 - 2q_1p_1 - 2q_1p_1 + 4q_1^2 = \boldsymbol{a}_1^{\mathrm{T}}\boldsymbol{a}_1 - 4q_1p_1 + 4q_1^2$$

where p_1 is the first element of $\boldsymbol{p} = \mathrm{A}\boldsymbol{\omega}$.
Further $q_1 = p_1 - k\omega_1 = p_1 \because \omega_1 = 0$.

$$\therefore \quad \boldsymbol{u}_1^{\mathrm{T}}\boldsymbol{u}_1 = \boldsymbol{a}_1^{\mathrm{T}}\boldsymbol{a}_1 - 4q_1^2 + 4q_1^2$$

$$= \boldsymbol{a}_1^{\mathrm{T}}\boldsymbol{a}_1.$$

Now let us describe the method for evaluating $\omega_2, \omega_3, \ldots \omega_n$.
We have proved that

$$\sum_{j=1}^{n} u_{1j}^2 = \sum_{j=1}^{n} a_{1j}^2.$$

As $u_{11} = a_{11}$ from (10.66) above relation becomes

$$\sum_{j=2}^{n} u_{1j}^2 = \sum_{j=2}^{n} a_{1j}^2 = s \text{ (say)}. \tag{10.71}$$

In order that first row of U conforms to the first row of tridiagonal matrix T, we should have

$$u_{1j} = 0, \; j = 3, 4, \ldots. n. \tag{10.72}$$

In that case we get form (10.71) using (10.72)

$$u_{12}^2 = s$$

or

$$u_{12} = \pm\sqrt{s}. \tag{10.73}$$

We can now write various elements of the first row of U, i.e., $\boldsymbol{u}_1^{\mathrm{T}}$ from (10.70b) and equate them appropriately as follows:

$$u_{11} = a_{11}. \tag{10.74}$$

$$u_{12} = a_{12} - 2q_1\omega_2 = \pm\sqrt{s}, \tag{10.75}$$

$$u_{1j} = a_{1j} - 2q_1\omega_j = 0, \; j = 3, 4, \ldots. n. \tag{10.76}$$

It may be noted that (10.74) is the same as (10.66) which we had obtained earlier giving different explanation. The vector ω satisfies the condition

$$\omega^{\mathrm{T}}\omega = \sum_{i=2}^{n} \omega_i^2 = 1, \quad \because \omega_1 = 0. \tag{10.77}$$

Multiplying (10.75) by ω_2 and (10.76) by ω_j, $j = 3(1)n$ and adding we get

$$\sum_{j=2}^{n} a_{1j}\omega_j - 2q_1 \sum_{j=2}^{n} \omega_j^2 = \pm\omega_2\sqrt{s}$$

or

$$p_1 - 2q_1 = \pm\omega_2\sqrt{s}, \text{ from (10.77)}.$$

or

$$q_1 = \mp\omega_2\sqrt{s}, \; \because q_1 = p_1. \tag{10.78}$$

Putting value of q_1 from (10.78) in (10.75) gives

$$a_{12} \pm 2\omega_2^2\sqrt{s} = \pm\sqrt{s}$$

or

$$\omega_2{}^2 = \frac{1}{2}\left(1 \mp \frac{a_{12}}{\sqrt{s}}\right)$$

or

$$\omega_2 = \pm \left\{ \frac{1}{2} \left(1 \mp \frac{a_{12}}{\sqrt{s}} \right) \right\}^{1/2}. \tag{10.79}$$

The $\pm$ sign outside the bracket in (10.79) may be taken always as positive since it will not change the character of vector ω. Hence

$$\omega_2 = \left\{ \frac{1}{2} \left(1 \mp \frac{a_{12}}{\sqrt{s}} \right) \right\}^{1/2}. \tag{10.80}$$

Further, from (10.76) and (10.78)

$$\omega_j = a_{1j}/2q_1 = \mp \frac{1}{2} \left\{ a_{1j} / \left(\sqrt{s} \omega_2 \right) \right\} \cdot j = 3,\ 4,\ \ldots\ n. \tag{10.81}$$

In order to decide the sign of $\sqrt{s}$, we see that ω_2 occurs in the denominator in computing ω_j from (10.81). We want to choose that sign which makes ω_2 larger in magnitude so that the components ω_j do not get larger. Thus we can write (10.80) and (10.81) in the following modified forms respectively,

$$\omega_2 = \left\{ \frac{1}{2} \left(1 + \frac{a_{12}}{(\text{sign of } a_{12})\sqrt{s}} \right) \right\}^{1/2} \tag{10.82}$$

$$\omega_j = \frac{1}{2} \left\{ \frac{a_{1j}(\text{sign of } a_{12})}{\sqrt{s}\omega_2} \right\},\ j = 3,\ 4,\ \ldots.\ n. \tag{10.83}$$

Having determined the components ω_i, $i = 2(1)n$, the elements of the transformed matrix U can be evaluated from (10.62). As U is symmetric and has same e.values as A, the above procedure may be applied on U, making its second row conforming to tridiagonal form. It may be mentioned that first row and first column of U will not be affected at all in the next transformation.

10.14.5 General procedure

Let us describe the procedure for k^{th} transformation where

$$A_k = S_k^T A_{k-1} S_k,\ k = 1,\ 2,\ \ldots.\ n-2, \tag{10.84}$$

and $A_0 = A$.

It may be noted that A_{n-2} will be tridiagonal.

We assume that first $(k-1)$ rows have been transformed to tridiagonal form and k^{th} row is to be transformed through (10.84) for $k = 1(1)n-1$. We proceed as follows:

(*i*) Choose vector $\omega^{\mathrm{T}} = \left(\underbrace{0\ 0\ \cdots\ 0}_{k \text{ zeros}}\ \omega_{k+1}\ \omega_{k+2}\ \cdots\ \omega_n \right)$; $\mathrm{S}_k = \mathrm{I} - 2\omega\,\omega^{\mathrm{T}}$.

Then first k rows and k columns of S will form a $k \times k$ unit matrix. Hence elements of $k \times k$ submatrix of A_{k-1} consisting of first k rows and k columns will remain unaltered under the transformation.

(*ii*) Calculate $s = \sum_{j=k+1}^{n} \{a_{kj}^{(k-1)}\}^2$, where $a_{kj}^{(k-1)}$ denotes the elements in the kth row of A_{k-1}. Also find $\sqrt{s}$.

(*iii*) sign(s) = if $a_{k,k+1}^{(k-1)}$ is ≥ 0 then 1 (+ve) else -1 ($-$ve).

(*iv*) Compute $\omega_{k+1} = \left\{ \frac{1}{2}\left(1 + \frac{a_{k,\,k+1}^{(k-1)}}{\sqrt{s}\cdot \mathrm{sign(s)}}\right)\right\}^{1/2}$.

(*v*) Compute $\omega_j = \dfrac{a_{kj}^{(k-1)}}{2\omega_{k+1}\sqrt{s}\cdot\mathrm{sign(s)}}$, $j = k+2,\ k+3,\ \ldots.\ n$.

(*vi*) Compute vector $\boldsymbol{p} = \mathrm{A}_{k-1}\omega$; first $(k-1)$ elements of $\boldsymbol{p}$ will be zeros. Since first k elements of ω are zero and all the elements in $(k-1)^{\text{th}}$ row (and above) of A_{k-1} in the columns beyond k, i.e., from $(k+1)$ to n are zero; $p_i = \sum_{j=k+1}^{n} a_{ij}^{(k-1)}\omega_j$, $i = k(1)n$.

(*vii*) Compute scalar $c = \omega^{\mathrm{T}}\mathrm{A}_{k-1}\omega = \omega^{\mathrm{T}}p = \sum_{j=k+1}^{n} \omega_j p_j$.

(*viii*) Compute $\boldsymbol{q} = \boldsymbol{p} - c\omega$, first $(k-1)$ elements will be zero and $q_k = p_k$ since $\omega_k = 0$; $q_i = p_i - c\omega_i$, $i = k+1(1)n$.

(*ix*) Compute elements of A_k, i.e., $a_{i,\,j}^{(k)}$, $i = k(1)n$, $j = k(1)n$.

$$a_{kk}^{(k)} = a_{kk}^{(k-1)};\ a_{k,\,k+1}^{(k)} = \sqrt{s}.$$

$$a_{kj}^{(k)} = 0, j = k+2(1)n.$$

$$a_{ij}^{(k)} = a_{i,\,j}^{(k-1)} - 2q_i\omega_j - 2\omega_i q_j,\quad i = k+1(1)n,\quad j = i(1)n.$$

(x) Since A_k is symmetric, copy elements of upper triangular to lower triangular, i.e.,

$$a_{i,j}^{(k)} = a_{ji}^{(k)},\ i = k+1(1)n,\ j = k(1)i-1.$$

Householder's method is more efficient than the Givens method as the number of arithmetic operations required will be almost half.

Example 10.3

Reduce the following matrix to tridiagonal form by Householder's method:

$$A = \begin{bmatrix} 3 & 0 & 1 & 1 \\ 0 & 2 & 1 & 0 \\ 1 & 1 & 4 & 1 \\ 1 & 0 & 1 & 5 \end{bmatrix}.$$

Solution Let $A_1 = S_1^T A S_1$.

$$\omega^T = (0 \quad \omega_2 \quad \omega_3 \quad \omega_4)$$

$$s = a^2{}_{12} + a^2{}_{13} + a^2{}_{14} = 0 + 1^2 + 1^2 = 2$$

$$\sqrt{s} = \pm\sqrt{2}$$

Let us take $\sqrt{s} = \sqrt{2}$ (according to sign of a_{12}, say $+0$)

$$\omega_2{}^2 = \frac{1}{2}\left(1 + \frac{0}{\sqrt{2}}\right) \quad \text{or} \quad \omega_2 = \frac{1}{\sqrt{2}} = 0.7071$$

$$\omega_3 = \frac{a_{13}}{2\omega_2\sqrt{s}} = \frac{1}{2 \times \frac{1}{\sqrt{2}} \times \sqrt{2}} = \frac{1}{2}$$

$$\omega_4 = \frac{a_{14}}{2\omega_2\sqrt{s}} = \frac{1}{2 \times \frac{1}{\sqrt{2}} \times \sqrt{2}} = \frac{1}{2}$$

Check: $\omega_2{}^2 + \omega_3{}^2 + \omega_4{}^2 = \frac{1}{2} + \frac{1}{4} + \frac{1}{4} = 1.$

$$p = A\omega = \begin{bmatrix} 3 & 0 & 1 & 1 \\ 0 & 2 & 1 & 0 \\ 1 & 1 & 4 & 1 \\ 1 & 0 & 1 & 5 \end{bmatrix} \begin{bmatrix} 0 \\ 0.7071 \\ 0.5 \\ 0.5 \end{bmatrix} = \begin{bmatrix} 1 \\ 1.9142 \\ 3.2071 \\ 3 \end{bmatrix}$$

$$c = \boldsymbol{\omega}^{\mathrm{T}}\boldsymbol{p} = (0 \quad 0.7071 \quad 0.5 \quad 0.5)\begin{bmatrix} 1 \\ 1.9142 \\ 3.2071 \\ 3 \end{bmatrix} = 4.4571$$

$$\boldsymbol{q} = \boldsymbol{p} - c\boldsymbol{\omega} = \begin{bmatrix} 1 \\ 1.9142 \\ 3.2071 \\ 3 \end{bmatrix} - 4.4571\begin{bmatrix} 0 \\ 0.7071 \\ 0.5 \\ 0.5 \end{bmatrix} = \begin{bmatrix} 1 \\ -1.2374 \\ 0.9786 \\ 0.7714 \end{bmatrix}$$

$$a_{12}^{(1)} = \sqrt{s} = \sqrt{2} = 1.4142$$

$$a_{13}^{(1)} = 0,\ a_{14}^{(1)} = 0 \text{ (annihilated)}$$

$$a_{22}^{(1)} = a_{22} - 2\omega_2 q_2 - 2q_2\omega_2 = a_{22} - 4\omega_2 q_2$$
$$= 2 - 4 \times 0.7071 \times (-1.2374) = 5.4999$$

$$a_{23}^{(1)} = a_{23} - 2\omega_2 q_3 - 2q_2\omega_3$$
$$= 1 - 2\,(0.7071)\,(0.9786) - 2(-1.2374)(0.5) = 0.8535$$

$$a_{24}^{(1)} = a_{24} - 2\omega_2 q_4 - 2q_2\omega_4$$
$$= 0 - 2\,(0.7071)\,(0.7714) - 2(-1.2374)(0.5) = 0.1465$$

$$a_{33}^{(1)} = a_{33} - 2\omega_3 q_3 - 2q_3\omega_3 = a_{33} - 4\omega_3 q_3$$
$$= 4 - 4\left(\frac{1}{2}\right)(0.9786) = 2.0428$$

$$a_{34}^{(1)} = a_{34} - 2\omega_3 q_4 - 2q_3\omega_4$$
$$= 1 - 2\left(\frac{1}{2}\right)(0.7714) - 2\,(0.9786)\left(\frac{1}{2}\right) = -0.75$$

$$a_{44}^{(1)} = a_{44} - 2\omega_4 q_4 - 2q_4\omega_4$$
$$= 5 - 4 \times \left(\frac{1}{2}\right)(0.7714) = 3.4572$$

$$A_1 = \begin{bmatrix} 3 & 1.4142 & 0 & 0 \\ 1.4142 & 5.4999 & 0.8535 & 0.1465 \\ 0 & 0.8535 & 2.0428 & -0.75 \\ 0 & 0.1465 & -0.75 & 3.4572 \end{bmatrix}$$

Check: Tr $(A_1) = \text{Tr}(A) = 14.$

$$A_2 = S_2^T A_1 S_2$$

$$\omega^T = (0 \ \ 0 \ \ \omega_3 \ \ \omega_4)$$

$$s = a_{23}^{(1)2} + a_{24}^{(1)2} = 0.8535^2 + 0.1465^2 = 0.7499$$

$\sqrt{s} = \pm\, 0.866$; Take $\sqrt{s} = 0.866$ ($\because a_{23}^{(1)}$ is positive(+ ve))

$$\omega_3 = \left[\frac{1}{2}\left(1 + \frac{0.8535}{0.866}\right)\right]^{1/2} = 0.9964$$

$$\omega_4 = \frac{a_{24}}{2\sqrt{s}\omega_3} = \frac{0.1465}{2 \times 0.866 \times 0.9964} = 0.0849$$

Check: $\omega_3{}^2 + \omega_4{}^2 = 0.9964^2 + 0.0849^2 = 1.0000.$

$$\boldsymbol{p} = A_1\omega = \begin{bmatrix} 3 & 1.4142 & 0 & 0 \\ 1.4142 & 5.4999 & 0.8535 & 0.1465 \\ 0 & 0.8535 & 2.0428 & -0.75 \\ 0 & 0.1465 & -0.75 & 3.4572 \end{bmatrix} \begin{bmatrix} 0 \\ 0 \\ 0.9964 \\ 0.0849 \end{bmatrix} = \begin{bmatrix} 0 \\ 0.8629 \\ 1.9718 \\ -0.4538 \end{bmatrix}$$

$$c = \omega^T \boldsymbol{p} = 0.9964 \times 1.9718 + (0.0849)(-0.4538) = 1.9262$$

$$\boldsymbol{q} = \boldsymbol{p} - c\omega = \begin{bmatrix} 0 \\ 0.8629 \\ 1.9718 \\ -0.4538 \end{bmatrix} - 1.9262 \begin{bmatrix} 0 \\ 0 \\ 0.9964 \\ -0.0849 \end{bmatrix} = \begin{bmatrix} 0 \\ 0.8629 \\ 0.0525 \\ -0.6173 \end{bmatrix}$$

$a_{22}^{(2)} = a_{22}^{(1)}$ (does not change)

$$a_{23}^{(2)} = a_{23}^{(1)} - 2\omega_2 q_3 - 2q_2\omega_3$$

$$= 0.8535 - 2 \times 0.8629 \times 0.9964 = -0.8661$$

$$a_{24}^{(2)} = 0 \text{ (annihilated)}$$

$$a_{33}^{(2)} = a_{33}^{(1)} - 2\omega_3 q_3 - 2q_3\omega_3 = a_{33}^{(1)} - 4\omega_3 q_3$$

$$= 2.0428 - 4 \times 0.0525 \times 0.9964 = 1.8334$$

$$a_{34}^{(2)} = a_{34}^{(1)} - 2\omega_3 q_4 - 2q_3\omega_4$$

$$= -0.75 - 2 \times 0.9964 \times (-0.6173) - 2 \times 0.0525 \times 0.0849 = 0.4712$$

$$a_{44}^{(2)} = a_{44}^{(1)} - 2\omega_4 q_4 - 2q_4\omega_4 = a_{44}^{(1)} - 4\omega_4 q_4$$

$$= 3.4572 - 4 \times 0.0849 \times (-0.6173) = 3.6668$$

$$\mathrm{A}_2 = \begin{bmatrix} 3 & 1.4142 & 0 & 0 \\ 1.4142 & 5.4999 & -0.8661 & 0 \\ 0 & -0.8661 & 1.8334 & 0.4712 \\ 0 & 0 & 0.4712 & 3.6668 \end{bmatrix}$$

$$\mathrm{Tr}\,(\mathrm{A}_2) = 14.0001 \simeq 14.$$

10.15 Sturm Sequence and its Properties

In the foregoing two methods, viz., Givens and Householder's we reduce a symmetric matrix A of order n to a symmetric tridiagonal matrix T (say) of order n through a number of similarity transformations. Matrix T has same e.values as A which are given by the roots of the characteristic equation $|\mathrm{T} - \lambda \mathrm{I}| = 0$. But finding the roots of this equation will be as difficult as finding the roots of the characteristic equation of A, i.e., $|\mathrm{A} - \lambda \mathrm{I}| = 0$. There is however another simpler way to find the roots when the matrix is tridiagonal. Instead of expanding $|\mathrm{T} - \lambda \mathrm{I}|$ in powers of λ we take recourse to the Sturm sequence which is discussed in the following sections.

10.15.1 Sturm sequence

A Sturm sequence is a sequence of real functions $f_0(x), f_1(x), f_2(x) \ldots . f_n(x)$ defined for $a \leq x \leq b$ which satisfies the following conditions:

(*i*) functions $f_i(x)$, $i = 0(1)n$ are continuous.

(*ii*) $f_0(x)$ does not change sign in $[a,\ b]$.

(*iii*) if any of the inner functions $f_i(x),\ i = 1(1)n-1$ vanishes at some point $\overline{x}$ in $(a,\ b)$ then none of $f_{i-1}(x)$ and $f_{i+1}(x)$ can be zero there, i.e., if $f_i(\overline{x}) = 0$, then neither $f_{i-1}(\overline{x}) = 0$ nor $f_{i+1}(\overline{x}) = 0$. It implies that no two consecutive functions can have a common zero.

(*iv*) if $f_i(\overline{x}) = 0$ for $x = \overline{x}$ in $(a,\ b)$ then $f_{i-1}(\overline{x})$ and $f_{i+1}(\overline{x})$ have opposite signs, i.e.,

$$\text{sign } f_{i+1}(\overline{x}) = -\text{sign} f_{i-1}(\overline{x}),\ i = 1(1)n-1.$$

(*v*) if $f_n(\overline{x}) = 0$ for $x = \overline{x}$ in $(a,\ b)$ then for sufficiently small $h > 0$ following relations hold

$$\text{sign } f_n(\overline{x}-h) = -\text{sign } f_{n-1}(\overline{x}-h) \text{ and}$$
$$\text{sign } f_n(\overline{x}+h) = \text{sign } f_{n-1}(\overline{x}+h).$$

The Sturm sequence defined above can be used to find the broad intervals in which the zeros of the function $f_n(x)$ lie. After knowing the interval of a particular root of $f_n(x) = 0$ it can be determined to desired accuracy by method of bisection or any other method discussed in Chapter 3. Following theorem is used for finding the interval (s) in which the zero (s) of the function $f_n(x)$ lie.

10.15.2 Theorem

Let V (x) be a non-negative integer function denoting the number of sign changes in the Sturm sequence $f_0(x),\ f_1(x),\ \ldots.\ f_n(x)$ for any given value of x in $[a,\ b]$. Then the number of zeros of $f_n(x)$ in $[a,\ b]$ equals $\mathrm{V}(a) - \mathrm{V}(b)$.

Proof: Obviously there can not be more than n sign changes in the sequence of $(n+1)$ functions when each function assumes a different sign from its predecessor and minimum number of sign changes will be zero when all the functions have the same sign. In other words $0 \leq \mathrm{V}(x) \leq n$ for any x in $[a,\ b]$.

Let us suppose that for some value of x, the number of sign changes $\mathrm{V}(x) = m$. This value of $\mathrm{V}(x)$ will change with increasing the value of x only if one or more of functions $f_1(x),\ f_2(x),\ \ldots.\ f_n(x)$ change signs as $f_0(x)$ does not change its sign in $[a,\ b]$. On account of continuity, a function has to pass through its zero for change of sign. Let us examine the change in $\mathrm{V}(x)$ when a function $f_i(x)$ passes through its zero at $x = \overline{x}$ (say). For h small we want to see the change in the value of $\mathrm{V}(\overline{x}-h)$ and $\mathrm{V}(\overline{x}+h)$. First let us consider the case when $f_i(x)$ is one of the inner functions, i.e., $i = 1(1)n-1$. In view of the properties (*iii*) and (*iv*) the possibilities of signs of $f_{i-1}(x), f_i(x)$ and $f_{i+1}(x)$ are shown in Table 10.2.

Table 10.2. Sign changes when $f_i(\bar{x}) = 0,\ i = 1(1)n-1$

x	$f_{i-1}(x)$	$f_i(x)$	$f_{i+1}(x)$	*Number of sign changes*	$f_{i-1}(x)$	$f_i(x)$	$f_{i+1}(x)$	*Number of sign changes*
$\bar{x}-h$	$-$	$\pm$	$+$	1	$+$	$\pm$	$-$	1
$\bar{x}$	$-$	0	$+$	1	$+$	0	$-$	1
		$(+\text{ or }-)$				$(+\text{ or }-)$		
$\bar{x}+h$	$-$	$\mp$	$+$	1	$+$	$\mp$	$-$	1

Note: Any sign + or – may be given to $f_i(\bar{x}) = 0$. According to Wilkinson (ref [4]) $f_i(\bar{x})$ should be given sign opposite to $f_{i-1}(\bar{x})$.

It is clear from the above table that number of sign changes are not affected if any of the inner functions passes through its zero. That is, $\mathrm{V}(\bar{x}-h) = \mathrm{V}(\bar{x}) = \mathrm{V}(\bar{x}+h)$ for h small.

Now let us consider the case when $f_n(x)$ changes sign when it passes through its zero at $x = \bar{x}$. Again let $h > 0$ be sufficiently small such that no zero of $f_{n-1}(x)$ lies between $\bar{x}-h$ and $\bar{x}+h$. Keeping property (v) in view, the possibilities of sign changes are shown in Table 10.3.

Table 10.3. Sign changes when $f_n(\bar{x}) = 0$

x	$f_{n-1}(x)$	$f_n(x)$	*Number of sign changes*	$f_n - 1(x)$	$f_n(x)$	*Number of sign changes*
$\bar{x}-h$	$-$	$+$	1	$+$	$-$	1
$\bar{x}$	$-$	0	1 or 0	$+$	0	0 or 1
		$(+\text{ or }-)$			$(+\text{ or }-)$	
$\bar{x}+h$	$-$	$-$	0	$+$	$+$	0

We see from Table 10.3 that number of sign changes reduce by 1 as $f_n(x)$ passes through its zero. That is, $\mathrm{V}(\bar{x}-h) - \mathrm{V}(\bar{x}+h) = 1$. It may be concluded that every time $f_n(x)$ passes through its zero the number of sign changes will reduce by one. Thus, $\mathrm{V}(a) - \mathrm{V}(b)$ will be equal to the number of zeros of $f_n(x)$ in $[a,\ b]$.

Note: If $f_n(a) = 0$, we should give it a sign opposite to that of $f_{n-1}(a)$ and if$f_n(b) = 0$ then it should be given same sign as of $f_{n-1}(b)$.

10.16 Eigenvalues of Symmetric Tridiagonal Matrix

We shall now discuss how to find the eigenvalues of a symmetric tridiagonal matrix T, given by (10.55), using Sturm sequence. The eigenvalues of T are the roots of the equation $|T-\lambda I| = (-1)^n|\lambda I - T| = 0$, i.e.,

$$\begin{vmatrix} \lambda-\alpha_1 & -\beta_1 & 0 & 0 & \cdots & 0 \\ -\beta_1 & \lambda-\alpha_2 & -\beta_2 & 0 & \cdots & 0 \\ \vdots & \vdots & \ddots & & & \vdots \\ & & & & & 0 \\ \vdots & \vdots & & -\beta_{n-2} & (\lambda-\alpha_{n-1}) & -\beta_{n-1} \\ 0\cdots 0 & 0 & 0\cdots 0 & & -\beta_{n-1} & \lambda-\alpha_n \end{vmatrix} = 0. \tag{10.85}$$

Let $f_i(\lambda)$ denote the ith principal minor of $|\lambda I - T|$ and let

$f_0(\lambda) = 1$, then we have

$$f_1(\lambda) = |\lambda - \alpha_1| = \lambda - \alpha_1,$$

$$f_2(\lambda) = \begin{vmatrix} \lambda-\alpha_1 & -\beta_1 \\ -\beta_1 & \lambda-\alpha_2 \end{vmatrix} = \begin{vmatrix} f_1(\lambda) & -\beta_1 \\ -\beta_1 & \lambda-\alpha_2 \end{vmatrix},$$

$$= (\lambda-\alpha_2)f_1(\lambda) - {\beta_1}^2 = (\lambda-\alpha_2)f_1(\lambda) - {\beta_1}^2 f_0(\lambda),$$

$$f_3(\lambda) = \begin{vmatrix} \lambda-\alpha_1 & -\beta_1 & 0 \\ -\beta_1 & \lambda-\alpha_2 & -\beta_2 \\ 0 & -\beta_2 & \lambda-\alpha_3 \end{vmatrix} = (\lambda-\alpha_3)\begin{vmatrix} \lambda-\alpha_1 & -\beta_1 \\ -\beta_1 & \lambda-\alpha_2 \end{vmatrix} + \beta_2 \begin{vmatrix} \lambda-\alpha_1 & 0 \\ -\beta_1 & -\beta_2 \end{vmatrix}$$

$$= (\lambda-\alpha_3)f_2(\lambda) - {\beta_2}^2 f_1(\lambda).$$

In general we can show that the following recurrence relation holds

$$f_i(\lambda) = (\lambda-\alpha_i)f_{i-1}(\lambda) - \beta^2_{i-1}f_{i-2}(\lambda),\ i = 2,\ 3,\ \ldots.\ n. \tag{10.86}$$

Choosing $\beta_0 = 0$, $f_1(\lambda)$ can also be expressed in the form of recurrence relation (10.86) for $i = 1$.

It should be clear that $f_n(\lambda)$, i.e., the nth principal minor is nothing but the whole determinant $|\lambda I - T|$. Thus $f_n(\lambda)$ is the characteristic polynomial of matrix T. Hence we shall be interested in finding the roots of the equation $f_n(\lambda) = 0$. It can be shown that $f_0(\lambda)$, $f_1(\lambda)$, ... $f_n(\lambda)$ form a Sturm sequence. Therefore zeros of $f_n(\lambda)$ can be computed by the method described in Sec 10.15.2. If any of β_i, $i = 1(1)n-1$ is zero then the roots of $|\lambda\mathrm{I} - \mathrm{T}| = 0$ will be repeated. In that case matrix T can be broken into separate tridiagonal matrices (ref [4]). However, we shall assume that the roots are simple, not repeated. It may be emphasised that the characteristic polynomial is not to be expressed in powers of λ. Instead we make use of the recurrence relation (10.86) as shown in Example 10.4.

Example 10.4

Form Sturm sequence of functions $f_0(\lambda)$, $f_1(\lambda)$, $f_2(\lambda)$, $f_3(\lambda)$ using recurrence relation, where λ is an e.value of the symmetric tridiagonal matrix.

$$\mathrm{T} = \begin{bmatrix} 4 & 1 & 0 \\ 1 & 2 & 1 \\ 0 & 1 & 5 \end{bmatrix}.$$

Examine the sign changes in the sequence for $\lambda = 0(1)6$ taking $f_0(\lambda) = 1$ and identify the intervals separating the e.values.

Solution $f_0(\lambda) = 1, f_1(\lambda) = \lambda - 4, f_2(\lambda) = (\lambda - 2)f_1(\lambda) - 1,$

$f_3(\lambda) = (\lambda - 5)f_2(\lambda) - f_1(\lambda).$

λ	$f_0(\lambda)$	$f_1(\lambda)$	$f_2(\lambda)$	$f_3(\lambda)$	*Number of sign changes*
0	1	−4	7	−31	3
1	1	−3	2	−5	3
2	1	−2	−1	5	2
3	1	−1	−2	5	2
4	1	0 (+ or −)	−1	1	2
5	1	1	2	−1	1
6	1	2	7	5	0

The e.values lie in the intervals, (1, 2), (4, 5) and (5, 6).

Note: For $\lambda = 4$, the sign of $f_1(\lambda)$ should be taken usually opposite to its predecessor $f_0(\lambda)$, i.e., minus, although it will not alter the number of changes even if plus sign is taken.

10.17 Upper and Lower Bounds of Eigenvalues

In the foregoing section we have discussed a method by which we can determine the intervals for various e.values. However, as we do not have prior knowledge about the range of e.values we have to depend on some trial-and-error method for searching all e.values. Following theorem helps us that provides the upper and lower bounds of eigenvalues of a given matrix.

10.17.1 Gerschgorin's theorem

The modulus of the largest eigenvalue of a matrix can not exceed the largest sum of the moduli of its elements along any row. Since the e.values of a matrix and its transpose are same the theorem applies to its columns also.

Proof: Let A be an $n \times n$ matrix and λ_r be an e.value with corresponding e.vector $\boldsymbol{x}_r$ where $\boldsymbol{x}_r^{\mathrm{T}} = (x_{1r} \ \ x_{2r} \ \ \ldots. \ \ x_{nr})$. Further, let $|x_{sr}|$ be the largest component in magnitude, of the vector $\boldsymbol{x}_r$, $1 \le s \le n$. Writing $\mathrm{A}\boldsymbol{x}_r = \lambda_r \boldsymbol{x}_r$ in expanded form we have,

$$a_{i1}x_{1r} + a_{i2}x_{2r} + \ldots + a_{is}x_{sr} + \ldots + a_{in}x_{nr} = \lambda_r x_{ir}, \tag{10.87}$$

$$i = 1,\ 2,\ \ldots.\ n.$$

Dividing (10.87) by x_{sr} throughout and taking modulus gives

$$|a_{i1}| + |a_{i2}| + \ \ldots.\ |a_{is}| + \ \ldots.\ + |a_{in}| \ge |\lambda_r| \frac{|x_{ir}|}{|x_{sr}|}, \tag{10.88}$$

since $\dfrac{|x_{1r}|}{|x_{sr}|} \le 1$, $\dfrac{|x_{2r}|}{|x_{sr}|} \le 1 \cdots$, etc., as $|x_{sr}|$ is largest.

Inequality (10.88) is true for all rows and will also hold for the sth row when the right side is largest. It gives

$$|\lambda_r| \le |a_{s1}| + |a_{s2}| + \ldots. + |a_{ss}| + \ldots. |a_{sn}|.$$

$$= \sum_{j=1}^{n} |a_{sj}| \tag{10.89}$$

Replacing the right side of inequality (10.89) by the row having largest sum of the moduli of its elements, we can write

$$|\lambda_r| \le \max_i \left| \sum_{j=1}^{n} a_{ij} \right|, \ r = 1(1)n \tag{10.90}$$

As (10.90) is true for any $\lambda_r,\ r = 1(1)n$ we have

$$|\lambda_{\max}| \leq \max_i \left| \sum_{j=1}^{n} |a_{ij}| \right|. \tag{10.91}$$

10.17.2 Corollary

For each e.value λ there exists a circle with centre a_{ss} and radius $\sum_{j=1,\, j\neq s}^{n} |a_{sj}|$, i.e., $|\lambda - a_{ss}| = \Sigma |a_{sj}|,\ j \neq s$, for some $s(1 \leq s \leq n)$ such that it lies inside that circle or on its boundary. The inside of the circle along with its boundary is also called a 'disc'. Obviously this applies to complex e.values.

The proof of the above corollary immediately follows by writing s^{th} equation from (10.87),

$$a_{s1}x_{1r} + a_{s2}x_{2r} + \ \ldots\ + a_{ss}x_{sr} + \ldots + a_{sn}x_{nr} = \lambda_r x_{sr}.$$

Dividing throughout by x_{sr}, transposing a_{ss} on the right side and taking modulus on both sides,

$$|\lambda_r - a_{ss}| \leq |a_{s1}| + |a_{s2}| + \ \ldots\ + |a_{s,\ s-1}| + |a_{s,\ s+1}| + \ldots + |a_{sn}|$$

$$= \sum_{j=1,\ j\neq s}^{n} |a_{sj}| \tag{10.92}$$

10.17.3 Brauer's theorem

If P_s denotes sum of the moduli of the elements along the sth row excluding the diagonal element a_{ss} of matrix A, then every e.value λ of A will lie inside or on the boundary of at least one of the circles $|\lambda - a_{ss}| = P_s$.

For proof of the theorem we have from (10.92),

$$|\lambda_r - a_{ss}| \leq P_s. \tag{10.93}$$

Inequality (10.93) is true for any λ_r, for some value for $s,\ 1 \leq s \leq n$. Thus, we can also write (10.93) as

$$|\lambda - a_{ss}| \leq P_s, \text{ for some } s,\ 1 \leq s \leq n. \tag{10.94}$$

For finding the bounds for λ we get from (10.94),

$$-\mathrm{P}_s \leq \lambda - a_{ss} \leq \mathrm{P}_s \text{ giving}$$

$$a_{ss} - \mathrm{P}_s \leq \lambda \leq a_{ss} + \mathrm{P}_s,\ s = 1, 2, \ldots .\ n. \tag{10.95}$$

The lower bound for λ will be given by min $(a_{ss} - \mathrm{P}_s)$ and upper bound by

$$\max(a_{ss} + \mathrm{P}_{ss}),\ s = 1,\ 2,\ \ldots .\ n.$$

Hence we have $\min(a_{ss} - \mathrm{P}_s) \leq \lambda \leq \max(a_{ss} + \mathrm{P}_s),\ s = 1,\ 2,\ \ldots .\ n.$ (10.96)

Obviously, the interval (10.96) for λ is the union of all intervals defined by (10.95).

Example 10.5

Find the range of eigenvalues matrix A using Gerschgorin's/Brauer's theorem where

$$\mathrm{A} = \begin{bmatrix} 4 & 1 & 0 \\ 1 & 2 & 1 \\ 0 & 1 & 5 \end{bmatrix}$$

Solution

A

$$\begin{bmatrix} 4 & 1 & 0 \\ 1 & 2 & 1 \\ 0 & 1 & 5 \end{bmatrix}$$

s	a_{ss}	P_s	$(a_{ss} - \mathrm{P}_s)$	$(a_{ss} + \mathrm{P}_s)$
1	4	1	3	5
2	2	2	0	4
3	5	1	4	6

$$\min(a_{ss} - \mathrm{P}_s) = 0$$

$$\max(a_{ss} + \mathrm{P}_s) = 6$$

$$\therefore \quad 0 \leq \lambda \leq 6.$$

Note: Matrix A is same as in Example 10.4.

10.18 Determination of Eigenvectors

Let us suppose that in Jacobi's method, symmetric matrix A reduces to a diagonal matrix D after, say r transformations

$$S_r^T \ldots (S_2^T (S_1^T AS_1) S_2) \ldots S_r = D \tag{10.97}$$

where $S_1, S_2 \ldots S_r$ are orthogonal matrices.
The above can also be written as

$$S^T AS = D = S^{-1}AS \tag{10.98}$$

since $S = S_1, S_2 \ldots S_r$ is also an orthogonal matrix, i.e., $S^T = S^{-1}$.

We know that diagonal elements of D are the e.values of A. Comparing (10.98) with (10.19) we see that the columns of S should give e.vectors of A. However in practical computations, matrix D is almost diagonal, not exactly diagonal as its off-diagonal elements are made zero only to certain level of accuracy. Secondly the convergence of the method is generally very slow, hence the number of orthogonal matrices may be quite large. Even the e.vectors and e.values so obtained may not satisfy the original system (10.1) at all.

Further, suppose we try to compute the e.vector directly by solving the system of equations (10.1) or (10.3). Note that only $(n-1)$ equations are needed to solve homogeneous equations in n unknowns as arbitrary value may be assigned to any one of the variables. For example, we may select first $(n-1)$ equations of the system (10.1); assign $x_1 = 1$ and solve them for $x_2, x_3, \ldots x_n$. But when these values are put in the nth equation it may not be satisfied at all. The reason being that the value of λ is known only approximately and there may also be rounding errors in computation of $x_2, x_3, \ldots x_n$. This problem may be circumvented by adopting the procedure outlined in Chapter 2. If $\overline{\lambda}$ is an approximate e.value of A, then following iterative scheme of Power method may be followed:

$$\left.\begin{aligned}(A - \overline{\lambda} I) z_{k+1} &= \boldsymbol{x}_k \\ \boldsymbol{x}_{k+1} &= \frac{1}{\alpha_{k+1}} z_{k+1}, \quad k = 0, 1, 2, \ldots\end{aligned}\right\} \tag{10.99}$$

where α_{k+1} is the numerically largest element of z_{k+1}. See [2].

Starting from an arbitrary vector $\boldsymbol{x}_0$ which is generally taken to be having all 1's as its components, the iterative procedure (10.99) converges to the e.value closest to $\overline{\lambda}$ with $\boldsymbol{x}_k$, the corresponding e.vector as $k \to \infty$. It is easy to see for if $\overline{\lambda}$ is an e.value of A with corresponding e.vector $\boldsymbol{x}$, then,

$$A\boldsymbol{x} = \lambda \boldsymbol{x}$$

or

$$(A - \overline{\lambda} I)\boldsymbol{x} = (\lambda - \overline{\lambda})\boldsymbol{x}$$

or $$(A-\bar{\lambda}I)^{-1}x = \frac{1}{\lambda-\bar{\lambda}}x. \tag{10.100}$$

Equation (10.100) shows that e.value of $(A-\bar{\lambda}I)^{-1}$ is given by $\frac{1}{\lambda-\bar{\lambda}}$ while the e.vector x remains same. Applying power method to find the e.value closest to $\bar{\lambda}$ results in (10.99). For performing iterations, $(A-\bar{\lambda}I)$ in (10.99) may be decomposed to LU form where L is a unit lower triangular and U an upper triangular matrix (or LL^T form since $A-\bar{\lambda}I$ is symmetric, L is lower triangular) (See Chapter 2).

In the Givens and Householder's methods, matrix A is reduced to tridiagonal form T through similarity transformations so that e.values of A and T are same. Let us suppose a total m transformations are required, i.e.,

$$S_m^T \ldots S_2^T S_1^T AS_1S_2 \ldots S_m = T, \tag{10.101}$$

where $S_1, S_2 \ldots$, etc., are orthogonal matrices.

It may be noted that m is equal to $(n-2)(n-1)/2$ in Givens method and $(n-2)$ in Householder's method.

Again, we can write (10.101) as

$$S^TAS = T, \tag{10.102}$$

where $S = S_1, S_2, \ldots S_m$ is also an orthogonal matrix.

Also, if y is an e.vector of T, the e.vector x of A would be

$$x = S^Ty. \tag{10.103}$$

In order to find y we have to solve the system of equations $Ty = \lambda y$ as given below:

$$\left.\begin{array}{l}
\alpha_1y_1+\beta_1y_2 = \lambda y_1\\
\beta_1y_1+\alpha_2y_2+\beta_2y_3 = \lambda y_2\\
\cdots\cdots\cdots\cdots\cdots\cdots\\
\beta_{i-1}y_{i-1}+\alpha_iy_i+\beta_iy_{i+1} = \lambda y_i\\
\cdots\cdots\cdots\cdots\cdots\cdots\\
\beta_{n-1}y_{n-1}+\alpha_ny_n = \lambda y_n
\end{array}\right\} \tag{10.104}$$

The solution of (10.104) can be obtained using first $(n-1)$ equations. Choosing $y_1 = 1$ (say) gives $y_2 = -(\alpha_1-\lambda)/\beta_1$ and

$$y_{i+1} = -\{\beta_{i-1}y_{i-1}+(\alpha_i-\lambda)\}/\beta_i,\ i = 2, 3, \ldots n-1.$$

However, substituting these values in the last equation would not satisfy it due to error in the computation of λ. As explained before we can use iterative method,

$$(\mathrm{T}-\bar{\lambda}\mathrm{I})z_{i+1} = y_i, \; i = 0, 1, 2\ldots.$$

$$y_{i+1} = \frac{1}{c_{i+1}} z_{i+1}, \tag{10.105}$$

where $\bar{\lambda}$ is an approximate e.value and c_{i+1} is the numerically largest element of z_{i+1}.

Having got $\boldsymbol{y}$ we can compute $\boldsymbol{x}$ from (10.103). Alternatively $\boldsymbol{x}$ can be computed from the original matrix A as given by the iterative scheme (10.99).

10.19 LR Method

The LR method was suggested by H. Rutishauser in 1958 to find the e.values of an arbitrary matrix A (not necessarily symmetric). The method consists of two steps:

$$\left.\begin{aligned} \mathrm{A}_k &= \mathrm{L}_k\mathrm{R}_k, \\ \mathrm{A}_{k+1} &= \mathrm{R}_k\mathrm{L}_k \end{aligned}\right\} \tag{10.106}$$

$k = 1, 2, \ldots$. and $\mathrm{A}_1 = \mathrm{A}$; L_k is a unit lower triangular matrix and R_k an upper triangular. For large k, matrix A_{k+1} generally approaches to an upper triangular matrix. Since the determinant of an upper triangular matrix (or a lower triangular matrix) is simply the product of its diagonal elements, its e.values are given by the diagonal elements. Further, it is easy to see that e.values of A_k and A_{k+1} are same, for we can write from (10.106)

$$\mathrm{A}_{k+1} = \mathrm{R}_k\mathrm{A}_k\mathrm{R}_k^{-1} \tag{10.107}$$

which shows that right side of (10.107) is a similarity transformation on A_k. Thus, the e.values of the final upper triangular matrix will be same as those of A. If A is symmetric, then the iterative scheme may be modified as

$$\left.\begin{aligned} \mathrm{A}_k &= \mathrm{R}_k{}^{\mathrm{T}}\mathrm{R}_k \\ \mathrm{A}_{k+1} &= \mathrm{R}_k\mathrm{R}_k{}^{\mathrm{T}} \end{aligned}\right\} \tag{10.108}$$

where R_k is an upper triangular matrix.

If matrix A is symmetric and positive definite then A_k converges to a diagonal matrix as k increases (ref [3]). The decomposition LR or $\mathrm{R}^{\mathrm{T}}\mathrm{R}$(or LL^{T}) may be obtained by the methods discussed in Chapter 2.

10.20 QR Method

This method was developed by J.F.G. Francis in 1961 which is similar to LR in the sense that a matrix is decomposed as a product of two matrices and in the next step a new matrix is formed by multiplying them in reverse order. It may be described as

$$\left.\begin{aligned} A_k &= Q_k R_k \\ A_{k+1} &= R_k Q_k \end{aligned}\right\} \tag{10.109}$$

$k = 1, 2, \ldots.$ and $A_1 = A$. In the kth step, matrix A_k is decomposed as a product of an orthogonal matrix Q_k and an upper triangular matrix R_k. Then in the following step they are multiplied in the reverse order. The process is continued until A_{k+1} converges to an upper triangular matrix. We can see as before that e.values of A_k and A_{k+1} are same and hence the e.values of the final upper triangular matrix will be same as those of the original matrix A.

An arbitrary matrix A can be factorised into QR form by performing following operations

$$Q_r \ldots . \, Q_2^T \, Q_1^T \, A = R \tag{10.110}$$

where $Q_1, \; Q_2 \ldots,$ etc., are orthogonal matrices of Jacobi method. The operations in (10.110) are carried out such that the elements below diagonal are made equal to zero in a cyclic manner column-wise or row-wise. Note that one cycle will take $n(n-1)/2$ operations. This cycle has to be repeated (since the zeros produced do not persist) until all the elements below the diagonal are zero. Therefore number of operations r can not be known beforehand.

From (10.110) we can write

$$A = Q_1 \, Q_2 \, \ldots . \, Q_r R = QR \tag{10.111}$$

where $\quad Q = Q_1 \, Q_2 \ldots Q_r.$

It may be mentioned that reduction of A to QR is more stable in comparison to LR as the elements of L may become too large if pivotal condensation is not used in Gaussian elimination (See Chapter 2). However, the process of reduction of A to upper triangular matrix R through Jacobi rotations is extremely slow. In order to make reduction process (of an arbitrary matrix A to QR) faster we may first convert A to upper Hessenberg matrix through similarity transformations so that its e.values remain same as that of A. In an upper Hessenberg matrix all the elements below subdiagonal are zero. The form of a (5×5) upper Hessenberg matrix, say A, will be as given below by (10.112).

$$
\mathrm{A} = \begin{bmatrix} a_{11} & a_{12} & a_{13} & a_{14} & a_{15} \\ a_{21} & a_{22} & a_{23} & a_{24} & a_{25} \\ 0 & a_{32} & a_{33} & a_{34} & a_{35} \\ 0 & 0 & a_{43} & a_{44} & a_{45} \\ 0 & 0 & 0 & a_{54} & a_{55} \end{bmatrix}. \tag{10.112}
$$

That is, $a_{ij} = 0$ for $i = 3(1)5, j = 1(1)i-2$. In general for matrix of order n, $a_{ij} = 0$, $i = 3(1)n, j = 1(1)i-2$. Similarly a lower Hessenberg matrix is defined as

$$
\mathrm{A} = \begin{bmatrix} a_{11} & a_{12} & 0 & 0 & 0 \\ a_{21} & a_{22} & a_{23} & 0 & 0 \\ a_{31} & a_{32} & a_{33} & a_{34} & 0 \\ a_{41} & a_{42} & a_{43} & a_{44} & a_{45} \\ a_{51} & a_{52} & a_{53} & a_{54} & a_{55} \end{bmatrix}. \tag{10.113}
$$

In a lower Hessenberg matrix A of order n, $a_{ij} = 0$ for $i = 1(1)n-2, j = (i+2)(1)n$. When A is symmetric, a Hessenberg matrix is tridiagonal.

An arbitrary matrix can be transformed to a similar Hessenberg matrix by performing elementary row and column transformations. For example, consider a 5×5 arbitrary matrix A given by

$$
\mathrm{A} = \begin{bmatrix} a_{11} & a_{12} & a_{13} & a_{14} & a_{15} \\ a_{21} & a_{22} & a_{23} & a_{24} & a_{25} \\ a_{31} & a_{32} & a_{33} & a_{34} & a_{35} \\ a_{41} & a_{42} & a_{43} & a_{44} & a_{45} \\ a_{51} & a_{52} & a_{53} & a_{54} & a_{55} \end{bmatrix}. \tag{10.114}
$$

Choose a lower triangular matrix.

$$
\mathrm{L}_1 = \begin{bmatrix} 1 & 0 & 0 & 0 & 0 \\ 0 & 1 & 0 & 0 & 0 \\ 0 & l_{32} & 1 & 0 & 0 \\ 0 & l_{42} & 0 & 1 & 0 \\ 0 & l_{52} & 0 & 0 & 1 \end{bmatrix} \tag{10.115}
$$

where $l_{32} = -\frac{a_{31}}{a_{21}},\ l_{42} = -\frac{a_{41}}{a_{21}},\ l_{52} = -\frac{a_{51}}{a_{21}}.$ (10.116)

The inverse of matrix L_1 is given by

$$L_1^{-1} = \begin{bmatrix} 1 & 0 & 0 & 0 & 0 \\ 0 & 1 & 0 & 0 & 0 \\ 0 & -l_{32} & 1 & 0 & 0 \\ 0 & -l_{42} & 0 & 1 & 0 \\ 0 & -l_{52} & 0 & 0 & 1 \end{bmatrix} \tag{10.117}$$

Following transformation is performed

$$A_1 = L_1 A L_1^{-1} \tag{10.118}$$

so that matrix A_1 is similar to A. Postmultiplication by L_1^{-1} does not change the elements of first column of L_1A.

Matrix A_1 will have the form

$$A_1 = \begin{bmatrix} a_{11} & a_{12} & a_{13} & a_{14} & a_{15} \\ a_{21} & a_{22} & a_{23} & a_{24} & a_{25} \\ 0 & a_{32}^{(1)} & a_{33}^{(1)} & a_{34}^{(1)} & a_{35}^{(1)} \\ 0 & a_{42}^{(1)} & a_{43}^{(1)} & a_{44}^{(1)} & a_{45}^{(1)} \\ 0 & a_{52}^{(1)} & a_{53}^{(1)} & a_{54}^{(1)} & a_{55}^{(1)} \end{bmatrix}. \tag{10.119}$$

Next, choose L_2 as

$$L_2 = \begin{bmatrix} 1 & 0 & 0 & 0 & 0 \\ 0 & 1 & 0 & 0 & 0 \\ 0 & 0 & 1 & 0 & 0 \\ 0 & 0 & l_{43} & 1 & 0 \\ 0 & 0 & l_{53} & 0 & 1 \end{bmatrix}, \tag{10.120}$$

where $l_{43} = -a_{42}^{(1)}/a_{32}^{(1)},\ l_{53} = -a_{52}^{(1)}/a_{32}^{(1)}.$ (10.121)

After performing $A_2 = L_2A_1L_2^{-1}$ we get

$$A_2 = \begin{bmatrix} a_{11} & a_{12} & a_{13} & a_{14} & a_{15} \\ a_{21} & a_{22} & a_{23} & a_{24} & a_{25} \\ 0 & a_{32}^{(1)} & a_{33}^{(1)} & a_{34}^{(1)} & a_{35}^{(1)} \\ 0 & 0 & a_{43}^{(2)} & a_{44}^{(2)} & a_{45}^{(2)} \\ 0 & 0 & a_{53}^{(2)} & a_{54}^{(2)} & a_{55}^{(2)} \end{bmatrix}. \qquad (10.122)$$

Finally choose L_3 as

$$L_3 = \begin{bmatrix} 1 & 0 & 0 & 0 & 0 \\ 0 & 1 & 0 & 0 & 0 \\ 0 & 0 & 1 & 0 & 0 \\ 0 & 0 & 0 & 1 & 0 \\ 0 & 0 & 0 & l_{54} & 1 \end{bmatrix} \qquad (10.123)$$

where $l_{54} = -a_{53}^{(2)}/a_{43}^{(2)}$.

Performing $A_3 = L_3A_2L_3^{-1}$ gives

$$A_3 = \begin{bmatrix} a_{11} & a_{12} & a_{13} & a_{14} & a_{15} \\ a_{21} & a_{22} & a_{23} & a_{24} & a_{25} \\ 0 & a_{32}^{(1)} & a_{33}^{(1)} & a_{34}^{(1)} & a_{35}^{(1)} \\ 0 & 0 & a_{43}^{(2)} & a_{44}^{(2)} & a_{45}^{(2)} \\ 0 & 0 & 0 & a_{54}^{(3)} & a_{55}^{(3)} \end{bmatrix}. \qquad (10.124)$$

Matrix A_3 is the required Hessenberg matrix which has the same e.values as A. Now Jacobi rotations can be made on the upper Hessenberg matrix to reduce it to upper triangular. Alternatively, Householder's method can also be used to transform matrix A to upper Hessenberg form as discussed for symmetric matrix in Sec 10.14.

Let us suppose a non-symmetric (or Hessenberg) matrix B is to be reduced to an upper triangular matrix through r Jacobi rotations, i.e.,

$$S_r^T \ldots . S_2^T S_1^T B = R. \qquad (10.125)$$

We may perform the above transformations successively as

$$\mathrm{B}_m = \mathrm{S}_m^{\mathrm{T}}\,\mathrm{B}_{m-1},\ m = 1(1)r;\ \mathrm{B}_0 = \mathrm{B},\ \mathrm{B}_r = \mathrm{R}. \quad (10.126)$$

For computing elements of B_m, let us refer back to equations (10.32). Making $(q,\ p)$ element $(q > p)$ zero gives

$$\tan\theta \ = -b_{qp}^{(m-1)}/b_{pp}^{(m-1)}$$

which determines sign of $\theta; c = 1$ if $\tan\theta \geq 0$ and -1 if $\tan\theta < 0$. (10.127)

Compute $k = \sqrt{b_{qp}^{(m-1)2} + b_{pp}^{(m-1)2}}$ giving $\sin\theta = \dfrac{c|b_{qp}^{(m-1)}|}{k}$; $\cos\theta = \dfrac{|b_{pp}^{(m-1)}|}{k}$. (10.128)

The elements of B_m are given by

$$\left.\begin{aligned}
b_{ij}^{(m)} &= (\mathrm{S}_m{}^{\mathrm{T}}\,\mathrm{B}_{m-1})_{i,j} = b_{ij}^{(m-1)},\ i \neq p \text{ or } q,\\
b_{pj}^{(m)} &= (\mathrm{S}_m{}^{\mathrm{T}}\,\mathrm{B}_{m-1})_{pj} = b_{pj}^{(m-1)}\cos\theta - b_{qj}^{(m-1)}\sin\theta,\\
b_{qj}^{(m)} &= (\mathrm{S}_m{}^{\mathrm{T}}\,\mathrm{B}_{m-1})_{qj} = b_{pj}^{(m-1)}\sin\theta + b_{qj}^{(m-1)}\cos\theta,\ j \neq p,\\
b_{qp}^{(m)} &= 0.
\end{aligned}\right\} \quad (10.129)$$

Since $\mathrm{S}_1,\ \mathrm{S}_2,\ \ldots.\ \mathrm{S}_r$ are orthogonal matrices the reduction of A to QR may be obtained where

$$\mathrm{Q} = \mathrm{S}_1\ \mathrm{S}_2\ \ldots.\ \mathrm{S}_r. \quad (10.130)$$

Example 10.6

Find the e.values of the following matrix A using LR method,

$$\mathrm{A} = \begin{bmatrix} 3 & 2 \\ 1 & 4 \end{bmatrix}.$$

Compute up to four places of decimal and get the answer correct up to two decimals only.

Solution $\mathrm{A}_1 = \mathrm{A} = \begin{bmatrix} 3 & 2 \\ 1 & 4 \end{bmatrix} = \mathrm{L}_1\mathrm{R}_1;\ \mathrm{L}_1 = \begin{bmatrix} 1 & 0 \\ 0.3333 & 1 \end{bmatrix},\ \mathrm{R}_1 = \begin{bmatrix} 3 & 2 \\ 0 & 3.3333 \end{bmatrix}$

$$A_2 = R_1L_1 = \begin{bmatrix} 3 & 2 \\ 0 & 3.3333 \end{bmatrix}\begin{bmatrix} 1 & 0 \\ 0.3333 & 1 \end{bmatrix} = \begin{bmatrix} 3.6666 & 2 \\ 1.1101 & 3.3333 \end{bmatrix}$$

$$= L_2R_2;\;\; L_2 = \begin{bmatrix} 1 & 0 \\ 0.3028 & 1 \end{bmatrix},\; R_2 = \begin{bmatrix} 3.6666 & 2 \\ 0 & 2.7277 \end{bmatrix}$$

$$A_3 = R_2L_2 = \begin{bmatrix} 3.6666 & 2 \\ 0 & 2.7277 \end{bmatrix}\begin{bmatrix} 1 & 0 \\ 0.3028 & 1 \end{bmatrix} = \begin{bmatrix} 4.2722 & 2 \\ 0.8259 & 2.7277 \end{bmatrix}$$

$$= L_3R_3;\;\; L_3 = \begin{bmatrix} 1 & 0 \\ 0.1933 & 1 \end{bmatrix},\; R_3 = \begin{bmatrix} 4.2722 & 2 \\ 0 & 2.3411 \end{bmatrix}$$

$$A_4 = R_3L_3 = \begin{bmatrix} 4.2722 & 2 \\ 0 & 2.3411 \end{bmatrix}\begin{bmatrix} 1 & 0 \\ 0.1933 & 1 \end{bmatrix} = \begin{bmatrix} 4.6588 & 2 \\ 0.4525 & 2.3411 \end{bmatrix}$$

$$= L_4R_4;\;\; L_4 = \begin{bmatrix} 1 & 0 \\ 0.0971 & 1 \end{bmatrix},\; R_4 = \begin{bmatrix} 4.6588 & 2 \\ 0 & 2.1469 \end{bmatrix}$$

$$A_5 = R_4L_4 = \begin{bmatrix} 4.6588 & 2 \\ 0 & 2.1469 \end{bmatrix}\begin{bmatrix} 1 & 0 \\ 0.0971 & 1 \end{bmatrix} = \begin{bmatrix} 4.8530 & 2 \\ 0.2085 & 2.1469 \end{bmatrix}$$

$$= L_5R_5;\;\; L_5 = \begin{bmatrix} 1 & 0 \\ 0.0430 & 1 \end{bmatrix},\; R_5 = \begin{bmatrix} 4.8530 & 2 \\ 0 & 2.0609 \end{bmatrix}$$

$$A_6 = R_5L_5 = \begin{bmatrix} 4.8530 & 2 \\ 0 & 2.0609 \end{bmatrix}\begin{bmatrix} 1 & 0 \\ 0.0430 & 1 \end{bmatrix} = \begin{bmatrix} 4.9390 & 2 \\ 0.0886 & 2.0609 \end{bmatrix}$$

$$= L_6R_6;\;\; L_6 = \begin{bmatrix} 1 & 0 \\ 0.0179 & 1 \end{bmatrix},\; R_6 = \begin{bmatrix} 4.9390 & 2 \\ 0 & 2.0251 \end{bmatrix}.$$

$$A_7 = R_6L_6 = \begin{bmatrix} 4.9390 & 2 \\ 0 & 2.0251 \end{bmatrix}\begin{bmatrix} 1 & 0 \\ 0.0179 & 1 \end{bmatrix} = \begin{bmatrix} 4.9748 & 2 \\ 0.0362 & 2.0251 \end{bmatrix}$$

$$= L_7R_7;\ L_7 = \begin{bmatrix} 1 & 0 \\ 0.0073 & 1 \end{bmatrix},\ R_7 = \begin{bmatrix} 4.9748 & 2 \\ 0 & 2.0105 \end{bmatrix}$$

$$A_8 = R_7L_7 = \begin{bmatrix} 4.9748 & 2 \\ 0 & 2.0105 \end{bmatrix} \begin{bmatrix} 1 & 0 \\ 0.0073 & 1 \end{bmatrix} = \begin{bmatrix} 4.9894 & 2 \\ 0.0147 & 2.0105 \end{bmatrix}$$

$$= L_8R_8;\ L_8 = \begin{bmatrix} 1 & 0 \\ 0.0029 & 1 \end{bmatrix},\ R_8 = \begin{bmatrix} 4.9894 & 2 \\ 0 & 2.0047 \end{bmatrix}$$

$$A_9 = R_8L_8 = \begin{bmatrix} 4.9894 & 2 \\ 0 & 2.0047 \end{bmatrix} \begin{bmatrix} 1 & 0 \\ 0.0029 & 1 \end{bmatrix} = \begin{bmatrix} 4.9952 & 2 \\ 0.0058 & 2.0047 \end{bmatrix}$$

$$= L_9R_9;\ L_9 = \begin{bmatrix} 1 & 0 \\ 0.0012 & 1 \end{bmatrix},\ R_9 = \begin{bmatrix} 4.9952 & 2 \\ 0 & 2.0023 \end{bmatrix}$$

$$A_{10} = R_9L_9 = \begin{bmatrix} 4.9952 & 2 \\ 0 & 2.0023 \end{bmatrix} \begin{bmatrix} 1 & 0 \\ 0.0012 & 1 \end{bmatrix} = \begin{bmatrix} 4.9976 & 2 \\ 0.0024 & 2.0023 \end{bmatrix}$$

Comparing diagonal elements of A_9, and A_{10}; we find that they agree up to two places of decimal giving $\lambda_1 = 5.00,\ \lambda_2 = 2.00$. The difference between the corresponding diagonal elements of A_9 and A_{10} is less than 0.005.

Note: The actual e.values are $\lambda_1 = 5,\ \lambda_2 = 2$.

Example 10.7

Use QR method to find the e.values of matrix A correct to one place of decimal only where

$$A = \begin{bmatrix} 3 & 2 \\ 1 & 4 \end{bmatrix}.$$

(Matrix A is same as in Example 10.6).

Solution Element (2, 1) to be annihilated; $q = 2,\ p = 1$. Formulae used for $B = S_1^T A$.

$$\tan\theta = -\frac{a_{21}}{a_{11}} \text{ which gives sign of } \theta$$

$$b_{11} = a_{11}\cos\theta - a_{21}\sin\theta$$

$$b_{12} = a_{12}\cos\theta - a_{22}\sin\theta$$

$$b_{21} = 0$$

$$b_{22} = a_{12}\sin\theta + a_{22}\cos\theta$$

$$\tan\theta = -\frac{1}{3};\ c = -1;\ k = \sqrt{1^2+3^2} = \sqrt{10} = 3.1623$$

$$\sin\theta = -\frac{1}{\sqrt{10}};\ = -0.3162,\ \cos\theta = \frac{3}{\sqrt{10}} = 0.9487.$$

$$b_{11} = 3\times 0.9487 - 1\times(-0.3162) = 3.1623$$

$$b_{12} = 2\times 0.9487 - 4\times(-0.3162) = 3.1622$$

$$b_{22} = 2\times(-0.3162) + 4(0.9487) = 3.1624$$

$$\text{Check: } A_1 = Q_1R_1 = \begin{bmatrix} 0.9487 & -0.3162 \\ 0.3162 & 0.9487 \end{bmatrix}\begin{bmatrix} 3.1623 & 3.1622 \\ 0 & 3.1624 \end{bmatrix} = \begin{bmatrix} 3 & 2 \\ 0.9999 & 4 \end{bmatrix} \simeq A$$

$$A_2 = R_1Q_1 = \begin{bmatrix} 3.1623 & 3.1622 \\ 0 & 3.1624 \end{bmatrix}\begin{bmatrix} 0.9487 & -0.3162 \\ 0.3162 & 0.9487 \end{bmatrix} = \begin{bmatrix} 4.0000 & 2.0000 \\ 1.0000 & 3.0000 \end{bmatrix}$$

$$\tan\theta = -\frac{1}{4};\ \sin\theta = -\frac{1}{\sqrt{17}};\ = -0.2425,\ \cos\theta = \frac{4}{\sqrt{17}} = 0.97.$$

$$b_{11} = 4\times 0.97 - 1\times(-0.2425) = 4.1225$$

$$b_{12} = 2\times 0.97 - 3\times(-0.2425) = 2.6675$$

$$b_{22} = 2\times(-0.2425) + 3\times 0.97 = 2.4250$$

$$A_3 = R_2Q_2 = \begin{bmatrix} 4.1225 & 2.6675 \\ 0 & 2.425 \end{bmatrix}\begin{bmatrix} 0.97 & -0.2425 \\ 0.2425 & 0.97 \end{bmatrix} = \begin{bmatrix} 4.6457 & 1.5878 \\ 0.5881 & 2.3522 \end{bmatrix} = Q_3R_3$$

$$\tan\theta = -\frac{0.5881}{4.6457} = -0.1256;\ k = \sqrt{0.5881^2 + 4.6457^2} = 4.6828$$

$$\sin\theta = -\frac{0.5881}{4.6828} = -0.1256;\ \cos\theta = \frac{4.6457}{4.6828} = 0.9921$$

$$b_{11} = 4.6457\times(0.9921) - 0.5881\times(-0.1256) = 4.6829$$

$$b_{12} = 1.5878\times(0.9921) - 2.3522\times(-0.1256) = 1.8707$$

$$b_{22} = 1.5878(-0.1256) + 2.3522 \times (0.9921) = 2.1342$$

$$A_4 = R_3Q_3 = \begin{bmatrix} 4.6829 & 1.8707 \\ 0 & 2.1342 \end{bmatrix} \begin{bmatrix} 0.9921 & -0.1256 \\ 0.1256 & 0.9921 \end{bmatrix} = \begin{bmatrix} 4.8808 & 1.2677 \\ 0.2681 & 2.1173 \end{bmatrix} = Q_4R_4$$

$$\tan\theta = -\frac{0.2681}{4.8808};\ k = \sqrt{0.2681^2 + 4.8808^2} = 4.8882$$

$$\sin\theta = -\frac{0.2681}{4.8882} = -0.0548,\ \cos\theta = \frac{4.8808}{4.8882} = 0.9985.$$

$$b_{11} = 4.8808 \times 0.9985 - 0.2681 \times (-0.0548) = 4.8882$$

$$b_{12} = 1.2677 \times (0.9985) - 2.1173 \times (-0.0548) = 1.3814$$

$$b_{22} = 1.2677 \times (-0.0548) + 2.1173 \times 0.9985 = 2.0446.$$

$$A_5 = R_4Q_4 = \begin{bmatrix} 4.8882 & 1.3814 \\ 0 & 2.0446 \end{bmatrix} \begin{bmatrix} 0.9985 & -0.0548 \\ 0.0548 & 0.9985 \end{bmatrix} = \begin{bmatrix} 4.9566 & 1.0938 \\ 0.1120 & 2.0415 \end{bmatrix} = Q_5R_5$$

$$\tan\theta = -\frac{0.1120}{4.9566};\ k = \sqrt{0.1120^2 + 4.9566^2} = 4.9579$$

$$\sin\theta = -\frac{0.1120}{4.9579} = -0.0226,\ \cos\theta = \frac{4.9566}{4.9579} = 0.9997$$

$$b_{11} = 4.9566 \times 0.9997 - 0.1120 \times (-0.0226) = 4.9576$$

$$b_{12} = 1.0938 \times 0.9997 - 2.0415 \times (-0.0226) = 1.1396$$

$$b_{22} = 1.0938 \times (-0.0226) + 2.0415 \times 0.9997 = 2.0162$$

$$A_6 = \begin{bmatrix} 4.9576 & 1.1396 \\ 0 & 2.0162 \end{bmatrix} \begin{bmatrix} 0.9997 & -0.0226 \\ 0.0226 & 0.9997 \end{bmatrix} = \begin{bmatrix} 4.9819 & 1.0272 \\ 0.0456 & 2.0156 \end{bmatrix}$$

Difference between diagonal elements of A_5 and A_6.

$$|4.9819 - 4.9566| = 0.0253$$

$$|2.0156 - 2.0415| = 0.0259$$

Since the error is less than .05, we can say that for an accuracy of one place of decimal

$$\lambda_1 = 5.0,\ \lambda_2 = 2.0$$

Exercise 10

10.1 Perform only three rotations of classical Jacobi method on the matrix

$$A = \begin{bmatrix} 4 & 1 & 2 \\ 1 & 2 & 0 \\ 2 & 0 & 6 \end{bmatrix}.$$

Compute up to four places of decimal.

10.2 Use Jacobi method to compute e.values and e.vectors of the matrix

$$A = \begin{bmatrix} 4 & \sqrt{2} \\ \sqrt{2} & 3 \end{bmatrix}$$

Compute up to four decimal places.

10.3 Reduce the following matrix to tridiagonal form by Givens method

$$A = \begin{bmatrix} 5 & 0 & 3 \\ 0 & 2 & 1 \\ 3 & 1 & 4 \end{bmatrix}.$$

Compute up to four decimal places.

10.4 Reduce the matrix A of Exercise 10.3 to tridiagonal form by Householder's method.

10.5 Form Sturm sequence of functions of λ where λ is an e.value of the tridiagonal matrix

$$T = \begin{bmatrix} 5 & 3 & 0 \\ 3 & 4 & 1 \\ 0 & 1 & 2 \end{bmatrix}.$$

Show by Gerschgorin's theorem that all e.values lie between 0 and 8.
Determine sign changes in the sequence for $\lambda = 0\ (1)\ 8$. Hence mention the intervals separately for each e.value.

10.6 The LR method is defined by succession of two steps $A_k = L_k R_k$, $A_{k+1} = R_k L_k$, where L_k is a unit lower triangular matrix and R_k an upper triangular. Perform these steps twice on the following matrix

$$A = \begin{bmatrix} 8 & 4 & 4 \\ 0 & 4 & 2 \\ 2 & 2 & 5 \end{bmatrix} = A_1$$

Compute up to four decimal places only.

10.7 Find the e.values of the following matrix using LR method correct to one place of decimal only.

$$A = \begin{bmatrix} 4 & 3 \\ 1 & 2 \end{bmatrix}.$$

Stop the process when the difference between the corresponding diagonal elements in the successive transformation is less than or equal to 0.05.

References and Some Useful Related Books/Papers

1. Gerald, C.F., and Wheatley P.O., *Applied Numerical Analysis*, Pearson Education Asia.
2. Goodwin, E.T., *Modern Computing Methods*, HMSO, London.
3. Schwarz, H.R., Rutishauser, H., and Stiefel, E. (Translated by Hertlendy P.), *Numerical Analysis of Symmetric Matrices*, Prentice-Hall, Inc., N.J.
4. Wilkinson J.H., *The Algebraic Eigenvalue Problem*, Clarendon Press, Oxford.

11

Partial Differential Equations

11.1 Introduction

Many physical phenomena can be represented by a second order partial differential equation

$$a\frac{\partial^2 u}{\partial x^2}+b\frac{\partial^2 u}{\partial x\partial y}+c\frac{\partial^2 u}{\partial y^2}+e=0 \tag{11.1}$$

where $a,\ b,\ c$ and e may be functions of $x,\ y,\ u,\ u_x$ and u_y. If equation (11.1) is linear in u and its partial derivatives, it is called linear otherwise nonlinear. The p.d.e. (11.1) may be classified, depending on the sign of b^2-4ac as

$$\left.\begin{array}{lll}(i) & \text{Parabolic,} & \text{if}\quad b^2-4ac=0\\ (ii) & \text{Elliptic,} & \text{if}\quad b^2-4ac<0\\ (iii) & \text{Hyperbolic,} & \text{if}\quad b^2-4ac>0.\end{array}\right\} \tag{11.2}$$

The above nomenclature has a correspondence in coordinate geometry where a general second degree equation

$$ax^2+hxy+by^2+gx+fy+c=0 \tag{11.3}$$

represents a parabola, ellipse or hyperbola according as h^2-4ab being zero, negative or positive respectively. In fact a term involving h^2-4ab under radical sign determines the existence (or non-existence) of asymptotes. Similarly we will see in Sec. 11.7.5 that the parabolic equation has one characteristic, hyperbolic equation has two characteristics and an elliptic equation has none.

As stated above $a,\ b,\ c$ may be function of $x,\ y,\ u,\ u_x$ and u_y in general, therefore sign of b^2-4ac may be dependent on them. Hence the same equation may be parabolic in one domain, elliptic in other domain and hyperbolic in yet another domain. For example, consider a p.d.e.

$$(x+1)\frac{\partial^2 u}{\partial x^2}+2(x+y+1)\frac{\partial^2 u}{\partial x\partial y}+2(y+1)\frac{\partial^2 u}{\partial y^2}=0 \tag{11.4}$$

for which $b^2-4ac=4(x^2+y^2-1)$.

It is obvious that equation (11.4) is elliptic inside the circle $x^2+y^2=1$, i.e., for $x^2+y^2<1$, parabolic on the circle $x^2+y^2=1$ and hyperbolic outside the circle, i.e., for $x^2+y^2>1$.

11.2 Some Standard Forms

There are three most familiar forms of the linear second order p.d.e.'s. They are given below:

A. Parabolic Equation

A parabolic equation in one space dimension may be written as

$$\frac{\partial u}{\partial t}=k\frac{\partial^2 u}{\partial x^2}, \tag{11.5}$$

where k is a constant. The equation (11.5) represents conduction of heat in the x-direction with u denoting temperature at a point x in a homogeneous medium at time t. Besides heat conduction, eq. (11.5) also represents several other physical processes, like diffusion of gas, fluid flow, etc.

A parabolic equation in two space dimensions can be written analogously as

$$\frac{\partial u}{\partial t}=k\left(\frac{\partial^2 u}{\partial x^2}+\frac{\partial^2 u}{\partial y^2}\right)=k\nabla^2 u \tag{11.6}$$

where $\nabla^2 u=\nabla\cdot\nabla u$ and $\nabla\equiv i\frac{\partial}{\partial x}+j\frac{\partial}{\partial y}$. The operator ∇^2 is known as Laplacian operator or simply Laplacian. In case of heat conduction (diffusion) the parameter k is called coefficient of heat conduction (diffusion) and is equal to $k=\mathrm{K}/\rho c$ where K is conductivity, ρ the density and c is the specific heat of the medium.

As parabolic equations are time-dependent, they are known as 'transient problems'.

B. Elliptic Equation

With time increasing, all transient problems tend to reach steady state, i.e., when there is no more change in the value of u in spite of increase in time, which mathematically means $\frac{\partial u}{\partial t} = 0$. The elliptic equations describe steady state processes and can be represented as

$$\nabla^2 u \equiv \frac{\partial^2 u}{\partial x^2} + \frac{\partial^2 u}{\partial y^2} = 0, \text{ (Laplace equation)}, \tag{11.7}$$

$$\nabla^2 u \equiv \frac{\partial^2 u}{\partial x^2} + \frac{\partial^2 u}{\partial y^2} = f(x, y), \text{ (Poisson equation)}. \tag{11.8}$$

Since u may also represent voltage in a conductor or velocity potential of fluids or gravitational potential in space, the elliptic equations are generally referred to as potential problems.

C. Hyperbolic Equation

The most common example of a hyperbolic equation in one-space dimension is the wave equation,

$$\frac{\partial^2 u}{\partial t^2} = c^2 \frac{\partial^2 u}{\partial x^2}, \tag{11.9}$$

where c is constant.

Equation (11.9) represents the motion of a vibrating string stretched between two points where u denotes the displacement of a point on the string at a distance x, at any instant t while the string vibrates in the u-x plane. The string is assumed to be uniform and elastic and that $c^2 = \text{T}/m$, where T is the tension in the string and m is its mass per unit length. The equation may also represent the displacement of a longitudinally vibrating bar or of soundwaves in a pipe. In two-space dimension it may represent deflection of a membrane.

11.3 Boundary Conditions

A problem described by a p.d.e. is said to be well-posed if sufficient number of conditions are prescribed so that its solution can be determined uniquely. Otherwise the problem is said to be ill-defined or not well-posed. These conditions may be prescribed either on the boundary/surface, called boundary conditions or on time, called initial conditions. Broadly speaking both kinds of these conditions may be referred to as boundary conditions. The boundary conditions may be of the following types:

(*i*) Value of the function u is prescribed on the boundary. Such a boundary condition is known as Dirichlet condition.

(*ii*) Value of normal derivative (flux/flow rate) $\frac{\partial u}{\partial n}$ is prescribed on the boundary where n is the direction of the outward normal to the surface (in 2 or 3D) or in the direction of x increasing (in 1D). This condition is called Neumann condition.

(*iii*) The third type of condition may be a combination of u and $\frac{\partial u}{\partial n}$, given as

$$\alpha\frac{\partial u}{\partial n}+\beta u=\gamma,$$

where α, β and γ are constants. It is called mixed type condition.

It may be noted that since one-dimensional parabolic equation (11.5) contains first order derivative in t and second order in x, there should be prescribed one condition on t (initial) and two conditions on x (boundary) for the problem to be well-posed. Similarly, in case of hyperbolic equation (11.9) we need two initial conditions and two boundary conditions for a unique solution. The two conditions on t together are called 'Cauchy' conditions.

11.4 Finite Difference Approximations for Derivatives

Let us recollect that if $f(x)$ is a function of single variable x, then finite difference approximations for its first and second derivatives by forward difference (FD), backward difference (BD) and central difference (CD) are as given below:

$$f'(x)=\begin{cases}\dfrac{f(x+h)-f(x)}{h}+\mathrm{O}(h) & :\ \text{FD} \qquad (11.10a)\\[2ex] \dfrac{f(x)-f(x-h)}{h}+\mathrm{O}(h) & :\ \text{BD} \qquad (11.10b)\\[2ex] \dfrac{f(x+\frac{1}{2}h)-f(x-\frac{1}{2}h)}{h}+\mathrm{O}(h^2) & :\ \text{CD} \qquad (11.10c)\end{cases}$$

$$f''(x)=\begin{cases}\dfrac{f(x)-2f(x+h)+f(x+2h)}{h^2}+\mathrm{O}(h) & :\ \text{FD} \qquad (11.11a)\\[2ex] \dfrac{f(x)-2f(x-h)+f(x-2h)}{h^2}+\mathrm{O}(h) & :\ \text{BD} \qquad (11.11b)\\[2ex] \dfrac{f(x-h)-2f(x)+f(x+h)}{h^2}+\mathrm{O}(h^2) & :\ \text{CD} \qquad (11.11c)\end{cases}$$

$\mathrm{O}(h^r)$ means terms of h^r and higher powers, $r=1,\ 2\ldots$

Also, if the function values are given at unequal intervals, say at $x-h_2$, x and $x+h_1$, then central difference approximation for second derivative may be written as [See Chapter 5],

$$f''(x) = \frac{2}{h_1 h_2(h_1+h_2)}[h_1 f(x-h_2) - (h_1+h_2)f(x) + h_2 f(x+h_1)]$$

$$-\frac{1}{6}(h_1-h_2)f'''(x) - \frac{1}{6}\frac{h_1^3+h_2^3}{h_1+h_2}f^{iv}(\xi), \ x-h_2 \le \xi \le x+h_1. \tag{11.12}$$

We shall now discuss methods for solving p.d.e.'s.

11.5 Methods for Solving Parabolic Equations

We will discuss a few finite difference methods, their merits and demerits, for solving one-dimensional parabolic equation (11.5). Generally the equation is divided by the parameter k throughout which is absorbed in t. Hence the resulting equation is written in the normalised form as,

$$\frac{\partial u}{\partial t} = \frac{\partial^2 u}{\partial x^2}, \quad 0 \le x \le \text{L}, \ t > 0. \tag{11.13}$$

As can be seen, equation (11.13) is defined in the space domain $0 \le x \le \text{L}$. This domain can also be normalised varying from 0 to 1 by change of variable, if required. Let us suppose Dirichlet conditions are prescribed at both the ends $x = 0$ and $x = \text{L}$, (i.e., values of u are given) and an initial condition is prescribed at time $t = 0$ as given below:

$$u(0, t) = u_0, \qquad t > 0 \tag{11.13a}$$

$$u(\text{L}, t) = u_\text{L}, \qquad t > 0 \tag{11.13b}$$

$$u(x, 0) = f(x), \ 0 \le x \le \text{L}, \qquad t = 0. \tag{11.13c}$$

As must be clear, the domain of integration of the p.d.e. (11.13) or its solution domain is $\text{D} = [0 \le x \le \text{L}] \times [t \ge 0]$, as shown in Fig. 11.1.

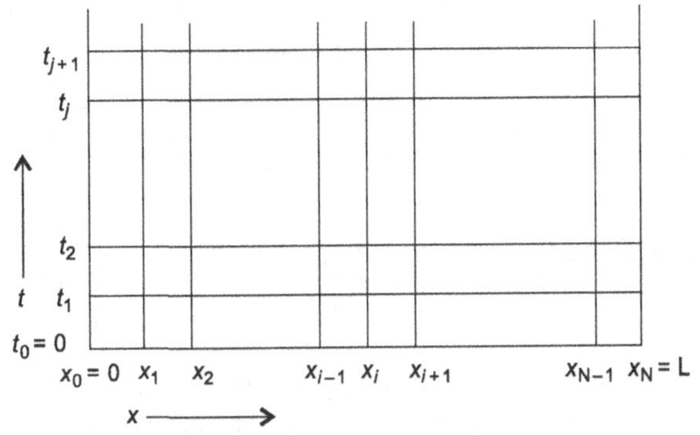

Figure 11.1 Subdivision of domain into rectangular mesh.

Let us consider $x - t$ plane (See Fig. 11.1) such that x is represented by horizontal axis and time t by vertical axis. A rectangular mesh (or grid) is formed in domain D by drawing lines parallel to the axes. We subdivide the interval $0 \leq x \leq \mathrm{L}$ into, say N subintervals each of width Δx such that $\mathrm{N}\Delta x = \mathrm{L}$ and mark the points on the x-axis as $x_i,\ i = 0(1)\ \mathrm{N}$ where $x_0 = 0,\ x_\mathrm{N} = \mathrm{L}$ and $x_i = i\Delta x$. We draw lines through these points parallel to t-axis and also draw lines parallel to x-axis at distances $t = \Delta t = t_1,\ 2\Delta t = t_2,\ j\Delta t = t_j$, etc. In this way the domain D is subdivided into rectangular meshes. The points of intersection of these lines are called mesh points, grid points and pivotal or nodal points. We find the solution at these mesh points in a step by step manner in t-direction. That is, if $u_{i,j}$ denotes the value of u at the mesh point $(i,\ j)$, then to start with we compute at $t = t_1 = \Delta t$ the values of $u_{i1},\ i = 1(1)\ \mathrm{N}{-}1$. Once the values are known at $t = t_1$, the process may be repeated to get the values at $t = t_2 = 2\Delta t$, i.e., $u_{i2},\ i = 1(1)\ \mathrm{N}{-}1$. In general when the values of $u_{i,j}$ have been computed up to $t = t_j = j\Delta t$, i.e., j^{th} time level, the values at the next time level $(j+1)^{\text{th}}$, are computed to give $u_{i,j+1},\ i = 1(1)\ \mathrm{N}{-}1$. It may be noted that the values of u at the boundaries $x_0 = 0$ and, $x_\mathrm{N} = \mathrm{L}$ are known as u_0 and u_L, by virtue of prescribed boundary conditions (11.13a) and (11.13b) respectively and values of u_{i0} on account of initial condition (11.13c) as $u_{i0} = f(x_i),\ i = 0(1)\ \mathrm{N},\ t = 0$.

11.5.1 Explicit method/scheme/formula

We discretize (approximate by finite differences) equation (11.13) at the point $(x_i,\ t_j)$ such that the time derivative $\dfrac{\partial u}{\partial t}$ is replaced by forward difference (FD) and space derivative $\dfrac{\partial^2 u}{\partial x^2}$ by central difference (CD). In operator form we can write,

$$\frac{1}{\Delta t}\Delta_t u(x_i,\ t_j) + \mathrm{O}(\Delta \mathrm{t}) = \frac{1}{(\Delta x)^2}\delta_x^2 u(x_i,\ t_j) + \mathrm{O}(\Delta x^2) \tag{11.14}$$

where Δ_t and δ_x^2 denote FD and CD in the directions of t and x respectively. The formulation (11.14) can be written in terms of u values as,

$$\frac{u(x_i, t_j + \Delta t) - u(x_i, t_j)}{\Delta t} = \frac{u(x_i - \Delta x, t_j) - 2u(x_i, t_j) + u(x_i + \Delta x, t_j)}{(\Delta x)^2} + \mathrm{O}(\Delta t) + \mathrm{O}(\Delta x^2)$$

or
$$\frac{u(x_i, t_{j+1}) - u(x_i, t_j)}{\Delta t} = \frac{u(x_{i-1}, t_j) - 2u(x_i, t_j) + u(x_{i+1}, t_j)}{(\Delta x)^2} + \mathrm{O}(\Delta t) + \mathrm{O}(\Delta x^2) \tag{11.15}$$

The error $\mathrm{O}(\Delta t) + \mathrm{O}(\Delta x^2)$ is called truncation error in the discretization of (11.13) by Explicit Scheme. Denoting the computed value at the mesh point $(i,\ j)$, i.e., at $(i\Delta x, j\Delta t)$ by $u_{i,j}$ etc. (11.15) may be written as

$$u_{i,j+1} = ru_{i-1,j} + (1-2r)u_{i,j} + ru_{i+1,j} + [O(\Delta t) + O(\Delta x^2)]\Delta t \tag{11.16}$$

where $r = \Delta t/\Delta x^2$.
Neglecting the error, formula (10.16) becomes

$$u_{i,j+1} = ru_{i-1,j} + (1-2r)u_{i,j} + ru_{i+1,j}. \tag{11.17}$$

Thus the value of $u_{i,j+1}$ at the time level $(j+1)$ can be computed explicitly in terms of the known values of u at the previous level j. As the values of u are known at the end points by virtue of the boundary conditions (11.13a and b) formula (11.17) may be applied at all the internal mesh points $i = 1(1)n-1$, at $(j+1)$ time level successively in a step by step manner for $j = 0, 1, \ldots$. As stated earlier, values of $u_{i,0}$ are known due to initial condition (11.13c). This method is also known as Schmidt's method.

Explicit method/formula (11.17) can be conveniently represented by a diagram, called 'molecule' (See Fig. 11.2). It shows the mesh points involved for computation of $u_{i,j+1}$ by the formula while the number written at the meshpoints indicate the coefficients of the u values corresponding to the respective mesh points.

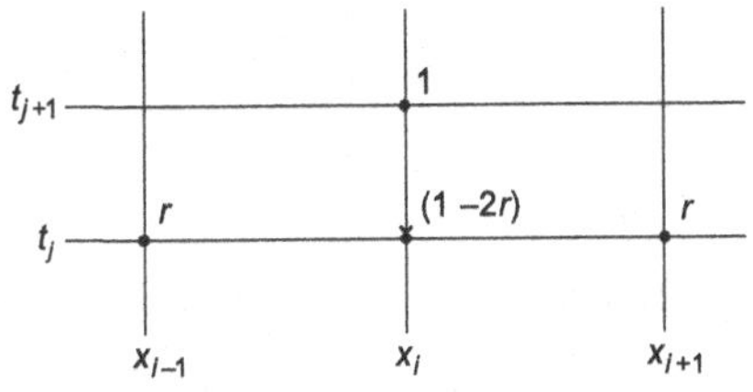

Figure 11.2 Molecule for Explicit method/scheme.

Note: $\times$ denotes the point at which the p.d.e. is discretized.

11.5.2 Fully Implicit scheme/method

In this scheme the differential equation is discretized at the meshpoint $(i, j+1)$ such that the time derivative is replaced by backward difference and the space derivative by central difference, i.e.,

$$\frac{1}{\Delta t}\nabla_t u_{i,j+1} + O(\Delta t) = \frac{1}{\Delta x^2}\delta_x^2 u_{i,j+1} + O(\Delta x^2)$$

or
$$\frac{u_{i,j+1} - u_{i,j}}{\Delta t} = \frac{u_{i-1,j+1} - 2u_{i,j+1} + u_{i+1,j+1}}{\Delta x^2} + O(\Delta t) + O(\Delta x^2) \tag{11.18}$$

The truncation error in the Fully Implicit Scheme is also $O(\Delta t) + O(\Delta x^2)$. Using $r = \Delta t/\Delta x^2$, it can be written as

$$-ru_{i-1,\,j+1} + (1+2r)u_{i,\,j+1} - ru_{i+1,\,j+1} = u_{i,\,j} + [O(\Delta t) + O(\Delta x^2)]\Delta t \tag{11.19}$$

Neglecting the error term, formula (11.19) may be written as,

$$-ru_{i-1,\,j+1} + (1+2r)u_{i,\,j+1} - ru_{i+1,\,j+1} = u_{i,\,j}. \tag{11.20}$$

Formula (11.20) is an implicit formula, i.e., the value of $u_{i,\,j+1}$ can not be computed straightaway, and is interrelated with other neighbouring values of u at the current level of time. In order to find the values at the $(j+1)$ time level, we apply formula (11.20) for $i = 1(1)\ \mathrm{N}-1$, giving (N−1) equations in $u_{i,\,j+1}$, $i = 1(1)\ \mathrm{N}-1$. These equations are to be solved to get the values of u at the time level $j+1$. This formula is also known as Laasonen's formula. The molecule for this formula is shown by Fig. 11.3.

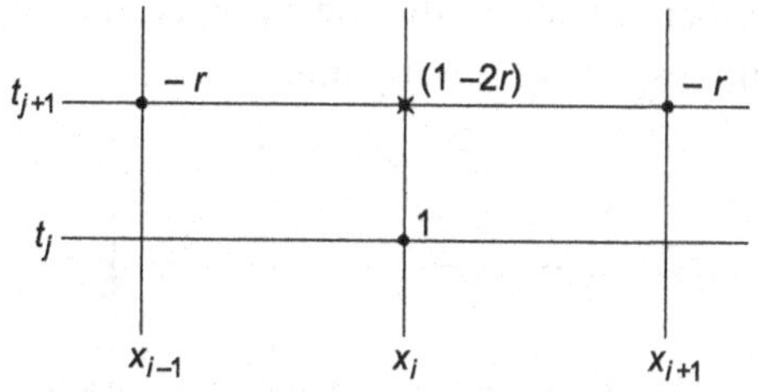

Figure 11.3 Molecule for Fully Implicit method/scheme.

Note: × denotes the point at which p.d.e. is discretized.

11.5.3 Crank–Nicolson's (C–N) scheme

The C–N scheme is also an implicit scheme. Instead of a meshpoint, the discretization is made at the midpoint of jth and $(j+1)^{\text{th}}$ levels, i.e., at $\left(x_i,\ t_j + \frac{1}{2}\Delta t\right) = (x_i,\ t_{j+1/2})$. The time derivative as well as the space derivative in (11.13), both are approximated by central differences, i.e.,

$$\frac{1}{\Delta t}\delta_t u_{i,\,j+\frac{1}{2}} + O(\Delta t^2) = \frac{1}{\Delta x^2}\delta_x^2 u_{i,\,j+\frac{1}{2}} + O(\Delta x^2). \tag{11.21}$$

Replacing the space derivative by the average value at the jth and the $(j+1)^{\text{th}}$ level, the above can be written as

$$\frac{u_{i,\,j+1} - u_{i,\,j}}{\Delta t} = \frac{1}{2}\left[\frac{1}{\Delta x^2}\delta_x^2 u_{i,\,j} + \frac{1}{\Delta x^2}\delta_x^2 u_{i,\,j+1}\right] + O(\Delta t^2) + O(\Delta x^2)$$

$$= \frac{1}{2}\left[\frac{u_{i-1,j} - 2u_{i,j} + u_{i+1,j}}{\Delta x^2} + \frac{u_{i-1,j+1} - 2u_{i,j+1} + u_{i+1,j+1}}{\Delta x^2}\right]$$

$$+ \text{O}(\Delta t^2) + \text{O}(\Delta x^2) \qquad (11.22)$$

It may be noted that since the values of u are not available at the midpoint, the second derivative at the point $\left(i,\ j+\frac{1}{2}\right)$ is replaced by the average values at the $(i,\ j)$ and $(i,\ j+1)$ points. It must also be noted that the truncation error in the C–N scheme is $\text{O}(\Delta t^2)+\text{O}(\Delta x^2)$ while it is $\text{O}(\Delta t)+\text{O}(\Delta x^2)$ in both the methods, namely Explicit and Implicit, discussed earlier. Rearranging various terms in (11.22) and putting $r = \Delta t/\Delta x^2$, we get,

$$-ru_{i-1,j+1} + 2(1+r)u_{i,j+1} - ru_{i+1,j+1} = ru_{i-1,j} + 2(1-r)u_{i,j} + ru_{i+1,j} + \Delta t[\text{O}(\Delta t^2) + \text{O}(\Delta x^2)]$$

Neglecting the error terms, the C–N scheme becomes

$$-ru_{i-1,j+1} + 2(1+r)u_{i,j+1} - ru_{i+1,j+1} = ru_{i-1,j} + 2(1-r)u_{i,j} + ru_{i+1,j}, \quad i = 1(1)N-1. \qquad (11.23)$$

As may be seen that C–N scheme (11.23) is an implicit scheme. In order to find the value of u at the $(j+1)^{\text{th}}$ level, we have to invoke it for $i = 1(1)$ N-1 which will result in N-1 equations in N-1 unknowns and their solution will give the value of u at at the $(j+1)^{\text{th}}$ time level. It may also be mentioned that the resulting system of equations is tridiagonal and can be very easily solved by Gaussian Elimination. The molecule for the C–N scheme is given by Fig. 11.4.

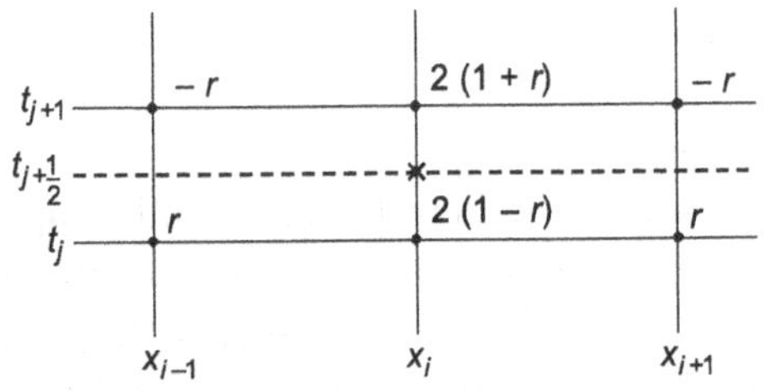

Figure 11.4 Molecule for C–N scheme.

Note: × denotes point of discretization

11.5.4 Comparison of three schemes

Criterion	*Explicit* *(Schmidt)*	*Fully Implicit* *(Laasonen)*	*Implicit* *(C–N)*
(*i*) Point of Discretization	(i, j)	$(i, j+1)$	$\left(i, j, \frac{1}{2}\right)$
(*ii*) Formula	$\frac{1}{\Delta t}\Delta_t u_{i,j} = \frac{1}{\Delta x^2}\delta_x^2 u_{i,j}$	$\frac{1}{\Delta t}\nabla_t u_{i,j+1} = \frac{1}{\Delta x^2}\delta_x^2 u_{i,j+1}$	$\frac{1}{\Delta t}\delta_t u_{i,j+\frac{1}{2}} = \frac{1}{\Delta x^2}\delta_x^2 u_{i,j+\frac{1}{2}}$
(*iii*) Truncation error	$O(\Delta t) + O(\Delta x^2)$	$O(\Delta t) + O(\Delta x^2)$	$O(\Delta t^2) + O(\Delta x^2)$
(*iv*) Complexity	Value of *u* obtained directly	Simultaneous equations to be solved.	Simultaneous equations to be solved.

We would now introduce some important concepts which will be useful in analysing the different finite difference schemes rigorously.

11.5.5 Compatibility, stability and convergence

Let us suppose that u represents the exact solution of the partial differential equation and u_D be the exact solution of the discretized equation, i.e., of the finite difference formula. The error $(u - u_D)$ is called 'discretization error'. Expressing equation (11.13) in operator form we can write

$$Lu = 0 = L_D(u) + T \tag{11.24}$$

where $L \equiv \frac{\partial}{\partial t} - \frac{\partial^2}{\partial x^2}$ and L_D denotes the discretizing operator which contains Δx and Δt as parameters and T is the truncation error. Let u_D be the exact solution of $L_D(u) = 0$. If the discretization error $(u - u_D)$ tends to zero as Δx and Δt tend to zero, then $u_D \to u$, i.e., the finite difference solution is said to converge to the exact solution.

If the finite difference approximation $L_D(u)$ tends to Lu as Δx and Δt tend to zero, i.e., T tends to zero as Δx and Δt tend to zero then the finite difference approximation is said to be compatible (consistent) with the original p.d.e.

While employing the discretized system to find solution u_D, other errors like rounding/truncation, etc., also creep in. Therefore, the final solution we get after computations is different from u_D which may be called u_C. The error $(u_D - u_C)$ is called 'stability error'. Hence the total error in u can be expressed as sum of the discretization error plus the stability error, i.e.,

$$\text{Total error} = u - u_C = (u - u_D) + (u_D - u_C). \tag{11.25}$$

Obviously the first term on the right side of (11.25) is the discretization error and the second is stability error.

With number of space intervals kept fixed, if the error $(u_D - u_C)$ tends to zero or it does not increase exponentially, at each of the meshpoints, with the increase in the number of time steps, then the finite difference scheme/method is said to be 'stable'.

As it happens, the convergence of the solution of a p.d.e. is more complex phenomenon to investigate. However, it has been observed that in case of the linear equations, compatibility and stability implies convergence. Therefore we shall examine these two aspects in respect of various methods. Let us discuss Explicit method first.

11.5.6 Compatibility of explicit scheme

As illustrated in Chapter 5, we can write using Taylor's expansion,

$$\frac{u(x,\, t+\Delta t) - u(x,\, t)}{\Delta t} = \frac{\partial u}{\partial t} + \frac{\Delta t}{2}\frac{\partial^2 u}{\partial t^2} + \frac{\Delta t^2}{6}\frac{\partial^3 u}{\partial t^3} + \frac{\Delta t^3}{24}\frac{\partial^4 u}{\partial t^4} + \cdots \tag{11.26}$$

$$\frac{u(x-\Delta x,\, t) - 2u(x,\, t) + u(x+\Delta x,\, t)}{\Delta x^2} = \frac{\partial^2 u}{\partial x^2} + \frac{\Delta x^2}{12}\frac{\partial^4 u}{\partial x^4} + \frac{\Delta x^4}{360}\frac{\partial^6 u}{\partial x^6} + \cdots \tag{11.27}$$

Subtracting (11.27) from (11.26) we get

$$\frac{\partial u}{\partial t} - \frac{\partial^2 u}{\partial x^2} = \frac{u(x,\, t+\Delta t) - u(x,\, t)}{\Delta t} - \frac{u(x-\Delta x,\, t) - 2u(x,\, t) + u(x+\Delta x,\, t)}{\Delta x^2} + \mathrm{T}$$

$$\text{where } \mathrm{T} = -\left(\frac{\Delta t}{2}\frac{\partial^2 u}{\partial t^2} + \frac{\Delta t^2}{6}\frac{\partial^3 u}{\partial t^3} + \cdots\right) + \left(\frac{\Delta x^2}{12}\frac{\partial^4 u}{\partial x^4} + \frac{\Delta x^4}{360}\frac{\partial^6 u}{\partial x^6} + \cdots\right) \tag{11.28}$$

The term T given by (11.28) is the truncation error in the Explicit scheme. As can be seen, $\mathrm{T} \to 0$ as Δt and Δx both tend to zero. Therefore we can say that the scheme is compatible or consistent with the partial differential equation.

Let us re-write the truncation error in the following form,

$$\mathrm{T} = -\left(\frac{\Delta t}{2}\frac{\partial^2 u}{\partial t^2} - \frac{\Delta x^2}{12}\frac{\partial^4 u}{\partial x^4}\right) - \left(\frac{\Delta t^2}{6}\frac{\partial^3 u}{\partial t^3} - \frac{\Delta x^4}{360}\frac{\partial^6 u}{\partial x^6}\right) - \cdots \tag{11.29}$$

From the original p.d.e. we have $\dfrac{\partial}{\partial t} = \dfrac{\partial^2}{\partial x^2}$ which gives

$$\frac{\partial^2 u}{\partial t^2} = \frac{\partial^4 u}{\partial x^4} \text{ and } \frac{\partial^3 u}{\partial t^3} = \frac{\partial^6 u}{\partial x^6}, \text{ etc.} \tag{11.30}$$

Using relations (11.30) in (11.29),

$$T = -\frac{\Delta x^2}{2}\left(r - \frac{1}{6}\right)\frac{\partial^4 u}{\partial x^4} - \frac{\Delta x^4}{6}\left(r^2 - \frac{1}{60}\right)\frac{\partial^6 u}{\partial x^6} - \cdots, \tag{11.31}$$

where $r = \Delta t/\Delta x^2$.
We see that if $r = 1/6$, the truncation error (11.31) reduces to

$$T = -\frac{\Delta x^4}{540}\frac{\partial^6 u}{\partial x^6} + \cdots \tag{11.32}$$

Thus for $r = 1/6$, the truncation error will be of highest order.

11.5.7 Stability of explicit scheme

There are two methods for analysing the stability of a linear finite difference schemes:
(*i*) Matrix method (*ii*) Neumann's method.
We illustrate these methods with regard to Explicit scheme.

(*i*) Matrix Method
In the Explicit scheme the values of u at the level of time $j+1$ are given explicitly in terms of the values of u at the previous level j, i.e.,

$$u_{i,j+1} = ru_{i-1,j} + (1-2r)u_{i,j} + ru_{i+1,j}, \quad i = 1(1)\,N-1. \tag{11.33}$$

Let us suppose that the values of u at the jth level are not the true values and involve certain errors. We have computed u at the $(j+1)^{th}$ level using these values which will also have error in it. Let us denote the true value by u and approximate (actually computed) value by u^*, then

$$u^*_{i,j+1} = ru^*_{i-1,j} + (1-2r)u^*_{i,j} + ru^*_{i+1,j}. \tag{11.34}$$

Subtracting (11.34) from (11.33), and denoting the error $u - u^*$ by e we get,

$$e_{i,j+1} = re_{i-1,j} + (1-2r)e_{i,j} + re_{i+1,j}, \quad i = 1(1)\,N-1 \tag{11.35}$$

where $e_{i,j}$ denotes the error in the exact value u and the computed value u^* at the $(i,\ j)$ mesh point, i.e., $e_{i,j} = (u_{i,j} - u^*_{i,j})$ etc. As the Explicit scheme is linear, the errors (11.35) follow the same formula as the scheme itself, i.e., (11.33).

It may be remembered that $u_{0,j+1} = u_0$ and $u_{N,j+1} = u_L$ on account of the boundary conditions and their values are not being computed at any time. Hence there is no question of errors in the values of u at the endpoints so that $e_{0,j} = e_{N,j} = 0$ for all j. We can now write (11.35) in matrix form as follows:

$$\begin{bmatrix} e_{1,j+1} \\ e_{2,j+1} \\ e_{3,j+1} \\ \vdots \\ e_{N-1,j+1} \end{bmatrix} \begin{bmatrix} 1-2r & r & 0 & 0 & \cdots & 0 \\ r & 1-2r & r & 0 & \cdots & 0 \\ 0 & 1 & 1-2r & r & \cdots & 0 \\ \vdots & & & \ddots & & \\ 0 & 0 & & & r & 1-2r \end{bmatrix} \begin{bmatrix} e_{1,j} \\ e_{2,j} \\ e_{3,j} \\ \vdots \\ e_{N-1,j} \end{bmatrix} \tag{11.36}$$

or

$$\boldsymbol{e}_{j+1} = \mathrm{A}\boldsymbol{e}_j \tag{11.37}$$

where

$$\boldsymbol{e}_{j+1}^{\mathrm{T}} = (e_{1,j+1} \;\; e_{2,j+1} \;\ldots.\; e_{N-1,j+1}), \tag{11.37a}$$

$$\boldsymbol{e}_j^{\mathrm{T}} = (e_{1,j} \;\; e_{2,j} \;\ldots.\; e_{N-1,j}) \tag{11.37b}$$

and

$$\mathrm{A} = \begin{bmatrix} 1-2r & r & 0 & \cdots\cdots & 0 \\ r & 1-2r & r & \cdots\cdots & 0 \\ & \ddots & & \ddots & \\ & & & r & 1-2r \end{bmatrix}. \tag{11.37c}$$

Let us suppose that at some point of time $t = t_0$, say, (t_0 not necessarily the initial time $t = 0$) the error introduced in the computed values are represented by error vector $\boldsymbol{e}_0$. Then according to formula (11.37) these errors will become $\mathrm{A}\boldsymbol{e}_0$ at the next step ignoring the computing errors introduced at that step, so that $\boldsymbol{e}_1 = \mathrm{A}\boldsymbol{e}_0$. Then at the following step it will become $\boldsymbol{e}_2 = \mathrm{A}\boldsymbol{e}_1 = \mathrm{A}^2\boldsymbol{e}_0$. Proceeding in the same manner, after k steps, the error $\boldsymbol{e}_0$ will contribute to overall error

$$\boldsymbol{e}_k = \mathrm{A}^k\boldsymbol{e}_0, \quad k = 1, 2, \ldots. \tag{11.38}$$

It should be noted that local computational errors have been neglected at each time level. We are interested in investigating the effect of $\boldsymbol{e}_0$ on the subsequent computations as how it behaves as k becomes larger and larger. We observe from (11.38) that $\boldsymbol{e}_k$ is dependent on A^k. If elements of matrix A^k become smaller and smaller tending to zero, or remain bounded, as k becomes infinitely large then the scheme will be stable. On the other hand, if they increase, then the scheme will be unstable. In order to examine the magnitudes of elements of A^k we proceed as follows.

Let us suppose that the matrix A $\{(N-1) \times (N-1)\}$ possesses e.values λ_s with corresponding e.vectors $\boldsymbol{v}_s, s = 1(1)\,N-1$ and that they are all distinct. They satisfy the equation,

$$A\boldsymbol{v} = \lambda \boldsymbol{v} \quad \text{or} \quad A\boldsymbol{v}_s = \lambda_s \boldsymbol{v}_s, \quad s = 1(1)N-1. \tag{11.39}$$

As the e.vectors are distinct they form a set of linearly independent vectors and we can express the error vector $\boldsymbol{e}_0$ by a linear combination of these vectors, say

$$\boldsymbol{e}_0 = \sum_{s=1}^{N-1} c_s \boldsymbol{v}_s \tag{11.40}$$

where $$\boldsymbol{v}_s^T = (v_{1s}\ v_{2s}\ \ldots.\ v_{N-1,\,s}). \tag{11.41}$$

It may be noted that the (N−1) coefficients, c_s can be uniquely determined from the system of equations (11.40).

From (11.40) we have,

$$A^k \boldsymbol{e}_0 = A^{k-1} \sum_{s=1}^{N-1} c_s A \boldsymbol{v}_s, \text{ from (11.40)}$$

$$= A^{k-1} \sum c_s \lambda_s \boldsymbol{v}_s, \text{ from (11.39)}$$

$$= \sum_{s=1}^{N-1} c_s \lambda_s^k \boldsymbol{v}_s. \tag{11.42}$$

The term $A^k \boldsymbol{e}_0$ on the left side of (11.42) denotes the propagation of error $\boldsymbol{e}_0$ at the kth step as shown in (11.38). If this error is not to increase indefinitely with k, then we must have

$$|\lambda_s| \le 1, s = 1(1)\,N-1$$

or $$|\lambda_{\max}| \le 1 \tag{11.43}$$

Thus for the Explicit scheme to be stable, the modulus of the largest e.value of matrix A, given by (11.37c), should not exceed unity.

We note that the elements of matrix A contain parameter $r(= \Delta t/\Delta x^2)$, therefore its e.values will be dependent on r. Let us derive condition as for what values of r, the e.values of A remain less than or equal to 1 in modulus. In order to achieve this we use Brauer's theorem which is an extension of the Gerschgorin's theorem. These theorems, discussed in Chapter 10, are restated as follows:

Gerschgorin's Theorem

Let λ be an eigenvalue of an arbitrary $(n \times n)$ matrix $A = [a_{ij}]$. Then for each λ there exists an inequality, for certain s, such that

$$|\lambda - a_{ss}| \leq \sum_{\substack{j=1 \\ j \neq s}}^{n} |a_{sj}|, \quad s = 1, 2, \ldots.. n. \tag{11.44}$$

That is, for each e.value there exists a circle with centre a_{ss} and radius $\sum_{j=1, j\neq s}^{n} |a_{sj}|$ for some $s(1 \leq s \leq n)$ such that it lies inside that circle or on its boundary.

In its weak form the theorem can also be stated giving the bound for numerically largest e.value of the matrix A, as follows:

The modulus of the largest e.value of the square matrix A can not exceed the largest sum of the moduli of the terms along any row, i.e.,

$$|\lambda_{\max}| \leq \max_{i} \sum_{j=1}^{n} |a_{i,\ j}|. \tag{11.45}$$

Since the e.values of A and A^T are same, the above also holds for columns of A, i.e.,

$$|\lambda_{\max}| \leq \max_{j} \sum_{i=1}^{n} |a_{i,\ j}| \tag{11.46}$$

We have to choose smallest of the right sides of (11.45) or (11.46) as an estimate of $|\lambda_{\max}|$.

Brauer's Theorem

Let P_s be the sum of moduli of the terms along the sth row of A excluding the diagonal element a_{ss}. Then every e.value of A lies inside or on the boundary of at least one of the circles $|\lambda - a_{ss}| = P_s$, i.e., a disc with centre a_{ss} and radius P_s, $s = 1(1)n$.

Following properties may also be observed with regard to e.values of a square matrix.

If λ is an eigenvalue of A, i.e., $A\boldsymbol{x} = \lambda \boldsymbol{x}$, then

(*i*) e.values of A^m will be λ^m for m being positive or negative integer.

(*ii*) e.values of matrix polynomial $P(A)$ will be $P(\lambda)$.

(*iii*) e.values of $\dfrac{P(A)}{Q(A)}$ will be $\dfrac{P(\lambda)}{Q(\lambda)}$ where $P(A)$ and $Q(A)$ are polynomials in A, and $Q(A)$ is non-singular.

We have to find condition on r such that the largest e.value λ of matrix A does not exceed unity in modulus. Let us use Brauer's theorem for determining the bounds of λ. Using Brauer's theorem we get from the first and last row of A (11.37c),

$$|\lambda-(1-2r)|\le r$$

or $$-r\le\lambda-1+2r\le r$$

or $$1-3r\le\lambda\le 1-r$$

For $|\lambda|\le 1$, the right inequality gives $0\le r\le 2$ and the left inequality gives $0\le r\le\frac{2}{3}$. From the intermediate rows, i.e., from 2nd to (N−2)th rows we have

$$|\lambda-(1-2r)\le 2r$$

or $$1-4r\le\lambda\le 1$$

For $|\lambda|\le 1$ the left inequality gives $0\le r\le\frac{1}{2}$.

There are three conditions on r for $|\lambda|\le 1$. We have to choose the value of r so that all the conditions are satisfied i.e., the smallest value of r. Thus we see that for the Explicit scheme to be stable $r\le\frac{1}{2}$.

(*ii*) Neumann's Method

When solving a general parabolic equation (11.13) by method of separation of variables we assume $u(x, t)$ as product of two functions X (x) and T(t) where X (x) is a function of variable x only and T(t) is a function of t only, i.e., $u(x, t)=\mathrm{X}(x)\cdot\mathrm{T}(t)$. After substituting in the p.d.e. the solution obtained is of the form

$$u(x, t)=e^{\alpha t}(\mathrm{A}e^{i\beta x}+\mathrm{B}e^{-i\beta x})$$

where A and B and α, β are constants. This concept is used in the method.

In this method since i will be used for complex number $(=\sqrt{-1})$ we shall use suffices p, q in place of i, j respectively in the error formula (11.35). Thus we write,

$$e_{p,\,q+1}=re_{p-1,\,q}+(1-2r)e_{p,\,q}+re_{p+1,\,q}. \tag{11.47}$$

Since the error (11.47) follows the same formula as u, the solution to error is assumed of the form (with an arbitrary constant A) as,

$$e_{p,\,q} = \text{A}e^{\alpha q \Delta t} \cdot e^{i\beta p \Delta x} = \text{A}\xi^{q} \cdot e^{i\beta p \Delta x} \tag{11.48}$$

where α and β may be complex constants in general, and $\xi = e^{\alpha \Delta t}$. For stability $|\xi| \leq 1$, since in that case error $e_{p,\,q}$ will tend to zero or will not increase with increasing q.

Substituting from (11.48) in (11.47) we get

$$\xi^{q+1} e^{i\beta p \Delta x} = r\xi^{q} e^{i\beta (p-1)\Delta x} + (1-2r)\xi^{q} e^{i\beta p \Delta x} + r\xi^{q} e^{i\beta (p+1)\Delta x}$$

or

$$\begin{aligned} \xi &= re^{-i\beta \Delta x} + (1-2r) + re^{i\beta \Delta x} \\ &= 2r\cos \beta \Delta x - 2r + 1 \\ &= 1 - 4r\sin^2 \frac{\beta \Delta x}{2} \end{aligned}$$

Since for stability $|\xi| \leq 1$,

$$\left| 1 - 4r\sin^2 \frac{\beta \Delta x}{2} \right| \leq 1$$

or

$$-1 \leq 1 - 4r\sin^2 \frac{\beta \Delta x}{2} \leq 1.$$

The right inequality gives $0 \leq r$ while the left inequality gives,

$$r \leq \frac{1}{2\sin^2 \dfrac{\beta \Delta x}{2}}$$

Since the value of $\sin^2 \dfrac{\beta \Delta x}{2}$ is always less than or equal to one, the minimum value of r should be $\dfrac{1}{2}$. The condition for stability of the Explicit scheme is $r \leq \dfrac{1}{2}$.

11.5.8 Stability of C–N scheme

It will be shown that C–N scheme is stable for all values of r. We will discuss both methods again, viz., Matrix method and Neumann's method.

(*i*) Matrix Method

We see that the C–N Scheme (11.23) is linear in u values defined at the mesh points. Hence the errors will also follow the same formula. Changing subscripts i and j by p and q respectively, the error formula may be written as,

$$-re_{p-1,\,q+1}+2(1+r)e_{p,\,q+1}-re_{p+1,\,q+1}=re_{p-1,\,q}+2(1-r)e_{p,\,q}+re_{p+1,\,q}. \quad (11.49)$$

Writing (11.49) for $p = 1(1)N-1$ in matrix form remembering that there is no error at $x = 0$ and $x = \mathrm{L}$,

$$\mathrm{P}\boldsymbol{e}_{q+1} = \mathrm{Q}\boldsymbol{e}_{\mathrm{q}}$$

or

$$\boldsymbol{e}_{q+1} = \mathrm{P}^{-1}\mathrm{Q}\boldsymbol{e}_{q} \quad (11.50)$$

where

$$\mathrm{P}=\begin{bmatrix} 2(1+r) & -r & 0 & \cdots & 0 \\ -r & 2(1+r) & -r & & \\ & & \ddots & & \\ & & & \ddots & \\ & & & -r & 2(1+r) \end{bmatrix},\ \mathrm{Q}=\begin{bmatrix} 2(1-r) & r & 0 & & \\ r & 2(1-r) & r & & \\ & & \ddots & & \\ & & & \ddots & \\ & & & r & 2(1-r) \end{bmatrix}$$

$$\boldsymbol{e}_{q+1}^{\mathrm{T}} = (e_{1,\,q+1}\ \ e_{2,\,q+1}\ \ldots.\ e_{\mathrm{N}-1,\,q+1})$$

$$\boldsymbol{e}_{q}^{\mathrm{T}} = (e_{1,\,q}\ \ e_{2,\,q}\ \ldots.\ e_{\mathrm{N}-1,\,q})$$

For stability the e.values of $\mathrm{P}^{-1}\mathrm{Q}$ should be less than or equal to one in modulus. Let us define a tridiagonal matrix T as

$$\mathrm{T}=\begin{bmatrix} 2 & -1 & & \\ -1 & 2 & -1 & \\ & & \ddots & \\ & & -1 & 2 \end{bmatrix}.$$

Then we can write (11.50) as

$$\boldsymbol{e}_{q+1} = (2\mathrm{I}+r\mathrm{T})^{-1}(2\mathrm{I}-r\mathrm{T})\boldsymbol{e}_q$$

$$= \mathrm{S}\boldsymbol{e}_q, \text{ where matrix } \mathrm{S} = (2\mathrm{I}+r\mathrm{T})^{-1}(2\mathrm{I}-r\mathrm{T}).$$

The e.values of matrix T are given by

$$\mu_s = 4\sin^2\left(\frac{s\pi}{2\mathrm{N}}\right), \quad s = 1(1)\ \mathrm{N}-1.$$

Hence, $$\lambda_s = \frac{1-2r\sin^2(s\pi/2N)}{1+2r\sin^2(s\pi/2N)}.$$

Obviously $|\lambda_s|$ will always be less than 1 for positive $r(=\Delta t/\Delta x^2)$.

Alternatively, if μ is an e.value of T then the e.value λ of matrix S is given by $\lambda = \frac{2-\mu r}{2+\mu r}$. For stability $|\lambda| \leq 1$, so that $-1 \leq \frac{2-\mu r}{2+\mu r} \leq 1$ which implies that $\mu \geq 0$. Using Brauer's theorem on matrix T, we see that $-2 \leq \mu - 2 \leq 2$ or $0 \leq \mu \leq 4$. Hence, C–N scheme is stable for all values of r, i.e., unconditionally stable.

***(ii)* Neumann's Method**

Substituting $e_{pq} = A\xi^q e^{i\beta p\Delta x}$ in (11.49), A being constant, we get

$$\xi(2+2r-2r\cos\beta\Delta x) = (2-2r+2r\cos\beta\Delta x)$$

or $$\xi = \frac{1-2r\sin^2\dfrac{\beta\Delta x}{2}}{1+2r\sin^2\dfrac{\beta\Delta x}{2}}.$$

For stability $|\xi| \leq 1$ which is true for all values of $r \geq 0$ since $0 \leq \sin^2\frac{\beta\Delta x}{2} \leq 1$.

It would be easy to show that the Fully Implicit (Laasonen's) scheme is also stable unconditionally. These two implicit schemes namely C–N and Laasonen's are compatible with the parabolic equation $\frac{\partial u}{\partial t} = \frac{\partial^2 u}{\partial x^2}$.

11.5.9 Further comparison of schemes

The drawback of the implicit schemes, namely, C–N scheme and Fully Implicit (Laasonen's) scheme is that a set of simultaneous equations has to be solved at each time level for computing the values of u, whereas they are computed directly in an explicit scheme. We have also seen that the Explicit scheme is stable when $r \leq \frac{1}{2}$ while the implicit schemes are stable unconditionally. It means that if $\Delta x = 0.1$, then for stability of the Explicit scheme the largest value for Δt can be chosen as $\Delta t = 0.01 \times 0.5 = 0.005$. Thus if the computations are to be carried up to $t = 1.00$, say, then we need 200 time steps. But in an implicit scheme, as there is no condition on r, we can take $\Delta t = 0.05$, i.e., ten times of 0.005 which will require one tenth number of time steps of the Explicit scheme. Further the truncation error in the Explicit scheme would be $O(\Delta t) + O(\Delta x^2) = O(0.005) + O(0.01)$; in the Laasonen's $O(\Delta t)$ $+O(\Delta x^2) = O(0.05) + O(0.01)$ while in the C–N scheme it would be $O(\Delta t^2) + O(\Delta x^2) =$

$O(0.0025)+O(0.01)$. Thus C–N scheme should provide a better estimate and at the same time requiring less number of time steps. It should also be realised that rounding error with each step goes on accumulating as number of time steps increase. For example, if the values are rounded to four decimal places at each step, then after 100 steps the error may accumulate to $0.00005 \times 100 = 0.005$. Thus the values may be trusted up to two decimal places only.

The three schemes can be represented by a single formula (See Exercise 11.3).

11.5.10 Derivative boundary conditions

We have considered so far only Dirichlet boundary condition which could be homogeneous, i.e., $u = 0$ or non-homogeneous when u is given not to be zero. Now we discuss as how other types of boundary conditions can be handled while using the foregoing methods.

Let us suppose following boundary conditions are prescribed

$$\alpha_0 \frac{\partial u}{\partial x} + \beta_0 u = \gamma_0 \text{ at } x = x_0 \tag{11.51}$$

$$\alpha_N \frac{\partial u}{\partial x} + \beta_N u = \gamma_N \text{ at } x = x_N, \tag{11.52}$$

where α_0, β_0, γ_0 and α_N, β_N, γ_N are constants including zero.

Let us further suppose that the values of u are known at the jth level, i.e., $u_{i,\ j}$, $i = 0(1)N$ are known and we are required to compute $u_{i,\ j+1}$. There are two ways for dealing with the boundary conditions when derivative is involved, namely *(a)* using forward difference, backward difference approximations *(b)* using central difference approximation. They are discussed as follows.

(a) Forward Difference and Backward Difference Approximation

(i) Write appropriate equations for internal meshpoints for $i = 1(1)\ N-1$.

(ii) Approximate the boundary condition at $x = x_0$, replacing the derivative by forward difference as $\alpha_0 \frac{u_{1,\ j+1} - u_{0,\ j+1}}{\Delta x} + \beta_0 u_{0,\ j+1} = \gamma_0$, which gives

$$u_{0,\ j+1} = (\alpha_0 u_{1,\ j+1} - \gamma_0 \Delta x)/(\alpha_0 - \beta_0 \Delta x). \tag{11.53}$$

(iii) Similarly to find $u_{N,\ j+1}$, approximate the boundary condition at $x = x_N$, replacing the derivative by backward difference as $\alpha_N \frac{u_{N,\ j+1} - u_{N-1,\ j+1}}{\Delta x} + \beta_N u_{N,\ j+1} = \gamma_N$, giving

$$u_{N,\ j+1} = (\alpha_N u_{N-1,\ j+1} + \gamma_N \Delta x)/(\alpha_N + \beta_N \Delta x). \tag{11.54}$$

(b) Central Difference Approximation

(*i*) Extend the domain to the left of $x = x_0$ by Δx, i.e., up to $x = x_0 - \Delta x = x_{-1}$ (say) and similarly to the right of $x = x_N$ up to $x = x_N + \Delta x = x_{N+1}$ (say). The points x_{-1} and x_{N+1} on the x-axis are called fictitious points (See Fig. 11.5)

(*ii*) Approximate the boundary condition (11.51) at $x = x_0$ replacing the derivative by central difference at the jth level as,

$$\alpha_0 \frac{u_{1,\, j} - u_{-1,\, j}}{2\Delta x} + \beta_0 u_{0,\, j} = \gamma_0 \text{ which gives}$$

$$u_{-1,\, j} = u_{1,\, j} + \frac{2\Delta x}{\alpha_0}(\beta_0 u_{0,\, j} - \gamma_0). \tag{11.55}$$

Analogously at the $(j+1)^{\text{th}}$ level as

$$u_{-1,\, j+1} = u_{1,\, j+1} + \frac{2\Delta x}{\alpha_0}(\beta_0 u_{0,\, j+1} - \gamma_0). \tag{11.56}$$

It should be borne in mind that value of $u_{-1,\, j}$ is known through (11.55) but $u_{-1,\, j+1}$ has to be computed.

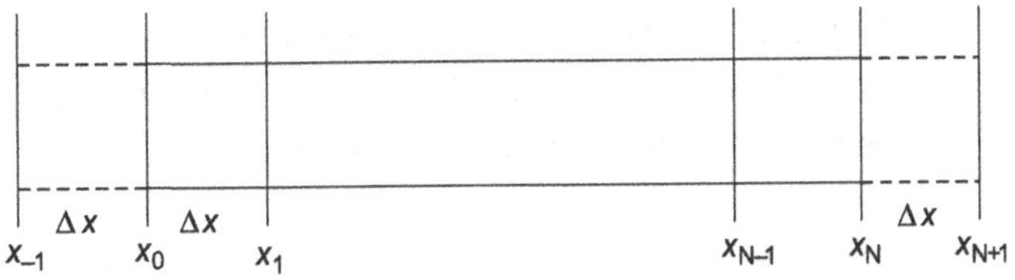

Figure 11.5 Fictitious points.

(*iii*) When the boundary condition at the other end $x = x_N$ also involves derivative, then following the procedure of (*ii*) above we get

$$u_{N+1,\, j} = u_{N-1,\, j} - \frac{2\Delta x}{\alpha_N}(\beta_N u_{N,\, j} - \gamma_N), \tag{11.57}$$

and $$u_{N+1,\, j+1} = u_{N-1,\, j+1} - \frac{2\Delta x}{\alpha_N}(\beta_N u_{N,\, j+1} - \gamma_N). \tag{11.58}$$

Here also value of $u_{N+1,\, j}$ is known while $u_{N+1,\, j+1}$ is not known.

(*iv*) Earlier we invoked the various formulae at the internal mesh points only, viz., for $i = 1(1)N-1$. Now we apply them at the end points also, i.e., for $i = 0$ and $i = N$.

While using Explicit formula (11.17) for $i = 0$, we get

$$u_{0,\ j+1} = ru_{-1,\ j} + (1-2r)u_{0,\ j} + ru_{1,\ j} \tag{11.59}$$

in which the values of $u_{-1,\ j}$ can be substituted from (11.55) while values of $u_{0,\ j}$ and $u_{1,\ j}$ are known. Similarly at the other end for $i = \mathrm{N}$,

$$u_{\mathrm{N},\ j+1} = ru_{\mathrm{N}-1,\ j} + (1-2r)u_{\mathrm{N},\ j} + ru_{\mathrm{N}+1,\ j} \tag{11.60}$$

in which value of $u_{\mathrm{N}+1,\ j}$ can be substituted from (11.57). If implicit scheme is being used, for example, C–N scheme (11.23), then, for $i = 0$,

$$-ru_{-1,\ j+1} + 2(1+r)u_{0,\ j+1} - ru_{1,\ j+1} = ru_{-1,\ j} + 2(1-r)u_{0,\ j} + ru_{1,\ j}. \tag{11.61}$$

The values of $u_{-1,\ j+1}$ and $u_{-1,\ j}$ can be substituted in (11.61) from (11.55) and (11.56) respectively, thus rendering an equation in $u_{0,\ j+1}$ and $u_{1,\ j+1}$ (unknowns). Similarly at the other end for $i = \mathrm{N}$, we have

$$-ru_{\mathrm{N}-1,\ j+1} + 2(1+r)u_{\mathrm{N},\ j+1} - ru_{\mathrm{N}+1,\ j+1} = ru_{\mathrm{N}-1,\ j} + 2(1-r)u_{\mathrm{N},\ j} + ru_{\mathrm{N}+1,\ j}. \tag{11.62}$$

The values of $u_{\mathrm{N}+1,\ j+1}$ and $u_{\mathrm{N}+1,\ j}$ can be used in (11.62) from (11.57) and (11.58) respectively reducing it to an equation in unknowns $u_{\mathrm{N}-1,\ j+1}$ and $u_{\mathrm{N},j+1}$.

Thus $(\mathrm{N}-1)$ equations at the internal meshpoints together with two equation (11.61) and (11.62) at the end points form a set of $(\mathrm{N}+1)$ equations in $(\mathrm{N}+1)$ unknowns, namely $u_{i,\ j+1},\ i = 0(1)\ \mathrm{N}$. These equations form a tridiagonal system which can be solved by Gaussian elimination.

Example 11.1

Find the numerical solution of the heat conduction equation

$$\frac{\partial u}{\partial t} = \frac{\partial^2 u}{\partial x^2};\ 0 \le x \le 1,\ t > 0$$

with boundary conditions

$$u(0,\ t) = u(1,\ t) = 1$$

and initial condition

$$u(x,\ 0) = \begin{cases} 1+2x, & 0 \le x \le \frac{1}{2} \\ 3-2x, & \frac{1}{2} \le x \le 1. \end{cases}$$

Use Explicit method taking $\Delta x = 0.2$, $\Delta t = 0.02$ and compute up to $t = 0.24$ up to six decimal places.

Solution We note that the problem is symmetric about $x = 0.5$ since the initial temperature $u(x, 0)$ is symmetric and the boundary conditions at $x = 0$ and $x = 1.0$ are also same. Therefore, the temperature at the subsequent times will remain also symmetric about $x = 0.5$.

The domain $0 \le x \le 1.0$ is subdivided into five sub-intervals each of width $\Delta x = 0.2$.

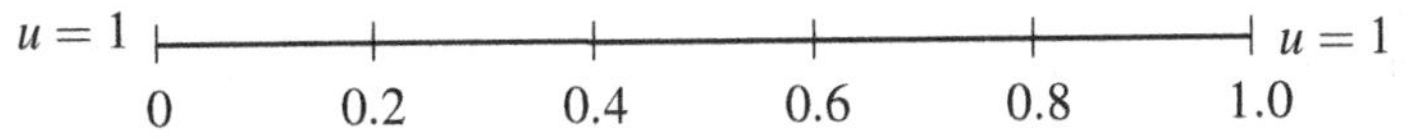

We have $x_i, = 0.2 \times i,\ \ i = 0(1)5$
Due to symmetry at $x = 0.5$, for all j

$$u_{0,\,j} = u_{5,\,j},\ u_{1,\,j} = u_{4,\,j},\ u_{2,\,j} = u_{3,\,j}$$

$$r = \Delta t/\Delta x^2 = 0.02/0.04 = 0.5.$$

Putting $r = 0.5$ in the Explicit formula

$$u_{i,\,j+1} = ru_{i-1,\,j} + (1-2r)u_{i,\,j} + ru_{i+1,\,j}$$

$$= \frac{1}{2}(u_{i-1,\,j} + u_{i+1,\,j}),\ i = 1,\ 2$$

The values for $i = 3$ and $i = 4$ can be written by symmetry. Computed values are as follows:

$x \to$	0	0.2	0.4	0.6	0.8	1.0
$t = 0.00$	1.0	1.4	1.8	1.8	1.4	1.0
0.02	1.0	1.4	1.6	1.6	1.4	1.0
0.04	1.0	1.3	1.5	1.5	1.3	1.0
0.06	1.0	1.25	1.4	1.4	1.25	1.0
0.08	1.0	1.20	1.325	1.325	1.20	1.0
0.10	1.0	1.1625	1.2625	1.2625	1.1625	1.0
0.12	1.0	1.13125	1.21250	1.21250	1.13125	1.0
0.14	1.0	1.10625	1.171875	1.171875	1.10625	1.0
0.16	1.0	1.085938	1.139062	1.139062	1.085938	1.0
0.18	1.0	1.069531	1.112500	1.112500	1.069531	1.0
0.20	1.0	1.056250	1.091016	1.091016	1.056250	1.0
0.22	1.0	1.045508	1.073633	1.073633	1.045508	1.0
0.24	1.0	1.036816	1.059570	1.059570	1.036816	1.0

Example 11.2

Show by solving the Example 11.1 taking $\Delta t = 0.04$ and $\Delta x = 0.2$ that the Explicit Scheme is unstable.

Solution $r = 0.04/0.04 = 1$

The Explicit formula becomes

$$u_{i,\,j+1} = u_{i-1,\,j} - u_{i,\,j} + u_{i+1,\,j}$$

Using the above formula we get the following values:

$x \rightarrow$	0	0.2	0.4	0.6	0.8	1.0
$t = 0.00$	1.0	1.4	1.8	1.8	1.4	1.0
0.04	1.0	1.4	1.4	1.4	1.4	1.0
0.08	1.0	1.0	1.4	1.4	1.0	1.0
0.12	1.0	1.4	1.0	1.0	1.4	1.0
0.16	1.0	0.6	1.4	1.4	0.6	1.0
0.20	1.0	1.8	0.6	0.6	1.8	1.0
0.24	1.0	–0.2	1.8	1.8	–0.2	1.0

The scheme starts behaving erratically from $t = 0.12$ onwards when the value of u increases from its previous value at $x = 0.2$. After that the value oscillates and later at $t = 0.24$ the temperature becomes negative also. It is impossible since the temperature at the boundary is unity and it should not go below it anywhere anytime.

Example 11.3

Solve the problem of Example 11.1 by C–N scheme taking $\Delta t = 0.08$ and $\Delta x = 0.2$.

Solution $r = \Delta t/\Delta x^2 = 0.08/0.04 = 2$

Putting $r = 2$ in C–N scheme

$$-ru_{i-1,\,j+1} + 2(1+r)u_{i,\,j+1} - ru_{i+1,\,j+1} = ru_{i-1,\,j} + 2(1-r)u_{i,\,j} + ru_{i+1,\,j}$$

$$-u_{i-1,\,j+1} + 3u_{i,\,j+1} - u_{i+1,\,j+1} = u_{i-1,\,j} - u_{i,\,j} + u_{i+1,\,j}$$

Applying the scheme for $i = 1, 2$

$$i = 1 \Rightarrow -u_{0,\,j+1} + 3u_{1,\,j+1} - u_{2,\,j+1} = u_{0,\,j} - u_{1,\,j} + u_{2,\,j}$$

or $3u_{1,\,j+1} - u_{2,\,j+1} = u_{0,\,j+1} + u_{0,\,j} - u_{1,\,j} + u_{2,\,j}$

or $$3u_{1,\,j+1} - u_{2,\,j+1} = 2 - u_{1,\,j} + u_{2,\,j}, \quad \because u_{0,\,j} = u_{0,\,j+1} = 1 \tag{i}$$

$i = 2 \Rightarrow -u_{1,\,j+1} + 3u_{2,\,j+1} - u_{3,\,j+1} = u_{1,\,j} - u_{2,\,j} + u_{3,\,j}$

or $$-u_{1,\,j+1} + 2u_{2,\,j+1} = u_{1,\,j}, \quad \because u_{2,\,j} = u_{3,\,j}, \text{ etc.} \qquad \text{...(ii)}$$

Computing the values for $j = 0,\ 1,\ 2,$

$j = 0 \Rightarrow 3u_{11} - u_{21} = 2.4;\ -u_{11} + 2u_{21} + 1.4$ giving $u_{11} = 1.24,\ u_{21} = 1.32.$

$j = 1 \Rightarrow 3u_{12} - u_{22} = 2.125;\ -u_{12} + 2u_{22} = 1.20$ giving $u_{12} = 1.090,\ u_{22} = 1.145.$

$j = 2 \Rightarrow 3u_{13} - u_{23} = 2.055;\ -u_{13} + 2u_{23} = 1.090$ giving $u_{13} = 1.040,\ u_{23} = 1.065.$

The values of u are shown in the table below:

$x \rightarrow$	*0*	*0.2*	*0.4*	*0.6*	*0.8*	*1.0*
$t = 0.00$	1.0	1.4	1.8	1.8	1.4	1.0
0.08	1.0	1.24	1.32	1.32	1.24	1.0
		(1.20)	(1.325)			
0.16	1.0	1.090	1.145	1.145	1.090	1.0
		(1.086)	(1.139)			
0.24	1.0	1.040	1.065	1.065	1.040	1.0
		(1.037)	(1.060)			

Note: The bracketed figures correspond to Explicit scheme with $\Delta t = 0.02$.

Example 11.4

Given the heat conduction equation

$$\frac{\partial u}{\partial t} = \frac{\partial^2 u}{\partial x^2}, \quad 0 \le x \le 0.5,\ t > 0$$

with boundary conditions

$$u(0,\,t) = 0 \quad \text{and} \quad \frac{\partial u}{\partial x} = 4 \text{ at } x = 0.5$$

and initial condition $u(x,\ 0) = 4x^2$.

Solve the above problem dividing the interval $[0,\ 0.5]$ into two equal parts, by Explicit method for $t = 0\ (0.025)\ 0.200$. Approximate the derivative boundary condition by backward difference.

Solution

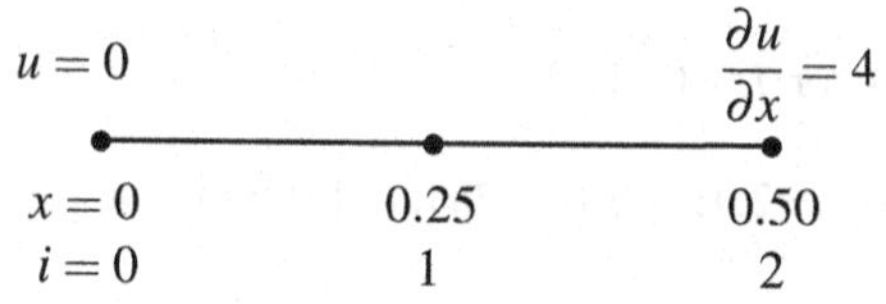

$$\Delta x = 0.25,\ \Delta t = 0.025. \quad r = \Delta t/\Delta x^2 = 0.025/0.0625 = 0.4.$$

$$u_{i,\ j+1} = 0.4u_{i-1,\ j} + (1 - 0.8)u_{i,\ j} + 0.4u_{i+1,\ j}$$

$$= \frac{1}{5}[2(u_{i-1,\ j} + u_{i+1,\ j}) + u_{i,\ j}]$$

For $i = 1 \Rightarrow \quad u_{1,\ j+1} = \dfrac{1}{5}(2u_{2,\ j} + u_{i,\ j}) \quad \because u_{0,\ j} = 0.$ ....(i)

At $i = 2$, approximating b.c. by backward difference

$$\frac{u_{2,\ j+1} - u_{1,\ j+1}}{0.25} = 4 \text{ gives } u_{2,\ j+1} = u_{1,\ j+1} + 1.$$ ....(ii)

Computed values are given below:

t \ x	0	0.25	0.50	t \ x	0	0.25	0.50
0.000	0	0.25	1.00	0.125	0	0.9287	1.9287
0.025	0	0.45	1.45	0.150	0	0.9572	1.9572
0.050	0	0.67	1.67	0.175	0	0.9743	1.9743
0.075	0	0.802	1.802	0.200	0	0.9846	1.9846
0.100	0	0.8812	1.8812				

Note: For simplification substitute $u_{2,\ j}$ from (*ii*) in (*i*) for $j = 1$ onwards which gives

$$u_{1,\ j+1} = \frac{1}{5}[3u_{1,\ j+2}],\quad j = 1,\ 2,\ \ldots. \qquad \text{....}(iii)$$

Example 11.5

Solve the problem of Example 11.4 replacing the boundary condition at $x = 0.5$ by central difference.

Solution $r = 0.4$

For $i = 1 \Rightarrow\ u_{1,\ j+1} = \frac{1}{5}(2u_{2,\ j} + u_{1,\ j})$ ....(*i*)

For $i = 2 \Rightarrow$ Replacing derivative by central difference

$$\frac{u_{3,\ j} - u_{1,\ j}}{0.5} = 4 \text{ giving } u_{3,\ j} = u_{1,\ j} + 2$$

where $u_{3,\ j}$ is a fictitious point.
Applying Explicit formula for $i = 2$

$$u_{2,j+1} = \frac{1}{5}(2u_{1,j} + u_{2,j} + 2u_{3,j}) = \frac{1}{5}(2u_{1,j} + u_{2,j} + 2u_{1,j} + 4)$$

$$= \frac{1}{5}[4(u_{1,j+1}) + u_{2,j}] \qquad \text{....}(ii)$$

Computed value from (*i*) and (*ii*) are given below:

t \ x	*0*	*0.25*	*0.50*	t \ x	*0*	*0.25*	*0.50*
0.000	0	0.25	1.00	0.125	0	0.8084	1.7286
0.025	0	0.45	1.20	0.150	0	0.8531	1.7924
0.050	0	0.57	1.40	0.175	0	0.8876	1.8410
0.075	0	0.674	1.536	0.200	0	0.9139	1.8783
0.100	0	0.7492	1.6464				

Example 11.6

Solve the problem of Example 11.4 by C–N method with $\Delta t = 0.1$ and approximating the derivative boundary condition by backward difference.

Solution $\Delta t = 0.1,\ \Delta x = 0.25;\ \ r = 1.6$
Putting value of r in C–N scheme

$$-ru_{i-1,j+1}+2(1+r)u_{i,j+1}-ru_{i+1,j+1}=ru_{i-1,j}+2(1-r)u_{i,j}+ru_{i+1,j}$$

$$-4u_{i-1,j+1}+13u_{i,j+1}-4u_{i+1,j+1}=4u_{i-1,j}-3u_{i,j}+4u_{i+1,j}$$

For $i=1$

$$-4u_{0,j+1}+13u_{1,j+1}-4u_{2,j+1}=4u_{0,j}-3u_{1,j}+4u_{2,j}$$

or

$$13u_{1,j+1}-4u_{2,j+1}=-3u_{1,j}+4u_{2,j} \quad ....(i)$$

$$\frac{u_{2,j+1}-u_{1,j+1}}{0.25}=4, \text{ giving } u_{2,j+1}=u_{1,j+1}+1, \text{ using BD.} \quad ....(ii)$$

Substituting from (*ii*) $u_{2,j+1}$ in (*i*)

$$9u_{1,j+1}=4-3u_{1,j}+4u_{2,j}$$

or

$$u_{1,j+1}=\frac{1}{9}[4(u_{2,j+1})-3u_{1,j}] \quad .... (iii)$$

We use (*iii*) for $j=0$. But for $j=1$ onwards we can use $u_{2,j}$ from (*ii*) in (*iii*) giving

$$u_{1,j+1}=\frac{1}{9}[u_{1,j}+8] \quad .... (iv)$$

Computed values are given below:

t \ x	*0*	*0.25*	*0.50*
0.000	0	0.25	1.00
0.100	0	0.8944	1.8944
		(0.8812)	
0.200	0	0.9883	1.9883
		(0.9846)	

Note: Bracketed values correspond to Explicit method (Example 11.4).

Example 11.7

Solve the problem of Example 11.4 by C–N scheme with $\Delta t=0.1$ and replacing the derivative at the boundary by central difference.

Solution $\Delta t = 0.1,\ \ \Delta x = 0.25;\ \ r = 1.6$

The C–N scheme as in Example 11.6 is given by

$$-4u_{i-1,j+1} + 13u_{i,j+1} - 4u_{i+1,j+1} = 4u_{i-1,j} - 3u_{i,j} + 4u_{i+1,j}$$

$$i = 1 \Rightarrow \qquad 13u_{1,j+1} - 4u_{2,j+1} = 4u_{2,j} - 3u_{1,j} \qquad \text{...}(i)$$

Approximating boundary condition by central difference,

$$\frac{u_{3,j+1} - u_{1,j+1}}{0.5} = 4 \ \text{ giving } \ u_{3,j+1} = u_{1,j+1}. \qquad \text{...}(ii)$$

$$i = 2 \Rightarrow -4u_{1,j+1} + 13u_{2,j+1} - 4u_{3,j+1} = 4u_{1,j} - 3u_{2,j} + 4u_{3,j}$$

Using (*ii*) for $j+1$ as well as for j in the above, we get

$$-8u_{1,j+1} + 13u_{2,j+1} = 8(u_{1,j} + 2) - 3u_{2,j}. \qquad \text{...}(iii)$$

For $j = 0$, i.e., for the first time level

$$13u_{11} - 4u_{21} = 3.250$$

$$-8u_{11} + 13u_{21} = 15.00$$

Solving these, $u_{11} = 0.7463,\ u_{21} = 1.6131$
For $j = 1$: for second time level

$$13u_{12} - 4u_{22} = 4.2135$$

$$-8u_{12} + 13u_{22} = 17.1311$$

Solving the above gives $u_{12} = 0.9000,\ u_{22} = 1.8716$
The values are given in tabular form below:

t \ x	*0*	*0.25*	*0.50*
0.000	0	0.25	1.00
0.100	0	0.7463	1.6131
		(0.7492)	(1.6464)
0.200	0	0.9000	1.8716
		(0.9139)	(1.8783)

Note: Bracketed figures correspond to Explicit scheme (Example 11.5).

11.5.11 Zero-time discontinuity at endpoints

In a parabolic equation an initial condition like, $u(x,\ 0) = f(x)$ is prescribed at $t = 0$. There are also imposed boundary conditions of the form $\alpha\frac{\partial u}{\partial x} + \beta u = \gamma,\ t > 0$ where $\alpha,\ \beta,\ \gamma$ are constants including zero at the endpoints. Usually the boundary condition does not match with the value of $f(x)$ and/or its derivative at the endpoint at $t = 0$. Thus there may appear a discontinuity at zero time at the endpoint. For instance, in Example 11.1, one or both ends could be subjected to boundary conditions different from unit temperature or in Example 11.4 there could have been prescribed at $x = 0.5$, $\frac{\partial u}{\partial x} = 2$ instead of $\frac{\partial u}{\partial x} = 4$ that matched with the initial condition (See Examples 11.8 and 11.9).

On account of discontinuity, the finite difference solution near the endpoint(s) gives rise to initial errors for short period of time which smooths out as computations proceed. However, this error may be quite significant depending on the magnitude of discontinuity. Further, as large Δt will produce larger errors, small value of Δt may be used for first few steps. In general C–N scheme should provide better result as compared to other methods since the information is used at two levels of time. We may also use the analytical solution for small time till the effect of discontinuity tapers off near the boundary. One may say that if the analytical solution is available then why not to use it for all times. In fact, the analytical solution is always expressed in terms of infinite series which involves exponential and trigonometric functions. It will require much greater efforts to evaluate the value of $u(x,\ t)$ for different values of x and t in comparison to that in a finite difference method.

Example 11.8

Given a parabolic equation

$$\frac{\partial u}{\partial t} = \frac{\partial^2 u}{\partial x^2},\quad 0 \le x \le 0.6,\ t > 0$$

with initial condition $u(x,\ 0) = \frac{1}{2}(1-x)^2,\ t = 0$. Following boundary conditions are imposed for $t > 0$ at the endpoints;

$$\frac{\partial u}{\partial x} = -0.5,\quad x = 0$$

and $\qquad u(0.6,\ t) = 0,\quad x = 0.6.$

Use C–N method to find the solution dividing the interval $[0,\ 0.6]$ into three equal parts and taking $\Delta t = 0.02$. Compute the values for $t = 0\ (0.02)\ 0.06$ approximating the derivative boundary condition by forward difference.

Solution

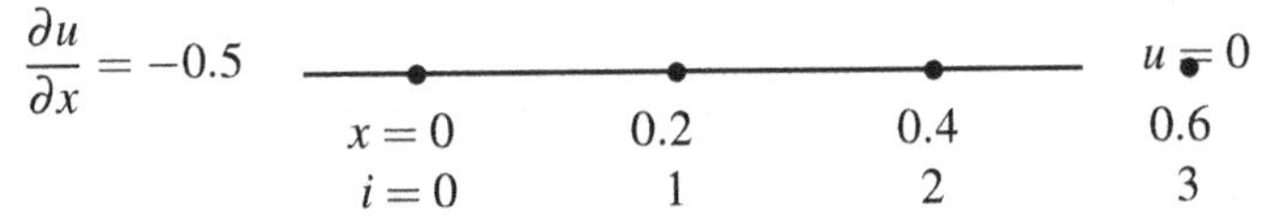

$\Delta t = 0.02, \;\; \Delta x = 0.2; \;\; r = 0.5.$

C–N scheme with $r = 0.5$ becomes

$$-u_{i-1,\,j+1} + 6u_{i,\,j+1} - u_{i+1,\,j+1} = u_{i-1,\,j} + 2u_{i,\,j} + u_{i+1,\,j} \qquad ....(i)$$

Replacing b.c. at $x = 0$ by forward difference

$$\frac{u_{1,\,j+1} - u_{0,\,j+1}}{0.2} = -0.5 \;\; \text{giving} \;\; u_{0,\,j+1} = u_{1,\,j+1} + 0.1, \;\; j \geq 0 \qquad ....(ii)$$

Writing (i) for $i = 1$:

$$-u_{0,\,j+1} + 6u_{1,\,j+1} - u_{2,\,j+1} = u_{0,\,j} + 2u_{1,\,j} + u_{2,\,j} \qquad ....(iii)$$

For $j = 0$ eqn. (iii) becomes

$$-u_{0,\,1} + 6u_{1,\,1} - u_{2,\,1} = u_{0,0} + 2u_{1,0} + u_{2,0}$$

or $$5u_{11} - u_{21} = 3u_{1,0} + u_{2,0} + 0.1, \;\; \text{using } (ii). \qquad .... (iv)$$

For $j > 0$ equation can be simplified on left and right side as

$$5u_{1,\,j+1} - u_{2,\,j+1} = 3u_{1,\,j} + u_{2,\,j} + 0.2 \qquad ....(v)$$

Writing (i) for $i = 2$:

$$-u_{1,\,j+1} + 6u_{2,\,j+1} - u_{3,\,j+1} = u_{1,\,j} + 2u_{2,\,j} + u_{3,\,j}, \qquad ....(vi)$$

For $j = 0$ equation (vi) becomes

$$-u_{1,1} + 6u_{2,1} = u_{1,0} + 2u_{2,0} + u_{3,0}, \;\; \because u_{3,\,j+1} = 0 \qquad ....(vii)$$

For $j > 0$ equation (vi) gives

$$-u_{1,\,j+1} + 6u_{2,\,j+1} = u_{1,\,j} + 2u_{2,\,j}, \;\; j > 0 \qquad ....(viii)$$

Computations are shown below:
For $j = 0$, i.e., for first time level; $t = 0.02$

$i = 0 \Rightarrow$ $$5u_{11} - u_{21} = 0.5 + 2 \times 0.32 + 0.18 + 0.1 = 1.42$$

$i = 1 \Rightarrow$ $$-u_{11} + 6u_{21} = 0.32 + 2 \times 0.18 + 0.08 = 0.76$$

Solving the above equations

$$u_{11} = 0.32,\ u_{21} = 0.18$$

For $j = 1$, i.e., second time level; $t = 0.04$

$i = 0 \Rightarrow$ $$5u_{12} - u_{22} = 3 \times 0.32 + 0.18 + 0.2 = 1.34$$

$i = 1 \Rightarrow$ $$-u_{12} + 6u_{22} = 0.32 + 2 \times 0.18 = 0.68$$

From above equations

$$u_{12} = 0.3004,\ u_{22} = 0.1634$$

For $j = 2$, i.e., $t = 0.06$

$i = 0 \Rightarrow$ $$5u_{13} - u_{23} = 3 \times 0.3004 + 0.1634 + 0.2 = 1.2644$$

$i = 1 \Rightarrow$ $$-u_{13} + 6u_{23} = 0.3004 + 2 \times 0.1634 = 0.6272$$

$$u_{13} = 0.2830,\ u_{23} = 0.1517$$

The values of u_{ij}, are tabulated below:

	$i = 0$	$i = 1$	$i = 2$	$i = 3$
t \ x	*0*	*0.2*	*0.4*	*0.6*
0.00	0.50	0.32	0.18	0.08
0.02	0.42	0.32	0.18	0
0.04	0.4004	0.3004	0.1634	0
0.06	0.3830	0.2830	0.1517	0

Note: There is discontinuity at zero time in the derivative at $x = 0$ and in the function value at $x = 0.6$.

Example 11.9

Solve the problem of Example 11.8 by Explicit method replacing the boundary condition at $x = 0$ by forward difference for $t = 0\ (0.02)\ 0.10$. Take the same value of Δt and Δx.

Solution $\Delta t = 0.02,\ \Delta x = 0.2;\ r = 0.5$

The Explicit scheme for $r = 0.5$ becomes

$$u_{i,j+1} = \frac{1}{2}(u_{i-1,j} + u_{i+1,j}). \qquad ....(i)$$

The forward difference replacement of $\frac{\partial u}{\partial x} = -0.5$ at $x = 0$ gives

$$\frac{u_{1,j+1} - u_{0,j+1}}{0.2} = 0.5 \text{ or } u_{0,j+1} = u_{1,j+1} + 0.10 \qquad ....(ii)$$

The computed values of $u_{i,j}$ from formulae (*i*) and (*ii*) are given below:

t \ x	*0*	*0.2*	*0.4*	*0.6*
0.00	0.50	0.32	0.18	0.08
0.02	0.44	0.34	0.20	0
0.04	0.42	0.32	0.17	0
0.06	0.395	0.295	0.16	0
0.08	0.3775	0.2775	0.1475	0
0.10	0.3625	0.2625	0.1388	0

11.5.12 Parabolic equation in two dimensions

A two-dimensional parabolic equation may be represented by partial differential equation

$$\frac{\partial u}{\partial t} = \frac{\partial^2 u}{\partial x^2} + \frac{\partial^2 u}{\partial y^2}\ ;\quad 0 \le x \le a,\quad 0 \le y \le b;\quad t > 0. \qquad (11.63)$$

An initial condition may be prescribed as

$$u(x,\ y,\ 0) = g(x,\ y),\quad t = 0. \qquad (11.64)$$

Some kind of boundary conditions are to be imposed on $x = 0,\ x = a$ and $y = 0,\ y = b$. Let us suppose Dirichlet conditions are prescribed on all these sides, i.e., value of u is given.

The solution domain of (11.63) is defined by $\text{D} = [0 \le x \le a] \times [0 \le y \le b] \times [t > 0]$.

To solve the problem we subdivide the intervals $[0 \le x \le a]$ and $[0 \le y \le b]$ into, say, M subintervals each of width Δx along x-axis and into N subintervals each of width Δy along y-axis respectively such that $\text{M}\Delta x = a$ and $\text{N}\Delta y = b$. (See Fig. 11.6). Let us denote the points of subdivision on x-axis by $x_i,\ i = 0(1)\text{M}$ where $x_0 = 0$ and $x_{\text{M}} = a$ and on y-axis by $y_j,\ j = 0\ (1)\ \text{N}$ where $y_0 = 0$ and $y_{\text{N}} = b$. By drawing lines $x = x_i,\ i = 1(1)\text{M}$ and $y = y_j,\ j = 1(1)\text{N}$, the region $[0 \le x \le a] \times [0 \le y \le b]$ is subdivided and we shall compute the value of $u(x_i,\ y_j,\ t),\ i = 1(1)\ \text{M}-1,\ j = 1(1)\ \text{N}-1$ for different t.

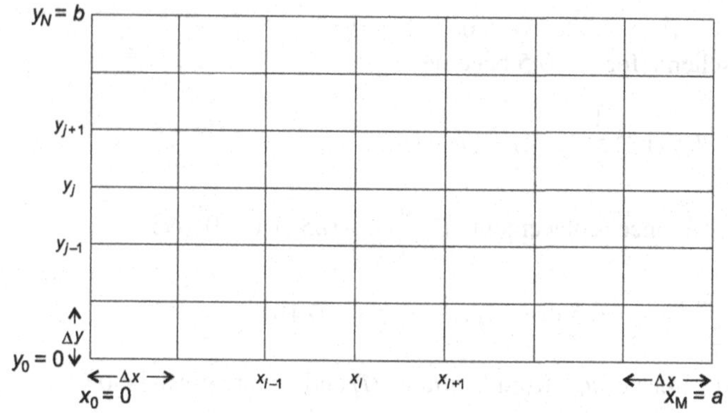

Figure 11.6 Subdivision of *x*-*y* region.

Let us represent time along an axis, say, t-axis which is normal to the x-y plane. The computations are carried in a step-by-step manner choosing a suitable step size Δt. Let us assume that values of u are known at the kth time level (step), i.e., $t = k\Delta t$ and we have to compute their values at $(k+1)^{\text{th}}$ level (step). Let us denote by $u_{i,j}^k$ the computed value of u at the mesh point $(i,\ j,\ k)$ in a three-dimensional domain where $x = i\Delta x = x_i,\ y = j\Delta y = y_j$ and $t = k\Delta t = t_k$ (See Fig. 11.7).

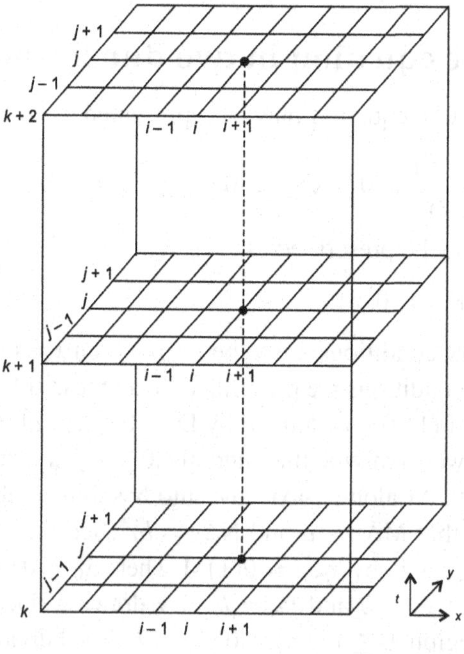

Figure 11.7 Subdivision of space and time domain.

The Explicit scheme and C–N scheme may be invoked for solving the two-dimensional (space) problem in the following way:

Explicit Scheme

We discretize the p.d.e. (11.63) at the meshpoint $(i,\ j,\ k)$ approximating time derivative by forward difference and the space derivative by central difference. Let it be denoted as

$$\left(\frac{\partial u}{\partial t}\right)_{i,\ j,\ k} = \left(\frac{\partial^2 u}{\partial x^2}\right)_{i,\ j,\ k} + \left(\frac{\partial^2 u}{\partial y^2}\right)_{i,\ j,\ k}$$

which after discretization gives

$$\frac{u_{i,j}^{k+1} - u_{i,j}^{k}}{\Delta t} + \mathrm{O}(\Delta t) = \frac{u_{i-1,j}^{k} - 2u_{i,j}^{k} + u_{i+1,j}^{k}}{\Delta x^2} + \frac{u_{i,j-1}^{k} - 2u_{i,j}^{k} + u_{i,j+1}^{k}}{\Delta y^2} + \mathrm{O}(\Delta x^2) + \mathrm{O}(\Delta y^2)$$

Neglecting the truncation error the above may be written as

$$u_{i,j}^{k+1} = u_{i,j}^{k} + \Delta t\left[\frac{u_{i-1,j}^{k} - 2u_{i,j}^{k} + u_{i+1,j}^{k}}{\Delta x^2} + \frac{u_{i,j-1}^{k} - 2u_{i,j}^{k} + u_{i,j+1}^{k}}{\Delta y^2}\right], \qquad (11.65)$$

$$i = 1(1)\,\mathrm{M}-1,\ j = 1(1)\,\mathrm{N}-1.$$

Formula (11.65) gives the value of u at the $(k+1)^{\text{th}}$ time level explicitly in terms of the known values at the kth level. It may be shown that the scheme (11.65) is stable only for $\Delta t\left\{\frac{1}{\Delta x^2} + \frac{1}{\Delta y^2}\right\} \leq 0.5$. If $\Delta x = \Delta y$, then (11.65) may be written as,

$$u_{i,j}^{k+1} = r(u_{i-1,j}^{k} + u_{i+1,j}^{k}) + (1-4r)u_{i,j}^{k} + r(u_{i,j-1}^{k} + u_{i,j+1}^{k}),\ r = \Delta t/\Delta x^2. \qquad (11.66)$$

C-N Scheme

The p.d.e. (11.63) is discretized at $\left(i,\ j,\ k+\frac{1}{2}\right)$ replacing the time derivative by central difference and the space derivatives by the average of the second derivatives at the kth and $(k+1)^{\text{th}}$ level replacing them by central difference. That is,

$$\left(\frac{\partial u}{\partial t}\right)_{i,j,k+\frac{1}{2}} = \left(\frac{\partial^2 u}{\partial x^2} + \frac{\partial^2 u}{\partial y^2}\right)_{i,j,k+\frac{1}{2}}$$

$$= \frac{1}{2}\left[\left(\frac{\partial^2 u}{\partial x^2} + \frac{\partial^2 u}{\partial y^2}\right)_{i,j,k} + \left(\frac{\partial^2 u}{\partial x^2} + \frac{\partial^2 u}{\partial y^2}\right)_{i,j,k+1}\right]$$

or $$\frac{u_{i,j}^{k+1}-u_{i,j}^{k}}{\Delta t}+O(\Delta t^2)=\frac{1}{2}\left[\frac{u_{i-1,j}^{k}-2u_{i,j}^{k}+u_{i+1,j}^{k}}{\Delta x^2}+\frac{u_{i,j-1}^{k}-2u_{i,j}^{k}+u_{i,j+1}^{k}}{\Delta y^2}\right]$$

$$+\frac{1}{2}\left[\frac{u_{i-1,j}^{k+1}-2u_{i,j}^{k+1}+u_{i+1,j}^{k+1}}{\Delta x^2}+\frac{u_{i,j-1}^{k+1}-2u_{i,j}^{k+1}+u_{i,j+1}^{k+1}}{\Delta y^2}\right]+O(\Delta x^2)+O(\Delta y^2)$$

Transposing the terms to be computed on the left and the terms which are known to the right, the above can be written after neglecting the truncation error as,

$$-\Delta t\left[\frac{u_{i-1,j}^{k+1}-2u_{i,j}^{k+1}+u_{i+1,j}^{k+1}}{\Delta x^2}+\frac{u_{i,j-1}^{k+1}-2u_{i,j}^{k+1}+u_{i,j+1}^{k+1}}{\Delta y^2}\right]+2u_{i,j}^{k+1}$$

$$=\Delta t\left[\frac{u_{i-1,j}^{k}-2u_{i,j}^{k}+u_{i+1,j}^{k}}{\Delta x^2}+\frac{u_{i,j-1}^{k}-2u_{i,j}^{k}+u_{i,j+1}^{k}}{\Delta y^2}\right]+2u_{i,j}^{k}, \tag{11.67}$$

$i=1(1)\,M-1,\ j=1(1)\,N-1$

If $\Delta x=\Delta y$ then after putting $r=\Delta t/\Delta x^2$, (11.67) reduces to

$$-r(u_{i-1,j}^{k+1}+u_{i+1,j}^{k+1})+2(1+2r)u_{i,j}^{k+1}-r(u_{i,j-1}^{k+1}+u_{i,j+1}^{k+1})$$

$$=r(u_{i-1,j}^{k}+u_{i+1,j}^{k})+2(1-2r)u_{i,j}^{k}+r(u_{i,j-1}^{k}+u_{i,j+1}^{k}), \tag{11.68}$$

$i=1(1)\,M-1,\ j=1(1)\,N-1.$

At each time step we will be required to solve $(M-1)\times(N-1)$ equations, while the scheme is stable for any value of $\Delta t\left(\frac{1}{\Delta x^2}+\frac{1}{\Delta y^2}\right)$.

11.5.13 Alternating Direction Implicit (ADI) method

We have seen that the Explicit method is stable only for $\Delta t\left(\frac{1}{\Delta x^2}+\frac{1}{\Delta y^2}\right)\le\frac{1}{2}$ which restricts the size of Δt to be very small. On the other hand C–N method although unconditionally stable, requires solving a large number of equations, $(M-1)\times(N-1)$, at each time step which is very time-consuming process. It was stated in Chapter 2 that in solving a system of $n\times n$ equations it involves $O\left(\frac{n^3}{3}\right)$ multiplications/divisions and $O\left(\frac{n^3}{6}\right)$ additions/subtractions. Thus even for a moderately large system, say $n=9\times 8=72$, $(M=10,\ N=9)$, the amount of computational work may be enormous.

The ADI method suggested by Peaceman and Rachford in 1955, is a combination of explicit and implicit methods in a sense. The method is unconditionally stable, i.e., for all values of $\Delta t/\Delta x^2$ and $\Delta t/\Delta y^2$ and at the same time the computations are reduced greatly. The method is performed in two stages; one, from k^{th} level to $(k+1)^{\text{th}}$ level and two, from $(k+1)^{\text{th}}$ level to $(k+2)^{\text{th}}$ level. These two stages together comprising a time interval of $2\Delta t$ may be considered as one operation of ADI method which is repeated for the next two intervals and so on. Let us describe the method below.

Stage-1. Like C–N scheme the discretization may be made at the mesh point $\left(i,\ j,\ k+\dfrac{1}{2}\right)$ but instead of taking the average of both the space derivatives, one of the derivatives is approximated at $(k+1)^{\text{th}}$ level and the other at the known level, i.e., k^{th} level. Suppose $\dfrac{\partial^2 u}{\partial x^2}$ is approximated at the unknown level and $\dfrac{\partial^2 u}{\partial y^2}$ at the known level, then in the first stage we have (neglecting error terms),

$$\frac{u_{i,j}^{k+1}-u_{i,j}^{k}}{\Delta t}=\frac{u_{i-1,j}^{k+1}-2u_{i,j}^{k+1}+u_{i+1,j}^{k+1}}{\Delta x^2}+\frac{u_{i,j-1}^{k}-2u_{i,j}^{k}+u_{i,j+1}^{k}}{\Delta y^2}$$

$$\text{or}\quad -\frac{\Delta t}{\Delta x^2}\left[u_{i-1,j}^{k+1}-2u_{i,j}^{k+1}+u_{i+1,j}^{k+1}\right]+u_{i,j}^{k+1}=\frac{\Delta t}{\Delta y^2}\left[u_{i,j-1}^{k}-2u_{i,j}^{k}+u_{i,j+1}^{k}\right]+u_{i,j}^{k},$$

$$\text{or}\quad -r_1u_{i-1,j}^{k+1}+(1+2r_1)u_{i,j}^{k+1}-r_1u_{i+1,j}^{k+1}=r_2u_{i,j-1}^{k}+(1-2r_2)u_{i,j}^{k}+r_2u_{i,j+1}^{k},\qquad (11.69)$$

where $r_1=\Delta t/\Delta x^2$, $r_2=\Delta t/\Delta y^2$; $i=1(1)\,\mathrm{M}-1$ for each $j=1(1)\,\mathrm{N}-1$.

Stage-2. Advancing from $(k+1)^{\text{th}}$ to $(k+2)^{\text{th}}$ level $\dfrac{\partial^2 u}{\partial y^2}$ is approximated at the unknown level $(k+2)$ and $\dfrac{\partial^2 u}{\partial x^2}$ at the known level, i.e., $(k+1)^{\text{th}}$. Thus, the direction of discretization of $\dfrac{\partial^2 u}{\partial x^2}$ and $\dfrac{\partial^2 u}{\partial y^2}$ are altered at the known and unknown levels. So we have,

$$\frac{u_{i,j}^{k+2}-u_{i,j}^{k+1}}{\Delta t}=\frac{u_{i-1,j}^{k+1}-2u_{i,j}^{k+1}+u_{i+1,j}^{k+1}}{\Delta x^2}+\frac{u_{i,j-1}^{k+2}-2u_{i,j}^{k+2}+u_{i,j+1}^{k+2}}{\Delta y^2}$$

$$\text{or}\quad -\frac{\Delta t}{\Delta y^2}\left[u_{i,j-1}^{k+2}-2u_{i,j}^{k+2}+u_{i,j+1}^{k+2}\right]+u_{i,j}^{k+2}=\frac{\Delta t}{\Delta x^2}\left[u_{i-1,j}^{k+1}-2u_{i,j}^{k+1}+u_{i+1,j}^{k+1}\right]+u_{i,j}^{k+1},$$

$$\text{or}\quad -r_2u_{i,j-1}^{k+2}+(1+2r_2)u_{i,j}^{k+2}-r_2u_{i,j+1}^{k+2}=r_1u_{i-1,j}^{k+1}+(1-2r_1)u_{i,j}^{k+1}+r_1u_{i+1,j}^{k+1}\qquad (11.70)$$

where $r_1 = \Delta t/\Delta x^2$, $r_2 = \Delta t/\Delta y^2$; $j = 1(1)\ \text{N}-1$ for each $i = 1(1)\ \text{M}-1$.

From $(k+2)^{\text{th}}$ to $(k+4)^{\text{th}}$ level again the stage-1 and stage-2 are repeated and so on. It may be noted that from k^{th} to $(k+1)^{\text{th}}$ level we have to solve (M−1) equations (N−1) times. For 9×8 system we have to solve 9 equations in nine unknowns 8 times rather than solving 72 equations in 72 unknowns. Similarly from $(k+1)$ to $(k+2)^{\text{th}}$ level we solve (N−1) equations (M−1) times which means for 9×8 system, solving 8 equations 9 times. Although we have considered Dirichlet condition other conditions may also be dealt with, easily.

Example 11.10

Given a heat conduction equation

$$\frac{\partial u}{\partial t} = \frac{\partial^2 u}{\partial x^2} + \frac{\partial^2 u}{\partial y^2},\ 0 \le x,\ y \le 0.75,\ t \ge 0$$

with initial temperature distribution

$$u(x,\ y,\ 0) = 10xy,\quad t = 0.$$

At $t \ge 0$, the sides $x = 0$ and $y = 0$ are kept at zero degree temperature while the side $x = 0.75$ is subjected to $\frac{\partial u}{\partial x} = -u$ and $y = 0.75$ is subjected to $\frac{\partial u}{\partial y} = -u$.

Use ADI method after subdividing the domain into square meshes of side 0.25 and approximating the derivative boundary conditions by forward and backward difference as required.

Take $\Delta t = 0.0625$ and compute for $t = 0\ (0.0625)\ 0.250$, i.e., four time steps only.

Solution $\Delta x = \Delta y = 0.25$ and $\Delta t = 0.0625$.

$$r_1 = \Delta t/\Delta x^2 = 1,\ r_2 = \Delta t/\Delta y^2 = 1,\ r_1 = r_2 = r = 1$$

Initial temperature distribution is given as follows:

$t = 0$

y	j				
0.75	3	0	1.0875	3.75	5.625
0.50	2	0	1.25	2.5	3.75
0.25	1	0	0.625	1.25	1.875
0	0	0	0	0	0
↑ y	↑ j	$i \rightarrow 0$	1	2	3
	x	0	0.25	0.50	0.75

Formulae used for $r = 1$

$$[-u_{i-1,j} + 3u_{i,j} - u_{i+1,j}]^{k+1} = [u_{i,j-1} - u_{i,j} + u_{i,j+1}]^{k},$$

$i = 1,\ 2$ for each $j = 1,\ 2$.

$$[-u_{i,j-1} + 3u_{i,j} - u_{i,j+1}]^{k+2} = [u_{i-1,j} - u_{i,j} + u_{i+1,j}]^{k+1},$$

$j = 1,\ 2$ for each $i = 1,\ 2$.

Approximation of derivative boundary conditions

$$\frac{u_{3j} - u_{2j}}{\Delta x} = -u_{3j} \text{ or } u_{3j} = \frac{u_{2j}}{1 + \Delta x} = 0.8u_{2j},\quad j = 1,\ 2$$

$$\frac{u_{i,3} - u_{i2}}{\Delta y} = -u_{i3} \text{ or } u_{i,3} = \frac{u_{i,2}}{1 + \Delta y} = 0.8u_{i,2},\quad i = 1,\ 2$$

$\boldsymbol{t = 0.0625}$: Superscripts are omitted

$\boldsymbol{j = 1}$

$i = 1 \Rightarrow \qquad -u_{01} + 3u_{11} - u_{21} = 0.625$ or $3u_{11} - u_{21} = 0.625$ ....(1)

$i = 2 \Rightarrow \qquad -u_{11} + 3u_{21} - u_{31} = 1.25$ or $-u_{11} + 2\cdot 2u_{21} = 1.25$ ....(2)

Solving (1) and (2)gives $\quad u_{11} = 0.4688,\ u_{21} = 0.7812$
Also from boundary condition we get, $u_{31} = 0.8u_{21} = 0.6250$

$\boldsymbol{j = 2}$

$i = 1 \Rightarrow \qquad -u_{02} + 3u_{12} - u_{22} = 1.25$ or $3u_{12} - u_{22} = 1.25$ ....(3)

$i = 2 \Rightarrow \qquad -u_{12} + 3u_{22} - u_{32} = 2.5$ or $-u_{12} + 2\cdot 2u_{22} = 2.5$ ....(4)

Solving (3) and (4) gives $\quad u_{12} = 0.9375,\ u_{22} = 1.5625$
From boundary condition, $u_{32} = 0.8u_{22} = 1.25$

$\boldsymbol{t = 0.125}$

$\boldsymbol{i = 1}$

$j = 1 \Rightarrow \qquad -u_{10} + 3u_{11} - u_{12} = 0.3124$ or $3u_{11} - u_{12} = 0.3124$ ....(5)

$j = 2 \Rightarrow \qquad -u_{11} + 3u_{12} - u_{13} = 0.625$ or $-u_{11} + 2\cdot 2u_{12} = 0.625$ ....(6)

From (5), (6) and boundary conditions,

$$u_{11} = 0.2343,\ u_{12} = 0.3906,\ u_{13} = 0.3125$$

$i = 2$

$j = 1 \Rightarrow \qquad -u_{20} + 3u_{21} - u_{22} = 0.3126$ or $3u_{21} - u_{22} = 0.3126$ ....(7)

$j = 2 \Rightarrow \qquad -u_{21} + 3u_{22} - u_{23} = 0.625$ or $-u_{21} + 2 \cdot 2u_{22} = 0.625$ ....(8)

From (7), (8) and boundary conditions,

$$u_{21} = 0.2344,\ u_{22} = 0.3906,\ u_{23} = 0.3125.$$

The values are given in tabular form as follows:

$t = 0.0625$

$j \uparrow$ / $i \rightarrow$	0	1	2	3
3	*	0.75	1.25	*
2	0	0.9375	1.5625	1.25
1	0	0.4688	0.7812	0.6250
0	0	0	0	*

$t = 0.1250$

$j \uparrow$ / $i \rightarrow$	0	1	2	3
3	*	0.3125	0.3125	*
2	0	0.3906	0.3906	0.3125
1	0	0.2343	0.2344	0.1875
0	0	0	0	*

Note: *denotes singularity in boundary condition or solution

$t = 0.1875$

$j = 1$

$i = 1 \Rightarrow \qquad 3u_{11} - u_{21} = 0.1563$ ....(9)

$i = 2 \Rightarrow \qquad -u_{11} + 2 \cdot 2u_{21} = 0.1562$ ....(10)

solving (9) and (10) and form boundary conditions,

$$u_{11} = 0.0891,\ u_{21} = 0.1116,\ u_{31} = 0.0893$$

$j = 2$

$i = 1 \Rightarrow \qquad 3u_{12} - u_{22} = 0.1562$ ....(11)

$i = 2 \Rightarrow \qquad -u_{12} + 2 \cdot 2u_{22} = 0.1563$ ....(12)

From (11), (12) and boundary conditions,

$$u_{12} = 0.0893,\ u_{22} = 0.1116,\ u_{32} = 0.0893$$

$t = 0.2500$

$i = 1$

$j = 1 \Rightarrow \qquad 3u_{11} - u_{12} = 0.0223$ ....(13)

$j = 2 \qquad -u_{11} + 2 \cdot 2u_{12} = 0.0223$ ....(14)

From (13), (14) and boundary conditions,

$$u_{11} = 0.0127,\ u_{12} = 0.0159,\ u_{13} = 0.0127.$$

$i = 2$

$j = 1 \Rightarrow \qquad 3u_{21} - u_{22} = 0.067$ ....(15)

$-u_{21} + 2 \cdot 2u_{22} = 0.067$ ....(16)

From (15), (16) and boundary conditions,

$$u_{21} = 0.0383,\ u_{22} = 0.0479,\ u_{23} = 0.0383$$

The above values are shown in tabular form as follows:

$t = 0.1875$

*	0.0714	0.0893	*
0	0.0893	0.1116	0.0893
0	0.0893	0.1116	0.0893
0	0	0	*

$t = 0.2500$

*	0.0127	0.0383	*
0	0.0159	0.0479	0.0383
0	0.0127	0.0383	0.0306
0	0	0	*

11.5.14 Non-rectangular space domains

When dealing with practical problems more often than not the geometrical shape of the space domain is not a square or rectangle. Thus when a rectangular or square mesh is fitted on such a domain some of the meshes near the boundary may remain incomplete. Hence approximation of derivatives at the mesh points near the boundary may involve unequal mesh lengths in x-direction or in y-direction or in both. For example, consider a semi-circular domain as shown in Fig. 11.8. Let us suppose that the initial distribution is symmetrical about y-axis and that Dirichlet conditions are prescribed on the circular boundary as well as on the x-axis which is also symmetrical about y-axis. Further, let the domain be subdivided taking $\Delta x = \Delta y$. Then we need to consider the solution of the problem at the six internal meshpoints numbered 1 to 6 only. At the meshpoints 1, 2 and 4 the discretization of the second derivatives can be made in the normal way but at the meshpoints 3, 5 and 6 formula (11.12) has to be used for unequal interval. Such problems will be dealt with under elliptic equations in Sec. 11.6.

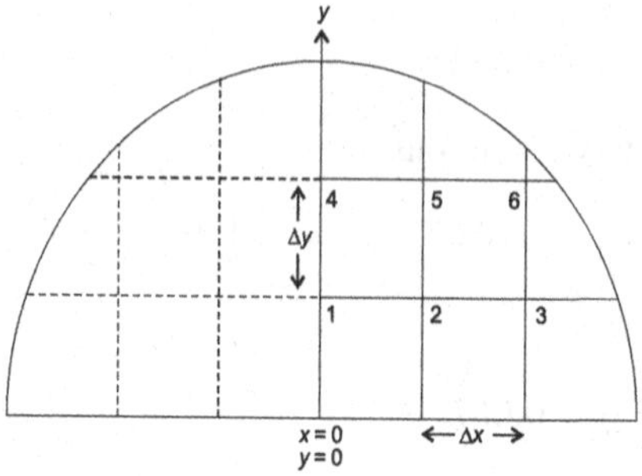

Figure 11.8 Subdivision of non-rectangular space domain.

11.6 Methods for Solving Elliptic Equations

When a system represented by a parabolic equation reaches steady state or state of equilibrium, that is, there is no further change in the value of the dependent variable anywhere in the medium as time progresses, then that ultimate state is defined by an elliptic equation. In fact some processes may attain their final state over a very short period of time so that their transient behaviour may not be of much consequence and the only information which may be of interest would be about their final state. Obviously, initial condition/state does not play any role in an elliptic equation as the solution is fully dependent on the boundary conditions. Let us write again the most familiar forms of elliptic equations:

$$\text{Laplace equation:} \quad \nabla^2 u \equiv \frac{\partial^2 u}{\partial x^2} + \frac{\partial^2 u}{\partial y^2} = 0, \tag{11.71}$$

$$\text{Poisson equation:} \quad \nabla^2 u \equiv \frac{\partial^2 u}{\partial x^2} + \frac{\partial^2 u}{\partial y^2} = f(x, y), \tag{11.72}$$

where equations (11.71) and (11.72) are defined over a domain D in the x-y plane and appropriate boundary conditions are prescribed on bounding curve C; $f(x, y)$ is a known function which denotes a source or a sink at point (x, y) in D.

As far as method for solving an elliptic equation is concerned we subdivide the domain into rectangular or square mesh and discretize the p.d.e. at the mesh points and also the derivative boundary conditions. This leads us to a set of linear system of equations which can be solved by Gaussian elimination or Gauss–Seidel method.

To illustrate the method, let us consider Poisson equation (11.72) defined over a rectangular domain $D = [0 \le x \le a] \times [0 \le y \le b]$. Let Dirichlet condition $u = u_c$ be prescribed on the three sides $x=0,\ y=0,\ y=b$ and a mixed condition $\dfrac{\partial u}{\partial x}=\alpha u+\beta$ be prescribed on $x=a$. Let us suppose the domain D is subdivided into square mesh with side h, i.e., $\Delta x=\Delta y=h$ so that $\mathrm{M}h=a$ and $\mathrm{N}h=b$. Further if $x_i=ih,\ i=0(1)\mathrm{M}$ and $y_j=jh,\ j=0(1)\mathrm{N}$, then the mesh point $(i,\ j)$ corresponds to the point $(x_i,\ y_j)$ in the domain D (See Fig. 11.9).

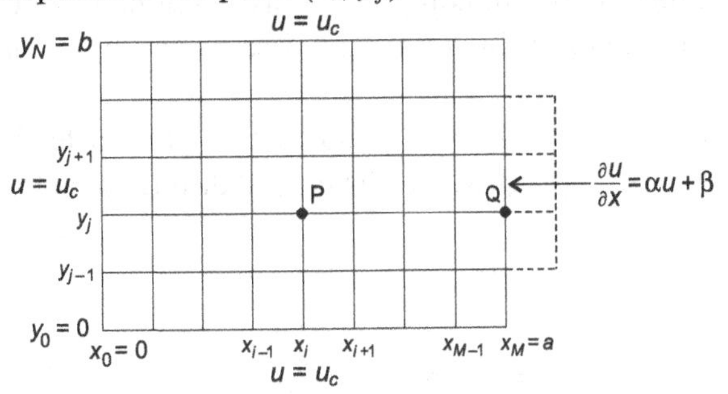

Figure 11.9 Subdivision of space domain showing fictitious points.

Discretizing (11.72) at the mesh point P$(i,\ j)$,

$$\frac{u_{i-1,j}-2u_{i,j}+u_{i+1,j}}{h^2}+\frac{u_{i,j-1}-2u_{i,j}+u_{i,j+1}}{h^2}=f_{i,j}$$

or $$u_{i-1,j}+u_{i,j-1}-4u_{i,j}+u_{i+1,j}+u_{i,j+1}=h^2 f_{i,j} \tag{11.73}$$

where $$f_{i,j}=f(x_i,y_j),\quad i=1(1)\mathrm{M},\quad j=1(1)\mathrm{N}-1.$$

11.6.1 Solution by Gauss–Seidel and Gaussian elimination

The set of simultaneous equations (11.73) can be solved by Gauss–Seidel method quite conveniently. The scheme for their solution may be written as,

$$u_{i,j}^{(n+1)}=\frac{1}{4}[u_{i-1,j}^{(n+1)}+u_{i,j-1}^{(n+1)}+u_{i+1,j}^{(n)}+u_{i,j+1}^{(n)}-h^2 f_{i,j}] \tag{11.74}$$

where $u_{i,j}^{(n+1)}$ denotes the value of $u_{i,j}$ at $(n+1)^{\text{th}}$ iteration. We may start computations taking initial values of u, i.e., $u_{i,j}^{(0)}=0$. The iterations are carried in the row-wise order, varying i in $(i,\ j), i=1(1)\mathrm{M}$ for fixed $j,\ j=1(1)\ \mathrm{N}-1$. As shown in the formula (11.74) the most recent values of u's are used as soon as they are computed. However the convergence would be very slow as the coefficient of $u_{i,j}^{n+1}$ is equal to sum of the coefficients of u's in absolute value.

However, if the number of mesh points is not large they may be numbered serially row-wise (or column-wise) and the system may be solved by Gaussian elimination method.

It is very important to note that there may be symmetry in the problem about a line or lines. Due to symmetry in the domain as well as in the boundary conditions and the function $f(x, y)$, the problem may reduce to computing only about half or quarter of the values. It must be used wherever required. In the present case, if $f(x, y) = \text{const.}$ then there will be symmetry about $y = b/2$.

The boundary condition prescribed at $x = a$ can be approximated by backward or central difference. For example at point Q(M, j), the central difference approximation will give,

$$\frac{u_{M+1,j} - u_{M-1,j}}{2h} = \alpha u_{M,j} + \beta$$

or $$u_{M+1,j} = u_{M-1,j} + 2h(\alpha u_{M,j} + \beta) \tag{11.75}$$

where $(M+1, j)$ is fictitious point.

Example 11.11

A p.d.e. $\dfrac{\partial^2 u}{\partial x^2} + \dfrac{\partial^2 u}{\partial y^2} = 0.5$ is defined over a rectangular domain $[0 \le x \le 0.6] \times [0 \le y \le 0.6]$ with following boundary conditions,

$u = 1$ on the three sides $y = 0$, $y = 0.6$ and $x = 0$;

$$\frac{\partial u}{\partial x} = u \text{ on } x = 0.6.$$

Solve the equation after subdividing the domain into square mesh of side 0.2 and approximating the derivative boundary condition by central difference. (Without using symmetry).

Solution $\Delta x = \Delta y = h = 0.2, \quad f(x, y) = 0.5$

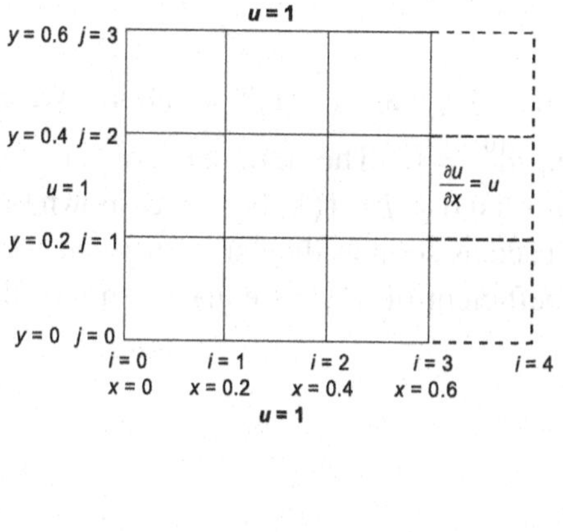

We have to compute $u_{i,j}$, $i = 1(1)3$, $j = 1(1)2$.
Approximating derivative boundary condition by central difference at the mesh point (3, 1)

$$\frac{u_{4,1} - u_{2,1}}{0.4} = u_{31} \quad \text{or} \quad u_{41} = u_{21} + 0.4u_{31} \qquad \text{....(1)}$$

Similarly at mesh point (3, 2)

$$u_{42} = u_{22} + 0.4u_{32}. \qquad \text{....(2)}$$

Discretizing the p.d.e. at various mesh points (i, j), $i = 1(1)3$, $j = 1(1)2$:

$$(1, 1) \Rightarrow \qquad 4u_{11} = u_{10} + u_{01} + u_{21} + u_{12} - h^2 f(x, y)$$

$$\text{or} \qquad u_{11}^{(n+1)} = \frac{1}{4}(u_{21}^{(n)} + u_{12}^{(n)} + 1.98) \qquad \text{....(3)}$$

$$(2, 1) \Rightarrow \qquad 4u_{21} = u_{11} + u_{20} + u_{31} + u_{22} - 0.04 \times 0.5$$

$$\text{or} \qquad u_{21}^{(n+1)} = \frac{1}{4}(u_{11}^{(n+1)} + u_{31}^{(n)} + u_{22}^{(n)} + 0.98) \qquad \text{....(4)}$$

$$(3, 1) \Rightarrow \qquad 4u_{31} = u_{21} + u_{30} + u_{41} + u_{32} - 0.02$$

$$= u_{21} + 1 + u_{21} + 0.4u_{31} + u_{32} - 0.02, \text{ using (1)}$$

$$\text{or} \qquad u_{31}^{(n+1)} = \frac{5}{18}(2u_{21}^{(n+1)} + u_{32}^{(n)} + 0.98) \qquad \text{....(5)}$$

$$(1, 2) \Rightarrow \qquad 4u_{12} = u_{02} + u_{11} + u_{22} + u_{13} - 0.02$$

$$\text{or} \qquad u_{12}^{(n+1)} = \frac{1}{4}(u_{11}^{(n+1)} + u_{22}^{(n)} + 1.98) \qquad \text{....(6)}$$

$$(2, 2) \Rightarrow \qquad 4u_{22} = u_{12} + u_{21} + u_{32} + u_{23} - 0.02$$

$$u_{22}^{(n+1)} = \frac{1}{4}(u_{12}^{(n+1)} + u_{21}^{(n+1)} + u_{32}^{(n)} + 0.98) \qquad \text{....(7)}$$

$$(3,2) \Rightarrow \qquad 4u_{32} = u_{22} + u_{31} + u_{42} + u_{33} - 0.02$$

$$= u_{22} + u_{31} + u_{22} + 0.4u_{32} + 0.98, \quad \text{using (2)}$$

$$\text{or} \qquad u_{32}^{(n+1)} = \frac{5}{18}(2u_{22}^{(n+1)} + u_{31}^{(n+1)} + 0.98). \qquad \text{....(8)}$$

We compute the value of $u_{i,j}$ using formulae (3) through (8) in that order substituting the most recent values of u. The convergence is achieved after 14 iterations. The iterated values are shown below at alternate iterations starting from first.

$j \downarrow$ / $i \rightarrow$	1	2	3
	0.495	0.3688	0.4771
	0.8642	0.8656	1.0141
	0.9756	1.0126	1.1486
1	1.0046	1.0503	1.1830
	1.0120	1.0599	1.1918
	1.0137	1.0622	1.1939
	1.0142	1.0628	1.1944
	0.6188	0.4919	0.6788
	0.9070	0.5230	1.0667
	0.9870	1.0274	1.1620
2	1.0089	1.0544	1.1866
	1.0126	1.0608	1.1926
	1.0139	1.0624	1.1941
	1.0142	1.0628	1.1944

Example 11.12

Solve the problem of Example 11.11 using its symmetry.

Solution The problem is symmetrical about $y = 0.3$ as the boundary conditions are same in the lower half and upper half of $y = 0.3$ and $f(x, y) = 0.5$ is also same (const). Therefore $u_{12} = u_{11}$, $u_{22} = u_{21}$ and $u_{32} = u_{31}$. Substituting these values in equations (3), (4) and (5) of Example 11.11 we get,

$$u_{11} = \frac{1}{3}(u_{21} + 1.98) \qquad \text{....(1)}$$

$$u_{21} = \frac{1}{3}(u_{11} + u_{31} + 0.98) \qquad \text{....(2)}$$

$$u_{31} = \frac{10}{13}(u_{21} + 0.49) \qquad \text{....(3)}$$

The solution of (1), (2) and (3) by Gauss–Seidel scheme is obtained in 11 iterations which are shown below ($j = 1$):

Iteration \ i	1	2	3	Iteration \ i	1	2	3
1	0.66	0.5467	0.7975	7	1.0132	1.0617	1.1936
2	0.8422	0.8732	1.0486	8	1.0139	1.0625	1.1942
3	0.9511	0.9932	1.1409	9	1.0142	1.0628	1.1944
4	0.9911	1.0373	1.1749	10	1.0143	1.0629	1.1945
5	1.0058	1.0536	1.1874	11	1.0143	1.0629	1.1945
6	1.0112	1.0595	1.1919	—	—	—	—

Equations can also be solved by Gaussian elimination. In that case the equations may be written, dropping suffix $j = 1$, as

$$3u_1 - u_2 = 1.98 \quad \text{....(4)}$$

$$u_1 - 3u_2 + u_3 = -0.98 \quad \text{....(5)}$$

$$u_2 - 1.3u_3 = -0.49 \quad \text{....(6)}$$

Solving (4), (5) and (6), we get

$$u_1 = 1.0143,\ u_2 = 1.0630,\ u_3 = 1.1946$$

Example 11.13

A Poisson equation $\frac{\partial^2 u}{\partial x^2} + \frac{\partial^2 u}{\partial y^2} = f(x, y)$ is defined over a semi-circular domain enclosed by the circle $x^2 + y^2 = 0.25$ and x-axis with $f(x, y) = 4(|x| + y)$. The boundary condition prescribed on the circle is $u = 1$ and on the x-axis, $\frac{\partial u}{\partial y} = 0$. Subdivide the domain into square mesh by drawing lines parallel to y-axis through $x = -0.25, 0, 0.25$ and a line parallel to x-axis through $y = 0.25$.

Solution

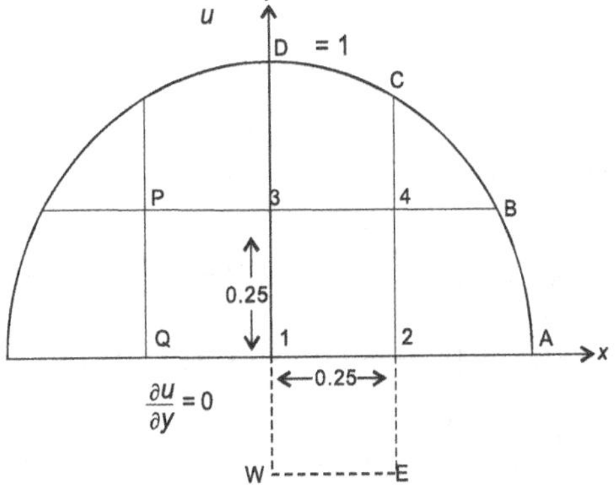

As the problem is symmetric about the y-axis we need find the solution in only quarter of the circle at the mesh points 1, 2, 3 and 4.

First let us find out the length of the sides B4 = C4. Putting $x = 0.25$ in the equation of the circle gives

$$y^2 = 0.25 - 0.0625 = 0.1875 \quad \text{or} \quad y = 0.433$$

$$C4 = 0.433 - 0.25 = 0.183 = B4.$$

The value of the function $f(x, y) = 4(|x| + y)$ at the mesh points 1, 2, 3 and 4 is given respectively as,

$$f_1 = 0,\ f_2 = 1,\ f_3 = 1 \text{ and } f_4 = 2 \qquad \text{....(1)}$$

Approximation of $\dfrac{\partial u}{\partial y} = 0$ on x-axis by central difference gives

$$u_4 = u_E \quad \text{and} \quad u_3 = u_W \qquad \text{....(2)}$$

Discretizing the p.d.e. at various mesh points:

at 1 ⇒ $u_Q + u_2 - 4u_1 + u_3 + u_W = h^2 f_1$

or $$2u_1 - u_2 - u_3 = 0 \qquad \text{....(3)}$$

at 2 ⇒ $u_1 + u_A - 4u_2 + u_4 + u_E = h^2 f_2$

or $$u_1 + 1 - 4u_2 + u_4 + u_4 = 0.0625$$

or $$u_1 - 4u_2 + 2u_4 = -0.9375 \qquad \text{....(4)}$$

at 3 ⇒ $u_P + u_4 - 4u_3 + u_1 + u_0 = h^2 f_3$

or $$2u_4 - 4u_3 + u_1 = -1 + 0.0625 \times 1$$

or $$u_1 - 4u_3 + 2u_4 = -0.9375 \qquad \text{....(5)}$$

at 4 ⇒ Using formula (11.12) for unequal interval

$$\frac{2}{0.25 \times 0.183(0.25 + 0.183)}[0.183u_3 - (0.25 + 0.183)u_4 + 0.25 \times u_B]$$

$$+ \frac{2}{0.25 \times 0.183(0.25 + 0.183)}[0.183u_2 - (0.25 + 0.183)u_4 + 0.25 \times u_C] = f_4$$

or $$0.183u_2 + 0.183u_3 - 0.866u_4 = -0.5 + 2 \times 0.25 \times 0.183 \times 0.433/2$$

or $$0.183u_2 + 0.183u_3 - 0.866u_4 = -0.4802 \qquad \text{....(6)}$$

From (4) and (5) we get $u_2 = u_3$

Using this in (1) gives $u_1 = u_2 = u_3$.

Simplifying (4) and (6) after substituting from above

$$3u_3 - 2u_4 = 0.9375 \qquad \text{....(7)}$$

$$0.366u_3 - 0.866u_4 = -0.4802 \qquad \text{....(8)}$$

Solving (7) and (8) gives,

$$u_3 = 0.9657 (= u_1 = u_2);\ u_4 = 0.9798.$$

Note:

(*i*) Although we have got $u_2 = u_3$ from equations (4) and (5), we can use symmetry of the problem about the radius joining the mesh points 1 and 4 and infer that $u_2 = u_3$.

(*ii*) The same problem can be defined over a quadrant of a circle with $\dfrac{\partial u}{\partial x} = 0$ on y-axis and $\dfrac{\partial u}{\partial y} = 0$ on x-axis; $f(x, y) = 4(x + y)$.

(*iii*) It can also be defined over the circular domain with $u = 1$ on the circle all around and $f(x, y) = 4(|x| + |y|)$.

11.6.2 Solution by SOR method

The SOR scheme for Gauss–Seidel method for computing the value of $u_{i,j}^{(n+1)}$ from (11.74) may be written following the discussion of SOR in Chapter 2, as

$$u_{i,j}^{(n+1)} = u_{i,j}^{(n)} + \omega \mathrm{R}_{i,j}$$

where $\mathrm{R}_{i,j} = \dfrac{1}{4}\left\{u_{i-1,j}^{(n+1)} + u_{i,j-1}^{(n+1)} + u_{i,j+1}^{(n)} + u_{i+1,j}^{(n)} - h^2 f_{i,j}\right\} - u_{i,j}^{(n)}$

$= \text{Value from Gauss–Seidel} - \text{Value at previous iteration.}$

or $$u_{i,j}^{(n+1)} = (1-\omega)u_{i,j}^{(n)} + \frac{\omega}{4}[u_{i-1,j}^{(n+1)} + u_{i,j-1}^{(n+1)} + u_{i,j+1}^{(n)} + u_{i+1,j}^{(n)} - h^2 f_{i,j}]. \qquad (11.76)$$

$= (1-\omega)u_{i,j}^{(n)} + \omega \times$ value computed by Gauss–Seidel.

$i = 1(1)\,\mathrm{M}-1,\ j = 1(1)\,\mathrm{N}-1.$

Let us suppose Dirichlet condition is prescribed on the boundary so that the values of u at the mesh points on the boundary are known, i.e., $u_{i,\,0}$, $u_{i,\mathrm{N}}$; $i=0(1)\mathrm{M}$ and $u_{0,\,j}$, $u_{\mathrm{M},j}$; $j=1\,(1)\,\mathrm{N}-1$ are known. It is assumed that there exists no singularity at the corners (although in finite difference method it will not pose any hindrance).

The optimum value of ω for fastest convergence of the SOR scheme (11.76) is given by

$$\omega_{\text{opt}} = 2-2\sqrt{1-4t^2} \tag{11.77}$$

where $$t = \cos\frac{\pi}{\mathrm{M}} + \cos\frac{\pi}{\mathrm{N}}.$$

For large M and N (11.77) may be approximated as

$$\omega_{\text{opt}} = 2-\sqrt{2}\pi\left(\frac{1}{\mathrm{M}^2}+\frac{1}{\mathrm{N}^2}\right)^{\frac{1}{2}} \tag{11.77a}$$

$$= 2\left(1-\frac{\pi}{\mathrm{M}}\right),\ \text{if M = N} \tag{11.77b}$$

Thus for M = N = 10, the value of ω may be chosen as 1.4 or slightly on higher side. For values of M and N less than 10, the value of ω may be chosen around 1.2. It may be remembered that for $\omega < 1$, the residual R_{ij} is under-relaxed. It may also be borne in mind that for too large a value ($\omega < 2$) the convergence may take much longer than the basic Gauss–Seidel method due to larger oscillations (See Example 11.15).

Example 11.14

Solve the problem of Example 11.11 by SOR method corresponding to Gauss–Seidel scheme taking $\omega = 1.2$.

Solution The SOR scheme at different points may be written as follows:

$$(1,1) \Rightarrow \quad u_{11}^{(n+1)} = (1-\omega)u_{11}^{(n)} + \frac{\omega}{4}[u_{21}^{(n)} + u_{12}^{(n)} + 1.98]$$

$$\text{or} \quad u_{11}^{(n+1)} = -0.2u_{11}^{(n)} + 0.3[u_{21}^{(n)} + u_{12}^{(n)} + 1.98]$$

$$(2,1) \Rightarrow \quad u_{21}^{(n+1)} = (1-\omega)u_{21}^{(n)} + \frac{\omega}{4}[u_{11}^{(n+1)} + u_{31}^{(n)} + u_{22}^{(n)} + 0.98]$$

$$\text{or} \quad u_{21}^{(n+1)} = -0.2u_{21}^{(n)} + 0.3[u_{11}^{(n+1)} + u_{31}^{(n)} + u_{22}^{(n)} + 0.98]$$

$$(3,1) \Rightarrow \quad u_{31}^{(n+1)} = (1-\omega)u_{31}^{(n)} + \frac{5\omega}{18}[2u_{21}^{(n+1)} + u_{32}^{(n)} + 0.98]$$

$$\text{or} \qquad u_{31}^{(n+1)} = -0.2u_{31}^{(n)} + \frac{1}{3}[2u_{21}^{(n+1)} + u_{32}^{(n)} + 0.98]$$

$$(1, 2) \Rightarrow \quad u_{12}^{(n+1)} = (1-\omega)u_{12}^{(n)} + \frac{\omega}{4}[u_{11}^{(n+1)} + u_{22}^{(n)} + 1.98]$$

$$\text{or} \qquad u_{12}^{(n+1)} = -0.2u_{12}^{(n)} + 0.3[u_{11}^{(n+1)} + u_{22}^{(n)} + 1.98]$$

$$(2, 2) \Rightarrow \quad u_{22}^{(n+1)} = (1-\omega)u_{22}^{(n)} + \frac{\omega}{4}[u_{12}^{(n+1)} + u_{21}^{(n+1)} + u_{32}^{(n)} + 0.98]$$

$$\text{or} \qquad u_{22}^{(n+1)} = -0.2u_{22}^{(n)} + 0.3[u_{12}^{(n+1)} + u_{21}^{(n+1)} + u_{32}^{(n)} + 0.98]$$

$$(3, 2) \Rightarrow \quad u_{32}^{(n+1)} = (1-\omega)u_{32}^{(n)} + \frac{5\omega}{18}[2u_{22}^{(n+1)} + u_{31}^{(n+1)} + 0.98]$$

$$\text{or} \qquad u_{32}^{(n+1)} = -0.2u_{32}^{(n)} + \frac{1}{3}[2u_{22}^{(n+1)} + u_{31}^{(n+1)} + 0.98]$$

Using the above scheme the convergence is obtained in 7 iterations which are half those in the basic Gauss–Seidel; they are shown as follows:

$j \downarrow$ / $i \rightarrow$	1	2	3
1	0.594	0.4722	0.6415
	0.8485	0.8467	1.0913
	0.9466	1.0296	1.1767
	1.0114	1.0591	1.1937
	1.0135	1.0632	1.1951
	1.0145	1.0632	1.1947
	1.0144	1.0630	1.1946
2	0.7722	0.6673	0.9854
	0.8943	0.9785	1.1457
	0.9927	1.0487	1.1889
	1.0135	1.0627	1.1952
	1.0142	1.0632	1.1948
	1.0145	1.0631	1.1947
	1.0144	1.0630	1.1946

Example 11.15

Solve the problem of Example 11.11 by Gauss–Seidel-SOR method taking $\omega = 1.6$.

Solution The equations corresponding to various mesh points are:

$$(1,1) \Rightarrow \quad u_{11}^{(n+1)} = -0.6u_{11}^{(n)} + 0.4[u_{21}^{(n)} + u_{12}^{(n)} + 1.98]$$

$$(2,1) \Rightarrow \quad u_{21}^{(n+1)} = -0.6u_{21}^{(n)} + 0.4[u_{21}^{(n+1)} + u_{31}^{(n)} + u_{22}^{(n)} + 0.98]$$

$$(3,1) \Rightarrow \quad u_{31}^{(n+1)} = -0.6u_{31}^{(n)} + \frac{4}{9}[2u_{21}^{(n+1)} + u_{32}^{(n)} + 0.98]$$

$$(1,2) \Rightarrow \quad u_{12}^{(n+1)} = -0.6u_{12}^{(n)} + 0.4[2u_{11}^{(n+1)} + u_{22}^{(n)} + 1.98]$$

$$(2,2) \Rightarrow \quad u_{22}^{(n+1)} = -0.6u_{22}^{(n)} + 0.4[u_{12}^{(n+1)} + u_{21}^{(n+1)} + u_{32}^{(n)} + 0.98]$$

$$(3,2) \Rightarrow \quad u_{32}^{(n+1)} = -0.6u_{32}^{(n)} + \frac{4}{9}[2u_{22}^{(n+1)} + u_{31}^{(n+1)} + 0.98]$$

Using the above equations the convergence is achieved in 19 iterations which is about one and a half times of the basic Gauss–Seidel method. The values are shown in the following table at alternate iterations starting from first:

0.7920	0.7088	1.0656
1.0657	1.1193	1.2348
0.9423	1.0767	1.2412
0.9884	1.0400	1.1308
1.0033	1.0610	1.1988
1.0213	1.0643	1.1928
1.0156	1.0633	1.1963
1.0150	1.0635	1.1945
1.0139	1.0629	1.1946
1.0143	1.0630	1.1945
1.0143	1.0629	1.1946
1.1088	1.1190	1.9038
1.1760	0.9132	1.2482
1.0390	1.0388	1.1873
1.0129	1.0470	1.1426
1.0043	1.0754	1.1985
1.0110	1.0619	1.1913
1.0141	1.0637	1.1972
1.0149	1.0627	1.1946
1.0143	1.0629	1.1947
1.0143	1.0629	1.1944
1.0143	1.0630	1.1946

11.6.3 Solution of elliptic equation by ADI method

Earlier, the ADI method was discussed in the context of parabolic equations (Sec. 11.5.13) where the Gaussian elimination is used row-wise in one time step and in the next it is used column-wise. The same approach is employed, with some modification, where time steps are replaced by iterations. For example, let us consider a more general elliptic equation

$$\frac{\partial^2 u}{\partial x^2} + \frac{\partial^2 u}{\partial y^2} - \alpha u = f(x, y), \quad \alpha \geq 0 \tag{11.78}$$

defined over a rectangular domain $\mathrm{D} = [0 \leq x \leq a] \times [0 \leq y \leq b]$ with Dirichlet condition prescribed on the boundary C. Let us suppose as before that the domain is subdivided into square mesh of side h such that $\mathrm{M}h = a$ and $\mathrm{N}h = b$. It is required to determine $u_{i,j}$ at $(\mathrm{M}-1) \times (\mathrm{N}-1)$ internal mesh points for $i = 1(1)\mathrm{M}-1$, $j = 1(1)\mathrm{N}-1$.

Discretizing (11.78) at the (i, j) mesh point we get

$$(u_{i-1,j} - 2u_{i,j} + u_{i+1,j}) + (u_{i,j-1} - 2u_{i,j} + u_{i,j+1}) - \alpha h^2 u_{i,j} = h^2 f_{i,j}.$$

Let us suppose, values of u are known at the nth iteration. The term containing u is broken into two equal parts; in the present case, $u = \frac{1}{2}u + \frac{1}{2}u$. One of them is used for the current iteration and the other for the previous iteration. Let ρ be the relaxation parameter.

Then ADI scheme from n^{th} iteration to $(n+1)^{\text{th}}$ iteration and then from $(n+1)^{\text{th}}$ to $(n+2)^{\text{th}}$ iteration would be as follows:

At first stage we write the terms corresponding to $\frac{\partial^2 u}{\partial x^2}$ for $(n+1)$th iteration and those to $\frac{\partial^2 u}{\partial y^2}$ for nth iteration. The term corresponding to u is written as $\frac{1}{2}u + \frac{1}{2}u$, one for $(n+1)$th iteration and the other half for nth iteration. Hence we get,

$$u_{i-1,j}^{(n+1)} - \left(2 + \frac{\alpha}{2}h^2 + \rho\right) u_{i,j}^{(n+1)} + u_{i+1,j}^{(n+1)}$$

$$= -\left\{u_{i,j-1}^{(n)} - \left(2 + \frac{\alpha}{2}h^2 + \rho\right) u_{i,j}^{(n)} + u_{i,j+1}^{(n)}\right\} + h^2 f_{i,j}, \tag{11.79a}$$

$$i = 1(1)\ \mathrm{M}-1 \text{ for each } j = 1(1)\ \mathrm{N}-1.$$

At the second stage the process is reversed from $(n+1)$ to $(n+2)$ iteration which provides,

$$u_{i,j-1}^{(n+2)} - \left(2 + \frac{\alpha}{2}h^2 + \rho\right) u_{i,j}^{(n+2)} + u_{i,j+1}^{(n+2)}$$

$$= -\left\{u_{i-1,j}^{(n+1)} - \left(2 + \frac{\alpha}{2}h^2 + \rho\right) u_{i,j}^{(n+1)} + u_{i+1,j}^{(n+1)}\right\} + h^2 f_{i,j}, \qquad (11.79b)$$

$$j = 1(1)\ \mathrm{N}-1 \text{ for each } i = 1(1)\ \mathrm{M}-1.$$

The equations (11.79a and b) constitute the ADI scheme; ρ is positive and its optimum value for maximum rate of convergence is given as,

$$\rho = \left\{\left(\frac{1}{2}\alpha h^2 + 4\sin^2\frac{\pi}{2\mathrm{R}}\right)\left(\frac{1}{2}\alpha h^2 + 4\cos^2\frac{\pi}{2\mathrm{R}}\right)\right\}^{1/2} \qquad (11.79c)$$

where R is larger of M and N. We will not deal with this method in detail, however reference may be made to [5]. The method can also be used without ρ, i.e., with $\rho = 0$.

It was originally suggested by D.M. Young in 1954.

11.7 Methods for Solving Hyperbolic Equations

For continuity let us write the hyperbolic equation (11.9) again,

$$\frac{\partial^2 u}{\partial t^2} = c^2 \frac{\partial^2 u}{\partial x^2}, \quad 0 \le x \le \mathrm{L}, \ t \ge 0 \qquad (11.80)$$

where c is a constant depending on the properties of the medium. Generally, constant c is absorbed in t by change of variable, therefore equation (11.80) may be expressed, without loss of generality, as

$$\frac{\partial^2 u}{\partial t^2} = \frac{\partial^2 u}{\partial x^2}, \ 0 \le x \le \mathrm{L}, \ t \ge 0. \qquad (11.81)$$

Appropriate boundary conditions will be prescribed at the endpoints $x = 0$ and $x = \mathrm{L}$. For example, if $u(x, t)$ represents displacement of a point at a distance x at time t of a vibrating string tied between two points $x = 0$ and $x = \mathrm{L}$, then the requisite boundary condition at these points will be $u = 0$ (no displacement).

There will also be two initial conditions (known as Cauchy conditions) prescribed at $t = 0$. One condition describes the position of each point of the string, i.e., shape (equation of curve) of the string in the x-u plane and the other specifying the velocity of each point in the plane at $t = 0$. Let these conditions be given as

$$u(x, 0) = f(x), \qquad (11.82)$$

and

$$\frac{\partial u}{\partial t} = g(x), \ t = 0. \qquad (11.83)$$

The hyperbolic equations can be solved by finite difference methods similar to parabolic equations. However, there is one method, known as 'method of characteristics' which is typically suited to hyperbolic equations only. We shall now discuss the methods for their solution.

11.7.1 Finite difference methods

Like parabolic equation we subdivide the domain $D = [0 \leq x \leq L] \times [t \geq 0]$ into rectangular meshes such that $N\Delta x = L$ and x_i denotes a point on the x-axis $x_i = i\Delta x,\ i = 0(1)N,\ x_0 = 0,\ x_N = L$. Let Δt be the size of the time step so that $u_{i,\,j}$ denotes the value of u at the mesh point $(i,\,j)$ where $x_i = i\Delta x,\ t_j = j\Delta t$ (See Fig. 11.10). Let us suppose Dirichlet condition is prescribed at both the ends $x = 0$ and $x = L$, although other boundary conditions can be easily dealt with like parabolic equations.

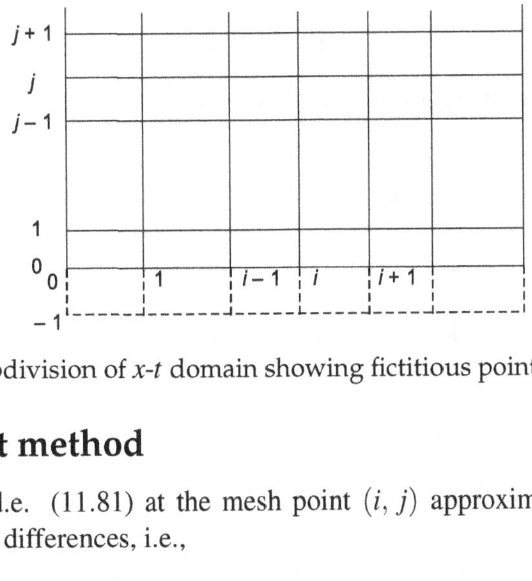

Figure 11.10 Subdivision of x-t domain showing fictitious points in time direction.

11.7.2 Explicit method

We discretize the p.d.e. (11.81) at the mesh point $(i,\,j)$ approximating both the partial derivatives by central differences, i.e.,

$$\frac{u_{i,\,j-1} - 2u_{i,\,j} + u_{i,\,j+1}}{\Delta t^2} + O(\Delta t^2) = \frac{u_{i-1,\,j} - 2u_{i,\,j} + u_{i+1,\,j}}{\Delta x^2} + O(\Delta x^2) \tag{11.84}$$

Neglecting the truncation error, we can write from (11.84)

$$u_{i,\,j+1} = r^2(u_{i-1,\,j} + u_{i+1,\,j}) + 2(1 - r^2)u_{i,\,j} - u_{i,\,j-1}, \tag{11.85}$$

$$i = 1(1)\ N-1$$

where $$r = \Delta t/\Delta x.$$

Assuming that the values of u have been computed upto j^{th} level $(j \geq 1)$ formula (11.85) computes the value of u at the $(j+1)^{\text{th}}$ level explicitly. It may be noted that, to start with, for the second time level, i.e., for $j = 1$, formula (11.85) becomes

$$u_{i,\,2} = r^2(u_{i-1,\,1} + u_{i+1,\,1}) + 2(1 - r^2)u_{i,\,1} - f_i \tag{11.86}$$

as from (11.82) $u_{i,\,0} = f(x_i) = f_i$.

The formula (11.85) can not be used directly to compute the values of u at the first level of time, i.e., for $j = 0$, since it involves u_{i-1}. To deal with it, we extend our grid in the negative time direction by one step and approximate the second Cauchy condition (11.83) by central difference at the mesh point $(i, 0)$ giving

$$\frac{u_{i,1} - u_{i,-1}}{2\Delta t} = g(x_i) = g_i$$

or

$$u_{i,-1} = u_{i,1} - 2\Delta t g_i. \tag{11.87}$$

Using (11.87) for $j = 0$, the formula (11.85) becomes,

$$u_{i,\,1} = \frac{r^2}{2}(u_{i-1,\,0} + u_{i+1,\,0}) + (1 - r^2)u_{i,\,0} + \Delta t \cdot g_i, \tag{11.88}$$

$$i = 1(1)\,\text{N}-1.$$

The molecule for the explicit scheme is shown in Fig. 11.11.

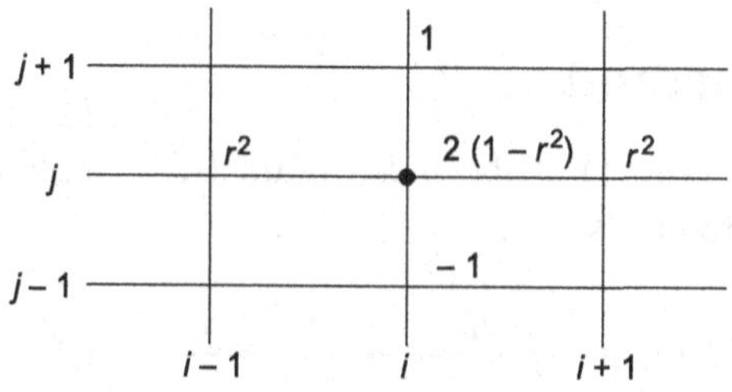

Figure 11.11 Molecule for explicit scheme for hyperbolic equation.

11.7.3 Implicit method

Discretizing p.d.e. at (i, j) mesh point such that time derivative is replaced by central difference and space derivative by the average of central difference approximations at $(i, j+1)$ and $(i,\, j-1)$, i.e.,

$$\frac{u_{i,\,j-1}-2u_{i,\,j}+u_{i,\,j+1}}{\Delta t^2}=\frac{1}{2}\left[\frac{u_{i-1,\,j+1}-2u_{i,\,j+1}+u_{i+1,\,j+1}}{\Delta x^2}+\frac{u_{i-1,\,j-1}-2u_{i,\,j-1}+u_{i+1,\,j-1}}{\Delta x^2}\right]$$

After transposing terms we get,

$$-r^2u_{i-1,\,j+1}+2(1+r^2)u_{i,\,j+1}-r^2u_{i+1,\,j+1}=r^2u_{i-1,\,j-1}-2(1+r^2)u_{i,\,j-1}+r^2u_{i+1,\,j-1}+4u_{i,\,j},\quad(11.89)$$

$$i=1(1)\ \mathrm{N}-1.$$

The equations (11.89) will have to be solved at each level of time. The molecule for the implicit scheme (11.89) is shown in Fig. 11.12.

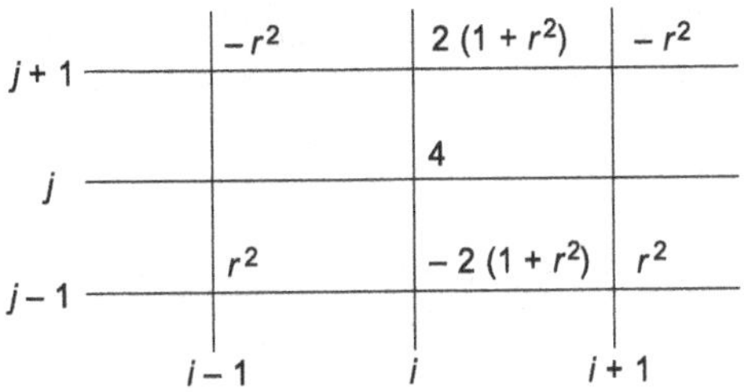

Figure 11.12 Molecule for implicit scheme for hyperbolic equation.

It may also be stated that a variety of implicit schemes may be formulated depending on the weightage given to the approximation of $\dfrac{\partial^2 u}{\partial x^2}$ at the $(j-1)^{\text{th}}$, j^{th} and $(j+1)^{\text{th}}$ levels.

11.7.4 Stability analysis

The explicit scheme (11.85) is linear in u, therefore the formula for the errors can be easily written. Further since we use the complex number $i=\sqrt{-1}$, let us replace i and j by p and q respectively. Thus, the formula for propagation of errors may be written as,

$$e_{p,\,q+1}=r^2(e_{p-1,\,q}+e_{p+1,\,q})+2(1-r^2)e_{p,\,q}-e_{p,\,q-1}\qquad(11.90)$$

Like in parabolic case we express

$$e_{p,q}=Ae^{\alpha q\Delta t}\cdot e^{i\beta p\Delta x}=A\xi^q\cdot e^{i\beta p\Delta x},\ \text{A constant.}\qquad(11.91)$$

where α and β may be complex constants and $\xi=e^{\alpha\Delta t}$. Substituting from (11.91), the equation (11.90), becomes

$$\xi^{q+1}=[r^2(e^{-i\beta\Delta x}+e^{i\beta\Delta x})+2(1-r^2)]\xi^q-\xi^{q-1}$$

or $$\xi^2 = 2(r^2\cos\beta\Delta x + 1 - r^2)\xi - 1$$

or $$\xi^2 - 2\left(1 - 2r^2\sin^2\frac{\beta\Delta x}{2}\right)\xi + 1 = 0. \tag{11.92}$$

If ξ_1 and ξ_2 are roots of the equation (11.92) then

$$\xi_1 + \xi_2 = 2\left(1 - 2r^2\sin^2\frac{\beta\Delta x}{2}\right) \tag{11.93}$$

and $$\xi_1\xi_2 = 1. \tag{11.94}$$

From (11.94) we have $|\xi_1| = \frac{1}{|\xi_2|}$, implying that if $|\xi_2| < 1$ then $|\xi_1| > 1$ and vice-versa. But for stability $|\xi_1| \le 1$ and $|\xi_2| \le 1$. Therefore, for stability of the scheme we must have $|\xi_1| = |\xi_2| = 1$. Now from (11.93),

$$|\xi_1 + \xi_2| = 2\left|1 - 2r^2\sin^2\frac{\beta\Delta x}{2}\right|$$

or $$|\xi_1| + |\xi_2| \ge 2\left|1 - 2r^2\sin^2\frac{\beta\Delta x}{2}\right|$$

or $\left|1 - 2r^2\sin^2\frac{\beta\Delta x}{2}\right| \le 1$ or $-1 \le 1 - 2r^2\sin^2\frac{\beta\Delta x}{2} \le 1$

As the right inequality gives no information we get from left inequality,

$$r^2 \le 1 \Big/ \sin^2\frac{\beta\Delta x}{2} \quad \text{or} \quad r \le 1, \text{ where } r = \Delta t/\Delta x. \tag{11.95}$$

It may be shown that implicit scheme is unconditionally stable.

Example 11.16

A function u satisfies the equation

$$\frac{\partial^2 u}{\partial t^2} = \frac{\partial^2 u}{\partial x^2}, \quad 0 \le x \le 1, \ t \ge 0$$

with boundary condition $u = 0$ at $x = 0$ and $x = 1$, $t \ge 0$; and initial conditions

$$u(x, 0) = 2x(1-x) \text{ and } \frac{\partial u}{\partial t} = 0,\ 0 \le x \le 1,\ t = 0.$$

Use explicit method to find the solution for $t = 0\ (0.1)\ 1.0$ taking $\Delta x = 0.1$ and $\Delta t = 0.1$, approximating the derivative initial condition by central difference.

Solution $\Delta t = 0.1,\ \Delta x = 0.1,\ r = \Delta t/\Delta x = 1.$

Explicit formula becomes for $r = 1$,

$$u_{i,j+1} = u_{i-1,j} + u_{i+1,j} - u_{i,j-1},\ j \ge 1,\ i = 1(1)9. \qquad .....(1)$$

After approximating the initial condition by central difference, the relevant formula becomes for $j = 0$ as

$$u_{i,\,1} = \frac{1}{2}(u_{i-1,\,0} + u_{i+1,\,0}),\ i = 1(1)9, \text{ since } g(x) = 0. \qquad .....(2)$$

The problem is symmetric about the mid-point $x = 0.5$, i.e., $u(x, 0)$ is symmetric and boundary conditions $u = 0$ as well as $\frac{\partial u}{\partial t}$ are symmetric; we need consider the solution for $i = 1(1)5$ only. The values of u computed from (2) and (1) are as follows:

t \ x	0	*0.1*	*0.2*	*0.3*	*0.4*	*0.5*	*0.6*
0.0	0	0.18	0.32	0.42	0.48	0.50	0.48
0.1	0	0.16	0.30	0.40	0.46	0.48	0.46
0.2	0	0.12	0.24	0.34	0.40	0.42	0.40
0.3	0	0.08	0.16	0.24	0.30	0.32	0.30
0.4	0	0.04	0.08	0.12	0.16	0.18	0.16
0.5	0	0	0	0	0	0	0
0.6	0	−0.04	−0.08	−0.12	−0.16	−0.18	−0.16
0.7	0	−0.08	−0.16	−0.24	−0.30	−0.32	−0.30
0.8	0	−0.12	−0.24	−0.34	−0.40	−0.42	−0.40
0.9	0	−0.16	−0.30	−0.40	−0.46	−0.48	−0.46
1.0	0	−0.18	−0.32	−0.42	−0.48	−0.50	−0.48

Example 11.17

A tightly stretched string with fixed end points $x = 0$ and $x = 1.0$ is at rest in its equilibrium position. At $t = 0$, each point of the string is given a velocity $20x(1-x)$. Find displacement of the string at $x = 0\,(0.1)\,1.0$ for $t = 0\,(0.1)\,1.0$ by finite differences using explicit method. Consider the normal form $\frac{\partial^2 u}{\partial t^2} = \frac{\partial^2 u}{\partial x^2}$ for vibration of string.

Solution Equation is given

$$\frac{\partial^2 u}{\partial t^2} = \frac{\partial^2 u}{\partial x^2}, \quad 0 \le x \le 1.0, \ t \ge 0.$$

Boundary conditions are $u = 0$ at $x = 0$ and $x = 1$ since string is fixed at the ends. As string is at rest in equilibrium position

$$u(x, 0) = 0 = f(x), \quad t = 0.$$

Another condition is the velocity given to each point, i.e.,

$$\frac{\partial u}{\partial t} = 20x(1 - x) = g(x), \ t = 0.$$

$$\Delta t = 0.1, \quad \Delta x = 0.1, \ r = \Delta t / \Delta x = 1.$$

Explicit formula for $r = 1$ becomes

$$u_{i,j+1} = u_{i-1,j} + u_{i+1,j} - u_{i,j-1}, \quad j \ge 1, \ i = 1(1)9 \qquad .....(1)$$

For $j = 0$, approximating the velocity condition by central difference

$$u_{i,1} = \frac{r^2}{2}(u_{i-1,0} + u_{i+1,0}) + (1 - r^2)u_{i,0} + \Delta t g(x_i)$$

$$= 0.1 \times 20x_i(1 - x_i) = 2x_i(1 - x_i) \qquad .....(2)$$

Using (2) and (1) we get the following result (symmetry about $x = 0.5$):

t \ x	0	0.1	0.2	0.3	0.4	0.5	0.6
0.0	0	0	0	0	0	0	0
0.1	0	0.18	0.32	0.42	0.48	0.50	0.48
0.2	0	0.32	0.60	0.80	0.92	0.96	0.92
0.3	0	0.42	0.80	1.10	1.28	1.34	1.28
0.4	0	0.48	0.92	1.28	1.52	1.60	1.52
0.5	0	0.50	0.96	1.34	1.60	1.70	1.60
0.6	0	0.48	0.92	1.28	1.52	1.60	1.52
0.7	0	0.42	0.80	1.10	1.28	1.34	1.28
0.8	0	0.32	0.60	0.80	0.92	0.96	0.92
0.9	0	0.18	0.32	0.42	0.48	0.50	0.48
1.0	0	0	0	0	0	0	0

11.7.5 Characteristics of a partial differential equation

Let us consider a general second order (quasi-linear) p.d.e.

$$a\frac{\partial^2 u}{\partial x^2}+b\frac{\partial^2 u}{\partial x\partial y}+c\frac{\partial^2 u}{\partial y^2}+e=0 \tag{10.96}$$

defined in a domain D, where $a,\ b,\ c$ and e are functions of $x,\ y,\ u,\ \frac{\partial u}{\partial x},\ \frac{\partial u}{\partial y}$ in general, but not of second order partial derivatives. Let us introduce the following notations:

$$p=\frac{\partial u}{\partial x}=u_x,\quad q=\frac{\partial u}{\partial y}=u_y$$

$$r=\frac{\partial^2 u}{\partial x^2}=\frac{\partial p}{\partial x},\quad s=\frac{\partial^2 u}{\partial x\partial y}=\frac{\partial^2 u}{\partial y\partial x}=\frac{\partial p}{\partial y}=\frac{\partial q}{\partial x},\quad t=\frac{\partial^2 u}{\partial y^2}=\frac{\partial q}{\partial y}.$$

Equation (11.96) can be written in terms of above notations as

$$ar+bs+ct+e=0. \tag{11.97}$$

Let G be a curve in D where p.d.e. holds. Assuming continuity of $u,\ u_x$ and u_y in D we have on curve G,

$$\begin{aligned}dp&=\frac{\partial p}{\partial x}\cdot dx+\frac{\partial p}{\partial y}\cdot dy\\&=rdx+sdy,\end{aligned} \tag{11.98}$$

$$\begin{aligned}dq&=\frac{\partial q}{\partial x}\cdot dx+\frac{\partial q}{\partial y}\cdot dy\\&=sdx+tdy.\end{aligned} \tag{11.99}$$

Since p.d.e. is satisfied at each point of G, we put in (11.97), the values of r and t from (11.98) and (11.99) respectively, getting

$$\left(a\frac{dp}{dx}+c\frac{dq}{dy}+e\right)-s\left(a\frac{dy}{dx}-b+c\frac{dx}{dy}\right)=0$$

or $$\frac{dy}{dx}\left(a\frac{dp}{dx}+c\frac{dq}{dy}+e\right)-s\left\{a\left(\frac{dy}{dx}\right)^2-b\frac{dy}{dx}+c\right\}=0. \tag{11.100}$$

If curve G is described by the solution of the differential equation

$$a\left(\frac{dy}{dx}\right)^2 - b\frac{dy}{dx} + c = 0, \tag{11.101}$$

then on curve G, from eq. (11.100) we have

$$a\frac{dy}{dx}\cdot\frac{dp}{dx} + c\frac{dq}{dx} + e\frac{dy}{dx} = 0 = a\frac{dy}{dx}dp + cdq + edy. \tag{11.102}$$

Equation (11.101) is quadratic in $\frac{dy}{dx}$ whose roots are given by

$$\frac{dy}{dx} = \frac{b \pm \sqrt{b^2 - 4ac}}{2a}. \tag{11.103}$$

The solution of (11.103) defines curve (s) G in the x-y plane called characteristic (s) or characteristic curve (s). If a, b and c are constant or functions of x, y then the equation of curve (s) can be obtained explicitly otherwise they may be dependent on u, u_x and u_y. Obviously (11.103) gives the directions (of tangents) of characteristics at the point (x, y) of G.

Further, if $b^2 - 4ac < 0$ in (11.103), there exist no real characteristic and this case corresponds to elliptic equation; if $b^2 - 4ac = 0$, then there will be only one characteristic which corresponds to parabolic equation; if $b^2 - 4ac > 0$, then equation (11.103) gives two real characteristics and the case corresponds to hyperbolic equation.

11.7.6 Significance of characteristics

The importance of the characteristics lies in the fact that the solution of a linear partial differential equation is unique at any point in the region enclosed by an initial curve C and the characteristics. However, if the p.d.e. is quasi-linear, i.e., coefficients contain u, u_x or u_y, then the region of determinacy gets narrower.

In the case of a parabolic equation (11.13) the characteristic directions are obtained by putting $a = 1$, $b = 0$ and $c = 0$ in (11.103), which gives after changing y to t,

$$\left(\frac{dt}{dx}\right)^2 = 0 \quad \text{or} \quad \frac{dt}{dx} = 0. \tag{11.104}$$

From the above, we get the equation of the only characteristic as $t =$ const. Thus the region of determinacy of solution in the case of parabolic equation is the region enclosed by the initial line, $t = t_0$ (say) and the characteristic $t = t_C$ (const.) in the x-t plane.

In the case of hyperbolic equation (11.81) the characteristic directions may be obtained by substituting $a = 1$, $b = 0$ and $c = -1$ in (11.103). After changing y to t we can write their equations as

$$\left(\frac{dt}{dx}\right)^2 = 1 \text{ or } \frac{dt}{dx} = \pm 1. \tag{11.105}$$

The equations of characteristics are the straight lines in the x-t plane, given by

$$t = x + \text{const.} \tag{11.106a}$$

$$t = -x + \text{const.} \tag{11.106b}$$

Referring to Fig. 11.13, if OA denotes initial curve C on which appropriate conditions are prescribed, then region of determinacy is given by OBAB′O. There are two characteristics at each of the points O and A.

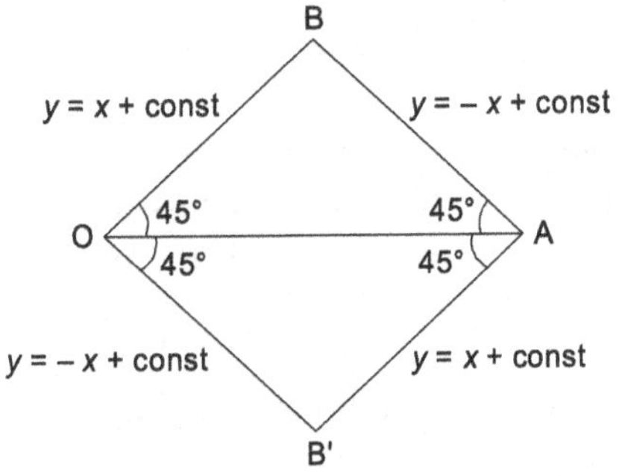

Figure 11.13 Region of determinacy of hyperbolic equation.

Further, let us consider rectangular mesh in context of a finite difference method (See Fig. 11.14). Let P, Q and R be three consecutive points in the x-direction. Then region of determinacy is given by PTRQP. The point A where the value of u has to be computed should lie within this area, i.e., below point T and not above it. That is $QA = \Delta t \leq QT \leq PQ = \Delta x$ or $\Delta t / \Delta x \leq 1$ which is the required stability condition for the explicit scheme.

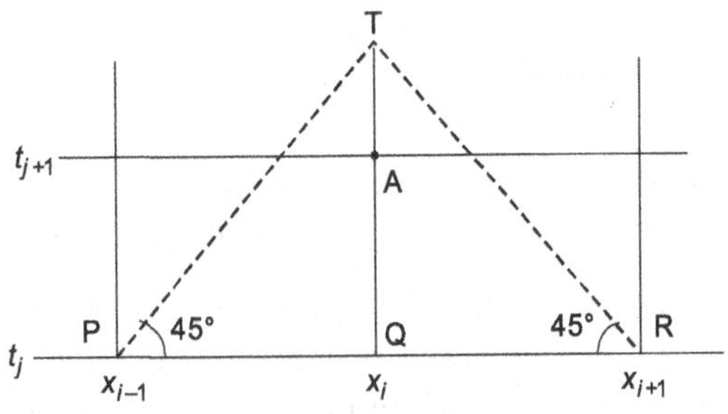

Figure 11.14 Stability domain for hyperbolic equation.

11.7.7 Method of characteristics for solving hyperbolic equations

So far we have developed finite difference methods with respect to rectangular grids only, *i.e.* dividing the domain by drawing lines parallel to the axes. However when an equation happens to be hyperbolic, there are two distinct directions of characteristics at every point that enables us to form a grid based on characteristics.

Referring to general equation (11.96) and assuming it to be hyperbolic let us denote the two characteristic directions as

$$\frac{dy}{dx} = \frac{b + \sqrt{b^2 - 4ac}}{2a} = f \tag{11.107}$$

and

$$\frac{dy}{dx} = \frac{b - \sqrt{b^2 - 4ac}}{2a} = g, \tag{11.108}$$

where f and g are functions of x, y, u, u_x and u_y in general.

Now, leaving the boundary conditions aside or assuming that the hyperbolic domain extends between $-\infty < x, y < \infty$ in the x-y plane, the problem may be posed as follows: Given the values of u, $\frac{\partial u}{\partial x}(= p)$, $\frac{\partial u}{\partial y}(= q)$ on a curve Γ (should not be a characteristic), find $u(x, y)$ at different points in the hyperbolic domain.

We proceed as follows (See Fig. 11.15)

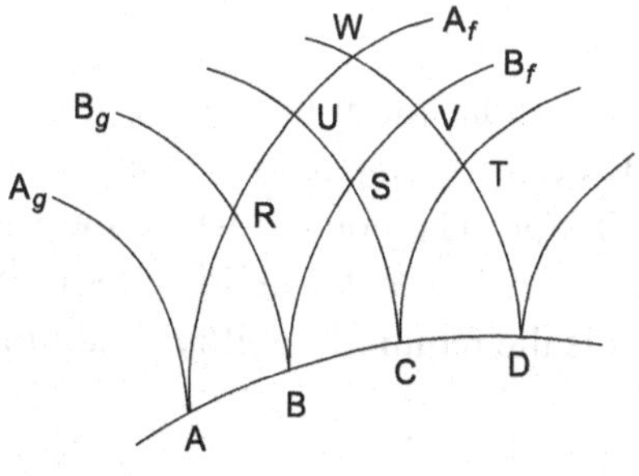

Figure 11.15 Characteristic grid.

Step-1. First of all we choose convenient points, namely, ... A, B, C, D .... (not necessarily equidistant) on the given curve Γ and denote their positions by $\mathrm{A}(x_\mathrm{A}, y_\mathrm{A}), \mathrm{B}(x_\mathrm{B}, y_\mathrm{B})$, etc.

Step-2. we intend to draw characteristics through A, B, C, D etc. If f and g in (11.107) and (11.108) involve only x and y then equations of characteristics can be found and a grid based on characteristics can be formed as shown in Fig. 11.15; A_f is the f characteristic and A_g is g-characteristic through A and so on. However, if f and g involve u and/or its derivatives, then we have to determine the location of the points of intersection of the characteristics, viz. R, S, T, etc., and compute the values of u there. In that case characteristic grid can not be drawn beforehand. It has rather to be generated point-by-point and value of u computed simultaneously. Let us consider this general case when the position of the grid point R is to be determined and value of u is to be computed there. Same procedure will be adopted for the points S and T also.

Integrating (11.107) from A to R and (11.108) from B to R we get the respective approximations as

$$y_R - y_A = f_A(x_R - x_A) \tag{11.109a}$$

$$y_R - y_B = g_B(x_R - x_B) \tag{11.109b}$$

Solving (11.109a and b) we get a crude approximation for the coordinates (x, y) of the grid point R.

Step-3. For advancing the solution to grid points U, V the values of u, u_x, u_y will be required at the grid points R, S, T. The values of $u_x(=p)$ and $u_y(=q)$ at R, i.e., p_R and q_R are computed from (11.102) as follows:

$$a\frac{dy}{dx}dp + cdq + edy = 0 \tag{11.110}$$

Integration of (11.110) from A to R along the characteristic f gives

$$a_A f_A(p_R - p_A) + c_A(q_R - q_A) + e_A(y_R - y_A) = 0 \tag{11.111a}$$

Similarly integration from B to R along g-characteristic gives

$$a_B f_B(p_R - p_B) + c_B(q_R - q_B) + e_B(y_R - y_B) = 0 \tag{11.111b}$$

Solution of (11.111a and b) will provide the values of p_R and q_R since all the other values are known as they are defined on the initial curve. It may be noted that during integration the only values which are known on the characteristics have been used in (11.111a and b).

Step-4. In order to calculate u at R, i.e., u_R, we make use of the formula

$$\begin{aligned} du &= \frac{\partial u}{\partial x}dx + \frac{\partial u}{\partial y}dy \\ &= pdx + qdx. \end{aligned} \tag{11.112}$$

Integrating along either of the characteristics, say f, from A to R (or along g-characteristic from B to R) we get

$$u_R = u_A + \frac{p_R + p_A}{2}(x_R - x_A) + \frac{q_R + q_A}{2}(y_R - y_A). \tag{11.113}$$

After getting the approximations for the position of the grid point (x_R, y_R) and, the values of u, u_x and u_y there, we improve upon them by modifying steps 2 to 4 as follows.

Step-5. Since values of f and g can be computed at the point R also, we modify step-2 by using their average values, i.e.,

$$y_R - y_A = \frac{f_R + f_A}{2}(x_R - x_A), \tag{11.114a}$$

and $$y_R - y_B = \frac{g_R + g_A}{2}(x_R - x_B). \tag{11.114b}$$

Solution of eqns (11.114a and b) gives improved (x_R, y_R).

Step-6. Similarly we modify step-3 by using average values of all the parameters, thus getting

$$\left(\frac{a_A + a_R}{2}\right)\left(\frac{f_A + f_R}{2}\right)(p_R - p_A) + \frac{c_A + c_R}{2}(q_R - q_A) + \frac{e_A + e_R}{2}(y_R - y_A) = 0. \tag{11.115a}$$

$$\left(\frac{a_B + a_R}{2}\right)\left(\frac{g_B + g_R}{2}\right)(p_R - p_B) + \frac{c_B + c_R}{2}(q_R - q_B) + \frac{e_B + e_R}{2}(y_R - y_B) = 0. \tag{11.115b}$$

After solving (11.115a and b) we get improved estimates for p_R and q_R.

Step-7. The value of u may now also be modified by using the latest values of (x_R, y_R) and p_R, q_R, i.e.,

$$u_R = u_A + \frac{p_A + p_R^*}{2}(x_R^* - x_A) + \frac{q_A + q_R^*}{2}(y_R^* - y_A). \tag{11.116}$$

where (*) denotes improved values obtained in steps 5 and 6. To get further refined estimates we can iterate the procedure by repeating the cycle of steps 5, 6 and 7.

Looking cumbersome, the method of characteristics has certain advantages over the finite difference methods. Firstly, non-linear equations can be solved in a routine manner. Secondly, the discontinuities on the initial curve travel along the characteristics thereby not affecting the solution adversely. Thirdly, the method can be used effectively for solving simultaneous first order p.d.e.'s. See Sec. 11.8.5.

Example 11.18

A function u is a solution of the equation

$$\frac{\partial^2 u}{\partial x^2} + (1-2x)\frac{\partial^2 u}{\partial x \partial y} - 2x\frac{\partial^2 u}{\partial y^2} + x = 0$$

and satisfies the initial conditions

$$\left.\begin{aligned} u &= x \\ \frac{\partial u}{\partial y} &= 0 \end{aligned}\right\} y = 0, \ -\infty < x < \infty.$$

Calculate u at the point R which is the intersection of the characteristics through the points A (0.2, 0) and B (0.4, 0) on the initial curve $y = 0$.

Solution

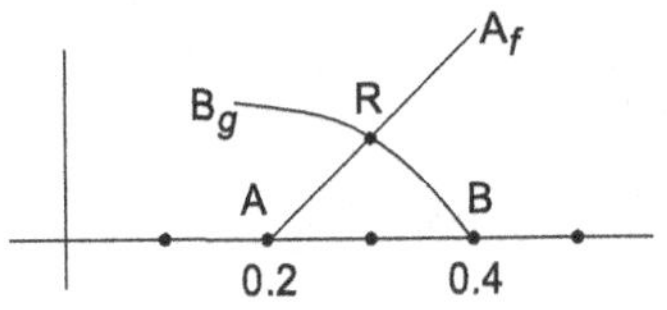

Here, $a = 1,\ b = 1-2x,\ c = -2x,\ e = x;\ u = x,\ u_x = 1,\ u_y = 0.$

The characteristic directions are given by

$$\left(\frac{dy}{dx}\right)^2 - (1-2x)\frac{dy}{dx} - 2x = 0$$

$$\frac{dy}{dx} = \frac{1-2x \pm \sqrt{(1-2x)^2 + 8x}}{2} = \frac{1-2x \pm (1+2x)}{2} = 1 \quad \text{or} \quad -2x.$$

Let $$\frac{dy}{dx} = 1 = f \qquad \text{....(1)}$$

and $$\frac{dy}{dx} = -2x = g \qquad \text{....(2)}$$

f-characteristic from (1) at A (0.2, 0) is given as

$$y = x - 0.2 \qquad \text{....(3)}$$

g-characteristic from (2) at B (0.4, 0) is

$$y = -x^2 + 0.16 \qquad \text{....(4)}$$

Differential equation satisfied along the characteristics

$$\frac{dy}{dx}dp - 2xdq + xdy = 0 \qquad \text{....(5)}$$

Step-1. Points A (0.2, 0), B (0.4, 0) on the initial curve already specified.

Step-2. Point of intersection of f-characteristic through A and g-characteristic through B is found by solving (3) and (4) which is given by $R(x_R, y_R)$ where $x_R = 0.281,\ y_R = 0.081$. Negative value of x is discarded.

Step-3. Equation (5) along f-characteristic after using (1) becomes

$$dp - 2xdq + xdy = 0 \qquad \text{....(6)}$$

Integrating (6) from A to R

$$(p_R - p_A) - 2\frac{x_R + x_A}{2}(q_R - q_A) + \frac{x_R + x_A}{2}(y_R - y_A) = 0$$

or $$(p_R - 1) - (0.281 + 0.2)(q_R - 0) + \frac{0.281 + 0.2}{2}(0.081 - 0) = 0$$

or $$p_R - 0.481q_R = 0.9805. \qquad \text{....(7)}$$

Equation (5) along g-characteristic, using (2), becomes

$$-2xdp - 2xdq + xdy = 0$$

or $$2dp + 2dq - dy = 0, \quad \because x \neq 0 \qquad \text{....(8)}$$

Integrating (8) from B to R,

$$2(p_R - p_B) + 2(q_R - q_B) - (y_R - y_B) = 0$$

or $$2\ (p_R - 1) + 2(q_R - 0) - (0.081 - 0) = 0$$

or $$p_R + q_R = 1.0405 \qquad \text{....(9)}$$

Solving (7) and (9) gives $p_R = 1.0,\ q_R = 0.0405$

Step-4. Integrating $du = pdx + qdy$ from A to R along f-characteristic

$$u_R = u_A + \frac{p_R + p_A}{2}(x_R - x_A) + \frac{q_R + q_A}{2}(y_R - y_A)$$

$$= 0.2 + \frac{1+1}{2}(0.281 - 0.2) + \frac{0.0405 + 0}{2}(0.081 - 0)$$

$$= 0.2826$$

Note: If integration is done along g-characteristic from B to R, then

$$u_R = u_B + \frac{p_R + p_B}{2}(x_R - x_B) + \frac{q_R + q_B}{2}(y_R - y_B)$$

$$= 0.4 + \frac{1+1}{2}(0.281 - 0.4) + \frac{0.0405}{2}(0.081 - 0) = 0.2826$$

Steps 5, 6 and 7 are not required as all parameters are independent of u, u_x and u_y.

Example 11.19

Use the method of characteristics to find the solution of the non-linear (quasi-linear) equation,

$$\frac{\partial^2 u}{\partial x^2} - u^2 \frac{\partial^2 u}{\partial y^2} = 0$$

at the first characteristic grid point between $x = 0.2$ and 0.3, $y > 0$ where u satisfies the conditions

$u = 5x^2$ and $\frac{\partial u}{\partial y} = 2x$ along the initial line $y = 0$, $0 \le x \le 1$. Also compute the coordinates of the grid point and values of u_x, u_y there correct up to two decimals.

Solution

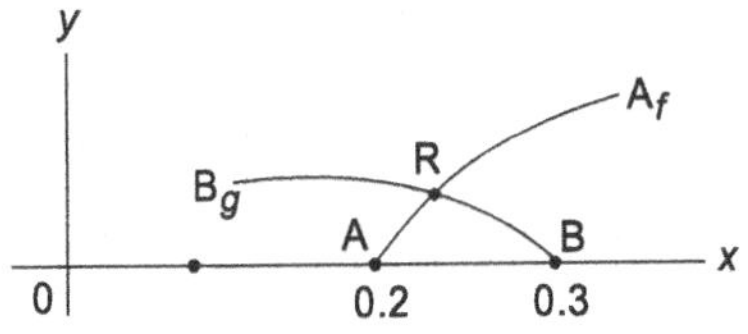

$$a = 1,\ b = 0,\ c = -u^2,\ e = 0.$$

Characteristic directions are given by

$$\left(\frac{dy}{dx}\right)^2 - u^2 = 0 \quad \text{or} \quad \frac{dy}{dx} = \pm u$$

Let $$\frac{dy}{dx} = u = f \qquad \text{....(1)}$$

$$\frac{dy}{dx} = -u = g \qquad \text{.....(2)}$$

Differential equation along characteristics

$$\frac{dy}{dx} dp - u^2 dq = 0 \qquad \text{....(3)}$$

Step-1. Points on the initial curve $y = 0$ are given A (0.2, 0), B (0.3, 0).

Step-2. Integrating (1) from A to R

$$y_R - y_A = u_A(x_R - x_A). \quad \text{.....(4)}$$

Substituting the values

$$y_R - 0 = 0.2\,(x_R - 0.2)$$

or $$y_R - 0.2x_R = -0.04 \quad \text{....(5)}$$

Integrating (2) from B to R

$$y_R - y_B = -u_B(x_R - x_B) \quad \text{....(6)}$$

Substituting the values

$$y_R - 0 = -0.45(x_R - 0.3)$$

$$y_R + 0.45x_R = 0.135 \quad \text{....(7)}$$

Solving (5) and (7) gives $x_R = 0.2692,\ y_R = 0.0138$.

Step-3. Differential equation along f-characteristic (3) becomes after using (1)

$$dp - udq = 0 \quad \text{....(8)}$$

Integrating (8) from A to R.

$$p_R - p_A - u_A(q_R - q_A) = 0 \quad \text{or} \quad p_R - 2 - 0.2(q_R - 0.4) = 0$$

or $$p_R - 0.2q_R = 1.92 \quad \text{....(9)}$$

Differential equation along g-characteristic (3) becomes after using (2)

$$dp + udq = 0 \quad \text{.....(10)}$$

Integrating (10) from B to R

$$p_R - p_B + u_B(q_R - q_B) = 0 \quad \text{or} \quad p_R - 3 + 0.45(q_R - 0.6) = 0$$

or $$p_R + 0.45q_R = 3.27 \quad \text{....(11)}$$

Solving (9) and (11) gives $p_R = 2.3354,\ q_R = 2.0769$.

Step-4. $$du = pdx + qdy$$

Integrating (12) from A to R

$$u_R = u_A + \frac{p_R + p_A}{2}(x_R - x_A) + \frac{q_R + q_A}{2}(y_R - y_A)$$

$$u_R = 0.2 + \frac{2.3354 + 2}{2}(0.2692 - 0.2) + \frac{2.0769 + 0.45}{2}(0.0138 - 0)$$

$$= 0.2 + 0.1500 + 0.0174 = 0.35 \qquad \text{....(13)}$$

Step-5. Modification of (4)

$$y_R - 0 = \frac{0.2 + 0.35}{2}(x_R - 0.2)$$

or $$y_R - 0.275x_R = -0.055 \qquad \text{....(14)}$$

Modification of (6)

$$y_R - 0 = -\frac{0.45 + 0.35}{2}(x_R - 0.3)$$

or $$y_R + 0.4x_R = 0.12 \qquad \text{....(15)}$$

Solving (14) and (15) gives $x_R = 0.2592,\ y_R = 0.0163$ ....(16)

Step-6. Modification of (9)

$$p_R - 2 - \frac{0.2 + 0.35}{2}(q_R - 0.4) = 0$$

or $$p_R - 0.275q_R = 1.89 \qquad \text{....(17)}$$

Modification of (11)

$$p_R - 3 + \frac{0.45 + 0.35}{2}(q_R - 0.6) = 0$$

or $$p_R + 0.4q_R = 3.24 \qquad \text{....(18)}$$

Solving (17) and (18) gives $p_R = 2.44,\ q_R = 2.00$ ....(19)

Step-7. Modification of (13)

$$u_R = 0.2 + \frac{2.44 + 2}{2}(0.2592 - 0.2) + \frac{2 + 0.45}{2}(0.0163 - 0)$$

$$= 0.2 + 0.1314 + 0.0200 = 0.3514 \qquad \text{....(20)}$$

value of u_R agrees up to two significant figures. Hence

$$u = 0.3514,\ \mathrm{R} = \mathrm{R}\,(0.2592,\ 0.0163),\ p_R = 2.44,\ q_R = 2.00$$

If required, the process may be iterated as follows:

Iteration-1.

Step-5. $$y_R - 0 = \frac{0.2+0.3514}{2}(x_R - 0.2)$$

or $$y_R - 0.2757x_R = -0.0551 \qquad ....(21)$$

and $$y_R - 0 = -\frac{0.45+0.3514}{2}(x_R - 0.3)$$

$$y_R + 0.4007x_R = 0.1202 \qquad ....(22)$$

Solution of (21) and (22) gives $x_R = 0.2592, y_R = 0.01635 = 0.0164$

Step-6. $$p_R - 2 - \frac{0.2+0.3514}{2}(q_R + 0.4) = 0$$

or $$p_R - 0.2757q_R = 1.8897 \qquad ....(23)$$

$$p_R - 3 - \frac{0.45+0.3514}{2}(q_R - 0.6) = 0$$

or $$p_R + 0.4007q_R = 3.2404 \qquad ....(24)$$

Solution of (23) and (24) gives $p_R = 2.4402,\ q_R = 1.9967$.

Step-7. $$u_R = 0.2 + \frac{2.4402+2}{2}(0.2592 - 0.2) + \frac{1.9967+0.45}{2}(0.0164)$$

$$= 2 + 0.1314 + 0.0201 = 0.35146 = 0.3515$$

Note: If we are rounding the numbers to four figures, there can be an error of ± 0.00005. Therefore in adding three numbers, the rounding error may be ± 0.00015. Therefore a difference of 0.0002 between two values may be due to rounding. Hence we should not expect an accuracy in the result beyond it, i.e., a number $\text{N} = \text{N} \pm .0002$.

11.8 Hyperbolic Equation of First Order

We have discussed in Sec. 11.7 finite difference methods and method of characteristics for solving hyperbolic (wave) equations of second order. It may be observed that the vibration/wave Eq. (11.9) is satisfied by a general solution (known as D'Alembert's solution) given by,

$$u(x,\ t) = f_1(x - ct) + f_2(x + ct) \qquad (11.117)$$

where f_1 and f_2 are functions of $(x - ct)$ and $(x + ct)$ respectively, c being a constant. Physically $f_1\ (x - ct)$ represents a wave travelling in the positive direction of x with velocity c and $f_2\ (x + ct)$, a wave travelling in the opposite direction with same velocity c.

Analogously, a first order hyperbolic equation (wave equation) is represented by a p.d.e.,

$$\frac{\partial u}{\partial t}+c\frac{\partial u}{\partial x}=0;\ \ x>0,\ t>0 \tag{11.118}$$

where c may be positive or negative. It may be verified that a general solution of (11.118) can be expressed as,

$$u(x,\ t)=f(x-ct) \tag{11.119}$$

where $f(x-ct)$ represents a wave travelling in positive x-direction if c is positive (+ve) and in the negative direction if c is negative (-ve).

Since Eq. (11.118) is of order one in both x and t, only one condition is required on x and t each, for the problem to be well-posed. Thus the problem is an initial value problem in x as well as in t. However, in general, we can refer these conditions as 'boundary conditions'. For instance, let us consider the following (boundary/initial) conditions,

$$u(x,\ 0)=x, \tag{11.120a}$$

and

$$u(0,\ t)=t. \tag{11.120b}$$

Before discussing numerical methods let us examine briefly, the analytical solution of (11.118) subject to boundary conditions (11.120a) and (11.120b) in order to have some physical idea. We will use Laplace Transform (L.T.) to get the solution. However, those who are not familiar with the technique may skip the analysis and go straight to the solution (11.126). Applying L.T. on (11.118) and using (11.120a) we get,

$$\frac{d\bar{u}}{dx}+\frac{s}{c}\bar{u}=\frac{x}{c} \tag{11.121}$$

where $\bar{u}=L(u)=\int_0^\infty u(x,\ t)e^{-st}dt$, L denotes L.T.

The solution of (11.121) gives,

$$\bar{u}=\frac{x}{s}-\frac{c}{s^2}+Ae^{-\frac{xs}{c}}, \tag{11.122}$$

where A is an arbitrary constant.

After using second condition (11.120b) in (11.122) we get,

$$\bar{u}=\frac{x}{s}-\frac{c}{s^2}+\frac{1+c}{s^2}.e^{-\frac{xs}{c}}. \tag{11.123}$$

Finally, the inverse L.T. gives

$$u = L^{-1}(\bar{u}) = x - ct + \frac{1+c}{c}(ct - x)U\left(t - \frac{x}{c}\right) \tag{11.124}$$

where U is unit step function defined as,

$$U(t-a) = \begin{cases} 0 \ , & t < a \\ 1 \ , & t \geq a \quad a \text{ is } +\text{ve} \end{cases} \tag{11.125}$$

The solution (11.124) can also be expressed as

$$u(x,\, t) = \begin{cases} x - ct, & \text{for } t < \dfrac{x}{c} \quad \text{or} \quad x > ct \\ \dfrac{1}{c}(ct - x), & \text{for } t \geq \dfrac{x}{c} \quad \text{or} \quad x \leq ct \end{cases} \tag{11.126}$$

The graph of the solution (11.126) at t=1.0 for $0 \leq x \leq 2.0$ when $c = 1$ is shown in Fig. 11.16. Note that AB corresponds to input (11.120b) and BC to initial condition (11.120a).

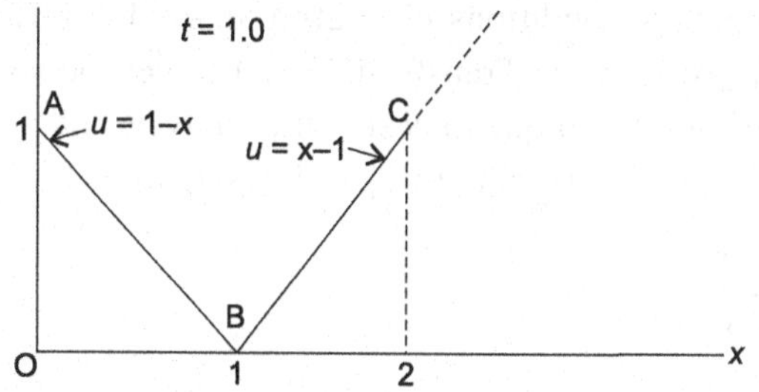

Figure 11.16 Graph of (11.126) when $t = 1.0$, $0 \leq x \leq 2.0$ with $c = 1$.

11.8.1 Finite difference methods

Let us now discuss finite difference methods for solving Eq. (11.118) with prescribed initial/boundary conditions at $t = 0$ and $x = 0$, namely (11.120a) and (11.120b) respectively. Suppose the solution is required over a space domain $0 \leq x \leq \mathrm{L}$ for $t > 0$. As before, we subdivide the interval [0, L] into n sub-intervals such that $\Delta x = \mathrm{L}/n$ and denote the points along x-axis as $x = x_i = i\Delta x,\ i = 0(1)n;\ x_0 = 0$ and $x_n = \mathrm{L}$. Choose a suitable step size Δt along t-direction and subdivide the domain $\mathrm{D} = [0 \leq x \leq L] \times [t \geq 0]$ into rectangular meshes of equal sizes as shown in Fig. 11.17.

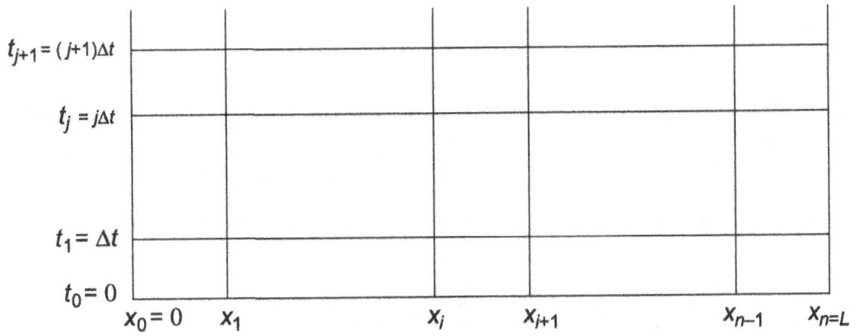

Figure 11.17 Subdivision of domain into rectangular meshes.

Let us denote the computed value of u at $x = x_i$, $t = t_j$ by u_{ij} and the exact value by $u(x_i, t_j)$. Suppose that the values of u have been computed up to j^{th} time level and we are required to compute the values at $(j+1)^{th}$ level. In other words, values of $u_{i,j}$ are known and values of $u_{i,j+1}$ are to be computed for $i = 1(1)n$, $j = 0, 1, 2, \ldots$. The values $u_{0,j}$ are known by virtue of boundary condition prescribed at $x = 0$ and those of $u_{i,0}$ due to boundary condition at $t = 0$.

We shall discuss various schemes in respect of their implementation and stability, denoting $r = \Delta t/\Delta x$. In case of analysing stability of a method, the mesh point (p, q) will be considered instead of (i, j) since i will be used for complex number $i = \sqrt{-1}$.

11.8.2 Lax–Wendroff's method

Discretizing p.d.e. (11.118) at mesh point (i, j) replacing time derivative by FD and the space derivative by CD, we get

$$\frac{u_{i,j+1} - u_{i,j}}{\Delta t} + c\frac{u_{i+1,j} - u_{i-1,j}}{2\Delta x} + O(\Delta t) + O\left(\Delta x^2\right) = 0,$$

and after neglecting error term,

$$u_{i,j+1} = u_{i,j} - \frac{cr}{2}\left(u_{i+1,j} - u_{i-1,j}\right), \quad r = \frac{\Delta t}{\Delta x} \tag{11.127}$$

Replacing (i, j) by (p, q), a relation for the error can be obtained as,

$$e_{p,q+1} = e_{p,q} - \frac{cr}{2}\left(e_{p+1,q} - e_{p-1,q}\right). \tag{11.128}$$

Substituting $e_{p,q} = Ae^{i\beta\Delta x}.\xi^q$ in (11.128) gives

$$\xi = 1 - \frac{cr}{2}\left(e^{i\beta\Delta x} - e^{-i\beta\Delta x}\right) = 1 - icr\sin\beta\Delta x$$

$$|\xi| = \left(1 + c^2r^2\sin^2\beta\Delta x\right)^{1/2}. \tag{11.129}$$

We see from (11.129) that $|\xi| > 1$ for all r; hence the scheme is unstable irrespective of c being positive (+ve) or negative (−ve).

Lax and Wendroff modified the above scheme by incorporating an extra term in the approximation of $\frac{\partial u}{\partial t}$ making it $O(\Delta t^2)$ in the following manner:

Write Taylors' expansion of $u(x_i,\ t_{j+1}) = u(x_i,\ t_j + \Delta t)$about $(x_i,\ t_j)$ as

$$u(x_i,\ t_{j+1}) = u(x_i,\ t_j) + \Delta t\frac{\partial u}{\partial t} + \frac{\Delta t^2}{2}\frac{\partial^2 u}{\partial t^2} + \frac{\Delta t^3}{6}\frac{\partial^3 u}{\partial t^3} + \ldots$$

or
$$\frac{\partial u}{\partial t} = \frac{u_{i,\ j+1} - u_{i,\ j}}{\Delta t} - \frac{\Delta t}{2}\frac{\partial^2 u}{\partial t^2} - \frac{\Delta t^2}{6}\frac{\partial^3 u}{\partial t^3} - \cdots \tag{11.130}$$

where values of u are replaced by computed values and partial derivatives taken at $(x_i,\ t_j)$.

From original Eq. (11.118) we have a relationship

$$\frac{\partial}{\partial t} = -c\frac{\partial}{\partial x} \text{ and } \frac{\partial^2 u}{\partial t^2} = c^2\frac{\partial^2 u}{\partial x^2}. \tag{11.131}$$

Using (11.131) in (11.130) and approximating $\frac{\partial^2 u}{\partial x^2}$ by CD at $(i,\ j)$,

$$\frac{\partial u}{\partial t} = \frac{u_{i,j+1} - u_{i,j}}{\Delta t} - \frac{c^2\Delta t}{2}\cdot\frac{u_{i-1,j} - 2u_{i,j} + u_{i+1,j}}{\Delta x^2} + O\left(\Delta t^2\right). \tag{11.132}$$

Now again discretizing (11.118) at $(x_i,\ t_j)$ such that $\frac{\partial u}{\partial t}$is approximated by (11.132) and $\frac{\partial u}{\partial x}$ by CD, we get

$$\frac{u_{i,\ j+1} - u_{i,j}}{\Delta t} - \frac{c^2\Delta t}{2}\cdot\frac{u_{i-1,j} - 2u_{i,j} + u_{i,\ j+1}}{\Delta x^2} + c\frac{u_{i+1,j} - u_{i-1,j}}{2\Delta x} + O\left(\Delta t^2\right) + O\left(\Delta x^2\right) = 0$$

or
$$u_{i,\ j+1} = u_{i,\ j} + \frac{c^2 r^2}{2}\left(u_{i-1,\ j} - 2u_{i,\ j} + u_{i+1,\ j}\right) - \frac{cr}{2}\left(u_{i+1,\ j} - u_{i-1,\ j}\right)$$

$$= \left(1 - c^2 r^2\right)u_{i,\ j} + \frac{1}{2}cr\left(1 + cr\right)u_{i-1,\ j} - \frac{1}{2}cr\left(1 - cr\right)u_{i+1,\ j}, \quad i = 1(1)n. \tag{11.133}$$

The formula given by (11.133) is the Lax–Wendroff's formula which is $O(\Delta t^2) + O(\Delta x^2)$. It will now be shown that it is stable for $cr \le 1$. As before we can obtain,

$$\xi = 1 + \frac{c^2 r^2}{2}\left(e^{-i\beta\Delta x} - 2 + e^{i\beta\Delta x}\right) - \frac{cr}{2}\left(e^{i\beta\Delta x} - e^{-i\beta\Delta x}\right)$$

$$= 1 + c^2r^2(\cos\beta\Delta x - 1) - icr\sin\beta\Delta x$$

$$= 1 - 2c^2r^2\sin^2\frac{\beta\Delta x}{2} - icr\sin\beta\Delta x$$

$$|\xi|^2 = 1 - 4c^2r^2\sin^2\frac{\beta\Delta x}{2} + 4c^4r^4\sin^4\frac{\beta\Delta x}{2} + 4c^2r^2\sin^2\frac{\beta\Delta x}{2}(1 - \sin^2\frac{\beta\Delta x}{2})$$

$$= 1 + 4c^4r^4\sin^4\frac{\beta\Delta x}{2} - 4c^2r^2\sin^4\frac{\beta\Delta x}{2}$$

$$= 1 - 4c^2r^2\left(1 - c^2r^2\right)\sin^4\frac{\beta\Delta x}{2}$$

For $|\xi| \leq 1$, we must have

$$0 \leq 4c^2r^2(1 - c^2r^2) \leq 1,$$

since largest value of $\sin^4\frac{\beta\Delta x}{2}$ cannot exceed unity.

Right inequality is satisfied for any value of cr since

$$4c^4r^4 - 4c^2r^2 + 1 \geq 0$$

or $$\left(2c^2r^2 - 1\right)^2 \geq 0.$$

The left inequality gives

$$c^2r^2(1 - c^2r^2) \geq 0.$$

This will be satisfied if $c^2r^2 \leq 1$. Hence the Lax–Wendroff scheme is stable for $cr \leq 1$ irrespective of the sign of c.

Note: i) For computation of $u_{n,\,j+1}$ from (11.133), the value of $u_{n+1,j}$ is required which is not available. Therefore, some special procedure is required for it. See Example 11.20.

ii)From (11.131) we see that equation (11.118) can be converted to second order hyperbolic equation but we do not solve it since there would not be available enough boundary conditions. If we do, it will involve extraneous solution $f(x+ct)$ which is not required and will interfere with the required solution.

iii)If derivative boundary conditions are prescribed at both ends then FD and BD replacements can be used.

11.8.3 Wendroff's method

The Wendroff's formula can be obtained by discretizing the p.d.e. (11.118) at a point halfway between x_i, x_{i+1} and t_j, t_{j+1} i.e., at a point $P\left(x_{i+\frac{1}{2}}, t_{j+\frac{1}{2}}\right)$ (similar to CN scheme). See Fig. 11.8.

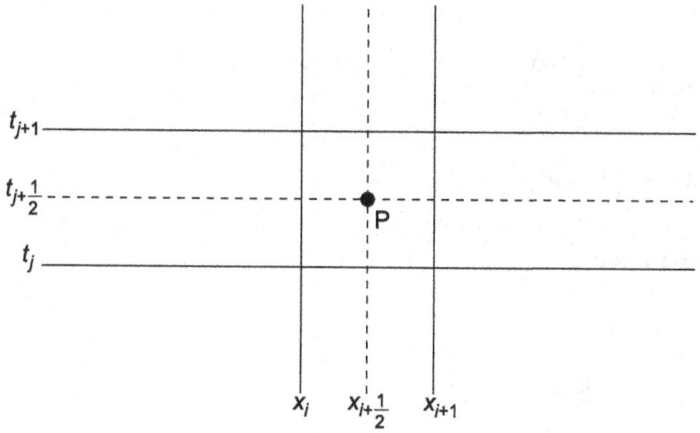

Figure 11.18 Point P halfway between x_i, x_{i+1} and t_j, t_{j+1}.

Discretizing the time and space derivatives both by CD, we get

$$\frac{u_{i+\frac{1}{2},j+1}-u_{i+\frac{1}{2},j}}{\Delta t}+c\frac{u_{i+1,j+\frac{1}{2}}-u_{i,j+\frac{1}{2}}}{\Delta x}+O\left(\Delta t^2\right)+O\left(\Delta x^2\right)=0.$$

Replacing the halfway values (as they are not available) by the corresponding average values, it becomes after neglecting error terms,

$$\frac{u_{i+1,j+1}-u_{i,j+1}}{2}-\frac{u_{i+1,j}+u_{i,j}}{2}+cr\left[\frac{u_{i+1,j+1}+u_{i+1,j}}{2}-\frac{u_{i,j+1}+u_{i,j}}{2}\right]=0.$$

Rearranging terms we get,

$$(1+cr)\,u_{i+1,j+1}+(1-cr)\,u_{i,j+1}-(1-cr)\,u_{i+1,j}-(1+cr)\,u_{i,j}=0$$

or $$u_{i+1,j+1}=u_{i,j}+\frac{1-cr}{1+cr}\left(u_{i+1,j}-u_{i,j+1}\right),\quad i=0(1)n-1. \tag{11.134}$$

Although two values of u i.e., $u_{i+1,j+1}$ and $u_{i,j+1}$ are involved at $(j+1)$th level in (11.134), it is an explicit formula since for $i=0$, $u_{i,j+1}$ can be computed as $u_{0,j+1}$ is known on account of boundary condition $u(0,t)$; then $u_{2,j+1}$ can be computed after $u_{1,j+1}$ has been computed and so on until $u_{n,j+1}$, $(i=n-1)$. Note that no special procedure is required to compute the value of u at $x=x_n$ as in the Lax–Wendroff's formula.

For analysing stability, we have the relation

$$\left\{(1+cr)\,e^{i\beta\Delta x}+(1-cr)\right\}\xi=(1+cr)+(1-cr)\,e^{i\beta\Delta x}$$

or $$\left\{(1+cr)\,e^{i\beta\Delta x/2}+(1-cr)\,e^{-i\beta\Delta x/2}\right\}\xi=(1+cr)\,e^{-i\beta\Delta x/2}+(1-cr)\,e^{i\beta\Delta x/2}$$

or $$\xi=\frac{\cos\beta\Delta x/2-icr\sin\beta\Delta x/2}{\cos\beta\Delta x/2+icr\sin\beta\Delta x/2}$$

or $|\xi|^2=1$ and $|\xi|=1$.

Hence the scheme is stable unconditionally for any value of r, irrespective of c being positive (+ve) or negative (−ve).

This scheme should be better compared to Lax–Wendroff's since this is a natural extension to C–N scheme.

11.8.4 Other explicit/implicit methods

Discretising time derivative by FD and space derivative by BD at mesh point $(i,\ j)$ gives

$$\frac{u_{i,j+1}-u_{i,j}}{\Delta t}+c\frac{u_{i,j}-u_{i-1,j}}{\Delta x}+O(\Delta t)+O(\Delta x)=0$$

or $$u_{i,j+1}=u_{i,j}-cr\left(u_{i,j}-u_{i-1,j}\right). \tag{11.135}$$

For stability,

$$|\xi|=\left\{1-4cr\,(1-cr)\sin^2\beta\Delta x/2\right\}^{1/2}$$

It can be shown, as before that (11.135) is stable for $cr\le 1$ when c is positive.

When both time and space derivatives are discretized by FD at $(i,\ j)$, we get

$$\frac{u_{i,j+1}-u_{i,j}}{\Delta t}+c\frac{u_{i+1,j}-u_{i,j}}{\Delta x}+O(\Delta t)+O(\Delta x)=0$$

or $$u_{i,j+1}=u_{i,j}-cr\left(u_{i+1,j}-u_{i,j}\right) \tag{11.136}$$

The stability condition gives

$$|\xi|=\left\{1+4cr\,(1+cr)\sin^2\beta\Delta x/2\right\}^{1/2}.$$

Obviously, it is unstable for any value of r, for c positive (+ve). But when c is negative, then (11.136) will be stable for $cr \le 1$ while formula (11.135) unstable.

For implicit schemes we can develop Crank–Nicolson type formula by discretization at point $(x_i, t_{j+\frac{1}{2}})$. Similarly a fully implicit scheme may be easily constructed. Both of these schemes are unconditionally stable. They may be employed when an extra condition is to be satisfied at the end point x_n; although in that case the system would be over-determined (not well-posed) i.e., more conditions prescribed than order of derivative in the p.d.e.

Example 11.20

Given a p.d.e. $\frac{\partial u}{\partial t} + \frac{\partial u}{\partial x} = 0$ with boundary conditions $u(x, 0) = x$ and $u(0, t) = t$, Find its solution by Lax–Wendroff's method over $0 \le x \le 2.0$ for $0 \le t \le 1.0$ at the mesh points $x = 0(0.25)2.0$ and $t = 0(0.125)1.0$. Plot the graph for $t = 0.5$ and 1.0.

Solution

$$\Delta t = 0.125,\ \ \Delta x = 0.25;\ \ r = \frac{\Delta t}{\Delta x} = 0.5;\ \ i = 0(1)8,\ \ j = 0(1)8.$$

Lax–Wendroff formula after putting $r = 0.5$,

$$u_{i,j+1} = \frac{1}{8}\left(6u_{i,j} + 3u_{i-1,j} - u_{i+1,j}\right);\ \ i = 1(1)8,\ \ j = 0(1)8.$$

The computations are shown in the following table:

$j \downarrow$ \ $i \rightarrow$		0	1	2	3	4	5	6	7	8
	t \ x	0	0.25	0.50	0.75	1.00	1.25	1.50	1.75	2.00
0	0.000	0	0.2500	0.5000	0.7500	1.0000	1.2500	1.5000	1.7500	2.0000
1	0.125	0.125	0.1250	0.3750	0.6250	0.8750	1.1250	1.3750	1.6250	1.8750
2	0.250	0.250	0.0938	0.2500	0.5000	0.7500	1.0000	1.2500	1.5000	1.7500
3	0.375	0.375	0.1328	0.1602	0.3750	0.6250	0.8750	1.1250	1.3750	1.6250
4	0.500	0.500	0.2202	0.1231	0.2632	0.5000	0.7500	1.0000	1.2500	1.5000
5	0.625	0.625	0.3373	0.1420	0.1811	0.3800	0.6250	0.8750	1.1250	1.3750
6	0.750	0.750	0.4696	0.2104	0.1416	0.2748	0.5012	0.7500	1.0000	1.2500
7	0.875	0.875	0.6072	0.3162	0.1508	0.1966	0.3852	0.6254	0.8750	1.1250
8	1.000	1.000	0.7440	0.4460	0.2071	0.1558	0.2844	0.5041	0.7502	1.0002

For instance: $u_{3,4} = \frac{1}{8}(6u_{3,3} + 3u_{2,3} - u_{4,3})$

$$= \frac{1}{8}(6 \times 0.3750 + 3 \times 0.1602 - 0.6250) = 0.2632$$

$$u_{5,7} = \frac{1}{8}(6u_{5,6} + 3u_{4,6} - u_{6,6})$$

$$= \frac{1}{8}(6 \times 0.5012 + 3 \times 0.2748 - 0.7500) = 0.3852 \text{ etc.}$$

In computing $u_{8,\,j+1}$ we have assumed that the velocity $\frac{\partial u}{\partial t}$ in the last column is same as in the column one before it i.e.,

$$\frac{u_{8,\,j+1} - u_{8,\,j}}{\Delta t} = \frac{u_{7,\,j+1} - u_{7,\,j}}{\Delta t}$$

or $$u_{8,\,j+1} = u_{8,\,j} + (u_{7,\,j+1} - u_{7,\,j})$$

For instance, $u_{8,5} = u_{8,4} + (u_{7,5} - u_{7,4})$

$$= 1.5 + (1.125 - 1.250) = 1.375 \text{ etc.}$$

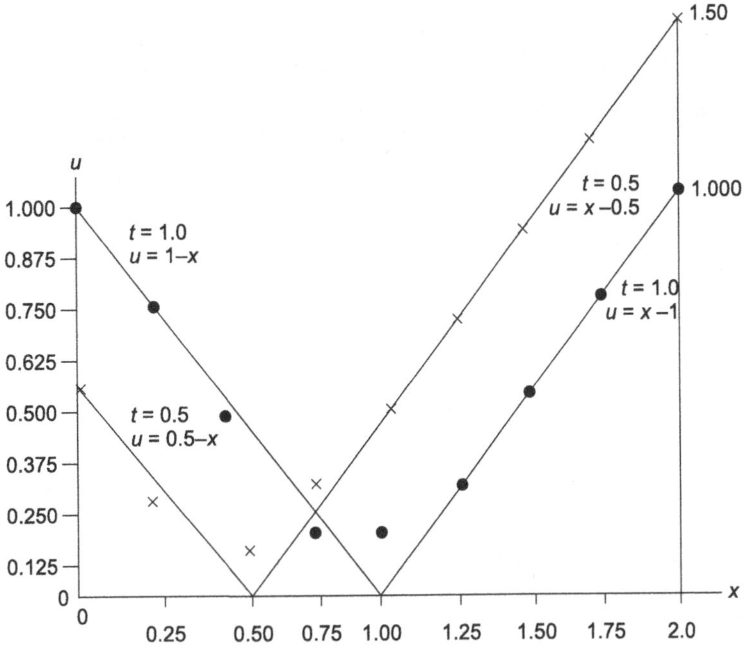

Note: Graphs with solid lines are due to analytical solution (11.126). The cross($\times$) corresponds to computed values for $t = 0.5$ and dot ($\cdot$) corresponds to $t = 1.0$.

Example 11.21

Solve the problem of Example 11.20 by Wendroff's method taking the same values of Δx and Δt.

Solution

Wendroff's formula when $c = 1$ as from (11.134),

$$u_{i+1,j+1} = u_{i,j} + \frac{1-r}{1+r}\left(u_{i+1,j} - u_{i,j+1}\right), \quad i = 0(1)n-1.$$

For r = 0.5, formula becomes

$$u_{i+1,j+1} = u_{i,j} + \frac{1}{3}\left(u_{i+1,j} - u_{i,j+1}\right), \quad i = 0(1)7.$$

The computations are shown in the following table:

j ↓	i →	0	1	2	3	4	5	6	7	8
	t \ x	0	0.25	0.50	0.75	1.00	1.25	1.50	1.75	2.00
0	0	0	0.2500	0.5000	0.7500	1.0000	1.2500	1.5000	1.7500	2.0000
1	0.125	0.125	0.0417	0.4028	0.6157	0.8781	1.1240	1.3753	1.6249	1.8750
2	0.250	0.250	0.0556	0.1574	0.5556	0.7232	1.0117	1.2452	1.5019	1.7493
3	0.375	0.375	0.1435	0.0602	0.3225	0.6892	0.8307	1.1499	1.3625	1.6308
4	0.500	0.500	0.2562 0.2202	0.0782 0.1231	0.1416 0.2632	0.5050 0.5000	0.7978 0.7500	0.9481 1.0000	1.2880 1.2500	1.4768 1.5000
5	0.625	0.625	0.3771	0.1566	0.0732	0.2855	0.6758	0.8856	1.0822	1.4195
6	0.750	0.750	0.5007	0.2624	0.0935	0.1375	0.4649	0.8160	0.9743	1.2306
7	0.875	0.875	0.6252	0.3798	0.1670	0.0837	0.2646	0.6487	0.9245	1.0763
8	1.000	1.000	0.7501 0.7440	0.5018 0.4460	0.2682 0.2071	0.1055 0.1558	0.1367 0.2844	0.4353 0.5041	0.8118 0.7502	1.0127 1.0002

Note : For t = 0.5 and 1.0, the lower entries correspond to Lax–Wendroff method in Example 11.20.

For instance: $u_{5,4} = u_{4,3} + \frac{1}{3}\left(u_{5,3} - u_{4,4}\right)$

$$= 0.6892 + \frac{1}{3}(0.8307 - 0.5050) = 0.7978$$

$$u_{7,8} = u_{6,7} + \frac{1}{3}\left(u_{7,7} - u_{6,8}\right)$$

$$= 0.6487 + \frac{1}{3}(0.9245 - 0.4353) = 0.8118 \text{ etc.}$$

11.8.5 Solving second order equation by simultaneous equations of first order

Let us consider a second order hyperbolic equation

$$\frac{\partial^2 u}{\partial t^2} = \frac{\partial^2 u}{\partial x^2}, \quad 0 \le x \le 1, \; t \ge 0 \tag{11.137}$$

with initial conditions

$$u(x, 0) = f(x) \text{ and } \frac{\partial u}{\partial t} = g(x), \; t = 0; \tag{11.138}$$

and boundary conditions

$$u(0, t) = \phi(t) \text{ and } u(1, t) = \psi(t) \tag{11.139}$$

Putting $\dfrac{\partial u}{\partial t} = v$ and $\dfrac{\partial u}{\partial x} = w$, eq (11.137) may be written as

$$\frac{\partial v}{\partial t} = \frac{\partial w}{\partial x}. \tag{11.140a}$$

Assuming that $u(x, t)$ is analytic, so that $\dfrac{\partial^2 u}{\partial x \partial t} = \dfrac{\partial^2 u}{\partial t \partial x}$, we also have the relation

$$\frac{\partial w}{\partial t} = \frac{\partial v}{\partial x}. \tag{11.140b}$$

Now we have two simultaneous equations viz. (11.140a) and (11.140b) to be solved for $v(x, t)$ and $w(x, t)$ over the domain $\mathrm{D} = [0 \le x \le 1] \times [t \ge 0]$. The corresponding boundary conditions on v and w would be as follows,

$$v(x, 0) = g(x), \; v(0, t) = \dot{\phi}(t) \text{ and } v(1, t) = \dot{\psi}(t), \tag{11.141}$$

$$w(x, 0) = f'(x), \tag{11.142}$$

where dot denotes differentiation w.r.t. t and dash w.r.t. x. To solve the system by finite difference method let the interval [0, 1] be subdivided into n sub-intervals each of width Δx such that $n\Delta x = 1$ and $x_i = \mathrm{i}\Delta\mathrm{x}$ with $x_0 = 0$ and $x_n = 1$. Choosing suitable Δt and denoting $t_j = j\Delta t$, the discretization of derivatives by CD in (11.140a) and (11.140b) at mesh point (i, j) gives $(r = \Delta t/\Delta x)$,

$$v_{i,\, j+1} = v_{i,\, j-1} + r(w_{i+1,\, j} - w_{i-1,\, j}), \tag{11.143}$$

$$w_{i,\, j+1} = w_{i,\, j-1} \text{ etc.} + r(v_{i+1,\, j} - v_{i-1,\, j}), \quad i = 1(1)n-1, \; j \ge 1. \tag{11.144}$$

Let us suppose that the values of v and w have already been computed upto j^{th} time level i.e., $t = t_j$ and we are required to compute at the $(j+1)^{\text{th}}$ time level. At the internal mesh points values of $v_{i,\,j+1}$ and $w_{i,\,j}$ can be computed from (11.143) and (11.144) respectively. At the end points we have from 2nd and 3rd conditions of (11.141)

$$v_{0,\,j+1} = \dot{\phi}(t_{j+1}) = \dot{\phi}(j+1) \text{ and } v_{n,\,j+1} = \dot{\psi}(t_{j+1}) = \dot{\psi}_{j+1} \tag{11.145}$$

For computing $w_{0,\,j+1}$ and $w_{n,\,j+1}$ we can use respectively FD and BD formulae in either (11.140a) or (11.140b). For example in case of (11.140b), the replacement of derivatives at $x = 0$ may be

$$\frac{w_{0,\,j+1} - w_{0,\,j}}{\Delta t} = \frac{v_{1,\,j+1} - v_{0,\,j+1}}{\Delta x}$$

or
$$w_{0,\,j+1} = w_{0,\,j} + r(v_{1,\,j+1} - v_{0,\,j+1}). \tag{11.146a}$$

Similarly at the other end $x = 1$, we may have

$$w_{n,\,j+1} = w_{n,\,j} + r(v_{n,\,j+1} - v_{n-1,\,j+1}).$$

However, to start the process, we need the values of v and w at the first level of time $t_1\,(j=0)$ which ought to be obtained by some other technique. As explained in solving hyperbolic equation of second order we make use of the initial conditions as follows:

From (11.140a) and (11.142)

$$\frac{v_{i,1} - v_{i,0}}{\Delta t} = f''(x_i) = f_i''$$

or
$$v_{i,1} = v_{i,0} + \Delta t f_i'' = g_i + \Delta t f_i'', \quad i = 1(1)n-1 \tag{11.147a}$$

Also
$$v_{0,1} = \dot{\phi}_1 \text{ and } v_{n,1} = \dot{\psi}_1. \tag{11.147b}$$

Similarly from (11.140b) and (11.141)

$$\frac{w_{i,1} - w_{i,0}}{\Delta t} = g'(x_i) = g_i', \quad i = 1(1)n-1$$

or
$$w_{i,1} = w_{i,0} + \Delta t g_i' = f_i' + \Delta t g_i'.$$

From (11.146a and b) we can get

$$w_{0,1} = w_{0,0} + r(v_{1,1} - v_{0,1}) = f_0' + r(v_{1,1} - v_{0,1}) \tag{11.148a}$$

and
$$w_{n,1} = w_{n,0} + r(v_{n,1} - v_{n-1,1}) = f_n' + r(v_{n,1} - v_{n-1,1}) \tag{11.148b}$$

Having got the values of $v_{i,j}, w_{i,j}$ for $i = 0(1)n$ at the jth time level, the values of $u_{i,j}$ may be computed in various ways. For example.

$$w_{i,j} = \left.\frac{\partial u}{\partial x}\right|_{\substack{x = x_i \\ t = t_j}} = \frac{u_{i,j} - u_{i-1,j}}{\Delta x}$$

or
$$u_{i,j} = u_{i-1,j} + \Delta x \cdot w_{i,j}, \quad i = 1(1)n - 1 \tag{11.149a}$$

Alternatively, if the replacement is made at the midpoint $(x_{i+\frac{1}{2}}, t_j)$ we can have,

$$\frac{w_{i,j} + w_{i-1,j}}{2} = \frac{u_{i,j} - u_{i-1,j}}{\Delta x}$$

giving
$$u_{i,j} = u_{i-1,j} + \frac{\Delta x}{2}\left(w_{i-1,j} + w_{i,j}\right), \quad i = 1(1)n - 1. \tag{11.149b}$$

The relations similar to (11.149a and b) can be derived in terms of v also.

In order to analyse stability, we replace (i, j) by (p, q) and express the errors in $v_{p,q}$ and $w_{p,q}$ by $\mathrm{A}e^{i\beta p\Delta x}\xi q$ and $\mathrm{B}e^{i\beta p\Delta x}\xi q$ respectively where A and B are arbitrary constants.

Hence from (11.143) and (11.144) we get,

$$A\left(\xi - \xi^{-1}\right) = Br\left(e^{i\beta\Delta x} - e^{-i\beta\Delta x}\right)$$

and $B\left(\xi - \xi^{-1}\right) = Ar\left(e^{i\beta\Delta x} - e^{-i\beta\Delta x}\right)$

Eliminating A and B,

$$\left(\xi - \xi^{-1}\right)^2 = r^2\left(e^{i\beta\Delta x} - e^{-i\beta\Delta x}\right)^2$$

or
$$\xi - \frac{1}{\xi} = \pm 2ir\sin\beta\Delta x$$

or
$$\xi^2 \pm 2ir\sin\beta\Delta x\xi - 1 = 0.$$

This equation has two roots ξ_1 and ξ_2; for stability we should have $|\xi_1| \leq 1$ and $|\xi_2| \leq 1$. It can be shown as discussed in Sec. 11.7.4 that $r(= \Delta t/\Delta x) \leq 1$ for the system to be stable.

11.8.6 Solution of first order hyperbolic equation by method of characteristics

Let us consider a first order hyperbolic equation

$$a\frac{\partial u}{\partial x}+b\frac{\partial u}{\partial y}+c=0 \tag{11.150}$$

where $a,\ b$ and c may be functions of $x,\ y$ and u only.

Denoting, $\frac{\partial u}{\partial x}=p$ and $\frac{\partial u}{\partial y}=q$, eq. (11.150) may be written as

$$ap+bq+c=0 \tag{11.151}$$

Also, we do have

$$\begin{aligned} du &= \frac{\partial u}{\partial x}dx+\frac{\partial u}{\partial y}dy \\ &= pdx+qdy. \end{aligned} \tag{11.152}$$

Eliminating p between (11.151) and (11.152) gives

$$q(bdx-ady)+adu+cdx=0. \tag{11.153}$$

Let us choose

$$bdx-ady=0 \quad \text{or} \quad \frac{dy}{dx}=\frac{b}{a}. \tag{11.154}$$

The equation (11.154) defines a characteristic (curve). If a and b do not contain u, then the characteristic may be found explicitly in the form $y=f(x)$. However, if it does involve u then we have to proceed point by point determining the position of mesh point as well as the value of u there, like in the case of second order equation. See Sec. 11.7.7.

From (11.153), it is clear that along the characteristic defined by (11.154) following relation holds,

$$adu+cdx=0 \quad \text{or} \quad du=-\frac{c}{a}dx. \tag{11.155a}$$

After using (11.154) we can also obtain,

$$\frac{du}{-c}=\frac{dx}{a}=\frac{dy}{b} \quad \text{or} \quad du=-\frac{c}{b}dy \tag{11.155b}$$

For the problem to be well-posed value of u should be provided on an initial curve (non-characteristic). We proceed to find the solution in a manner as explained in Sec. 11.7.7. It may be noted that any of the relations (11.155a) or (11.155b) may be used whichever is convenient, to find u. See Examples 11.22 and 11.23 for computational procedure.

Example 11.22

Given a p.d.e.

$$\frac{\partial u}{\partial x} + 2(x-2)\frac{\partial u}{\partial y} = x + 3y$$

with value of u prescribed on the x-axis as $u = 0$. Find the characteristic passing through the point A (2, 0). Hence find the value of u at the point B where it intersects the line $y = 0.25$ and also at point C where it intersects $y = 1.0$. Perform the integration along the characteristic analytically.

Solution

$$a = 1, \ \ b = 2\,(x-2), \ \ c = -(x+3y); \ \ du = (x+3y)dx.$$

The characteristic is given by

$$\frac{dy}{dx} = 2(x-2) \ \text{ or } \ y = (x-2)^2 + \text{const.}$$

Since it passes through (2, 0), the equation of characteristic is

$$y = (x-2)^2 \tag{1}$$

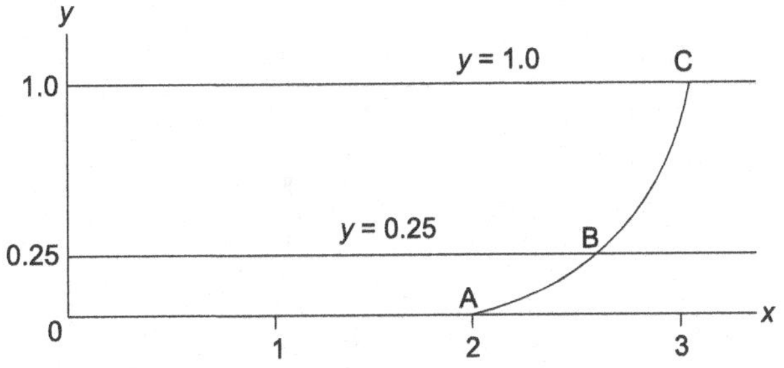

From (1) the coordinates of B and C are B (2.50, 0.25) and C(3.0, 1.0). Integrating the differential equation $du = (x+3y)dx$, along AB from A to B, we get using (1):

$$u_B - u_A = \int_A^B (x+3y)dx = \int_{2.0}^{2.5} \left\{x + 3(x-2)^2\right\} dx = \left[\frac{x^2}{2} + (x-2)^3\right]_{x=2.0}^{x=2.50}$$

or $$u_B = \left(\frac{6.25}{2} + 0.125 - 2\right) = 1.25, \quad \because u_A = 0$$

Integrating along BC from B to C,

$$u_C - u_B = \left[\frac{x^2}{2} + (x-2)^3\right]_{x=2.5}^{x=3.0} = (4.5 + 1.0 - 3.125 - 0.125) = 2.25.$$

or $$u_C = u_B + 2.25 = 3.50.$$

Example 11.23

Show that the p.d.e. $x\frac{\partial u}{\partial x} + 2u\frac{\partial u}{\partial y} = x + y$ reduces to an o.d.e. $xdu - (x+y)dx = 0$ along the characteristic defined by $xdy - 2udx = 0$.

Let it be given that $u = 1.0$ on the x-axis and let the characteristic pass through a point A (1.0, 0). Find approximately up to 4 decimals, the position of the point B on the characteristic where it meets the line $x = 1.25$ and also the value of u at B.

Solution

$$a = x,\ \ b = 2u \ \text{ and } \ c = -(x+y). \ \text{ Then show,}$$

$$\text{Characteristic} : xdy - 2udx = 0. \tag{1}$$

$$\text{o.d.e.} : xdu - (x+y)dx = 0. \tag{2}$$

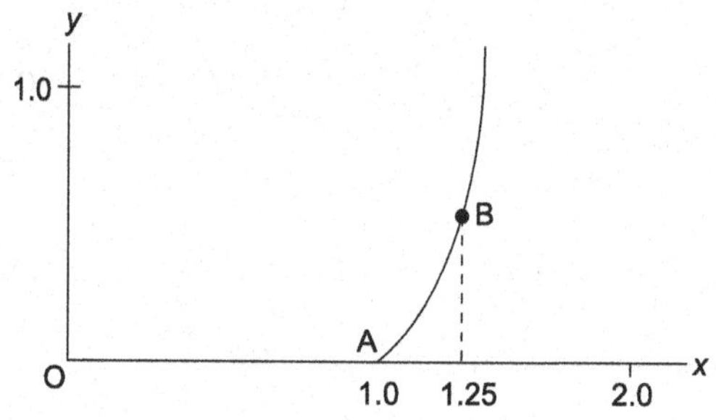

Integrating (1) along AB from A to B,

$$\frac{1.25+1.0}{2}(y_B - y_A) - 2\left(\frac{u_B + u_A}{2}\right)(1.25 - 1.0) = 0$$

or $$1.125(y_B - 0) - (u_B + 1.0)(0.25) = 0, \quad \because u_A = 1.0.$$

or $$1.125y_B - 0.25u_B = 0.25$$

or $$4.5y_B - u_B = 1.0. \tag{3}$$

Integrating (2) along AB from A to B,

$$\frac{1.25+1.0}{2}(u_B - u_A) - \left(\frac{1.25 + y_B + 1.0 + 0}{2}\right)(1.25 - 1.0) = 0$$

or $2.25u_B - 0.25y_B = 2.25 + 0.5625 = 2.8125, \quad \because \; u_A = 1.0.$

or $$9u_B - y_B = 11.25. \tag{4}$$

Solving (3) and (4) gives

$$y_B = 0.5126; \; u_B = 1.3070.$$

Exercise 11

11.1 Richardson's explicit scheme for solving parabolic equation $\dfrac{\partial u}{\partial t} = \dfrac{\partial^2 u}{\partial x^2}$, is given by discretizing the time derivative and space derivative both by central difference at the mesh point (i, j) as follows:

$$\frac{u_{i,\,j+1} - u_{i,\,j-1}}{2\Delta t} = \frac{u_{i-1,\,j} - 2u_{i,\,j} + u_{i+1,\,j}}{\Delta x^2}.$$

Show by Neumann's method that the scheme is unstable for all values of $r(= \Delta t/\Delta x^2)$.

11.2 Du-Fort and Frankel modified the Richardson's scheme (Exercise 11.1) replacing the term $u_{i,j}$ by average value at the $(j-1)^{\text{th}}$ and $(j+1)^{\text{th}}$ levels, obtaining

$$\frac{u_{i,\,j+1} - u_{i,\,j-1}}{2\Delta t} = \frac{u_{i-1,\,j} - (u_{i,\,j-1} + u_{i,\,j+1}) + u_{i+1,\,j}}{\Delta x^2}.$$

Show by Neumann's method that this scheme is unconditionally stable, i.e., stable for all values of r. Also show that the scheme is not compatible with the parabolic equation and approaches hyperbolic equation $c^2 \dfrac{\partial^2 u}{\partial t^2} + \dfrac{\partial u}{\partial t} = \dfrac{\partial^2 u}{\partial x^2} (c = \Delta t/\Delta x)$ as $\Delta t, \, \Delta x \to 0$.

Hint: Obtain the following expansions

$$\frac{u(x,\, t+\Delta t) - u(x,\, t-\Delta t)}{2\Delta t} = \frac{\partial u}{\partial t} + \frac{\Delta t^2}{6}\frac{\partial^3 u}{\partial t^3} + \frac{\Delta t^4}{120}\frac{\partial^5 u}{\partial t^5} + \cdots$$

$$\frac{u(x,\, t+\Delta t) + u(x,\, t-\Delta t)}{\Delta x^2} = \frac{2u}{\Delta x^2} + \frac{\Delta t^2}{\Delta x^2}\frac{\partial^2 u}{\partial x^2} + \frac{1}{12}\frac{\Delta t^4}{\Delta x^2}\frac{\partial^4 u}{\partial t^4} + \cdots$$

$$\frac{u(x+\Delta x,\, t) + u(x-\Delta x,\, t)}{\Delta x^2} = \frac{2u}{\Delta x^2} + \frac{\partial^2 u}{\partial x^2} + \frac{\Delta x^2}{12}\frac{\partial^4 u}{\partial x^4} + \cdots$$

Find truncation error T in the formula when

$$\frac{u_{i,\,j+1} - u_{i,\,j-1}}{2\Delta t} - \frac{u_{i-1,j} - (u_{i,j-1} + u_{i,j+1}) + u_{i+1,j}}{\Delta x^2} + \mathrm{T} = \frac{\partial u}{\partial t} - \frac{\partial^2 u}{\partial x^2}$$

Show $T \to -c^2 \dfrac{\partial^2 u}{\partial t^2}$ as $\Delta t,\ \Delta x \to 0$.

11.3 The parabolic equation $\dfrac{\partial u}{\partial t} = \dfrac{\partial^2 u}{\partial x^2}$ is approximated by the finite difference scheme

$$u_{p,q+1} - u_{p,q} = r\{\theta(u_{p-1,q+1} - 2u_{p,q+1} + u_{p+1,q+1})$$
$$+(1-\theta)(u_{p-1,q} - 2u_{p,q} + u_{p+1,q})$$

where $r = \Delta t/\Delta x^2$ and θ is a parameter $0 \le \theta \le 1$.
Prove by means of (*i*) Matrix method using Brauer's theorem (*ii*) Neumann's method, that the scheme is stable for $r \le 1/2(1-2\theta)$ when $0 \le \theta \le \frac{1}{2}$ and is unconditionally stable when $\frac{1}{2} \le \theta \le 1$.
It may be assumed that Dirichlet condition is prescribed at the end points.

Hint: (*i*) $(\mathrm{I} + r\theta \mathrm{T})\boldsymbol{e}_{q+1} = \{\mathrm{I} - (1-\theta)r\mathrm{T}\}\boldsymbol{e}_q$

error propagation matrix $\mathrm{S} = (\mathrm{I} + r\theta\mathrm{T})^{-1}\{\mathrm{I} - (1-\theta)r\mathrm{T}\}$. If μ is an e.value of T, then

$$\left|\frac{1-(1-\theta)r\mu}{1+\theta r\mu}\right| \le 1 \text{ giving } \mu \le \frac{2}{r(1-2\theta)}.$$

But $0 \le \mu \le 4$ gives $r \le \dfrac{1}{2(1-2\theta)},\ 0 \le \theta \le \dfrac{1}{2}$ or $0 \le 1-2\theta \le 1$.

When $\frac{1}{2} \le \theta < 1$, then $\dfrac{1-(1-\theta)r\mu}{1+\theta r\mu} \le 1$, irrespective of value of r, since $0 \le 1-\theta \le \frac{1}{2}$.

Alternatively e values of matrix T can be used

$$(ii)\ \xi = \frac{1 - 4r(1-\theta)\sin\dfrac{\beta\Delta x}{2}}{1 + 4r\theta\sin^2\dfrac{\beta\Delta x}{2}}$$

11.4 Given the p.d.e.

$$\frac{\partial u}{\partial t} = \frac{\partial^2 u}{\partial x^2},\quad 0 \le x \le 1.0,\ t \ge 0$$

with boundary condition $u = 0$ at $x = 0$ and $x = 1$ and initial condition $u(x, 0) = \sin \pi x$, $t = 0$. Compute $u(x, t)$ for $x = 0\,(0.1)\,1.0$, for t up to 0.02 by Explicit method taking $r = 1/2$.
Hint: $\Delta t = 0.005$. The problem is symmetric about the midpoint $x = 0.5$, therefore values should be computed for $x = 0(0.1)0.5$. Use the following values:

x	0	0.1	0.2	0.3	0.4	0.5	0.6
$\sin \pi x$	0	0.3091	0.5877	0.8090	0.9510	1.0	0.9510

11.5 Solve the parabolic equation

$$\frac{\partial u}{\partial t} = \frac{\partial^2 u}{\partial x^2}, \ 0 \le x \le 0.5, \ t \ge 0$$

by Explicit method for x =0 (0.1) 0.5 for t = 0 (0.005)0.020 when the boundary conditions are given as

$$u(0, t) = 0, \ x = 0 \text{ and } \frac{\partial u}{\partial x} = 0 \text{ at } x = 0.5.$$

At $t = 0$, $u(x, 0) = \sin \pi x$.
Approximate the derivative boundary condition by central difference. (Take π = 3.1416).

or alternatively

The boundary conditions are

$$\frac{\partial u}{\partial x} = 0, \ x = 0; \ u = 0 \text{ at } x = 0.5$$

when the initial condition is given as $u(x, 0) = \cos \pi x$.

11.6 Given the parabolic equation

$$\frac{\partial u}{\partial t} = \frac{\partial^2 u}{\partial x^2}, \ 0 \le x \le 1.0, \ t \ge 0$$

with $u = 0$ as boundary conditions and $u(x, 0) = \sin \pi x$.

Solve the problem by

(*i*) Explicit method (*ii*) C–N method
taking $\Delta x = 0.2$ and $\Delta t = 0.02$. Show only one step of each method. Following values may be taken for π =3.1416

x	0	0.2	0.4	0.6	0.8	1.0
$\sin \pi x$	0	0.5877	0.9510	0.9510	0.5877	0

Hint: Use symmetry about $x = 0.5$, i.e., values of u will be same at x and $(1-x)$.

11.7 Given a parabolic equation $\dfrac{\partial u}{\partial t} = \dfrac{\partial^2 u}{\partial x^2}$, $0 \le x \le 1.0$, $t \ge 0$ with boundary conditions $\dfrac{\partial u}{\partial x} = 0$, at $x = 0$ and $u = 0$ at $x = 1.0$ and initial condition $u(x, 0) = 1 - x^2$.
Solve the problem for $x = 0\ (0.25)\ 1.0$ by
(*i*) Explicit method taking $r = 1/2$.
(*ii*) C–N method taking $r = 1$.

Compare results at $t = 0.0625$. Approximate $\dfrac{\partial u}{\partial x}$ at $x = 0$ by central difference. Alternatively the problem may be defined for $-1 \le x \le 1$ with boundary conditions $u = 0$ at both ends and same initial condition $u(x, 0) = 1 - x^2$. Problem is symmetrical about y-axis.

11.8 Given a two-dimensional parabolic equation

$$\frac{\partial u}{\partial t} = \frac{\partial^2 u}{\partial x^2} + \frac{\partial^2 u}{\partial y^2}, \quad 0 \le x, y \le 1.0, \ t \ge 0$$

with initial condition $u(x, y, 0) = x + y$, $t = 0$. Then following boundary conditions are imposed for $t > 0$,

$$\frac{\partial u}{\partial x} = u \quad \text{at} \quad x = 0 \text{ and } u = 1 \text{ at } x = 1.$$

$$u = 1 \text{ at } y = 0 \text{ and } \frac{\partial u}{\partial y} = -u \text{ at } y = 1.$$

Solve the problem by Explicit method for only two time steps taking $\Delta x = \Delta y = 0.25$ and $\Delta t = 0.01$. Approximate the derivative boundary condition by forward and backward differences as required.

11.9 Solve by ADI scheme the parabolic equation

$$\frac{\partial u}{\partial t} = \frac{\partial^2 u}{\partial x^2} + \frac{\partial^2 u}{\partial y^2}, \quad 0 \le x, y \le 0.6, \ t \ge 0$$

with boundary conditions $u = 1$ on $x = 0$, $y = 0$, $y = 0.6$ and $\dfrac{\partial u}{\partial x} = u$ on $x = 0.6$. The initial distribution of u is given by $u(x, y, 0) = 5(x^2 + y^2)$, $t = 0$.

Solve for $t = 0.02$ and $t = 0.04$ only taking $\Delta t = 0.02$ and $\Delta x = \Delta y = 0.2$. Approximate the derivative boundary condition by backward difference.

11.10 Find the solution of the Dirichlet problem

$$\frac{\partial^2 u}{\partial x^2} + \frac{\partial^2 u}{\partial y^2} = 0,\ 0 \le x,\ 1.0,\ 0 \le y \le 0.75$$

when the sides $x = 0$ and $x = 1.0$ are subjected to boundary conditions $\dfrac{\partial u}{\partial x} = 5$ and $\dfrac{\partial u}{\partial x} = -3$ respectively and $u = 5$ is imposed on $y = 0$ and $y = 0.75$.
Subdivide the domain into square meshes of side 0.25 and replace the boundary conditions by forward and backward differences.
Hint: Use symmetry about the line $y = 0.375$.

11.11 Find the solution of the Poisson equation

$$\frac{\partial^2 u}{\partial x^2} + \frac{\partial^2 u}{\partial y^2} = 1$$

by the method of finite differences. The domain and the prescribed boundary conditions are given below:

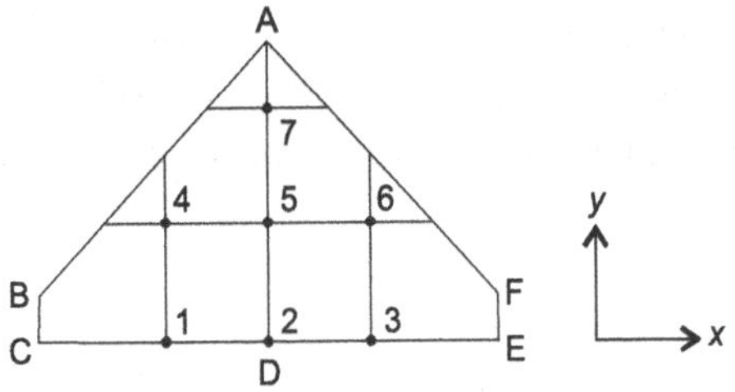

The domain ABCDEFA is subdivided into square mesh of side 0.2, while each of the incomplete side is of length 0.1. The boundary condition $u = 4$ is prescribed on the sides CBAFE while $\dfrac{\partial u}{\partial y} = 5$ on the side CDE. Approximate the derivative boundary condition by forward difference.
Hint: Use symmetry about AD.

11.12 Consider a long uniform hollow cylinder whose cross-section is as shown below:

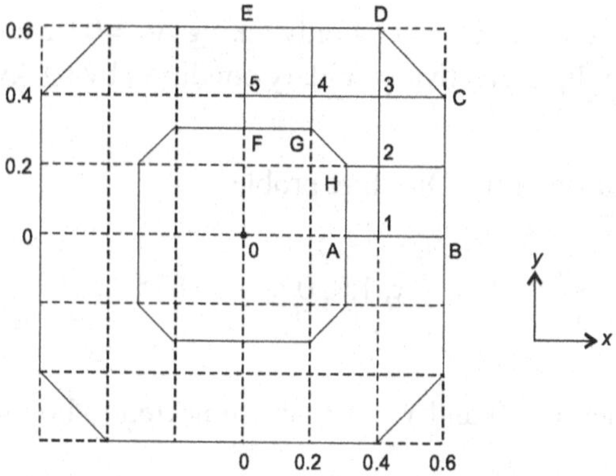

The Poisson equation $\frac{\partial^2 u}{\partial x^2} + \frac{\partial^2 u}{\partial y^2} = 10|xy|$ holds in the solid region. Boundary condition $u = 5$ is prescribed on the outer wall while $u = 1$ is prescribed on the inner wall of the cylinder. Find the solution in the region ABCDEFGHA at the points 1, 2, 3, 4 and 5 remembering that there is symmetry about x-axis and y-axis while O is the origin. The region is subdivided into square mesh of side 0.2 and points F, G, H and A are midpoints.
Hint: Also use symmetry about the diagonals $y = x$ and $y = -x$.

11.13 An elliptic equation

$$\frac{\partial^2 u}{\partial x^2} + \frac{\partial^2 u}{\partial y^2} = 0$$

is defined over a domain enclosed by x-axis, y-axis and the straight line $x + y = 0.8$. Subdivide the domain into square mesh of side 0.2 by drawing lines parallel to the axes. Following boundary conditions are prescribed on the three sides of the domain:

$$\frac{\partial u}{\partial x} = -2 \text{ at } x = 0,$$

$$\frac{\partial u}{\partial y} = 4 \text{ at } y = 0$$

and $\quad u = 10(x^2 + y^2)$ on the line $x + y = 0.8$.

Find the solution at the internal mesh points approximating the derivative boundary conditions by forward difference.

11.14 A Poisson problem is defined as

$$\frac{\partial^2 u}{\partial x^2} + \frac{\partial^2 u}{\partial y^2} = x + y$$

over a square domain $0 \le x,\ y \le 0.6$. The boundary conditions on the four sides are prescribed as follows:

$$\frac{\partial u}{\partial x} = 2 \text{ at } x = 0, \quad \frac{\partial u}{\partial x} = 4 \text{ at } x = 0.6;$$

$$u = 2 \text{ at } y = 0, \quad u = 8 \text{ at } y = 0.6.$$

Divide the domain into square mesh of side 0.2 and find the solution using Gauss–Seidel method at the internal points approximating the boundary conditions by forward and backward differences.
Compute up to an accuracy of three decimal places.

11.15 Solve Exercise 11.14 by Gauss–Seidel SOR method using $\omega = 1.2$.

11.16 A function u is the solution of the equation

$$\frac{\partial^2 u}{\partial x^2} + 2(1-2x)\frac{\partial^2 u}{\partial x \partial y} - 8x\frac{\partial^2 u}{\partial y^2} = 0$$

and satisfies the initial conditions

$$u = x^2, \quad \frac{\partial u}{\partial y} = 0 \text{ on the initial curve as } x\text{-axis.}$$

Calculate u at the point $R(x_R, y_R)$ which is intersection of the characteristics through the points A (0.2, 0) and B(0.4,0).

References and Some Useful Related Books/Papers

1. Ames, W.F., *Numerical Methods for Partial Differential Equations*, Academic Press.
2. Forsythe, G.E. and Wasow, W.R., *Finite Difference methods for Partial Differential Equations*, John Wiley.
3. Fox, L., *Numerical Solution of Ordinary and Partial Differential Equations*, Pergamon.
4. Jain, M.K., *Numerical Solution of Differential Equations*, Wiley Eastern Ltd.
5. Smith, G.D., *Numerical Solution of Partial Differential Equations*, Oxford University Press.

12

Finite Element Method

12.1 Introduction

We discussed boundary value problems (bvp's) in one-dimension in Chapter 7 and in two-dimensions (elliptic equations) in Chapter 11. In case of ordinary differential equation the domain of the problem was a finite interval along, say x-axis. Some nodal points or nodes were selected over the interval by subdividing it into uniform (evenly-spaced) sub-intervals/subdomains. Similarly an elliptic equation was defined over a domain in the x-y plane which was subdivided into rectangular/square meshes. In both cases the sizes of the subdomains were of uniform size except perhaps near the boundary. The solution was obtained by approximating the derivatives by finite differences (forward, backward or central). These methods are called Finite Difference Methods (FDMs).

The Finite Element Method (FEM) in this chapter, will be discussed mainly in respect of boundary value problems in one and two dimensions only. It can be used for solving transient (time-dependent) problems also and in that case, as one of the methods, the time derivative will be approximated by the finite difference techniques. They are also discussed briefly. The FEM approach for solving a bvp differs from FDM in mainly two aspects. First, in the FEM, the domain may be subdivided in an arbitrary manner. For example, in two-dimension it is usually subdivided into triangles which are not necessarily of equal sizes – on the other hand they are often of unequal sizes. This gives us the freedom to choose larger triangles in the part of the domain where variation in the solution is expected to be not too large and choose smaller triangles where solution may vary too sharply or it may be too sensitive over some part of the domain. These triangles (or polygons in general) are called 'elements' and vertices of the triangles (or polygons) as 'nodes'. We are required to find the solution at these nodes. The second difference is with regard to the mathematical strategy adopted in FEM which is totally different from FDM. As the nodes are not positioned along the axes in FEM, the derivatives can not be approximated by finite differences. Instead, the value of some kind of integral is minimised over the entire domain and the process of minimisation is carried out element-wise (or node-wise) to get an over-all minimum.

As stated, some kind of integral is to be minimised in FEM. This situation arises on account of the mathematical strategy adopted that uses either of the two approaches, *viz.*

(*i*) Galerkin's method or

(*ii*) Rayleigh–Ritz (or Ritz) method.

The first of the above methods falls under the broad category of 'weighted residual methods' while the second under 'variational methods'. Therefore we shall give brief outlines of these methods before embarking upon FEM. It will be shown, however, that the two approaches are equivalent in case of linear equations.

12.2 Weighted Residual Methods

Consider a boundary value problem

$$\mathrm{L}u = g \tag{12.1}$$

defined in region R bounded by S; L is a differential operator and g is a known function. In one dimension, S will consist of two points; in two dimensions it will represent a curve and in three dimensions, a surface. Let us assume that appropriate boundary conditions are prescribed on S and we are required to find the solution in domain D = R∪S.

The underlying principle of a 'weighted residual' method is that instead of finding the solution u which satisfies (12.1) at every-point in D, we seek for a solution $\overline{u}$ which satisfies it in a weaker form. This point will be explained in the following analysis.

Let $\Phi = \{\phi_1, \phi_2, \ldots.\}$ be a set of linearly independent functions satisfying the boundary condition $u = 0$ on S; obviously there can be infinite number of such functions. These functions form a basis for approximating u in D. Let us assume an approximate solution that satisfies the boundary condition as

$$\overline{u} = \sum_{i=1}^{n} a_i \phi_i, \tag{12.2}$$

where a_i's are unknown parameters and are to be determined. We substitute the assumed solution (12.2) in (12.1) and obtain a 'residual' Re which is a function of the space variables and contains the parameters a_i's, i.e.,

$$Re = \mathrm{L}\overline{u} - g. \tag{12.3}$$

In general, the assumed solution $\overline{u}$ hardly ever satisfies the equation (12.1) exactly.

In the 'weighted residual method' the integral of the residual Re alongwith certain weights, is minimised over R. There are various methods which differ from each other in the way the integral is minimised by choosing weight functions (weights) to determine the parameters $a_1, a_2, \dots a_n$ in (12.2).

12.2.1 Galerkin's method

It is a widely used method to find approximate solution of ordinary and partial differential equations. The residual Re is minimised such that it is orthogonal to all the basis function ϕ_i, $i = 1(1)n$ over the region R, i.e.

$$\iint_R Re\phi_j d\sigma = \iint_R (L\bar{u} - g)\phi_j d\sigma = 0, \; j = 1(1)n. \tag{12.4}$$

We get from (12.4) a set of n linear simultaneous equations in $a_1, a_2, \dots a_n$. After solving these equations and putting the values a_i's in (12.2), the approximate solution $\bar{u}$ may be obtained. This method will be our basis in the discussion of the Finite Element Method.

12.2.2 Least squares method

This method has been discussed in respect of discrete data in Chapter 9. It can also be used for finding approximate solution to ordinary and partial differential equations. We minimise the integral of square of the residual with respect to the unknown parameters a_i's over the region R, i.e.,

$$\frac{\partial}{\partial a_i} \iint_R R^2 e \, d\sigma = 0, \; i = 1(1)n,$$

giving

$$2 \iint Re \frac{\partial Re}{\partial a_i} d\sigma = 0 \quad \text{or} \quad \iint_R Re \frac{\partial Re}{\partial a_i} d\sigma = 0.$$

12.2.3 Subdomain method

In this method, domain D is subdivided into as many subdomains as the number of unknowns and the residual is then zeroised over each subdomain, e.g.,

$$\iint_{D_i} Re d\sigma = 0, \; i = 1(1)n$$

12.2.4 Collocation method

In this method, the residual *Re* is made zero at as many points (known as collocation points) of the domain as there are unknown parameters. It can be expressed, in one dimension, as

$$\int_a^b Re\delta\ (x-x_i)dx = 0,\ i = 1(1)n$$

where $\delta(x-x_i)$ is Dirac's delta function defined for $-\infty < x < \infty$ such that $\delta(x-x_i) = 0$ at all points except at $x = x_i$ where it tends to infinity such that

$$\int_{-\infty}^{\infty} Re\delta\ (x-x_i)dx = Re(x_i)$$

Obviously, (12.9) can be written over the interval $(a,\ b)$ as

$$\int_a^b Re\delta\ (x-x_i)dx = Re(x_i),\ i = 1(1)n.$$

For further details of residual methods, see [2].

Example 12.1

Find an approximate solution by Galerkin's method, of the differential equation

$$\frac{d^2y}{dx^2} - u = x,\ 0 \le x \le 1,$$

with boundary conditions $u(0) = u(1) = 0$.
Use only two basis functions.

Solution A set of functions satisfying the boundary conditions may be

$$\Phi = \{x\ (x-1),\ x^2(x-1),\ x^2(x-1)^2,\ \ldots.\}$$

Let us choose

$$\bar{u} = a_1 x\ (x-1) + a_2 x^2 (x-1)$$

Differentiating it twice gives

$$\bar{u}'' = 2a_1 + a_2\ (6x-2).$$

Substituting $\overline{u}$ and $\overline{u}''$ in the differential equation gives

$$Re = 2a_1 + a_2(6x - 2) - a_1x(x - 1) - a_2x^2(x - 1) - x$$
$$= a_1\{2 - x(x - 1)\} + a_2\{6x - 2 - x^2(x - 1)\} - x$$

Applying Galerkin's method

$$(i) \int_0^1 Re\, x(x-1)dx = \int_0^1 [a_1\{2 - x(x-1)\} + a_2\{6x - 2 - x^2(x-1)\} - x]x(x-1)dx = 0$$

gives $-\dfrac{11}{30}a_1 - \dfrac{11}{60}a_2 + \dfrac{1}{12} = 0$

or $\qquad 11a_1 + 5.5a_2 = 2.5. \qquad$ .... (1)

$$(ii) \int_0^1 Re\, x^2(x-1)dx = \int_0^1 [a_1\{2 - x(x - 1)\} + a_2\{6x - 2 - x^2(x - 1)\} - x]x^2(x - 1)dx = 0$$

gives $-\dfrac{11}{60}a_1 - \dfrac{1}{7}a_2 + \dfrac{1}{20} = 0$

or $\qquad 11a_1 + \dfrac{60}{7}a_2 = 3. \qquad$ .... (2)

By solving (1) and (2) we get

$$a_1 = 0.1460, \quad a_2 = 0.1626$$

The approximate solution is

$$\overline{u} = 0.1460x(x - 1) + 0.1626x^2(x - 1)$$
$$= -0.1460x - 0.0166x^2 + 0.1626x^3.$$

Note: For comparison with exact solution see Example 12.2.

Example 12.2

Find the approximate solution of the differential equation of Example 12.1 by method of Least Squares.

Solution The residual is given by

$$Re = a_1\{2 - x(x - 1)\} + a_2\{6x - 2 - x^2(x - 1)\} - x$$

Applying method of Least Square

$$(i) \qquad \frac{\partial}{\partial a_1}\int_0^1 Re^2 dx = 2\int_0^1 Re\frac{\partial Re}{\partial a_1}dx = 0$$

i.e., $\int_0^1 [a_1\{2-x\,(x-1)\}+a_2\{6x-2-x^2(x-1)\}-x]\{2-x(x-1)\}dx = 0$, giving

$$\frac{47}{10}a_1+\frac{141}{60}a_2=\frac{13}{12}$$

or $\qquad 282a_1+141a_2=65 \qquad$ .... (1)

$$(ii) \qquad \frac{\partial}{\partial a_2}\int_0^1 Re^2 dx = 2\int_0^1 Re\frac{\partial Re}{\partial a_2}dx = 0$$

i.e., $\int_0^1 [a_1\{2-x\,(x-1)\}+a_2\{6x-2-x^2(x-1)\}-x]\{6x-2-x^2(x-1)\}dx = 0$, giving

$$\frac{141}{60}a_1+\frac{449}{105}a_2=\frac{21}{20}$$

or $\qquad 141a_1+256.5714a_2=63 \qquad$ .... (2)

Solving (1) and (2) gives

$$a_1 = 0.1485,\ \ a_2 = 0.1639$$

$$\bar{u} = 0.1485x(x-1)+0.1639x^2(x-1)$$

$$= -0.1485x\ -0.0154x^2+0.1639x^3$$

Note: Analytical solution is $u=\dfrac{e^x-e^{-x}}{e-e^{-1}}-x = 0.4254(e^x-e^{-x})-x$

$$\text{or} = \frac{\sin hx}{\sin h1}-x = 0.8509\sin hx - x$$

$$\simeq 0.1492x+0.1418x^3+\ldots.$$

Values of u at some selected points

	Method \ x	0	0.25	0.50	0.75	1.0
(i)	Analytical	0	– 0.0351	– 0.0569	– 0.0521	0
(ii)	Galerkin	0	– 0.0350	– 0.0568	– 0.0502	0
(iii)	Least Square	0	– 0.0355	– 0.0576	– 0.0509	0

Example 12.3

Solve the problem of Example 12.1 by subdomain method dividing the interval $0 \leq x \leq 1$ into two equal intervals and using the same basis functions.

Solution

$$Re = a_1\{2 - x(x-1)\} + a_2\{6x - 2 - x^2(x-1)\} - x$$

$$\int_0^{1/2} Re\, dx = 0 \text{ gives}$$

$$\frac{26}{24}a_1 - \frac{43}{64 \times 3}a_2 - \frac{1}{8} = 0$$

or

$$208a_1 - 43a_2 = 24 \qquad \text{.... (1)}$$

$$\int_{1/2}^{1} Redx = 0 \text{ gives}$$

$$\frac{26}{24}a_1 + \frac{251}{64 \times 3}a_2 - \frac{3}{8} = 0$$

or

$$208a_1 + 251a_2 = 72. \qquad \text{.... (2)}$$

Solving (1) and (2) we get

$$a_1 = 0.1491, \quad a_2 = 0.1633$$

Solution is:

$$\bar{u} = 0.1491x(x-1) + 0.1633x^2(x-1).$$

Example 12.4

Solve the differential equation of Example 12.1 by collocation method taking the collocation points which divide the interval in three equal parts; use same basis functions.

Solution

$$Re(x) = a_1\{2 - x(x-1)\} + a_2\{6x - 2 - x^2(x-1)\} - x$$

Collocation points are $x = \frac{1}{3}, \frac{2}{3}$

$$Re\left(\frac{1}{3}\right) \equiv \frac{20}{9}a_1 + \frac{2}{27}a_2 - \frac{1}{3} = 0$$

or $$60a_1 + 2a_2 = 9 \qquad \text{.... (1)}$$

$$Re\left(\frac{2}{3}\right) \equiv \frac{20}{9}a_1 + \frac{58}{27}a_2 - \frac{2}{3} = 0$$

or $$60a_1 + 58a_2 = 18 \qquad \text{.... (2)}$$

Solving (1) and (2) gives

$$a_1 = 0.1446, \;\; a_2 = 0.1607$$

Solution is: $$\overline{u} = 0.1446x(x-1) + 0.1607x^2(x-1).$$

Example 12.5

Find an approximate solution by Galerkin's method, of the Poisson equation

$$\frac{\partial^2 u}{\partial x^2} + \frac{\partial^2 u}{\partial y^2} = -1$$

defined in domain D where D $= \{x, y| -1 \leq x, y \leq 1\}$ and homogeneous Dirichlet boundary conditions are prescribed on the boundary, i.e., $u = 0$ on $x = \pm 1$ and $y = \pm 1$.

Use only one basis function.

Solution The problem is symmetrical about x and y axes as well as about the diagonals $y = \pm x$ of the square domain D. Therefore the basis functions satisfying the boundary conditions may be of the form as given in the set

$$\Phi = \{(1-x^2)(1-y^2), \; x^2y^2(1-x^2)(1-y^2), \; \ldots .\}$$

Choosing $\overline{u} = a(1-x^2)(1-y^2)$, we get

$$\frac{\partial^2 \overline{u}}{\partial x^2} = -2a(1-y^2), \;\; \frac{\partial^2 \overline{u}}{\partial y^2} = -2a\,(1-x^2)$$

Putting these values in p.d.e. gives

$$Re(x,y) = -2a\,\{1 - x^2 + 1 - y^2\} + 1$$

Using Galerkin's method

$$\int_{-1}^{1}\int_{-1}^{1} \{-2a\,(1-x^2+1-y^2)+1\}(1-x^2)(1-y^2)dxdy = 0$$

or $\quad 8a\int_0^1\int_0^1(1-x^2+1-y^2)(1-x^2)(1-y^2)dxdy = 4\int_0^1\int_0^1(1-x^2)(1-y^2)dxdy$

or $\quad \frac{256}{45}a = \frac{16}{9}$ or $a = \frac{5}{16}$ and hence $\overline{u} = \frac{5}{16}(1-x^2)(1-y^2)$.

Example 12.6

Use method of Least Squares in problem of Example 12.5.

Solution $\qquad Re = -2a\{1-x^2+1-y^2\}+1$

$$\int_{-1}^{1}\int_{-1}^{1} Re\frac{\partial Re}{\partial a}dxdy = 0 \text{ gives}$$

$$\int_{-1}^{1}\int_{-1}^{1}\{-2a\,(1-x^2+1-y^2)+1\}\{-2(1-x^2+1-y^2)\}dxdy = 0.$$

We get $a = \frac{15}{44}$; hence $\overline{u} = \frac{15}{44}(1-x^2)(1-y^2)$.

12.3 Non-homogeneous Boundary Conditions

In the preceding section we have considered problems with homogeneous boundary conditions, i.e., when the value of u is given to be zero at the boundary. The residual methods can be applied to solve problems having non-homogeneous boundary conditions also. In the case of ordinary differential equation we can assume a solution of the form $\overline{u} = \phi_0(x) + \sum_{i=1}^{n} a_i\phi_i(x)$ where $\phi_0(x)$ is a function which satisfies the given boundary conditions while basis functions ϕ_i's $i = 1(1)n$ may be chosen to have zero values there. Following example will illustrate the procedure.

Example 12.7

Find an approximate solution to the given differential equation by Galerkin's method:

$$\frac{d^2u}{dx^2} - u = x, \quad 0 \le x \le 1$$

$$x\,(0) = 1, \quad x\,(1) = 3.$$

Use basis functions $\phi_1 = x(x-1)$ and $\phi_2 = x^2(x-1)$.

Solution Let us assume the solution satisfying the boundary conditions as

$$\bar{u} = 2x+1+a_1x(x-1)+a_2x^2(x-1) \text{ where } \phi_0 = 2x+1$$

$$Re = a_1\{2-x(x-1)\}+a_2\{6x-2-x^2(x-1)\}-(2x+1+x)$$

Applying Galerkin's method

$$\int_0^1 Re\phi_1(x)dx = 0 \text{ gives}$$

$$\int_0^1 a_1\{2-x(x-1)\}x(x-1)dx + a_2\int_0^1 \{6x-2-x^2(x-1)\}x(x-1)dx$$

$$= \int_0^1 (3x+1)x(x-1)dx \quad \text{or} \quad 22a_1+11a_2 = 25 \qquad \text{...(1)}$$

Similarly $\int_0^1 Re\phi_1(x)dx = 0$ gives

$$22a_1 + \frac{120}{7}a_2 = 28 \qquad \text{....(2)}$$

On solving (1) and (2): $a_1 = 0.8922, \;\; a_2 = 0.4884$

$$\bar{u} = 2x+1+0.8922x(x-1)+0.4884x^2(x-1)$$

Note: Analytical solution is: $u = 1.5453e^x - 0.5453e^{-x} - x$

At $x = 0.5$ we have $u(0.5) = 1.7170$ and $\bar{u}(0.5) = 1.7159$.

12.4 Variational Methods

In the variational methods, the problem is solved rather indirectly in the sense that the given differential equation is reformulated. It is converted into a problem for finding an extremum (generally minimum) of a definite integral over the domain of the given differential equation with respect to a function. The function for which the integral attains its minimum value is the solution of the given differential equation. We discuss the method in respect of some typical problems assuming the existence of the solution and fulfilment of continuity and differentiability conditions. Nevertheless our treatment of FEM will be based on Galerkin technique.

12.4.1 Functional and its variation

Let $F(x, u, u')$ be a function of x, $u(x)$ and $u'(x)$ $(= \frac{du}{dx})$ while $u(x)$ belongs to a class of functions, defined in the interval $a \le x \le b$, satisfying the boundary conditions

$$u(a) = A, \quad u(b) = B. \tag{12.5}$$

We define an integral

$$I\,[u(x)] = I\,[u] = \int_a^b F(x, u, u')dx. \tag{12.6}$$

The value of the integral $I\,[u]$ does vary with the choice of the function $u(x)$. Thus $I[u]$ is a function of a function which acquires a definite value for each function $u(x)$ satisfying conditions (12.5); it is called 'Functional'.

We now derive an expression for the corresponding change in the value of the functional $I[u]$ when the function $u(x)$ changes to $u(x) + \delta u(x)$ where $\delta u(x)$ is an arbitrary function such that

$$\delta u(a) = 0 \text{ and } \delta u(b) = 0. \tag{12.7}$$

It may be noted that the conditions (12.7) ensure that the new functions $u(x) + \delta u(x)$ belongs to the class of functions satisfying conditions (12.5).

The variation in the value of the functional may be computed using Taylor's expansion noting that $\delta u'$ is the corresponding change in $u'(x)$ as follows:

$$\begin{aligned}
\delta I &= I[u(x) + \delta u(x)] - I[u(x)] \\
&= \int_a^b \{F(x, u + \delta u, u' + \delta u') - F(x, u, u')\}dx \\
&= \int_a^b \left\{\frac{\partial F}{\partial u} \cdot \delta u + \frac{\partial F}{\partial u'}\delta u'\right\} dx + \frac{1}{2}\int_a^b \left\{\delta^2 u \frac{\partial^2 F}{\partial u^2} + 2\delta u \cdot \delta u' \frac{\partial^2 F}{\partial u \partial u'} + \delta u'^2 \frac{\partial^2 F}{\partial u'^2}\right\} dx + \cdots
\end{aligned} \tag{12.8}$$

Assuming that second and higher order partial derivatives of $F\,(x, u, u')$ remain bounded over $[a, b]$, the first order variation in $I\,[u]$ may be written as

$$\delta I = \int_a^b \left\{\frac{\partial F}{\partial u}\delta u + \frac{\partial F}{\partial u'}\delta u'\right\} dx. \tag{12.9}$$

Integration of second term in (12.9) gives

$$\int_a^b \frac{\partial F}{\partial u'}\delta u' \cdot dx = \frac{\partial F}{\partial u'} \cdot \delta u \Big|_a^b - \int_a^b \frac{d}{dx}\left(\frac{\partial F}{\partial u'}\right) \cdot \delta u \cdot dx$$

$$= -\int_a^b \frac{d}{dx}\left(\frac{\partial F}{\partial u'}\right) \cdot \delta u \cdot dx \qquad (12.10)$$

since $\delta u(a) = 0$ and $\delta u(b) = 0$ due to (12.7).
Using (12.10) in (12.9) we get

$$\delta I = \int_a^b \left\{\frac{\partial F}{\partial u} - \frac{d}{dx}\left(\frac{\partial F}{\partial u'}\right)\right\} \delta u \cdot dx \qquad (12.11)$$

For the functional $I[u]$ defined by (12.6) to attain its extremum (stationary value) for a given function $u(x)$, the necessary condition would be $\delta I = 0$, i.e.,

$$\frac{\partial F}{\partial u} - \frac{d}{dx}\left(\frac{\partial F}{\partial u'}\right) = 0. \qquad (12.12)$$

The condition (12.12) is known as Euler's (or Euler-Lagrange) equation. Thus the extremal of the functional (12.6) provides the solution of the boundary value problem

$$\frac{\partial F}{\partial u} - \frac{d}{dx}\left(\frac{\partial F}{\partial u'}\right) = 0 \qquad (12.13)$$

with $\qquad u(a) = A,\ u(b) = B.$

Note: We can arrive at the result (12.12) by choosing $\delta u = \varepsilon\eta(x),\ \delta u' = \varepsilon\eta'(x);\ \eta(a) = \eta(b) = 0$ and get $\frac{dI}{d\varepsilon} = 0$ evaluated as $\varepsilon \to 0$.

12.4.2 Rayleigh–Ritz (or Ritz) method

In this method, first of all an appropriate form of a functional is sought corresponding to the given differential equation. Then like residual method, an approximate solution is assumed which is a linear combination of basis functions involving unknown parameters. This solution is substituted in the functional which is extremised with respect to the unknown parameters. As a result, as many linear simultaneous equations are obtained as there are unknowns. The solution of simultaneous equations gives the values of unknowns which are put in the assumed solution to provide the final solution of the given differential equation.

Example 12.8

Solve the problem given in Example 12.1 by Ritz method assuming the same approximate solution as in the Galerkin's method.

Solution The given differential equation is

$$\frac{d^2u}{dx^2} - u - x = 0, \quad 0 \le x \le 1,$$

with boundary conditions

$$u(0) = 0, \quad u(1) = 0.$$

The integrand corresponding to the given differential equation is

$$\mathrm{F}(x, u, u') = -\frac{1}{2}(u^2 + 2ux + u'^2) \qquad ....(1)$$

since from Euler's equation

$$\frac{\partial \mathrm{F}}{\partial u} - \frac{d}{dx}\left(\frac{\partial \mathrm{F}}{\partial u'}\right) = -\frac{1}{2}\{2u + 2x - 2u''\} = u'' - u - x = 0$$

which is the same differential equation as given.
We may choose the above integrand or a more convenient form

$$\mathrm{F}(x, u, u') = u^2 + 2xu + u'^2, \qquad ....(2)$$

since it will give the same Euler's eqn. except for a constant factor $(-\frac{1}{2})$.
Let us suppose that for

$$u = a_1x(x-1) + a_2x^2(x-1) \qquad .... (3)$$

the functional $\mathrm{I}[u]$ attains its extermum value where

$$\mathrm{I}[u] = \int_0^1 (u^2 + 2xu + u'^2)dx. \qquad .... (4)$$

We get the following,

$$u = a_1x(x-1) + a_2x^2(x-1)$$

$$u^2 = a_1^2x^2(x-1)^2 + a_2^2x^4(x-1)^2 + 2a_1a_2x^3(x-1)^2$$

$$u' = a_1(2x-1) + a_2(3x^2-2x)$$

$$u'^2 = a_1^2(2x-1)^2 + a_2^2(3x^2-2x)^2 + 2a_1a_2(2x-1)(3x^2-2x)$$

$$2ux = 2a_1(x^3-x^2) + 2a_2(x^4-x^3)$$

$$\begin{aligned} u^2+u'^2+2ux &= a_1^2\,(x^4-2x^3+5x^2-4x\,+1) + a_2^2(x^6-2x^5+10x^4-12x^3 \\ &\quad +4x^2) + 2a_1a_2(x^5-2x^4+7x^3-7x^2+2x) + 2a_1(x^3-x^2) \\ &\quad +2a_2(x^4-x^3) \end{aligned}$$

$$\text{I}\,[u] = \int_0^1 (u^2+u'^2+2ux)dx$$

Find $\dfrac{\partial \text{I}}{\partial a_1} = 0,$

$$\int_0^1 [2a_1(x^4-2x^3+5x^2-4x+1) + 2a_2(x^5-2x^4+7x^3-7x^2+2x) + 2(x^3-x^2)]dx = 0$$

or $\quad \dfrac{22}{60}a_1 + \dfrac{11}{60}a_2 = \dfrac{1}{12}$

or $\quad 22a_1 + 11a_2 = 5$ .... (5)

Find $\dfrac{\partial \text{I}}{\partial a_2} = 0,$

$$\int_0^1 [2a_2(x^6-2x^5+10x^4-12x^3+4x^2) + 2a_1(x^5-2x^4+7x^3-7x^2+2x) + 2(x^4-x^3)]dx = 0$$

or $\quad \dfrac{1}{7}a_2 + \dfrac{11}{60}a_1 = \dfrac{1}{20} \quad$ or $\quad 11a_1 + \dfrac{60}{7}a_2 = 3.$ .... (6)

Solving (5) and (6) gives

$$a_1 = 0.1460, \;\; a_2 = 0.1626$$

$$u = 0.1460x(x-1) + 0.1626x^2(x-1)$$

Note: We have got the same simultaneous equations as in Galerkin's method.

12.5 Equivalence of Rayleigh–Ritz and Galerkin Methods (1–D)

We have seen in Example 12.8 that Rayleigh–Ritz and Galerkin methods lead to the same simultaneous equations for homogeneous boundary conditions. Let us show they are same for the non-homogeneous boundary conditions also if same basis functions are chosen. We are required to solve

$$\frac{d^2u}{dx^2} - u - x = 0, \quad 0 \leq x \leq 1$$

with non-homogeneous boundary conditions at $x = 0$ and $x = 1$ (See Example 12.7). Let us assume a solution with two basis functions as

$$u = \phi_0(x) + a_1\phi_1(x) + a_2\phi_2(x)$$

where $\phi_0(x)$ satisfies the prescribed boundary conditions while $\phi_1(x)$ and $\phi_2(x)$ assume zero values at the end points.

Rayleigh–Ritz Method

$$u = \phi_0 + a_1\phi_1 + a_2\phi_2; \; u' = \phi_0' + a_1\phi_1' + a_2\phi_2'$$

$$\mathrm{I}[u] = \int_0^1 (u^2 + u'^2 + 2ux)dx$$

$$\frac{\partial \mathrm{I}}{\partial a_1} = \int_0^1 [2a_1(\phi_1^2 + \phi_1'^2) + 2a_2(\phi_1\phi_2 + \phi_1'\phi_2') + 2x\phi_1 + 2\phi_0\phi_1]dx = 0$$

or $$a_1\int_0^1 (\phi_1^2 + \phi_1'^2)dx + a_2\int_0^1 (\phi_1\phi_2 + \phi_1'\phi_2')dx = -\int_0^1 (\phi_0 + x)\phi_1 dx \quad (12.14)$$

Similarly, $\dfrac{\partial \mathrm{I}}{\partial a_2} = 0$ gives

$$a_1\int_0^1 (\phi_1\phi_2 + \phi_1'\phi_2')dx + a_2\int_0^1 (\phi_2^2 + \phi_2'^2)dx = -\int_0^1 (\phi_0 + x)\phi_2 dx \quad (12.15)$$

Galerkin's method

After substituting u, u'' in the original differential equation, the first equation is given by

$$\int_0^1 \{a_1\phi_1'' + a_2\phi_2'' - (\phi_0 + a_1\phi_1 + a_2\phi_2) - x\}\phi_1 dx = 0$$

or $$a_1 \int_0^1 (\phi_1\phi_1'' - \phi_1^2)dx + a_2 \int_0^1 (\phi_1\phi_2'' - \phi_1\phi_2)dx = \int_0^1 (\phi_0 + x)\phi_1 dx. \tag{12.16}$$

We have $$\int_0^1 \phi_1\phi_1'' dx = \phi_1\phi_1'\Big|_0^1 - \int_0^1 \phi_1'^2 dx = -\int_0^1 \phi_0'^2 dx,$$

and $$\int_0^1 \phi_1\phi_2'' dx = \phi_1\phi_2'\Big|_0^1 - \int_0^1 \phi_1'\phi_2' dx = -\int_0^1 \phi_1'\phi_2' dx$$

since ϕ_1 and ϕ_2 have zero values at the end points.
Hence (12.16) becomes

$$a_1 \int_0^1 (\phi_1^2 + \phi_1'^2)dx + a_2 \int_0^1 (\phi_1\phi_2 + \phi_1'\phi_2')dx = -\int_0^1 (\phi_0 + x)\phi_1 dx. \tag{12.17}$$

In the same manner the second equation can be written as

$$a_1 \int_0^1 (\phi_1\phi_2 + \phi_1'\phi_2')dx + a_2 \int_0^1 (\phi_2^2 + \phi_2'^2)dx = -\int_0^1 (\phi_0 + x)\phi_2 dx. \tag{12.18}$$

Equations (12.17) and (12.18) are same as eqs (12.14) and (12.15). The result can be generalised for n basis functions easily.

12.6 Construction of Functional

In the foregoing section we have defined a functional whose argument is a function of one variable only. It is expressed in the form of an integral, the extremal of which is the solution of a differential equation. The outcome of the condition for minimality of the functional is Euler's equation. We can define functional whose argument is a function of two variables expressing it in the form of double integral and derive the relevant Euler's equation for realising the minimum. Its minimal will obviously correspond to the solution of a partial differential equation in two dimensions. Similarly, a functional may be represented by triple integral when its argument is a function of three variables. However the problem is the other way round. That is, we are given a differential equation to solve and are interested to construct the corresponding functional so that its minimal provides the solution to the given problem. Here we shall deal with this aspect confining ourselves to two-dimensional boundary value problem, particularly having elliptic p.d.e. in mind. See [1], [4].

Let a two-dimensional Poisson equation be given

$$\nabla^2 u + g(x,y) = 0$$

$$\text{or} \qquad -\nabla^2 u = g(x,\ y) \tag{12.19}$$

which is defined over a region R enclosed by a curve S and we are required to find its solution in domain D = R ∪ S. Four types of boundary conditions associated with (12.19) will be considered on S:

(*i*) homogeneous Dirichlet condition, $u = 0$.

(*ii*) non-homogeneous Dirichlet condition, $u = u_1$ (constant)

(*iii*) non-homogeneous Dirichlet condition,
$u = f(x,\ y)$ on boundary S.

(*iv*) mixed condition, $\alpha u + \dfrac{\partial u}{\partial n} = 0$, where n is the outward normal to S.

12.6.1 Preliminaries from vector calculus

Let us note the following results from vector calculus in two dimensions:

If $u(x,\ y)$ and $v(x,\ y)$ are scalar fields like temperature/potential, defined in certain domain, then

grad $u \equiv \nabla u = \boldsymbol{i}\dfrac{\partial u}{\partial x} + \boldsymbol{j}\dfrac{\partial u}{\partial y} = \dfrac{\partial u}{\partial n}\boldsymbol{n}$ where $\boldsymbol{n}$ is the unit vector in the direction of the normal to the curve $u =$ const.

$$\text{div.grad } u = \nabla \cdot \nabla u = \nabla^2 u = \frac{\partial^2 u}{\partial x^2} + \frac{\partial^2 u}{\partial y^2}.$$

If $\boldsymbol{A}$ is a vector field defined in domain D = R ∪ S, i.e., region R with bounding curve S then from divergence theorem in two dimensions (Green's theorem),

$$\iint_R \nabla.\ \mathrm{A}dxdy = \int_S \mathrm{A}\cdot\boldsymbol{n}\,ds \tag{12.20}$$

where $\boldsymbol{n}$ is the outward unit normal to S and s is length of curve along S.

Following equalities will also be used where u and v are scalar fields:

$$\nabla \cdot (u\nabla v) = u\nabla^2 v + \nabla u \cdot \nabla v \tag{12.21a}$$

$$\nabla \cdot (v\nabla u) = v\nabla^2 u + \nabla v \cdot \nabla u \tag{12.21b}$$

$$\nabla \cdot (u\nabla u) = u\nabla^2 u + \nabla u \cdot \nabla u \tag{12.21c}$$

Integration of (12.21c) over R and use of (12.20) and also $\boldsymbol{n} \cdot \boldsymbol{n} = 1$ gives

$$\iint\limits_{R} u\nabla^2 u \, dxdy = \int\limits_{S} u\frac{\partial u}{\partial n} ds - \iint\limits_{R} \left\{ \left(\frac{\partial u}{\partial x}\right)^2 + \left(\frac{\partial u}{\partial y}\right)^2 \right\} dxdy \tag{12.22}$$

Subtracting (12.21a) from (12.21b) and integrating over R, we get,

$$\iint\limits_{R} (v\nabla^2 u - u\nabla^2 u) dxdy = \int\limits_{S} \left(v\frac{\partial u}{\partial n} - u\frac{\partial u}{\partial n} \right) ds, \quad \text{since } \nabla u \cdot \nabla v = \nabla v \cdot \nabla u. \tag{12.23}$$

12.6.2 Minimum Functional Theorem (MFT)

Before stating the theorem we give a few simple definitions.

The inner product of two functions $u(x, y)$ and $v(x, y)$, each defined over a region R, is denoted and defined as

$$< u, v > = \iint\limits_{R} u(x, y) \cdot v(x, y) dxdy. \tag{12.24}$$

Further, if L is an operator which operates on u, v, w, etc., in domain D$(= \text{R} \cup \text{S}$, curve S encloses region R) then it is called:

(*i*) Linear, if

$$\text{L}(\alpha u + \beta v) = \alpha \text{L}u + \beta \text{L}v, \tag{12.25a}$$

(*ii*) Symmetric, if

$$< \text{L}u, v > = < \text{L}v, u > \tag{12.25b}$$

(*iii*) Positive Definite, if

$$< \text{L}u, u > \geq 0, \tag{12.25c}$$

equality occuring only if $u = 0$ everywhere in D.

Theorem: The Minimum Functional Theorem (MFT) states that if an equation $Lu = g$ is defined in domain D where the operator L is linear, symmetric and positive definite, then its solution u and u only provides the minimum value to the functional

$$I[u] = < Lu, u > -2 < g, u > \tag{12.26}$$

We assume that solution to $Lu = g$ exists and is unique.

Conversely, if the functional $I(u)$ given by (12.26) attains its minimum value for a function $u(x, y)$ then $u(x, y)$ is the solution to the equation $Lu = g$ in D where the operator L is linear, symmetric and positive definite.

For proof of the above reference may be made to [3].

Example 12.9

Find the functional for the Poisson equation

$$-\nabla^2 u = g(x, y) \text{ in R}$$

when $u = 0$ on S, enclosing the region R.

Solution We have to check if operator $L \equiv -\nabla^2$ is linear, symmetric and positive definite in $D = R \cup S$.

(*i*) Linear

$$L(\alpha u + \beta v) = \alpha Lu + \beta Lv$$

$$-\nabla^2(\alpha u + \beta v) = -\alpha\nabla^2 u - \beta\nabla^2 u = \alpha(-\nabla^2 u) + \beta(-\nabla^2 v)$$

(*ii*) Symmetric

$$< Lu, v > = < Lv, u > \text{ or } < Lu, v > - < Lv, u >= 0$$

$$\iint_R (-v\nabla^2 u + u\nabla^2 v)dxdy = \int_S \left(u\frac{\partial v}{\partial n} - v\frac{\partial u}{\partial n}\right) ds = 0, \text{ from (12.23)}$$

since $u = 0,\ v = 0$ on S.

$$\therefore \quad < Lu, v >=< Lv, u >$$

(*iii*) Positive Definite

$$< Lu, u > \geq 0$$

$$\iint_R (-u\nabla^2 u)dxdy = \iint_R \left\{\left(\frac{\partial u}{\partial x}\right)^2 + \left(\frac{\partial u}{\partial y}\right)^2\right\} dxdy - \int_S u\frac{\partial u}{\partial n}ds \text{ from (12.22)}$$

$$= \iint_R \left\{\left(\frac{\partial u}{\partial x}\right)^2 + \left(\frac{\partial u}{\partial y}\right)^2\right\} dxdy \qquad \because \; u = 0 \text{ on S.}$$

The integral on the right is always positive. It will be zero only when $u = \text{const}$ in $\text{D}(=\text{R} \cup \text{S})$. But $u = 0$ on the boundary; hence it should be zero everywhere. Therefore $-\nabla^2$ is positive definite.

Since $-\nabla^2$ is linear, symmetric and positive definite, applying Minimum Functional Theorem, the required functional is given by

$$\text{I}[u] = <\text{L}u, u> -2 <g, u>$$

$$= \iint_R \left\{\left(\frac{\partial u}{\partial x}\right)^2 + \left(\frac{\partial u}{\partial y}\right)^2 - 2gu\right\} dxdy.$$

Example 12.10

Find the variational form for solving

$$-\nabla^2 u = g(x, y) \text{ in R}$$

where $u = u_0$ (const) on the curve S enclosing R.

Solution Changing the variable by putting $v = u - u_0$. The problem is redefined as

$$-\nabla^2 v = g(x, y) \text{ in R}$$

$$v = 0 \text{ on boundary S.}$$

Applying MFT

$$\text{I}[v] = <\text{L}v, v> -2 <g, v>$$

$$= \iint_R -v\nabla^2 v \, dxdy \; -2 <g, v>$$

$$= \iint_R \nabla v \cdot \nabla v \, dxdy \; - \iint_R \nabla \cdot v\nabla v \, dxdy - 2 \; <g, v>$$

$$= \iint_R \left\{\left(\frac{\partial v}{\partial x}\right)^2 + \left(\frac{\partial v}{\partial y}\right)^2\right\} dxdy - \int_S v\frac{\partial u}{\partial n}ds - 2 <g, v>$$

$$= \iint_R \left\{ \left(\frac{\partial v}{\partial x}\right)^2 + \left(\frac{\partial v}{\partial y}\right)^2 \right\} dxdy - 2 < g, v >$$

Putting $v = u - u_0$ in the above

$$\mathrm{I}[u - u_0] = \iint_R \left\{ \left(\frac{\partial u}{\partial x}\right)^2 + \left(\frac{\partial u}{\partial y}\right)^2 \right\} dxdy - 2 < g, u > + 2 < g, u_0 >$$

Since u_0 does not vary, the term $2 < g, u_0 >$ is also fixed and its variation w.r.t. u will be zero. Therefore, we minimise the functional

$$\mathrm{I}[u] = \iint_R \left\{ \left(\frac{\partial u}{\partial x}\right)^2 + \left(\frac{\partial u}{\partial y}\right)^2 \right\} dxdy - 2 \iint_R gudxdy$$

taking $u = u_0 + v$ where u_0 is the given boundary condition and v vanishes on the boundary.

Example 12.11

Find the variational form for the solution of

$$-\nabla^2 u = g(x, y) \text{ in R}$$

where R is enclosed by a curve S and it is given that

$$u = f_0(x, y) \text{ on S.}$$

Solution Let us change the variable as

$$v = u - \psi$$

where $\psi(x, y)$ is a chosen fixed function such that

$$\psi(x, y) = f_0(x, y) \text{ on the boundary S.}$$

The problem now is redefined as

$$-\nabla^2 v = g(x, y) + \nabla^2 \psi = g_1(x, y), \text{ say,}$$

with $v = 0$ on the boundary S.

Applying MFT,

$$\mathrm{I}[v] = < \mathrm{L}v, v > -2 < g_1, v >$$

$$= \iint -v\nabla^2 v \, dxdy \; - 2 < g_1, v >$$

$$= \iint \nabla v \cdot \nabla v \; dxdy - \iint v \cdot \frac{\partial v}{\partial n} ds - 2 < g_1, v >$$

$$\mathrm{I}\,[u-\psi] = \iint (\nabla u - \nabla \psi) \cdot (\nabla u - \nabla \psi) \; dxdy - 2 < g + \nabla^2 \psi, \; u - \psi >$$

$$= \iint \nabla u \cdot \nabla u \; dxdy - 2 \iint \nabla u \cdot \nabla \psi dxdy + \iint \nabla \psi \cdot \nabla \psi dxdy$$

$$-2 < g, u > +2 < g, \psi > -2 < \nabla^2 \psi, u > +2 < \nabla^2 \psi, \psi >$$

$$= \iint \nabla u \cdot \nabla u \; dxdy - 2 \iint (\nabla u \cdot \nabla \psi + u \nabla^2 \psi) dxdy - 2 < g, \; u >$$

$$+2 \iint (\nabla \psi \cdot \nabla \psi + g\psi + \psi \nabla^2 \psi) \; dxdy$$

$$= \iint \nabla u \cdot \nabla u \; dxdy - 2 \iint \nabla \cdot (u \nabla \psi) \; dxdy - 2 \; < g, u >$$

$$+2 \iint (\nabla \psi \cdot \nabla \psi + \psi \nabla^2 \psi) dxdy + 2 < g, \psi >$$

$$= \iint_R \nabla u \cdot \nabla u \; dxdy - 2 \int_S u \frac{\partial \psi}{\partial n} ds - 2 < g, u > +2 \int_S \psi \frac{\partial \psi}{\partial n} dxdy$$

$$+2 < g, \psi >,$$

$$= \iint_R \left\{ \left(\frac{\partial u}{\partial x} \right)^2 + \left(\frac{\partial u}{\partial y} \right)^2 - 2gu \right\} dxdy + 2 < g, \psi >,$$

$\because \; u = \psi = f_0$ on S.

Since the second term on the right side is independent of u

$$\mathrm{I}[u] = \iint_R \left\{ \left(\frac{\partial u}{\partial x} \right)^2 + \left(\frac{\partial u}{\partial y} \right)^2 - 2gu \right\} dxdy$$

u is chosen in the form $u = \psi + v$ where ψ satisfies the boundary condition and v is zero on S.

Example 12.12

Find the functional for solving equation

$$\frac{\partial^2 u}{\partial x^2} + \frac{\partial^2 u}{\partial y^2} - g(x, y) = 0 \text{ in R}$$

with boundary condition $\alpha u + \frac{\partial u}{\partial n} = 0$ on S where n is the outward normal to S and α is known function of x and y.

Solution Express the problem as

$$-\nabla^2 u = g \text{ where } \nabla^2 u = \frac{\partial^2 u}{\partial x^2} + \frac{\partial^2 u}{\partial y^2}; \ \mathrm{L} \equiv -\nabla^2.$$

(i) $-\nabla^2$ is linear since $-\nabla^2(au + bv) = a(-\nabla^2 u) + b(-\nabla^2 v)$

where a and b are scalars.

(ii) Symmetric

$$< \mathrm{L}u, v > - < \mathrm{L}v, u >$$

$$= \iint_R (-v\nabla^2 u + u\nabla^2 v)\, dxdy$$

$$= \int_S \left(u\frac{\partial v}{\partial n} - v\frac{\partial u}{\partial n}\right) ds$$

since u and v satisfy the same boundary condition on S, i.e., $\alpha u + \frac{\partial u}{\partial n} = 0$ and $\alpha v + \frac{\partial v}{\partial n} = 0$, we get after multiplying first by v and second by u and subtracting and integrating over S,

$$u\frac{\partial v}{\partial n} - v\frac{\partial u}{\partial n} = 0 \quad \therefore \quad \int_S \left(u\frac{\partial v}{\partial n} - v\frac{\partial u}{\partial n}\right) ds = 0.$$

Hence $< \mathrm{L}u, v >=< \mathrm{L}v, u >$.

(iii) Positive Definite

$$< \mathrm{L}u, u > = \iint_R -u\nabla^2 u du = \iint \left\{\left(\frac{\partial u}{\partial x}\right)^2 + \left(\frac{\partial u}{\partial y}\right)^2\right\} dxdy - \int_S u\frac{\partial u}{\partial n} ds$$

$$= \iint_R \left\{\left(\frac{\partial u}{\partial x}\right)^2 + \left(\frac{\partial u}{\partial y}\right)^2\right\} dxdy + \int_S \alpha u^2 ds. \qquad \text{....(1)}$$

For the function $\alpha(x, y) > 0$, the right side is positive since integrands under both of the integrals are positive.

$< \mathrm{L}u, u >$ can be zero only if both the integrals are zero, i.e.,

$$(i) \iint\limits_{R} \left\{ \left(\frac{\partial u}{\partial x}\right)^2 + \left(\frac{\partial u}{\partial y}\right)^2 \right\} dxdy = 0 \text{ and } (ii) \int\limits_{S} \alpha u^2 ds = 0.$$

First integral (*i*) suggests $u = \text{C}$ (cosnt) everywhere.

$\therefore$ From second integral (*ii*) $\int\limits_{S} \alpha u^2 = \text{C}^2 \int\limits_{S} \alpha ds = 0$

But the value of the integral (*ii*) can not be zero since α is positive (+ve).

Hence C should be zero everywhere. So $< \text{L}u, u >$ is positive (+ve) def. Thus, the functional is given as

$$\text{I}[u] = < \text{L}u, u > -2 < g, u >$$

$$= \iint\limits_{R} \left\{ \left(\frac{\partial u}{\partial x}\right)^2 + \left(\frac{\partial u}{\partial y}\right)^2 - 2gu \right\} dxdy + \int\limits_{S} \alpha u^2 du. \tag{2}$$

The minimum of functional (2) will provide the solution to the given equation with boundary condition being satisfied automatically. Hence given boundary condition is called 'natural' boundary condition for functional $\text{I}[u]$.

12.6.3 Application of MFT to one-dimension problem

The minimum functional theorem discussed for partial differential equation in two dimensions has straight forward application in solving ordinary differential equation of boundary value type. The only difference will be that the double integral will be replaced by single integral and line integral by the end points. Following example will make it clear.

Example 12.13

Find the functional for solving the differential equation

$$\frac{d^2u}{dx^2} - u = x, \quad 0 \le x \le 1$$

with boundary conditions $u = 0$ on $x = 0$ and $x = 1$.

Solution Write the equation as $\text{L}u = g$, i.e.,

$$-\left(\frac{d^2}{dx^2} - 1\right) u = -x$$

where $$\mathrm{L} \equiv -\left(\frac{d^2}{dx^2} - 1\right) = -\frac{d^2}{dx^2} + 1$$

The functional is given as

$$\begin{aligned}
\mathrm{I}[u] &= <\mathrm{L}u, u> -2 <g, u> \\
&= \int_0^1 u\left(-\frac{d^2u}{dx^2} + u\right) dx + 2\int_0^1 xudx \\
&= \int_0^1 -u\frac{d^2u}{dx^2} dx + \int_0^1 u^2 dx + 2\int_0^1 xudx \\
&= -u\frac{du}{dx}\bigg|_0^1 + \int_0^1 \frac{du}{dx}\cdot\frac{du}{dx}\cdot dx + \int_0^1 u^2 dx + \int_0^1 2uxdx \\
&= \int_0^1 \left\{\left(\frac{du}{dx}\right)^2 + u^2 + 2ux\right\} dx, \text{ since } u(0) = u(1) = 0.
\end{aligned}$$

Note: It is same as given in Example 12.8.

12.7 Equivalence of Rayleigh–Ritz and Galerkin Methods (2–D)

In Sec 12.5 we have shown the equivalence of Rayleigh–Ritz and Galerkin methods in regard to one-dimensional boundary value problems. Here we show their equivalence in two dimensions while the earlier case may be considered as a particular case of the present analysis. Let us consider a two-dimensional elliptic equation

$$\nabla^2 u = g \text{ in R}$$

with a Dirichlet boundary condition $u = f(x, y)$ on S which encloses R. We have to solve it over domain $\mathrm{D} = \mathrm{R} \cup \mathrm{S}$.

Rayleigh–Ritz Method

Let us assume a solution as

$$u = \phi_0(x, y) + a_1\phi_1(x, y) + a_2\phi_2(x, y)$$

where $\phi_0 = f$ on S and ϕ_1 and ϕ_2 have zero value there; a_1 and a_2 are unknown parameters which are to be determined. The functional to be minimised is given by

$$\mathrm{I}[u] = \iint_R \left\{ \left(\frac{\partial u}{\partial x}\right)^2 + \left(\frac{\partial u}{\partial y}\right)^2 - 2gu \right\} dxdy$$

$$\frac{\partial u}{\partial x} = \frac{\partial \phi_0}{\partial x} + a_1 \frac{\partial \phi_1}{\partial x} + a_2 \frac{\partial \phi_2}{\partial x}; \frac{\partial u}{\partial y} = \frac{\partial \phi_0}{\partial y} + a_1 \frac{\partial \phi_1}{\partial y} + a_2 \frac{\partial \phi_2}{\partial y}.$$

$$\left(\frac{\partial u}{\partial x}\right)^2 = \left(\frac{\partial \phi_0}{\partial x}\right)^2 + a_1^2 \left(\frac{\partial \phi_1}{\partial x}\right)^2 + a_2^2 \left(\frac{\partial \phi_2}{\partial x}\right)^2 + 2a_1 \frac{\partial \phi_0}{\partial x} \cdot \frac{\partial \phi_1}{\partial x} + 2a_2 \frac{\partial \phi_0}{\partial x} \cdot \frac{\partial \phi_2}{\partial x}$$
$$+ 2a_1 a_2 \frac{\partial \phi_1}{\partial x} \cdot \frac{\partial \phi_2}{\partial x}$$

$$\left(\frac{\partial u}{\partial y}\right)^2 = \left(\frac{\partial \phi_0}{\partial y}\right)^2 + a_1^2 \left(\frac{\partial \phi_1}{\partial y}\right)^2 + a_2^2 \left(\frac{\partial \phi_2}{\partial y}\right)^2 + 2a_1 \frac{\partial \phi_0}{\partial y} \cdot \frac{\partial \phi_1}{\partial y} + 2a_2 \frac{\partial \phi_0}{\partial y} \cdot \frac{\partial \phi_2}{\partial y}$$
$$+ 2a_1 a_2 \frac{\partial \phi_1}{\partial y} \cdot \frac{\partial \phi_2}{\partial y}$$

$$-2gu = -2g(\phi_0 + a_1\phi_1 + a_2\phi_2)$$

$\dfrac{\partial I}{\partial a_1} = 0$ gives

$$\iint_R \left[a_1 \left\{ \left(\frac{\partial \phi_1}{\partial x}\right)^2 + \left(\frac{\partial \phi_1}{\partial y}\right)^2 \right\} + a_2 \left\{ \frac{\partial \phi_1}{\partial x} \cdot \frac{\partial \phi_2}{\partial x} + \frac{\partial \phi_1}{\partial y} \cdot \frac{\partial \phi_2}{\partial y} \right\} \right.$$
$$\left. + \left\{ \frac{\partial \phi_0}{\partial x} \cdot \frac{\partial \phi_1}{\partial x} + \frac{\partial \phi_0}{\partial y} \cdot \frac{\partial \phi_1}{\partial y} \right\} - g\phi_1 \right] dxdy = 0 \quad (12.27)$$

Similarly from $\dfrac{\partial \mathrm{I}}{\partial a_2} = 0$ we get

$$\iint_R \left[a_1 \left\{ \frac{\partial \phi_1}{\partial x} \cdot \frac{\partial \phi_2}{\partial x} + \frac{\partial \phi_1}{\partial y} \cdot \frac{\partial \phi_2}{\partial y} \right\} + a_2 \left\{ \left(\frac{\partial \phi_2}{\partial x}\right)^2 + \left(\frac{\partial \phi_2}{\partial y}\right)^2 \right\} \right.$$
$$\left. + \left\{ \frac{\partial \phi_0}{\partial x} \cdot \frac{\partial \phi_2}{\partial x} + \frac{\partial \phi_0}{\partial y} \cdot \frac{\partial \phi_2}{\partial y} \right\} - g\phi_2 \right] dxdy = 0 \quad (12.28)$$

Galerkin's method

Substituting $u = \phi_0 + a_1\phi_1 + a_2\phi_2$ in the given equation $-\nabla^2 u - g = 0$, we get

$$Re = -\nabla^2(u_0 + a_1\phi_1 + a_2\phi_2) - g$$

Multiplying *Re* by ϕ_1 and integrating we get the equation

$$\iint_R [-(\nabla^2\phi_0 + a_1\nabla^2\phi_1 + a_2\nabla^2\phi_2 + g)\phi_1]dxdy = 0$$

$$\text{or} \quad \iint_R [-\phi_1\nabla^2\phi_0 - a_1\phi_1\nabla^2\phi_1 - a_2\phi_1\nabla^2\phi_2 - g\phi_1]dxdy = 0 \tag{12.29}$$

We have, after using (12.21a, b, c)

$$\iint_R -\phi_1\nabla^2\phi_0\, dxdy = \iint_R \nabla\cdot(\phi_1\nabla\phi_0)\, dxdy - \iint \nabla\cdot\phi_1\nabla\phi_0\, dxdy$$

$$= -\int_S \phi_1\frac{\partial\phi_0}{\partial n}ds + \iint\left(\frac{\partial\phi_0}{\partial x}\cdot\frac{\partial\phi_1}{\partial x} + \frac{\partial\phi_0}{\partial y}\cdot\frac{\partial\phi_1}{\partial y}\right)dxdy$$

$$= \iint_R\left(\frac{\partial\phi_0}{\partial x}\cdot\frac{\partial\phi_1}{\partial x} + \frac{\partial\phi_0}{\partial y}\cdot\frac{\partial\phi_1}{\partial y}\right)dxdy \text{ since } \phi_1 = 0 \text{ on S.}$$

$$\iint_R -\phi_1\nabla^2\phi_1 dxdy = \iint_R\left[\left(\frac{\partial\phi_1}{\partial x}\right)^2 + \left(\frac{\partial\phi_1}{\partial y}\right)^2\right]dxdy$$

$$\iint -\phi_1\nabla^2\phi_2 dxdy = \iint_R\left(\frac{\partial\phi_1}{\partial x}\cdot\frac{\partial\phi_2}{\partial x} + \frac{\partial\phi_1}{\partial y}\cdot\frac{\partial\phi_2}{\partial y}\right)dxdy$$

Substituting these values in (12.29) gives

$$\iint_R\left[a_1\left\{\left(\frac{\partial\phi_1}{\partial x}\right)^2 + \left(\frac{\partial\phi_1}{\partial y}\right)^2\right\} + a_2\left\{\frac{\partial\phi_1}{\partial x}\cdot\frac{\partial\phi_2}{\partial x} + \frac{\partial\phi_1}{\partial y}\cdot\frac{\partial\phi_2}{\partial y}\right\}\right.$$
$$\left. + \left\{\frac{\partial\phi_0}{\partial x}\cdot\frac{\partial\phi_1}{\partial x} + \frac{\partial\phi_0}{\partial y}\cdot\frac{\partial\phi_1}{\partial y}\right\} - g\phi_1\right]dxdy = 0 \tag{12.30}$$

Similarly multiplying R by ϕ_2 and integrating we can obtain

$$\iint\limits_R \left[a_1 \left\{ \frac{\partial \phi_1}{\partial x} \cdot \frac{\partial \phi_2}{\partial x} + \frac{\partial \phi_1}{\partial y} \cdot \frac{\partial \phi_2}{\partial y} \right\} + a_2 \left\{ \left(\frac{\partial \phi_2}{\partial x} \right)^2 + \left(\frac{\partial \phi_2}{\partial y} \right)^2 \right\} \right.$$

$$\left. + \left\{ \frac{\partial \phi_0}{\partial x} \cdot \frac{\partial \phi_2}{\partial x} + \frac{\partial \phi_0}{\partial y} \cdot \frac{\partial \phi_2}{\partial y} \right\} - g\phi_2 \right] dxdy = 0. \quad (12.31)$$

The equations (12.30) and (12.31) are same as (12.27) and (12.28) respectively. The result can be easily generalised for n parameters.

12.8 Pre-requisites for Finite Element Method

As stated earlier, there may be mainly two approaches for the formulation of finite element method – one based on weighted residual methods, particularly Galerkin's method and the other based on variational principle. The Rayleigh–Ritz method requires continuity of lower order derivative as compared to Galerkin's but a functional has to be obtained for each problem before applying it while Galerkin's method can be used directly even in the non-linear problems. Further, as we have seen that the Galerkin's method finally arrives at the same set of equations as in the Rayleigh–Ritz method. Therefore, we will describe the FEM adopting the Galerkin's criterion. But prior to that some terms and techniques will be discussed as they are pre-requisites for the development of FEM.

12.8.1 Shape functions

Let us suppose we are interested in solving a boundary value problem $\mathrm{L}u = g$ over a domain D. As stated earlier, the domain D is subdivided into subdomains called 'elements' not necessarily of same sizes and shapes. The elements are formed by joining the conveniently selected points called 'nodes' through straight lines or curves (in our case, only straight lines).

In case of a one-dimensional problem the domain is simply a straight line, say x-axis. Let the domain be subdivided into (N−1) elements at the nodes numbering 1, 2, 3, ... N and the coordinate of the rth node be x_r, $r = 1(1)$ N. Thus the ith element e_i will be defined by the interval $[x_i, x_{i+1}]$. If the solution $u(x)$ is approximated by a straight line over this element then

$$u(x) = \phi_i(x)u_i + \phi_{i+1}(x)u_{i+1}, \quad x_i \le x \le x_{i+1} \quad (12.32)$$

where $\phi_i(x)$ and $\phi_{i+1}(x)$ are Lagrange's functions given as

$$\phi_i(x) = \frac{x - x_{i+1}}{x_i - x_{i+1}}, \quad \phi_{i+1}(x) = \frac{x - x_i}{x_{i+1} - x_i} \quad (12.33)$$

and u_i and u_{i+1} are the approximate values of u at the nodes i and $(i+1)$ respectively.

Similarly the linear approximation for $u(x)$ over the element e_{i-1} may be written as

$$u(x) = \phi_{i-1}(x)u_{i-1} + \phi_i(x)u_i, \quad x_{i-1} \le x \le x_i \tag{12.34}$$

$$\text{where} \quad \phi_{i-1}(x) = \frac{x - x_i}{x_{i-1} - x_i}, \quad \phi_i(x) = \frac{x - x_{i-1}}{x_i - x_{i-1}} \tag{12.35}$$

We can represent the two straight lines (12.32) and (12.34) by a single expression

$$f(x) = \phi_{i-1}(x)u_{i-1} + \phi_i(x)u_i + \phi_{i+1}(x)u_{i+1}, \quad x_{i-1} \le x \le x_{i+1} \tag{12.36}$$

when the functions $\phi_{i-1}(x)$, $\phi_i(x)$ and $\phi_{i+1}(x)$ are defined as follows:

$$\phi_{i-1}(x) = \begin{cases} \dfrac{x - x_i}{x_{i-1} - x_i}, & x_{i-1} \le x \le x_i \\ 0 & x_i \le x \le x_{i+1} \end{cases} \tag{12.37a}$$

$$\phi_i(x) = \begin{cases} \dfrac{x - x_{i-1}}{x_i - x_{i-1}}, & x_{i-1} \le x \le x_i \\ \dfrac{x - x_{i+1}}{x_i - x_{i+1}}, & x_i \le x \le x_{i+1} \end{cases} \tag{12.37b}$$

$$\phi_{i+1}(x) = \begin{cases} 0 & x_{i-1} \le x \le x_i \\ \dfrac{x - x_i}{x_{i+1} - x_i}, & x_i \le x \le x_{i+1} \end{cases} \tag{12.37c}$$

The functions $\phi_{i-1}(x)$, $\phi_i(x)$ and $\phi_{i+1}(x)$ defined by (12.37a), (12.37b) and (12.37c) respectively are called 'shape functions' over domain $[x_{i-1}, x_{i+1}]$. The most important property of a shape function $\phi_r(x)$ is that its value is 1 at the node r and 0 at the other nodes of the domain. Keeping this definition in mind we can represent the approximation of $u(x)$ over the entire domain $[x_1, x_\text{N}]$ by a single expression

$$\begin{aligned} u(x) &= \phi_1(x)\, u_1 + \phi_2(x)\, u_2 + \ldots. + \phi_\text{N}(x)\, u_\text{N} \\ &= \sum_{r=1}^{\text{N}} \phi_r(x)u_r, \end{aligned} \tag{12.38}$$

where $\phi_1(x)$, $\phi_2(x)\ldots$, etc. are the shape functions defined over the domain $[x_1, x_\text{N}]$. We can also represent $\phi_r(x_s) = \delta_{rs}$ where δ_{rs} is Kronecker's delta. The graphs of $\phi_1(x)$, $\phi_r(x)$ for $r = 2(1)\,\text{N}-1$ and $\phi_\text{N}(x)$ are shown in Fig. 12.1a, 12.1b and 12.1c respectively by ABCDEFG.

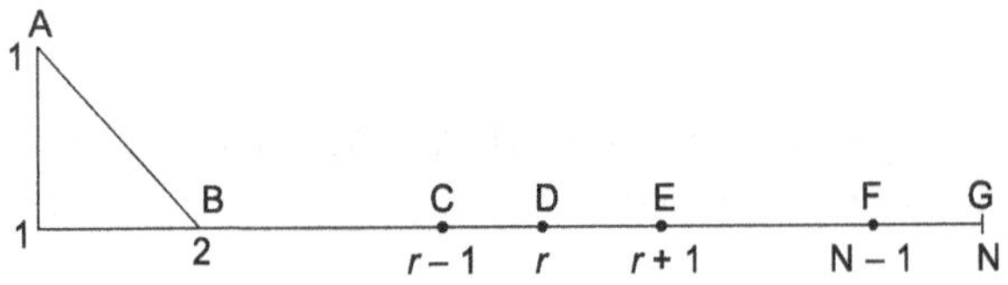

Figure 12.1a Graph of $\phi_1(x)$

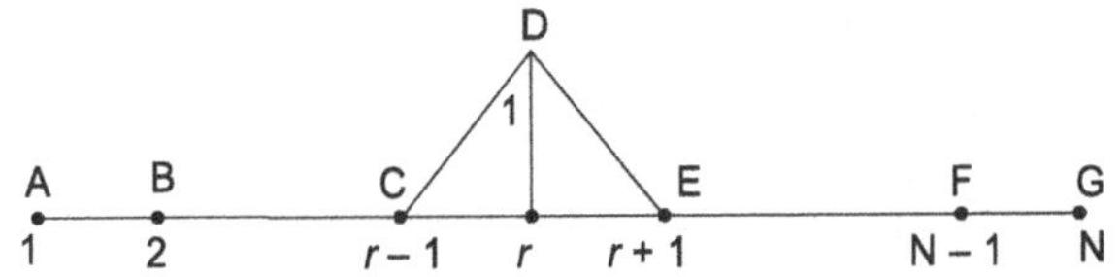

Figure 12.1b Graph of $\phi_r(x)$

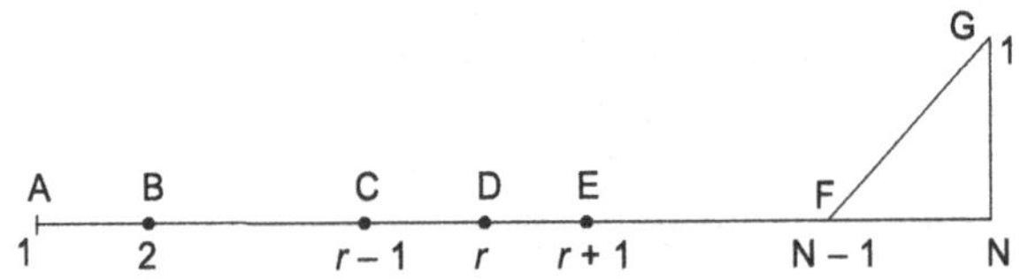

Figure 12.1c Graph of $\phi_N(x)$

Let us now consider a two-dimensional problem defined over a domain D in the *x*-*y* plane. Suppose that domain D has been subdivided into triangular elements. Consider an element *e* having nodes (vertices) 1, 2 and 3, i.e., $\Delta 123$ as shown in Fig. 12.2.

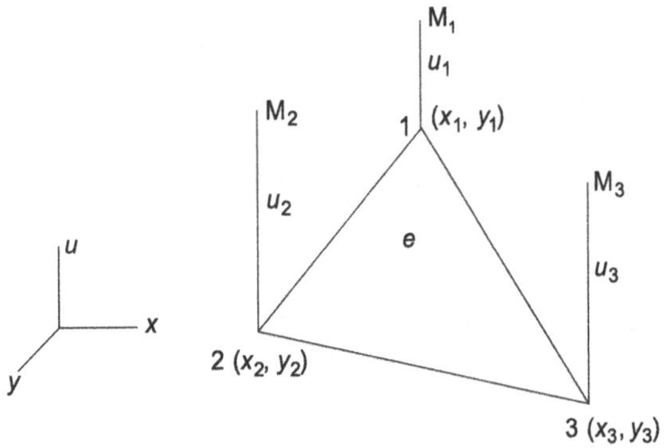

Figure 12.2 Triangular element

Let the value of *u* at the nodes 1, 2 and 3 be denoted by u_1, u_2 and u_3 respectively and shown by $1M_1, 2M_2$ and $3M_3$ respectively in the direction perpendicular to the *x*-*y* plane. Let us approximate $u(x, y)$ over element *e* by

$$u = \alpha x + \beta y + \gamma \tag{12.39}$$

i.e., a plane passing through the points M_1, M_2 and M_3. The values of α, β and γ may be determined from the following equations,

$$\begin{aligned} u_1 &= \alpha x_1 + \beta y_1 + \gamma \\ u_2 &= \alpha x_2 + \beta y_2 + \gamma \\ u_3 &= \alpha x_3 + \beta y_3 + \gamma \end{aligned} \tag{12.40}$$

We can write (12.40) in matrix form as

$$\begin{bmatrix} u_1 \\ u_2 \\ u_3 \end{bmatrix} = \begin{bmatrix} x_1 & y_1 & 1 \\ x_2 & y_2 & 1 \\ x_3 & y_3 & 1 \end{bmatrix} \begin{bmatrix} \alpha \\ \beta \\ \gamma \end{bmatrix}. \tag{12.41}$$

From (12.41) we get the values of α, β, γ as

$$\begin{bmatrix} \alpha \\ \beta \\ \gamma \end{bmatrix} = \begin{bmatrix} x_1 & y_1 & 1 \\ x_2 & y_2 & 1 \\ x_3 & y_3 & 1 \end{bmatrix}^{-1} \begin{bmatrix} u_1 \\ u_2 \\ u_3 \end{bmatrix}$$

$$= \frac{1}{2A} \begin{bmatrix} y_2 - y_3 & y_3 - y_1 & y_1 - y_2 \\ x_3 - x_2 & x_1 - x_3 & x_2 - x_1 \\ x_2y_3 - y_2x_3 & x_3y_1 - y_3x_1 & x_1y_2 - y_1x_2 \end{bmatrix} \begin{bmatrix} u_1 \\ u_2 \\ u_3 \end{bmatrix}, \tag{12.42}$$

where A denotes the area of the element e, given by

$$A = \frac{1}{2} \begin{bmatrix} x_1 & y_1 & 1 \\ x_2 & y_2 & 1 \\ x_3 & y_3 & 1 \end{bmatrix}. \tag{12.43}$$

Substituting the values of α, β, γ from (12.42) in (12.39) and rearranging terms, we get

$$u(x, y) = \frac{1}{2A}\{(y_2 - y_3)x + (x_3 - x_2)y + x_2y_3 - y_2x_3\}u_1$$

$$+ \frac{1}{2A}\{(y_3 - y_1)x + (x_1 - x_3)y + x_3y_1 - y_3x_1\}u_2$$

$$+\frac{1}{2A}\{(y_1-y_2)x+(x_2-x_1)y+x_1y_2-y_1x_2\}u_3. \tag{12.44}$$

Equation (12.44) can be written in concise form as

$$u(x,y)=\sum_{i=1}^{3}(a_ix+b_iy+c_i)u_i \tag{12.45}$$

where

$$\left.\begin{aligned} a_1&=\frac{1}{2A}(y_2-y_3),\ b_1=\frac{1}{2A}(x_3-x_2),\ c_1=\frac{1}{2A}(x_2y_3-y_2x_3),\\ a_2&=\frac{1}{2A}(y_3-y_1),\ b_2=\frac{1}{2A}(x_1-x_3),\ c_2=\frac{1}{2A}(x_3y_1-y_3x_1),\\ a_3&=\frac{1}{2A}(y_1-y_2),\ b_3=\frac{1}{2A}(x_2-x_1),\ c_3=\frac{1}{2A}(x_1y_2-y_1x_2),\end{aligned}\right\} \tag{12.46}$$

A being area of the element e.

We can also write equation of the approximating function (12.45) in the form

$$u(x,y)=\phi_1(x,y)u_1+\phi_2(x,y)u_2+\phi_3(x,y)u_3 \tag{12.47}$$

where $\phi_i(x,y)=a_ix+b_1y+c_i,\ i=1,\ 2,\ 3$
and a_i, b_i and c_i are given as (12.46).

The functions $\phi_1(x,y)$, $\phi_2(x,y)$ and $\phi_3(x,y)$ are the shape functions defined over the triangular element e. It can be verified also that $\phi_1(x,y)$ has got a value 1 at the node 1 and 0 at the other two nodes *viz.* node 2 and node 3. Similarly $\phi_2(x,y)$ has a value 1 at node 2 and zero at node 1 and node 3; $\phi_3(x,y)$ attains a value 1 at node 3 and 0 at node 1 and node 2. The graphs of $\phi_1(x,y)$, $\phi_2(x,y)$ and $\phi_3(x,y)$ are triangular planes A_1 23, A_2 13 and A_3 12 as shown in Fig. 12.3a, 12.3b and 12.3c, respectively.

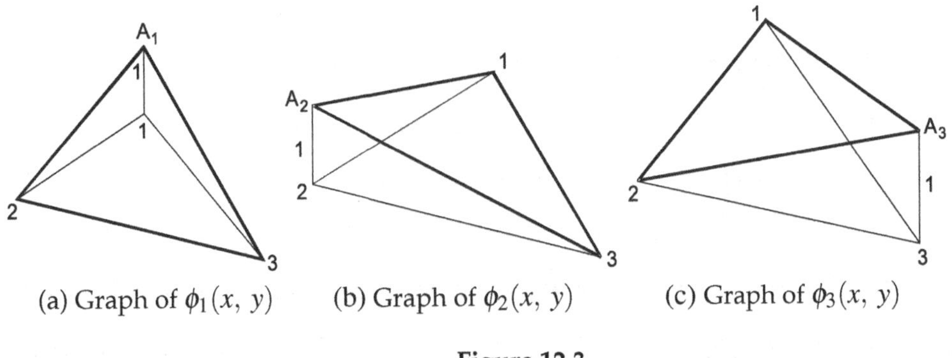

(a) Graph of $\phi_1(x,y)$ (b) Graph of $\phi_2(x,y)$ (c) Graph of $\phi_3(x,y)$

Figure 12.3

If domain D has been subdivided into say, total M triangular elements having N nodes the function $u(x, y)$ may be approximated over the whole domain as

$$u\,(x, y) = \phi_1(x, y)u_1 + \phi_2(x, y)u_2 + \ldots. + \phi_N(x, y)\,u_N$$

$$= \sum_{r=1}^{N} \phi_r(x, y)u_r \tag{12.48}$$

where the shape function $\phi_r(x, y)$ will have value 1 at the node r having coordinates $(x_r,\ y_r)$ and 0 at all the other nodes.

It may be mentioned that the element may have a different shape but the basic property of the shape function will remain same, i.e., $\phi_r(x,\ y) = 1$ at node r and zero at the other nodes. Secondly it may also be stated that a straight line (in one dimension) or a plane (in two dimensions) are the simplest form of approximations. There can be other forms of approximations also.

12.8.2 Normalised/natural coordinates

Let us refer again to the problem in one dimension where an element e_r has nodes r and $r+1$ with coordinates x_r and x_{r+1} respectively as shown in Fig. 12.4 and P is any point between nodes r and $r+1$.

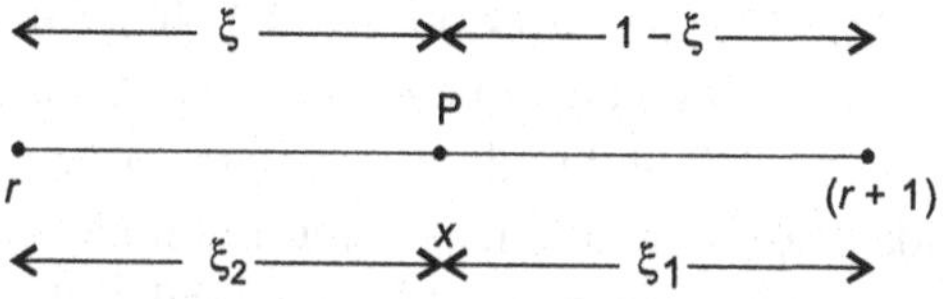

Figure 12.4 Element in one dimension

We can approximate u over element e_r linearly as

$$u(x) = \phi_r(x)u_r + \phi_{r+1}(x)u_{r+1},\ \ x_r \leq x \leq x_{r+1}, \tag{12.49}$$

where $\phi_1(x) = \dfrac{x - x_{r+1}}{x_r - x_{r+1}},\ \ \phi_{r+1}(x) = \dfrac{x - x_r}{x_{r+1} - x_r}.$

The interval $x_r \leq x \leq x_{r+1}$ may be written in normalised form as $0 \leq \dfrac{x - x_r}{x_{r+1} - x_r} \leq 1.$

Changing the variable by putting $\xi = \dfrac{x - x_r}{x_{r+1} - x_r}$, shifts the origin to x_r and the length of the interval $x_r \le x \le x_{r+1}$ has been normalised to unity since $0 \le \xi \le 1$. The approximation for $u(x)$ in the interval $x_1 \le x \le x_{r+1}$ converts to $u(\xi)$ in terms of ξ as

$$u(\xi) = (1-\xi)u_r + \xi u_{r+1}, \quad 0 \le \xi \le 1. \tag{12.50}$$

$$= \phi_r(\xi)u_r + \phi_{r+1}(\xi)u_{r+1}, \tag{12.50a}$$

$$\text{where } \phi_r(\xi) = 1-\xi \text{ and } \phi_{r+1}(\xi) = \xi \tag{12.50b}$$

Expression (12.50) represents u in terms of normalised length variable ξ which varies from 0 to 1; it means that coordinate of node r is $\xi = 0$ and that of node $r+1$ is $\xi = 1$. If we want to differentiate or integrate u, then following relation may be used over the element e_r,

$$\mathrm{L}_r d\xi = dx, \text{ where } \mathrm{L}_r = x_{r+1} - x_r. \tag{12.51}$$

Expression (12.50) may be obtained independently using local (natural) coordinate system. Let a point P(x) in $[x_r, x_{r+1}]$ be represented by P(ξ_1, ξ_2) where ξ_1 is the normalised length from P to node $r+1$ and ξ_2 is the normalised length from node r to P, so that

$$\xi_1 + \xi_2 = 1 \tag{12.52}$$

The coordinates of node r in terms of (ξ_1, ξ_2) will be $(1,0)$ and that of node $(r+1)$ will be $(0,1)$. Following relation will hold

$$x_r\xi_1 + x_{r+1}\xi_2 = x \tag{12.53}$$

Solving (12.52) and (12.53) we get

$$\xi_2 = \frac{x - x_r}{x_{r+1} - x_r} = \xi \text{ and } \xi_1 = 1 - \xi_2 = 1 - \xi, \tag{12.54}$$

which is same as in (12.50).

Hence $u(x)$ can be represented as

$$u(\xi_1, \xi_2) = \xi_1 u_r + \xi_2 u_{r+1}, \quad 0 \le \xi_1, \ \xi_2 \le 1, \quad \xi_1 + \xi_2 = 1. \tag{12.55}$$

Consider a two dimensional domain when it has been subdivided into triangular elements. Let one of its elements e has nodes 1, 2 and 3 with coordinates (x_1, y_1), (x_2, y_2) and (x_3, y_3) respectively. Let P(x, y) be any point inside the element or on its boundary (see Fig. 12.5).

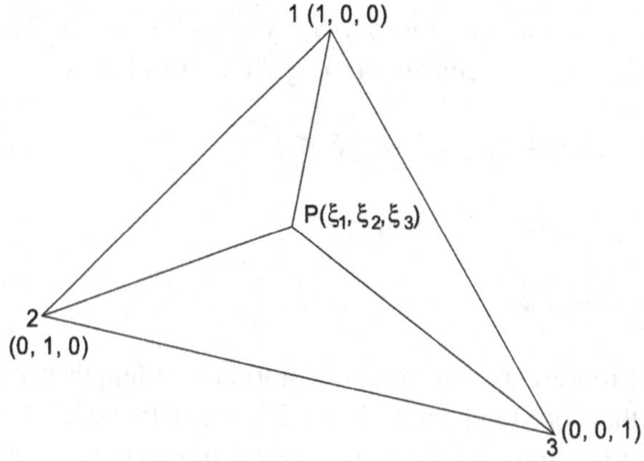

Figure 12.5 Triangular element *e*.

Let us suppose ξ_1, ξ_2, ξ_3 are shape functions defined over the element *e*; then we can express the position of the point P(x, y) in terms of ξ_1, ξ_2, ξ_3. It should be clear that since ξ_1, ξ_2 and ξ_3 are shape functions, the coordinates of the nodes 1, 2 and 3 should be (1, 0, 0), (0, 1, 0) and (0, 0, 1) respectively. Further, assuming a linear relationship between them, following equations should be satisfied:

$$\left.\begin{aligned} \xi_1 + \xi_2 + \xi_3 &= 1 \\ x_1\xi_1 + x_2\xi_2 + x_3\xi_3 &= x \\ y_1\xi_1 + y_2\xi_2 + y_3\xi_3 &= y \end{aligned}\right\} \tag{12.56}$$

The equations (12.56) may be written in matrix form as

$$\begin{bmatrix} 1 & 1 & 1 \\ x_1 & x_2 & x_3 \\ y_1 & y_2 & y_3 \end{bmatrix} \begin{bmatrix} \xi_1 \\ \xi_2 \\ \xi_3 \end{bmatrix} = \begin{bmatrix} 1 \\ x \\ y \end{bmatrix} \tag{12.57}$$

Using Cramer's rule we get from (12.57)

$$\xi_1 = \frac{\Delta_1}{\Delta}, \ \xi_2 = \frac{\Delta_2}{\Delta} \text{ and } \xi_3 = \frac{\Delta_3}{\Delta} \tag{12.57a}$$

where

$$\Delta_1 = \begin{vmatrix} 1 & 1 & 1 \\ x & x_2 & x_3 \\ y & y_2 & y_3 \end{vmatrix}, \quad \Delta_2 = \begin{vmatrix} 1 & 1 & 1 \\ x_1 & x & x_3 \\ y_1 & y & y_3 \end{vmatrix}$$

$$\Delta_3 = \begin{vmatrix} 1 & 1 & 1 \\ x_1 & x_2 & x \\ y_1 & y_2 & y \end{vmatrix}, \quad \text{and } \Delta = \begin{vmatrix} 1 & 1 & 1 \\ x_1 & x_2 & x_3 \\ y_1 & y_2 & y_3 \end{vmatrix}. \tag{12.58}$$

The approximation $u(\xi_1, \xi_2, \xi_3)$ may be written as

$$u(\xi_1, \xi_2, \xi_3,) = \xi_1 u_1 + \xi_2 u_2 + \xi_3 u_3, \tag{12.59}$$

which is equivalent to

$$u(x, y) = \left(\frac{\Delta_1}{\Delta} u_1 + \frac{\Delta_2}{\Delta} u_2 + \frac{\Delta_3}{\Delta} u_3 \right) \tag{12.60a}$$

$$= \frac{1}{\Delta} (\Delta_1 u_1 + \Delta_2 u_2 + \Delta_3 u_3). \tag{12.60b}$$

We see that (12.60a) is same as obtained in (12.44). We must also note that the natural (normalised) coordinates

$$\xi_1 = \frac{\Delta_1}{\Delta} = \frac{\Delta_1/2}{\Delta/2} = \frac{\text{Area } \Delta\text{P23}}{\text{Area } \Delta 123}$$

$$\xi_2 = \frac{\text{Area } \Delta\text{P13}}{\text{Area } \Delta 123} \text{ and } \xi_3 = \frac{\text{Area } \Delta\text{P12}}{\text{Area}\Delta 123}.$$

Thus ξ_1, ξ_2, ξ_3 show the fractions of the area of the element e. Hence they are also called 'area coordinates'.

12.9 Finite Element Method

The finite element method was originally suggested by the famous mathematician Courant in 1943. But it was later exploited extensively by the engineers since Argyris used it in 1954. Now there exists a vast amount of literature on the subject dealing with all kinds of problems, steady state and transient, with various types of elements. The objective of this chapter is to provide basic fundamentals of the method through the illustration of problems in one and two dimensions only. Galerkin's criterion will be followed for solution of the problems.

12.9.1 Ordinary differential equation

Consider a one-dimensional steady state heat conduction (fluid flow/gas diffusion) equation represented by an ordinary differential equation

$$\frac{d}{dx}\left(k_x\frac{du}{dx}\right)+q(x)=0,\ \ a\le x\le b \tag{12.61}$$

alongwith the associated boundary conditions:

$$x=a, \qquad u=u_a \text{ (Dirichlet condition)} \tag{12.62a}$$

$$x=b, \qquad -k_b\frac{du}{dx}=\text{Q}_b \text{ (Neumann's condition)} \tag{12.62b}$$

where
$$\text{Q}=-k_x\frac{\partial u}{\partial x}. \tag{12.62c}$$

Physically, in context of heat conduction, $u(x)$ denotes the temperature at a distance x; k_x is thermal conductivity depending on the properties of the medium; $\text{Q}=-k_x\frac{du}{dx}$, known as flux, is the amount of heat flow at a distance x, per unit area per unit time in the direction of x increasing; the term $q(x)$ denotes the amount of heat being generated (if positive) or lost (if negative) at a distance x (in fluid flow called 'source' or 'sink').

Let the domain $a\le x\le b$ be subdivided into N$-$1 elements, not necessarily of equal lengths with nodes from 1 to N. Let the coordinate of node r be x_r such that $x_1=a<x_2<x_3 \ldots <x_\text{N}=b$. Clearly the domain of the rth element e_r is $x_r\le x\le x_{r+1},\ r=1(1)\text{N}-1$. Let the approximate solution over the entire domain $a\le x\le b$ be represented by

$$\begin{aligned}\bar{u} &= \phi_1(x)\,u_1+\phi_2(x)\,u_2+\ldots.+\phi_\text{N}(x)\,u_\text{N}\\ &=\sum_{j=1}^{\text{N}}\phi_j(x)u_j\end{aligned} \tag{12.63}$$

where $\phi_j(x)$ is the shape function with the property that it has got a value 1 at $x=x_j$ and 0 at all the other nodes; u_j denotes the approximate value of u at node x_j which is to be computed for $i=1(1)$ N. In the present case u_1 is given as $u_1=u_a$ from the boundary condition (12.62a).

Using Galerkin's criterion we form equations

$$\int_{x_1=a}^{x_\text{N}=b}\left[\frac{d}{dx}\left(k_x\frac{d\bar{u}}{dx}\right)+q(x)\right]\phi_i(x)dx=0,\ i=1(1)\,\text{N}. \tag{12.64}$$

From (12.64) we get N equations in $u_1,\ u_2,\ \ldots.\ u_\text{N}$, one each for $i=1(1)$ N. The solution of these equations provides the solution to the problem.

Now, rewriting (12.64) and simplifying it we get

$$\int_a^b \phi_i(x)\frac{d}{dx}\left(k_x\frac{d\bar{u}}{dx}\right)dx+\int_a^b q(x)\phi_i(x)dx=0$$

or $$\left[\phi_i(x)k_x\frac{d\bar{u}}{dx}\right]_{x=a}^{b}-\int k_x\frac{d\phi_i}{dx}\cdot\frac{d\bar{u}}{dx}dx+\int_a^b q(x)\phi_i(x)d_x=0$$

or $$\int_a^b k_x\frac{d\phi_i}{dx}\cdot\frac{d\bar{u}}{dx}dx=\int_a^b q(x)\phi_i(x)dx-[\phi_i(x)Q(x)]_{x=a}^{b}, \tag{12.65}$$

due to (12.62c), $i=1(1)\,N$.

The integral (12.65) is evaluated element-wise over each elements e_r, $r=1(1)\,N{-}1$ and therefore may be expressed as

$$\sum_{r=1}^{N-1}\int_{e_r} k_x\frac{d\phi_i}{dx}\cdot\frac{d\bar{u}}{dx}dx=\sum_{r=1}^{N-1}\int_{e_r} q(x)\phi_i(x)dx-[\phi_i(x)Q(x)]_{x=a}^{b};\quad i=1(1)\,N. \tag{12.66}$$

Further, the second term on the right side of (12.66)

$$\begin{aligned}[\phi_i(x)Q(x)]_{x=a}^{b}&=\phi_i(b)Q(b)-\phi_i(a)\,Q(a)\\&=\phi_i(x_N)Q_b-\phi_i(x_1)Q_a\end{aligned} \tag{12.67}$$

where Q_b is known from boundary condition (12.62b) and Q_a, the flux at $x=a$ is not known since Dirichlet condition $u=u_a$ is prescribed there as per (12.62a).

Using (12.67), we can write (12.66) as

$$\sum_{r=1}^{N-1}\int_{e_r} k_x\frac{d\phi_i}{dx}\cdot\frac{d\bar{u}}{dx}dx=\sum_{r=1}^{N-1}\int_{e_r} q(x)\phi_i(x)dx-[\phi_i(x_N)Q_b-\phi_i(x_1)Q_a],\quad i=1(1)\,N. \tag{12.68}$$

Now, let us substitute $\bar{u}$ from (12.63) in the integral on the left side of (12.68), giving

$$\sum_{r=1}^{N-1}\int_{e_r} k_x\frac{d\phi_i}{dx}\cdot\frac{d\bar{u}}{dx}dx=\sum_{r=1}^{N-1}\int_{x_r}^{x_{r+1}} k_x\frac{d\phi_i}{dx}\cdot\frac{d\bar{u}}{dx}dx$$

$$= \sum_{r=1}^{N-1} k_r \int_{x_r}^{x_{r+1}} \frac{d\phi_i}{dx} \cdot \frac{d}{dx}\left(\sum_{j=1}^{N} \phi_j u_j\right) dx$$

k_x is assumed to be constant over the rth element e_r as $k_r = k(x_r)$ or $k\left(\frac{x_r + x_{r+1}}{2}\right)$.

$$= \sum_{r=1}^{N-1} k_r \int_{x_r}^{x_{r+1}} \frac{d\phi_i}{dx}\left(\sum_{j=1}^{N} \frac{d\phi_j}{dx} u_j\right) dx$$

$$= \sum_{r=1}^{N-1} k_r \sum_{j=1}^{N} u_j \int_{x_r}^{x_{r+1}} \frac{d\phi_i}{dx} \cdot \frac{d\phi_j}{dx} dx, \tag{12.69}$$

$$i = 1(1)\,\text{N}.$$

Assuming that $q(x)$ varies smoothly we take its value as constant over the element e_r, say, $q_r = q\left(\frac{x_r + x_{r+1}}{2}\right)$. Hence we can write (12.68) after using (12.69) as,

$$\sum_{r=1}^{n-1} k_r \sum_{j=1}^{n} u_j \int_{x_r}^{x_{r+1}} \frac{d\phi_i}{dx} \cdot \frac{d\phi_j}{dx} dx = \sum_{r=1}^{N-1} q_r \int_{x_r}^{x_{r+1}} \phi_i(x)dx - [\phi_i(x_N)Q_b - \phi_i(x_1)Q_a] \quad i = 1(1)\,\text{N}. \tag{12.70}$$

Now we describe the method for computing various terms in (12.70). The expression on the left side in (12.70) is evaluated for $i = 1(1)$N over each element e_r, $r = 1(1)\,\text{N}-1$. The corresponding coefficients of u_j obtained from various elements, for particular ϕ_i, are added up. Thus the final form of the expression in matrix form would be PU where P is an $\text{N} \times \text{N}$ matrix and U is a column vector $\text{U}^\text{T} = (u_1, u_2, u_3 \ldots. u_\text{N})$. Let us suppose we get, corresponding to the rth element, P_rU where P_r is an $\text{N} \times \text{N}$ matrix, then

$$\text{PU} = \text{P}_1\text{U} + \text{P}_2\text{U} + \ldots. + \text{P}_{\text{N}-1}\text{U}$$

$$= (\text{P}_1 + \text{P}_2 + \ldots. + \text{P}_{\text{N}-1})\,\text{U}. \tag{12.71}$$

The terms on the right side do not contain u_j in the present case, hence both of them can be expressed by the column vectors having constant terms. Let the first and second terms be denoted by R and T column vectors; then (12.70) may be finally represented by the system of equations

$$\text{PU} = \text{R} - \text{T}. \tag{12.72}$$

After solving (12.72) for U we get the solution of the problem. If mixed type boundary condition is prescribed then vector T will involve term(s) containing u which may be transferred to the left side of (12.72) and P will be modified accordingly.

A. Computation of PU

We compute PU, element by element i.e., first computing P_1, then P_2 and so on up to P_{N-1}; finally make P by adding them up as shown in (12.71). We have

$$\text{PU} = \sum_{r=1}^{N-1} k_r \sum_{j=1}^{n} u_j \int_{x_r}^{x_{r+1}} \frac{d\phi_i}{dx} \cdot \frac{d\phi_j}{dx} dx, \quad i = 1(1)\text{N}. \tag{12.73}$$

Let us consider evaluation of (12.73) element-wise.

Element $e_1 (r = 1)$

Over first element e_1, $\bar{u}$ is approximated as

$$\bar{u} = \phi_1(x)\, u_1 + \phi_2(x)\, u_2$$

$$\text{where} \quad \phi_1(x) = \frac{x - x_2}{x_1 - x_2}, \quad \phi_2(x) = \frac{x - x_1}{x_2 - x_1}. \tag{12.74}$$

All the other shape functions $\phi_3, \phi_4, \ldots \phi_N$ will be zero in e_1, i.e., for $x_1 \leq x \leq x_2$. Therefore we will get only u_1 and u_2 terms for first row $(i = 1)$ and for the second row $(i = 2)$.

Now let us compute $P_1(1, 1)$, i.e., in element e_1 $(r = 1)$, corresponding to $i = 1$, the coefficient of $u_1 (j = 1)$

$$\begin{aligned} P_1(1, 1) &= k_1 \int_{x_1}^{x_2} \frac{d\phi_1}{dx} \cdot \frac{d\phi_1}{dx} dx = k_1 \int_{x_1}^{x_2} \left(\frac{1}{x_1 - x_2}\right)\left(\frac{1}{x_1 - x_2}\right) dx \\ &= \frac{k_1}{L_1} \end{aligned} \tag{12.75a}$$

where L_1 is the length of the first element, i.e., $L_1 = x_2 - x_1$. The coefficient of $u_1 (j = 1)$ in the second row $(i = 2)$ from the first element is given by

$$P_1(2, 1) = k_1 \int_{x_1}^{x_2} \frac{d\phi_2}{dx} \cdot \frac{d\phi_1}{dx} dx = k_1 \int_{x_1}^{x_2} \frac{1}{x_2 - x_1} \cdot \frac{1}{x_1 - x_2} dx$$

$$= -\frac{k_1}{L_1}. \tag{12.75b}$$

Similarly we can compute coefficients of $u_2(j = 2)$ corresponding to node 2 in the first element e_1 and for $i = 2$ and $i = 1$. It gives

$$P_1(2, 2) = k_1 \int_{x_1}^{x_2} \frac{d\phi_2}{dx} \cdot \frac{d\phi_2}{dx} \cdot dx = k_1 \int_{x_1}^{x_2} \frac{1}{x_2 - x_1} \cdot \frac{1}{x_2 - x_1} dx$$

$$= \frac{k_1}{L_1}. \tag{12.75c}$$

$$P_1(1, 2) = k_1 \int_{x_1}^{x_2} \frac{d\phi_1}{dx} \cdot \frac{d\phi_2}{dx} dx$$

$$= -\frac{k_1}{L_1}, \tag{12.75d}$$

which is equal to P(2, 1) showing that P_1 is symmetrical.

The rest of the terms in P_1 will be zero. Thus P_1 provides terms in the 1st and 2nd rows and 1st and 2nd columns of P.

Element $e_2(r = 2)$

Over element e_2,

$$\bar{u} = \phi_2 u_2 + \phi_3 u_3$$

$$\text{where} \quad \phi_2(x) = \frac{x - x_3}{x_2 - x_3}, \quad \phi_3(x) = \frac{x - x_2}{x_3 - x_2}. \tag{12.76}$$

Also, $\phi_1 = 0$ and $\phi_4 = \phi_5 = \ldots = \phi_N = 0$ over e_2. If $L_2 = x_3 - x_2$, then

$$P_2(2, 2) = k_2 \int_{x_2}^{x_3} \frac{d\phi_2}{dx} \cdot \frac{d\phi_2}{dx} dx = \frac{k_2}{L_2} \tag{12.77a}$$

$$P_2(3, 2) = k_2 \int_{x_2}^{x_3} \frac{d\phi_3}{dx} \cdot \frac{d\phi_2}{dx} dx = -\frac{k_2}{L_2} \tag{12.77b}$$

$$P_2(2, 3) = k_2 \int_{x_2}^{x_3} \frac{d\phi_2}{dx} \cdot \frac{d\phi_3}{dx} dx = -\frac{k_2}{L_2} \text{ (by symmetry)} \tag{12.77c}$$

$$P_2(3,3) = k_2 \int_{x_2} \frac{d\phi_3}{dx} \cdot \frac{d\phi_3}{dx} dx = \frac{k_2}{L_2}. \tag{12.77d}$$

Element e_2 provides terms in the 2nd and 3rd rows and columns of P.

Let us add matrices P_1 and P_2; it will provide terms in the first three rows and columns of P. Further, since P_1 and P_2 are symmetric, $P_1 + P_2$ will also be symmetric, given as

$$P_1 + P_2 = \begin{bmatrix} \frac{k_1}{L_1} & -\frac{k_1}{L_1} & 0 & \cdots & 0 \\ -\frac{k_1}{L_1} & \frac{k_1}{L_1} + \frac{k_2}{L_2} & -\frac{k_2}{L_2} & \cdots & 0 \\ 0 & -\frac{k_2}{L_2} & \frac{k_2}{L_2} & & \vdots \\ \vdots & \vdots & \vdots & & \vdots \\ 0 & 0 & 0 & & 0 \end{bmatrix}. \tag{12.78}$$

In practice we compute in terms of local coordinates as follows:

Element e_r

We compute various terms over an element in terms of local coordinates. For example, while dealing with element e_r, node r may be treated as node 1 and node $r+1$ as node 2. A record of the local coordinates vis-a-vis the global coordinates for each element is kept through the Connectivity Table (See Table 12.1).

Table 12.1. Connectivity table

Element \ *Local nodes*	*1*	*2*
e_1	1	2
e_2	2	3
⋮	⋮	⋮
e_r	r	$r+1$
⋮	⋮	⋮
e_{N-1}	N − 1	N

We can compute various terms pertaining to an element, say e using normalised/natural coordinates as follows:

Express u in accordance with (12.50a, b) as

$$u^{(e)}(\xi) = \phi_1^{(e)}(\xi)u_1^{(e)} + \phi_2^{(e)}(\xi)u_2^{(e)}, \quad 0 \le \xi \le 1 \tag{12.79a}$$

where
$$\phi_1^{(e)}(\xi) = 1 - \xi \text{ and } \phi_2^{(e)}(\xi) = \xi. \tag{12.79b}$$

Also from $\xi = \dfrac{x - x_1}{x_2 - x_1}$ we have, $\mathrm{L}d\xi = dx$, $\mathrm{L} = x_2 - x_1$. (12.80)

From the above we get

$$\frac{d\phi_1^{(e)}}{dx} = \frac{d\phi_1^{(e)}}{d\xi} \cdot \frac{d\xi}{dx} = -\frac{1}{\mathrm{L}}; \quad \frac{d\phi_2^{(e)}}{dx} = \frac{d\phi_2^{(e)}}{d\xi} \cdot \frac{d\xi}{dx} = \frac{1}{\mathrm{L}}$$

Hence, we can compute

$$\left.\begin{aligned}
\mathrm{P}^{(e)}(1, 1) &= k\int_e \frac{d\phi_1^{(e)}}{dx} \cdot \frac{d\phi_1^{(e)}}{dx} dx = k\int_0^1 \left(-\frac{1}{\mathrm{L}}\right)\left(-\frac{1}{\mathrm{L}}\right)\mathrm{L}d\xi = \frac{k}{\mathrm{L}} \\
\mathrm{P}^{(e)}(1, 2) &= k\int_e \frac{d\phi_1^{(e)}}{dx} \cdot \frac{d\phi_2^{(e)}}{dx} dx = k\int_0^1 \left(-\frac{1}{\mathrm{L}}\right)\left(\frac{1}{\mathrm{L}}\right)\mathrm{L}d\xi = -\frac{k}{\mathrm{L}} \\
\mathrm{P}^{(e)}(2, 2) &= k\int_e \frac{d\phi_2^{(e)}}{dx} \cdot \frac{d\phi_2^{(e)}}{dx} dx = k\int_0^1 \left(\frac{1}{\mathrm{L}}\right)\left(\frac{1}{\mathrm{L}}\right)\mathrm{L}d\xi = \frac{k}{\mathrm{L}}
\end{aligned}\right\} \tag{12.81}$$

$\mathrm{P}^{(e)}(2, 1) = \mathrm{P}^{(e)}(1, 2) = -\dfrac{k}{\mathrm{L}}$, by symmetry; k and L pertain to e.

It must be borne in mind that the shape functions $\phi_1^{(e)}(\xi)$, $\phi_2^{(e)}(\xi)$ and the field (solution) variable $u_1^{(e)}$, $u_2^{(e)}$ are defined only over the element e and have no meaning outside it. For any element e_r, the global nodes r and $r+1$ corresponding to local nodes 1 and 2 can be decoded from the Connectivity Table 12.1 and put at appropriate places as shown below:

	r^{th} **column**	$(r+1)^{\text{th}}$ **column**
r^{th} **row**	$\mathrm{P}_r(r, r) = \dfrac{k_r}{\mathrm{L}_r}$	$\mathrm{P}_r(r, r+1) = \dfrac{k_r}{\mathrm{L}_r}$
$(r+1)^{\text{th}}$ **row**	$\mathrm{P}_r(r+1, r) = \dfrac{k_r}{\mathrm{L}_r}$	$\mathrm{P}_r(r+1, r+1) = \dfrac{k_r}{\mathrm{L}_r}$

(12.82)

Practically we do not store all the terms in matrix form. Instead the term $\mathrm{P}_r(i, j)$ is added to $\mathrm{P}(i, j)$ as soon as it is computed. Assembling the corresponding terms from all the elements we obtain the matrix P.

Putting $\beta_r = k_r/\mathrm{L}_r$, the matrix P may be expressed as

$$\mathrm{P} = \begin{bmatrix} \beta_1 & -\beta_1 & 0 & 0 & \cdots & 0 & 0 \\ -\beta_1 & \beta_1+\beta_2 & -\beta_2 & & & \vdots & \vdots \\ 0 & -\beta_2 & \beta_2+\beta_3 & & & \vdots & \vdots \\ \vdots & \vdots & \vdots & \ddots & & \vdots & \vdots \\ \vdots & \vdots & \vdots & & & \vdots & \vdots \\ \vdots & \vdots & \vdots & & & \vdots & -\beta_{n-1} \\ 0 & 0 & 0 & \cdots & \cdots & -\beta_{n-1} & \beta_{n-1}+\beta_n \end{bmatrix} \tag{12.83}$$

The matrix P is symmetric and tridiagonal and is generally diagonally dominant.

Note: Separate treatment for element-1 and element-2 was given earlier to make method understand easily. They are also particular cases of element e_r, $r = 1,\ 2$.

B. Computation of R

The first term on the right side of eqn. (12.70) may be expressed by a column vector R whose i^{th} term is given by

$$\mathrm{R}_i = \sum_{r=1}^{\mathrm{N}-1} \int_{e_r} q_r(x)\phi_i(x)\,dx,\ \ i = 1(1)\,\mathrm{N}. \tag{12.84}$$

In one dimension the integral in (12.84) may also be evaluated exactly over each element depending on the function $q(x)$. But in two dimensions it may not be easier to compute it. In that case we can assume the value of $q(x)$ to be constant over the element and may chose it suitably, e.g., in case of triangular element its value may be taken at the centroid. Now, considering the evaluation of (12.84) over an element e locally with nodes 1 and 2, having length $\mathrm{L} = x_2 - x_1$,

$$\mathrm{R}_i^{(e)} = \int_{e_r} q_e(x)\ \phi_i(x)\,dx,\ \ i = 1, 2.$$

$$= q_e \int_{x_1}^{x_2} \phi_i(x)\,dx,\ \text{ where } q_e = q\left(\frac{x_1+x_2}{2}\right),\ \text{say.}$$

$$\mathrm{R}_1^{(e)} = q_e \int_{x_1}^{x_2} \phi_1(x)dx = q_e \int_0^1 (1-\xi)\,\mathrm{L}d\xi = \frac{1}{2}q_e\mathrm{L}. \qquad \leftarrow \text{first row}$$

$$R_2^{(e)} = q_e \int_{x_1}^{x_2} \phi_2(x)dx = q_e \int_0^1 \xi L d\xi = \frac{1}{2}q_e L \qquad \leftarrow \text{second row}$$

Assembling for all the elements $e_r,\ r = 1(1)\ N-1$, we get

$$R = \begin{bmatrix} \frac{1}{2}q_1L_1 \\ \frac{1}{2}(q_1L_1 + q_2L_2) \\ \cdots \\ \frac{1}{2}(q_{i-1}L_{i-1} + q_iL_i) \\ \cdots \\ \frac{1}{2}(q_{N-2}L_{N-2} + q_{N-1}L_{N-1}) \\ \frac{1}{2}q_{N-1}L_{N-1} \end{bmatrix} \tag{12.85}$$

C. Computation of T

The ith term of vector T is given by

$$T_i = \phi_i(x_N)Q_b - \phi_i(x_1)Q_a, \quad i = 1(1)\ N. \tag{12.86}$$

From the definition $\phi_r(x_s) = 1$ if $r = s$ and 0 otherwise, we have

$$T_1 = -Q_a$$

$$T_i = 0,\ i = 2(1)\ N-1$$

$$T_N = +Q_b.$$

Thus the vector T may be written as

$$T = \begin{bmatrix} -Q_a \\ 0 \\ \vdots \\ \vdots \\ 0 \\ Q_b \end{bmatrix} \tag{12.87}$$

It may be noted that at node 1, $x = x_1 = a$, Dirichlet boundary condition $u = u_a$ is prescribed, so that Q_a is not known; while at node N, $x = x_N = b$, flux Q_b is known as per boundary condition $(12.62b)$.

Solution of PU = R–T

We get $P(N \times N)$, a symmetric tridiagonal matrix (12.83), $R(N \times 1)$ matrix as given by (12.85) and $T(N \times 1)$ as per (12.87) giving rise to system of equations

$$PU = R - T, \tag{12.88}$$

where we have to find $U^T = (u_1 \ \ u_2 \ \dots \ u_N)$.

Now since Q_a is not known, first equation can not be used. But u_1 is given; therefore we can use remaining $(N-1)$ equations to compute $(N-1)$ values of u, namely $u_2, u_3, \dots u_N$. The system of equations may be solved by the method described in Chapter 2, for symmetric tridiagonal matrix. Flux Q_a can be then computed from the first equation, if required.

Example 12.14

Using FEM solve the differential equation

$$\frac{d^2u}{dx^2} - u + x = 0, \quad 0 \le x \le 1.0$$

with boundary conditions

$$x = 0, \ u = 1 \text{ and } x = 1, \ \frac{du}{dx} = 1.$$

Take the nodes at $x = 0, \ 0.2, \ 0.5, \ 0.8, \ 1.0.$

Solution

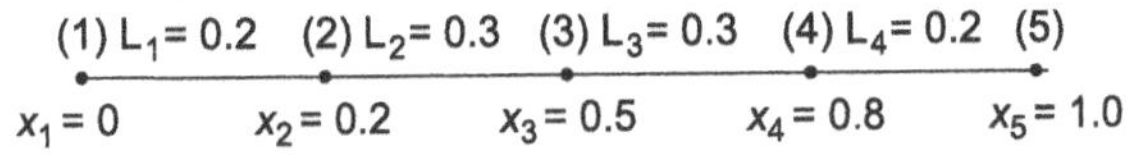

$N = 5.$

Let the solution be

$$\bar{u} = \sum_{j=1}^{5} \phi_j(x) u_j, \quad j = 1(1)5. \qquad \dots(1)$$

Substituting (1) in the equation and applying Galerkin's method

$$\int_0^1 \left(\frac{d^2\bar{u}}{dx^2} - \bar{u} + x\right)\phi_i(x)dx = 0,\ i = 1(1)5 \qquad \text{....(2)}$$

or $$\int_0^1 \phi_i(x)\frac{d^2\bar{u}}{dx^2}dx - \int_0^1 \phi_i(x)\,\bar{u}\,dx + \int_0^1 x\phi_i(x)dx = 0 \qquad \text{....(3)}$$

$$\int_0^1 \phi_i(x)\frac{d^2\bar{u}}{dx^2}dx = \phi_i(x)\frac{d\bar{u}}{dx}\bigg|_{x=0}^{1} - \int_0^1 \frac{d\phi_i}{dx}\cdot\frac{d\bar{u}}{dx}dx$$

$$= -\phi_i(x_5)Q_5 + \phi_i(x_1)Q_1 - \int_0^1 \frac{d\phi_i}{dx}\cdot\frac{d\bar{u}}{dx}dx$$

where $\quad Q = -\dfrac{du}{dx}.$

Equation (3) becomes

$$\int_0^1 \frac{d\phi_i}{dx}\cdot\frac{d\bar{u}}{dx}dx + \int_0^1 \phi_i(x)\,\bar{u}\,dx = \int_0^1 x\phi_i(x)dx - \phi_i(x_5)Q_5 + \phi_i(x_1)Q_1$$

or $$\sum_{j=1}^{5} u_j \int_0^1 \frac{d\phi_i}{dx}\cdot\frac{d\phi_j}{dx}dx + \sum_{j=1}^{5} u_j \int_0^1 \phi_i(x)\cdot\phi_j(x)dx = \int_0^1 x\phi_i(x)dx - [\phi_i(x_5)Q_5 - \phi_i(x_1)Q_1],$$

$$i = 1(1)5.$$

Integrating element-wise

$$\sum_{r=1}^{4}\sum_{j=1}^{5} u_j \int_{e_r} \frac{d\phi_i}{dx}\cdot\frac{d\phi_j}{dx}dx + \sum_{r=1}^{4}\sum_{j=1}^{5} u_j \int_{e_r} \phi_i(x)\phi_j(x)dx = R - T,\ \ i = 1(1)5, \qquad \text{.... (4)}$$

where $$R = \sum_{r=1}^{4}\int_{e_r} x\phi_i\,dx \qquad \text{.... (5)}$$

$$T = \phi_i(x_5)Q_5 - \phi_i(x_1)Q_1. \qquad \text{....(6)}$$

In this problem there are two terms on the left side of eqn. (4) containing u.

Consider first term on the left side of eqn (4), and denote it in matrix form as

$$\mathrm{AU} = \sum_{r=1}^{4} \sum_{j=1}^{5} u_j \int_{e_r} \frac{d\phi_i}{dx} \cdot \frac{d\phi_j}{dx} dx, \quad i = 1(1)5. \qquad \text{....(7)}$$

First let us evaluate the integral over an element e having local nodes 1 and 2, i.e.,

$$\int_e \frac{d\phi_i^{(e)}}{dx} \cdot \frac{d\phi_j^{(e)}}{dx} dx \text{ for } i = 1,\ 2 \text{ and } j = 1,\ 2.$$

In terms of normalised coordinates

$$\phi_1^{(e)} = 1 - \xi, \ \phi_2^{(e)} = \xi$$

where
$$\xi = \frac{x - x_1}{x_2 - x_1} = \frac{1}{\mathrm{L}}(x - x_1) \text{ and } \mathrm{L}d\xi = dx.$$

$$\frac{d\phi^{(e)}}{dx} = \frac{d\phi^{(e)}}{d\xi} \cdot \frac{d\xi}{dx} = \frac{1}{\mathrm{L}} \frac{d\phi^{(e)}}{d\xi}$$

$$\mathrm{A}_{11}^{(e)} = \int_{x_1}^{x_2} \frac{d\phi_1^{(e)}}{dx} \cdot \frac{d\phi_1^{(e)}}{dx} dx = \int_0^1 \left(-\frac{1}{\mathrm{L}} \frac{d\phi_1^{(e)}}{d\xi} \right) \left(-\frac{1}{\mathrm{L}} \frac{d\phi_1^{(e)}}{d\xi} \right) \mathrm{L}_1 d\xi = \frac{1}{\mathrm{L}}$$

$$\mathrm{A}_{12}^{(e)} = \int_{x_1}^{x_2} \frac{d\phi_1^{(e)}}{dx} \cdot \frac{d\phi_2^{(e)}}{dx} dx = \int_0^1 \left(-\frac{1}{L} \frac{d\phi_1^{(e)}}{d\xi} \right) \left(\frac{1}{L} \frac{d\phi_2^{(e)}}{d\xi} \right) \mathrm{L}_1 d\xi = -\frac{1}{\mathrm{L}}$$

$$\mathrm{A}_{22}^{(e)} = \int_{x_1}^{x_2} \frac{d\phi_2^{(e)}}{dx} \cdot \frac{d\phi_2^{(e)}}{dx} = \int_0^1 \left(\frac{1}{\mathrm{L}} \frac{d\phi_2^{(e)}}{d\xi} \right) \left(\frac{1}{\mathrm{L}} \frac{d\phi_2^{(e)}}{d\xi} \right) \mathrm{L}_1 d\xi = \frac{1}{L}$$

$\mathrm{A}_{21}^{(e)} = a_{12}^{(e)}$ by symmetry.

For an element e_r, the subscripts 1 and 2 will be replaced by r and $r+1$ respectively, $r = 1(1)4$. The coefficients computed above will be inserted in the rth, $(r+1)^{\text{th}}$ rows under r^{th}, $(r+1)^{\text{th}}$ columns of the Assumbley Table. The matrix $\mathrm{A}^{(e)}$ is given as

$$A^{(e)} = \begin{bmatrix} \frac{1}{L} & -\frac{1}{L} \\ -\frac{1}{L} & \frac{1}{L} \end{bmatrix}$$

The second term on the left side of eqn. (4) may be denoted as

$$BU = \sum_{r=1}^{4} \sum_{j=1}^{5} u_j \int_{e_r} \phi_i(x)\phi_j(x)\,dx, \quad i = 1(1)5.$$

Again over the element e

$$B_{11}^{(e)} = \int_{x_1}^{x_2} \phi^{(e)}(x) \cdot \phi^{(e)}(x)dx = \int_0^1 (1-\xi)^2 \cdot Ld\xi = \frac{L}{3}$$

$$B_{12}^{(e)} = \int_{x_1}^{x_2} \phi_1^{(e)}(x) \cdot \phi_2^{(e)}(x)dx = \int_0^1 (1-\xi)\xi Ld\xi = \frac{L}{6}$$

$$B_{22}^{(e)} = \int_{x_1}^{x_2} \phi_2^{(e)}(x) \cdot \phi_2^{(e)}(x)dx = \int_0^1 \xi^2 Ld\xi = \frac{L}{3}$$

$$B_{21}^{(e)} = B_{12}^{(e)}.$$

$$B^{(e)} = \begin{bmatrix} \frac{L}{3} & \frac{L}{6} \\ \frac{L}{6} & \frac{L}{3} \end{bmatrix}$$

Adding $A^{(e)}$ and $B^{(e)}$

$$P^{(e)} = \begin{bmatrix} \frac{1}{L} & -\frac{1}{L} \\ -\frac{1}{L} & \frac{1}{L} \end{bmatrix} + \begin{bmatrix} \frac{L}{3} & \frac{L}{6} \\ \frac{L}{6} & \frac{L}{3} \end{bmatrix} = \begin{bmatrix} \frac{3+L^2}{3L} & -\frac{6-L^2}{6L} \\ -\frac{6-L^2}{6L} & \frac{3+L^2}{3L} \end{bmatrix}$$

Substituting the values $L_1 = 0.2,\ L_2 = 0.3,\ L_3 = 0.3,\ L_4 = 0.2$

$$P_1 = \begin{bmatrix} \dfrac{3+0.04}{0.6} & -\dfrac{6-0.04}{1.2} \\[2ex] -\dfrac{6-0.04}{1.2} & \dfrac{3+0.04}{0.6} \end{bmatrix} = \begin{bmatrix} \dfrac{15.2}{3} & -\dfrac{14.9}{3} \\[2ex] -\dfrac{14.9}{3} & \dfrac{15.2}{3} \end{bmatrix} \begin{matrix} \leftarrow 1^{\text{st}} \text{ row} \\[2ex] \leftarrow 2^{\text{nd}} \text{ row} \end{matrix}$$

(columns of the right-hand matrix: 1st col, 2nd col)

$$P_2 = \begin{bmatrix} \dfrac{3+0.09}{0.9} & -\dfrac{6-0.09}{1.8} \\[2ex] -\dfrac{6-0.09}{1.8} & \dfrac{3+0.09}{0.9} \end{bmatrix} = \begin{bmatrix} \dfrac{10.3}{3} & -\dfrac{19.7}{6} \\[2ex] -\dfrac{19.7}{6} & \dfrac{10.3}{3} \end{bmatrix} \begin{matrix} \leftarrow 2^{\text{nd}} \text{ row} \\[2ex] \leftarrow 3^{\text{rd}} \text{ row} \end{matrix}$$

(columns of the right-hand matrix: 2nd col, 3rd col)

$P_3 = P_2$ and $P_4 = P_1$ since $L_3 = L_2$ and $L_4 = L_1$.

Values of P_1, P_2, P_3 and P_4 to be inserted in appropriate rows and columns.

The ith term of the column vector R of eq (4) is given by

$$R_i = \int_0^1 x\phi_i(x)dx, \quad i = 1(1)5.$$

$$R_i^{(e)} = \sum_{r=1}^{4} \int_{e_r} x\phi_i(x)dx.$$

Let us evaluate the above integral over an element e having nodes 1 and 2.

$$R_1^{(e)} = \int_{x_1}^{x_2} x\phi_1(x)dx = \int_{x_1}^{x_2} x\frac{(x-x_2)}{x_1-x_2}dx = -\frac{1}{L}\int_{x_1}^{x_2} x(x-x_2)dx$$

$$= -\frac{1}{L}\left\{x \cdot \frac{(x-x_2)^2}{2} - \frac{(x-x_2)^3}{6}\right\}_{x_1}^{x_2} = -\frac{1}{L}\left\{-x_1\frac{L^2}{2} - \frac{L^3}{6}\right\} = \frac{1}{2}x_1L + \frac{L^2}{6}$$

$$R_2^{(e)} = \int_{x_1}^{x_2} x\phi_2(x)dx = \int_{x_1}^{x_2} x\,\frac{(x-x_1)}{x_2-x_1}dx = \frac{1}{L}\left[x\,\frac{(x-x_1)^2}{2} - \frac{(x-x_1)^3}{6}\right]_{x_1}^{x_2} = \frac{1}{2}x_2L - \frac{L^2}{6}$$

$$\mathrm{R} = \begin{bmatrix} \frac{1}{2}x_1\mathrm{L}_1 - \frac{\mathrm{L}_1^2}{6} \\ \frac{1}{2}x_2\mathrm{L}_1 - \frac{\mathrm{L}_1^2}{6} + \frac{1}{2}x_2\mathrm{L}_2 - \frac{\mathrm{L}_2^2}{6} \\ \frac{1}{2}x_3\mathrm{L}_2 - \frac{\mathrm{L}_2^2}{6} + \frac{1}{2}x_3\mathrm{L}_3 - \frac{\mathrm{L}_3^2}{6} \\ \frac{1}{2}x_4\mathrm{L}_3 - \frac{\mathrm{L}_3^2}{6} + \frac{1}{2}x_4\mathrm{L}_4 - \frac{\mathrm{L}_4^2}{6} \\ \frac{1}{2}x_5\mathrm{L}_4 - \frac{\mathrm{L}_4^2}{6} \end{bmatrix} = \begin{bmatrix} \frac{0.02}{3} \\ \frac{0.35}{6} \\ \frac{0.90}{6} \\ \frac{1.15}{6} \\ \frac{0.56}{6} \end{bmatrix}$$

$x_1 = 0,\ x_2 = 0.2,\ x_3 = 0.5,\ x_4 = 0.8,\ x_5 = 1.0.$

$\mathrm{T}_i = -\phi_i(x_5)\mathrm{Q}_5 + \phi_i(x_1)\mathrm{Q}_1$, given that $\mathrm{Q}_5 = -1.$

$$\mathrm{T} = \begin{bmatrix} \mathrm{Q}_1 \\ 0 \\ 0 \\ 0 \\ 1 \end{bmatrix}$$

Insert the values of R and T in the Assembly Table.

Assembly Table

$\phi_i \backslash u_j$	u_1	u_2	u_3	u_4	u_5	R	T
ϕ_1	$\frac{15.2}{3}$	$-\frac{14.9}{3}$	0	0	0	$\frac{0.02}{3}$	$-Q_1$
ϕ_2	$-\frac{14.9}{3}$	$\frac{15.2}{3}+\frac{10.3}{3}=\frac{25.5}{3}$	$-\frac{19.7}{6}$	0	0	$\frac{0.35}{6}$	0
ϕ_3	0	$\frac{-19.7}{6}$	$\frac{10.3}{3}+\frac{10.3}{3}=\frac{20.6}{3}$	$-\frac{19.7}{6}$	0	$\frac{0.90}{6}$	0
ϕ_4	0	0	$-\frac{19.7}{6}$	$\frac{10.3}{3}+\frac{15.2}{3}=\frac{25.5}{3}$	$-\frac{14.9}{3}$	$\frac{1.15}{6}$	0
ϕ_5	0	0	0	$-\frac{14.9}{3}$	$\frac{15.2}{3}$	$\frac{0.56}{6}$	1

In the above table Q_1 is not known; therefore neglect first equation and consider 2nd to 4th equations. Further since $u_1 = 1$ (given), these equations may be written as

u_2	u_3	u_4	u_5	
25.5	−9.85	0	0	15.075
−9.85	20.6	−9.85	0	0.45
0	−9.85	25.5	−14.9	0.575
0	0	−14.9	15.2	3.28

Multipliers					
	25.5	−9.85	0	0	15.075
0.3863	0	16.7949	−9.85	0	6.2735
0.5865	0	0	19.7230	−14.9	4.2544
0.7555	0	0	0	3.9430	6.4942

$u_5 = 1.6470,\ u_4 = 1.4600,\ u_3 = 1.2298,\ u_2 = 1.0662,\ u_1 = 1$ (Given)

Analytical solution is $u = \dfrac{e^{1-x} + e^{-(1-x)}}{e^1 + e^{-1}} + x = \dfrac{\cosh(1-x)}{\cosh 1} + x$

The corresponding values are:

$u(1.0) = 1.6480,\ u(0.8) = 1.4610,\ u(0.5) = 1.2307,\ u(0.2) = 1.0667,\ u(0) = 1$

We can compute Q_1 from first eqn., $Q_1 = 0.2355$

Note: We can compute R_i by taking the value of x as constant over the element, e.g.,

$$\int_{x_{i-1}}^{x_i} x\phi_i(x)dx + \int_{x_i}^{x_{i+1}} x\phi_i(x)dx = \frac{x_{i-1}+x_i}{2}\int_{x_{i-1}}^{x_i} \phi_i(x)dx + \frac{x_i+x_{i+1}}{2}\int_{x_i}^{x_{i+1}} \phi_i(x)dx$$

But the accuracy may not be as good.

12.9.2 Elliptic equation

The versatility of the FEM will be better appreciated in two dimensions. If $u(x, y)$ represents the temperature at a point (x, y), the partial differential equation governing the steady state heat flow in a non-homogeneous medium, can be expressed by an elliptic equation

$$\frac{\partial}{\partial x}\left(k_1\frac{\partial u}{\partial x}\right) + \frac{\partial}{\partial y}\left(k_2\frac{\partial u}{\partial y}\right) + q(x, y) = 0, \text{ in R} \tag{12.89}$$

where region R is enclosed by a curve S; $q(x, y)$ is sink/source term; $k_1(x, y)$ and $k_2(x, y)$ are conductivities depending on the properties of the medium. All these terms, namely q, k_1 and k_2 are known and we are interested in finding u in domain $D = R \cup S$ with given boundary conditions on S. Let us assume that different boundary conditions are prescribed on different segments of S. For example, let $S = S_1 \cup S_2 \cup S_3 \cup S_4$ and the conditions prescribed on S_1, S_2, S_3 and S_4 are as follows (See Fig. 12.6).

(*i*) on $S_1 = AB$, Dirichlet condition is prescribed, i.e., u is given (12.90a)

(*ii*) on $S_2 = BC$, Neumann condition is prescribed, i.e., flux Q is given (12.90b)

(*iii*) on $S_3 = CD$, Dirichlet condition (12.90c)

(*iv*) on $S_4 = DA$, Neumann condition. (12.90d)

The normal flux Q is given by,

$$Q = -\left(k_1\frac{\partial u}{\partial x} + k_2\frac{\partial u}{\partial y}\right) = -\left(k_1\frac{\partial u}{\partial x}\boldsymbol{i} + k_2\frac{\partial u}{\partial y}\boldsymbol{j}\right)\cdot \boldsymbol{n} \tag{12.91}$$

in the direction of the outward unit normal $\boldsymbol{n}$ to S.

We subdivide the domain into elements of any shapes or sizes; in the present case triangular. Subdivision is made according to convenience in the choice of nodes and the nodes may be selected keeping in view the boundary conditions and the desirability of accuracy of the solution in portions of region R.

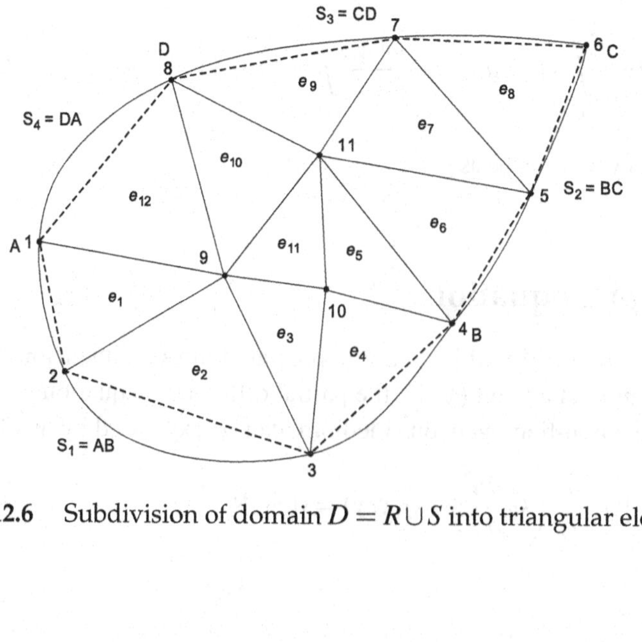

Figure 12.6 Subdivision of domain $D = R \cup S$ into triangular elements

As shown in Fig. 12.6, the domain D has been subdivided into twelve elements with total eleven nodes. Various segments of S consist of the nodes as shown against each below:

$$S_1 = AB \rightarrow 1,\ 2,\ 3,\ 4$$

$$S_2 = BC \rightarrow 4,\ 5,\ 6$$

$$S_3 = CD \rightarrow 6,\ 7,\ 8$$

$$S_4 = DA \rightarrow 8,\ 1$$

Since Dirichlet conditions are prescribed on S_1 and S_3 the values of u at the nodes 1, 2, 3, 4, 6, 7, 8 are known and hence we are required to compute the values at the nodes 5, 9, 10 and 11. However, we will discuss the method in general for M elements, say $e_r,\ r = 1$ (1) M and N nodes with coordinates $(x_i,\ y_i),\ i = 1$ (1) N.

Let the approximate solution of (12.89) be expressed as

$$\bar{u} = \sum_{j=1}^{N} \phi_j(x,\ y)u_j, \tag{12.92}$$

where ϕ_j's are the global shape functions such that $\phi_j = 1$ at node j and zero at all the other nodes. They are also called 'pyramid' shape functions because of their shape having a tip of height one at node j and zero at all the neighbouring nodes; u_j is the value of u at node $j-$ to be computed or given.

Following the Galerkin's method we evaluate

$$\iint_R \left\{ \frac{\partial}{\partial x}\left(k_1 \frac{\partial \bar{u}}{\partial x}\right) + \frac{\partial}{\partial y}\left(k_2 \frac{\partial \bar{u}}{\partial y}\right) + q(x,\ y) \right\} \phi_i(x,\ y)\ dxdy = 0 \tag{12.93}$$

$$i = 1(1)\ N.$$

Using vector calculus, let us express a vector field V as

$$V = k_1 \frac{\partial \bar{u}}{\partial x} \boldsymbol{i} + k_2 \frac{\partial \bar{u}}{\partial y} \boldsymbol{j}$$

so that
$$\nabla \cdot V = \frac{\partial}{\partial x}\left(k_1 \frac{\partial \bar{u}}{\partial x}\right) + \frac{\partial}{\partial y}\left(k_2 \frac{\partial u}{\partial y}\right) \tag{12.94}$$

where
$$\nabla = \boldsymbol{i}\frac{\partial}{\partial x} + \boldsymbol{j}\frac{\partial}{\partial y}.$$

Putting from (12.94) into (12.93) we get

$$\iint_R \phi_i \cdot \nabla \cdot \mathrm{V} dxdy + \iint_R q(x, y)\phi_i \cdot dxdy = 0$$

or $$\iint_R \{\nabla \cdot (\mathrm{V}\phi_i) - \mathrm{V} \cdot \nabla\phi_i\}\, dxdy + \iint_R \phi_i q(x, y)\, dxdy = 0$$

Using result from vector calculus,

$\nabla \cdot (\mathrm{V}\phi_i) = \mathrm{V} \cdot \nabla\phi_i + \phi_i \nabla \cdot \mathrm{V}$. Also see 12.21$a$ when $\mathrm{V} = \nabla v$.

Now, $$\iint_R \nabla \cdot (\mathrm{V}\phi_i)\, dxdy - \iint_R \mathrm{V} \cdot \nabla\phi_i\, dxdy + \iint_R \phi_i q dxdy = 0, \quad q = q(x, y).$$

or $$\int_S \phi_i \mathrm{V} \cdot \boldsymbol{n} ds - \iint_R \left(k_1 \frac{\partial \bar{u}}{\partial x}\boldsymbol{i} + k_2 \frac{\partial \bar{u}}{\partial y}\boldsymbol{j}\right) \cdot \left(\frac{\partial \phi_i}{\partial x}\boldsymbol{i} + \frac{\partial \phi_i}{\partial y}\boldsymbol{j}\right) dxdy + \iint \phi_i q dxdy = 0.$$

or $$\iint_R \left(k_1 \frac{\partial \bar{u}}{\partial x} \cdot \frac{\partial \phi_i}{\partial x} + k_2 \frac{\partial \bar{u}}{\partial y} \cdot \frac{\partial \phi_i}{\partial y}\right) dxdy - \int_S \phi_i \left(k_1 \frac{\partial \bar{u}}{\partial x}\boldsymbol{i} + k_2 \frac{\partial \bar{u}}{\partial y}\boldsymbol{j}\right) \cdot \boldsymbol{n}\, dS.$$

$$= \iint_R \phi_i q dxdy, \quad i = 1(1)\,\mathrm{N}.$$

or $$\iint_R \left(k_1 \frac{\partial \bar{u}}{\partial x} \cdot \frac{\partial \phi_i}{\partial x} + k_2 \frac{\partial \bar{u}}{\partial y} \cdot \frac{\partial \phi_i}{\partial y}\right) dxdy = \iint_R \phi_i q dxdy - \int_S \phi_i \mathrm{Q}\, ds, \tag{12.95}$$

due to (12.91), $i = 1(1)$ N.

The expression (12.95) is analogous to (12.65) in one dimension. The integrals in expression (12.95) are evaluated element-wise over M elements so that it may be written as

$$\sum_{r=1}^{\mathrm{M}} \iint_{e_r} \left(k_1 \frac{\partial \bar{u}}{\partial x} \cdot \frac{\partial \phi_i}{\partial x} + k_2 \frac{\partial \bar{u}}{\partial y} \cdot \frac{\partial \phi_i}{\partial y}\right) dxdy = \sum_{r=1}^{\mathrm{M}} \iint_{e_r} \phi_i q dxdy - \int_{\mathrm{S}} \phi_i Q ds, \tag{12.96}$$

$$i = 1(1)\mathrm{N}$$

where e_r denotes the rth element and the line integral is taken along the sides of the boundary elements approximating S. We can assume that the diffusivity does not vary over an element, so that we can consider $k_1(x, y) = k_2(x, y) = k_r$ (const) where k_r is the diffusivity in element e_r which may be taken as average of k_1 and k_2 at the centroid of the triangular element or average value at the three nodes of e_r. Substituting $\overline{u}$ from (12.92), the left side of (12.96) may be written as

$$\sum_{r=1}^{M} \iint_{e_r} \left(k_1 \frac{\partial \overline{u}}{\partial x} \cdot \frac{\partial \phi_i}{\partial x} + k_2 \frac{\partial \overline{u}}{\partial y} \cdot \frac{\partial \phi_i}{\partial y} \right) dxdy$$

$$= \sum_{r=1}^{M} k_r \sum_{j=1}^{N} u_j \iint_{e_r} \left(\frac{\partial \phi_i}{\partial x} \cdot \frac{\partial \phi_j}{\partial x} + \frac{\partial \phi_i}{\partial y} \cdot \frac{\partial \phi_j}{\partial y} \right) dxdy, \tag{12.97}$$

$$i = 1(1)\,\mathrm{N}.$$

Similarly the value of sink/source term $q(x, y)$ may also be considered as constant over the element e_r and may be taken as q_r. Thus (12.96) may be written after using (12.97) as

$$\sum_{r=1}^{M} k_r \sum_{j=1}^{N} u_j \iint_{e_r} \left(\frac{\partial \phi_i}{\partial x} \cdot \frac{\partial \phi_j}{\partial x} + \frac{\partial \phi_i}{\partial y} \cdot \frac{\partial \phi_j}{\partial y} \right) dxdy = \sum_{r=1}^{M} q_r \iint_{e_r} \phi_i dxdy - \int_{S} \phi_i Q ds, \tag{12.98}$$

$$i = 1(1)\,\mathrm{N}.$$

We may note that (12.98) represents a system of N equations, one each for $i = 1(1)\mathrm{N}$ and each equation being linear in u_j, $j = 1(1)$ N. This system of linear equations may be represented in matrix form as

$$\mathrm{PU} = \mathrm{R} - \mathrm{T}, \tag{12.99}$$

where $\quad \mathrm{U}^{\mathrm{T}} = (u_1 \quad u_2 \quad \ldots. \quad u_{\mathrm{N}});$

P is a $(\mathrm{N} \times \mathrm{N})$ coefficient matrix obtained from left side of (12.98); $\mathrm{R}(\mathrm{N} \times 1)$ and $\mathrm{T}(\mathrm{N} \times 1)$ are column vectors corresponding to the first and the second terms respectively on the right side of (12.98) for $i = 1(1)$ N. Further, if $\mathrm{P}_r\mathrm{U}$ denotes the computation of the term on the left side of (12.98) over element e_r, then

$$\mathrm{P} = \mathrm{P}_1 + \mathrm{P}_2 + \ldots. + \mathrm{P}_{\mathrm{M}} = \sum_{r=1}^{M} \mathrm{P}_r. \tag{12.100}$$

Let us suppose that l, m and n are the nodes/vertices of the triangular element e_r with global coordinates (x_l, y_l), (x_m, y_m) and (x_n, y_n) respectively. As stated earlier we express the concerned formulae in terms local coordinates which may be used over all the elements globally. A connectivity table is prepared connecting the global nodes with local nodes as shown in Table 12.2.

Table 12.2 Connectivity table

Elements \ Local nodes	Node 1	Node 2	Node 3
1	⋮	⋮	⋮
2	⋮	⋮	⋮
⋮	⋮	⋮	⋮
r	l	m	n
⋮	⋮	⋮	⋮
M	⋮	⋮	⋮

A. Computation of PU

Let us compute the relevant terms of the coefficient matrix P_r. We should note that all the global shape functions except ϕ_l, ϕ_m and ϕ_n will have zero value over element e_r, so that we will get terms corresponding $j = l,\ m,\ n$ and $i = l,\ m,\ n$ only. Further, since the expression on the left side of (12.98) is symmetrical w.r.t i and j we need compute only following terms and the remaining may be written by symmetry (See Table 12.3).

Table 12.3

i \ j	1	2	3
1	$P_r(1, 1)$	$P_r(1, 2)$	$P_r(1, 3)$
2		$P_r(2, 2)$	$P_r(2, 3)$
3	symmetric		$P_r(3, 3)$

After computing in terms of local nodes the values are inserted in the Assembly Table decoding the global nodes from the Connectivity Table 12.2.

Let $u(x, y)$ be represented over the element e locally as

$$u^{(e)} = \phi_1^{(e)}(x, y)u_1^{(e)} + \phi_2^{(e)}(x, y)u_2^{(e)} + \phi_3^{(e)}(x, y)u_3^{(e)}, \tag{12.101}$$

where $\phi_1^{(e)}$, $\phi_2^{(e)}$ and $\phi_3^{(e)}$ are defined in terms of local coordinates. From (12.47) and (12.46) we can express them as

$$\left.\begin{aligned} \phi_1^{(e)} &= a_1x + b_1y + c_1 \\ \phi_2^{(e)} &= a_2x + b_2y + c_2 \\ \phi_3^{(e)} &= a_3x + b_3y + c_3 \end{aligned}\right\} \tag{12.102}$$

where

$$\left.\begin{aligned}
a_1 &= \frac{1}{2A_e}(y_2 - y_3),\ b_1 = \frac{1}{2A_e}(x_3 - x_2),\ c_1 = \frac{1}{2A_e}(x_2y_3 - y_2x_3);\\
a_2 &= \frac{1}{2A_e}(y_3 - y_1),\ b_2 = \frac{1}{2A_e}(x_1 - x_3),\ c_2 = \frac{1}{2A_e}(x_3y_1 - y_3x_1);\\
a_3 &= \frac{1}{2A_e}(y_1 - y_2),\ b_3 = \frac{1}{2A_e}(x_2 - x_1),\ b_3 = \frac{1}{2A_e}(x_1y_2 - y_1x_2);
\end{aligned}\right\} \tag{12.103}$$

A_e being the area of the element e. (check: $\Sigma a_i = 0,\ \Sigma b_i = 0,\ \Sigma c_i = 1$). We can now find various terms corresponding to $P^{(e)}$:

$$P^{(e)}(1,\ 1) = k_e \iint_e \left(\frac{\partial \phi_1^{(e)}}{\partial x} \cdot \frac{\partial \phi_1^{(e)}}{\partial x} + \frac{\partial \phi_1^{(e)}}{\partial y} \cdot \frac{\partial \phi_1^{(e)}}{\partial y}\right) dxdy$$

$$= k_e \iint_e (a_1^2 + b_1^2)dxdy = A_e k_e (a_1^2 + b_1^2). \tag{12.104a}$$

$$P^{(e)}(1,\ 2) = k_e \iint_e \left(\frac{\partial \phi_1^{(e)}}{\partial x} \cdot \frac{\partial \phi_2^{(e)}}{\partial x} + \frac{\partial \phi_1^{(e)}}{\partial y} \cdot \frac{\partial \phi_2^{(e)}}{\partial y}\right) dxdy$$

$$= k_e \iint_e (a_1a_2 + b_1b_2)dxdy = A_e k_e (a_1a_2 + b_1b_2). \tag{12.104b}$$

$$P^{(e)}(1,\ 3) = k_e \iint_e \left(\frac{\partial \phi_1^{(e)}}{\partial x} \cdot \frac{\partial \phi_3^{(e)}}{\partial x} + \frac{\partial \phi_1^{(e)}}{\partial y} \cdot \frac{\partial \phi_3^{(e)}}{\partial y}\right) dxdy$$

$$= A_e k_e (a_1a_3 + b_1b_3). \tag{12.104c}$$

$$P^{(e)}(2,\ 2) = k_e \iint_e \left(\frac{\partial \phi_2^{(e)}}{\partial x} \cdot \frac{\partial \phi_2^{(e)}}{\partial x} + \frac{\partial \phi_2^{(e)}}{\partial y} \cdot \frac{\partial \phi_2^{(e)}}{\partial y}\right) dxdy$$

$$= A_e k_e (a_2^2 + b_2^2). \tag{12.104d}$$

$$P^{(e)}(2,\ 3) = k_e \iint_e \left(\frac{\partial \phi_2^{(e)}}{\partial x} \cdot \frac{\partial \phi_3^{(e)}}{\partial x} + \frac{\partial \phi_2^{(e)}}{\partial y} \cdot \frac{\partial \phi_3^{(e)}}{\partial y}\right) dxdy$$

$$= A_e k_e (a_2a_3 + b_2b_3). \tag{12.104e}$$

$$\mathrm{P}^{(e)}(3,\ 1) = k_e \iint\limits_{e} \left(\frac{\partial \phi_3^{(e)}}{\partial x} \cdot \frac{\partial \phi_3^{(e)}}{\partial x} + \frac{\partial \phi_3^{(e)}}{\partial y} \cdot \frac{\partial \phi_3^{(e)}}{\partial y} \right) dxdy$$

$$= \mathrm{A}_e k_e (a_3^2 + b_3^2). \qquad (12.104\text{f})$$

The terms below the diagonal may be written by symmetry.
Putting in matrix form

$$\mathrm{P}^{(e)} = \mathrm{A}_e k_e \begin{bmatrix} a_1^2 + b_1^2 & a_1 a_2 + b_1 b_2 & a_1 a_3 + b_1 b_3 \\ & a_2^2 + b_2^2 & a_2 a_3 + b_2 b_3 \\ \text{Symmetric} & & a_3^2 + b_3^2 \end{bmatrix}. \qquad (12.105)$$

It may be understood that $\mathrm{P}^{(e)}$ has been computed locally for element e. While computing over element e_r the terms are computed with nodes $l,\ m,\ n$ corresponding to local nodes 1, 2 and 3. The various terms in (12.105) are inserted in Assembly Table at appropriate places in P_r corresponding to Table 12.3 which may be decoded from the Connectivity Table 12.2. The other terms of matrix P_r over the element e_r are zero. The values of various terms in (12.105) may be computed as shown in (12.104a-f).

B. Computation of R

As explained, the first term on the right side of (12.98) is a $(\mathrm{N} \times 1)$ column vector whose i^{th} term is given by

$$\mathrm{R}_i = \sum_{r=1}^{\mathrm{M}} q_r \iint\limits_{e_r} \phi_i dxdy,\ i = 1(1)\mathrm{N} \qquad (12.106)$$

where ϕ_i is the global shape function.

If $i \neq l,\ m,\ n$ then $\phi_i(x,\ y)$ will be zero over the element e_r and it will not contribute anything to the summation in (12.106). For an element e, the nodes $l,\ m,\ n$ correspond to nodes 1, 2 and 3 respectively. Then for one of these nodes, say node 2,

$$\mathrm{R}_2^{(e)} = q_e \iint\limits_{e} \phi_2 dxdy$$

Instead of evaluating ϕ_2 over e we can take its average value at the nodes of element i.e., $\dfrac{0+1+0}{3} = \dfrac{1}{3}$. Hence we may approximate,

$$R_2^{(e)} = \frac{1}{3} q_e A_e,$$

where A_e is the area of element e.

Thus, we have

$$R_1^{(e)} = R_2^{(e)} = R_3^{(e)} = \frac{1}{3} q_e A_e. \quad (12.107)$$

The values of $R_1^{(e)}$, $R_2^{(e)}$ and $R_3^{(e)}$ may be inserted in the lth, m^{th} and n^{th} components of R in the Assembly Table. Referring to Fig. 12.6, it may be noted that

$$R_1 = \frac{1}{3} A_1 q_1 + \frac{1}{3} A_{12} q_{12}$$

$$R_3 = \frac{1}{3} A_2 q_2 + \frac{1}{3} A_3 q_3 + \frac{1}{3} A_4 q_4$$

$$R_9 = \frac{1}{3} A_1 q_1 + \frac{1}{3} A_2 q_2 + \frac{1}{3} A_3 q_3 + \frac{1}{3} A_{10} q_{10} + \frac{1}{3} A_{11} q_{11} + \frac{1}{3} A_{12} q_{12}$$

etc.

The number of terms to be added in the ith term of R, i.e., R_i depends on node i as how many elements it is common to. The common node i is also called 'apex' node for these elements. For example, node 9, is the 'apex' node for elements e_1, e_2, e_3, e_{10}, e_{11} and e_{12}.

C. Computation of T

The second term on the right side of (12.98) is the line integral

$$T_i = \int_S Q\phi_i ds, \ i = 1(1)N \quad (12.108)$$

where T is a column vector of size N; ds is the elemental length along S. The line integral along a curve is evaluated tracing the curve in an anti-clockwise direction, generally.

Referring to Fig. 12.7 let us suppose that nodes m and n (global) of element e_r or nodes 2 and 3 (local) lie on the bounding curve S. The local nodes are shown in Fig. 12.7.

Instead of evaluating the integral (12.108) along curve S, say S_{23} we evaluate it along the side of the element, 2-3 which approximates S_{23} as

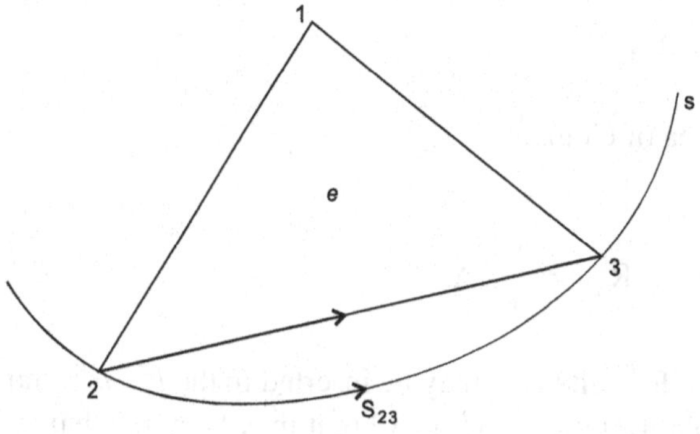

Figure 12.7 Element with one side approximating the boundary.

$$T_i^{(e)} = \int_\sigma Q\phi_i d\sigma,\ i = 1(1)\,N$$

where σ denotes the length along side 2-3 of element e. Let us suppose that its length is σ_{23} so that at node 2, $\sigma = 0$ and at 3, $\sigma = \sigma_{23}$ and we integrate from $\sigma = 0$ to $\sigma = \sigma_{23}$, i.e.,

$$T_i^{(e)} = \int_0^{\sigma_{23}} Q\phi_i d\sigma. \tag{12.109}$$

In terms of normalised/natural coordinates u may be represented along 2-3 as

$$u^{(e)} = \phi_2^{(e)} u_2^{(e)} + \phi_3^{(e)} u_3^{(e)}$$

where $\phi_2^{(e)} = \dfrac{\sigma - \sigma_{23}}{0 - \sigma_{23}},\ \phi_3^{(e)} = \dfrac{\sigma - 0}{\sigma_{23}}.$

Putting $\dfrac{\sigma}{\sigma_{23}} = \xi$, gives

$$\phi_2^{(e)} = 1 - \xi \text{ and } \phi_3^{(e)} = \xi,\ 0 \le \xi \le 1.$$

and $\quad d\sigma = \sigma_{23} d\xi.$

Let us suppose that flux Q varies linearly from Q_2 to Q_3 along side 2-3 of e, so that it may be expressed as

$$Q^{(e)} = (1 - \xi)Q_2^{(e)} + \xi Q_3^{(e)},\ 0 \le \xi \le 1. \tag{12.110}$$

Now line integral (12.109) for $\phi_i = \phi_2^{(e)}$ gives

$$T_2^{(e)} = \int_0^1 \{(1 - \xi)Q_2^{(e)} + \xi Q_3^{(e)}\}(1 - \xi)\sigma_{23} d\xi$$

$$= \frac{\sigma_{23}(2Q_2^{(e)} + Q_3^{(e)})}{6}. \quad (12.111a)$$

Similarly, we can get for $\phi_i = \phi_3^{(e)}$

$$T_3^{(e)} = \frac{\sigma_{23}(Q_2^{(e)} + 2Q_3^{(e)})}{6}. \quad (12.111b)$$

In the Assembly Table the value of $T_2^{(e)}$ will be put in T_m and that of $T_3^{(e)}$ in T_n. If there are two sides of an element which approximate S, like in element e_8 in Fig. 12.6, then values of components of T may be computed independently along both sides and put at appropriate places. If all the three sides approximate S, then it should be a single element in R. If no side of an element forms boundary or part of boundary S, then its contribution to T will be naught, like e_3, e_5, e_7, e_{10} and e_{11} in Fig. 12.6.

Solution of PU = R−T

As explained in the case of one-dimensional problem, the coefficient matrix P is usually symmetric and diagonally dominant. Some equations will involve the unknown flux Q on the right side; solution may be found omitting those equations. The equations may be solved by a suitable method (direct or iterative) described in Chapter 2.

Example 12.15

Given a steady state equation in two dimensions

$$\nabla^2 u + x + y = 0 \text{ in R}$$

where R is bounded by $S = S_1 \cup S_2 \cup S_3$. The boundary conditions on the curves S_1, S_2 and S_3 are prescribed as follows:

on S_1 : $y^2 = 4x, \quad u = 2x + y$

on S_2 : $y = 0, \quad \dfrac{\partial u}{\partial y} = -u$

on S_3 : $x = 1, \quad \dfrac{\partial u}{\partial x} = -1$

Find the solution by FEM over $D = R \cup S$ when domain D has been subdivided into three elements as shown in the figure :

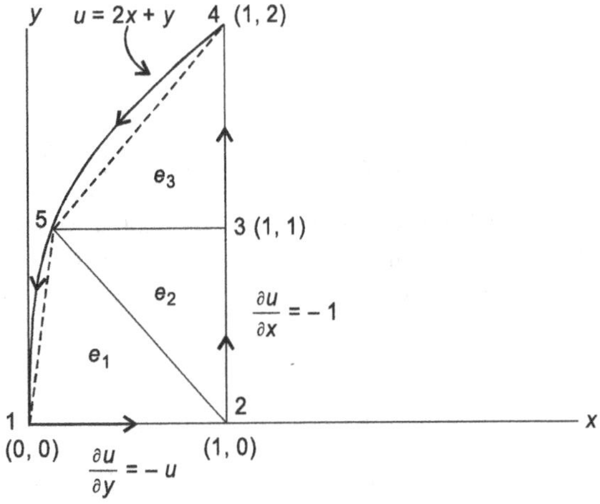

Solution

Connectivity table

Element \ Node	1	2	3
e_1	1	2	5
e_2	2	3	5
e_3	3	4	5

A. Computation of PU

To find x-coordinate of node 5, put $y = 1$ in the equation of the parabola, giving $x = 0.25$.

element e_1

Local Nodes	1	2	3
Coordinates	(0, 0)	(1, 0)	(0.25, 1)

$$2A = \begin{vmatrix} 1 & 1 & 1 \\ 0 & 1 & 0.25 \\ 0 & 0 & 1 \end{vmatrix} = 1, \ A = 0.5$$

$$a_1 = \frac{1}{2A}(y_2 - y_3) = -1 \quad ; \quad b_1 = \frac{1}{2A}(x_3 - x_2) = (0.25 - 1) = -0.75$$

$$a_2 = \frac{1}{2A}(y_3 - y_1) = 1 \quad ; \quad b_2 = \frac{1}{2A}(x_1 - x_3) = (0 - 0.25) = -0.25$$

$$a_3 = \frac{1}{2A}(y_1 - y_2) = 0 \quad ; \quad b_3 = \frac{1}{2A}(x_2 - x_1) = (1 - 0) = 1$$

$$P_1 = Ak_1 \begin{bmatrix} a_1^2 + b_1^2 & a_1a_2 + b_1b_2 & a_1a_3 + b_1b_3 \\ & a_2^2 + b_2^2 & a_2a_3 + b_2b_3 \\ \text{symmetric} & & a_3^2 + b_3^2 \end{bmatrix} = \begin{bmatrix} 0.78125 & 0.40625 & -0.375 \\ & 0.53125 & -0.125 \\ \text{symmetric} & & 0.5 \end{bmatrix}$$

Note: For simplification in computation we can compute $2Aa_1$, $2Ab_1$, etc., and substitute in P_1 to get a multiplying factor $k_1/4A$ instead of Ak_1. Here we are not using it for showing the complete details.

element e_2

1	2	3
(1, 0)	(1, 1)	(0.25, 1)

$$2A = 0.75, \quad A = 0.375$$

$$P_2 = 0.375\begin{bmatrix} 1 & -1 & 0 \\ -1 & \frac{25}{9} & -\frac{4}{3} \\ 0 & -\frac{4}{3} & 1 \end{bmatrix} = \begin{bmatrix} 0.375 & -0.375 & 0 \\ & 1.04167 & -0.5 \\ \text{symmetric} & & 0.375 \end{bmatrix}$$

element e_3

1	2	3
(1, 1)	(1, 2)	(0.25, 1)

$$2A = 0.75, \quad A = 0.375$$

$$P_3 = 0.375\begin{bmatrix} \frac{25}{9} & -1 & -\frac{16}{9} \\ & 1 & 0 \\ \text{symmetric} & & \frac{16}{9} \end{bmatrix} = \begin{bmatrix} 1.04167 & -0.375 & -0.66667 \\ & 0.375 & 0 \\ & & 0.66667 \end{bmatrix}$$

B. Computation of R

$$q(x, y) = x + y, \; R_i^{(e)} = \frac{1}{3}q_e A_e = \frac{1}{3}q\left(\frac{x_1 + x_2 + x_3}{3}, \frac{y_1 + y_2 + y_3}{3}\right) A$$

element e_1

$$x = \frac{0 + 1 + 0.25}{3} = \frac{1.25}{3}; \; y = \frac{0 + 0 + 1}{3} = \frac{1}{3}; \; A = 0.5$$

$$R_1^{(e)} = R_2^{(e)} = R_3^{(e)} = \frac{1}{3}\left(\frac{2.25}{3}\right) \times 0.5 = 0.125$$

element e_2

$$x = \frac{1 + 1 + 0.25}{3} = \frac{2.25}{3}; \; y = \frac{0 + 1 + 1}{3} = \frac{2}{3}; \; A = 0.375$$

$$R_1^{(e)} = R_2^{(e)} = R_3^{(e)} = \frac{1}{3} \cdot \frac{4.25}{3} \times 0.375 = 0.17708$$

element e_3

$$x = \frac{1+1+0.25}{3} = \frac{2.25}{3};\ y = \frac{1+2+1}{3} = \frac{4}{3};\ A = 0.375$$

$$R_1^{(e)} = R_2^{(e)} = R_3^{(e)} = \frac{1}{3}\cdot\frac{6.25}{3}\times 0.375 = 0.26402$$

C. Computation of T

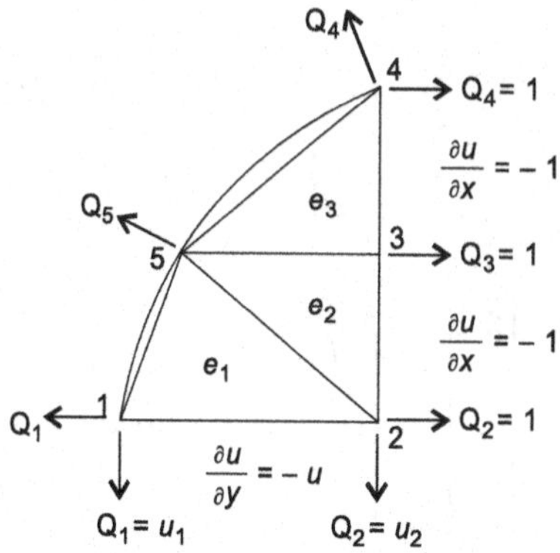

Computing T_i directly in terms of global coordinates:

element e_1

(*i*) along σ_{12}: $\quad T_1^{(e)} = \frac{\sigma_{12}}{6}(2Q_1 + Q_2) = \frac{1}{6}(2u_1 + u_2)$

$$T_2^{(e)} = \frac{\sigma_{12}}{6}(Q_1 + 2Q_2) = \frac{1}{6}(u_1 + 2u_2)$$

(*ii*) along σ_{51}: $\quad T_1^{(e)} = \frac{\sigma_{51}}{6}(2Q_1 + Q_5) = \frac{1.03075}{6}(2Q_1 + Q_5) = 0.17180\ (2Q_1 + Q_5)$

$$T_5^{(e)} = \frac{\sigma_{51}}{6}(Q_1 + 2Q_5) = \frac{1.03075}{6}(Q_1 + 2Q_5) = 0.17180(Q_1 + 2Q_5),$$

$$\sigma_{51} = (1^2 + 0.25^2)^{1/2} = 1.03075.$$

element e_2

along σ_{23}: $\quad T_2^{(e)} = \frac{\sigma_{23}}{6}(2Q_2 + Q_3) = \frac{1}{6}(2+1) = \frac{1}{2}$

$$T_3^{(e)} = \frac{\sigma_{23}}{6}(Q_2 + 2Q_3) = \frac{1}{6}(1+2) = \frac{1}{2}$$

element e_3

(*i*) along σ_{34}: $\quad T_3^{(e)} = \dfrac{\sigma_{34}}{6}(2Q_3 + Q_4) = \dfrac{1}{6}(2+1) = \dfrac{1}{2}$

$$T_4^{(e)} = \frac{\sigma_{34}}{6}(Q_3 + 2Q_4) = \frac{1}{6}(1+2) = \frac{1}{2}$$

(*ii*) along σ_{45}: $\quad T_4^{(e)} = \dfrac{\sigma_{45}}{6}(2Q_4 + Q_5) = \dfrac{1.25}{6}(2Q_4 + Q_5) = 0.20833(2Q_4 + Q_5)$

$$T_5^{(e)} = \frac{\sigma_{45}}{6}(Q_4 + 2Q_5) = \frac{1.25}{6}(Q_4 + 2Q_5) = 0.20833(Q_4 + 2Q_5)$$

$$\sigma_{45} = \{(2-1)^2 + (1-0.25)^2\}^{1/2} = 1.25$$

Assembly table

$i \backslash j$	1	u_2	u_3	u_4	u_5	R	T
1	0.78125	0.40625			− 0.375	0.125	$\frac{1}{6}(2u_1+u_2)$ $0.17180(2Q_1+Q_5)$
2	0.40625	0.53125	− 0.375		− 0.125 0	0.125 + 0.17708	$\frac{1}{6}(u_1+2u_2)$ $+\frac{1}{2}$
3		− 0.375	1.04167 + 1.04167	− 0.375	− 0.5 − 0.66667	0.17708 + 0.26402	$\frac{1}{2}+\frac{1}{2}$
4			− 0.375	0.375	0	0.26402	$\frac{1}{2}$ $0.20833(2Q_4+Q_5)$
5	− 0.375	− 0.125 0	− 0.5 − 0.66667	0	0.5 + 0.375 + 0.66667	0.125 + 0.17708 + 0.26402	$0.17180(Q_1+2Q_5)$ $0.20833(Q_4+2Q_5)$

On the parabola $y^2 = 4x$, u is given as $u = 2x + y$

$\therefore$ at node 1, $u_1 = 0$; at node 4, $u_4 = 4$ and at node 5, $u_5 = 1.5$. $u_2 = ?$, $u_3 = ?$

Using equations 2 and 3 and putting the values u_1, u_4 and u_5, we get

$$0.90625u_2 - 0.375u_3 = 0.125 \times 1.5 - \frac{1}{6}(0 + 2u_2) - 0.5 + 0.30208$$

or $\quad 1.23958u_2 - 0.375u_3 = -0.01042 \qquad$ .... (1)

$$-0.375u_2+2.08334u_3=0.375\times 4+1.16667+0.4375-1$$

or $-0.375u_2+2.08334u_3=2.68750$ .... (2)

$$1.96990u_3=2.68435$$

$$u_3=1.3627,\ u_2=0.4038.$$

Note: (*i*) If required we can find Q_1, Q_4 and Q_5 from equations 1, 4, and 5.

(*ii*) There is singularity in the gradient $\frac{\partial u}{\partial n}$ at the corner nodes 1, 2 and 4.

12.9.3 Node-wise (point-wise) assembly

In the foregoing approach of FEM, the coefficient matrix P is constructed by assembling the terms computed from various elements. The elements are chosen sequentially starting from one and going up to the last. The computed terms over an element are inserted in the appropriate rows and columns of P. In the case of one-dimensional problem a submatrix of dimension (2×2) and in the case of two-dimensional problem a submatrix of dimension (3×3) is obtained from each element and the terms are inserted in the rows and columns in the Assembly Table.

There is however an alternative approach also in which the terms are computed node-wise (point-wise); that is, starting from node one and going up to the last node. The elements associated with the node (apex node) are identified and all the terms over these elements are computed and inserted in the column pertaining to that node and in various rows of the Assembly Table. For example, when dealing with node j, the terms are put in the jth column as coefficient of u_j. Obviously we have to maintain a Connectivity Table identifying the elements with the node. For example, in case of Example 12.15, the table may be as follows:

Table 12.5 Connectivity table

Node	*Elements*
1	e_1
2	e_1, e_2
3	e_2, e_3
4	e_3
5	e_1, e_2, e_3

Mathematically, we can write for one-dimensional problem from (12.70)

$$\text{PU} = \sum_{r=1}^{\text{N}-1} k_r \sum_{j=1}^{\text{N}} u_j \int_{e_r} \frac{\partial \phi_i}{\partial x} \cdot \frac{\partial \phi_j}{\partial x} \cdot dx, \quad i = 1(1)\text{N}$$

$$= \sum_{j=1}^{\text{N}} u_j \sum_{r=1}^{\text{N}-1} k_r \int_{e_r} \frac{\partial \phi_i}{\partial x} \cdot \frac{\partial \phi_j}{\partial x} \cdot dx. \tag{12.112}$$

Thus, the coefficient of u_j in the ith row is given by

$$\text{p}_{ij} = \sum_{r=1}^{\text{N}-1} k_r \int_{e_r} \frac{\partial \phi_i}{\partial x} \cdot \frac{\partial \phi_j}{\partial x} \cdot dx. \tag{12.113}$$

The summation in (12.113) will be taken over the elements to which node j is common. In case of one dimension it will consist of two elements only, viz., $(j-1)^{\text{th}}$ and j^{th} elements, i.e., e_{j-1} and e_j. It may be remembered that $\phi_j(x) = \dfrac{x - x_{j-1}}{x_j - x_{j-1}}$ over e_{j-1} and $\phi_j(x) = \dfrac{x - x_{j+1}}{x_j - x_{j+1}}$ over e_j.

Similarly, in case of two dimensions we can have from (12.98)

$$\text{PU} = \sum_{r=1}^{\text{M}} k_r \sum_{j=1}^{\text{N}} u_j \iint_{e_r} \left(\frac{\partial \phi_i}{\partial x} \cdot \frac{\partial \phi_j}{\partial x} + \frac{\partial \phi_i}{\partial y} \cdot \frac{\partial \phi_j}{\partial y} \right) dxdy, \; i = 1(1)\,\text{N}.$$

$$= \sum_{j=1}^{\text{N}} u_j \sum_{r=1}^{\text{M}} k_r \iint_{e_r} \left(\frac{\partial \phi_i}{\partial x} \cdot \frac{\partial \phi_j}{\partial x} + \frac{\partial \phi_i}{\partial y} \cdot \frac{\partial \phi_j}{\partial y} \right) dxdy \tag{12.114}$$

The coefficient of u_j in the ith row will be given by

$$\text{P}_{ij} = \sum_{r=1}^{\text{M}} k_r \iint_{e_r} \left(\frac{\partial \phi_i}{\partial x} \cdot \frac{\partial \phi_j}{\partial x} + \frac{\partial \phi_i}{\partial y} \cdot \frac{\partial \phi_j}{\partial y} \right) dxdy. \tag{12.115}$$

The summation in (12.115) has to be taken over the elements to which node j is common.

12.9.4 Higher order elements

An element is linear, quadratic or cubic, etc., according to the degree of the polynomial approximated over it. We have considered only linear elements so far. Let us give a brief description of the higher order elements.

(*i*) In One Dimension

Suppose we want to approximate the function $u(x)$ by a quadratic over the elements, then there should be available three points over an element. If there are N elements each containing 3 nodes then, there should be $(2N+1)$ nodes, say having coordinate x_i, $i = 1(1)$ $2N+1$. Each element $e_{r/2}$, $r = 2, 4, \ldots 2N$, contains three nodes $r-1$, r and $r+1$. The function that interpolates $u(x)$ at these nodes may be expressed as

$$u(x) = \phi_{r-1}(x)u_{r-1} + \phi_r(x)u_r + \phi_{r+1}(x)u_{r+1}, \quad x_{r-1} \le x \le x_{r+1} \tag{12.116}$$

where ϕ_{r-1}, ϕ_r and ϕ_{r+1} are Lagrange's coefficients given by

$$\phi_{r-1}(x) = \frac{(x-x_r)(x-x_{r+1})}{(x_{r-1}-x_r)(x_{r-1}-x_{r+1})}, \quad \phi_r(x) = \frac{(x-x_{r-1})(x-x_{r+1})}{(x_r-x_{r-1})(x_r-x_{r+1})}$$

$$\phi_{r+1}(x) = \frac{(x-x_{r-1})(x-x_r)}{(x_{r+1}-x_{r-1})(x_{r+1}-x_r)}. \tag{12.117}$$

Assume that node r is chosen as midpoint of x_{r-1} and x_{r+1} so that

$$x_{r+1} - x_r = x_r - x_{r-1} = h_r \text{ (say)}. \tag{12.118}$$

Then the Lagrange's coefficients (shape functions) (12.117) may be written as

$$\phi_{r-1} = \frac{(x-x_r)(x-x_{r+1})}{2h_r^2}, \quad \phi_r = \frac{(x-x_{r-1})(x-x_{r+1})}{-h_r^2}, \quad \phi_{r+1} = \frac{(x-x_{r-1})(x-x_r)}{2h_r^2}. \tag{12.119}$$

We can normalise the interval $x_{r-1} \le x \le x_{r+1}$ to either $[-1, 1]$ or to $[0, 1]$. For normalising to interval $[-1, 1]$, we make the following change of variable, shifting the origin to midpoint of x_{r-1} and x_{r+1}, i.e., x_r,

$$\xi = \frac{x-x_r}{x_{r+1}-x_r} = \frac{x-x_r}{h_r}, \quad -1 \le \xi \le 1. \tag{12.120}$$

The transformation (12.120) reduces the shape functions (12.119) to the normalised shape functions

$$\phi_1(\xi) = \frac{1}{2}\xi(\xi-1); \; \phi_2(\xi) = -(\xi+1)(\xi-1); \; \phi_3(\xi) = \frac{1}{2}\xi(\xi+1), \; -1 \le \xi \le 1. \tag{12.121}$$

For normalising the interval $[x_{r-1}, x_{r+1}]$ to $[0, 1]$ we shift the origin to x_{r-1} and divide by the total length, i.e., put

$$\xi = \frac{x-x_{r-1}}{x_{r+1}-x_{r-1}} = \frac{x-x_{r-1}}{2h_r}, \quad 0 \le \xi \le 1.$$

Consequently the shape functions become

$$\phi_1(\xi) = (2\xi - 1)(\xi - 1); \ \ \phi_2(\xi) = -4\xi(\xi - 1); \ \ \phi_3(\xi) = \xi(2\xi - 1), \ \ 0 \le \xi \le 1. \quad (12.122)$$

The normalised shape functions (12.121) may be converted in terms of natural coordinate (ξ_1, ξ_2). Assuming $\xi_1 + \xi_2 = 1$ when coordinates of x_{r-1} and x_{r+1} are respectively (1, 0) and (0, 1). We see that $-1 \le \xi \le 1$ and $0 \le \frac{\xi + 1}{2} < 1$ so let $\xi_2 = \frac{\xi + 1}{2}$ giving $\xi_1 = 1 - \xi_2 = \frac{1 - \xi}{2}$.

Putting these values transforms (12.121) to

$$\phi_1(\xi_1, \xi_2) = \xi_1(2\xi_1 - 1); \ \ \phi_2(\xi_1, \xi_2) = 4\xi_1\xi_2; \ \ \phi_3(\xi_1, \xi_2) = \xi_2(2\xi_2 - 1), \quad (12.123)$$
$$\xi_1 + \xi_2 = 1.$$

Similarly by putting $\xi_2 = \xi$ and $\xi_1 = 1 - \xi_2 = 1 - \xi$ converts (12.122) immediately to (12. 123) given above.

The graphs of ϕ_1, ϕ_2 and ϕ_3 are shown in Fig. 12.8.

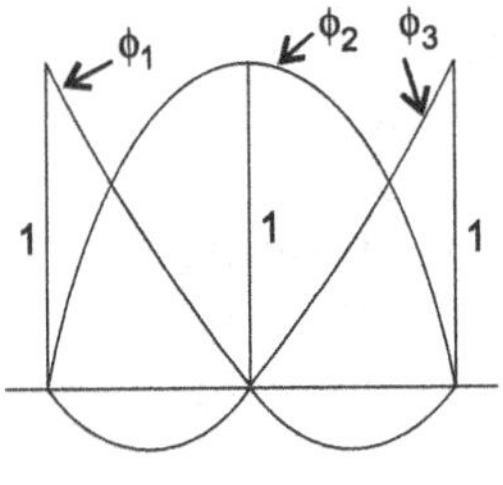

Figure 12.8 Quadratic shape functions in 1−D.

There would be no complexity in solving the problem except that submatrix $P_r^{(e)}$ will be of order 3 and column vectors R and T will have dimensions (3×1) each.

(*ii*) **In Two Dimensions**

We have considered linear variation of u over the triangular element in two dimensions also, i.e., $u(x, y) = ax + by + c$. A general second order approximation may be of the form

$$u(x, y) = \alpha x^2 + \beta y^2 + \gamma xy + ax + by + c. \quad (12.124)$$

As the quadratic expression (12.124) contains six unknowns, there must be available six nodes over the triangular element where function values must be satisfied. Thus in addition to three nodes at the vertices another three nodes are needed. An obvious choice would be the midpoints of the sides of the triangular element. The six nodes are shown over an element e in Fig. 12.9.

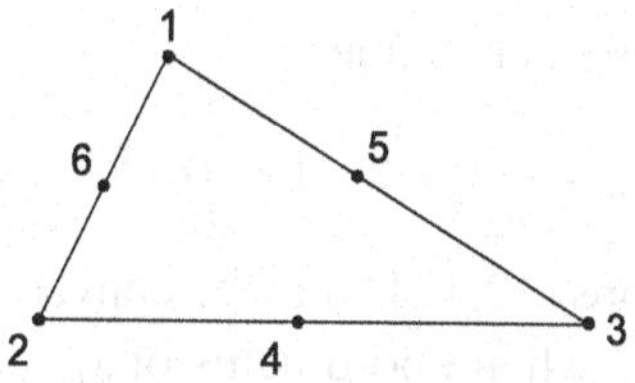

Figure 12.9 Triangular element with six nodes

In the above case function $u(x, y)$ may be approximated over the element e in terms of shape functions as

$$u(x, y) = \sum_{r=1}^{6} \phi_r(x, y) u_r. \qquad (12.125)$$

The shape function $\phi_i(x, y)$ fulfils the required property that its value is 1 at the node i and 0 at the other nodes for $i = 1(1)6$. For example $\phi_1(x, y)$ will vary quadratically along the sides 1–6–2 and 1–5–3 whereas it will have zero value along the side 2–4–3.

In terms of natural coordinates (ξ_1, ξ_2, ξ_3), we should note that ξ_3 will be zero along the side 1–6–2 and the coordinates of node 1 and node 2 will be (1, 0, 0) and (0, 1, 0) respectively. From (12.123) we can write

$$\phi_1(\xi_1, \xi_2) = \xi_1(2\xi_1 - 1) \text{ along side } 1\text{–}6\text{–}2$$

$$\phi_1(\xi_1, \xi_3) = \xi_1(2\xi_1 - 1) \text{ along side } 1\text{–}5\text{–}3$$

Thus we can express $\phi_1(\xi_1, \xi_2, \xi_3)$ over the element as

$$\phi_1(\xi_1, \xi_2, \xi_3) = \xi_1(2\xi_1 - 1) \qquad (12.126a)$$

Similarly ϕ_2 and ϕ_3 may be expressed as

$$\phi_2(\xi_1, \xi_2, \xi_3) = \xi_2(2\xi_2 - 1) \qquad (12.126b)$$

$$\phi_3(\xi_1, \xi_2, \xi_3) = \xi_3(2\xi_3 - 1). \qquad (12.126c)$$

The shape functions corresponding to the midpoint nodes 4, 5 and 6 may also be expressed using (12.123) as

$$\phi_4(\xi_1, \xi_2, \xi_3) = 4\xi_2\xi_3 \qquad (12.127a)$$

$$\phi_5(\xi_1, \xi_2, \xi_3) = 4\xi_1\xi_3 \qquad (12.127b)$$

$$\phi_6(\xi_1, \xi_2, \xi_3) = 4\xi_1\xi_2 \tag{12.127c}$$

In this case the submatrix $P_r^{(e)}$ will be of order (6×6) and R and T of order (6×1) each over each element.

This procedure can be further extended to cubic variation of u over a triangular element. Two extra points may be selected on each side of the triangle besides the vertices; thus making altogether nine nodes. But it should be noted that a general cubic expression in x and y will consist of 10 unknowns. Therefore one more node will be required to uniquely determine the 10 unknowns. This node may be chosen inside the triangle – may be centroid. However, it may be worth mentioning that a higher order approximate does not always guarantee better accuracy.

12.9.5 Element of rectangular shape

So far we have dealt with only two types of elements-one, linear (straight line) in one-dimensional problem and triangular in two-dimensional problem. Nothing can be done as regards the choice of the shape of the element as far as 1–D is concerned. However, in 2–D we could have chosen any shape of the element, other than a triangle, depending on the configuration of the domain or otherwise. As stated earlier, the freedom in selecting arbitrary shapes of the elements makes the FEM more competitive as compared to the conventional finite difference methods. The shapes may include curved boundaries in an element and of course three-dimensional element when dealing with 3–D problem. Here we will discuss element of rectangular shape whose sides are parallel to the axes.

As there are four nodes in a rectangular element (see Fig. 12.10) the interpolating function should contain only four unknowns.

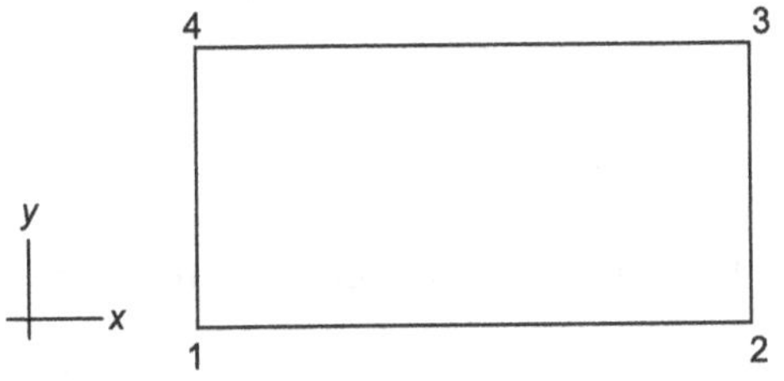

Figure 12.10 Rectangular element.

The most suitable representation would be

$$u(x, y) = ax + by + c + dxy, \tag{12.128}$$

where a, b, c, d are four parameters to be satisfied at the nodes.

The expression approximating u in (12.128) is known as bilinear form since it is linear in x (for constant y) and is also linear in y (for constant x). It is also a form which will be obtained by multiplying a linear function in x, say $\alpha_1 x+\beta_1$ and a linear function in y, say $\alpha_2 y+\beta_2$. Instead of substituting the coordinates of four nodes and solving for a, b, c and d we proceed by expressing u in terms of shape functions as

$$u(x,\ y)=\phi_1 u_1+\phi_2 u_2+\phi_3 u_3+\phi_4 u_4 \tag{12.129}$$

where $\phi_i=1$ at node i and zero at the other nodes, $i=1(1)4$.

We use normalised coordinates $(\xi,\ \eta)$ such that the origin is shifted to $(x_1,\ y_1)$ and the sides of the rectangle are transformed to that unit lengths. That is, we change the variable by putting

$$\xi=\frac{x-x_1}{x_2-x_1};\quad \eta=\frac{y-y_1}{y_4-y_1}. \tag{12.130}$$

Thus, the transformation (12.130) transforms the rectangle to a square of unit length in the ξ-η plane such that coordinates of the nodes 1, 2, 3 and 4 become $(0,\ 0)$, $(1,\ 0)$, $(1,\ 1)$ and $(0,\ 1)$ respectively. Hence along the side $1-2$, i.e., $\eta=0$

$$\phi_1(0,\ 0)=1 \text{ and } \phi_1(1,\ 0)=0$$

$$\phi_2(0,\ 0)=0 \text{ and } \phi_2(1,\ 0)=1$$

This suggests,

$$\phi_1(\xi,\ 0)=1-\xi \text{ and } \phi_2(\xi,\ 0)=\xi \text{ on side } 1-2. \tag{12.131}$$

Similarly along the side $1-4$

$$\phi_1(0,\ \eta)=1-\eta \text{ and } \phi_4(0,\ \eta)=\eta. \tag{12.132}$$

From (12.131) and (12.132), the bilinear expression for $\phi_1(\xi,\ \eta)$ will be given by

$$\phi_1(\xi,\ \eta)=(1-\xi)(1-\eta). \tag{12.133a}$$

The expressions for other shape functions may also be written as

$$\phi_2(\xi,\ \eta)=\xi(1-\eta);\quad \phi_3(\xi,\ \eta)=\xi\eta;\quad \phi_4(\xi,\ \eta)=(1-\xi)\eta. \tag{12.133b}$$

Alternatively the origin can be shifted to the centre of the rectangle and normalise the length to 2.

Since $u(x, y)$ varies linearly over the edges of the rectangle it can be adjoined with a triangular element such that one side is common in both. When two elements, same shape or different shapes, are adjoined having a common edge, they are called 'conforming' if the function is determined uniquely along the common edge, i.e., its value computed from the two elements remains same. As already mentioned there are various kinds of transformations and different types of elements which have been used in solving problems by FEM. The interested reader may refer to [7], [3].

12.9.6 Parabolic equation (one dimension)

Let us consider a general one-dimensional parabolic equation

$$\frac{\partial u}{\partial t} = \frac{\partial}{\partial x}\left(k_x \frac{\partial u}{\partial x}\right) + q(x), \quad a \le x \le b, \tag{12.134}$$

where $u(x, t)$ denotes temperature/concentration (in general potential) at a distance x at time t; $q(x)$ is a known function of x representing absorption/generation (sink/source) at x; k_x is taken to be conductivity of the medium which is also variable.

It is assumed that appropriate boundary conditions are prescribed at the end points $x = a$, $x = b$ and an initial condition at $t = 0$.

We subdivide the domain $a \le x \le b$ into say (N−1) elements, not necessarily of equal lengths by selecting the N number of nodes at $x = x_r$, $r = 1(1)\mathrm{N}$ such that $a = x_1 < x_2 \ldots. < x_\mathrm{N} = b$.

At any instant t, let the solution to (12.134) be approximated by

$$\bar{u} = \phi_1(x)u_1(t) + \phi_2(x)u_2(t) + \ldots. + \phi_\mathrm{N}(x)u_\mathrm{N}(t) \tag{12.135}$$

where $\phi_1(x)$, $\phi_2(x)$ etc are shape functions. Using Galerkin's criterion we form N equations as given below

$$\int_a^b \left\{\frac{\partial \bar{u}}{\partial t} - \frac{\partial}{\partial x}\left(k_x \frac{\partial \bar{u}}{\partial x}\right) - q(x)\right\} \phi_i(x)dx = 0, \; i = 1(1)\;\mathrm{N}. \tag{12.136a}$$

$$\text{or} \quad \int_a^b \frac{\partial \bar{u}}{\partial t} \cdot \phi_i(x)dx - \int_a^b \phi_i(x) \cdot \frac{\partial}{\partial x}\left(k_x \frac{\partial \bar{u}}{\partial x}\right) dx = \int_a^b q(x)\phi_i(x),$$

$$i = 1,\ 2,\ \ldots.\ \mathrm{N}. \tag{12.136b}$$

As per FEM procedure we perform the integration in (12.136b) element-wise over (N−1) elements then assemble the contributions to various terms. Consider the first term on the left side of (12.136b).

$$\int_a^b \phi_i(x)\frac{\partial \bar{u}}{\partial t}dx = \sum_{r=1}^{N-1}\int_{x_r}^{x_{r+1}} \phi_i(x)\frac{\partial}{\partial t}\sum_{j=1}^{N}\phi_j(x)u_j dx$$

$$= \sum_{r=1}^{N-1}\sum_{j=1}^{N}\dot{u}_j\int_{e_r}\phi_i(x)\phi_j(x)dx, \tag{12.137}$$

where $\dot{u}_j = \dfrac{\partial u_j}{\partial t}$ and e_r is the rth element extending between x_r and x_{r+1}.

The only contribution from the element e_r would be when $i = r,\ r+1$ and $j = r,\ r+1$ since other shape functions have zero value over e_r. We have for element e_r

$$\phi_r(x) = \frac{x - x_{r+1}}{x_r - x_{r+1}} = -\frac{1}{L_r}(x - x_{r+1})$$

$$\phi_{r+1}(x) = \frac{x - x_r}{x_{r+1} - x_r} = \frac{1}{L_r}(x - x_r),\ L_r = x_{r+1} - x_r.$$

$$\int_{x_r}^{x_{r+1}} \phi_r^2(x)dx = \frac{1}{L_r^2}\int_{x_r}^{x_{r+1}}(x - x_{r+1})^2 dx = \frac{L_r}{3}$$

$$\int_{x_r}^{x_{r+1}} \phi_r(x)\phi_{r+1}dx = -\frac{1}{L_r^2}\int_{x_r}^{x_{r+1}}(x - x_{r+1})(x - x_r)dx = \frac{L_r}{6}$$

$$\int_{x_r}^{x_{r+1}} \phi_{r+1}^2(x)dx = \frac{1}{L_r^2}\int_{x_r}^{x_{r+1}}(x - x_r)^2 dx = \frac{L_r}{3}.$$

Thus we get the coefficients of $\dot{u}_r$ and $\dot{u}_{r+1}$ in the rth, $(r+1)^{\text{th}}$ rows and r^{th} and $(r+1)^{\text{th}}$ columns as shown below:

	$\dot{u}_r$	$\dot{u}_{r+1}$
	r^{th} col.	**$(r+1)^{\text{th}}$ col.**
r^{th} row	$\frac{L_r}{3}$	$\frac{L_r}{6}$
$(r+1)^{\text{th}}$ row	$\frac{L_r}{6}$	$\frac{L_r}{3}$

(12.138)

For computing the second term on the left side of (12.136b), referring to Sec 12.9.1

$$\int_a^b \phi_i(x)\frac{\partial}{\partial x}\left(k_r\frac{\partial \bar{u}}{\partial x}\right)dx = \left[\phi_i(x)k_x\frac{\partial \bar{u}}{\partial x}\right]_{x=a}^{b} - \int_a^b k_x\frac{\partial \phi_i(x)}{\partial x}\cdot\frac{\partial \bar{u}}{\partial x}dx$$

$$= \phi_i(a)\mathrm{Q}_a - \phi_i(b)\mathrm{Q}_b - \int_a^b k_x\frac{\partial \phi_i(x)}{\partial x}\cdot\frac{\partial \bar{u}}{\partial x}dx$$

$$\text{where}\quad \mathrm{Q}_x = -k_x\frac{\partial u}{\partial x}.$$

$$= \phi_i(x_1)\mathrm{Q}_a - \phi_i(x_\mathrm{N})\mathrm{Q}_b - \sum_{r=1}^{\mathrm{N}-1} k_r \sum_{j=1}^{\mathrm{N}} u_j \int_{x_r}^{x_{r+1}} \frac{\partial \phi_i}{\partial x}\cdot\frac{\partial \phi_j}{\partial x}dx, \quad (12.139)$$

$i = 1(1)\ \mathrm{N}.$

As explained in Sec 12.9.1, the contribution of the summation of the integral over j from element e_r will be as follows.

$$\begin{array}{lcc} & \dot{u}_{\mathbf{r}} & \dot{u}_{\mathbf{r}+1} \\ & \boldsymbol{r^{\text{th}}}\ \textbf{col} & \boldsymbol{(r+1)^{\text{th}}}\ \textbf{col} \\ \boldsymbol{r^{\text{th}}}\ \textbf{row} & \dfrac{k_r}{\mathrm{L}_r} & -\dfrac{k_r}{\mathrm{L}_r} \\ \boldsymbol{(r+1)^{\text{th}}}\ \textbf{row} & -\dfrac{k_r}{\mathrm{L}_r} & \dfrac{k_r}{\mathrm{L}_r} \end{array} \quad (12.140)$$

The right side of (12.136b) is

$$\int_a^b q(x)\phi_i(x)dx = \sum_{r=1}^{\mathrm{N}-1} q_r \int_{x_r}^{x_{r+1}} \phi_i(x)dx, \quad i = 1(1)\ \mathrm{N}. \quad (12.141)$$

The contribution of the element e_r in the rth and $(r+1)^{\text{th}}$ row, after taking value of $q(x)$ to be constant over the element, (i.e., q_r) will be

$$\begin{array}{ll} \boldsymbol{r^{\text{th}}}\ \textbf{row} & \dfrac{1}{2}q_r\mathrm{L}_r \\ \boldsymbol{(r+1)^{\text{th}}}\ \textbf{row} & \dfrac{1}{2}q_r\mathrm{L}_r \end{array} \quad (12.142)$$

It may be noted that

$$\phi_i(x_1) = 1 \text{ for } i = 1 \text{ and } 0 \text{ otherwise.} \tag{12.143a}$$

$$\phi_i(x_N) = 1 \text{ for } i = N \text{ and } 0 \text{ otherwise.} \tag{12.143b}$$

Rewriting (12.136b) using summation over elements

$$\sum_{r=1}^{N-1}\sum_{j=1}^{N} \dot{u}_j \int_{e_r} \phi_i(x)\phi_j(x)dx + \sum_{r=1}^{N-1} k_r \sum_{j=1}^{N} u_j \int_{e_r} \frac{\partial \phi_i}{\partial x}\cdot\frac{\partial \phi_j}{\partial x}dx$$

$$= \sum_{r=1}^{N-1} q_r \int_{e_r} \phi_i(x)dx + \{\phi_i(x_1)Q_a - \phi_i(x_N)Q_b\}, \quad i = 1(1)N, \tag{12.144}$$

where element e_r extends from x_r to x_{r+1}, $r = 1, 2, \ldots$. N–1. We can express (12.144) in matrix form as

$$\mathbf{A}\dot{\boldsymbol{u}} + \mathbf{B}\boldsymbol{u} = \mathrm{R} + \mathrm{T} \tag{12.145}$$

where A and B are square matrices of order N and R and T are column matrices with N components. Let us write down the elements of these matrices for N = 4 using (12.138), (12.140), (12.142) and (12.143a and b),

$$A = \frac{1}{6}\begin{bmatrix} 2L_1 & L_1 & 0 & 0 \\ L_1 & 2(L_1+L_2) & L_2 & 0 \\ 0 & L_2 & 2(L_2+L_3) & L_3 \\ 0 & 0 & L_3 & 2L_3 \end{bmatrix} \tag{12.146a}$$

$$B = \begin{bmatrix} \dfrac{k_1}{L_1} & -\dfrac{k_1}{L_1} & 0 & 0 \\ -\dfrac{k_1}{L_1} & \dfrac{k_1}{L_1}+\dfrac{k_2}{L_2} & -\dfrac{k_2}{L_2} & 0 \\ 0 & -\dfrac{k_2}{L_2} & \dfrac{k_2}{L_2}+\dfrac{k_3}{L_3} & -\dfrac{k_3}{L_3} \\ 0 & 0 & -\dfrac{k_3}{L_3} & \dfrac{k_3}{L_3} \end{bmatrix} \tag{12.146b}$$

$$R = \frac{1}{2}\begin{bmatrix} q_1L_1 \\ q_1L_1 + q_2L_2 \\ q_2L_2 + q_3L_3 \\ q_3L_3 \end{bmatrix}, T = \begin{bmatrix} Q_a \\ 0 \\ 0 \\ -Q_b \end{bmatrix}. \tag{12.146c}$$

We find the solution in a step-by-step manner choosing a time step Δt, say. Let us denote by u_i^n, the value of u at node $i(x = x_i)$ at time $t = n\Delta t$. Let us suppose that the values are known at the nth time level, i.e., $t = n\Delta t$. Then we are required to compute the values of u at the $(n+1)^{\text{th}}$ level, i.e., u_i^{n+1}, $i = 1(1)$ N. We can replace the time derivative $\dot{\boldsymbol{u}}$ in (12.145) in any of the following manners:

(*i*) by forward difference at $(i,\ n)$

(*ii*) by backward difference at $(i,\ n+1)$

(*iii*) by central difference at $(i,\ n+\frac{1}{2})$—C–N Scheme.

However, even if we use (*i*) we can not compute u_i's explicitly—we shall be required to solve a set of simultaneous equations. Similarly we have to solve a system of equations in (*ii*) and (*iii*). But due to stability and smaller truncation error in central difference replacement we use (*iii*) Crank–Nicolson scheme. Hence using C–N scheme we get from (12.145).

$$\frac{1}{\Delta t}\, A\, (\boldsymbol{u}^{n+1} - \boldsymbol{u}^n) + \frac{1}{2} B(\boldsymbol{u}^{n+1} + \boldsymbol{u}^n) = R + T$$

$$\text{or} \quad (2A + \Delta t B)\boldsymbol{u}^{n+1} - (2A - \Delta t B)\boldsymbol{u}^n = 2\Delta t(R + T)$$

$$\text{or} \quad (2A + \Delta t B)\boldsymbol{u}^{n+1} = (2A - \Delta t B)\boldsymbol{u}^n + 2\Delta t(R + T). \tag{12.147}$$

With right side known, the system of equations (12.147) can be solved for $\boldsymbol{u}^{n+1}$, giving values of u at $(n+1)^{\text{th}}$ time level.

The equations (12.147) may be written in expanded form as follows:

$$\left(2L_1 + \frac{3k_1\Delta t}{L_1}\right) u_1^{n+1} + \left(L_1 - \frac{3k_1\Delta t}{L_1}\right) u_2^{n+1}$$

$$= \left(2L_1 - \frac{3k_1\Delta t}{L_1}\right) u_1^n + \left(L_1 + \frac{3k_1\Delta t}{L_1}\right) u_2^n + 6\Delta t(R_1 + Q_a)$$

$$\left(\mathrm{L}_{i-1} - \frac{3k_{i-1}\Delta t}{\mathrm{L}_{i-1}}\right) u_{i-1}^{n+1} + \left\{2(\mathrm{L}_{i-1}+\mathrm{L}_i) + 3\Delta t\left(\frac{k_{i-1}}{\mathrm{L}_{i-1}} + \frac{k_i}{\mathrm{L}_i}\right)\right\} u_i^{n+1} + \left(\mathrm{L}_i - \frac{3k_i\Delta t}{\mathrm{L}_i}\right) u_{i+1}^{n+1}$$

$$= \left(\mathrm{L}_{i-1} - \frac{3k_{i-1}\Delta t}{\mathrm{L}_{i-1}}\right) u_{i-1}^{n} + \left\{2(\mathrm{L}_{i-1}+\mathrm{L}_i) - 3\Delta t\left(\frac{k_{i-1}}{\mathrm{L}_{i-1}} + \frac{k_i}{\mathrm{L}_i}\right)\right\} u_i^{n} + \left(\mathrm{L}_i + \frac{3k_i\Delta t}{\mathrm{L}_i}\right) u_{i+1}^{n} +$$
$$6\Delta t\mathrm{R}_i \quad i = 2,\ 3,\ \ldots .\ \mathrm{N}-1.$$

$$\left(\mathrm{L}_{\mathrm{N}-1} - \frac{3k_{\mathrm{N}-1}\Delta t}{\mathrm{L}_{\mathrm{N}-1}}\right) u_{\mathrm{N}-1}^{n+1} + \left(2\mathrm{L}_{\mathrm{N}-1} + \frac{3k_{\mathrm{N}-1}\Delta t}{\mathrm{L}_{\mathrm{N}-1}}\right) u_{\mathrm{N}}^{n+1}$$

$$= \left(\mathrm{L}_{\mathrm{N}-1} + \frac{3k_{\mathrm{N}-1}\Delta t}{\mathrm{L}_{\mathrm{N}-1}}\right) u_{\mathrm{N}-1}^{n} + \left(2\mathrm{L}_{\mathrm{N}-1} + \frac{3k_{\mathrm{N}-1}\Delta t}{\mathrm{L}_{\mathrm{N}-1}}\right) u_{\mathrm{N}}^{n} + 6\Delta t(\mathrm{R}_{\mathrm{N}} - \mathrm{Q}_b) \tag{12.148}$$

where R_i and T_i, etc., are the ith elements of R and T respectively. Let us consider the equation in normalised form and without source/sink term, i.e.,

$$\frac{\partial u}{\partial t} = \frac{\partial^2 u}{\partial x^2}, \quad a \le x \le b, \quad t > 0 \tag{12.149}$$

If all the elements are taken of the same length Δx where $\Delta x = (b-a)/(\mathrm{N}-1)$ and put $r = \Delta t/\Delta x^2$, then above equations can be written as

$$\left.\begin{aligned}
&(2+3r)u_1^{n+1} + (1-3r)u_2^{n+1} = (2-3r)u_1^n + (1+3r)u_2^n + 6r\Delta x\mathrm{Q}_a \\
&(1-3r)u_{i-1}^{n+1} + 2(2+3r)u_i^{n+1} + (1-3r)u_{i+1}^{n+1} \\
&\qquad\qquad = (1+3r)u_{i-1}^n + 2(2-3r)u_i^n + (1+3r)u_{i+1}^n, \\
&\qquad\qquad\qquad i = 2,\ 3,\ \ldots .\ \mathrm{N}-1 \\
&(1-3r)u_{\mathrm{N}-1}^{n+1} + (2+3r)u_{\mathrm{N}}^{n+1} = (1+3r)u_{\mathrm{N}-1}^n + (2-3r)u_{\mathrm{N}}^n - 6r\Delta x\mathrm{Q}_b.
\end{aligned}\right\} \tag{12.150}$$

Example 12.16

Given a parabolic equation

$$\frac{\partial u}{\partial t} = \frac{\partial^2 u}{\partial x^2}, \quad 0 \le x \le 1.0, \quad t > 0$$

with boundary conditions $u(0, t) = 0$, $u(1, t) = 0$ and initial condition $u(x, 0) = 2x(1-x)$. Divide the interval into four elements of equal length. Taking $\Delta t = 0.0625$, write down the relevant finite element equations. Hence find the solution at first time level. Also compute flux at the end points $x = 0$ and $x = 1$.

Solution $\Delta x = \dfrac{1-0}{4} = 0.25$; $N = 5$; $r = \Delta t/\Delta x^2 = 1$.

Values of $u(x, t)$ given at $t = 0$

x	0	0.25	0.50	0.75	1.0
u	0	0.375	0.5	0.375	0
Node	1	2	3	4	5

Using equations (12.150)

$$\begin{aligned}
5u_1^1 - 2u_2^1 &= -u_1^0 + 4u_2^0 + 6 \times 0.25 \times Q_a \\
-2u_1^1 + 10u_2^1 - 2u_3^1 &= 4u_1^0 - 2u_2^0 + 4u_3^0 \\
-2u_2^1 + 10u_3^1 - 2u_4^1 &= 4u_2^0 - 2u_3^0 + 4u_4^0 \\
-2u_3^1 + 10u_4^1 - 2u_5^1 &= 4u_3^0 - 2u_4^0 + 4u_5^0 \\
-2u_4^1 + 5\,u_5^1 &= 4u_4^0 - u_5^0 - 6 \times 0.25 \times Q_b
\end{aligned}$$

Using $u_1 = u_5 = 0$ and also $u_2 = u_4$ due to symmetry we get from equations 2, 3 and 4

$$\begin{aligned}
5\,u_2^1 - u_3^1 &= 2 \times 0 - 0.375 + 2 \times 0.5 = 0.625 \\
-u_2^1 + 5u_3^1 - u_4^1 &= 2 \times 0.375 - 0.5 + 2 \times 0.375 = 1.0 \\
-u_3^1 + 5u_4^1 &= 2 \times 0.5 - 0.375 + 2 \times 0 = 0.625
\end{aligned}$$

On solving, $u_2 = u_4 = 0.1793$, $u_3 = 0.2717$
For computing flux at $x = 0$, use first equation

$$-2u_2^1 = 4 \times 0.375 + 1.5Q_a$$

or $$1.5Q_a = -2 \times 0.1793 - 1.5 \text{ giving } Q_a = -1.2391$$

For flux at $x = 1$ use last equation

$$-2u_4 = 4 \times 0.1793 - 6 \times 0.25 \times Q_b$$

or $$Q_b = 1.2391$$

Note: If finite differences are used, we get from C–N scheme

$$u_2 = u_4 = 0.1964, \; u_3 = 0.2857$$

Example 12.17

In Example 12.16, select the nodes at $x = 0,\ 0.2,\ 0.5,\ 0.8,\ 1.0$ and compute u with $\Delta t = 0.0625$.

Solution

Node	1	2	3	4	5
x	0	0.2	0.5	0.8	1.0
u	0	0.32	0.5	0.32	0

The problem is symmetrical about the midpoint as the location of the nodes is same and boundary conditions are same. Hence $u_2 = u_4$ for all times and we are required to compute u_2 and u_3 only. Using equations (12.148) with $L_1 = 0.2$, $L_2 = 0.3$, $L_3 = 0.3$ and $L_4 = 0.2$; $\Delta t = 0.0625$

For $i = 2$

$$\left(0.2 - \frac{3\times 0.0625}{0.2}\right)u_1 + \left\{2(0.2+0.3) + 3\times 0.0625\left(\frac{1}{0.2}+\frac{1}{0.3}\right)\right\}u_2$$

$$+\left(0.3 - \frac{3\times 0.0625}{0.3}\right)u_3$$

$$= \left(0.2 - \frac{3\times 0.0625}{0.2}\right)\times 0 + \left\{2(0.2+0.3) - 3\times 0.0625\left(\frac{1}{0.2}+\frac{1}{0.3}\right)\right\}\times 0.32$$

$$+\left(0.3 + \frac{3\times 0.0625}{0.3}\right)\times 0.5$$

or $\quad (0.2-0.9375)u_1 + (1+1.5625)u_2 + (0.3-0.625)u_3$

$$= (0.2+0.9375)\times 0 + (1-1.5625)\times 0.32 + (0.3+0.625)\times 0.5$$

or $\quad -0.7375u_1 + 2.5625u_2 - 0.325u_3 = 0.2825$ .... (*i*)

for $i = 3$

$$\left(0.3 - \frac{3\times 0.0625}{0.3}\right)u_2 + \left\{2(0.3+0.3) + 3\times 0.0625\left(\frac{1}{0.3}+\frac{1}{0.3}\right)\right\}u_3$$

$$+\left(0.3 - \frac{3\times 0.0625}{0.3}\right)u_4$$

$$= \left(0.3 + \frac{3\times 0.0625}{0.3}\right)\times 0.32 + \left\{2(0.3+0.3) - 3\times 0.0625\left(\frac{1}{0.3}+\frac{1}{0.3}\right)\right\}\times 0.5$$

$$+\left(0.3 - \frac{3\times 0.0625}{0.3}\right)\times 0.32$$

or $\quad -0.325u_2 + 2.45u_3 - 0.325u_4 = 0.567$ .... (*ii*)

Using $u_1 = 0$ and $u_2 = u_4$ equations (*i*) and (*ii*) become

$$2.5625\, u_2 - 0.325u_3 = 0.2825 \qquad \text{.... }(iii)$$

$$-0.65u_2 + 2.45u_3 = 0.567 \qquad \text{.... }(iv)$$

Solving (*iii*) and (*iv*)

$$u_2 = u_4 = 0.1654;\ u_3 = 0.2698.$$

Note: Q_a and Q_b can be computed from the first and the last equations of (12.148).

12.9.7 Parabolic equation (two dimensions)

Let us consider a normalised two-dimensional parabolic equation

$$\frac{\partial u}{\partial t} = \frac{\partial^2 u}{\partial x^2} + \frac{\partial^2 u}{\partial y^2} \tag{12.151}$$

defined over a domain R enclosed by a curve S. It is assumed that appropriate boundary conditions are prescribed on S and an initial condition is given at $t = 0$.

To apply FEM let us suppose that domain R has been subdivided into M triangular elements consisting of N number of nodes. We assume an approximate solution

$$\bar{u} = \phi_1(x,\ y)u_1(t) + \phi_2(x,\ y)u_2(t) + \ldots. + \phi_{\mathrm{N}}(x,\ y)u_{\mathrm{N}}(t) \tag{12.152}$$

where $\phi_1,\ \phi_2 \ldots$, etc. are shape functions defined at time t. Applying Galerkin's criterion,

$$\iint_{\mathrm{D}} \left\{ \frac{\partial \bar{u}}{\partial t} - \left(\frac{\partial^2 \bar{u}}{\partial x^2} + \frac{\partial^2 \bar{u}}{\partial y^2} \right) \right\} \phi_i(x,\ y)dxdy = 0,\ i = 1(1)\mathrm{N} \tag{12.153a}$$

or
$$\iint_{\mathrm{R}} \phi_i \frac{\partial \bar{u}}{\partial t} dxdy - \iint_{\mathrm{R}} \phi_i \nabla \cdot \nabla \bar{u}\, dxdy = 0 \tag{12.153b}$$

The second term in (12.153b) can be written as per (12.95)

$$-\iint_{\mathrm{R}} \phi_i \nabla \cdot \nabla \bar{u}\, dxdy = \sum_{r=1}^{\mathrm{M}} \sum_{j=1}^{\mathrm{N}} u_j \iint_{e_r} \left(\frac{\partial \phi_i}{\partial x} \cdot \frac{\partial \phi_j}{\partial x} + \frac{\partial \phi_i}{\partial y} \cdot \frac{\partial \phi_j}{\partial y} \right) dxdy + \int_{\mathrm{S}} \phi_i Q ds, \tag{12.154}$$

$$i = 1(1)\ \mathrm{N}.$$

Over a single element e with local nodes 1, 2, 3 we have from (12.105)

$$B^{(e)}\boldsymbol{u}^{(e)} = \sum_{j=1}^{3} u_j \iint_e \left(\frac{\partial \phi_i}{\partial x} \cdot \frac{\partial \phi_j}{\partial x} + \frac{\partial \phi_i}{\partial y} \cdot \frac{\partial \phi_j}{\partial y} \right) dxdy, \quad i = 1,\ 2,\ 3 \tag{12.155}$$

where $$B^{(e)} = \Delta_e \begin{bmatrix} a_1^2 + b_1^2 & a_1a_2 + b_1b_2 & a_1a_3 + b_1b_3 \\ & a_2^2 + b_2^2 & a_2a_3 + b_2b_3 \\ \text{symmetric} & & a_3^2 + b_3^2 \end{bmatrix}$$

and $\boldsymbol{u}^{\mathrm{T}} = (u_1 \ \ u_2 \ \ u_3)$; Δ_e denotes area of element e. The values of $a_i,\ b_i,\ c_i$ for $i = 1,\ 2,\ 3$ are as given in (12.105).

The first term in (12.153b) can be expressed as

$$\iint_R \phi_i \frac{\partial \bar{u}}{\partial t} dxdy = \sum_{r=1}^{M} \sum_{j=1}^{N} \dot{u}_j \iint_{e_r} \phi_i \phi_j dxdy, \ i = 1(1)\ \mathrm{N}. \tag{12.156}$$

Let us consider the triangular element e as shown in Fig. 12.11.

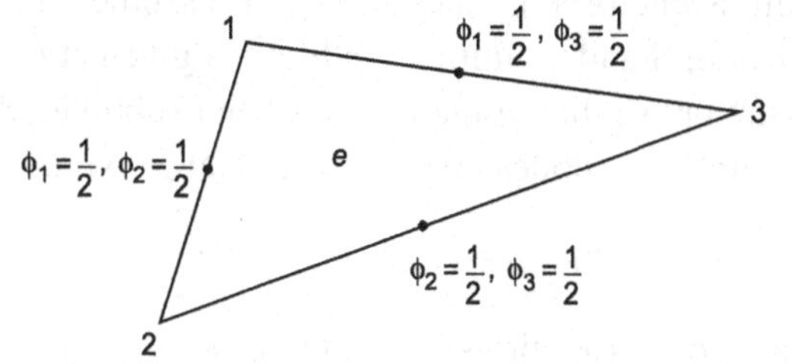

Figure 12.11 Values of shape functions at midpoints.

We want to compute $\iint_e \phi_i \phi_j dxdy, \ \ j = 1,\ 2,\ 3$ for $i = 1,\ 2,\ 3$.

Note that ϕ_1 varies over a plane having value 1 at node 1 and zero at nodes 2 and 3. It varies linearly along sides 1–2 and 1–3. Similar arguments may be given for ϕ_2 and ϕ_3. The values of the shape functions at the midpoints of the sides of the elements are shown in Fig. 12.11. We take the average value of $\phi_i \phi_j$ over the three sides of the element. Thus giving for $i = 1$,

$$\iint_e \phi_1^2 dxdy = \frac{1}{3}\left(\frac{1}{4} + \frac{1}{4} + 0\right)\Delta_e = \frac{1}{6}\Delta_e$$

$$\iint_e \phi_1\phi_2 dxdy = \frac{1}{3}\left(\frac{1}{4}+0+0\right)\Delta_e = \frac{1}{12}\Delta_e$$

$$\iint_e \phi_1\phi_3 dxdy = \frac{1}{3}\left(\frac{1}{4}+0+0\right)\Delta_e = \frac{1}{12}\Delta_e$$

Same expressions can be obtained for $i = 2$ and 3.

For element e we can write

$$\mathrm{A}^{(e)} \cdot \boldsymbol{u}^{(e)} = \sum_{j=1}^{3} \dot{u}_j \iint_e \phi_i\phi_j dxdy,\ i = 1,\ 2,\ 3 \tag{12.157}$$

where $$\mathrm{A}^{(e)} = \frac{\Delta_e}{12}\begin{bmatrix} 2 & 1 & 1 \\ 1 & 2 & 1 \\ 1 & 1 & 2 \end{bmatrix}$$

$$(\dot{\boldsymbol{u}}^{(e)})^{\mathrm{T}} = (\dot{u}_1\ \ \dot{u}_2\ \ \dot{u}_3).$$

After assembling the corresponding terms from all the elements we will get from (12.154) the equations of the form

$$\mathrm{A}\dot{\boldsymbol{u}} + \mathrm{B}\boldsymbol{u} = -\mathrm{T} \tag{12.158}$$

where A and B are $(\mathrm{N}\times\mathrm{N})$ matrices and T is a column vector $(\mathrm{N}\times 1)$ whose i^{th} element is

$$\mathrm{T}_i = \int_{\mathrm{S}} \phi_i \mathrm{Q} ds,\ \ i = 1(1)\ \mathrm{N}.$$

To find $\boldsymbol{u}^{n+1}$ when $\boldsymbol{u}^n$ is known we can use C–N scheme like one-dimensional problem and obtain from (12.158)

$$\frac{1}{\Delta t}\mathrm{A}\ (\boldsymbol{u}^{n+1} - \boldsymbol{u}^n) + \frac{1}{2}\mathrm{B}(\boldsymbol{u}^{n+1} + \boldsymbol{u}^n) = -\mathrm{T}$$

or $$(2\mathrm{A} + \mathrm{B}\Delta t)\boldsymbol{u}^{n+1} = (2\mathrm{A} - \mathrm{B}\Delta t)\boldsymbol{u}^n - 2\Delta t.\ \mathrm{T}. \tag{12.159}$$

The system of linear equations (12.159) can be solved by Gaussian elimination method.

12.9.8 Hyperbolic equation

Let us consider the hyperbolic equation

$$\frac{\partial^2 u}{\partial t^2} = \frac{\partial^2 u}{\partial x^2}, \quad a \leq x \leq b. \tag{12.160}$$

It is assumed that boundary conditions are prescribed at the end points $x = a$ and $x = b$. The initial conditions at $t = 0$ (Cauchy conditions) are also prescribed, viz., $u(x, 0)$ and $\frac{\partial u}{\partial t}(x, 0)$ are known. As discussed in Sec. 12.9.6 we can get

$$\mathrm{A}\ddot{\boldsymbol{u}} + \mathrm{B}\boldsymbol{u} = \mathrm{T}, \tag{12.161}$$

where matrices A, B and T have the forms as shown in (12.146 a, b and c); $\ddot{u}$ denotes $\frac{\partial^2 u}{\partial t^2}$.

Applying C–N scheme

$$\frac{1}{\Delta t^2}\mathrm{A}\,(\boldsymbol{u}^{n+1} - 2\boldsymbol{u}^n + \boldsymbol{u}^{n-1}) + \frac{1}{2}\mathrm{B}(\boldsymbol{u}^{n+1} + \boldsymbol{u}^{n-1}) = \mathrm{T},$$

or $$(2\mathrm{A} + \Delta t^2\mathrm{B})\boldsymbol{u}^{n+1} = 4\mathrm{A}\boldsymbol{u}^n - (2\mathrm{A} + \Delta t^2\mathrm{B})\boldsymbol{u}^{n-1} + 2\Delta t^2\mathrm{T}. \tag{12.162}$$

Assuming that computations have already been made up to $n^{\text{th}}(n \geq 1)$ level, the values of $\boldsymbol{u}^{n+1}$ can be computed from (12.162). For the first time level $(n = 0)$, values of u's may be obtained from the given initial condition as discussed in Chapter 11.

Further, the FEM formulation for the two space dimensional hyperbolic equation

$$\frac{\partial^2 u}{\partial t^2} = \frac{\partial^2 u}{\partial x^2} + \frac{\partial^2 u}{\partial y^2}$$

can be obtained as discussed in Sec. 12.9.7.

Exercise 12

12.1 Given a differential equation

$$\frac{d^2 u}{dx^2} + u = 1 + x, \quad 0 \leq x \leq 1$$

with boundary conditions $u(0) = 0,\ u(1) = 2$.

(a) Find an approximate solution by Galerkin's method in the form $u = \phi_0(x) + a_1\phi_1(x)$ where $\phi_0(x)$ is a linear function satisfying the given boundary conditions and $\phi_1(x)$ is quadratic in x which vanishes at the boundary points.

(b) Also find the solution by method of least squares with the same trial function.

(c) Compare the results with the analytical solution $u = 1 + x - \dfrac{\sin(1-x)}{\sin 1}$ at $x = 0.5$ by both the methods.

12.2 Given a differential equation

$$-\nabla^2 u + \mathrm{F}(x,\ y)u = g(x,\ y) \text{ in R}$$

with region R enclosed by curve S where $u = 0$; F and g are known. Show that the variational form for the solution of the above differential equation is given by

$$\mathrm{I}\,[u] = \iint\limits_{\mathrm{R}} \left\{ \left(\frac{\partial u}{\partial x}\right)^2 + \left(\frac{\partial u}{\partial y}\right)^2 + \mathrm{F}u^2 - 2gu \right\} dxdy$$

(Hint: $\mathrm{L} \equiv -\nabla^2 + \mathrm{F}$; Use MFT)

12.3 Find the approximate solution of the differential equation

$$\frac{d^2u}{dx^2} + u + 1 + x = 0,\quad 0 \le x \le 1,$$

$$u(0) = 0,\ u(1) = 0;$$

by Rayleigh–Ritz method taking basis functions $\phi_1 = x(1-x),\ \phi_2 = x^2(1-x)$. First derive the variational form

$$\mathrm{I}\,[u] = \int_0^1 \left\{ \left(\frac{d^2u}{dx^2}\right)^2 - u^2 - 2(1+x)\,u \right\} dx.$$

Also verify that the same solution is obtained by the Galerkin method.

12.4 Let $-\nabla^2 u + \mathrm{F}(x,\ y)u = g(x,\ y)$ in R with $u = f_0(x,\ y)$ on the curve S enclosing R. Show that the variational form is given by

$$\mathrm{I}\,[u] = \iint\limits_{\mathrm{R}} \left\{ \left(\frac{\partial u}{\partial x}\right)^2 + \left(\frac{\partial u}{\partial y}\right)^2 + \mathrm{F}u^2 - 2gu \right\} dxdy$$

where u has the form $u = f_0(x,\ y) + v(x,\ y)$; $v(x,\ y)$ vanishes on the boundary S.

12.5 Using the result of Exercise 12.4 solve the one-dimensional problem given in Exercise 12.1 by Rayleigh–Ritz method taking the same trial solution, i.e., $\bar{u} = 2x + a_1 x(1-x)$. Verify that the solution obtained in Exercise 12.1 by Galerkin method is same.

12.6 Solve by Galerkin method, the partial differential equation

$$\nabla^2 u = 10xy \text{ in R}$$

where R is a square bounded by $x = 0,\ x = 1$ and $y = 0,\ y = 1$. It is given that $u = 2$ on the boundary.
Take the trial solution $u = 2 + axy(1-x)(1-y)$.

12.7 Solve the problem of Exercise 12.6 by Rayleigh–Ritz method taking the same trial solution, $u = 2 + axy\,(1-x)(1-y)$.
(Hint: $-\nabla^2 u = -10xy$; $\mathrm{I}\,[u] = \iint \left\{ \left(\frac{\partial u}{\partial x}\right)^2 + \left(\frac{\partial u}{\partial y}\right)^2 - 2gu \right\} dxdy$)

12.8 Solve the problem of Exercise 12.6 by method of least squares taking the same trial solution.

12.9 Using FEM, find the solution of the differential equation given in Exercise 12.1 taking four nodes at 0, 0.3, 0.7 and 1.0.
Also compute $\frac{du}{dx}$ at $x = 0$ and $x = 1$.

12.10 A Poisson equation $\nabla^2 u + 9xy = 0$ is defined in R where R is enclosed by the curve $S = S_1 \cup S_2 \cup S_3$. The curves S_1, S_2 and S_3 are defined below and prescribed boundary conditions are shown against each;

S_1 : x-axis where $u = 1$

S_2 : y-axis where $u = 1$

S_3 : Straight line $x + y = 1$ where $\frac{\partial u}{\partial n} = -u$.

Divide the domain $D = R \cup S$ into two elements joining origin to the midpoint of S_3 and find the value of u there. Evaluate the function $9xy$ at the centroid of the element.

12.11 Solve the following Poisson equation by FEM,

$$\nabla^2 u = 3(x + y) \text{ in R}$$

where R is a triangle enclosed by the axes and a straight line $2x + y = 1$. The boundary conditions on the three sides are as given below:

(*i*) S_1 : x-axis, $\frac{\partial u}{\partial y} = -x$

(*ii*) S_2 : y-axis, $\frac{\partial u}{\partial x} = -y$

(*iii*) $S_3 : 2x+y=1,\ u=x+y.$

The domain has been subdivided into three triangular elements as shown below:

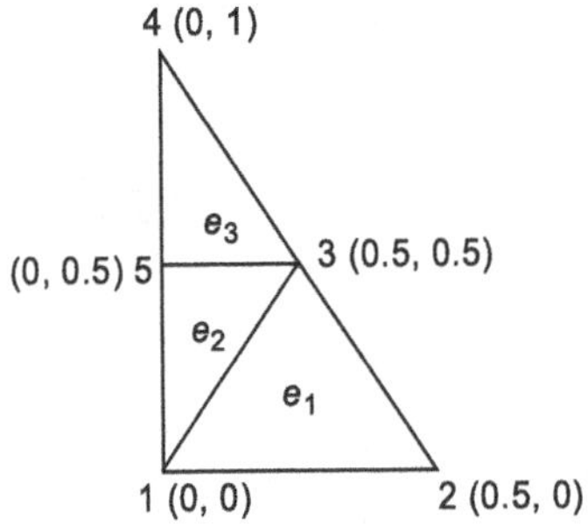

Values of u_2, u_3 and u_4 are given as in (*iii*), find u_1 and u_5.

References and Some Useful Related Books/Papers

1. Courant, R., Hilbert, D., *Methods of Mathematical Physics*, Interscience.
2. Finlayson, B.A., *The Method of Weighted Residuals and Variational Principles*, Academic Press.
3. Martin, H.C., *Introduction to Finite Element Analysis*, McGraw-Hill.
4. Mikhlin, S.G., *Variational Methods in Mathematical Physics*, Pergamon Press.
5. Strang, G. and Fix, G.J., A*n Analysis of the Finite Element Method*, Prentice Hall.
6. Whiteman, J.R., A*n Introduction to Variational and Finite Element Methods*, (*Lecture Notes, February 1972*) Dept. of Mathematics, Brunel University, U.K.
7. Zienkiewicz, O.C., *The Finite Element Method in Engineering Science*, McGraw-Hill.

13

Integral Equations

13.1 Introduction

In a differential equation, the unknown function occurs as differential coefficient while in an integral equation it appears under the sign of integration. An integral equation is called linear if the unknown function occurs linearly, otherwise non-linear. Like differential equations (o.d.e. or p.d.e.) in the integral equations also, the unknown function, i.e., the dependent variable may be a function of one or more variables. We shall be concerned with linear equations consisting of function of one variable only.

The integral equations may be classified, after the names of mathematicians, as Fredholm integral equations and Volterra integral equations. The Fredholm equations correspond to boundary value problem (bvp) and Volterra equations to initial value problems (ivp).

13.2 Fredholm Integral Equations

The Fredholm integral equations may be of two types;

(*i*) Fredholm Equation of First Kind

$$\int_a^b \mathrm{K}(x,\ \xi)y(\xi)\,d\xi = f(x),\ a \le x,\ \xi \le b. \tag{13.1}$$

(*ii*) Fredholm Equation of Second Kind

$$y(x) - \lambda \int_a^b \mathrm{K}(x,\ \xi)y(\xi)\,d\xi = f(x),\ a \le x,\ \xi \le b. \tag{13.2}$$

In equations (13.1) and (13.2), $y(x)$ is an unknown function which is to be determined over $a \le x \le b$; $\mathrm{K}(x,\ \xi)$ is a known real function of x and ξ which is called kernel (or nucleus) of the integral equation; λ is a parameter such that the solution becomes dependent on its value or alternatively it may be a constant; $f(x)$ is also a known function of x.

In the present chapter, it will be assumed that the kernel $K(x, \xi)$ and the function $f(x)$ are continuous in $[a, b]$ while the limits a and b are finite. However, there exist methods for dealing with the cases when they are discontinuous and the limits are infinite. For details reference may be made to [2], [4].

13.3 Volterra Integral Equations

Analogous to Fredholm's, the Volterra equations of first and second kinds are represented as follows:

(*i*) Volterra Equation of First Kind

$$\int_a^x K(x, \xi)y(\xi)\,d\xi = f(x),\ a \le \xi \le x. \tag{13.3}$$

(*ii*) Volterra Equation of Second Kind

$$y(x) - \lambda \int_a^x K(x, \xi)y(\xi)\,d\xi = f(x),\ a \le \xi \le x. \tag{13.4}$$

Obviously Volterra equations may be considered as particular cases of Fredholm equations when $K(x, \xi) = 0$ for $\xi > x$. Here also we will assume that kernel $K(x, \xi)$ and $f(x)$ have no discontinuities for $x \ge a$.

Let us show that, in general, a Volterra equation of the first kind can be converted to an equation of the second kind. First, note the important formula

$$\frac{d}{dx}\int_{a(x)}^{b(x)} F(x, \xi)\,d\xi = \int_{a(x)}^{b(x)} \frac{\partial F(x, \xi)}{\partial x}\,d\xi + F\{x, b(x)\}\frac{db}{dx} - F\{x, a(x)\}\frac{da}{dx}. \tag{13.5}$$

Now, differentiating (13.3) and using (13.5) we get

$$K(x, x)\,y(x) + \int_a^x \frac{\partial K(x, \xi)}{\partial x}\,y(\xi)\,d\xi = f'(x) \tag{13.6}$$

If $K(x, x) \ne 0$, we can divide equation (13.6) throughout by $K(x, x)$ and obtain

$$y(x) + \int_a^x K_1(x, \xi)\,y(\xi)\,d\xi = f_1(x) \tag{13.7}$$

where $K_1(x, \xi) = \frac{1}{K(x, x)} \frac{\partial K(x, \xi)}{\partial x}$ and $f_1(x) = \frac{f'(x)}{K(x, x)}$.

Equation (13.7) is the required Volterra equation of second kind. It may be noted that the new kernel K_1 and f_1 should have no discontinuities for $x \geq a$.

If, however $K(x, x)$ happens to be zero, we can differentiate equation (13.6) again. The process of differentiation may be continued until a non-vanishing derivative is found as coefficient of $y(\xi)$. It should be remembered that the equation of first kind does not always convert to equation of the second kind. It may convert to an equivalent differential equation.

13.4 Green's Function

The Green's function plays an important role in the solution of two-point boundary value problems. Let us suppose that a second order homogeneous differential equation is defined for $a \leq x \leq b$ with appropriate boundary conditions prescribed at $x = a$ and $x = b$, e.g.,

$$Ly(x) = 0 \tag{13.8}$$

$$\alpha_a y + \beta_a y' = 0, \quad x = a \tag{13.9a}$$

$$\alpha_b y + \beta_b y' = 0, \quad x = b \tag{13.9b}$$

where L is a linear differential operator and α_a, β_a, α_b, β_b are constants. If ξ is any point in the interval (a, b), the Green's function $G(x, \xi)$ is defined for $a \leq x \leq b$ as follows:

(i) $$G(x, \xi) = \begin{cases} G_1(x, \xi), \text{ such that } LG_1 = 0, & a \leq x < \xi \\ G_2(x, \xi), \text{ such that } LG_2 = 0, & \xi < x \leq b. \end{cases} \tag{13.10a}$$

That is, $G_1(x, \xi)$ satisfies (13.8) along with boundary condition (13.9a) in $a < x < \xi$ and $G_2(x, \xi)$ satisfies the differential equation (13.8) along with boundary condition (13.9b) in the interval $\xi < x < b$.

(ii) $$G_1(\xi, \xi) = G_2(\xi, \xi), \quad x = \xi. \tag{13.10b}$$

That is $G(x, \xi)$ is continuous at $x = \xi$.

(iii) $$G_2'(\xi, \xi) - G_1'(\xi, \xi) = \alpha(\xi), \quad x = \xi \text{ where } \alpha(\xi) \text{ is finite.} \tag{13.10c}$$

That is, $\frac{\partial}{\partial x}G(x,\ \xi)$ has a finite discontinuity at $x = \xi$. It can also be written as

$$G_2'(\xi + 0,\ \xi) - G_1'(\xi - 0,\ \xi) = \alpha(\xi),\ \ x = \xi.$$

The solution of a homogeneous differential equation can be expressed by Green's function as will be clear by the following simple example.

Example 13.1

Find in terms of Green's function the solution of the differential equation,

$$\frac{d^2y}{dx^2} + y = 0,\ \ a < x < b$$

with boundary condition $y(a) = 0,\ \ y(b) = 0$.

Solution The analytical solution of the differential equation is given by

$$y = \text{A}\cos x + \text{B}\sin x,\ \text{A and B being constants} \qquad \text{.... (1)}$$

Satisfying the boundary conditions in (1) gives

$$x = a: \qquad 0 = \text{A}\cos a + \text{B}\sin a \qquad \text{....(2)}$$

$$x = b: \qquad 0 = \text{A}\cos b + \text{B}\sin b \qquad \text{....(3)}$$

From (2) and (3) we get A = 0 and B = 0 giving $y = 0$ which is a trivial solution representing steady state or position of equilibrium.

If we use (2) in (1) we get

$$y = -\frac{\text{A}}{\sin a}\sin(x - a) \quad \text{or} \quad y = \frac{\text{B}\sin(x - a)}{\cos a}. \qquad \text{.... (4)}$$

Similarly, if we use (3) in (1) we get

$$y = -\frac{\text{A}}{\sin b}\sin(b - x) \quad \text{or} \quad y = -\frac{\text{B}}{\cos a}\sin(b - x). \qquad \text{....(5)}$$

It may be noted that solution (4) satisfies the boundary condition at $x = a$ and solution (5) at $x = b$.

Let ξ be a point in $[a,\ b]$ and assume the solution in accordance with (13.10a), as

$$y_1 = c_1\sin(x - a) = \text{G}_1(x,\ \xi);\ \ a \le x < \xi \qquad \text{....(6)}$$

$$y_2 = c_2\sin(b - x) = \text{G}_2(x,\ \xi);\ \ \xi < x \le b \qquad \text{....(7)}$$

where c_1 and c_2 are functions of ξ to be determined from (13.9b) and (13.9c), i.e.,

$$c_1 \sin(\xi - a) - c_2 \sin(b - \xi) = 0 \qquad \text{....(8)}$$

$$-c_2 \cos(b - \xi) - c_1 \cos(\xi - a) = \alpha(\xi)$$

or $$c_1 \cos(\xi - a) + c_2 \cos(b - \xi) = -\alpha(\xi) \qquad \text{....(9)}$$

Solving (8) and (9) we get

$$c_1(\xi) = -\frac{\alpha(\xi)\sin(b - \xi)}{\sin(b - a)}; \; c_2(\xi) = -\frac{\alpha(\xi)\sin(\xi - a)}{\sin(b - a)}. \qquad \text{....(10)}$$

Substituting $c_1(\xi)$ and $c_2(\xi)$ in (6) and (7) we get respectively

$$y_1 = -\frac{\alpha(\xi)\sin(b - \xi)\sin(x - a)}{\sin(b - a)} = G_1(x, \xi), \; a \le x \le \xi, \qquad \text{....(11a)}$$

$$y_2 = -\frac{\alpha(\xi)\sin(\xi - a)\sin(b - x)}{\sin(b - a)} = G_2(x, \xi), \; \xi \le x \le b. \qquad \text{....(11b)}$$

Thus we have the solution

$$y = \begin{cases} G_1(x, \xi), & a \le x \le \xi \\ G_2(x, \xi), & \xi \le x \le b. \end{cases} \qquad \text{....(12)}$$

Further, from the given differential equation we can write at $x = \xi$,

$$\left(\frac{d^2 y_2}{dx^2}\right)_{\xi+0} - \left(\frac{d^2 y_1}{dx^2}\right)_{\xi-0} = -\{y_2(\xi + 0) - y_1(\xi - 0)\}$$

or $$\frac{d}{dx}\left[\frac{dy_2}{dx} - \frac{dy_1}{dx}\right] = 0, \; \because \; y_2(\xi) = y_1(\xi).$$

or $$\frac{dy_2}{dx} - \frac{dy_1}{dx} = c \text{ (cosnt)}$$

or $$\alpha(\xi) = c, \text{ from property } (iii) \text{ of Green's function.} \qquad \text{....(13)}$$

Thus, the solution (11a) and (11b) after using (13), will become

$$y_1 = G_1(x, \xi) = -\frac{c}{\sin(b - a)}\sin(b - \xi)\sin(x - a), \; a \le x \le \xi \qquad \text{....(14a)}$$

$$y_2 = G_2(x, \xi) = -\frac{c}{\sin(b-a)}\sin(\xi - a)\sin(b-x), \quad \xi \le x \le b \qquad \text{....(14b)}$$

It may be seen from (14a) and (14b) that

$$G_1(x, \xi) = G_2(\xi, x), \text{ i.e., } G(x, \xi) = G(\xi, x). \qquad \text{....(15)}$$

If a Green's function satisfies condition (15), it is called symmetric.

13.5 Solution of Differential Equation Represented by Integral and Vice-Versa

Let us consider a second order linear differential equation

$$\frac{d^2y}{dx^2} = f(x), \quad a \le x \le b \qquad (13.11)$$

with homogeneous Dirichlet conditions prescribed at the boundary points, i.e.,

$$y(a) = 0, \ y(b) = 0 \qquad (13.12)$$

where $f(x)$ is a known function.
Integrating (13.11) w.r.t. x from a to x,

$$\frac{dy}{dx} = \int_a^x f(x)dx + c_1 \qquad (13.13)$$

where c_1 is a constant.
Integrating (13.13) again w.r.t. x from a to x,

$$y(x) = \int_a^x \int_a^x f(x)dx + c_1x + c_2, \qquad (13.14)$$

where c_2 is another constant. The constants c_1 and c_2 will be determined from the boundary conditions (13.12). Let us first reduce the double integration in (13.14) to single integration. Integrating by parts

$$\int_a^x \int_a^x f(x)dx\ dx = \int_{t=a}^{x} \left\{ \int_{\xi=a}^{t} f(\xi)d\xi \right\} dt$$

$$= \left[t \int_{\xi=a}^{t} f(\xi)d\xi \right]_{t=a}^{x} - \int_{t=a}^{x} f(t)\cdot t dt \quad \because \frac{d}{dt}\int_{\xi=a}^{t} f(\xi)d\xi = f(t)$$

$$= x \int_{\xi=a}^{x} f(\xi)d\xi - 0 - \int_{\xi=a}^{x} \xi f(\xi)d\xi$$

$$= \int_{a}^{x} (x-\xi)f(\xi)d\xi. \tag{13.15}$$

It is also an important formula which will be used in the future discussions. Using (13.15) in (13.14) gives

$$y(x) = \int_{a}^{x} (x-\xi)f(\xi)\,d\xi + c_1 x + c_2. \tag{13.16}$$

Now satisfying the boundary conditions (13.12) at $x = a$ and $x = b$, respectively, provides the equations

$$0 = c_1 a + c_2 \tag{13.17}$$

and $$0 = \int_{a}^{b} (b-\xi)f(\xi)\,d\xi + c_1 b + c_2 \tag{13.18}$$

Solving (13.17) and (13.18)

$$c_1 = -\frac{1}{b-a}\int_{a}^{b}(b-\xi)f(\xi)d\xi; \quad c_2 = \frac{a}{b-a}\int_{a}^{b}(b-\xi)f(\xi)d\xi. \tag{13.19}$$

Substituting the values of c_1 and c_2 from (13.19) in (13.16)

$$y(x) = \int_{a}^{x}(x-\xi)f(\xi)d\xi - \frac{x}{b-a}\int_{a}^{b}(b-\xi)f(\xi)d\xi + \frac{a}{b-a}\int_{a}^{b}(b-\xi)f(\xi)d\xi.$$

$$= \int_{a}^{x}(x-\xi)f(\xi)d\xi - \int_{a}^{b}\frac{(x-a)(b-\xi)f(\xi)}{b-a}d\xi. \tag{13.20}$$

This is the required integral form of d.e. (13.11).
Since $f(x)$ is a known function we can evaluate the right side of (13.20).
Further, we can also express (13.20) as follows:

$$y(x) = \int_a^x (x-\xi)f(\xi)\,d\xi - \int_a^x \frac{(x-a)(b-\xi)f(\xi)}{b-a}dx - \int_x^b \frac{(x-a)(b-\xi)f(\xi)}{b-a}d\xi$$

or $$y(x) = \int_a^x \frac{(x-b)(\xi-a)f(\xi)}{b-a}d\xi + \int_x^b \frac{(x-a)(\xi-b)f(\xi)d\xi}{b-a}. \tag{13.21}$$

To revert back to differential equation (13.11) we differentiate equation (13.21). Making use of the formula (13.5) we get

$$y'(x) = \int_a^x \frac{(\xi-a)f(\xi)d\xi}{b-a} + \frac{(x-b)(x-a)f(x)}{b-a} + \int_x^b \frac{(\xi-b)f(\xi)d\xi}{b-a} - \frac{(x-a)(x-b)f(x)}{b-a}.$$

Cancelling 2nd and 4th terms and differentiating again, we get

$$y''(x) = \frac{(x-a)f(x)}{b-a} - \frac{(x-b)f(x)}{b-a} = f(x).$$

It is the given differential equation (13.11).
In this case Green's function may be expressed as

$$\left.\begin{aligned} G_2(x,\,\xi) &= \frac{(x-a)(\xi-b)f(x)}{b-a}, \quad x \le \xi \le b \\ G_1(x,\,\xi) &= \frac{(x-b)(\xi-a)f(x)}{b-a}, \quad a \le \xi \le x. \end{aligned}\right\} \tag{13.22}$$

13.6 Reduction of Differential Equation to Integral Equation

A linear differential equation can be transformed to an equivalent integral equation by integrating it. We illustrate the method in respect of a differential equation having constant coefficients and simple boundary conditions; however it can be extended easily to incorporate more general conditions.

13.6.1 Reduction of a BVP to Fredholm equation

Let us consider the differential equation

$$y'' + y' + y = f(x), \quad a \leq x \leq b \tag{13.23}$$

with boundary conditions

$$y(a) = 0 \text{ and } y(b) = 0 \tag{13.24}$$

Integrating (13.23) w.r.t. x from a to x

$$y'(x) + y(x) + \int_a^x y(x)dx = \int_a^x f(x)dx + c_1 \tag{13.25}$$

where c_1 is a constant.
Integrating (13.25) again w.r.t. x from a to x

$$y(x) + \int_a^x y(x)dx + \int_a^x (x - \xi)y(\xi)d\xi = \int_a^x (x - \xi)f(\xi)d\xi + c_1 x + c_2 \tag{13.26}$$

where c_2 is another constant.
Eq. (13.26) can be written as

$$y(x) = -\int_a^x (x - \xi + 1)y(\xi)d\xi + \int_a^x (x - \xi)f(\xi)d\xi + c_1 x + c_2 \tag{13.27}$$

Satisfying boundary conditions (13.24) we get from (13.27)

$$0 = c_1 a + c_2 \tag{13.28a}$$

$$0 = -\int_a^b (b - \xi + 1)y(\xi)d\xi + \int_a^b (b - \xi)f(\xi)d\xi + c_1 b + c_2 \tag{13.28b}$$

From (13.28a and b) we get

$$c_1 = \frac{1}{b-a}\left[\int_a^b (b - \xi + 1)y(\xi)d\xi + \int_a^b (b - \xi)f(\xi)d\xi\right]$$

$$c_2 = -\frac{a}{b-a}\left[\int_a^b (b-\xi+1)\,y(\xi)\,d\xi - \int_a^b (b-\xi)\,f(\xi)\,d\xi\right]$$

Substituting the values of c_1 and c_2 in (13.27),

$$y(x) = -\int_a^x (x-\xi+1)\,y(\xi)\,d\xi + \frac{x-a}{b-a}\left[\int_a^b (b-\xi+1)\,y(\xi)\,d\xi - \int_a^b (b-\xi)\,f(\xi)\,d\xi\right]$$

$$+\int_a^x (x-\xi)\,f(\xi)\,d\xi.$$

or $$y(x)+\int_a^x (x-\xi+1)\,y(\xi)\,d\xi - \int_a^x \frac{(x-a)(b-\xi+1)}{b-a}\,y(\xi)\,d\xi - \int_x^b \frac{(x-a)(b-\xi+1)}{b-a}\,y(\xi)\,d\xi$$

$$= \int_a^x (x-\xi)\,f(\xi)\,d\xi - \int_a^x \frac{(x-a)(b-\xi)}{b-a}\,f(\xi)\,d\xi - \int_x^b \frac{(x-a)(b-\xi)}{b-a}\,f(\xi)\,d\xi$$

or $$y(x)+\int_a^x (x-\xi)\,y(\xi)\,d\xi - \int_a^x \frac{(x-a)(b-\xi)}{b-a}\,y(\xi)\,d\xi + \int_a^x \frac{b-x}{b-a}\,y(\xi)\,d\xi$$

$$+\int_x^b \frac{(x-a)(\xi-b-1)}{b-a}\,y(\xi)\,d\xi$$

$$= \int_a^x \frac{(x-b)(\xi-a)}{b-a}\,f(\xi)\,d\xi + \int_x^b \frac{(x-a)(\xi-b)}{b-a}\,f(\xi)\,d\xi$$

or $$y(x)+\int_a^x \frac{(x-b)(\xi-a)}{b-a}\,y(\xi)\,d\xi - \int_a^x \frac{(x-b)}{b-a}\,y(\xi)\,d\xi + \int_x^b \frac{(x-a)(\xi-b-1)}{b-a}\,y(\xi)\,d\xi$$

$$= \int_a^x \frac{(x-b)(\xi-a)}{b-a}\,f(\xi)\,d\xi + \int_x^b \frac{(x-a)(\xi-b)}{b-a}\,f(\xi)\,d\xi$$

$$\text{or} \quad y(x) + \int_a^x \frac{(x-b)(\xi-a-1)}{b-a} y(\xi)\,d\xi + \int_x^b \frac{(x-a)(\xi-b-1)}{b-a} y(\xi)\,d\xi$$

$$= \int_a^x \frac{(x-b)(\xi-a)}{b-a} f(\xi)\,d\xi + \int_x^b \frac{(x-a)(\xi-b)}{b-a} f(\xi)\,d\xi$$

$$\text{or} \quad y(x) + \int_a^x \mathrm{K}_1(x,\ \xi)\,y(\xi)\,d\xi + \int_x^b \mathrm{K}_2(x,\ \xi)\,y(\xi)\,d\xi$$

$$= \int_a^x \frac{(x-b)(\xi-a)}{b-a} f(\xi)\,d\xi + \int_x^b \frac{(x-a)(\xi-b)}{b-a} f(\xi)\,d\xi. \tag{13.29}$$

Equation (13.29) is a Fredholm's equation of second kind with kernel

$$\mathrm{K}(x,\ \xi) = \left\{ \begin{array}{ll} \mathrm{K}_1(x,\ \xi) = \dfrac{(x-b)(\xi-a-1)}{b-a}, & a \le \xi < x \\ \mathrm{K}_2(x,\ \xi) = \dfrac{(x-a)(\xi-b-1)}{b-a}, & x < \xi \le b \end{array} \right\}. \tag{13.30}$$

13.6.2 Reduction of IVP to Volterra equation

Let us consider an initial value problem

$$y'' + P(x)y' + Q(x)y = f(x), \quad x \ge \mathrm{a} \tag{13.31}$$

with conditions prescribed at $x = a$, as

$$y(a) = y_a \text{ and } y'(a) = y'_a; \tag{13.32}$$

$P(x)$, $Q(x)$ and $f(x)$ are known functions of x.
Integrating (13.31) w.r.t. x from a to x and using initial conditions (13.32), we get

$$y'(x) - y'(a) + P(x)\,y(x) - P(a)\,y(a) - \int_a^x P'(x)y(x)dx + \int_a^x Q(x)y(x)dx = \int_a^x f(x)dx$$

$$\text{or} \quad y'(x) + P(x)y(x) - \int_a^x \{P'(x) - Q(x)\}y(x)dx = \int_a^x f(x)dx + y'_a + P(a)y_a. \tag{13.33}$$

Integrating (13.33) w.r.t. x again from a to x gives

$$y(x)-y(a)+\int_a^x P(\xi)y(\xi)d\xi-\int_a^x (x-\xi)\{P'(\xi)-Q(\xi)\}y(\xi)d\xi=\int_a^x (x-\xi)f(\xi)d\xi$$

$$+\{y'_a+P(a)\,y_a\}(x-a)$$

or $$y(x)+\int_a^x [P(\xi)-(x-\xi)\{P'(\xi)-Q(\xi)\}]y(\xi)d\xi=\int_a^x (x-\xi)f(\xi)d\xi$$

$$+\{y'_a+P(a)\,y_a\}(x-a)+y_a. \tag{13.34}$$

Equation (13.34) is a Volterra equation of second kind with kernel

$$\text{K}(x,\ \xi)=P(\xi)-(x-\xi)\{P'(\xi)-Q(\xi)\} \tag{13.35}$$

For example, the differential equation

$$\frac{d^2y}{dx^2}-\frac{dy}{dx}-2y=4x$$

with boundary conditions $y(0)=0,\ y'(0)=1$ reduces to the Volterra integral equation

$$y(x)-\int_0^x \{1+2(x-\xi)\}y(\xi)d\xi=\frac{2}{3}x^3+x.$$

13.7 Methods for Solving Fredholm Equations

Let us rewrite the Fredholm equation of second kind as

$$y(x)=\lambda\int_a^x \text{K}(x,\ \xi)y(\xi)d\xi+f(x) \tag{13.36}$$

If $f(x)=0$, the equation (13.36) reduces to an e.value problem when λ is a parameter to be determind for the equation to have a non-trivial solution. We shall discuss three types of methods which may be suitable under different situations.

13.7.1 Analytical method

An analytical solution of (13.36) may be found if the kernel $K(x, \xi)$ is a degenerate kernel, i.e., it can be expressed as sum of the products of functions $u_i(x)$ and $v_i(\xi)$, $i = 1(1)n$, as

$$K(x, \xi) = \sum_{i=1}^{n} u_i(x)v_i(\xi) \tag{13.37}$$

where $u_i(x)$ is a function of x alone and $v_i(\xi)$, is a function of ξ alone. It may be mentioned that an arbitrary kernel may be converted to a degenerate kernel by various mathematical techniques, like, expanding it by Maclaurin's series or approximating it by orthogonal polynomials or through Fourier series expansion. Let us suppose that the given kernel is degenerate and is expressed as in (13.37). Then replacing $K(x, \xi)$ in (13.36) yields

$$y(x) = \lambda \int_a^b \sum_{i=1}^{n} u_i(x)v_i(\xi) \cdot y(\xi)\, d\xi + f(x)$$

$$= \lambda \sum_{i=1}^{n} u_i(x) \int_a^b v_i(\xi)y(\xi)\, d\xi + f(x)$$

$$= \lambda \sum_{i=1}^{n} u_i(x)\, c_i + f(x), \tag{13.38}$$

where $$c_i = \int_a^b v_i(\xi)y(\xi)\, d\xi, \quad i = 1, 2, \ldots. n. \tag{13.39}$$

It may be noted that c_i's are constants but their values are still undetermined since $y(\xi)$ is not known in (13.39). We substitute $y(\xi)$ from (13.38) in (13.39) to give

$$c_i = \int_a^b v_i(\xi) \left\{ \lambda \sum_{j=1}^{n} u_j(\xi)c_j + f(\xi) \right\} d\xi$$

$$= \lambda \sum_{j=1}^{n} c_j \int_a^b v_i(\xi)u_j(\xi)d\xi + \int_a^b v_i(\xi)f(\xi)d\xi,$$
$$i = 1(1)n.$$

In short, we can write

$$c_i = \lambda \sum_{j=1}^{n} \alpha_{ij} c_j + \beta_i, \quad i = 1(1)n \tag{13.40}$$

$$\text{where } \alpha_{ij} = \int_a^b v_i(\xi) u_j(\xi)\, d\xi \quad \text{and} \quad \beta_i = \int_a^b v_i(\xi) f(\xi)\, d\xi. \tag{13.41}$$

Equations (13.40) represent n equations for $i = 1(1)n$ in n unknowns c_j, $j = 1(1)n$. In matrix from, they can be written as

$$\boldsymbol{c} = \lambda \mathrm{A}\boldsymbol{c} + \boldsymbol{\beta} \tag{13.42}$$

$$\text{where} \quad \boldsymbol{c}^{\mathrm{T}} = (\boldsymbol{c}_1 \;\; \boldsymbol{c}_2 \;\; \ldots. \;\; \boldsymbol{c}_n), \; \boldsymbol{\beta}^{\mathrm{T}} = (\beta_1 \;\; \beta_2 \;\; \ldots \;\; \beta_n) \tag{13.43}$$

$$\text{and} \qquad \mathrm{A} = \begin{bmatrix} \alpha_{11} & \alpha_{12} & \cdots & \alpha_{1n} \\ \alpha_{21} & \alpha_{22} & \cdots & \alpha_{2n} \\ \vdots & \vdots & & \vdots \\ \alpha_{n1} & \alpha_{n2} & \cdots & \alpha_{nn} \end{bmatrix}. \tag{13.44}$$

From (13.42) we can write

$$(\mathrm{I} - \lambda \mathrm{A})\boldsymbol{c} = \boldsymbol{\beta}. \tag{13.45}$$

If $\boldsymbol{\beta} \neq \boldsymbol{0}$, we get vector $\boldsymbol{c}$ as

$$\boldsymbol{c} = (\mathrm{I} - \lambda \mathrm{A})^{-1}\boldsymbol{\beta}, \text{ provided } |\mathrm{I} - \lambda \mathrm{A}| \neq 0. \tag{13.46}$$

The values of c_i, $i = 1(1)n$ are substituted in (13.38) to get the solution.

If $\boldsymbol{\beta}$ is a null vector, i.e., $\boldsymbol{\beta} = \boldsymbol{0}$, as it will be a case when $f(x) = 0$ (See 13.42) then $|\mathrm{I} - \lambda \mathrm{A}| = 0$ is a polynomial equation of degree n in λ and will have n roots, say $\lambda = \lambda_1, \lambda_2, \ldots. \lambda_n$. If these roots are real then we solve the homogeneous system of equations

$$(\mathrm{I} - \lambda_i \mathrm{A})\boldsymbol{c} = \boldsymbol{0}, \quad i = 1(1)n. \tag{13.47}$$

For each real root λ_i we will have a solution $\boldsymbol{c}$. Of course $\boldsymbol{c}$ vector will not be unique and can be a normalised vector. Substituting the value of $\boldsymbol{c}$ in (13.38) will provide a solution corresponding to the particular value of λ. Thus for n real roots λ_i, $i = 1(1)n$ there will be n different solutions when $f(x) = 0$.

Again, when the integral equation (13.36) is homogeneous i.e., $f(x) = 0$, then we can write from (13.42) after putting $\boldsymbol{\beta} = \mathbf{0}$,

$$\lambda \mathrm{A}\boldsymbol{c} = \boldsymbol{c}$$

or
$$\mathrm{A}\boldsymbol{c} = \frac{1}{\lambda}\boldsymbol{c} = \mu\boldsymbol{c}, \text{ say.} \tag{13.48}$$

Equation (13.48) is an eigenvalue problem showing that e.values μ of A are given by $\mu_i = \dfrac{1}{\lambda_i}$ with the corresponding e.vector $\boldsymbol{c}_i$.

Example 13.2

Find the analytical solution of the integral equation

$$y(x) = x^2 + \lambda \int_0^1 (2x + 3\xi)\, y(\xi)\, d\xi.$$

Also find the solution when the equation is homogeneous.

Solution Given equation is

$$y(x) = x^2 + 2\lambda x \int_0^1 y(\xi)\, d\xi + 3\lambda \int_0^1 \xi y(\xi)\, d\xi. \qquad \text{....(1)}$$

Let
$$c_1 = \int_0^1 y(\xi)\, d\xi; \; c_2 = \int_0^1 \xi y(\xi)\, d\xi. \qquad \text{....(2)}$$

Put c_1 and c_2 in (1)

$$y(x) = x^2 + 2\lambda x c_1 + 3\lambda c_2. \qquad \text{....(3)}$$

Substituting $y(\xi)$ from (3) in eqns. (2) gives

$$c_1 = \int_0^1 (\xi^2 + 2\lambda\xi c_1 + 3\lambda c_2)\, d\xi = \frac{1}{3} + \lambda c_1 + 3\lambda c_2$$

or
$$(1-\lambda)c_1 - 3\lambda c_2 = \frac{1}{3} \qquad \text{....(4)}$$

$$c_2 = \int_0^1 (\xi^3 + 2\lambda\xi^2 c_1 + 3\lambda\xi)\, d\xi = \frac{1}{4} + \frac{2}{3}\lambda c_1 + \frac{3}{2}\lambda c_2 \qquad \text{....(5)}$$

Equations (4) and (5) can be written as

$$\begin{bmatrix} 1-\lambda & -3\lambda \\ -\frac{2}{3}\lambda & 1-\frac{3}{2}\lambda \end{bmatrix}\begin{bmatrix} c_1 \\ c_2 \end{bmatrix} = \begin{bmatrix} 1/3 \\ 1/4 \end{bmatrix}. \qquad ....(6)$$

Solving (6)

$$c_1 = -\frac{1}{6}\frac{3\lambda+4}{\lambda^2+5\lambda-2}; \; c_2 = \frac{1}{18}\frac{\lambda-9}{\lambda^2+5\lambda-2}. \qquad ....(7)$$

The solution is given by (3) as

$$y(x) = x^2 - \frac{1}{3}\cdot\frac{\lambda(3\lambda+4)}{\lambda^2+5\lambda-2}x + \frac{1}{6}\cdot\frac{\lambda(\lambda-9)}{\lambda^2+5\lambda-2}, \; \lambda^2+5\lambda-2 \neq 0. \qquad ....(8)$$

When equation is homogeneous it reduces to eigenvalue problem; the e.values are given by

$$\begin{vmatrix} 1-\lambda & -3\lambda \\ -\frac{2}{3}\lambda & 1-\frac{3}{2}\lambda \end{vmatrix} = 0$$

or $$\lambda^2+5\lambda-2=0; \; \lambda_1 = \frac{\sqrt{33}-5}{2}, \; \lambda_2 = -\frac{\sqrt{33}+5}{2}$$

For $\lambda = \lambda_1 = \frac{\sqrt{33}-5}{2}$, solution of (6) with right side zero, gives $c_2 = -\frac{\sqrt{33}+1}{12}c_1$. Hence the solution from (3), without x^2 term,

$$y(x) = c\left\{2x - \frac{\sqrt{33}+1}{4}\right\}, \text{ where } c \text{ is a constant.}$$

Similarly for $\lambda = \lambda_2 = -\frac{\sqrt{33}+5}{2}, \; c_2 = -\frac{\sqrt{33}-1}{12}c_1$

$$y(x) = c\left\{2x - \frac{\sqrt{33}-1}{4}\right\}, \text{ where } c \text{ is an arbitrary const.}$$

Example 13.3

Find the analytical solution of the integral equation

$$y(x) = f(x) + \lambda\int_0^1 \log(1+x\xi)\, y(\xi)\, d\xi \text{ for } f(x) = 1.$$

Also find the solution of the eigenvalue problem when $f(x) = 0$.

Transform the kernel to 'degenerate' kernel by expanding it by Maclaurin's series and consider first two terms only.

Solution
$$y(x) = 1 + \lambda \int_0^1 \log(1 + x\xi) y(\xi) d\xi$$

$$= 1 + \lambda \int_0^1 \left(x\xi - \frac{x^2\xi^2}{2} + \cdots \right) y(\xi) d\xi$$

$$= 1 + x\lambda \int_0^1 \xi y(\xi) d\xi - \frac{\lambda x^2}{2} \int_0^1 \xi^2 y(\xi) d\xi. \qquad \text{....(1)}$$

Let
$$c_1 = \int_0^1 \xi y(\xi) d\xi; \; c_2 = \int_0^1 \xi^2 y(\xi) d\xi. \qquad \text{....(2)}$$

Putting c_1 and c_2 in (1)

$$y(x) = 1 + \lambda x c_1 - \frac{\lambda x^2}{2} c_2 \qquad \text{....(3)}$$

From (3), substitute $y(\xi)$ in (2) to get

$$c_1 = \int_0^1 \xi \left\{ 1 + \lambda \xi c_1 - \frac{\lambda \xi^2}{2} c_2 \right\} d\xi = \frac{1}{2} + \frac{\lambda}{3} c_1 - \frac{\lambda}{8} c_2$$

or
$$\left(1 - \frac{\lambda}{3} \right) c_1 + \frac{\lambda}{8} c_2 = \frac{1}{2} \qquad \text{....(4)}$$

and
$$c_2 = \int_0^1 \xi^2 \left\{ 1 + \lambda \xi c_1 - \frac{\lambda \xi^2}{2} c_2 \right\} d\xi = \frac{1}{3} + \frac{\lambda}{4} c_1 - \frac{\lambda}{10} c_2$$

or
$$-\frac{\lambda}{4} c_1 + \left(1 + \frac{\lambda}{10} \right) c_2 = \frac{1}{3} \qquad \text{....(5)}$$

From (4) and (5)

$$\begin{bmatrix} 1 - \frac{\lambda}{3} & \frac{\lambda}{8} \\ -\frac{\lambda}{4} & 1 + \frac{\lambda}{10} \end{bmatrix} \begin{bmatrix} c_1 \\ c_2 \end{bmatrix} = \begin{bmatrix} \frac{1}{2} \\ \frac{1}{3} \end{bmatrix} \qquad \text{....(6)}$$

$$c_1 = -\frac{4(60+\lambda)}{\lambda^2+112\lambda-480};\; c_2 = -\frac{24+\lambda}{\lambda^2+112\lambda-480},\; \lambda^2+112\lambda-480 \neq 0. \qquad ....(7)$$

Substituting the values of c_1 and c_2 in (3) provides the solution.

When $f(x) = 0$, we solve (6) with right side equal to zero. To get a non-trivial solution we should have

$$\begin{vmatrix} 1-\frac{\lambda}{3} & \frac{\lambda}{8} \\ -\frac{\lambda}{4} & 1+\frac{\lambda}{10} \end{vmatrix} = 0$$

which gives $\lambda^2 + 112\lambda - 480 = 0$.

The roots are $\lambda_1 = -56 + 4\sqrt{226}$ and $\lambda_2 = -(56 + 4\sqrt{226})$.

For $\lambda = \lambda_1$ we get $c_2 = \frac{1}{15}(26 - \sqrt{226})c_1$

and for $\lambda = \lambda_2$, $c_2 = \frac{1}{15}(26 + \sqrt{226})c_1$

Substituting the above values in (3) with $f(x) = 0$ we can get two solutions corresponding to λ_1 and λ_2.

13.7.2 Classical iterative method

In this method also the solution is obtained in closed form. We start from an approximate solution which is improved successively in an iterative manner. Hence it is also called method of successive approximations. Let us consider the equation

$$y(x) = f(x) + \lambda \int_a^b \mathrm{K}(x, \xi)\, y(\xi)\, d\xi, \qquad (13.49)$$

where $\mathrm{K}(x, \xi)$ is an arbitrary kernel and λ is a parameter. Start from the initial approximation

$$y_0(x) = f(x).$$

Substituting $y_0(x)$ in (13.49) gives first iterated solution

$$\begin{aligned} y_1(x) &= f(x) + \lambda \int_a^b \mathrm{K}(x, \xi)\, f(\xi)\, d\xi, \\ &= f(x) + \lambda \int_a^b \mathrm{K}_1(x, \xi)\, f(\xi)\, d\xi, \end{aligned}$$

where $K_1(x, \xi) = K(x, \xi)$.
Substituting $y_1(x)$ in (13.49) renders the second approximation

$$y_2(x) = f(x) + \lambda \int_a^b K_1(x, \xi) \left\{ f(\xi) + \lambda \int_a^b K_1(\xi, t) f(t)\, dt \right\} d\xi$$

$$= f(x) + \lambda \int_a^b K_1(x, \xi) f(\xi)\, d\xi + \lambda^2 \int_a^b K_1(x, \xi) \int_a^b K_1(\xi, t) f(t)\, dt \cdot d\xi$$

$$= f(x) + \lambda \int_a^b K_1(x, \xi) f(\xi)\, d\xi + \lambda^2 \int_a^b \int_a^b K_1(x, \xi) \cdot K_1(\xi, t) f(t)\, dt\, d\xi.$$

Changing the order of integration in the second integral and interchanging ξ and t,

$$y_2(x) = f(x) + \lambda \int_a^b K_1(x, \xi) f(\xi)\, d\xi + \lambda^2 \int_a^b \left\{ \int_a^b K_1(x, t) K_1(t, \xi)\, dt \right\} f(\xi)\, d\xi$$

$$= f(x) + \lambda \int_a^b K_1(x, \xi) f(\xi)\, d\xi + \lambda^2 \int_a^b K_2(x, \xi) f(\xi)\, d\xi$$

where $\quad K_2(x, \xi) = \int_a^b K_1(x, t) \cdot K_1(t, \xi) dt.$

Proceeding in the same manner the nth approximation would be

$$y_n(x) = f(x) + \lambda \int_a^b K_1(x, \xi) f(\xi)\, d\xi + \lambda^2 \int_a^b K_2(x, \xi) f(\xi)\, d\xi + \cdots + \lambda^n \int_a^b K_n(x, \xi) f(\xi)\, d\xi, \qquad (13.50)$$

where $\quad K_m(x, \xi) = \int_a^b K_{m-1}(x, t) \cdot K_{m-1}(t, \xi) dt, \; m = 2, 3, 4, \ldots. n.$ $\qquad$ (13.51)

The classical iterative method converges to the true solution if $|\lambda| < \dfrac{1}{B}$ where

$$B^2 = \int_a^b \int_a^b K^2(x, \xi) dx\, d\xi. \qquad (13.52)$$

It is a sufficient condition, not necessary. For details see ref. [4].

This method is more of theoretical interest as the evaluation of the integral may not be possible. However, it has got an advantage in that it can be applied to nonlinear integral equations also.

Example 13.4

Using classical iterative method find the solution up to second iteration $y_2(x)$ only, to the integral equation

$$y(x) = x^2 + \lambda \int_0^1 (2x+3\xi)y(\xi)d\xi.$$

Find the sufficient condition on λ for convergence of the solution. Hence obtain $y_2(x)$ for $\lambda = 0.1$.

Solution

$$y(x) = x^2 + \lambda \int_0^1 (2x+3\xi)y(\xi)d\xi.$$

$$y_0(x) = x^2 = f(x)$$

$$y_1(x) = x^2 + \lambda \int_0^1 (2x+3\xi)y_0(\xi)d\xi = x^2 + \lambda \int_0^1 (2x+3\xi)f(\xi)d\xi$$

$$= x^2 + \lambda \int_0^1 (2x+3\xi)\xi^2 d\xi$$

$$= x^2 + \lambda\left(\frac{2}{3}x + \frac{3}{4}\right)$$

$$y_2(x) = x^2 + \lambda \int_0^1 (2x+3\xi)f(\xi)d\xi + \lambda^2 \int_0^1 \mathrm{K}_2(x,\xi)f(\xi)d\xi$$

$$\mathrm{K}_2(x,\xi) = \int_0^1 \mathrm{K}_1(x,t)\mathrm{K}_1(t,\xi)dt$$

$$= \int_0^1 (2x+3t)(2t+3\xi)dt = \int_0^1 (4xt+6x\xi+6t^2+9\xi t)dt$$

$$= 2x + 6x\xi + 2 + \frac{9}{2}\xi$$

$$\int_0^1 K_2(x, \xi)f(\xi)d\xi = \int_0^1 \left(2x + 6x\xi + 2 + \frac{9}{2}\xi\right)\xi^2 d\xi$$

$$= \frac{2}{3}x + \frac{3}{2}x + \frac{2}{3} + \frac{9}{8} = \frac{13}{6}x + \frac{43}{24}.$$

$$y_2(x) = x^2 + \lambda\left(\frac{2}{3}x + \frac{3}{4}\right) + \lambda^2\left(\frac{13}{6}x + \frac{43}{24}\right).$$

$$B^2 = \int_0^1\int_0^1 (2x + 3\xi)^2 dx d\xi = \int_0^1\int_0^1 (4x^2 + 12x\xi + 9\xi^2) dx d\xi = \frac{22}{3}$$

$1/B = 0.37$; for $\lambda = 0.1$ condition of convergence (13.52) is satisfied. Hence for $\lambda = 0.1$

$$y_2(x) = x^2 + 0.1\left(\frac{2}{3}x + \frac{3}{4}\right) + 0.01\left(\frac{13}{6}x + \frac{43}{24}\right)$$

$$= x^2 + 0.0883x + 0.0929.$$

Note: From Example 13.2 we get the analytical solution for $\lambda = 0.1$ as

$$y(x) = x^2 + 0.0962x + 0.0996.$$

13.7.3 Numerical method

In a numerical method the solution is obtained at the specific points in the domain $a \leq x \leq b$ of the integral equation

$$y(x) = \lambda \int_a^b K(x, \xi)y(\xi)d\xi + f(x). \qquad (13.53)$$

The integral in (13.53) is approximated by any of the integration formulas discussed in Chapter 6. If Cotes formulas are used then the points at which the solution is required are selected in advance, usually equi-spaced while if Gaussian type formulas are employed, the points happen to be zeros of some orthogonal polynomials, known as quadrature points. Let us first discuss the method when the nodal points are chosen in advance.

A. Using Cotes Integration Formula

The interval $[a, b]$ is subdivided into say $(n-1)$ subintervals such that $\dfrac{b-a}{n-1} = \Delta x = \Delta\xi$ and let the nodal points be denoted by $a = x_1, x_2 \ldots . x_n = b$ where $x_i = a + (i-1)\Delta x$, and we are required to find the solution at $x = x_i$ denoted by y_i, $i = 1(1)n$.

At $x = x_i$ eqn. (13.53) may be written as

$$y_i = \lambda \int_a^b \mathrm{K}(x_i, \xi) y(\xi) d\xi + f_i \tag{13.54}$$

where $f_i = f(x_i)$.
Replacing the integral in (13.54) by a Cotes formula taking abscissas $\xi_j = x_j$, $j = 1(1)n$

$$y_i - \lambda \sum_{j=1}^{n} \omega_j \mathrm{K}(x_i, \xi_j) y_j = f_i,$$

where $y_j = y(\xi_j)$ and ω_j are weights.
Again,

$$y_i - \lambda \sum_{j=1}^{n} \omega_j \mathrm{K}_{ij} y_j = f_i, \tag{13.55}$$

where $\mathrm{K}_{ij} = \mathrm{K}(x_i, \xi_j)$.
Writing the equation (13.55) for $i = 1(1)n$ in matrix form

$$\boldsymbol{y} - \lambda \mathrm{KD} \boldsymbol{y} = \boldsymbol{f},$$

or $$(\mathrm{I} - \lambda \mathrm{KD}) \boldsymbol{y} = \boldsymbol{f}, \tag{13.56}$$

where

$$\mathrm{K} = \begin{bmatrix} \mathrm{K}_{11} & \mathrm{K}_{12} & \cdots & \mathrm{K}_{1n} \\ \mathrm{K}_{21} & \mathrm{K}_{22} & \cdots & \mathrm{K}_{2n} \\ \vdots & \vdots & & \vdots \\ \mathrm{K}_{n1} & \mathrm{K}_{n2} & \cdots & \mathrm{K}_{nn} \end{bmatrix}, \ \mathrm{D} = \begin{bmatrix} \omega_1 & 0 & \cdots & 0 \\ 0 & \omega_2 & \cdots & 0 \\ \vdots & \vdots & \ddots & \vdots \\ 0 & 0 & \cdots & \omega_n \end{bmatrix}$$

$$\boldsymbol{y}^{\mathrm{T}} = (y_1 \ \ y_2 \ \ \ldots \ \ y_n), \ \boldsymbol{f}^{\mathrm{T}} = (f_1 \ \ f_2 \ \ \cdots \ \ f_n).$$

If matrix $\mathrm{I} - \lambda \mathrm{KD}$ is not singular, i.e., $|\mathrm{I} - \lambda \mathrm{KD}| \neq 0$, then system of equations (13.56) may be solved for $\boldsymbol{y}$ giving

$$\boldsymbol{y} = (\mathrm{I} - \lambda \mathrm{KD})^{-1} \boldsymbol{f}. \tag{13.57}$$

If $f(x) = 0$, then vector $\boldsymbol{f} = \mathbf{0}$ and (13.56) reduces to an eigenvalue problem. That is, for the system (13.56) to have a non-trivial solution (with $\boldsymbol{f} = \mathbf{0}$)

$$|\mathrm{I} - \lambda \mathrm{KD}| = 0. \tag{13.58}$$

Eqn (13.58) is a polynomial equation of degree n in λ and should have n roots of λ, say $\lambda = \lambda_1, \lambda_2, \ldots. \lambda_n$. If the roots are real, a solution $\boldsymbol{y}$ will be obtained for each value of λ by solving

$$(\mathrm{I} - \lambda \mathrm{KD})\boldsymbol{y} = \mathbf{0}. \tag{13.59}$$

The homogeneous system (13.59) can also be written as

$$\lambda \mathrm{KD}\boldsymbol{y} = \boldsymbol{y}$$

or

$$\mathrm{KD}\boldsymbol{y} = \frac{1}{\lambda}\boldsymbol{y} = \mu \boldsymbol{y} \tag{13.60}$$

where μ is an e.value of matrix KD with the corresponding vector $\boldsymbol{y}$ and that $\lambda = 1/\mu$.

If kernel is symmetric, i.e., $\mathrm{K}(x, \xi) = \mathrm{K}(\xi, x)$ then the system (13.60) may be reduced to finding the e.values of symmetric matrix as we can write

$$\mathrm{D}^{1/2}\mathrm{KD}^{1/2}\mathrm{D}^{1/2}\boldsymbol{y} = \mu \mathrm{D}^{1/2}\boldsymbol{y}$$

or

$$\mathrm{D}^{1/2}\mathrm{KD}^{1/2}z = \mu z, \tag{13.61}$$

where

$$\mathrm{D}^{1/2}\boldsymbol{y} = z. \tag{13.62}$$

It may be noted that D is a diagonal matrix, hence symmetric. Therefore $\mathrm{D}^{1/2}\mathrm{KD}^{1/2}$ is also symmetric having the same e.values as KD but e.vectors hold a relation given by (13.62) and can be obtained from the relation.

$$\boldsymbol{y} = (\mathrm{D}^{1/2})^{-1}z. \tag{13.63}$$

The methods discussed in Chapter 10 can be employed in this case.

B. Using Gauss–Legendre Quadrature Formula

First of all we must change the limits of integration from $[a, b]$ to $[-1, 1]$ through the transformation

$$\left.\begin{aligned} t &= \frac{2x-a-b}{(b-a)} \text{ or } x = \frac{1}{2}\{(b+a)+(b-a)t\} \\ \text{with} \quad dx &= \frac{b-a}{2}dt. \end{aligned}\right\} \tag{13.64}$$

The eqn (13.53) after transformation can be written as

$$y(t) - \frac{b-a}{2}\lambda\int_{-1}^{1} \mathrm{K}\left\{\frac{b+a+(b-a)t}{2},\ \xi\right\}y(\xi)d\xi = f\left\{\frac{b+a+(b-a)t}{2}\right\}. \tag{13.65a}$$

For brevity writing $\tau = \dfrac{b+a+(b-a)t}{2}$, the above eqn. (13.65a) becomes

$$y(t) - \frac{b-a}{2}\lambda\int_{-1}^{1} \mathrm{K}(\tau,\ \xi)y(\xi)d\xi = f(\tau). \tag{13.65b}$$

If n^{th} order formula is to be used to approximate the integral in (13.65a), then the abscissas $t = t_i,\ i = 1(1)n$ are given by the zero of the Legendre polynomial $\mathrm{P}_n(t)$. We shall illustrate the method for $n = 3$.

For $t = t_i$, eqn. (13.65b) can be written as

$$y(t_i) - \frac{b-a}{2}\lambda\int_{-1}^{1} \mathrm{K}(\tau_i,\ \xi)y(\xi)\,d\xi = f(\tau_i). \tag{13.66}$$

Taking $\xi_j = \tau_j$ for evaluating the integral in (13.66) by Gaussian quadrature we can write

$$y_i - \frac{b-a}{2}\lambda\sum_{j=1}^{3} \mathrm{K}(\tau_i,\ \xi_j)\omega_j y_j = f(\tau_i)$$

$$\text{or} \qquad y_i - \frac{b-a}{2}\lambda\sum_{j=1}^{3} \mathrm{K}_{ij}\omega_j y_j = f_i. \tag{13.67}$$

Writing (13.67) for $i = 1, 2, 3$ we can express them in matrix form as

$$\begin{bmatrix} y_1 \\ y_2 \\ y_3 \end{bmatrix} - \frac{b-a}{2}\lambda\begin{bmatrix} \mathrm{K}_{11}\omega_1 & \mathrm{K}_{12}\omega_2 & \mathrm{K}_{13}\omega_3 \\ \mathrm{K}_{21}\omega_1 & \mathrm{K}_{22}\omega_2 & \mathrm{K}_{23}\omega_3 \\ \mathrm{K}_{31}\omega_1 & \mathrm{K}_{32}\omega_2 & \mathrm{K}_{33}\omega_3 \end{bmatrix}\begin{bmatrix} y_1 \\ y_2 \\ y_3 \end{bmatrix} = \begin{bmatrix} f_1 \\ f_2 \\ f_3 \end{bmatrix}. \tag{13.68}$$

For $n=3$, the quadrature points t_i and the weights $\omega_i,\ i=1,\ 2,\ 3$ are given below:

i	$=$	1	2	3
t_i	$=$	$-\sqrt{\frac{3}{5}}$	0	$\sqrt{\frac{3}{5}}$
ω_i	$=$	$5/9$	$8/9$	$5/9.$

Solving (13.68) we get the values of $y(t)$ at $t_1,\ t_2$ and t_3. The values at the corresponding point x in $[a,\ b]$ can be found from the relation (13.64).

C. Using Chebyshev's Formula

As stated in Chapter 6, the Chebyshev's n^{th} order quadrature formula is given as

$$\int_{-1}^{1} f(x)dx = \frac{2}{n}\sum_{r=1}^{n} f(x_r), \tag{13.69}$$

where the quadrature points x_r are given by the zeros of the polynomial of degre n obtained from the expansion

$$\mathrm{F}(x) = x^n - \frac{n}{3!}x^{n-2} + \frac{n}{5!}\left(\frac{5n}{3} - 6\right)x^{n-4} - \frac{n}{7!}\left(\frac{35n^2}{9} - 42n + 120\right)x^{n-6} + \cdots \tag{13.70}$$

It may also be mentioned that for $n \le 7$, the zeros of the corresponding polynomials are real. But for $n=8$, the imaginary zeros occur. Hence due to its simple form, formula (13.69) may be used for $n \le 7$. It was also stated that this formula is exact only up to polynomial of degree n while Gauss–Legendre formula is exact for polynomials of degree up to $2n-1$.

In order to use Chebyshev's formula (13.69) we should change the limits of integration as in the Gauss quadrature formula. We arrive at the same equations as given by (13.68). In the present case for $n=3$, the quadrature points and weights would be

$$t_1 = -\sqrt{\frac{1}{2}},\ t_2 = 0,\ t_3 = \sqrt{\frac{1}{2}}$$

$$\omega_1 = \frac{2}{3},\ \omega_2 = \frac{2}{3},\ \omega_3 = \frac{2}{3}.$$

Example 13.5

Set up the relevant equations for solving the integral equation

$$y(x) - \lambda \int_a^b (x+\xi)y(\xi)\,d\xi = x$$

approximating the integral by trapezoidal rule after dividing the interval [0, 1] into two equal parts.

Also find the solution at the nodal points for $\lambda = 1/2$.

Solution Length of the subinterval $h = \dfrac{1-0}{2} = 0.5$

Nodal points (abscissas) are $x_1 = 0,\ x_2 = 0.5,\ x_3 = 1.0$.

At $x = x_i$, the integral equation can be written as follows:

Approximating the integral by trapezoidal rule with abscissas $\xi_j = x_j,\ \ j = 1,\ 2,\ 3$ we get

$$y_i - \frac{\lambda h}{2}[(x_i+\xi_1)y_1 + 2(x_i+\xi_2)y_2 + (x_i+\xi_3)y_3] = x_i$$

or $$y_i - \frac{\lambda}{4}[(x_i+0)y_1 + 2(x_i+0.5)y_2 + (x_i+1.0)y_3] = x_i$$

Writing the equations for $i = 1,\ 2,\ 3$ we get

$$y_1 - \frac{\lambda}{4}y_2 - \frac{\lambda}{4}y_3 = 0$$

$$-\frac{\lambda}{8}y_1 + \left(1 - \frac{\lambda}{2}\right)y_2 - \frac{3}{8}\lambda y_3 = 0.5$$

$$-\frac{\lambda}{4}y_1 - \frac{3\lambda}{4}y_2 + \left(1 - \frac{\lambda}{2}\right)y_3 = 1.0$$

For $\lambda = \dfrac{1}{2}$, the above equations may be written as

$$8y_1 - y_2 - y_3 = 0$$

$$-y_1 + 12y_2 - 3y_3 = 8$$

$$-y_1 - 3y_2 + 6y_3 = 8$$

Solving these equations exactly (in fractions rather than decimal)

$$y_1 = \frac{2}{5},\ y_2 = \frac{6}{5},\ y_3 = 2.$$

Example 13.6

Set up the relevant equations for solving the integral equation

$$y(x) - \lambda \int_0^1 K(x,\ \xi) y(\xi) d\xi = x$$

approximating the integral by 3-point Gauss quadrature formula. What is the value of y at $x = 0.5$ for $\lambda = 1/2$, when $K(x,\ \xi) = x + \xi$.

Solution Changing the variable from x to t so that limits of integration are $(-1,\ 1)$, put

$$t = \frac{2x-1}{1} \text{ or } x = \frac{t+1}{2};\ dx = \frac{dt}{2}.$$

The quadrature points are

$$t_1 = -\sqrt{\frac{3}{5}} = -0.7746,\ t_2 = 0,\ t_3 = \sqrt{\frac{3}{5}} = 0.7746$$

$$\tau = \frac{1+t}{2};\ \tau_1 = \frac{1-0.7746}{2} = 0.1127,\ \tau_2 = \frac{1+0}{2} = 0.5,\ \tau_3 = \frac{1+0.7746}{2} = 0.8873.$$

The relevant equations are

$$\begin{bmatrix} y_1 \\ y_2 \\ y_3 \end{bmatrix} - \frac{\lambda}{2} \begin{bmatrix} (0.1127+0.1127)\times\frac{5}{9} & (0.1127+0.5)\times\frac{8}{9} & (0.1127+0.8873)\times\frac{5}{9} \\ (0.5+0.1127)\times\frac{5}{9} & (0.5+0.5)\times\frac{8}{9} & (0.5+0.8873)\times\frac{5}{9} \\ (0.8873+0.1127)\times\frac{5}{9} & (0.8873+0.5)\times\frac{8}{9} & (0.8873+0.8873)\times\frac{5}{9} \end{bmatrix} \begin{bmatrix} y_1 \\ y_2 \\ y_3 \end{bmatrix}$$

$$= \begin{bmatrix} 0.1127 \\ 0.5 \\ 0.8873 \end{bmatrix}$$

For $\lambda = \frac{1}{2}$, the above equations become

$$0.9687y_1 - 0.1362y_2 - 0.1389y_3 = 0.1127$$

$$-0.0851y_1 + 0.7778y_2 - 0.1927y_3 = 0.5$$

$$-0.1389y_1 - 0.3083y_2 + 0.7535y_3 = 0.8873$$

Solving the above system of equations

$y_3 = 1.7366,\ y_2 = 1.305, y_1 = 0.5243$

When $x = 0.5,\ t = 2x - 1 = 0$
The value corresponding to $t = 0$ is $y_2 = 1.1305$

Note: Analytical solution is $y(0.5) = 1.1304$.

y_1 corresponds to the value of the y at $x = 0.1127$ and

y_2 corresponds to the value at $x = 0.8873$

The values at $x = 0$ and $x = 1.0$ are not available.

13.8 Methods for Solving Volterra Equation

As stated before the Volterra equation represents an initial value problem (IVP). Also that a Volterra equation of the first kind could be converted to an equation of the second kind, if conditions satisfied. We discuss below methods for solving Volterra equation, namely, Numerical, Taylor's Series and Iterative methods.

13.8.1 Numerical method

Let us suppose the given integral equation is

$$y(x) - \lambda \int_a^x \mathrm{K}(x,\ \xi)\, y(\xi)\, d\xi = f(x). \tag{13.71}$$

Like a differential equation initial value type we solve the integral equation (13.71) in a step-by-step manner. A small step-size h is chosen and the values of $y(x)$ are computed at $x = x_r = a + rh$ successively for $r = 1,\ 2,\ \ldots$. The integral in (13.71) is approximated by a suitable quadrature formula between $x = a$ and $x = a + rh$. Let us suppose trapezoidal formula is used. Denoting $y(x_i) = y_i, f(x_i) = f_i$ and $\mathrm{K}(x_i,\ \xi_j) = \mathrm{K}_{i,j}$ we proceed as follows:

Put $x = a$ in (13.71) giving $y(a) = f(a)$. Let $x_0 = a$ so that $y(x_0) = f(x_0)$ or $y_0 = f_0$. Taking quadrature points as $\xi_i = x_i$ and approximating the integral in (13.71) by trapezoidal rule we can write the relevant equations for $x = x_r,\ r = 1(1)n$ as

$$\left.\begin{aligned}
y_1 - \lambda \frac{h}{2}\mathrm{K}_{10}y_0 - \lambda \frac{h}{2}\mathrm{K}_{11}y_1 &= f_1 \\
y_2 - \lambda \frac{h}{2}\mathrm{K}_{20}y_0 - \lambda h\mathrm{K}_{21}y_1 - \frac{\lambda h}{2}\mathrm{K}_{22}y_2 &= f_2 \\
\vdots \qquad\qquad &\qquad \vdots \\
y_n - \lambda \frac{h}{2}\mathrm{K}_{n0}y_0 - \lambda h\mathrm{K}_{n1}y_1 - \cdots - \lambda h\mathrm{K}_{n,\ n-1} - \frac{\lambda h}{2}\mathrm{K}_{nn}y_n &= f_n
\end{aligned}\right\} \tag{13.72}$$

After rearrangement, the equations in (13.72) along with $y_0 = f_0$ can be written as

$$\left.\begin{array}{lll} y_0 & = & f_0 \\ -\frac{\lambda h}{2}K_{10}y_0 + \left(1 - \frac{\lambda h}{2}K_{11}\right) y_1 & = & f_1 \\ \dots\dots\dots & \dots & \dots \\ -\frac{\lambda h}{2}K_{n0}y_0 - \lambda h K_{n1}y_1 - \cdots - \lambda h K_{n,\, n-1}y_{n-1} + \left(1 - \frac{\lambda h}{2}K_{nn}\right) y_n & = & f_n \end{array}\right\} \quad (13.73)$$

Clearly, by putting the value of y_0 in the second equation of (13.73), y_1 can be computed; then putting the values of y_0 and y_1 in the third equation, y_2 can be computed and so on. It may be important to note that Volterra equation does not lead to an eigenvalue problem when $f(x) = 0$ since, in that case the only solution available will be a trivial solution, i.e., $y_0 = y_1 = \ldots = y_n = 0$.

For a general quadrature formula

$$\begin{bmatrix} 1 & & & & 0 \\ -\lambda\omega_0 K_{10} & (1-\lambda\omega_1 K_{11}) & & & \\ -\lambda\omega_0 K_{20} & -\lambda\omega_1 K_{21} & (1-\lambda\omega_2 K_{22}) & & \\ \vdots & & & \ddots & \\ -\lambda\omega_0 K_{n0} & -\lambda\omega_1 K_{n1} & \cdots & \cdots & (1-\lambda\omega_n K_{nn}) \end{bmatrix} \begin{bmatrix} y_0 \\ y_1 \\ y_2 \\ \vdots \\ y_n \end{bmatrix} = \begin{bmatrix} f_0 \\ f_1 \\ f_2 \\ \vdots \\ f_n \end{bmatrix} \quad (13.74)$$

The determinant of the matrix on the left side of (13.74) is given by

$$\Pi = (1 - \lambda\omega_1 K_{11})(1 - \lambda\omega_2 K_{22}) \ldots (1 - \lambda\omega_n K_{nn}).$$
$$\geq (1 - |\lambda|\omega_1 K_{11})(1 - |\lambda|\omega_2 K_{22}) \ldots (1 - |\lambda|\omega_n K_{nn}).$$

It can be shown that if the weights become smaller and smaller, i.e $h \to 0$, the determinant Π can not become zero as the value of the kernel remains bounded; hence no e.values will exist, even when $f(x) = 0$.

13.8.2 Taylor's series method

The Taylor's expansion of $y(x)$ about the point $x = a$ can be written (for $x \geq a$) as

$$y(x) = y(a) + (x-a)y'(a) + \frac{(x-a)^2}{2!}y''(a) + \cdots + \frac{(x-a)^n}{n!}y^n(a) + R_{n+1} \tag{13.75a}$$

where the truncation error

$$R_{n+1}(x) = \frac{(x-a)^{n+1}}{(n+1)!}y^{n+1}(\xi),\ a \leq \xi \leq a+x \tag{13.75b}$$

or equivalently $\xi = a + \theta x,\ 0 \leq \theta \leq 1$.

We get the values of $y(a),\ y'(a),\ y''(a)$, etc., by differentiating the integral equation successively and putting $x = a$ in it. Since no eigenvalues are involved in the Volterra equation, let us consider the equation

$$y(x) - \int_a^x K(x, \xi)y(\xi)d\xi = f(x). \tag{13.76}$$

Putting $x = a$ in (13.76) gives $y(a) = f(a)$.
Differentiating (13.76) once w.r.t. x,

$$y'(x) - \left\{ \int_a^x \frac{\partial K(x, \xi)}{\partial x} \cdot y(\xi)d\xi + K(x, x)y(x) \right\} = f'(x) \tag{13.77}$$

Putting $x = a$ in (13.77), we get

$$\begin{aligned} y'(a) &= K(a, a)\, y(a) + f'(a) \\ &= K(a, a)f(a) + f'(a). \end{aligned} \tag{13.78}$$

We can differentiate (13.77) again w.r.t. x and put $x = a$ to get $y''(a)$ and so on. Obviously function value and its derivative should remain bounded at $x = a$.

As should be clear from the truncation error (13.75b) we should use the formula (13.75a), without R_{n+1}, for the values of x in the neighbourhood of $x = a$, i.e., $(x-a)$ should not be too large. The formula can be used to compute only a first few values of $y(x)$, say $a+h,\ a+2h,\ a+3h$, for h small. Then again a new Taylor's series has to be constructed expanding $y(x)$ about $x = a+3h$ or so.

Note that Volterra equation of first kind should be changed to an equation of the second kind, as suggested in Sec 13.3, for applying this method.

13.8.3 Iterative method

An approximate solution in closed form can be obtained by applying an iterative scheme. For the equation (13.76) the scheme may be written as

$$y_{n+1}(x) = \int_a^x K(x, \xi) y_n(\xi) d\xi + f(x), \tag{13.79}$$

where $y_{n+1}(x)$ denotes $(n+1)^{\text{th}}$ iteration, $n = 0, 1, 2, \ldots$.

We can start by taking $y_0(x) = f(a)$ which after substituting in (13.79) produces $y_1(x)$; again putting $y_1(\xi)$ in (13.79) yields the second approximation $y_2(x)$ and so on.

In this method also like Taylor's series method, the Volterra equation of the first kind has to be converted to the equation of the second kind before this method can be applied. But as stated before the Volterra equation of first kind may not be converted to an equation of the second kind always.

The process of iteration may be carried until desired number of terms agree in the two successive iterations. Then the solution may be represented by the final iteration, retaining one extra term than the number of terms that agree.

Example 13.7

Find the solution of the Volterra equation

$$y(x) - \int_a^x (1 + 2x - 2\xi) y(\xi) d\xi + f(x) = x + \frac{2}{3}x^3$$

for $x = 0(0.1)0.6$ approximating the integral by trapezoidal rule.

Solution Putting $x = 0$ in the equation gives

$$y(0) = y_0 = 0.$$

$$h = 0.1; \; y_i = y(x_i), \; \text{for } x_i = ih, \; i = 0(1)6$$
$$K(x, \xi) = 1 + 2x - 2\xi.$$

Replacing the integral by trapezoidal formula for x_i, the above equation can be written as

$$y_i - \frac{h}{2}\left[K(x_i, \xi_0)y_0 + 2\sum_{j=1}^{i-1} K(x_i, \xi_j)y_j + K(x_i, \xi_i)y_i\right] = x_i + \frac{2}{3}x_i^3$$

or $$\left(1-\frac{h}{2}K_{ii}\right)y_i = h\sum_{j=1}^{i-1}K_{ij} + x_i + \frac{2}{3}x_i^3, \quad \because y_0 = 0.$$

In the present problem $K_{ii} = 1+2x_i-2\xi_i = 1,\ \xi_i = x_i$.

$$0.95y_i = 0.1\times\sum_{j=1}^{i-1}K_{i,\,j}y_j + x_i + \frac{2}{3}x_i^3,\ \ i>1 \tag{1}$$

$\boldsymbol{x = x_1 = 0.1}$

$$y_1 - \frac{0.1}{2}[K_{10}y_0 + K_{11}y_1] = 0.1 + \frac{2}{3}\times 0.001$$

$$0.95y_1 = 0.100667 \ \text{ or } \ y_1 = 0.105965$$

$\boldsymbol{x = x_2 = 0.2}$

$$0.95y_2 = 0.1\times K_{21}y_1 + 0.2 + \frac{0.016}{3} = 0.1(1+0.4-2\times 0.1)\times 0.105965 + 0.2 + \frac{0.016}{3}$$

$$y_2 = 0.229525$$

$\boldsymbol{x = x_3 = 0.3}$

$$0.95y_3 = 0.1[K_{31}y_1 + K_{32}y_2] + 0.3 + \frac{2}{3}\times 0.027$$

$$= 0.1[1.4\times 0.105965 + 1.2\times 0.229525] + 0.3 + 0.018$$

$$y_3 = 0.379345$$

$\boldsymbol{x = x_4 = 0.4}$

$$0.95y_4 = 0.1[1.6\times 0.105965 + 1.4\times 0.229525 + 1.2\times 0.379345] + 0.4 + \frac{2\times 0.064}{3}$$

$$y_4 = 0.565553$$

$\boldsymbol{x = x_5 = 0.5}$

$$0.95y_5 = 0.1[1.8\times 0.105965 + 1.6\times 0.229525 + 1.4\times 0.379345 + 1.2\times 0.56553]$$
$$+0.5 + \frac{2\times 0.125}{3}$$

$$y_5 = 0.800111$$

$\boldsymbol{x = x_6 = 0.6}$

$$0.95y_6 = 0.1[2 \times 0.105965 + 1.8 \times 0.229525 + 1.6 \times 0.379345 + 1.4 \times 0.565553 + 1.2 \times 0.800111] + 0.6 + \frac{2 \times 0.216}{3}$$

$$y_6 = 1.097256.$$

Example 13.8

Find the solution of the integral equation given in Example 13.7 for $x = 0(0.2)0.6$ using trapezoidal rule for approximating the integral.

Compare your result with that of Example 13.7.

Solution $x = x_0 = 0, \ y(0) = y_0 = 0; \ h = 0.2; \ x_1 = 0.2, \ x_2 = 0.4, \ x_3 = 0.6$

$\boldsymbol{x = x_1 = 0.2}$

$$y_1 - \frac{0.2}{2}[K_{10}y_0 + K_{11}y_1] = 0.2 + \frac{2 \times 0.008}{3}$$

$$y_1 - 0.1[1.4y_0 + 1 \times y_1] = 0.20533$$

$$0.9y_1 = 0.20533$$

$$y_1 = 0.22814$$

$\boldsymbol{x = x_2 = 0.4}$

$$0.9y_2 = 0.1 \times 2 \times 1.4 \times 0.22814 + 0.4 + 0.042667$$

$$y_2 = 0.56283$$

$\boldsymbol{x = x_3 = 0.6}$

$$0.9y_3 = 0.1[3.6y_1 + 2.8y_2] + 0.6 + \frac{2 \times 0.216}{3}$$

$$y_3 = 0.98372$$

Comparison of results

Method \ *x*	*0*	*0.2*	*0.4*	*0.6*
$h = .2$	*0*	0.22814	0.56283	0.98372
$h = .1$	*0*	0.22952	0.56555	1.09726
Analytical	*0*	0.22995	0.56647	1.09872

The corresponding differential equation is

$$y'' - y' - 2y = 4x, \ \ y(0) = 0, \ \ y'(0) = 1.$$

The analytical solution is $y = \frac{2}{3}e^{2x} + \frac{5}{3}e^{-x} - 2x + 1.$

Example 13.9

Given a Volterra equation

$$y(x) = \int_0^x (3 - 2x + 2\xi) y(\xi) d\xi + x + \frac{1}{3}x^3.$$

Find Taylor's series solution up to x^5.

Solution

$$y(x) = y(0) + xy'(0) + \frac{x^2}{2}y''(0) + \frac{x^3}{6}y'''(0) + \frac{x^4}{24}y^{iv}(0) + \frac{x^5}{120}y^{v}(0).$$

$$y(0) = 0$$

$$y'(x) = -2\int_0^x y(\xi)d\xi + 3y(x) + 1 + x^2$$

$$y'(0) = 3 \times y(0) + 1 = 1$$

$$y''(x) = -2y(x) + 3y'(x) + 2x$$

$$y''(0) = 3 \times 1 = 3$$

$$y'''(x) = -2y'(x) + 3y''(x) + 2$$

$$y'''(0) = -2 \times 1 + 3 \times 3 + 2 = 9$$

$$y^{iv}(x) = -2y''(x) + 3y'''(x)$$

$$y^{iv}(0) = -2 \times 3 + 3 \times 9 = 21$$

$$y^{v}(x) = -2y'''(x) + 3y^{iv}(x)$$

$$y^{v}(0) = -2 \times 9 + 3 \times 21 = 45$$

Taylor's series solution is

$$y(x) = 0 + x + \frac{x^2}{2} \times 3 + \frac{x^3}{6} \times 9 + \frac{x^4}{24} \times 21 + \frac{x^5}{120} \times 45$$

$$= x + \frac{3}{2}x^2 + \frac{3}{2}x^3 + \frac{7}{8}x^4 + \frac{3}{8}x^5$$

Note: The corresponding differential equations is

$$y'' - 3y' + 2y = 2x; \;\; y(0) = 0, \;\; y'(0) = 1.$$

The analytical solution is

$$y = \frac{3}{2}e^{2x} - 3e^x + x + \frac{3}{2}.$$

The series solution is

$$y(x) \simeq x + \frac{3}{2}x^2 + \frac{3}{2}x^3 + \frac{7}{8}x^4 + \frac{3}{8}x^5 + \cdots.$$

Example 13.10

Using iterative method find the solution of the Volterra equation

$$y(x) = \int_0^x (3 - 2x + 2\xi)y(\xi)d\xi + x + \frac{1}{3}x^3.$$

Perform the iterations until at least three terms agree. Express the final solution containing four terms.

Solution The iterative scheme is

$$y_{n+1}(x) = \int_0^x (3 - 2x + 2\xi)y_n(\xi)d\xi + x + \frac{1}{3}x^3.$$

$$y(0) = 0$$

Let $y_0(x) = 0$.

$$y_1(x) = \int_0^x (3 - 2x + 2\xi)y_0(\xi)d\xi + x + \frac{1}{3}x^3$$

$$= x + \frac{1}{3}x^3$$

$$y_2(x) = \int_0^x (3 - 2x + 2\xi)y_1(\xi)d\xi + x + \frac{1}{3}x^3$$

$$= \int_0^x (3 - 2x + 2\xi)\left(\xi + \frac{1}{3}\xi^3\right)d\xi + x + \frac{1}{3}x^3$$

$$= (3-2x)\left(\frac{x^2}{2}+\frac{x^4}{12}\right)+2\left(\frac{x^3}{3}+\frac{x^5}{15}\right)+x+\frac{1}{3}x^3$$

$$= x+\frac{3}{2}x^2+\frac{x^4}{4}-\frac{x^5}{30}$$

$$y_3(x) = \int_0^x (3-2x+2\xi)\left(\xi+\frac{3}{2}\xi^2+\frac{\xi^4}{4}-\frac{\xi^5}{30}\right)d\xi + x+\frac{1}{3}x^3$$

$$= (3-2x)\left(\frac{x^2}{2}+\frac{x^3}{2}+\frac{x^5}{20}-\frac{x^6}{180}\right)+2\left(\frac{x^3}{3}+\frac{3}{8}x^4+\frac{x^6}{24}\right)+x+\frac{1}{3}x^3$$

$$= x+\frac{3}{2}x^2+\frac{3}{2}x^3-\frac{x^4}{4}+\frac{3}{20}x^5+\frac{x^6}{15}$$

$$y_4(x) = \int_0^x (3-2x+2\xi)\left(\xi+\frac{3}{2}\xi^2+\frac{3}{2}\xi^3-\frac{\xi^4}{4}+\frac{3}{20}\xi^5\right)d\xi + x+\frac{1}{3}x^3$$

$$= (3-2x)\left(\frac{x^2}{2}+\frac{x^3}{2}+\frac{3}{8}x^4-\frac{x^5}{20}\right)+2\left(\frac{x^3}{3}+\frac{3}{8}x^4+\frac{3}{10}x^5\right)+x+\frac{1}{3}x^3$$

$$= x+\frac{3}{2}x^2+\frac{3}{2}x^3+\frac{7}{8}x^4-\frac{3}{10}x^5$$

Solution is

$$y(x) \simeq x+\frac{3}{2}x^2+\frac{3}{2}x^3+\frac{7}{8}x^4$$

Example 13.11

Find the Taylor's series solution of the Volterra equation of first kind

$$\int_0^x (x+2\xi)y(\xi)d\xi = 2x^2-\frac{5}{12}x^4+\frac{x^6}{45}$$

Compute up to x^4 only.

Solution Differentiating the integral eqn. w.r.t. x

$$\int_0^x y(\xi)d\xi + 3xy(x) = 4x-\frac{5}{3}x^3+\frac{2}{15}x^5 \qquad ....(1)$$

By putting $x=0$ in the above equation value of $y(0)$ can not be obtained as we get $0=0$.

Therefore, differentiating the above equation again, we get

$$y(x)\ +3y(x)\ +3xy'(x)\ =4-5x^2+\frac{2}{3}x^4$$

or $$4y(x)+3xy'(x)=4-5x^2+\frac{2}{3}x^4 \qquad .....(2)$$

Putting $x=0$ in eqn. (2) gives

$$4y(0)=4 \text{ or } y(0)=1 \text{ since } y'(0) \text{ is to be bounded.}$$

Differentiating (2) further

$$4y'(x)+3y'(x)+3xy''(x)=-10x+\frac{8}{3}x^3$$

or $$7y'(x)+3xy''(x)=-10x+\frac{8}{3}x^3 \qquad ....(3)$$

$$y'(0)=0, \ \because\ y''(0) \text{ is bounded}$$

Differentiating (3) gives

$$10y''(x)+3xy'''(x)=-10+8x^2 \qquad ....(4)$$

$$y''(0)=-1.$$

Differentiating (4)

$$13y'''(x)+3xy^{iv}(x)=16x \qquad ....(5)$$

and $$y'''(0)=0.$$

Again after differentiating (5) and putting $x=0$

$$y^{iv}(0)=1.$$

Hence the Taylor's series solution is

$$y(x)=1-\frac{x^2}{2}+\frac{x^4}{24}.$$

Exercise 13

13.1 Find the analytical solution of the integral equation

$$y(x) - \lambda \int_0^1 (x+\xi)y(\xi)d\xi = x.$$

What is the solution for $\lambda = \frac{1}{2}$?
Also find solution (s) when the right side is zero.

13.2 Using Simpson's $\frac{1}{3}$ rule, set up the relevant equations for solving the integral equation of Exercise 13.1.
Divide the integral $[0, 1]$ into two equal parts. Also find the solution for $\lambda = \frac{1}{2}$.

13.3 Find the solution of the Volterra equation

$$y(x) - \int_0^x (1+2x-2\xi)y(\xi)d\xi = x + \frac{2}{3}x^3$$

in terms of Taylor's series up to powers of x^5.

13.4 Using iterative method find the solution of the Volterra equation of Exercise 13.3.

13.5 Find the solution $y(x)$ for $x = 0(0.2)0.6$ numerically of the Volterra equation of the first kind $\int_0^x (x+2\xi)y(\xi)d\xi = 2x^2 - \frac{5}{12}x^4 + \frac{1}{45}x^6$
Use trapezoidal rule for approximating the integral.
Hint: Differentiate the given eqn w.r.t. x and obtain

$$3xy(x) + \int_0^x y(\xi)d\xi = 4x - \frac{5}{3}x^3 + \frac{2}{15}x^5.$$

Use it to compute $y(x)$. To obtain $y(0) = 1$, differentiate it once more; (See Example 13.11).

References and Some Useful Related Books/Papers

1. Kantorovich, L.V. and Krylov, V.I., *Approximate Methods of Higher Analysis,* Interscience Publishers.
2. Mikhlin, S.G., *Integral Equation,* Pergamon.
3. Myskis, A.D., (Translated by Volosov V.M. and Volosova I.G.) *Advanced Mathematics for Engineers: Special Courses,* Mir Publishers.
4. Tricomi, F.G., *Integral Equations,* Interscience Publishers.

14

Difference Equations

14.1 Introduction

A difference equation of order r has the form

$$a_0y(n)+a_1y(n-1)+a_2y(n-2)+\ldots.+a_ry(n-r)=f(n), \tag{14.1a}$$

where $a_0 \neq 0,\ a_r \neq 0$; r and n are positive integers such that $1 \leq r < n$.
Another way of expressing an rth order difference equation can be

$$a_0y(n+r)+a_1y(n+r-1)+\ldots.+a_ry(n)=f(n), \tag{14.1b}$$

where r is a positive integer $r \geq 1$.
In eqns (14.1a) or (14.1b), y is a dependent variable and n an independent variable (argument) which can assume non-negative integral values only; the coefficients a's can be function of n or constants. The order of a difference equation is obtained by subtracting the lowest argument of y from its highest. Obviously, above equations are linear in y. If the function $f(n)$ on the right side is zero, the equation is said to be homogeneous, otherwise non-homogeneous. The solution of a difference equation means to express $y(n)$ in terms of n. The general solution of an rth order difference equation contains r arbitrary constants. Therefore r conditions on y should be prescribed to determine these constants. Replacing arguments by suffices, the nth order difference equation from (14.1a) may be written as

$$a_0y_n+a_1y_{n-1}+a_2y_{n-2}+\ldots.+a_ny_0=f(n). \tag{14.2}$$

It should be noted that if the values of $y_0,\ y_1,\ \ldots.\ y_{n-1}$ are known, the value of y_n can be computed from (14.2) straightaway. Hence, a difference equation is basically a recurrence relation.

A few examples of difference equations are

$$y_n - 6y_{n-1} + 11y_{n-2} - 6y_{n-3} = 0 \tag{14.3a}$$

or $$y_{n+3} - 6y_{n+2} + 11y_{n+1} - 6y_n = 0$$

$$y_n - 3y_{n-1} + 2y_{n-2} = n \tag{14.3b}$$

or $$y_{n+2} - 3y_{n+1} + 2y_n = n+2$$

$$y_{n+1} - y_{n-1} = (a - by_n)y_n, \text{ } a \text{ and } b \text{ are constants.} \tag{14.3c}$$

Eqn. (14.3a) is a homogeneous equation of order 3; eqn (14.3b) is non-homogeneous equation of order 2; eqn (14.3c) is a nonlinear equation of order 2 and degree 2 due to y_n^2 term.

It may be noted that as the argument varies at a unit step, we can make use of difference operators, e.g. $\mathrm{E}^r y_n = y(n+r) = y_{n+r}$, $\mathrm{E}(1) = 1$; $\Delta y(n) = y(n+1) - y(n)$ or $\Delta y_n = y_{n+1} - y_n$ and $\nabla y_n = y_n - y_{n-1}$, etc.

14.2 Method of Solution

We will be concerned mainly with equations having constant coefficients. It will be assumed that the solution exists and is unique. The technique for solving a difference equation, i.e. to express y_n in terms of n is similar to finding a solution of a linear differential equation with constant coefficients which is explained as follows:

(*i*) Find the general solution of the associated homogeneous equation putting right side equal to zero, say y_n^{H}. It contains arbitrary constants.

(*ii*) Find a particular solution of the complete equation, say y_n^{P} which has no arbitrary constants.

(*iii*) The general solution of the complete equation is given by $y_n = y_n^{\mathrm{H}} + y_n^{\mathrm{P}}$.

(*iv*) Use given conditions to determine arbitrary constants in y_n to obtain a particular solution of the complete equation.

14.2.1 To find $\mathbf{y^H}$

Let us consider an rth order difference equation with constant coefficients

$$y_n + a_1 y_{n-1} + a_2 y_{n-2} + \ldots\ldots + a_r y_{n-r} = f(n). \tag{14.4}$$

The homogeneous equation corresponding to (14.4) is

$$y_n + a_1 y_{n-1} + a_2 y_{n-2} + \dots + a_r y_{n-r} = 0 \tag{14.5}$$

Assume a solution of (14.5) as

$$y_n = \xi^n. \tag{14.6}$$

Substituting (14.6) in (14.5) gives

$$(\xi^r + a_1 \xi^{r-1} + a_2 \xi^{r-2} + \dots + a_r)\xi^{n-r} = 0. \tag{14.7}$$

The value $\xi = 0$ from (14.7) gives a trivial solution $y_n = 0$. For a non-trivial solution we should have

$$\xi^r + a_1 \xi^{r-1} + a_2 \xi^{r-2} + \dots + a_r = 0. \tag{14.8}$$

Eqn. (14.8) is a polynomial equation of degree r, called 'characteristic' or 'auxiliary' equation. It will have r roots which may be real or complex and distinct or repeated. The general solution of the homogeneous equation (14.5), y^{H} may be written for the cases of distinct and repeated roots as follows:

(*a*) When the roots are distinct

Let us suppose that the roots of eqn. (14.8) are distinct, say $\xi_1, \xi_2 \dots \xi_r$. Then $y_n = \xi_i^n$, for $i = 1, 2, \dots r$ is a solution (particular) of (14.5). Further, since each of ξ_i^n is a solution, their linear combination is also a solution of the homogeneous equation (14.8). Thus the general solution of the homogeneous equation (14.5) is given by

$$y_n^{\mathrm{H}} = c_1 {\xi_1}^n + c_2 \xi_2^n + \dots + c_r \xi_r^n \tag{14.9}$$

where $c_1, c_2, \dots c_r$ are arbitrary constants, not all of them equal to zero.

(*b*) When the roots are repeated

Let us suppose that ξ_1 is a double root and let $\xi_2 = \xi_1$ so that roots may be written as $\xi = \xi_1, \xi_1, \xi_3, \xi_4 \dots \xi_r$. In that case characteristic polynomial as well as its first derivative will vanish for $\xi = \xi_1$. It can be shown that $y_n = \xi_1^n$ and $y_n = n{\xi_1}^n$ will be solutions of (14.5). Hence the general solution can be written as linear combinations

$$y_n^{\mathrm{H}} = (c_1 + c_2 \cdot n)\xi_1^n + c_3 \xi_3^n + c_4 \xi_4^n + \dots c_r \xi_r^n. \tag{14.10}$$

If root ξ_1 is repeated thrice so that the roots are $\xi = \xi_1,\ \xi_1,\ \xi_1,\ \xi_4,\ \xi_5\ \ldots\ \xi_r$ then three solutions will be available corresponding to $\xi = \xi_1$ viz $y_n = {\xi_1}^n,\ n{\xi_1}^n$ and $n^2{\xi_1}^n$. In that case, general solution of the homogeneous equation will be

$$y_n^{\mathrm{H}} = (c_1 + c_2 \cdot n + c_3 n^2)\xi_1^n + c_4\xi_4^n + \ldots . + c_r\xi_r^n \tag{14.11}$$

The above analysis can be generalised when the root is repeated k times as

$$y_n^{\mathrm{H}} = (c_1 + c_2 \cdot n + c_3 n^2 + \ \ldots .\ c_k n^{k-1})\xi_1^n + c_{k+1}\xi_{k+1}^n + c_{k+2}\xi_{k+2}^n + \ \ldots .\ + c_r\xi_r^n. \tag{14.12}$$

If the characteristic equation has a complex root, say $\xi_1 = \alpha + i\beta$, then its conjugate will also be a root (since a's are real); let $\xi_2 = \alpha - i\beta$. Using the polar form $\alpha \pm i\beta = \rho e^{\pm i\theta}$, we can write the corresponding solution as $\rho^n(c_1 \cos\ n\theta\ \pm c_2 \sin\ n\theta)$ where $\rho = \sqrt{\alpha^2 + \beta^2}$ and $\cos\theta\ = \alpha/\rho,\ \sin\theta\ = \beta/\rho$.

14.2.2 To find $\mathbf{y}^{\mathbf{P}}$

To find particular solution of the complete equation (14.4) we assume a solution, containing unknown parameters, in accordance with the form of the function $f(n)$. The unknown parameters are determined by substituting the assumed (trial) solution in the left side of (14.4) and comparing with $f(n)$. A few standard forms of $f(n)$ and corresponding choice of trial solutions are given in Table 14.1. (A, B and $\alpha,\ \beta$, etc., are constants). Broadly speaking same functions are chosen what $f(n)$ contains but with different coefficients.

Table 14.1. Forms of $f(n)$ and trial solutions

Sl. No.	$f(n)$	*Trial solution*
(i)	Exponential: $\mathrm{B}\alpha^{\beta n}$ or $\mathrm{B}e^{\beta n}$	$\mathrm{A}\,(\alpha)^{\beta n}$ or $\mathrm{A}e^{\beta n}$
(ii)	$\mathrm{B}n^k$	$\mathrm{A}n^k$
(iii)	$\mathrm{B}\alpha^n \cdot n^k$	$\mathrm{A}\alpha^n \cdot n^k$
(iv)	$\mathrm{B}\sin\beta n$ or $\mathrm{B}\cos\beta n$	$\mathrm{A}_1 \sin\beta n + \mathrm{A}_2 \cos\beta n$
(v)	$\mathrm{B}\alpha^n \sin\beta n$ or $\mathrm{B}\alpha^n \cos\beta n$	$\alpha^n(\mathrm{A}_1 \sin\beta n + \mathrm{A}_2 \cos\beta n)$

If $f(n)$ is linear combination of above functions, the trial solution may also be taken as linear combination of respective trial solutions with different parameters.

Alternatively, we can also use operator method to find a particular solution. Let us consider the difference equation of rth order in the form,

$$y_{n+r} + a_1 y_{n+r-1} + a_2 y_{n+r-2} + \ldots. + a_{r-1} y_{n+1} + a_r y_n = \mathrm{F}(n). \tag{14.13}$$

If equation is given in the form of eqn. (14.4), it can be converted to (14.13) replacing n by $(n+r)$ on both sides. In terms of operator E, (14.13) can be written as

$$(\mathrm{E}^r + a_1 \mathrm{E}^{r-1} + a_2 \mathrm{E}^{r-2} + \ldots. + a_{r-1}\mathrm{E} + a_r) y_n = \mathrm{F}(n) \tag{14.14a}$$

or $$\phi(\mathrm{E}) y_n = \mathrm{F}(n) \tag{14.14b}$$

where $$\phi(\mathrm{E}) = \mathrm{E}^r + a_1 \mathrm{E}^{r-1} + a_2 \mathrm{E}^{r-2} + \ldots. + a_{r-1}\mathrm{E} + a_r \tag{14.14c}$$

From (14.14b) we can obtain y_n from operation

$$y_n = \frac{1}{\phi(\mathrm{E})} \mathrm{F}(n). \tag{14.15}$$

Let us consider some particular cases:

(*i*) When $\mathrm{F}(n) = \mathrm{B}$ (*a* cosnt.)

Taking a trial solution $y_n = \mathrm{A}$ and substituting it in (14.14b)

$$\phi(\mathrm{E})\ \mathrm{A} = \mathrm{B} \text{ or } \mathrm{A}\ \phi(\mathrm{E})1 = \mathrm{B}$$

or $$\mathrm{A}\phi\ (1) = \mathrm{B} \text{ or } \mathrm{A} = \mathrm{B}/\phi(1), \text{ since } \mathrm{E}^p(1) = 1.$$

Hence $$y_n = \frac{\mathrm{B}}{\phi(1)}, \text{ provided } \phi\ (1) \neq 0. \tag{14.16}$$

If $\phi\ (1) = 0$, try An, An^2, etc.

(*ii*) When $\mathrm{F}(n) = \mathrm{B}(\alpha)^{\beta n}$; B, α and β are constants.

Assuming $y_n = \mathrm{A}\alpha^{\beta n}$ and substituting in (14.14a)

$$(\mathrm{E}^r + a_1 \mathrm{E}^{r-1} + a_2 \mathrm{E}^{r-2} + \ldots. + a_{r-1}\mathrm{E} + a_r)\mathrm{A}\alpha^{\beta n} = \mathrm{B}\alpha^{\beta n}$$

or $$\mathrm{A}\ (\alpha^{\beta r} + a_1 \alpha^{\beta(r-1)} + a_2 \alpha^{\beta(r-2)} + \ldots. + a_{r-1}\alpha^{\beta} + a_r)\alpha^{\beta n} = \mathrm{B}\alpha^{\beta n}$$

or $$\mathrm{A}\phi(\alpha^{\beta}) = \mathrm{B} \text{ giving } \mathrm{A} = \mathrm{B}/\phi(\alpha^{\beta}), \text{ since } \mathrm{E}^p \alpha^{\beta n} = \alpha^{\beta(n+p)} = \alpha^p.\alpha^{\beta n}.$$

Hence, $$y_n = \frac{B\alpha^{\beta n}}{\phi(\alpha^{\beta})}. \tag{14.17}$$

Solution (14.17) is true for $e^{\beta n}$ also, i.e. $\alpha = e$.

(iii) When $F(n) = B_0 + B_1 n + B_2 n^2 + \ldots. + B_k n^k$: polynomial of degree k.

In that case $y_n = \dfrac{1}{\phi(E)} F(n) = \{\phi(1+\Delta)\}^{-1} F(n)$.

Expand $\{\phi(1+\Delta)\}^{-1}$ by binomial series up to powers Δ^k since all higher order differences will be zero.

Important: If any trial solution provides a particular solution which is same as one of the solutions of y^H, then multiply the trial solution by n; if this also provides one of the solutions of y^H then multiply the trial solution by n^2 and so on until a different solution is obtained.

Example 14.1

(Fibonacci Sequence)
Any term in the Fibonacci sequence, is equal to the sum of preceding two terms while first and the second terms in the sequence are 0, 1. Find the nth term.

Solution Let n^{th} term of the sequence be t_n.

Then $t_n = t_{n-1} + t_{n-2}$ or $t_n - t_{n-1} - t_{n-2} = 0$.

It is given that $t_1 = 0$ and $t_2 = 1$.
The characteristic (auxiliary) equation is

$$\xi^2 - \xi - 1 = 0 \text{ giving } \xi_1 = \frac{1+\sqrt{5}}{2}, \; \xi_2 = \frac{1-\sqrt{5}}{2}.$$

Hence the general solution is

$$t_n = c_1\left(\frac{1+\sqrt{5}}{2}\right)^n + c_2\left(\frac{1-\sqrt{5}}{2}\right)^n.$$

Using the conditions $t_1 = 0$ and $t_2 = 1$, we get

$$0 = \frac{c_1}{2}(1+\sqrt{5}) + \frac{c_2}{2}(1-\sqrt{5}) \text{ or } (c_1+c_2) + \sqrt{5}(c_1-c_2) = 0.$$

$$1 = \frac{c_1}{2}(3+\sqrt{5}) + \frac{c_2}{2}(3-\sqrt{5}) \text{ or } 3\,(c_1+c_2) + \sqrt{5}(c_1-c_2) = 2.$$

From above equations we get

$c_1 + c_2 = 1$ and $c_1 - c_2 = -1/\sqrt{5}$, giving

$$c_1 = \frac{1}{2\sqrt{5}}(\sqrt{5}-1), \quad c_2 = \frac{1}{2\sqrt{5}}(\sqrt{5}+1).$$

Hence $$t_n = \frac{1}{2\sqrt{5}}(\sqrt{5}-1)\left(\frac{1+\sqrt{5}}{2}\right)^n + \frac{\sqrt{5}+1}{2\sqrt{5}}\left(\frac{1-\sqrt{5}}{2}\right)^n$$

$$= \frac{1}{\sqrt{5}}\frac{1}{2^{n-1}}\left[\left(1+\sqrt{5}\right)^{n-1} - \left(1-\sqrt{5}\right)^{n-1}\right]$$

Using binomial expansion on the braketed terms gives

$$t_n = \frac{1}{2^{n-1}}\left[\binom{n-1}{1} + \binom{n-1}{3}\cdot 5 + \binom{n-1}{5}\cdot 5^2 + \cdots + \binom{n-1}{2k+1}\cdot 5^k\right]$$

where $2k+1 = (n-1)$ if n is even and $(n-2)$ if n is odd.

Note: The problem can be put in an alternative manner.

The Fibonacci numbers follow the formula

$t_n = t_{n-1} + t_{n-2}$, $t_0 = 0$ and $t_1 = 1$; find t_n.

We will get $t_n = \dfrac{1}{\sqrt{5}}\left(\dfrac{1+\sqrt{5}}{2}\right)^n - \dfrac{1}{\sqrt{5}}\left(\dfrac{1-\sqrt{5}}{2}\right)^n.$

In the expansion $n-1$ will be replaced by n.

Example 14.2

(Sum of Natural numbers)
If s_n denotes the sum of first n natural numbers, then $s_n = s_{n-1} + n$, $s_0 = 0$. Find s_n.

Solution The difference equation is $s_n - s_{n-1} = n$, $s_0 = 0$.

$$s_n^{\mathrm{H}} = c \text{ (const.)}$$

Let particular solution be $s_n = \mathrm{A}_1 n + \mathrm{A}_2$.

Then $\quad s_n - s_{n-1} = \mathrm{A}_1$ (const.)

It is solution of the homogeneous equation. Modify particular solution.

Taking trial solution $s_n = n(\mathrm{A}_1 n + \mathrm{A}_2)$.

Then $\qquad s_n - s_{n-1} = (2n-1)A_1 + A_2 = n$

Comparing coefficients, $2A_1 = 1, -A_1 + A_2 = 0$ giving $A_1 = A_2 = \frac{1}{2}$.

Hence $\qquad s_n = c + \frac{1}{2}n(n+1).$

Using condition $s_0 = 0$ gives $c = 0$, so that

$$s_n = \frac{1}{2}n(n+1).$$

Example 14.3

(Sum of Geometric Series)
The difference equation $S_n = S_{n-1} + r^{n-1}$ with $S_1 = 1$ defines the sum S_n of n numbers of a geometric series with common ratio r and first term being 1. Find S_n.

Solution $\quad S_n - S_{n-1} = r^{n-1}, \; S_1 = 1.$

$$S_n^H = c \text{ (const.)}$$

Assuming trial solution as $S_n^P = Ar^{n-1}$ and substituting in the equation gives $A = r/(r-1)$. Hence the particular solution is

$$S_n^P = \frac{r^n}{r-1}.$$

The general solution is given by

$$S_n = c + \frac{r^n}{r-1}.$$

The condition $S_1 = 1$ gives $c = -\frac{1}{r-1}$.

Hence the particular solution satisfying the condition is

$$S_n = -\frac{1}{r-1} + \frac{r^n}{r-1} = \frac{r^n - 1}{r-1}.$$

Example 14.4

(Number of Comparisions in Merge-Sort Algorithm)
The time-complexity (number of maximum comparisions) in Merge-Sort algorithm for sorting n elements denoted by $T(n)$, is expressed by the relation

$$T(n) = 2T\left(\frac{n}{2}\right) + n - 1, \; T(2) = 1.$$

Find $T(n)$, assuming n as power of 2.

Solution Let $n = 2^k$. The difference equation transforms to

$$T(2^k) = 2T(2^{k-1}) + 2^k - 1$$

or $$y_k - 2y_{k-1} = 2^k - 1, \text{ where } y_k = T(2^k).$$

$$y_k^H = c \cdot 2^k.$$

Since 2^k is already a solution of y_k^H, we choose trial solution for determining particular solution, as

$$y_k^P = Ak \cdot 2^k + B.$$

Substituting in the difference equation

$$(A \cdot k \cdot 2^k + B) - 2\{A(k-1)2^{k-1} + B\} = 2^k - 1$$

or $A(k \cdot 2^k - k \cdot 2^k + 2^k) - B = 2^k - 1$, giving $A = 1$, $B = 1$.

Hence the particular solution is

$$y_k^P = k \cdot 2^k + 1.$$

The general solution is given by

$$y_k = c \cdot 2^k + k \cdot 2^k + 1 \text{ or } T(n) = c \cdot n + n \cdot \log_2 n + 1$$

Using the condition $T(2) = 1$ gives $c = 0$.
Hence the particular solution is

$$T(n) = n \log_2 n + 1.$$

Example 14.5

Find the general solution of the difference equation

$$y_n - 4y_{n-1} + 4y_{n-2} = n - 4 + 2^{n-1}.$$

Solution The characteristic (auxiliary) equation is

$$\xi^2 - 4\xi + 4 = 0$$

$\xi = 2$ is the double root, i.e., $\xi_1 = 2$, $\xi_2 = 2$.

Hence $$y_n^H = c_1 \cdot 2^n + c_2 \cdot n \cdot 2^n.$$

To find a particular solution we choose $A_1 n + A_2$ corresponding to $n - 4$ and $B \cdot n^2 \cdot 2^{n-1}$ corresponding to 2^{n-1} since $c_1 2^n$ and $c_2 \cdot n \cdot 2^n$ are already solution of the homogeneous equation. Hence select trial solution as

$$y_n^P = A_1 n + A_2 + B \cdot n^2 \cdot 2^{n-1}.$$

We can not choose $\mathrm{B} \cdot 2^{n-1}$ or $\mathrm{B} \cdot n \cdot 2^{n-1}$ as they are already solutions in y_n^{H} (See Note).

Substituting it in the equation

$$\mathrm{A}_1 \cdot n + \mathrm{A}_2 + \mathrm{B} \cdot n^2 \cdot 2^{n-1} - 4\{\mathrm{A}_1(n-1) + \mathrm{A}_2 + \mathrm{B}(n-1)^2 \cdot 2^{n-2}\}$$

$$+4\{\mathrm{A}_1(n-2) + \mathrm{A}_2 + \mathrm{B}(n-2)^2 \cdot 2^{n-3}\} = n - 4 + 2^{n-1}$$

or $\quad \mathrm{A}_1 n + \mathrm{A}_2 + \mathrm{B}n^2 \cdot 2^{n-1} + 4\{-\mathrm{A}_1 + 2^{n-3} \cdot \mathrm{B}(-n^2+2)\} = n - 4 + 2^{n-1}$

or $\quad \mathrm{A}_1 n + (\mathrm{A}_2 - 4\mathrm{A}_1) + \mathrm{B} \cdot 2^{n-1} = n - 4 + 2^{n-1}$

On comparing coefficients, $\mathrm{A}_1 = 1,\ \mathrm{A}_2 = 0$ and $\mathrm{B} = \dfrac{1}{2}$.
Hence the general solution is

$$y_n = y_n^{\mathrm{H}} + y_n^{\mathrm{P}} = c_1 \cdot 2^n + c_2 \cdot n \cdot 2^n + n + n^2 \cdot 2^{n-2}.$$

Note: We can put $c_1 2^n = 2c_1 \cdot 2^{n-1} = c_1' \cdot 2^{n-1}$ where c_1' is again a constant; similarly $c_2 \cdot n \cdot 2^n = 2c_2 n \cdot 2^{n-1} = c_2' n \cdot 2^{n-1}$ where $c_2' = 2c_2$. In Exercise 14.3 we shall have the solution of homogeneous equation as $c4^k = c \cdot 2^{2k}$. This can not be put in the form $c' \cdot 2^k$ since $c' = c/2^k$ which is not a constant. Hence a particular solution can be chosen as $\mathrm{A} \cdot 2^k$ if required.

14.3 Simultaneous Difference Equations and Exponentiation of Matrix

To illustrate the problem and method of solution, let us consider a system of three linear difference equations with constant coefficients of order one:

$$\left.\begin{aligned} x_1(t+1) &= \alpha_{11}x_1(t) + \alpha_{12}x_2(t) + \alpha_{13}x_3(t) \\ x_2(t+1) &= \alpha_{21}x_1(t) + \alpha_{22}x_2(t) + a_{23}x_3(t) \\ x_3(t+1) &= \alpha_{31}x_1(t) + \alpha_{32}x_2(t) + \alpha_{33}x_3(t) \end{aligned}\right\} \tag{14.18}$$

In (14.18) $x_1,\ x_2,\ x_3$ are dependent variables and t an independent variable which can assume only non-negative integral values. The coefficients α_{rs}, r and $s = 1,\ 2,\ 3$ are known. It may be mentioned that if the original equations involve more than one unknowns of $x_1(t+1),\ x_2(t+1),\ x_3(t+1)$ in one or more equations, then they be expressed in the form (14.18) by solving them. The variable t has been chosen (instead of n) simply because such problems commonly arise in the study of growth models wherein growth of population of various species is measured over fixed intervals of time.

The system of equations (14.18) can be written in the matrix form as

$$\mathrm{X}(t+1) = \mathrm{AX}(t) \tag{14.19}$$

where $X^T(t) = \{x_1(t) \quad x_2(t) \quad x_3(t)\}$ etc. (14.20)

and
$$A = \begin{bmatrix} \alpha_{11} & \alpha_{12} & \alpha_{13} \\ \alpha_{21} & \alpha_{22} & \alpha_{23} \\ \alpha_{31} & \alpha_{32} & \alpha_{33} \end{bmatrix}. \tag{14.21}$$

Let us suppose that the values $x_1(0)$, $x_2(0)$, $x_3(0)$ are known initially, i.e., X (0) is known. Then we can get successively X (1) = AX (0), X (2) = $A^2X(0)$ and so on and finally

$$X(t) = A^t X(0). \tag{14.22}$$

Our problem is to find X(t) in (14.22) which can be determined if A^t can be determined since X(0) is known. We shall discuss now a method as how the elements of A^t can be expressed in terms of t.

Let λ be an e.value of matrix A and $p_3(\lambda)$, the characteristic polynomial which is of degree 3. The characteristic equation is given by

$$|A - \lambda I| \equiv p_3(\lambda) = 0. \tag{14.23}$$

We know from Cayley-Hamilton theorem that a square matrix satisfies its own characteristic equation; hence from (14.23) we have

$$p_3(A) = O, \tag{14.24}$$

where O is a null matrix.

We can express the e.values of A^t, i.e., λ^t in terms of the characteristic polynomial $p_3(\lambda)$ of A, as

$$\lambda^t = p_3(\lambda)q_{t-3}(\lambda) + r_2(\lambda) \tag{14.25}$$

where p_3, q_{t-3} and r_2 are polynomials of degree 3, $t-3$ and 2 respectively. The polynomials q_{t-3} will have $t-2$ unknowns and r_2 will have 3 unknowns; thus there will be in all $t+1$ unknowns since coefficients of p_3 are known. These $t+1$ unknowns can be determined by comparing the coefficients of various powers of λ from 0 to t on both sides of (14.25). Here $q_{t-3}(\lambda)$ is the quotient and $r_2(\lambda)$ is the remainder when λ^t is divided by $p_3(\lambda)$. Hence degree of $r_2(\lambda)$ is lower than the degree of $p_3(\lambda)$.

In the above case since $p_3(\lambda) = 0$, (14.25) becomes

$$\lambda^t = r_2(\lambda) = a_0 + a_1\lambda + a_2\lambda^2. \tag{14.26}$$

where a_0, a_1 and a_2 are unknowns to be determined.

Further, we can also write a relation similar to (14.25) by the same logic as given above for matrix A, i.e.

$$\mathrm{A}^t = p_3(\mathrm{A}) \cdot q_{t-3}(\mathrm{A}) + r_2(\mathrm{A}) \tag{14.27}$$

But again, as $p_3(\mathrm{A}) = \mathrm{O}$ due to (14.24), eqn. (14.27) becomes

$$\mathrm{A}^t = r_2(\mathrm{A}) = a_0\mathrm{I} + a_1\mathrm{A} + a_2\mathrm{A}^2, \tag{14.28}$$

while $a_0,\ a_1$ and a_2 satisfy (14.26).

Let $\lambda_1,\ \lambda_2$ and λ_3 be e.values of A. Since they satisfy (14.26) we get

$$\left.\begin{aligned} {\lambda_1}^t &= a_0 + a_1\lambda_1 + a_2\lambda_1^2 \\ {\lambda_2}^t &= a_0 + a_1\lambda_2 + a_2\lambda_2^2 \\ {\lambda_3}^t &= a_0 + a_1\lambda_3 + a_2\lambda_3^2 \end{aligned}\right\}. \tag{14.29}$$

Writing (14.29) in matrix form

$$\begin{bmatrix} 1 & \lambda_1 & \lambda_1^2 \\ 1 & \lambda_2 & \lambda_2^2 \\ 1 & \lambda_3 & \lambda_3^2 \end{bmatrix} \begin{bmatrix} a_0 \\ a_1 \\ a_2 \end{bmatrix} = \begin{bmatrix} \lambda_1^t \\ \lambda_2^t \\ \lambda_3^t \end{bmatrix} \tag{14.30}$$

After finding the e.values of A and substituting them in eqn. (14.30), the values of $a_0,\ a_1,\ a_2$ can be computed by solving the system of equations. It may be noted that the e.values should be distinct otherwise two or more rows may be identical and matrix will be singular. Putting the values of $a_0,\ a_1,\ a_2$ in (14.27), the matrix A^t can be obtained and finally its value may be inserted in (14.22) to get X (t) which is the required solution.

Note: The result (14.27) can be extended to a function F(A) which can be expanded in powers of A and is convergent, if infinite series. That is, for a matrix of order k

$$F(\mathrm{A}) = p_k(\mathrm{A})q(\mathrm{A}) + r_{k-1}(\mathrm{A}) = r_{k-1}(\mathrm{A}). \tag{14.31}$$

If A is a 2×2 matrix, then,

$$F(\mathrm{A}) = a_0\mathrm{I} + a_1\mathrm{A}$$

and the coefficients $a_0,\ a_1$ will be determined from

$$F(\lambda) = a_0 + a_1\lambda \tag{14.32}$$

after substituting the e.values $\lambda_1,\ \lambda_2$ of matrix A.

Example 14.6

Develop a mathematical model in terms of difference equations for the growth of population group-wise when the groups are as given below:

x_1: Pre-productive (People below reproductive age)

x_2: Reproductive (People in reproductive age group)

x_3: Post-productive (People above reproductive age)

Incorporate the following parameters which are given over a unit interval of time:
a_{ij} = rate at which population in group i increases due to movement from group j to group i due to birth or age.
b_i = rate at which the population in group i decreases due to either movement or death
Express the equations in matrix form as $\mathrm{X}(t) = \mathrm{A}^t\mathrm{X}(0)$.

Solution Assuming population is known group-wise at time t, i.e., $x_1(t),\ x_2(t),\ x_3(t)$ known, we have to find $x_1(t+1),\ x_2(t+1),\ x_3(t+1)$.

$$x_1(t+1) = x_1(t) + a_{12}x_2(t) - b_1x_1(t) \qquad \ldots(1)$$

where a_{12} is the birth-rate in group x_2; $a_{13} = 0$.

$$x_2(t+1) = x_2(t) + a_{21}x_1(t) - b_2x_2(t) \qquad \ldots(2)$$

where a_{21} is the rate at which people from group x_1 move to group x_2; $a_{23} = 0$.

$$x_3(t+1) = x_3(t) + a_{32}x_2(t) - b_3x_3(t) \qquad \ldots(3)$$

where a_{32} is the rate at which people from group x_2 move to group x_3. Writing (1), (2) and (3) matrix of form

$$\begin{bmatrix} x_1(t+1) \\ x_2(t+1) \\ x_3(t+1) \end{bmatrix} = \begin{bmatrix} 1-b_1 & a_{12} & 0 \\ a_{21} & 1-b_2 & 0 \\ 0 & a_{32} & 1-b_3 \end{bmatrix} \begin{bmatrix} x_1(t) \\ x_2(t) \\ x_3(t) \end{bmatrix}$$

or $\qquad \mathrm{X}(t) = \mathrm{A}^t\mathrm{X}(0)$

where $\qquad \mathrm{X}^{\mathrm{T}}(t) = (x_1(t)\ \ x_2(t)\ \ x_3(t))$

and

$$\mathrm{A} = \begin{bmatrix} 1-b_1 & a_{12} & 0 \\ a_{21} & 1-b_2 & 0 \\ 0 & a_{32} & 1-b_3 \end{bmatrix}$$

Example 14.7

Form a matrix $B = A^3 + A^5$ using e.values of matrix A where

$$A = \begin{bmatrix} -2 & -3 \\ 4 & 5 \end{bmatrix}.$$

Also compute B directly evaluating A^3 and A^5.

Solution If λ is an e.value of A, then characteristic polynomial is given by

$$p(\lambda) = |\lambda I - A| = \lambda^2 - 3\lambda + 2 = (\lambda - 1)(\lambda - 2).$$

The e.values are $\lambda_1 = 1,\ \lambda_2 = 2$.
The e.values of $A^3 + A^5$ can be expressed as

$$\lambda^3 + \lambda^5 = p(\lambda)q(\lambda) + r(\lambda) = r(\lambda)$$

where $r(\lambda)$ is a polynomial of degree 1.
Let $\quad \lambda^3 + \lambda^5 = a_0 + a_1\lambda \qquad \ldots(1)$

Also we have

$$A^3 + A^5 = p(A).\, q(A) + r(A) = r(A)$$

or $$B = A^3 + A^5 = a_0 I + a_1 A. \qquad \ldots(2)$$

Substituting λ_1 and λ_2 in (1) we get

$$a_0 + a_1 = 2$$

$$a_0 + 2a_1 = 40 \qquad \ldots(3)$$

Solving (3) gives $a_0 = -36, a_1 = 38$. Hence from (2)

$$B = -36 \begin{bmatrix} 1 & 0 \\ 0 & 1 \end{bmatrix} + 38 \begin{bmatrix} -2 & -3 \\ 4 & 5 \end{bmatrix} = \begin{bmatrix} -112 & -114 \\ 152 & 154 \end{bmatrix}$$

From direct computation

$$A^2 = \begin{bmatrix} -8 & -9 \\ 12 & 13 \end{bmatrix},\ A^3 = \begin{bmatrix} -20 & -21 \\ 28 & 29 \end{bmatrix},\ A^5 = A^2.A^3 = \begin{bmatrix} -92 & -93 \\ 124 & 125 \end{bmatrix}$$

$$B = A^3 + A^5 = \begin{bmatrix} -112 & -114 \\ 152 & 154 \end{bmatrix}.$$

14.3.1 Property of constant Row-sum (Column-sum)

If two square matrices have constant row-sums then the row-sum of their product is also constant and is equal to the product of the row-sums of the matrices.

The property is true for constant column-sum also.

Proof. Let A and B be two $n \times n$ square matrices and let their respective row-sums be a and b, i.e.,

$$\sum_{j=1}^{n} a_{ij} = a; \ \sum_{j=1}^{n} b_{ij} = j; \ i = 1(1)n.$$

Let C = A.B, then sum of elements of C along i^{th} row, $i = 1(1)n$ is

$$\begin{aligned}
\sum_{j=1}^{n} c_{ij} &= \sum_{j=1}^{n}\sum_{k=1}^{n} a_{ik}b_{kj} \\
&= \sum_{k=1}^{n} a_{ik} \sum_{j=1}^{n} b_{kj} \\
&= \sum_{k=1}^{n} a_{ik}.b, \qquad \because \ \sum_{j=1}^{n} b_{kj} = b, \text{ for any } k \text{ (given)} \\
&= a.\, b \qquad \because \ \sum_{k=1}^{n} a_{ik} = a, \text{ for any } i \text{ (given)}
\end{aligned}$$

Similarly, if the column-sums of A and B are constant, say,

$$\sum_{i=1}^{n} a_{ij} = a'; \ \sum_{i=1}^{n} b_{i,\,j} = b'; \ \ j = 1(1)n.$$

Let C = A.B, then sum of elements of C along j^{th} col, $j = 1(1)n$,

$$\sum_{i=1}^{n} c_{ij} = \sum_{i=1}^{n}\sum_{k=1}^{n} a_{ik}b_{kj}$$

$$= \sum_{k=1}^{n} b_{kj} \sum_{i=1}^{n} a_{ik}$$

$$= a' \sum_{k=1}^{n} b_{kj}, \qquad \because \sum_{i=1}^{n} a_{ik} = a', \text{ for any } k \text{ (given)}$$

$$= a'.b', \qquad \because \sum_{k=1}^{n} b_{kj} = b', \text{ for any } j \text{ (given)}$$

Example 14.8

A set of simultaneous difference equations satisfy $\boldsymbol{x}(t+1) = \mathrm{A}\boldsymbol{x}(t)$ where

$$\mathrm{A} = \begin{bmatrix} 3 & 3 & -4 \\ 0 & -4 & 6 \\ -7 & 0 & 9 \end{bmatrix}$$

Find $\boldsymbol{x}(10)$, using property of Sec 14.3.1 when it is given that $\boldsymbol{x}^{\mathrm{T}}(0) = (5 \;\; 5 \;\; 5)$

Solution $\boldsymbol{x}(t+1) = \mathrm{A}\boldsymbol{x}(t)$

$$\boldsymbol{x}(1) = \mathrm{A}\,\boldsymbol{x}(0),\ \boldsymbol{x}(2) = \mathrm{A}\boldsymbol{x}(1) = \mathrm{A}^2\boldsymbol{x}(0).$$

Hence, $\boldsymbol{x}(10) = \mathrm{A}^{10}\boldsymbol{x}(0)$.

The matrix A has constant row-sum 2, hence the row sum of A^{10} will also be constant and will be equal to $2^{10} = 1024$. Since elements of $x^{\mathrm{T}}(0)$ are same, each element of $\boldsymbol{x}(10)$ will be equal to $1024 \times 5 = 5120$.

Thus, $$\boldsymbol{x}(10) = \mathrm{A}^{10}\boldsymbol{x}(0) = \begin{bmatrix} 5120 \\ 5120 \\ 5120 \end{bmatrix}$$

Exercise 14

14.1 Let S_n denote sum of n terms of an arithmetic series whose first term is 'a' and common difference is 'd'. Set up a difference equation for $n > 1$ and solve for S_n.

[Hint: $\mathrm{S}_n = \mathrm{S}_{n-1} + a + (n-1)d,\ \ \mathrm{S}_1 = a$.]

14.2 The difference scheme $T(n) = T(n-1) + n - 1$, $T(1) = 0$ arises when computing the time-complexity (max. comparisons) of Selection-sort and Bubble-sort for sorting n elements. Find $T(n)$.

14.3 Find the solution of the difference equation

$$T(n) = 4T\left(\frac{n}{2}\right) + n, \quad T(1) = 1$$

assuming n as power of 2.

14.4 The time-complexity of best case of Quick-sort (min. comparisions) is determined from the difference equation

$$T(n) = \begin{cases} 1, & n = 2 \\ 2T\left(\frac{n}{2}\right) + n, & n > 2. \end{cases}$$

Assume n as power of 2; find $T(n)$.

14.5 Solve the difference equation

$$T(n) = \begin{cases} 1, & n = 1 \\ 2T\left(\frac{n}{2}\right) + \log_2 n, & \text{otherwise} \end{cases}$$

Assume $n = 2^k$, where k is a positive integer.

14.6 Given the simultaneous difference equations

$$x_1(t+1) = 0 + 3x_2(t) + 0$$
$$x_2(t+1) = 0 + 4x_2(t) - x_3(t)$$
$$x_3(t+1) = -2x_1(t) + 3x_2(t) + 2x_3(t)$$

Express the equations in the matrix form $x(t+1) = Ax(t)$. Compute A^4 using the e.values of matrix A. Also compute $x(4)$ when $x^T(0) = (1\ \ 1\ \ 1)$.

Reference and Some Useful Related Books/Papers

1. Froberg, C.E., *Introduction to Numerical Analysis,* Addison Wesley.

2. Kapur, J.N., *Insight into Mathematical Modelling,* Mathematical Sciences Trust Society, New Delhi (India).

15

Fourier Series, Discrete Fourier Transform and Fast Fourier Transform

15.1 Introduction

A function $f(x)$ is said to be periodic if it satisfies the property $f(x) = f(x \pm T)$ where T is a positive number and the smallest value of T for which this property holds is called period of function $f(x)$. For example, $\sin x$ and $\cos x$ are periodic functions having a period 2π and $\sin nx$ and $\cos nx$ are periodic with period $2\pi/n$. Jean-Baptiste J. Fourier, a French physicist/mathematician, represented a periodic function by a series containing only sine and cosine terms which is known as Fourier series. Although Fourier series has wide applications in solving ordinary and partial differential equations, we will discuss its basic fundamentals only. We shall also discuss Fourier Transform, in particular Discrete Fourier Transform (DFT) which is a numerical version of Fourier Transform and is frequently used in electrical engineering in transforming time-domain to frequency-domain and vice-versa. The computational aspects of DFT, in terms of number of arithmetic operations $(+, -, \times, \div)$ involved in its evaluation, will be discussed leading to an computationally efficient algorithm known as Fast Fourier Transform (FFT).

15.2 Fourier Series

A periodic function $f(x)$ of period 2π defined in the interval $-\pi < x < \pi$ may be represented by a Fourier series as

$$f(x) = a_0 + \sum_{n=1}^{\infty} (a_n \cos nx + b_n \sin nx) \tag{15.1}$$

where a_0, a_n, b_n known as Fourier's coefficients are determined by multiplying both sides of (15.1) by 1, $\cos nx$ and $\sin nx$ respectively and integrating for $-\pi \leq x \leq \pi$, getting:

$$a_0 = \frac{1}{2\pi}\int_{-\pi}^{\pi} f(x)\,dx, \tag{15.2a}$$

$$a_n = \frac{1}{\pi}\int_{-\pi}^{\pi} f(x)\cos nx\,dx, \tag{15.2b}$$

$$b_n = \frac{1}{\pi}\int_{-\pi}^{\pi} f(x)\sin nx\,dx. \tag{15.2c}$$

Following points regarding Fourier series and its coefficients should be noted:

(i) For the Fourier series to be convergent, the function $f(x)$ should be piece-wise continuous and should possess left and right first derivatives at each point in $(-\pi, \pi)$. That is, the function $f(x)$ may contain a finite number of finite discontinuities in $(-\pi, \pi)$.

At the point of discontinuity, say $x = x_0$, the Fourier series gives the average value of the function to the left and to the right of x_0, i.e., $f(x_0) = \{f(x_0+) + f(x_0-)\}/2$. It is an important property from the point of view of application in electrical engineering when the sharp corners in the input signals, such as step and ramp functions, are smoothed out in a Fourier series representation.

(ii) Since $f(x)$ is periodic it repeats itself over the entire real line after each period.

(iii) The sum of periodic functions is itself a periodic function; its period is the L.C.M. of the periods of various functions. For example, cos mx, sin mx are periodic functions of periods $2\pi/m$, $m = 1, 2, 3, \ldots$. Hence, their sum will also be a periodic function with period 2π.

(iv) Let S be *a* set of functions given as,

$$S = \{1, \cos x, \sin x, \cos 2x, \sin 2x, \ldots, \cos mx, \sin mx, \ldots\}.$$

It may be verified that S is an orthogonal set w.r.t. integral over the interval $\alpha \leq x \leq \alpha + 2\pi$, where α is a constant. That is, the product of any two functions integrated over an interval of 2π will give a zero value e.g., for $m \neq n$ we have,

$$\int_{\alpha}^{\alpha+2\pi} \sin mx.\sin nx dx = \int_{\alpha}^{\alpha+2\pi} \cos mx.\cos nx dx = 0 = \int_{\alpha}^{\alpha+2\pi} \sin mx dx \text{ etc.}$$

Further the integration of any function of the set multiplied by itself provides a non-zero value e.g., for $m = n$

$$\int_{\alpha}^{\alpha+2\pi} \sin^2 mx dx = \int_{\alpha}^{\alpha+2\pi} \cos^2 mx dx = \pi \text{ and } \int_{\alpha}^{\alpha+2\pi} 1.dx = 2\pi$$

15.3 Fourier Series with Other Intervals

When the function $f(x)$ is defined for an interval of length other than 2π, we can convert it to 2π by changing the variable. Following are most commonly used intervals:

(i) $f(x), -c \leq x \leq c.$

Put $t = x\pi/c$ or $x = ct/\pi$ so that $-\pi \leq t \leq \pi$.

Let $f(x) = f\left(\frac{ct}{\pi}\right) = \phi(t)$, then $\phi(t)$ is defined over an interval $-\pi \leq t \leq \pi$ and hence can be represented by Fourier series as follows,

$$\phi(t) = a_0 + \sum_{n=1}^{\infty} (a_n \cos nt + b_n \sin nt) \tag{15.3}$$

where

$$a_0 = \frac{1}{2\pi} \int_{-\pi}^{\pi} \phi(t) dt, \tag{15.4a}$$

$$a_n = \frac{1}{\pi} \int_{-\pi}^{\pi} \phi(t) \cos nt dt, \tag{15.4b}$$

$$b_n = \frac{1}{\pi} \int_{-\pi}^{\pi} \phi(t) \sin nt dt. \tag{15.4c}$$

We can express the above in terms of the original function $f(x)$ by putting $t = \pi x/c$ and $dt = \pi dx/c$, getting

$$f(x) = a_o + \sum_{n=1}^{\infty} \left(a_n \cos \frac{n\pi x}{c} + b_n \sin \frac{n\pi x}{c}\right) \tag{15.5}$$

where

$$a_0 = \frac{1}{2c}\int_{-c}^{c} f(x)\,dx, \tag{15.6a}$$

$$a_n = \frac{1}{c}\int_{-c}^{c} f(x)\cos\frac{n\pi x}{c}\,dx, \tag{15.6b}$$

$$b_n = \frac{1}{c}\int_{-c}^{c} f(x)\sin\frac{n\pi x}{c}\,dx. \tag{15.6c}$$

(ii) $f(x), 0 \le x \le c$.

By putting $t = 2\pi x/c$ we can convert the interval to $0 \le t \le 2\pi$. The Fourier series may be written as,

$$f(x) = a_0 + \sum_{n=1}^{\infty}\left(a_n\cos\frac{2n\pi x}{c} + b_n\sin\frac{2n\pi x}{c}\right) \tag{15.7}$$

where

$$a_0 = \frac{1}{c}\int_{0}^{c} f(x)\,dx, \tag{15.8a}$$

$$a_n = \frac{2}{c}\int_{0}^{c} f(x)\cos\frac{2n\pi x}{c}\,dx, \tag{15.8b}$$

$$b_n = \frac{2}{c}\int_{0}^{c} f(x)\sin\frac{2n\pi x}{c}\,dx. \tag{15.8c}$$

15.4 Half-Range Fourier Series

In Sec. 15.3 (ii) we suggested that when the function $f(x)$ is defined in an interval $0 \le x \le c$, we can convert it to $0 \le t \le 2\pi$ by putting $t = 2\pi x/c$ and express $f(x)$ by appropriate Fourier series. However, there is an smarter way of changing the interval to a length of 2π. We extend the domain of function $f(x)$ from $0 \le x \le c$ to $-c \le x \le c$ defining the function in the extended domain $-c \le x \le 0$ in a suitable manner. The function $f(x)$ can be made even or odd in the interval $-c \le x \le c$, i.e., symmetric $f(x) = f(-x)$ or anti-symmetric $f(x) = -f(x)$ about y-axis. Then by putting $t = \pi x/c$, the interval can be converted to $-\pi \le$

$t \leq \pi$, as discussed in Sec. 15.3 (i). When the function $f(x)$ is represented by a Fourier series, it will contain either cosine terms or the sine terms only depending on whether $f(x)$ is made even or odd respectively. When $f(x)$ is made even, the integral (15.6c) vanishes since integrand is an odd function and when $f(x)$ is made an odd function, then integrals (15.6a) and (15.6b) vanish. Hence only cosine terms are left when $f(x)$ is even and only sine terms are left when the function is odd. These series are known as Fourier's Half-Range cosine series and Half-Range sine series. They can be expressed as follows when $f(x)$ is defined for $0 \leq x \leq c$:

(i) Half-Range cosine series

$$f(x) = a_0 + \sum_{n=1}^{\infty} a_n \cos \frac{n\pi x}{c} \tag{15.9}$$

where

$$a_0 = \frac{1}{c} \int_0^c f(x)\, dx, \tag{15.10a}$$

$$a_n = \frac{2}{c} \int_0^c f(x) \cos \frac{n\pi x}{c} dx. \tag{15.10b}$$

(ii) Half-Range sine series

$$f(x) = \sum_{n=1}^{\infty} b_n \sin \frac{n\pi x}{c} \tag{15.11}$$

where

$$b_n = \frac{2}{c} \int_0^c f(x) \sin \frac{n\pi x}{c} dx \tag{15.12}$$

Example 15.1

Express the function $f(x) = x$, $0 < x < c$ by a Fourier half-range cosine series. Also plot the graph represented by the series.

Solution

We extend the function in the interval $(-c, 0)$ to make it even i.e., $f(x) = f(-x)$ in $-c < x < c$. Using (15.10a) and (15.10b) we get

$$a_0 = \frac{1}{c}\int_0^c f(x)\,dx = \frac{1}{c}\int_0^c x\,dx = \frac{c}{2}$$

$$a_n = \frac{2}{c}\int_0^c x\cos\frac{n\pi x}{c}dx$$

$$= \frac{2}{c}\left[x\sin\frac{n\pi x}{c}.\frac{c}{n\pi} + \frac{c^2}{n^2\pi^2}\cos\frac{n\pi x}{c}\right]_0^c$$

$$= \frac{2c}{n^2\pi^2}\{(-1)^n - 1\} = -\frac{4c}{n^2\pi^2}, n = 1,\ 3,\ 5,\ldots$$

Hence, half-range Fourier cosine series is given by

$$f(x) = \frac{c}{2} - \frac{4c}{\pi^2}\sum_{m=0}^{\infty}\frac{1}{(2m+1)^2}.\cos\frac{(2m+1)\pi x}{c}.$$

The graph represented by the series is as given below :

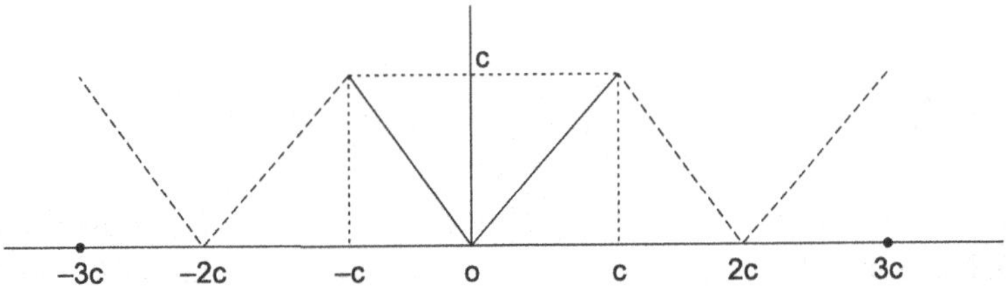

Figure 15.1 Graph of $f(x) = x, -c \le x \le c$ represented by Fourier series

Note: If $c = \pi$, then series may be written as

$$f(x) = \frac{\pi}{2} - \frac{4}{\pi}\left(\cos x + \frac{1}{9}\cos 3x + \frac{1}{25}\cos 5x + \ldots\right)$$

15.5 Fourier Series for Discrete Data

In many practical problems the function is not defined analytically; instead it is given in the form of numerical data. In such cases the integrals may be evaluated numerically using integration formulae, like rectangular and trapezoidal rules discussed in Chapter 6. Further, as stated in Sec. 15.2(i) that under given conditions, Fourier series would be convergent, we need take only a first few terms of the series to provide a reasonably good approximation, within a desired accuracy. Let us suppose that values of a periodic function $f(x)$ are known over the period $0 \leq x \leq c$, at N+1 equi-distant points $x = x_k,\ k = 0(1)N$ such that $c/N = h,\ x_k = kh, k = 0(1)N;\ x_0 = 0$ and $x_N = c$. Let the function value at $x = x_k$ be denoted by $f_k,\ k = 0(1)N$. If the function $f(x)$ is to be represented by a general form of Fourier series, we convert the interval from [0, c] to [0, 2π]. Using rectangular rule for evaluation of the integrals, the Fourier series, corresponding to (15.7), (15.8a, b, c) may be written as follows:

$$f(x) = a_0 + \sum_{n=1}^{K}\left(a_n \cos\frac{2n\pi x}{c} + b_n \sin\frac{2n\pi x}{c}\right) \tag{15.13}$$

where,

$$a_0 = \frac{1}{N}\sum_{k=0}^{N-1} f_k, \tag{15.14a}$$

$$a_n = \frac{2}{N}\sum_{k=0}^{N-1} f_k \cos\frac{2n\pi x_k}{c} = \frac{2}{N}\sum_{k=0}^{N-1} f_k.\cos\frac{2n\pi k}{N},\quad n = 1(1)K, \tag{15.14b}$$

$$b_n = \frac{2}{N}\sum_{k=0}^{N-1} f_k \sin\frac{2n\pi x_k}{c} = \frac{2}{N}\sum_{k=0}^{N-1} f_k.\sin\frac{2n\pi k}{N},\quad n = 1(1)K. \tag{15.14c}$$

where K denotes number of terms to be kept in the summation, also called terms upto K^{th} harmonic. A half-range Fourier series can also be written analogously when $x_k = kh = kc/N$. See Example 15.2.

Example 15.2

Represent the function $f(x)$ by Fourier half-range cosine series when its values are given as follows :

x	0	T/6	2T/6	3T/6	4T/6	5T/6	T
$f(x)$	1.0000	0.5924	0.3510	0.2079	0.1232	0.0730	0.0428

Use rectangular rule and compute upto second harmonic only (i.e., only two terms in the summation).

Solution

We have $N = 6;\ h = T/6;\ Nh = T;\ x_k = kh,\ k = 0(1)6;\ K = 2.$

The half-range cosine series upto second harmonic i.e., two terms in the summation can be written as follows:

$$f(x) = a_0 + a_1 \cos\frac{\pi x}{T} + a_2 \cos\frac{2\pi x}{T}$$

where

$$a_0 = \frac{1}{T}\int_0^T f(x)\,dx = \frac{1}{N}\sum_{k=0}^{N-1} f_k = \frac{1}{6}\sum_{k=0}^{5} f_k$$

$$a_n = \frac{2}{T}\int_0^T f(x)\cos\frac{n\pi x}{T}dx = \frac{2}{N}\sum_{k=0}^{N-1} f_k \cos\frac{n\pi k}{N}$$

$$= \frac{1}{3}\sum_{k=0}^{5} f_k \cos\frac{n\pi k}{6}, n = 1,2.$$

Note: We will also try to compute using Trapezoidal rule for the sake of comparision. That is, when

$$a_0 = \frac{1}{T}\cdot\frac{h}{2}\{f_0 + 2(f_1 + f_2 + f_3 + f_4 + f_5) + f_6\}$$

$$= \frac{1}{12}\{f_0 + 2(f_1 + f_2 + f_3 + f_4 + f_5) + f_6\}$$

$$a_n = \frac{2}{T}\cdot\frac{h}{2}\left\{f_0 + 2\sum_{k=0}^{5} f_k.\cos\frac{n\pi k}{6} + f_6 \cos n\pi\right\}$$

$$= \frac{1}{6}\left\{f_0 + 2\sum_{k=0}^{5} f_k.\cos\frac{n\pi k}{6} + f_6 \cos n\pi\right\}, n = 1,2.$$

(i) Rectangular Rule

$$a_0 = \frac{1}{6}\sum_{k=0}^{5} f_k = \frac{2.3475}{6} = 0.3912$$

$$a_1 = \frac{1}{3}\sum_{k=0}^{5} f_k.\cos k\pi/6 = \frac{1.5637}{3} = 0.5212$$

$$a_2 = \frac{1}{3}\sum_{k=0}^{5} f_k.\cos\frac{2\pi k}{6} = \frac{0.8877}{3} = 0.2959$$

For computations refer Table 15.1.

Table 15.1

k	x_k	f_k	$k\pi/6$	cos $k\pi/6$	f_k cos $k\pi/6$	2 $k\pi/6$	cos $2\pi k/6$	f_k cos $2\pi k/6$
0	0	1.0000	0	1.0000	1.0000	0	1.0000	1.0000
1	T/6	0.5924	$\pi/6$	0.8660	0.5130	$\pi/3$	0.50000	0.2962
2	T/3	0.3510	$\pi/3$	0.5000	0.1755	$2\pi/3$	−0.5000	−0.1755
3	T/2	0.2079	$\pi/2$	0	0	π	−1.0000	−0.2079
4	2T/3	0.1232	$2\pi/3$	−0.5000	−0.0616	$4\pi/3$	−0.5000	−0.0616
5	5T/6	0.0730	$5\pi/6$	−0.8660	−0.0632	$5\pi/3$	0.5000	0.0365
6	T	0.0428	π	−1.0000	−0.0428	2π	1.0000	0.0428

(ii) Trapezoidal Rule

$$a_0 = \frac{1}{12}\left[f_0 + 2\left(f_1 + f_2 + f_3 + f_4 + f_5\right) + f_6\right]$$

$$= \frac{1}{12}[1 + 2\times 1.3475 + 0.0428] = \frac{3.7378}{12} = 0.3115$$

$$a_1 = \frac{1}{6}\left[f_0\cos 0 + 2\sum_{i=1}^{5} f_k\cos\frac{k\pi}{6} + f_6\cos\pi\right]$$

$$= \frac{1}{6}[1 + 2\times(0.5637) - 0.0428] = \frac{2.0846}{6} = 0.3474$$

$$a_2 = \frac{1}{6}\left[f_0\cos 0 + 2\sum_{i=1}^{5} f_k\cos\frac{2k\pi}{6} + f_6\cos 2\pi\right]$$

$$= \frac{1}{6}[1 + 2\times(-0.1123) + 0.0428] = \frac{0.8182}{6} = 0.1364$$

Note: The data has been taken from the function $f(x) = e^{-x}$, $0 \le x \le 1.0$.

The analytical half-range cosine series representation for $f(x) = e^{-x}$, $0 \le x \le \pi$ is given by $f(x) = a_0 + \sum_{n=1}^{\infty} a_n \cos nx$ where

$$a_0 = \frac{1}{\pi}\left(e^{-x}\right)_0^{\pi} = \frac{1}{\pi}(1 - 0.0428) = 0.3046$$

$$a_n = \frac{2}{\pi}\cdot\frac{1}{n^2+1}\left[1 - e^{-\pi}(-1)^n\right] = \frac{2}{\pi}\cdot\frac{1}{n^2+1}[1 - 0.0428(-1)^n].$$

$$a_1 = 1.0428/\pi = 0.3319;\ a_2 = \frac{2}{5\pi}(0.9572) = 0.1219.\ \text{(See Exercise 15.3)}$$

15.6 Fourier Transform

Let $f(x)$ be a periodic function of period T extending from –T/2 to T/2, i.e., –T/2 $< x <$ T/2. If conditions satisfied, it can be expressed by Fourier series, in accordance with Sec. 15.3(i), as

$$f(x) = a_0 \sum_{n=1}^{\infty} \left(a_n \cos \frac{2n\pi x}{T} + b_n \sin \frac{2n\pi x}{T} \right) \tag{15.15}$$

where

$$a_0 = \frac{1}{T} \int_{-T/2}^{T/2} f(t)\, dt \tag{15.16a}$$

$$a_n = \frac{2}{T} \int_{-T/2}^{T/2} f(t) . \cos \frac{2n\pi t}{T} dt \tag{15.16b}$$

$$b_n = \frac{2}{T} \int_{-T/2}^{T/2} f(t) . \sin \frac{2n\pi x}{T} dt. \tag{15.16c}$$

Considering T to be a large positive integer i.e., T>>2π, let us denote $\delta\xi = 2\pi/T$ and $\xi_n = n\delta\xi,\ n = 0,\ 1,\ 2,\ 3,\ \ldots$T. Thus (15.15) may be expressed after using (15.16 a, b, c) in summation form as,

$$f(x) = \frac{1}{T} \int_{-T/2}^{T/2} f(t)\, dt + \frac{1}{\pi} \sum_{n=1}^{\infty} \left[\left\{ \int_{-T/2}^{T/2} f(t) . \cos \xi_n t\, dt \right\} \cos \xi_n x \right.$$

$$\left. + \left\{ \int_{-T/2}^{T/2} f(t) . \sin \xi_n t\, dt \right\} \sin \xi_n x \right] \delta\xi. \tag{15.17}$$

Let T $\to \infty$, when the function $f(x)$ should be defined in the interval $-\infty < x < \infty$. Further, since $\delta\xi\ (= 2\pi/T) \to 0$, we may replace $\delta\xi$ by $d\xi$ and remembering that $\xi_0 = 0$, the summation in (15.17) may be replaced by integration with respect to ξ under limits 0 to ∞, as follows,

$$f(x)=\frac{1}{T}\int_{-\infty}^{\infty} f(t)\,dt+\frac{1}{\pi}\int_{\xi=0}^{\infty}\left[\cos\xi x\int_{-\infty}^{\infty} f(t)\cos\xi t dt\right.$$

$$\left.+\sin\xi x\int_{-\infty}^{\infty} f(t)\sin\xi t dt\right]d\xi. \tag{15.18}$$

If $\int_{-\infty}^{\infty}|f(t)|\,dt$ remains bounded, the first term in (15.18) tends to zero as T tends to ∞. Hence, we can write,

$$f(x)=\frac{1}{\pi}\int_{\xi=0}^{\infty}\left[\cos\xi x\int_{-\infty}^{\infty} f(t)\cos\xi t dt+\sin\xi x\int_{-\infty}^{\infty} f(t)\sin\xi t dt\right]d\xi. \tag{15.19}$$

The expression (15.19) is known as Fourier Integral of the function $f(x)$ defined for $-\infty<x<\infty$. It may be pointed out that above is only a working proof, not a rigorous one. It must also be noted that if $f(x)$ is an even function then second term in (15.19) vanishes and if odd then first term vanishes.

Further, it is easy to see that (15.19) can be written as,

$$f(x)=\frac{1}{\pi}\int_{\xi=0}^{\infty}\left\{\int_{-\infty}^{\infty} f(t).\cos\xi(x-t)\,dt\right\}d\xi \tag{15.20}$$

As limits of ξ and t vary independently, the integrand in (15.20) which is an even function of ξ can be written as,

$$f(x)=\frac{1}{2\pi}\int_{\xi=-\infty}^{\infty}\left\{\int_{-\infty}^{\infty} f(t)\cos\xi(x-t)\,dt\right\}d\xi \tag{15.21}$$

Further, since $f(t)\sin\xi(x-t)$ is an odd function of ξ, we also have,

$$0=\frac{i}{2\pi}\int_{\xi=-\infty}^{\infty}\left\{\int_{-\infty}^{\infty} f(t)\sin\xi(x-t)\,dt\right\}d\xi, i=\sqrt{-1}. \tag{15.22}$$

Adding (15.21) and (15.22) we get,

$$f(x) = \frac{1}{2\pi} \int_{\xi=-\infty}^{\infty} \left[\int_{-\infty}^{\infty} f(t).e^{i\xi(x-t)} dt \right] d\xi$$

$$= \frac{1}{2\pi} \int_{\xi=-\infty}^{\infty} e^{i\xi x} \int_{-\infty}^{\infty} f(t) e^{-i\xi t} dt.d\xi. \tag{15.23}$$

Expression (15.23) is the complex form of Fourier Integral. It may be expressed by a pair of (reciprocal) functions, viz.,

$$F(\xi) = \frac{1}{2\pi} \int_{-\infty}^{\infty} f(t).e^{-i\xi t} dt, \tag{15.24a}$$

$$f(x) = \int_{-\infty}^{\infty} F(\xi).e^{i\xi t} d\xi. \tag{15.24b}$$

The first integral (15.24a) which carries the original function $f(x)$ to an imaginary (complex) function $F(\xi)$ is called Fourier Transform of $f(x)$ and the second integral (15.24b) returning the original function $f(x)$ from $F(\xi)$ is called Inverse Fourier Transform of $F(\xi)$. It may be mentioned that constant factor $1/2\pi$ in (15.23) may be distributed in (15.24a) and (15.24b) as desired.

It must also be observed that since in (15.21) the integrand $f(t)\cos\xi(x-t)$ is an even function of ξ, then proceeding as above we can get an alternative expression for the complex form of Fourier Integral as,

$$f(x) = \frac{1}{2\pi} \int_{\xi=-\infty}^{\infty} e^{-i\xi x} \int_{-\infty}^{\infty} f(t)\, e^{i\xi t} dt\, d\xi. \tag{15.25}$$

As before (15.25) may be expressed as a pair of (reciprocal) functions,

$$F(\xi) = \frac{1}{2\pi} \int_{-\infty}^{\infty} f(t)\, e^{-i\xi t} dt, \tag{15.26a}$$

$$f(x) = \int_{-\infty}^{\infty} F(\xi)\, e^{-i\xi x} d\xi. \tag{15.26b}$$

At some places (15.26a) and (15.26b) are defined as Fourier Transform and Inverse Fourier Transform respectively instead of (15.24 a, b). It should not be matter of concern anyway. Further, we can also write Fourier cosine transform and Fourier sine transform depending on whether $f(x)$ is an even or odd function respectively.

The Fourier Transform is of great practical importance in electrical engineering in that it transforms the input signals from time domain (real function) to frequency domain (complex function) defining amplitude and phase angle of frequency.

15.7 Discrete Fourier Transform (DFT)

As stated, Fourier Transform plays an important role in electrical engineering. However, from the viewpoint of practical applications we have to generally deal with numerical values of the function at discrete points, rather than its analytical, from defined over a finite interval (not infinite). We shall now discuss the Fourier Transform of a function which is given in the form of discrete data i.e., whose values are known at finite number of distinct points. Such a transform is known as Discrete Fourier Transform (DFT). Let us assume that a function $f(x)$ is defined numerically at a finite number of points over a normalised interval $0 \leq x \leq 1$. We subdivide the interval of unit length into N subintervals each of width 1/N at the points $x = x_k, k = 0(1)N$, such that $x_k = k/N$, $x_0 = 0$ and $x_N = 1$. Denoting $f(x_k)$ by f_k, the N-point Discrete Fourier Transform (DFT) may be defined as,

$$F_j = \frac{1}{N}\sum_{k=0}^{N-1} f(x_k).e^{-i2\pi jx_k}$$

$$= \frac{1}{N}\sum_{k=0}^{N-1} f_k.e^{-\frac{i2\pi j.k}{N}}, \quad j = 0(1)N-1. \tag{15.27a}$$

The Inverse DFT is defined as,

$$f_k = \sum_{j=0}^{N-1} F_j e^{i2\pi x_k.j}$$

$$= \sum_{j=0}^{N-1} F_j.e^{\frac{i.2\pi k.j}{N}}, \quad k = 0(1)N-1. \tag{15.27b}$$

The expressions (15.27a) and (15.27b) are analogous to (15.24a) and (15.24b) respectively for transform of a discrete function defined over an interval of unit length.

It would be pertinent to give some explanation regarding the formulation of the transforms (15.27a) and (15.27b). Let us recall that a function $f(x)$, defined for $0 \leq x \leq 1$ can be represented by a Fourier series containing terms $\sum(a_n \cos 2n\pi x + b_n \sin 2n\pi x)$ where a_n

and b_n are determined using orthogonal property of $\cos 2n\pi x$ and $\sin 2n\pi x$ with respect to an integral having limits from 0 to 1. Likewise, in the discrete transform, $f(x_k)$ is represented by (15.27b) containing f_j's while f_j's are determined by (15.27a). It will be shown that $\exp\left(i2\pi j.k/N\right)$ and $\exp\left(-i2\pi j.k/N\right), j = 0(1)N-1$ constitute sets of biorthogonal functions with respect to summation over k from 0 to N–1. To understand this point let us consider the sum over k = 0(1) N–1, of the product of these functions taking $j = m$ in $\exp\left(i2\pi j.k/N\right)$ and $j = n$ in $\exp\left(-i2\pi j.k/N\right)$. Denoting the sum by S, we have

$$S = \sum_{k=0}^{N-1} e^{\frac{i.2\pi mk}{N}} . e^{\frac{-i2\pi nk}{N}} = \sum_{k=0}^{N-1} e^{\frac{i.2\pi (m-n).k}{N}} \tag{15.28a}$$

$$= \frac{\exp\{i2\pi(m-n)\}-1}{1-\exp\{i2\pi(m-n)/N\}}, \tag{15.28b}$$

where m, n = 0(1)N–1.

It should be clear that S = 0 when $m \neq n$ from (15.28b) and $S = N$ when $m = n$ from (15.28a) i.e.,

$$S = \begin{cases} 0, & m \neq n \\ N, & m = n \end{cases} \tag{15.29}$$

Remember that (15.28a) represents the sum of N terms of a geometric series whose common ratio is $\exp\left(i2\pi(m-n)/N\right)$ and first term unity.

We shall now verify (15.27a) for computation of F_j when f_k has been represented by (15.27b). To find F_j, we multiply both sides of (15.27b) by $\exp\left(-i.2\pi j.k/N\right)$ to get,

$$f_k . e^{\frac{-i2\pi j.k}{N}} = \sum_{m=0}^{N-1} F_m e^{\frac{i2\pi(m-j)k}{N}}, \quad k = 0,\ 1,\ 2, \ldots\ N-1. \tag{15.30}$$

Now adding up all the terms on the left and right sides of N equations (15.30) for k = 0(1) N–1, we get

$$\sum_{k=0}^{N-1} f_k e^{-\frac{i2\pi jk}{N}} = \sum_{k=0}^{N-1}\sum_{m=0}^{N-1} F_m e^{\frac{i2\pi(m-j)k}{N}}$$

$$= \sum_{m=0}^{N-1} F_m \sum_{k=0}^{N-1} e^{\frac{i2\pi(m-j)k}{N}}$$

$= F_j.N$ due to (15.29).

or $$F_j = \frac{1}{N}\sum_{k=0}^{N-1} f_k e^{-\frac{i2\pi jk}{N}}, j = 0, 1, 2, \ldots N-1. \quad (15.31)$$

Thus (15.27b) can be viewed as another way of expressing the discrete function by a terminating Fourier series in complex form while the coefficients are determined by (15.27a). It may also be added that as rectangular rule has been used in approximating the integral over 0 to 1 summing for $k = 0(1)N-1$, a larger value of N will provide better results.

15.8 Representation of Transforms in Matrix Form

Putting $w = e^{i2\pi/N}$ we can express N-Point DFT (15.27a) and Inverse DFT (15.27b) respectively as,

$$\boldsymbol{F}_j = \frac{1}{N}\sum_{k=0}^{N-1} f_k.w^{-j.k}, \quad j = 0, 1, 2, \ldots (N-1) \quad (15.32a)$$

$$f_k = \sum_{j=0}^{N-1} F_j.w^{k.j}, \quad k = 0, 1, 2, \ldots (N-1) \quad (15.32b)$$

The transforms (15.32a) and (15.32b) both are systems of N simultaneous linear equations in N unknowns, which can be expressed in matrix form. Let $\boldsymbol{F}$ and $\boldsymbol{f}$ be two vectors such that

$$\boldsymbol{F}^T = (F_0, F_1, F_2, \ldots, F_{N-1}) \quad (15.33a)$$

$$\boldsymbol{f}^T = (f_0, f_1, f_2, \ldots, f_{N-1}) \quad (15.33b)$$

Further let $\boldsymbol{W}$ be a $(N \times N)$ matrix as follows:

$$\boldsymbol{W} = \begin{bmatrix} w^{0.0} & w^{0.1} & w^{0.2} & \ldots & w^{0.(N-1)} \\ w^{1.0} & w^{1.1} & w^{1.2} & \ldots & w^{1.(N-1)} \\ w^{k.0} & w^{k.1} & w^{k.2} & \ldots & w^{k.(N-1)} \\ \vdots & \vdots & \vdots & & \vdots \\ w^{(N-1).0} & w^{(N-1).1} & w^{(N-1).2} & \ldots & w^{(N-1).(N-1)} \end{bmatrix} \quad (15.34)$$

where dot (.) in the exponent indicates multiplication. It must be noted that $\boldsymbol{W}$ is a symmetric matrix i.e.,

$$\boldsymbol{W} = \boldsymbol{W}^T. \tag{15.35}$$

Let $\bar{\boldsymbol{W}}$ be a matrix which is obtained by conjugating each element of $\boldsymbol{W}$ i.e., changing $w^{k.j}$ to w^{-kj}. Matrix $\bar{\boldsymbol{W}}$ will also be symmetric i.e.,

$$\bar{\boldsymbol{W}} = \bar{\boldsymbol{W}}^T \tag{15.36}$$

Due to property (15.29) it is easy to see that

$$\boldsymbol{W}.\bar{\boldsymbol{W}} = N.I \text{ or } \frac{1}{N}.\boldsymbol{W}.\bar{\boldsymbol{W}} = I, \tag{15.37}$$

where I is a Unit/Identity matrix.

Thus using (15.35) and (15.36) we can write (15.32a) and (15.32b) respectively in matrix form as,

$$\boldsymbol{F} = \frac{1}{N}.\bar{\boldsymbol{W}}\boldsymbol{f} \tag{15.38a}$$

$$\boldsymbol{f} = \boldsymbol{W}\boldsymbol{F} \tag{15.38b}$$

Obviously we can derive in case of above equations one from the other by using relation (15.37).

15.9 Complex Roots of Unity

There is a direct relationship between the variable $w = e^{i2\pi/N}$ and complex roots of unity, as discussed below.

The N complex roots of unity are given by the solution of the complex equation $z^N = 1$. If $w_0, w_1, w_2, \ldots, w_{N-1}$ denote the N complex roots then,

$$w_j = e^{i2\pi j/N}, j = 0, 1, 2, \ldots, (N-1), i = \sqrt{-1} \tag{15.39}$$

It may be noted that $w_1 = e^{i2\pi/N} = w = w^1$ and in general $w_j = w^j$. Hence w_j and w^jcan be used interchangeably. The root w_1 is called 'primitive' root and all the roots are its various powers from 0 to N–1. Also note that $w_0 = w^0 = 1$.

We should also know that these N roots keep repeating in that order for j increasing. It means that if $m = qN + r$, where m, q and r are positive integers ($r < N$) including zero, then

$$w^m = w^r, \quad r = m\,(\text{modulus})\,N. \tag{15.40}$$

To compute N-point DFT of a function $f(t)$, say defined in an interval $0 \le t \le a$ we may proceed as follows:

(i) Convert the interval 0 to 1 by changing the variable
$x = t/a$ or $t = ax$; define $f(ax) = f(x), 0 \le x \le 1$.

(ii) Choose N points $x_k = k/N, k = 0(1)N-1$. Find function values $f(x_k) = f_k$ at these points.

(iii) Find N complex roots of unity $w^0 = w_0 = 1$ and $w^j = w_j = e^{i2\pi j/N}, j = 1(1)N-1$.

(iv) Use (15.38a) to compute $F_j, j = 0(1)N-1$. To compute elements of $\bar{\mathbf{W}}$ we need $w^{-j.k}$ for $j = 0(1)N-1$ and $k = 0(1)N-1$. Since $\bar{\mathbf{W}}$ is a symmetric matrix, compute only elements in the lower triangular (or in the upper triangular). Use (15.40) when $j.k \ge N$. We need powers of w from 0 to $(N-1) \times (N-1)$. Get complex conjugates by changing i to $-i$.

Example 15.3

Find 4-point Discrete Fourier Transform (DFT) of the function $f(t) = e^{-t}$, $0 \le t \le 4.0$. Check Inverse DFT.

Solution.

Put $x = t/4$ or $t = 4x$.

Hence we find DFT of the function

$$f(x) = e^{-4x}, 0 \le x \le 1.0$$

$N = 4$

$k \rightarrow$	0	1	2	3	4
$x_k \rightarrow$	0	$1/4$	$2/4$	$3/4$	
$4x_k \rightarrow$	0	1.0	2.0	3.0	
$f = e^{-4x_k}$	1.0000	0.3679	0.1353	0.0498	

Four roots of unity are given by $w^j = e^{\frac{i2\pi j}{4}}, j = 0, 1, 2, 3.$

$$w^0 = 1,\ w^1 = i,\ w^2 = -1,\ w^3 = -i$$

$$w^4 = w^0 = 1,\ w^5 = w^1 = i,\ w^6 = w^2 = -1,\ w^7 = w^3 = -i$$

$$w^8 = w^0 = 1,\ w^9 = w^1 = i$$

$$F = \frac{1}{N}\bar{\mathbf{W}}f = \frac{1}{4}\begin{bmatrix} 1 & 1 & 1 & 1 \\ 1 & -i & -1 & i \\ 1 & -1 & 1 & -1 \\ 1 & i & -1 & -i \end{bmatrix}\begin{bmatrix} 1.0000 \\ 0.3679 \\ 0.1353 \\ 0.0498 \end{bmatrix} = \begin{bmatrix} 0.3882 \\ 0.2162 - 0.0795i \\ 0.1794 \\ 0.2162 + 0.0795i \end{bmatrix}$$

To check Inverse DFT

$$f = \mathbf{W}F = \begin{bmatrix} 1 & 1 & 1 & 1 \\ 1 & i & -1 & -i \\ 1 & -1 & 1 & -1 \\ 1 & -i & -1 & i \end{bmatrix}\begin{bmatrix} 0.3882 \\ 0.2162 - 0.0795i \\ 0.1794 \\ 0.2162 + 0.0795i \end{bmatrix} = \begin{bmatrix} 1.0000 \\ 0.3678 \\ 0.1352 \\ 0.0498 \end{bmatrix}$$

Note: The error of $\pm.0001$ in 0.3678 and 0.1352 is due to rounding.

Example 15.4

Find 8-point DFT of the function $f(t) = e^{-t},\ 0 \le t \le 4.0$

Solution

As in Example 15.3 we find DFT of function $f(x) = e^{-4x},\ 0 \le x \le 1.0$. $N = 8$. Eight roots of unity are given by $e^{i2\pi j/8},\ j = 0(1)7$.

$$w^0 = 1,\ w^1 = \frac{1+i}{\sqrt{2}},\ w^2 = i,\ w^3 = -\frac{1-i}{\sqrt{2}}$$

$$w^4 = -1,\ w^5 = -\frac{1+i}{\sqrt{2}},\ w^6 = -i,\ w^7 = \frac{1-i}{\sqrt{2}}$$

Writing other powers of w

$$w^8 = w^0,\ w^9 = w^1,\ w^{10} = w^2,\ w^{11} = w^3 \text{ and so on.}$$

$$\bar{W} = \begin{bmatrix} 1 & & & & & & & \\ 1 & w^{-1} & & & & & & \\ 1 & w^{-2} & w^{-4} & & \text{symmetric} & & & \\ 1 & w^{-3} & w^{-6} & w^{-1} & & & & \\ 1 & w^{-4} & w^{0} & w^{-4} & w^{0} & & & \\ 1 & w^{-5} & w^{-2} & w^{-7} & w^{-4} & w^{-1} & & \\ 1 & w^{-6} & w^{-4} & w^{-2} & w^{0} & w^{-6} & w^{-4} & \\ 1 & w^{-7} & w^{-6} & w^{-5} & w^{-4} & w^{-3} & w^{-2} & w^{-1} \end{bmatrix}$$

$k \rightarrow$	0	1	2	3	4	5	6	7	8
$x_k \rightarrow$	0	1/8	2/8	3/8	4/8	5/8	6/8	7/8	
$4x_k \rightarrow$	0	1/2	1.0	1.5	2.0	2.5	3.0	3.5	
$f_k = e^{-4x_k}$	1.0000	0.6065	0.3679	0.2231	0.1353	0.0821	0.0498	0.0302	

$$F_0 = \frac{1}{8}[2.4949] = 0.3119$$

$$F_1 = \frac{1}{8}\left[f_0 + w^{-1}f_1 + w^{-2}f_2 + w^{-3}f_3 + w^{-4}f_4 + w^{-5}f_5 + w^{-6}f_6 + w^{-7}f_7\right]$$

$$= \frac{1}{8}\left[1.0000 + 0.6065 \times \frac{(1-i)}{\sqrt{2}} + 0.3679 \times (-i) + 0.2231 \times \left(-\frac{1+i}{\sqrt{2}}\right)\right.$$

$$\left. +0.1353 \times (-1) + 0.0821 \times \left(-\frac{1-i}{\sqrt{2}}\right) + 0.0498 \times (i) + 0.0302 \times \left(\frac{1+i}{\sqrt{2}}\right)\right]$$

$$= \frac{1}{8}\left[0.8647 + \frac{1}{\sqrt{2}}(0.3315) - \frac{1}{\sqrt{2}}(0.7173)i - 0.3181i\right]$$

$$= 0.1374 - 0.1032i$$

$$F_2 = \frac{1}{8}\left[f_0 + w^{-2}f_1 + w^{-4}f_2 + w^{-6}f_3 + w^{0}f_4 + w^{-2}f_5 + w^{-4}f_6 + w^{-6}f_7\right]$$

$$= \frac{1}{8}[1.0000 + 0.6065 \times (-i) + 0.3679 \times (-1) + 0.2231 \times (i)$$

$$+0.1353 \times (1) + 0.0821 \times (-i) + 0.0498 \times (-1) + 0.0302 \times (i)]$$

$$= 0.0897 - 0.0544i$$

$$F_3 = \frac{1}{8}\left[f_0 + w^{-3}f_1 + w^{-6}f_2 + w^{-1}f_3 + w^{-4}f_4 + w^{-7}f_5 + w^{-2}f_6 + w^{-5}f_7\right]$$

$$= \frac{1}{8}\left[1.0000 + 0.6065 \times \left(-\frac{1+i}{\sqrt{2}}\right) + 0.3679 \times (i) + 0.2231 \times \left(\frac{1-i}{\sqrt{2}}\right)\right.$$

$$\left. + 0.1353 \times (-1) + 0.0821 \times \left(\frac{1+i}{\sqrt{2}}\right) + 0.0498 \times (-i) + 0.0302 \times \left(-\frac{1-i}{\sqrt{2}}\right)\right]$$

$$= \frac{1}{8}\left[0.8647 - \frac{1}{\sqrt{2}}(0.3315) - \frac{1}{\sqrt{2}}(0.7173)\,i + 0.3181i\right] = \frac{1}{8}[0.6303 - 0.1891i]$$

$$= 0.0788 - 0.0236i$$

$$F_4 = \frac{1}{8}\left[f_0 + w^{-4}f_1 + w^{0}f_2 + w^{-4}f_3 + w^{0}f_4 + w^{-4}f_5 + w^{0}f_6 + w^{-4}f_7\right]$$

$$= \frac{1}{8}[1.0000 - 0.6065 + 0.3679 - 0.2231 + 0.1353 - 0.0821 + 0.0498 - 0.0302]$$

$$= 0.0764$$

$$F_5 = 0.0788 + 0.0236i$$

$$F_6 = 0.0897 + 0.0544i$$

$$F_7 = 0.1374 + 0.1032i$$

Note: We need compute F_j for $j = 0$ to $\frac{N}{2}$ only; the remaining can be written from the relation $F_{N-j} = \bar{F}_j,\ j = 0(1)\frac{N}{2}$, $\bar{F}$ denoting complex conjugate of F, since

$$F_{N-j} = \frac{1}{N}\sum_{k=0}^{N-1} f_k.e^{-i.2\pi(N-j)k/N}$$

$$= \frac{1}{N}\sum_{k=0}^{N-1} f_k.e^{i.2\pi j.k/N}.e^{-i.2\pi k}$$

$$= \frac{1}{N}\sum_{k=0}^{N-1} f_k.e^{i.2\pi j.k/N} = \bar{F}_j, j = 0, 1, 2, \ldots \frac{N}{2}$$

It should also be clear that $F_{N/2}$ should be real since $F_{N/2} = \bar{F}_{N/2}$.

15.10 Fast Fourier Transform (FFT)

The Fast Fourier Transform (FFT) is a procedure to compute DFT faster in terms of reduction in the number of arithmetic operations (multiplication/division, addition/subtraction), hence saving in computer time. A large amount of work has been done on FFT since Cooley and Tukey [3], first published their paper on the topic in 1965. We give below their version of FFT algorithm.

Let us rewrite the N-point DFT,

$$F_j = \frac{1}{N}\sum_{k=0}^{N-1} f_k e^{\frac{-i2\pi jk}{N}} = \frac{1}{N}\sum_{k=0}^{N-1} f_k w^{-j.k} = \frac{1}{N}\sum_{k=0}^{N-1} f_k.\left(w^{-j}\right)^k, j=0,1,2,\ldots,N-1 \quad (15.41)$$

where $w = e^{\frac{i2\pi}{N}} = w^1$

Further, putting $z_j = w^{-j}$, which is complex conjugate of jth root of unity, we can write (15.41) as

$$F_j = F(z_j) = \frac{1}{N}\sum_{k=0}^{N-1} f_k.z_j^k, \quad j = 0(1)N-1. \quad (15.42)$$

We see that right side of (15.42) is a complex polynomial in z of degree (N–1) evaluated for $z = z_j$. This requires multiplications and additions O(N). Such evaluations for $j = 0(1)N-1$ will require O(N^2) multiplications/additions (complex operations). It will be seen that FFT involves only O(NlogN) arithmetic operations. But there should be a condition on N, in that it should be a power of 2 i.e., $N = 2^p$, say, where p is an integer. This enables us to divide N successively by 2 while $p = \log_2 N$.

Cooley and Tukey [3] make use of the following properties in their algorithm:

Let $w_N^j, j = 0(1)N-1$ denote N^{th} roots of unity. Then $(N/2)^{th}$, roots of unity can be represented as $w_{N/2}^j$, $j = 0(1)\frac{N}{2}-1$. We have,

$$\text{(i) } w_{N/2}^j = \left(w_N^j\right)^2 = w_N^{2j}, \; j = 0(1)\tfrac{N}{2}-1, \quad (15.43)$$

since $w_{N/2}^j = \exp\left(\frac{i.2\pi j}{N/2}\right) = \exp\left(i2\pi.2j/N\right) = w_N^{2j}$, $j = 0(1)\frac{N}{2}-1$.

That is, (N/2)th roots of unity, $w_{N/2}^j$, $j = 0(1)\frac{N}{2}-1$ are squares of $w_N^j, j = 0(1)\frac{N}{2}-1$; or we can say that the roots $w_{N/2}^j$, $j = 0(1)\frac{N}{2}-1$ are simply even-numbered roots w_N^j, $j = 0(2)N-2$.

(ii) $w_N^{N/2} = -1$ (15.44)

for $w_N^{N/2} = \exp\left(\frac{i.2\pi}{N}.\frac{N}{2}\right) = \exp(i\pi) = -1.$

The basic concept in the algorithm is to break an N-point DFT into sum of two $\frac{N}{2}$-point DFT's. These two DFT's can be again broken into $\frac{N}{4}$-point DFT's each and so on until we reach total N/2 DFT's each a 2-point DFT. In general after m^{th} break-up there will be 2^m number of DFT's each of them being $\frac{N}{2^m}$-point DFT, $m = 1, 2, \ldots .p-1$, when $N = 2^p$. The computations can be carried by using a recursive procedure. There are two approaches which we shall discuss below.

1. From (15.42) we can write

$$F_j = \frac{1}{N}\sum_{k=0}^{N-1} f_k.z_j^k = \frac{1}{N}\sum_{k=0}^{\frac{N}{2}-1} f_k.z_j^k + \frac{1}{N}\sum_{k=\frac{N}{2}}^{N-1} f_k.z_j^k$$

$$= \frac{1}{N}\sum_{k=0}^{\frac{N}{2}-1} f_k.z_j^k + \frac{1}{N}\sum_{k=0}^{\frac{N}{2}-1} f_{k+\frac{N}{2}}.z_j^{k+\frac{N}{2}}$$

$$= \frac{1}{N}\sum_{k=0}^{\frac{N}{2}-1} f_k.z_j^k + z_j^{\frac{N}{2}}.\frac{1}{N}\sum_{k=0}^{\frac{N}{2}-1} f_{k+\frac{N}{2}}.z_j^k$$

$$= \frac{1}{N}\sum_{k=0}^{\frac{N}{2}-1} f_k.w_N^{-j.k} + w_N^{j.\frac{N}{2}}.\frac{1}{N}\sum_{k=0}^{\frac{N}{2}-1} f_{k+\frac{N}{2}}.w_N^{-j.k}$$

$$= \frac{1}{N}\sum_{k=0}^{\frac{N}{2}-1} f_k.w_N^{-j.k} + (-1)^j.\frac{1}{N}\sum_{k=0}^{\frac{N}{2}-1} f_{k+\frac{N}{2}}.w_N^{-j.k}, j = 0,1,2,\ldots,N-1. \text{ due to (15.44).} \quad (15.45)$$

For j even or odd (15.45) can be written as follows for $r = 0, 1, 2, \ldots, \frac{N}{2}-1$,

$$j = 2r \Rightarrow F_{2r} = \frac{1}{N}\sum_{k=0}^{\frac{N}{2}-1} f_k.w_N^{-2r.k} + \frac{1}{N}\sum_{k=0}^{\frac{N}{2}-1} f_{k+\frac{N}{2}}.w_N^{-2r.k} \quad (15.46)$$

$$= \frac{1}{N}\sum_{k=0}^{\frac{N}{2}-1} \left(f_k + f_{k+\frac{N}{2}}\right) w_N^{-2r.k} \quad (15.46a)$$

$$j = 2r+1 \Rightarrow F_{2r+1} = \frac{1}{N}\sum_{k=0}^{\frac{N}{2}-1} f_k.w_N^{-(2r+1).k} - \frac{1}{N}\sum_{k=0}^{\frac{N}{2}-1} f_{k+\frac{N}{2}}.w_N^{-(2r+1).k}$$

$$= \frac{1}{N}\sum_{k=0}^{\frac{N}{2}-1} \left\{\left(f_k - f_{k+\frac{N}{2}}\right).w_N^{-k}\right\} w_N^{-2r.k} \tag{15.46b}$$

It must be noted that (15.45a) and (15.45b) are $\frac{N}{2}$–point DFT's since $w_N^{2r} = w_{N/2}^{r}, r = 0(1)\frac{N}{2}-1$ (see property 15.43). Thus an N–point DFT (15.41) has been reduced to two summations (15.45). It is further shown that for j even and odd (15.45) is equivalent to $\frac{N}{2}$–point DFT (15.46a) and (15.46b) respectively, adjusting 1/N as 2/2N.

2. Alternatively we can express (15.42) in the following form,

$$F_j = \frac{1}{N}\sum_{k=0}^{N-1} f_k z_j^k$$

$$= \frac{1}{N}\left(f_0 z_j^0 + f_1 z_j^1 + f_2 z_j^2 + \ldots + f_{N-2} z_j^{N-2} + f_{N-1}.z_j^{N-1}\right)$$

$$= \frac{1}{N}\left(f_0 z_j^0 + f_1 z_j^2 + \ldots + f_{N-2} z_j^{N-2}\right) + \frac{1}{N}\left(f_1 z_j^1 + f_3 z_j^3 + \ldots + f_{N-1} z_j^{N-1}\right)$$

$$= \frac{1}{N}\sum_{r=0}^{\frac{N}{2}-1} f_{2r}.z_j^{2r} + z_j.\frac{1}{N}\sum_{r=0}^{\frac{N}{2}-1} f_{2r+1}.z_j^{2r}$$

$$= \frac{1}{N}\sum_{r=0}^{\frac{N}{2}-1} f_{2r}.w_N^{-2j.r} + w_N^{-j}.\frac{1}{N}\sum_{r=0}^{\frac{N}{2}-1} f_{2r+1}.w_N^{-2j.r} \tag{15.47}$$

Again since $w_N^{-2j} = w_{N/2}^{-j}$ from (15.43) the break-up (15.47) can be considered as sum of two N/2–point DFT's. In the language of analysis of algorithm if T(N) denotes the time-complexity of N–point DFT i.e., the number of arithmetic operations required in computing $F_j,\ j = 0(1)N-1$, then we have a recurrence relation in the Cooley and Tukey algorithm,

$$T(N) = 2T(N/2) + c.N, \tag{15.48}$$

with some initial condition like T(2) = 3 and c being a constant.

The solution of (15.48) can be found as T(N) = O (N log N); see Chapter 14 of the book. Thus we see that the number of arithmetic operations in Cooley and Tukey algorithm are O(N log N) vis-a-vis O(N^2) in the direct evaluation of DFT; hence the algorithm is called

Fast Fourier Transform i.e., FFT. There are some other versions of FFT also; see references [1] to [5].

15.11 Fast Fourier Transform via Inverse Transform (Author's Comments)

In this method we start off from (15.32b) which is a system of N simultaneous equations in N unknown $F_j, j = 0(1)\ N\text{–}1$ and known right sides $f_j, j = 0(1)\ N\text{–}1$. By using (15.32b) we arrive at (15.32a) which is the required N–point DFT. It will be seen that the algorithm is also O(N log N). In designing the algorithm we shall make use of some more properties of Nth roots of unity. They are discussed below:

(i) $w^{\frac{N}{2}+j} = -w^j, j = 0(1)\frac{N}{2} - 1,$ (15.49)

since $\exp\{\frac{i2\pi}{N}\left(\frac{N}{2} + j\right)\} = \exp(i\pi)\exp\left(\frac{i2\pi j}{N}\right) = -\exp\left(\frac{i2\pi j}{N}\right)$

It means that the first N/2 roots repeat with their signs changed, in that order, i.e.,

$$w^{\frac{N}{2}} = -w^0 = -1, w^{\frac{N}{2}+1} = -w^1, \ldots, w^{N-1} = -w^{\frac{N}{2}-1}. \qquad (15.50)$$

(ii) If $a + b = N/2$, a and b being positive integers, then $w^a = -w^{-b}$ or vice-versa since $w^{\frac{N}{2}} = -1$.

(iii) For any positive integer k,

$$w^{\left(\frac{N}{2}+j\right)k} = \begin{cases} w^{jk}, & \text{if } k \text{ is even} \\ -w^{jk}, & \text{if } k \text{ is odd} \end{cases} \qquad (15.51)$$

(iv) We rewrite (15.40) here to make this section complete in that the N roots repeat themselves for $j > N\text{–}1$. That is, $w^N = w^0 = 1$, $w^{N+1} = w^1$, $w^{N=2} = w^2$ and so on. In general, for any positive integer m,

$$w^m = w^r, \text{ where } r = m \text{ (modulus) } N. \qquad (15.52)$$

Instead of discussing the method in its generality we would like to explain its working through an example with $N = 8 = 2^3$, as follows:

For an 8-point FFT we are required to solve a system of simultaneous equations (15.38b) for eight unknowns $F_0, F_1, \ldots F_7$ when the function values $f_k = f(x_k)$, $k = 0(1)7$ are known. The coefficient matrix W is given by

$$W = \begin{bmatrix} 1 & 1 & 1 & 1 & 1 & 1 & 1 & 1 \\ 1 & w^{1\cdot 1} & w^{1\cdot 2} & w^{1\cdot 3} & w^{1\cdot 4} & w^{1\cdot 5} & w^{1\cdot 6} & w^{1\cdot 7} \\ 1 & w^{2\cdot 1} & w^{2\cdot 2} & w^{2\cdot 3} & w^{2\cdot 4} & w^{2\cdot 5} & w^{2\cdot 6} & w^{2\cdot 7} \\ 1 & w^{3\cdot 1} & w^{3\cdot 2} & w^{3\cdot 3} & w^{3\cdot 4} & w^{3\cdot 5} & w^{3\cdot 6} & w^{3\cdot 7} \\ 1 & w^{4\cdot 1} & w^{4\cdot 2} & w^{4\cdot 3} & w^{4\cdot 4} & w^{4\cdot 5} & w^{4\cdot 6} & w^{4\cdot 7} \\ 1 & w^{5\cdot 1} & w^{5\cdot 2} & w^{5\cdot 3} & w^{5\cdot 4} & w^{5\cdot 5} & w^{5\cdot 6} & w^{5\cdot 7} \\ 1 & w^{6\cdot 1} & w^{6\cdot 2} & w^{6\cdot 3} & w^{6\cdot 4} & w^{6\cdot 5} & w^{6\cdot 6} & w^{6\cdot 7} \\ 1 & w^{7\cdot 1} & w^{7\cdot 2} & w^{7\cdot 3} & w^{7\cdot 4} & w^{7\cdot 5} & w^{7\cdot 6} & w^{7\cdot 7} \end{bmatrix} \tag{15.53}$$

where dot $(\cdot)$ in the exponentiation of w stands for multiplication. We can express powers of w in W matrix (15.53) in its lowest degree using property 15.52 i.e., from w^0 to w^7. Further, due to property 15.49 we have

$$w^4 = -w^0 = -1;\; w^5 = -w^1;\; w^6 = -w^2;\; w^7 = -w^3 \tag{15.54}$$

Hence, matrix 15.53 can be expressed in terms of the elements w^0 (=1), w^1, w^2 and w^3 only. Let us write the system of equations (15.38b) in the following form (with unknowns at the top).

(15.55)

	F_0	F_1	F_2	F_3	F_4	F_5	F_6	F_7	f
$R1:$	1	1	1	1	1	1	1	1	f_0
$R2:$	1	w^1	w^2	w^3	-1	$-w^1$	$-w^2$	$-w^3$	f_1
$R3:$	1	w^2	-1	$-w^2$	1	w^2	-1	$-w^2$	f_2
$R4:$	1	w^3	$-w^2$	w^1	-1	$-w^3$	w^2	$-w^1$	f_3
$R5:$	1	-1	1	-1	1	-1	1	-1	f_4
$R6:$	1	$-w^1$	w^2	$-w^3$	-1	w^1	$-w^2$	w^3	f_5
$R7:$	1	$-w^2$	-1	w^2	1	$-w^2$	-1	w^2	f_6
$R8:$	1	$-w^3$	$-w^2$	$-w^1$	-1	w^3	w^2	w^1	f_7

where $R1, R2, \ldots R8$ denote row numbers.

It would be important to notice a special property satisfied by the elements of the coefficient matrix above and below the dotted line in (15.55) that is, the column elements in the first four rows, $R1$-$R4$ and in the last four rows, $R5$-$R8$. According to property (15.51) the elements in the odd columns 1, 3, 5 and 7 in rows $R5$-$R8$ are same as in rows $R1$-$R4$ while in the even columns 2, 4, 6 and 8 they are same but have opposite signs. Let us call this property Symmetry – Asymmetry – Column (SAC) property. There will be three stages in the procedure i.e., equivalent to 2's exponent in $N = 2^3$. Let us describe them as follows:

Stage 1: Reduction of (8×8) system to two (4×4) systems

Adding row $R5$ to $R1$, $R6$ to $R2$, $R7$ to $R3$ and $R8$ to $R4$ we get from (15.55),

	$2F_0$	$2F_2$	$2F_4$	$2F_6$	
$R1 \leftarrow R1+R5$:	1	1	1	1	$f_0 + f_4$
$R2 \leftarrow R2+R6$:	1	w^2	-1	$-w^2$	$f_1 + f_5$
$R3 \leftarrow R3+R7$:	1	-1	1	-1	$f_2 + f_6$
$R4 \leftarrow R4+R8$:	1	$-w^2$	-1	w^2	$f_3 + f_7$

(15.56)

Similarly, subtracting rows $R5$, $R6$, $R7$, $R8$ from $R1$, $R2$, $R3$, $R4$ respectively in (15.55) we get,

	$2F_1$	$2F_3$	$2F_5$	$2F_7$	
$R1 \leftarrow R1+R5$:	1	1	1	1	$f_0 - f_4$
$R2 \leftarrow R2-R6$:	w^1	w^3	$-w^1$	$-w^3$	$f_1 - f_5$
$R3 \leftarrow R3-R7$:	w^2	$-w^2$	w^2	$-w^2$	$f_2 - f_6$
$R4 \leftarrow R4-R8$:	w^3	w^1	$-w^3$	$-w^1$	$f_3 - f_7$

(15.57)

In (15.57) we divide each row by the corresponding element in the first column of that row. Further, also using property (15.50) i.e., replacing w^{-2} by $-w^2$ in the coefficient matrix, we get

	$2F_1$	$2F_3$	$2F_5$	$2F_7$	
$R1 \leftarrow 1 \times R1$:	1	1	1	1	$w^0\,(f_0 - f_4)$
$R2 \leftarrow w^{-1} \times R2$:	1	w^2	-1	$-w^2$	$w^{-1}\,(f_1 - f_5)$
$R3 \leftarrow w^{-2} \times R3$:	1	-1	1	-1	$w^{-2}\,(f_2 - f_6)$
$R4 \leftarrow w^{-3} \times R4$:	1	$-w^2$	-1	w^2	$w^{-3}\,(f_3 - f_7)$

(15.58)

It may be seen that the system of equations (15.55) containing 8 unknowns F_j, $j = 0(1)7$ has been reduced to 2 sets of simultaneous equations (15.56) and (15.58), each containing 4 unknowns. The set (15.56) contains $2F_0$, $2F_2$, $2F_4$, $2F_6$ as unknowns and set (15.58), $2F_1$,

$2F_3$, $2F_5$, $2F_7$.

Two things must be noted here; one, that the coefficient matrices in (15.56) and (15.58) are same (symmetric) and two, that they satisfy SAC property about the dotted line. Hence we can repeat the process of reducing them each to two matrices.

Stage 2: Reduction of (4×4) systems to (2×2) systems.

By performing the operation of addition on (15.56) we get

	$4F_0$	$4F_4$		
$R1 \leftarrow R1+R3$:	1	1	$(f_0+f_4)+(f_2+f_6)$	(15.59)
$R2 \leftarrow R2+R4$:	1	−1	$(f_1+f_5)+(f_3+f_7)$	

The operation of subtraction gives,

	$4F_2$	$4F_6$		
$R1 \leftarrow R1-R3$:	1	1	$(f_0+f_4)-(f_2+f_6)$	(15.60)
$R2 \leftarrow R2-R4$:	w^2	$-w^2$	$(f_1+f_5)-(f_3+f_7)$	

After subtraction we divide each row by the corresponding element in the first column. Hence from (15.60) we get

	$4F_2$	$4F_6$		
$R1 \leftarrow 1 \times R1$:	1	1	$w^0\{(f_0+f_4)-(f_2+f_6)\}$	(15.61)
$R2 \leftarrow w^{-2} \times R2$:	1	−1	$w^{-2}\{(f_1+f_5)-(f_3+f_7)\}$	

Again, the coefficient matrices in (15.59) and (15.61) are same (symmetric) and satisfy SAC property about the dotted line. Since coefficient matrix is same we can do same operations on (15.58) as in (15.56). Thus, by adding relevant rows we get from (15.58),

	$4F_1$	$4F_5$		
$R1 \leftarrow R1+R3$:	1	1	$w^0(f_0-f_4)+w^{-2}(f_2-f_6)$	(15.62)
$R2 \leftarrow R2+R4$:	1	−1	$w^{-1}(f_1-f_5)+w^{-3}(f_3-f_7)$	

By subtraction we get,

	$4F_3$	$4F_7$		
$R1 \leftarrow R1 - R3$:	1	1	$w^0(f_0 - f_4) - w^{-2}(f_2 - f_6)$	(15.63)
$R2 \leftarrow R2 - R4$:	w^2	$-w^2$	$w^{-1}(f_1 - f_5) - w^{-3}(f_3 - f_7)$	

Dividing the rows in (15.63) by the element in the first column gives,

	$4F_3$	$4F_7$		
$R1 \leftarrow 1 \times R1$:	1	1	$w^0\{w^0(f_0 - f_4) - w^{-2}(f_2 - f_6)\}$	(15.64)
$R2 \leftarrow w^{-2} \times R2$:	1	–1	$w^{-2}\{w^{-1}(f_1 - f_5) - w^{-3}(f_3 - f_7)\}$	

Now, the coefficient matrix in (15.62) and (15.64) is same (symmetric) and satisfies the SAC property. Hence we can again apply the foregoing procedure on them.

Stage 3: Obtaining solution from (2 × 2) system of equations.

Performing operations of addition and subtraction / division we immediately get the following solution i.e., the values of the transformed variable:

From (15.59) :-

$$8F_0 = \{(f_0 + f_4) + (f_2 + f_6)\} + \{(f_1 + f_5) + (f_3 + f_7)\} \tag{15.65.1}$$

$$8F_4 = w^0 [\{(f_0 + f_4) + (f_2 + f_6)\} - \{(f_1 + f_5) + (f_3 + f_7)\}] \tag{15.65.2}$$

From (15.61) :-

$$8F_2 = w^0 \{(f_0 + f_4) - (f_2 + f_6)\} + w^{-2} \{(f_1 + f_5) - (f_3 + f_7)\} \tag{15.65.3}$$

$$8F_6 = w^0 \left[w^0 \{(f_0 + f_4) - (f_2 + f_6)\} - w^{-2} \{(f_1 + f_5) - (f_3 + f_7)\}\right] \tag{15.65.4}$$

From (15.62) :-

$$8F_1 = \left\{w^0 (f_0 - f_4) + w^{-2} (f_2 - f_6)\right\} + \left\{w^{-1} (f_1 - f_5) + w^{-3} (f_3 - f_7)\right\} \tag{15.65.5}$$

$$8F_5 = w^0 \left[\left\{w^0 (f_0 - f_4) + w^{-2} (f_2 - f_6)\right\} - \left\{w^{-1} (f_1 - f_5) + w^{-3} (f_3 - f_7)\right\}\right] \tag{15.65.6}$$

From (15.64) :-

$$8F_3 = w^0 \left\{w^0 (f_0 - f_4) - w^{-2} (f_2 - f_6)\right\} + w^{-2} \left\{w^{-1} (f_1 - f_5) - w^{-3} (f_3 - f_7)\right\} \tag{15.65.7}$$

$$8F_7 = w^0\left[w^0\left\{w^0(f_0 - f_4) - w^{-2}(f_2 - f_6)\right\} - w^{-2}\left\{w^{-1}(f_1 - f_5) - w^{-3}(f_3 - f_7)\right\}\right] \quad (15.65.8)$$

Following points may be noted:

(a) Only four roots, out of eight roots of unity, are required, namely, $w^0(=1)$, w^1, w^2, w^3; the negative powers are simply their complex conjugates.

(b) In the 8-point FFT, only 12 multiplications are required, 4 in each stage and in general N/2 $\log_2$ N. The number of additions/subtractions required is 24 i.e., 8 in each stage and in general N $\log_2$ N. Thus a total of 3/2 N $\log_2$ N arithmetic operations are required.

Mathematically, in the language of Analysis of Algorithms, if T(N) denotes time complexity (number of arithmetic operations) of the present algorithm for computing N-point FFT, we can express it by a recurrence relation,

$$T(N) = 2T(N/2) + 3N/2,\ T(2) = 3, \quad (15.66)$$

where N is a power of 2.

It may be verified that the solution to (15.66) is $T(N) = \dfrac{3}{2}N\log_2 N = O(N\log N)$;

(c) Matrix $\boldsymbol{W}$ has been used for designing the algorithm; there is no need to compute its elements. The solution is obtained by making relevant operations on the right sides only.

For an N-point FFT where $N = 2^p$, following information is required:

(i) Values of N and p.

(ii) Values of complex conjugates of the first $N/2$ roots of unity i.e.,

$$w^0(=1), w^{-1}, w^{-2}, w^{-3}, \ldots, w^{-\left(\frac{N}{2}-1\right)}$$

(iii) N number of function values i.e.,

$$f_0,\ f_1,\ f_2, \ldots,\ f_{N-1}.$$

We have to compute N values of the transformed variable viz.

$$F_0,\ F_1,\ F_2,\ \ldots,\ F_{N-1}.$$

There will be p stages in the complete procedure. At any given stage s, $0 \le s \le p$ we have 2^s systems of simultaneous equations each containing $N/2^s = m$ (say) unknowns. The state zero ($s = 0$) refers to the original ($N \times N$) system of simultaneous equations. Assuming that values have been computed upto stage (s–1), the operations of addition and subtraction / multiplication at the next stage s may be performed as follows :

Note that there will be 2^{s-1} systems of equations at stage (s–1) and each system contains $N/2^{s-1} = 2m$ unknowns with $2m$ elements on the right side. The coefficient matrices ($2m \times 2m$), in all the systems will be same; they will be symmetric and satisfy SAC property. The right sides will be different in all of them and will have $2m$ elements each. Let the $2m$ elements in a system be given as,

$$t_0,\ t_1,\ t_2,\ \ldots t_{m-1},\ t_m,\ t_{m+1},\ \ldots\ t_{2m-1}.$$

This system will be reduced to two ($m \times m$) systems. Let g_j denote the right side element after addition operation and h_j, after subtraction / multiplication operations, $j = 0(1)m-1$. Then for s=1(1) p where p= $\log_2 N$, we compute for j = 0(1)m–1, as follows:

1. for addition operation

   $$g_j \leftarrow t_j + t_{j+m}$$

2. for subtraction and multiplication (division) operation

   $$d \leftarrow j * 2^{s-1};$$

   $$h_j \leftarrow w^{-d} * (t_j - t_{j+m});$$

This procedure has to be executed for all the $N/2m$ systems giving rise to N/m reduced systems.

Example 15.5

Using the method discussed in Sec.15.11, find 8-point FFT of the function $f(t) = e^{-t}$, $0 = t = 4.0$ (see Example 15.4)

Solution

$$N = 8$$

Eight roots of unity

$$w^0 = 1,\ w^1 = \frac{1+i}{\sqrt{2}},\ w^2 = i,\ w^3 = -\frac{1-i}{\sqrt{2}},\ w^4 = -1,\ w^5 = -\frac{1+i}{\sqrt{2}},\ w^6 = -i,\ w^7 = \frac{1-i}{\sqrt{2}}$$

The roots to be used : $w^0 = 1,\ w^{-1} = \frac{1-i}{\sqrt{2}},\ w^{-2} = -i,\ w^{-3} = -\frac{1+i}{\sqrt{2}}.$

$$f_0 = 1.0000,\ f_1 = 0.6065,\ f_2 = 0.3679,\ f_3 = 0.2231$$

$$f_4 = 0.1353,\ f_5 = 0.0821,\ f_6 = 0.0498,\ f_7 = 0.0302.$$

Stage 1

$$2F_0 \Rightarrow 1.0000 + 0.1353 = 1.1353$$

$$2F_2 \Rightarrow 0.6065 + 0.0821 = 06886$$

$$2F_4 \Rightarrow 0.3679 + 0.0498 = 0.4177$$

$$2F_6 \Rightarrow 0.2231 + 0.0302 = 0.2533$$

$$2F_1 \Rightarrow w^0\,(1.0000 - 0.1353) = 0.8647$$

$$2F_3 \Rightarrow w^{-1}\,(0.6065 - 0.0821) = \frac{1-i}{\sqrt{2}}(0.5244)$$

$$2F_5 \Rightarrow w^{-2}\,(0.3679 - 0.0498) = -i\,(0.3181)$$

$$2F_7 \Rightarrow w^{-3}\,(0.2231 - 0.0302) = -\frac{1+i}{\sqrt{2}}(0.1929)$$

Stage 2

$$4F_0 \Rightarrow 1.1353 + 0.4177 = 1.5530$$

$$4F_4 \Rightarrow 0.6886 + 0.2533 = 0.9419$$

$$4F_2 \Rightarrow w^0\,(1.1353 - 0.4177) = 0.7176$$

$$4F_6 \Rightarrow w^{-2}\,(0.6886 - 0.2533) = -i\,(0.4353) = 0.4353i$$

$$4F_1 \Rightarrow 0.8647 - 0.3181i$$

$$4F_5 \Rightarrow \frac{1-i}{\sqrt{2}}(0.5244) - \frac{1+j}{\sqrt{2}}(0.1929) = \frac{1}{\sqrt{2}}(0.3315 - 0.7173i)$$

$$4F_3 \Rightarrow w^0\,(0.8647 + 0.3181i) = 0.8647 + 0.3181i$$

$$4F_7 \Rightarrow w^{-2}\left\{\frac{1-i}{\sqrt{2}}(0.5244) + \frac{1+j}{\sqrt{2}}(0.1929)\right\} = \frac{1}{\sqrt{2}}(-0.3315 - 0.7173i)$$

Stage 3

$$8F_0 = 1.5530 + 0.9419 = 2.4949$$

$$8F_4 = w^0\,(1.5530 - 0.9419) = 0.6111$$

$$8F_2 = 0.7176 - 0.4353i$$

$$8F_6 = w^0\{0.7176+0.4353i\} = 0.7176+0.4353i$$

$$8F_1 = (0.8647-0.3181i)+\frac{1}{\sqrt{2}}(0.3315-0.7173i)$$

$$= 1.0991-0.8253i$$

$$8F_5 = w^0\left\{(0.8647-0.3181i)-\frac{1}{\sqrt{2}}(0.3315-0.7173i)\right\}$$

$$= 0.6303+0.1891i$$

$$8F_3 \Rightarrow (0.8647+0.3181i)+\frac{1}{\sqrt{2}}(-0.3315-0.7173i)$$

$$= 0.6303-0.1891i$$

$$8F_7 = w^0\left\{(0.8647+0.3181i)-\frac{1}{\sqrt{2}}(-0.3315-0.7173i)\right\}$$

$$= 1.0991+0.8253i$$

Note: Check with answers of Example 15.4.

Exercise 15

15.1 Express the following function (step function) by Fourier series,

$$f(x) = \begin{cases} k, & 0 < x < \pi \\ -k, & \pi < x < 2\pi \end{cases} \quad \text{and } f(x) = f(x \pm 2\pi).$$

Also plot a graph of $f(x)$ between -2π and 2π.

15.2 Represent the function $f(x) = x$, $0 < x < c$ by a half-range sine series. Plot the graph represented by the series. Also write the series for $c = \pi$ and prove that $\frac{\pi}{4} = 1 - \frac{1}{3} + \frac{1}{5} - \frac{1}{7} + \ldots$

15.3 Express the function $f(x) = e^{-x}, 0 < x < \pi$ by a half-range cosine series. You can use the integral, $\int e^{ax}.\cos bx dx = \frac{1}{a^2+b^2}e^{ax}\,(a\cos bx + b\sin bx)$

References and Some Useful Related Books/Papers

1. Aho A.V., Hopcroft J.E. and Ullman J.D., The Design and Analysis of Computer Algorithms, Pearson Education India

2. Brigham E.O. and Morrow R.E., The Fast Fourier Transform, IEEE Spectrum-4, 1967, pp.63–70.

3. Cooley J.W. and Tukey J.W., An algorithm for the machine calculations of complex Fourier series, Math. Comp., Vol 10, April 1965, pp 297–301.

4. Coremen T.H., Leiserson C.E. and Rivest R.L., Introduction to Algorithms, Prentice Hall of India.

5. Kreyszig E., Advanced Engineering Mathematics, Wiley Eastern Limited

6. Yakowitz S. and Szidarovszky F., An Introduction to Numerical Computations, Macmillan Publishing Co., New York.

16

Free and Moving Boundary Problems: A Brief Introduction

16.1 Introduction

In Chapter 11, we have dealt with parabolic equations defined over certain domain D with some kind of conditions prescribed on the boundary S. In these problems the boundary remains fixed for all times. However, there are very many problems of great practical importance where the boundary does not remain stationary; it moves. The boundary (whole or part) changes its position, shape and/or size with time. That is, S becomes a function of time in addition to the space variables. Considering Cartesian coordinates, the boundary can move along the x-axis only in case of one-dimensional problem, in the x-y plane in case of two-dimensional problem and in space in the case of three-dimensional problem. These problems are called moving boundary problems (mbp's). The most common example of a mbp is found in heat flow when a solid undergoes melting or a liquid changes its state under the process of solidification. In the melting or solidification problems, the solid/liquid interface (moving boundary) separating the solid and the liquid phases is a function of space and time. Another example from heat flow may be that of ablation, i.e., removal of the material from the surface of a solid body due to excessive heating – a practical example may be that of ablation from the space capsule at the time of its re-entry into the earth's atmosphere. Other familiar examples may be like consolidation of earth dam or diffusion of a gas in an absorbing medium. Besides, many more examples of moving boundary problems may be found in reference [1].

The free boundary problems (fbp's) are elliptic equations representing special type of steady state problems. They are defined over a fixed domain but location of the boundary is not known, *a priori*. A practical example of a fbp may be given from the context of 'fluid flow through porous media' is, the seepage of water into an earth dam. Assuming the ground to be impervious and water level known in the dam (reservoir) we are required to find the position of the seepage line/surface in the dam as well as pressures inside the dam.

Some authors refer mbp's and fbp's both as fbp's. However, we prefer to use different nomenclatures to distinguish them from each other. J. Stefan had published the first paper on mbp in 1889 concerning melting of polar ice; therefore mbp's are also known as Stefan problems. It may be mentioned here that due to mathematical complexity, only a few analytical solutions are available to moving boundary problems and those too for particular cases of one-dimensional problems. Therefore recourse is generally made to numerical methods. Various surveys on mbp's have been published by different authors from time to time. The interested readers may refer to them given in [1] and [2] or mentioned in the research papers by the author listed in the chapter.

We shall describe a couple of problems below as illustrations, of moving and free boundary problems.

16.2 Moving Boundary Problems

(i) Melting of Ice

Let us consider an ice slab of unit length extending from $x = 0$ to $x = 1$ which is at a temperature of $-2°$C initially. Then the surface $x = 0$ is raised to a temperature of 10°C while the other end $x = 1$ is maintained at a temperature of 0°C for all times. Assuming that ice melts at 0°C, the melting of the slab starts from $x = 0$ and the water/ice interface starts moving towards $x = 1$. The temperature distributions in the water and the ice phases along with the position of the interface (moving boundary) are shown in Fig. 16.1. Let $u_1(x, t)$ denote the temperature at a distance x at time t in the water phase and $u_2(x, t)$ that in the ice phase.

The temperature distributions in the water and ice are shown by the curves AS and SBC respectively, while S denotes the position of the moving boundary $s(t)$. The discontinuity in the gradients $\frac{\partial u_1}{\partial x}$ and $\frac{\partial u_2}{\partial x}$ at the interface $x = s(t)$ is important to note which needs some explanation.

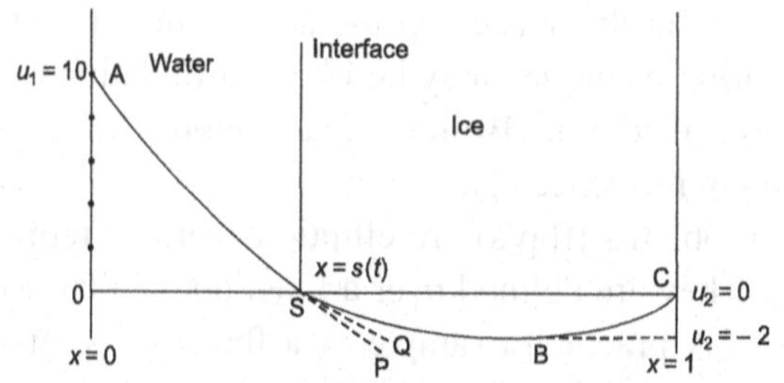

Figure 16.1 Temperature distribution in water/ice.

The amount of heat flowing (per unit area per unit time) from water to the ice across the interface is $-\kappa_1 \frac{\partial u_1}{\partial x}$ (i.e., flux) where κ_1 is the thermal conductivity of water. But in the ice phase the amount of heat flow would be $-\kappa_2 \frac{\partial u_2}{\partial x}$ at the interface where κ_2 is the conductivity of ice. This difference accrues since the heat transferred from water has been used to change the form of ice at 0°C to water at 0°C, i.e., without increasing the temperature. Let us suppose that ρ is density which does not change on its transformation to ice or vice-versa. This assumption is also necessary as we are not considering the volume change under the phase transformation. If L is the latent heat of ice, then it would be easy to see that

$$\kappa_2 \frac{\partial u_2}{\partial x} - \kappa_1 \frac{\partial u_1}{\partial x} = \mathrm{L}\rho \frac{ds}{dt}, \quad x = s(t) \tag{16.1}$$

for if the interface $s(t)$ moves a distance δs in time δt; then $\left(\kappa_2 \frac{\partial u_2}{\partial x} - \kappa_1 \frac{\partial u_1}{\partial x}\right) \delta t$ would be the heat given per unit area to the ice and $\mathrm{L}\rho\delta s$ will be the heat gained by the ice thickness δs for changing it to water. In fact, this is the governing condition which is used in determining the movement of the interface. It is known as Stefan condition.

A simplification/variation of the problem may be when the ice is initially at the melting temperature 0°C (instead of −2°C) and the end $x = 1$ is maintained at 0°C or the ice slab extends to infinity, $x \geq 0$. In this case the temperature in the ice will always remain zero and it will have to be computed in water only. Let us write down the relevant equation for this one-phase problem.

The equation in non-dimensional form can be expressed as follows:

$$\frac{\partial u}{\partial t} = \frac{\partial^2 u}{\partial x^2}, \quad 0 < x < s(t), \ t > 0 \tag{16.2}$$

with initial condition

$$u(x, 0) = 0, \ 0 \leq x \leq 1, \ t = 0 \tag{16.3}$$

and

$$s(0) = 0. \tag{16.4}$$

The boundary conditions at the fixed ends are

$$u(0, t) = 10 \tag{16.5a}$$

$$u(1, t) = 0, \ t \geq 0. \tag{16.5b}$$

The conditions to be satisfied at the interface (moving boundary) $x = s(t)$ are

$$u(s, t) = 0 \tag{16.6}$$

and the Stefan condition

$$-\frac{\partial u}{\partial x} = \lambda \frac{\partial s}{\partial t} \tag{16.7}$$

where λ is a known parameter depending on the properties of ice and other normalising factors.

By differentiating $u(x, t) = 0$ at $x = s$ and using (16.7), it can be shown that

$$\frac{\partial u}{\partial t} = \frac{1}{\lambda}\left(\frac{\partial u}{\partial x}\right)^2, \quad x = s(t) \tag{16.8}$$

proving that the equation is non-linear at the moving boundary.

(ii) Diffusion of oxygen in an absorbing medium

The problem we are going to discuss came up from a hospital in London in 1969 during the course of determining an optimum radiation dose for eradication of cancerous tumour. Physically, oxygen is allowed to diffuse in a body tissue which absorbs oxygen at a constant rate everywhere. The oxygen concentration at the surface say $x = 0$ is kept constant and process of diffusion/absorption continues till it reaches steady state, i.e., there is no further change in the oxygen concentration inside the tissue. In the steady state situation there will be constant concentration at the surface $x = 0$ which decreases continuously as x increases (due to absorption) until it becomes zero at some point, say $x = s_0$. It would be important to note that not only concentration is zero at $x = s_0$, the flux will also be zero there since the oxygen does not diffuse beyond this point. If m is the constant rate of absorption and $u(x, t)$ denotes concentration at a distance x at time t in the tissue, then the governing p.d.e. would be

$$\frac{\partial u}{\partial t} = k\frac{\partial^2 u}{\partial x^2} - m, \quad x > 0 \tag{16.9}$$

where k is coefficient of diffusion.

To find steady state solution of (16.9) we solve

$$k\frac{\partial^2 u}{\partial x^2} - m = 0 \tag{16.10}$$

with following boundary conditions

$$u = u_0 \text{ (const) at } x = 0 \tag{16.11}$$

$$u = 0 \text{ and } \frac{\partial u}{\partial x} = 0 \text{ at } x = s_0. \tag{16.12}$$

The solution to (16.10) with boundary conditions (16.11) and (16.12) can be easily obtained as

$$u(x) = \frac{m}{2k}(s_0 - x)^2, \tag{16.13}$$

where
$$s_0 = \sqrt{\frac{2ku_0}{m}}. \tag{16.14}$$

As can be seen (16.13) satisfies the conditions at $x = 0$ and $x = s_0$ while distance s_0 is dependent on the parameters k, u_0 and m.

Having achieved steady state, the surface $x = 0$ is sealed $\left(\frac{\partial u}{\partial x} = 0\right)$ so that no oxygen can get into or come out from the tissue. As the tissue still absorbs oxygen, the process of diffusion with absorption starts taking place under the new condition at the surface $x = 0$. Since the supply oxygen is cut off, the oxygen concentration begins to drop everywhere. Consequently, the position of the point of zero concentration, i.e., $x = s_0$ starts shifting towards the surface $x = 0$. That is, the boundary $x = s$ becomes $x = s(t)$ with $s(0) = s_0$. Thus we have to solve a moving boundary problem determining position $s(t)$ at time t and compute the concentration in the medium simultaneously. Mathematically, the problem in non-dimensional form, is defined as follows:

$$\frac{\partial u}{\partial t} = \frac{\partial^2 u}{\partial x^2} - 1, \quad 0 \le x \le s(t) \tag{16.15}$$

with initial condition

$$u(x,\, 0) = \frac{1}{2}(1 - x)^2, \quad 0 \le x \le 1, \quad t = 0. \tag{16.16}$$

The boundary condition at the surface is

$$\frac{\partial u}{\partial x} = 0, \quad x = 0, \quad t \ge 0. \tag{16.17}$$

The conditions to be satisfied at the moving boundary are as follows:

$$u = 0 \tag{16.18}$$

and
$$\frac{\partial u}{\partial x} = 0 \tag{16.19}$$

with
$$s(0) = 1.$$

It must be observed that in the oxygen diffusion problem there is no explicit relationship governing movement of the moving boundary like Stefan condition (16.1) or (16.7) in the ice melting problem. Therefore a condition has to be derived for determining the position of the moving boundary at any instant of time t. We express the concentration at a point in the neighbourhood of the moving boundary by a Taylor's series expansion about the moving boundary $x = s(t)$. At a point $(x - p\delta x)$, $p > 0$ in the neighbourhood of x, we have

$$u(x - p\delta x) = u(x) - p\delta x\frac{\partial u}{\partial x} + \frac{p^2\delta x^2}{2}\cdot\frac{\partial^2 u}{\partial x^2} - \frac{p^3\delta x^3}{6}\frac{\partial^3 u}{\partial x^3} + \cdots \tag{16.20}$$

At the moving boundary $x = s$, $u = 0$ and $\frac{\partial u}{\partial x} = 0$. To find $\frac{\partial^2 u}{\partial x^2}$ we differentiate $u(x, t) = 0$ w.r.t. t at $x = s$ giving

$$0 = \frac{\partial u}{\partial t} + \frac{\partial u}{\partial x}\cdot\frac{ds}{dt}$$

or $\frac{\partial u}{\partial t} = 0$ which gives from eqn (16.15)

$$\frac{\partial^2 u}{\partial x^2} = 1, \quad x = s. \tag{16.21}$$

Similarly, differentiating $\frac{\partial u}{\partial x} = 0$ w.r.t. t at $x = s$ gives

$$\frac{\partial^3 u}{\partial x^3} = -\frac{ds}{dt}, \quad x = s, \tag{16.22}$$

and so on,

Now we can use (16.21) and (16.22) to compute $p\delta x$ in terms of $u(x - p\delta x)$ or $u(s - p\delta x)$, i.e., the concentration in the neighbourhood of the moving boundary. If the terms only upto second derivatives are used in (16.20), then we get

$$p\delta x = \sqrt{2u(s - p\delta x)} \quad \text{or} \quad p = \frac{\sqrt{2u(s - p\delta x)}}{\delta x}, \tag{16.23}$$

where $u(s - p\delta x)$ denotes the concentration in the neighbourhood of the moving boundary s.

This problem was solved first time by the author. In a simple way, the procedure may be explained as follows:

Divide the zero time domain $0 \le x \le 1.0$ into sub-intervals of equal size of width δx, say. Choose a time step δt at which position of the moving boundary as well as the concentrations are to be computed. Suppose the computations have already been done up to time $t_j = j\delta t$ when the moving boundary lies somewhere in the $(r+1)^{\text{th}}$ interval, i.e., $x_r < s(t_j) < x_{r+1}$, x_0 represents $x = 0$. Assume that the moving boundary remains in the same interval at time t_{j+1}, i.e., $x_r < s(t_{j+1}) < x_{r+1}$. Referring to Fig. 16.2, we compute the concentrations at all the mesh points $x_0, x_1, \ldots. x_{r-1}$ using explicit formula with equal space intervals. At the mesh point x_r, explicit formula with unequal interval is used giving $u_{r,\, j+1}$. To get the position of the moving boundary, formula (16.23) is used when $p\delta x = \sqrt{2u_{r,\, j+1}}$ and $s(t_{j+1}) = s_{j+1} = r\delta x + p\delta x = (r+p)\delta x$. If the moving boundary crosses the meshpoint x_r or p is too small at t_{j+1}, we move to the next point x_{r-1} and omit the computation of u_r for future times. For complete details reference may be made to [1].

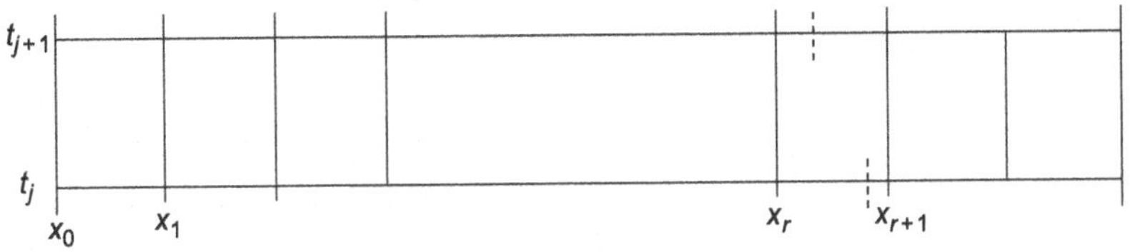

Figure 16.2 Dotted line shows position of moving boundary

As explained, we compute at successive time steps, the concentration at the mesh point nearest to the moving boundary and then the distance of the moving boundary $p\delta x$ from it in accordance with the formula (16.23). When $p\delta x$ becomes too small we shift the process of computing the distance of the moving boundary to the next point neglecting the earlier one. However, there is some drawback in this method in that at the time of changing the mesh point there appears an irregularity in the motion of the moving boundary. There are two reasons for this irregular behavior. One, there is a sudden change in the size of the last interval and two, the values of concentration are very small near the moving boundary which may give rise to larger computational errors. To circumvent this problem and maintain smoothness in the movement of the boundary throughout we describe below a method called 'Moving Grid Method'.

16.3 Moving Grid Method (MGM)

In the MGM we move the entire grid towards the fixed end $x = 0$ by a distance as traversed by the moving boundary at each time step, thus always keeping the unequal interval adjacent to the fixed end where concentration is higher. Again, when first interval becomes too small we merge it with the second and continue the process with bigger first interval. We will discuss two versions of the method, viz. (i). MGM With Interpolations, (ii). MGM Without Interpolations.

16.3.1 MGM with interpolations

Let us suppose that computations have already been made up to j^{th} level i.e., $t = t_j$ while the mesh points are located at $x = x_i,\ i = 0(1)r$ such that $x_i - x_{i-1} = \delta x,\ i = 2(1)r$ and $x_1 - x_0 = h_1$. Let the concentration be known at these mesh points with the rightmost mesh point x_r, coinciding with the moving boundary where concentration is zero. It may be borne in mind that position of the mesh points changes at each time step. We have to compute at the next time step $t_{j+1} = t_j + \delta t$, the position of the moving boundary and concentrations at the new mesh points. See Fig. 16.3. We proceed as follows:

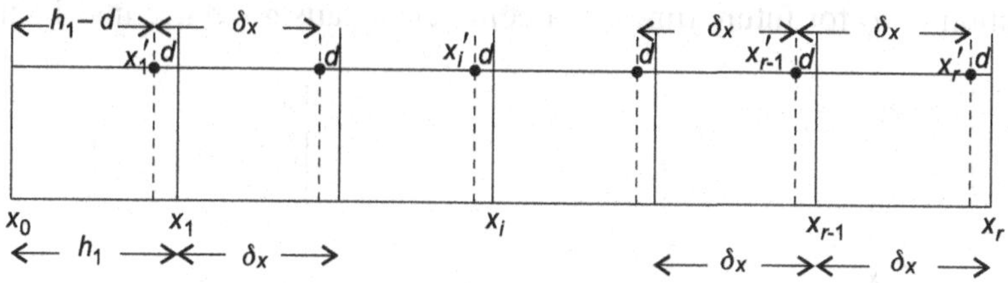

Figure 16.3 Mesh points x and x' before and after moving grid

(i) Compute $u(x_{r-1},\ t_{j+1})$ by discretizing (16.15) at $(x_{r-1},\ t_j)$,

$$\frac{u(x_{r-1},\ t_{j+1}) - u(x_{r-1},\ t_j)}{\delta t} = \frac{u(x_{r-2},\ t_j) - 2u(x_{r-1},\ t_j)}{\delta x^2} - 1, \ \because\ u(x_r,\ t_j) = 0. \quad (16.24)$$

(ii) Distance of moving boundary from x_{r-1} at t_{j+1} is given by

$$p\delta x = \sqrt{2u(x_{r-1},\ t_{j+1})}.$$

(iii) Displacement of moving boundary in time δt from t_j to t_{j+1} would be, $d = \delta x - p\delta x$.

(iv) Shift the grid towards the fixed end $x = x_0 = 0$ by a distance d. The new mesh points at t_{j+1} are shown by dots (.); let these be denoted by $x'_i = x_i - d,\ i = 1(1)r$. The distance between the points $x = x_i$ and $x = x'_i$ shows the distance traversed by the moving boundary from $t = t_j$ to $t = t_{j+1}$. We want to compute concentration at $x'_i, i = 1(1)r - 1$ and x_0 at $t = t_{j+1}$. In order to do that we need to compute second derivatives at these points at t_j which are obtained through interpolation using (a) Cubic Splines and (b) Cubic Polynomials.

(a) Interpolation by Cubic Spline

We fit a cubic spline $s(x)$ at t_j with knots $x_i,\ i = 0(1)r$ where respective concentrations are known as $u(x_i,\ t_j) = u_{i,\ j} = u_i$ (say). If $s''(x_i)$ denotes second derivative (approximation to $\frac{\partial^2 u}{\partial x^2}$) at $x = x_i$ i.e., $s''(x_i) = u''_i$ and that $s(x_i) = u_i$, the cubic spline in the interval $[x_{i-1},\ x_i]$

can be written by (8.7) as,

$$s(x) = s(x_{i-1}) + (x - x_{i-1})\left[\frac{s(x_i) - s(x_{i-1})}{x_i - x_{i-1}} - \frac{x_i - x_{i-1}}{6}\left\{s''(x_i) + 2s''(x_{i-1})\right\}\right]$$

$$+\frac{(x - x_{i-1})^2}{2}s''(x_{i-1}) + \frac{(x - x_{i-1})^3}{6}\left\{\frac{s''(x_i) - s''(x_{i-1})}{x_i - x_{i-1}}\right\}, \quad x_{i-1} \le x \le x_i \;\; i = 1(1)r \quad (16.25)$$

The second derivatives in (16.25) are determined from the following tri-diagonal system of equations arising due to continuity of the first derivative at $x = x_i$, See (8.12)

$$h_i s''(x_{i-1}) + 2(h_i + h_{i+1})s''(x_i) + h_{i+1}s''(x_{i+1}) = 6\left[\frac{u_{i+1} - u_i}{h_{i+1}} - \frac{u_i - u_{i-1}}{h_i}\right], \; i = 1(1)r - 1, \quad (16.26)$$

where $h_i = x_i - x_{i-1}$ etc.

In (16.26) there are (r+1) unknowns, namely $s''(x_i)$ $i = 0, 1, \ldots r$ while the number of equations is only r–1; hence we need two more conditions. They are obtained as follows :

From (16.21) we know that

$$s''(x_r) = 1. \quad (16.27)$$

Differentiating (16.25) w.r.t. x and putting $s'(x) = 0$, $x = 0$ in accordance with (16.17) we get (also see (8.9)),

$$2s''(x_0) + s''(x_1) = \frac{6(u_i - u_0)}{h_1^2}. \quad (16.28)$$

Now, from (16.26), (16.27) and (16.28) we can get values of $s''(x_i)$, $i = 0, 1, 2, \ldots r$. Thus, a cubic spline is completely defined in the entire interval $[x_0, x_r]$. We can find an estimate of concentration at $x = x_i'$ at $t = t_j$ by putting $x = x_i'$ in (16.25) i.e.,

$$u(x_i', t_j) = s(x_i'). \quad (16.29)$$

Similarly, the second derivative at $x = x_i'$ is given by,

$$u''(x_i', t_j) = s''(x_i') = s''(x_{i-1}') + (x_i' - x_{i-1})\frac{s''(x_i) - s''(x_{i-1})}{x_i - x_{i-1}}. \quad (16.30)$$

Finally, the concentration at $t = t_{j+1}$ at the new mesh point $x = x_i'$ may be obtained by discretizing (16.15) at (x_i', t_j) as

$$\frac{u(x_i', t_{j+1}) - u(x_i', t_j)}{\delta t} = s''(x_i') - 1, \; i = 0(1)r-1, \tag{16.31}$$

while values of $u(x_i', t_j)$ and $s''(x_i'')$may be substituted from (16.29) and (16.30) respectively.

As mentioned above, when first interval becomes too small we merge it with the second interval δx thus reducing the number of mesh points by one.

(b) Interpolation by Cubic Polynomial

Instead of a spline we can use an ordinary cubic polynomial to represent the concentration in the interval (x_{i-1}, x_i) at time t_j, by say,

$$P(x) = a_0 + a_1 x + a_2 x^2 + a_3 x^3, \quad x_{i-1} \le x \le x_i, \tag{16.32}$$

where a_0, a_1, a_2, a_3 are constants. We need four conditions to determine these constants, out of which two are readily available as concentrations are known at $x = x_{i-1}$ and $x = x_i$, giving

$$P(x_{i-1}) \equiv a_0 + a_1 x_{i-1} + a_2 x_{i-1}^2 + a_3 x_{i-1}^3 = u(x_{i-1}, t_j) \tag{16.33a}$$

$$P(x_i) \equiv a_0 + a_1 x_i + a_2 x_i^2 + a_3 x_i^3 = u(x_i, t_j) \tag{16.33b}$$

For other two conditions we compute the second derivatives at x_{i-1} and x_i by a 3-point formula and let the polynomial (16.32) satisfy them i.e.,

$$P''(x_{i-1}) \equiv 2a_2 + 6a_3 x_{i-1} = u''(x_{i-1}, t_j) \tag{16.33c}$$

$$P''(x_i) \equiv 2a_2 + 6a_3 x_i = u''(x_i, t_j) \tag{16.33d}$$

where $u''(x_i, t_j) = \dfrac{u(x_{i-1}, t_j) - 2u(x_i, t_j) + u(x_{i+1}, t_j)}{\delta x^2}, \quad i = 2(1)r-1,$

$$u''(x_1, t_j) = \frac{2}{h_1 \delta x (h_1 + \delta x)} \{\delta x u(x_0, t_j) - (h_1 + \delta x) u(x_1, t_j) + h_1 u(x_2, t_j)\}$$

and $u''(x_0, t_j) = \dfrac{2}{h_1^2} \{u(x_1, t_j) - u(x_0, t_j)\}, \; x_0 = 0.$

From eqs. (16.33a, b, c, d) we can find the values of the constants a_0, a_1, a_2 and a_3, thus defining the polynomial completely in the interval (x_{i-1}, x_i). The concentration and the second derivative at the new mesh point $x = x_i'$ at $t = t_j$ are given by,

$$u(x_i', t_j) = P(x_i') = a_0 + a_1 x_i' + a_2 x_i'^2 + a_3 x_i'^3 \tag{16.34}$$

$$u''(x_i', t_j) = P''(x_i') = 2a_2 + 6a_3 x_i'. \tag{16.35}$$

Having got the values of u and u'' from (16.34) and (16.35) respectively, we can proceed to compute the concentrations at t_{j+1} at the new mesh point x_i' as discussed in (a) above. For numerical details, see author's paper [2].

16.3.2 MGM without interpolations

In this method, same idea of moving the grid towards fixed surface $x = 0$, a distance as traversed by the moving boundary at each time step is maintained but interpolations are avoided by using a two-dimensional Taylor's series. Referring Fig. 16.3, let x_i be a mesh point at t which has moved to a new position $x_i' = x_i - d$ at time $t_{j+1} = t_j + \delta t$ where d is the distance traversed by the moving boundary in time δt. To compute concentration at mesh point x_i' at time t_{j+1} i.e., $u(x_i', t_{j+1})$ we express it by Taylor's expansion about $x = x_i$, $t = t_j$ as follows :

$$\begin{aligned} u\left(x_i', t_{j+1}\right) &= u(x-d, t+\delta t) \\ &= u(x, t) - d.\frac{\partial u}{\partial x} + \delta t.\frac{\partial u}{\partial t} + \frac{1}{2}\left\{d^2\frac{\partial^2 u}{\partial x^2} - 2d.\delta t\frac{\partial^2 u}{\partial x.\partial t} + \delta t^2\frac{\partial^2 u}{\partial t^2}\right\} + \ldots \\ &= u(x, t) - d.\frac{\partial u}{\partial t} + \delta t\left(\frac{\partial^2 u}{\partial x^2} - 1\right) + \frac{1}{2}d^2.\frac{\partial^2 u}{\partial x^2} \\ &\quad - d.\delta t\frac{\partial^3 u}{\partial x^3} + \frac{1}{2}\cdot\delta t^2\cdot\frac{\partial^4 u}{\partial x^4} + \ldots \text{ using (16.15)} \\ &= u(x, t) - d\cdot\frac{\partial u}{\partial x} + \delta t\left(\frac{\partial^2 u}{\partial x^2} - 1\right) + \frac{1}{2}d^2\cdot\frac{\partial^2 u}{\partial x^2}, \end{aligned} \tag{16.36}$$

neglecting higher order terms.

We can use formula (16.36) straightaway to compute $u(x_i', t_{j+1})$ approximating the first and second derivatives, on the right side by central differences as concentrations are known at all the mesh points at $t = t_j$. Further, if d is small i.e., boundary is moving slowly, the last term in (16.36) may be neglected. For numerical details reference may be made to author's paper [3].

It may be mentioned that several authors have solved this problem using other techniques. The author and associates have suggested various methods, from time to time, for solving moving boundary problems in one as well as in two or three dimensions. A list of publications by the author and associates on this topic is given at the end of this chapter.

16.4 Free Boundary Problem

The most familiar example of a Free Boundary Problem is the classical problem of seepage of water through rectangular earth dam. In this problem the boundary marking the seepage line does not move (is stationary) but its position is not known, 'a priori'. The problem is elliptic in nature and the water pressure is to be computed inside the dam along with the location of the free boundary. See Fig. 16.4.

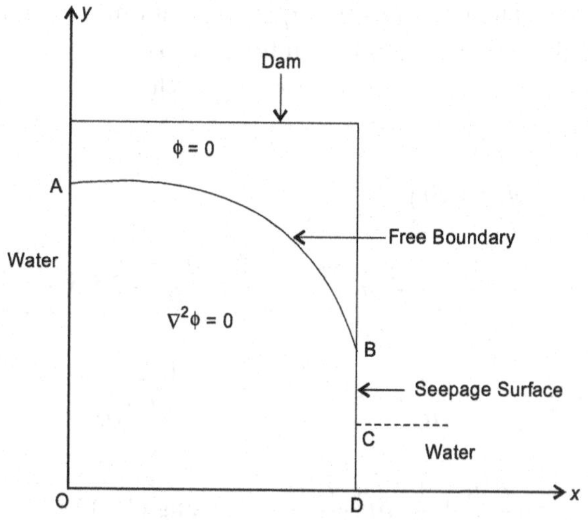

Figure 16.4 Rectangular earth dam (cross-section)

Let $\phi(x, y)$ denote water pressure, then we have to solve the following elliptic equation in the domain OABCDO and find the position of AB in the dam at the same time:

$$\nabla^2\phi = \frac{\partial^2\phi}{\partial x^2} + \frac{\partial^2\phi}{\partial y^2} = 0 \tag{16.37}$$

The following boundary conditions are to be satisfied:

$$(i)\ \phi = y_1,\ x = 0 \text{ (on OA)} \tag{16.38}$$

(*ii*) $\dfrac{\partial \phi}{\partial n} = 0,\ y = 0$ (on OD) (16.39)

(*iii*) $\phi = y_2,\ x = \text{L}$ (on CD) (16.40)

(*iv*) $\phi = y,\ x = \text{L}$ (on BC) : B is not known (16.41)

(*v*) $\phi = y$ and $\dfrac{\partial \phi}{\partial n} = 0$ on AB (Free Boundary) (16.42)

where n denotes the outward normal to the surface.
For further details reference may be made to [1].

References And Some Useful Related Books/Papers

1. Crank J., *Free and Moving Boundary problems*, Clarendon Press, Oxford

2. Rubinstein L.I., *The Stefan Problem*, Trans. Math. Monograph, Am. Math. Soc., Vol 27 (1971).

Papers published by the author and co-workers on moving boundary problems

1. Crank J. and Gupta R.S., A moving boundary problem arising from the diffusion of oxygen in absorbing tissue, J. Inst. Maths. Applics., Vol 10 (1972) pp. 19–33.

2. ———, A method for solving moving boundary problems in heat flow using cubic splines or polynomials, J. Inst. Math. Applics, Vol 10 (1972), pp. 296–304.

3. ———, Isotherm Migration Method in two dimensions, Int. J. Heat & Mass Transfer, vol. 18 (1975), pp. 1101–1107.

4. Gupta R.S., Moving Grid Method without interpolations, Comp. Methods in App. Mech & Engng, Vol 4 (1974), pp. 143–152.

5. Gupta R.S. and Kumar D., A modified variable time step method for one dimensional diffusion problem, Comp. Mthds. App. Mech. Engng, Vol 23 (1980), pp. 101–109.

6. ———, Complete numerical solution of oxygen diffusion problem involving moving boundary, Comp. Mthds. App. Mech. Engng, Vol 29 (1981), pp. 233–239.

7. ———, Variable time step methods for one dimensional Stefan problem with mixed boundary condition, Int. J. Heat & Mass Transfer Vol 24 (1981), pp. 251–259.

8. ———, Variable time step methods for the dissolution of a gas bubble in liquid, Comp. and Fluids, vol 11 (1983), pp. 341–349.

9. ———, Solution of phase change problem with non-uniform initial temperature, Comp. Methods Appd. Mech. & Engng, vol 37 (1983), pp. 139–150.

10. ———, Treatment of solidification problem inside and outside of cylinders by variable time methods, Int. J. Heat & Mass Transfer, vol 26 (1983), pp. 313–315.

11. Gupta R.S. and Kumar A., An efficient approach to isotherm migration method in two Dimensions, Int. J. Heat and Mass Transfer, vol 27 (1984), pp. 1939–1942.

12. ———, Variable time step method with coordinate transformation, Comp. Mthds Appd. Mech. Engng, vol 84 (1984), pp. 91–103.

13. ———, Treatment of multidimensional Stefan problems by coordinate transformation, Int. J. Heat & Mass Transfer, vol 28 (1985), pp. 1355–1366.

14. ———, Approximate analytical methods for multidimensional Stefan problems, Comp. Mthds. Appd. Mech. Engng, vol 56 (1986), pp. 127–138.

15. ———, Treatment of alloy solidification by fixed domain variable time step method, Int. Conf. on modelling, Simulation and Control, held at Naples, Italy, Sept 29 to Oct 1, 1986. Proc. ASME Press vol 11 (1987) pp. 11–23.

16. ———, Isotherm migration method applied to fusion problems with convective boundary conditions, Int. J. Num. Methods in Engng, vol 26 (1988) pp. 2547–2556.

17. Gupta R.S. and Banik N.C., Constrained integral method for solving moving boundary problems, Comp. Mthds. Appd. Mech Engng, vol 67 (1988) pp. 211–221.

18. ———, Approximate method for the oxygen diffusion problem, Int. J. Heat & Mass Transfer, vol 32 (1989) pp. 781–783.

19. ———, Diffusion of oxygen in a sphere with simultaneous absorption, Appd. Math. Modelling, vol 14 (1990) pp. 114–121.

20. ———, Solution weakly two-dimensional melting problem by an approximate method, J. Computational & Appd. Maths, vol 31 (1990) pp. 351–356.

Appendices

Appendix A: Some Theorems and Formulae

A.1. Binomial Theorem and Related Formulae

$$(p+q)^n = \binom{n}{0} p^n q^0 + \binom{n}{1} p^{n-1} q + \cdots + \binom{n}{r} p^{n-r} q^r + \cdots + \binom{n}{n} p^0 q^n$$

or $$\binom{n}{0} p^0 q^n + \binom{n}{1} p q^{n-1} + \cdots + \binom{n}{r} p^r q^{n-r} + \cdots + \binom{n}{n} p^n q^0$$

(*i*) $$\binom{n}{0} = \binom{n}{n} = 1$$

(*ii*) $$\binom{n}{r} = \binom{n}{n-r} = \frac{n!}{r!(n-r)!} = \frac{n(n-1)(n-2)\cdots(n-r+1)}{1\cdot 2\cdot 3\cdots r}$$

(*iii*) $$\binom{n}{r} + \binom{n}{n-r} = \binom{n+1}{r}$$

(*iv*) $$\binom{n}{r} - \binom{n-1}{r-1} = \binom{n-1}{r}$$

A.2. Rolle's Theorem

If $f(x)$ is a function which is continuous on a closed interval $[a, b]$ and differentiable in the open interval (a, b) and $f(a) = f(b) = 0$, then there is at least one point c in (a, b) at which the derivative of $f(x)$ is zero *i.e.* $f'(c) = 0$.

A.3. Mean Value Theorem

If a function $f(x)$ is continuous on a closed interval $[a,\ b]$ and differentiable in the open interval $(a,\ b)$, then there is at least one point c in $(a,\ b)$ at which

$$f(b)-f(a) = (b-a)f'(c),\ a<c<b.$$

A.4. Mean Value Theorem of Integrals

Let $f(x)$ be a function defined in a closed interval $[a,\ b]$. If the function is integrable in $[a,\ b]$, then there exists a point c in $(a,\ b)$ such that

$$\int_a^b f(x)dx = (b-a)f(c),\ a<c<b.$$

A.5. Taylor's Theorem/Formula

(*i*) Let $f(x)$ be a function which possesses continuous derivative upto order $(n+1)$ in an interval I. If a is a point in I, then for each point x in I, $f(x)$ can be expressed in powers of $(x-a)$, as

$$f(x) = f(a)+(x-a)f'(a)+\frac{(x-a)^2}{2!}f''(a)+\cdots+\frac{(x-a)^n}{n!}f^n(a)+\mathrm{R}_{n+1}(x),$$

where $\mathrm{R}_{n+1}(x)$, called remainder term is given by $\mathrm{R}_{n+1}(x) = \dfrac{(x-a)^{n+1}}{(n+1)!}f^{n+1}(\xi),\ a \leq \xi \leq x.$

If $\mathrm{R}_{n+1}(x) \to 0$ as $n \to \infty$, for all x in I, then $f(x)$ can be represented by a convergent series,

$$f(x) = f(a)+(x-a)f'(a)+\frac{(x-a)^2}{2!}f''(a)+\frac{(x-a)^3}{3!}f'''(a)+\cdots$$

If $x = a+h$, then Taylor's series may be written as

$$f(a+h) = f(a)+hf'(a)+\frac{h^2}{2!}f''(a)+\frac{h^3}{3!}f'''(a)+\cdots$$

(ii) If $f(x,\ y)$ is a function of two variables which is differentiable in a domain D containing a point $(a,\ b)$, then $f(x,\ y)$ can be expressed at any point $(x,\ y)$ or $(a+h,\ b+k)$ in D as,

$$f(a+h,\ b+k) = f(a,\ b) + \left(h\frac{\partial f}{\partial x} + k\frac{\partial f}{\partial y}\right) + \frac{1}{2!}\left(h^2\frac{\partial^2 f}{\partial x^2} + 2hk\frac{\partial^2 f}{\partial x\partial y} + k^2\frac{\partial^2 f}{\partial y^2}\right) +$$

$$\ldots + \frac{1}{n!}\left(h\frac{\partial}{\partial x} + k\frac{\partial}{\partial y}\right)^n + \cdots$$

where partial derivatives are taken at point $(a,\ b)$.

A.6. Leibniz Formula for Derivative of Integral

Let $f(t)$ be a continuous function of t defined in the interval $[\mathrm{a},\ b]$. If $u(x)$ and $v(x)$ are differentiable functions in $[a,\ b]$ and whose values correspond to some value of t in $[a,\ b]$, then,

(i) $$\frac{d}{dx}\int_{u(x)}^{v(x)} f(t)dt = f(v)\frac{dv}{dx} - f(u)\frac{du}{dx}$$

(ii) $$\frac{\partial}{\partial x}\int_{u(x)}^{v(x)} f(x,\ t)dt = f(x, v(x))\frac{dv}{dx} - f(x,\ u(x))\frac{du}{dx} + \int_{u(x)}^{v(x)} \frac{\partial f}{\partial x}dt$$

A.7. $$\int_a^x\int_a^x f(x)dxdx = \int_a^x (x-t)f(t)dt$$

A.8. (*i*) $e^{ix} = \cos x + i\sin x;\ e^{-ix} = \cos x - i\sin x$

(ii) $$\cosh x = \cos ix = \frac{e^x + e^{-x}}{2}$$

(*iii*) $$\sinh x = \frac{\sin ix}{i} = \frac{e^x - e^{-x}}{2}$$

Appendix B: Expansions of Some Functions

B.1. $e^x = 1 + x + \frac{x^2}{2!} + \frac{x^3}{3!} + \frac{x^4}{4!} + \cdots$, $-\infty < x < \infty$

B.2. $\log(1+x) = x - \frac{x^2}{2} + \frac{x^3}{3} - \frac{x^4}{4} + \frac{x^5}{5} - \cdots$, $-1 < x \le 1$

B.3. $\sin x = x - \frac{x^3}{3!} + \frac{x^5}{5!} - \frac{x^7}{7!} + \cdots$, $-\infty < x < \infty$

(x in radian)

B.4. $\cos x = 1 - \frac{x^2}{2!} + \frac{x^4}{4!} - \frac{x^6}{5!} \cdots\cdots$, $-\infty < x < \infty$

(x in radian)

B.5. $\sin^{-1} x = x + \frac{1 \cdot x^3}{2 \cdot 3} + \frac{1 \cdot 3x^5}{2 \cdot 4 \cdot 5} + \frac{1 \cdot 3 \cdot 5x^7}{2 \cdot 4 \cdot 6 \cdot 7} + \cdots$, $-1 \le x \le 1$

B.6. $\cos^{-1} x = \frac{\pi}{2} - \sin^{-1} x$, $-1 \le x \le 1$

B.7. $(1+x)^n = 1 + nx + \frac{n(n-1)x^2}{1 \cdot 2} + \frac{n(n-1)(n-2)x^3}{1 \cdot 2 \cdot 3} + \cdots$, $-1 < x < 1$

B.8. $(1-x)^{-1} = 1 + x + x^2 + x^3 \ldots.$, $-1 < x < 1$

B.9. $x^n - a^n = (x - a)(x^{n-1} + x^{n-2} \cdot a + x^{n-3} \cdot a^2 + \ldots. + xa^{n-2} + a^{n-1})$

B.10. $x^n + a^n = (x+a)\{x^{n-1} - x^{n-2}a + x^{n-3}a^2 \ldots. + (-1)^{n-2}xa^{n-2} + (-1)^{n-1}a^{n-1}\}$

Appendix C: Graphs of Some Functions

C.1. Exponential

(a) $y = e^x$ (Fig. C-1a)

x	$-\infty$	0	∞	$-\infty < x \leq 0$	$0 \leq x < \infty$
e^x	0	1	∞	increases from 0 to 1	increases from 1 to ∞

(b)$y = e^{-x}$ (Fig C-1b)

x	$-\infty$	0	∞	$-\infty < x \leq 0$	$0 \leq x < \infty$
e^{-x}	∞	1	0	decreases from ∞ to 1	decreases from 1 to 0

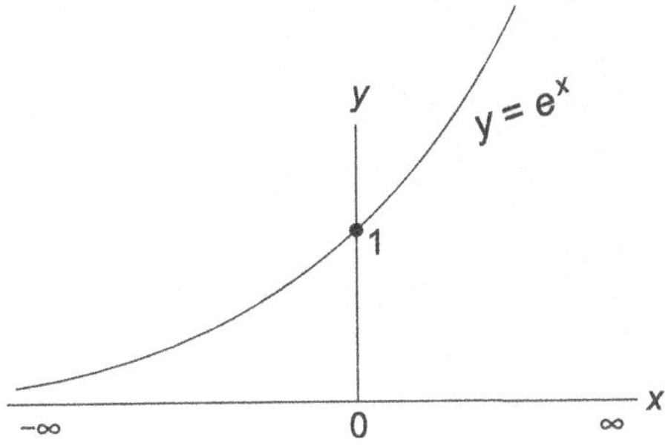

Figure C.1a $y = e^x$.

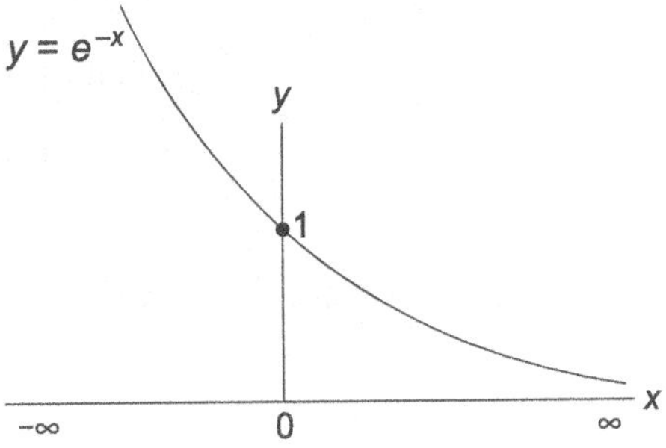

Figure C.1b $y = e^{-x}$.

C.2. Logarithmic

$y = \log_e x$ or $\ln x$ (Fig. C-2)

x	0	1	∞	$x < 0$	$0 < x < 1$	$1 \leq x < \infty$
$\ln x$	$-\infty$	0	∞	Not defined	increases $-\infty$ to 0	increases from 0 to ∞

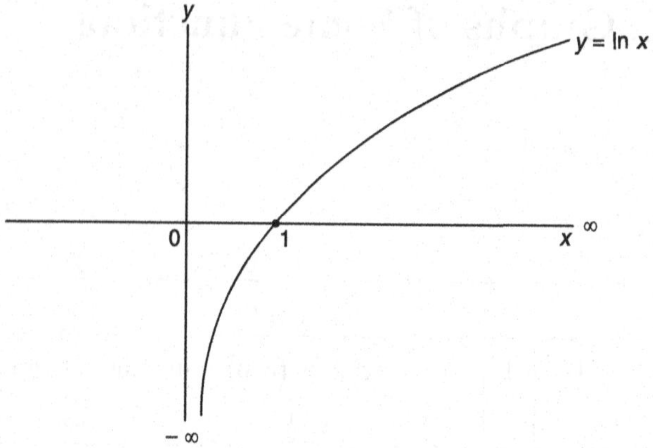

Figure C.2 $y = \ln x$.

C.3. Trigonometric functions

(**a**) $y = \sin x$. (Fig. C.3a)

x in radians	0	$\frac{\pi}{2}$	π	$\frac{3}{2}\pi$	2π
$\sin x$	0	1	0	−1	0

It is a periodic function with period 2π, i.e. repeats itself after an interval of 2π.

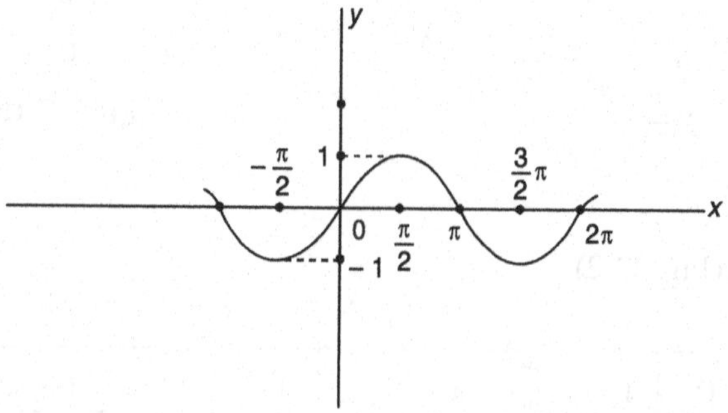

Figure C.3a $y = \sin x$.

(**b**) $y = \cos x$. (Fig. C-3b)

x in radians	0	$\frac{\pi}{2}$	π	$\frac{3}{2}\pi$	2π
$\cos x$	1	0	-1	0	1

It is a periodic function with period 2π.

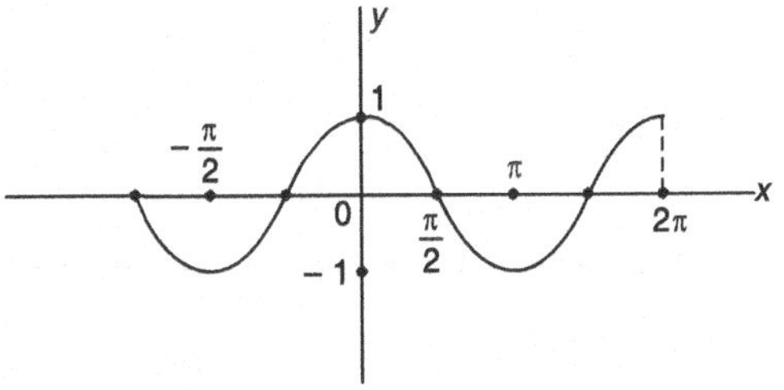

Figure C.3b $y = \cos x$.

(**c**) $y = \tan x$. (Fig. C-3c)

x	0	$\frac{\pi}{2}-$	$\frac{\pi}{2}+$	π	$\frac{3}{2}\pi-$	$\frac{3\pi}{2}+$	2π
$\tan x$	0	∞	$-\infty$	0	∞	$-\infty$	0

It is periodic function with period π and discontinuous at $x = \frac{\pi}{2}, \frac{3}{2}\pi$

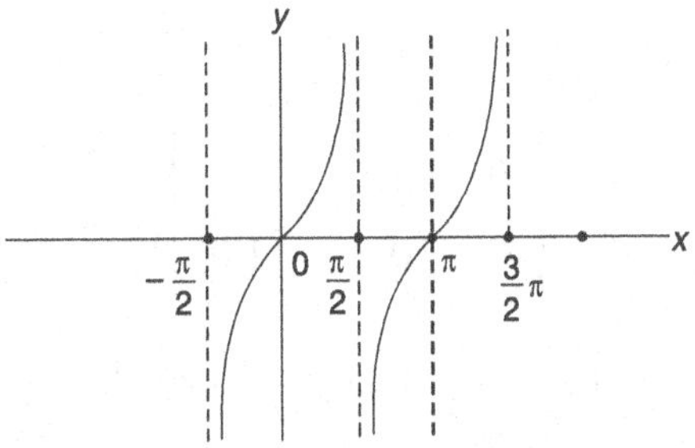

Figure C.3c $y = \tan x$.

Answers to Exercises

Exercise 1

1.1 (*i*) 0.005 (*ii*) 0.010 1.2 0.7856×10^{-2}, 0.78×10^{-2}, 0.56×10^{-4}

1.3 $9.025 < x < 9.975$ 1.4 0.2384 1.5 7.536 c.cms

1.6 $0 \le x \le 0.3$ 1.7 10.01

Exercise 2

2.1 $x_1 = 2,\ x_2 = 5,\ x_3 = 1$ 2.2 $x_1 = 3,\ x_2 = -2,\ x_3 = 2$ 2.3 $x_1 = 1.2,\ x_2 = 3.4,\ x_3 = 2.5$

2.4 $A^{-1} = \begin{bmatrix} 0.42 & -0.09 & -0.63 \\ -0.27 & 0.35 & 0.29 \\ 0.05 & -0.24 & 0.22 \end{bmatrix}$,

$x^T = (1.08 \quad -1.03 \quad 2.03)$

2.5 $A^{-1} = \begin{bmatrix} -0.63 & -0.04 & 0.42 \\ 0.29 & 0.32 & -0.28 \\ 0.22 & -0.24 & 0.05 \end{bmatrix}$

$x^T = (1.18 \quad -1.24 \quad 2.03)$

2.6 $L = \begin{bmatrix} 1 & 0 & 0 \\ 1.6 & 1 & 0 \\ 0.4 & -1 & 1 \end{bmatrix}$; $U = \begin{bmatrix} 5 & 4 & -2 \\ 0 & -3.4 & 4.2 \\ 0 & 0 & 11.0 \end{bmatrix}$, $y^T = (21 \quad 7.4 \quad 55)$
$x^T = (3 \quad 4 \quad 5)$

2.7 $L = \begin{bmatrix} 12 & 0 & 0 \\ 8 & 10.36 & 0 \\ 10 & 0.2 & 10.67 \end{bmatrix}$; $U = \begin{bmatrix} 1 & 0.58 & 0.42 \\ 0 & 1 & 0.64 \\ 0 & 0 & 1 \end{bmatrix}$, $y^T = (2.17 \quad 0.35 \quad -1.01)$
$x^T = (2.01, \quad 1.00, \quad -1.01)$

2.8 $LL^T = \begin{bmatrix} 2 & 0 & 0 & 0 \\ 2.5 & 1.32 & 0 & 0 \\ 1.5 & 0.19 & 1.65 & 0 \\ 2 & 0.76 & 2.34 & 1.40 \end{bmatrix} \begin{bmatrix} 2 & 2.5 & 1.5 & 2 \\ 0 & 1.32 & 0.19 & 0.76 \\ 0 & 0 & 1.65 & 2.34 \\ 0 & 0 & 0 & 1.40 \end{bmatrix}$

2.9 $P = \begin{bmatrix} 0 & 1 & 0 & 0 \\ 0 & 0 & 0 & 1 \\ 1 & 0 & 0 & 0 \\ 0 & 0 & 1 & 0 \end{bmatrix}$; $L = \begin{bmatrix} 1 & 0 & 0 & 0 \\ 0.50 & 1 & 0 & 0 \\ 0.67 & -0.38 & 1 & 0 \\ 0.50 & 0.71 & -0.87 & 1 \end{bmatrix}$; $U = \begin{bmatrix} 6 & 5 & 4 & 5 \\ 0 & 3.5 & 6 & 0.5 \\ 0 & 0 & 2.60 & 0.84 \\ 0 & 0 & 0 & 4.8 \end{bmatrix}$

Exact solution is $\boldsymbol{x}^T = (1 \quad -1 \quad 1 \quad 1)$

2.10 (*i*) Gauss–Jacobi

x \ n	0	1	2	3	4	5	6	7	8
x_1	0	2.80	2.20	1.88	1.98	2.02	2.00	2.00	2.00
x_2	0	−0.88	−1.31	−1.01	−0.96	−1.00	−1.01	−1.00	−1.00
x_3	0	2.1	3.31	3.11	2.95	2.99	3.01	3.00	3.00

(*ii*) Gauss–Seidel

x \ n	0	1	2	3	4	5	6
x_1	0	2.80	1.81	2.04	1.99	2.00	2.00
x_2	0	−1.58	−0.90	−1.02	−1.00	−1.00	−1.00
x_3	0	3.38	2.91	3.02	3.00	3.00	3.00

2.11 $\lambda_1 = 10$, $\boldsymbol{x}_1^T = (1 \quad 0.5)$; $\lambda_2 = 1$, $\boldsymbol{x}_2^T = (1 \quad -0.4)$

Exercise 3

3.1 (3.5625, 3.6250)

3.2 $x_{n+1} = e^{-x_n}$, 0.567

3.3 Min at $x = 0.739$; Min Value $= -0.801$; $f''(x) > 0$

3.4. 2.359

3.5. 0.53

3.6 $x_{n+1} = \dfrac{2x_n^3 + 100}{3x_n^2}, 4.642$

3.7 $x_1 = 0.776,\ y_1 = 0.683$

3.8 $x_1 = 1.5276,\ y_1 = 1.2002$

3.9 $x_1 = 0.312,\ y_1 = 0.168$

Exercise 4

4.3 (*i*) $(\mathrm{E} - 2 + \mathrm{E}^{-1})x^3 = 6h^2x$ (*ii*) $a_1b_2c_36!h^6$

4.4 32 should be 30

4.5. 5,185

4.6 $y_2 = 55.73,\ y_4 = 67.94$

4.7 2.3619, 2.8844

4.8 Stirling: 0.5306, Bessel: 0.7654

4.9 Everett : 0.7655, Steffensen : 0.5304

4.10 32

4.11 $f(2) = 12$

4.12 $f(6) = 6.56$; error $= -80$

Exercise 5

5.1 $x = 1.5$: $y' = 0.4060$, error $= 0.00135$

$y'' = -0.1168$, error $= -0.0099$

$x = 1.7$: $y' = 0.3834$, error $= 0.00004$

$y'' = -0.10624$, error $= -0.00428$

5.2 $x = 3.5$: $y' = 0.2681$, error $= -0.00135$

$y'' = -0.0328$, error $= -0.0099$

$x = 3.25$: $y' = 0.2851$, error $= -0.00495$

$y'' = -0.0406$, error $= -0.00135$

5.3 Stirling's: $x = 32^\circ,\ p = 0.2$

$y' = 1.4032,\ y'' = 1.8006$

Bessel's: $x = 36^o,\ p = 0.6$

$$y' = 1.5306,\ y'' = 2.4259$$

5.4 $t_0 = 3,\ p = t - 3,\ h = 1.$

$21p^2 + 23p - 11 = 0;\ p = .360;\ t = 3.36;\ u = 16.57$

5.5 Stirling's: $t_0 = 3$

$3p^2 + 8p - 3 = 0;\ p = \frac{1}{3};\ t = 3.333;\ \mathrm{H} = 21.52$

5.6 $f'(x_0) = \frac{1}{6h}[-5f_0 + 9f_2 - 4f_3]$; error $= h^2 f'''(\xi),\ x_0 \le \xi \le x_3.$

Exercise 6

6.1 (*i*) 0.9333 (*ii*) 0.9203; exact 0.9184

6.2 $h \le 0.42;\ h = 0.25,\ \mathrm{I} = 0.9445$; Series 0.9461

6.3 $h \le 0.242;\ h = 0.2,\ \mathrm{I} = 0.6832$

6.4. $h \le 0.2449;\ h = 0.25,\ \mathrm{I} = 0.8795$; exact 0.8814.

6.5 $\mathrm{T}_0^{(0)} = 0.75;\ \mathrm{T}_1^{(0)} = 0.775,\ \mathrm{T}_1^{(1)} = 0.7833;\ \mathrm{T}_2^{(0)} = 0.7828,\ \mathrm{T}_2^{(1)} = 0.7854,\ \mathrm{T}_2^{(2)} = 0.7855;$
$\tan^{-1} 1 = \frac{\pi}{4} = 0.7854$

6.6 $\mathrm{T}_0^{(0)} = 0.7854;\ \mathrm{T}_1^{(0)} = 0.7038,\ \mathrm{T}_1^{(1)} = 0.6766;\ \mathrm{T}_2^{(0)} = 0.7426,\ \mathrm{T}_2^{(1)} = 0.7555,$
$\mathrm{T}_2^{(2)} = 0.7608.$

6.7 (*i*) 23.1521 (*ii*) 23.6049; exact 23.6044

6.8 (*i*) 1.3991 (*ii*) 1.3987; exact 1.3987.

6.9 (*i*) 1.6988 (*ii*) 1.6317; value of fourth derivative is too large.

6.10 0.25

6.11 57.8051

6.12 $a = 0,\ b = 3/4,\ c = 1/4$; error $-\frac{9}{4}h^4 f'''(\xi).$

6.13 $a = \frac{7}{15}$, $b = \frac{16}{15}$, $c = -\frac{1}{15}$; error $\frac{h^7}{4725} f^{iv}(\xi)$

6.14 (*i*) 0.85926 (*ii*) 0.86908; exact 0.87636

6.15 (*i*) 1.02928 (*ii*) 1.02926.

Exercise 7

7.1 $y_1 = 1 - x^2$; $y_2 = 1 - x^2 + \frac{x^4}{2}$; $y_3 = 1 - x^2 + \frac{x^4}{2} - \frac{x^6}{12}$;

error $= \frac{x^6}{12} = 0.00034$ (at $x = 0.4$); (anal. soln. $y = e^{-x^2}$.)

7.2 $y_2 = x + \frac{x^3}{3}$; $y_4 = x + \frac{x^3}{3} + \frac{2}{15}x^5 + \frac{17}{7 \times 45}x^7 + \cdots$

at $x = 0.2$ error $= \frac{17x^7}{7 \times 45} \simeq 0.7 \times 10^{-6}$; (anal. soln. $y = \tan x$)

7.3 $y_3 = 1 + \frac{x^2}{2} + \frac{3}{8}x^4 + \frac{5}{16}x^6$; (anal. soln. $y = (1 - x^2)^{-\frac{1}{2}}$.)

7.4 $y'(1) = 1$, $y''(1) = 2$, $y'''(1) = 2$, $y^{iv}(1) = 2 \ldots$.

$$y = (x-1) + (x-1)^2 + \frac{(x-1)^3}{3} + \frac{(x-1)^4}{12}; \; y(1.5) = 0.7969$$

approx error $= \frac{(x-1)^5}{60} \simeq 0.26 \times 10^{-3}$ at $x = 1.5$

(anal. soln. $y = 2e^{(x-1)} - (x+1)$, at $x = 1.5$ $y = 0.7974$.)

7.5 $y''(0) = 3$, $y'''(0) = -8$, $y^{iv}(0) = 34$, $y^{v}(0) = -186$

$$y = 1 - x + \frac{3}{2}x^2 - \frac{4}{3}x^3 + \frac{17}{12}x^4; \; y(0.2) = 0.8516$$

error $= \frac{x^5}{120}y^{v} = 0.000496$ (approx).

7.6 $y'''(0) = -4,\ y^{iv}(0) = 7,\ y^{v}(0) = -9$

$y(x) = x + x^2 - \frac{2}{3}x^3 + \frac{7}{24}x^4 + \mathrm{R}(x);\ \mathrm{R}(x) = -\frac{3}{40}x^5$

$y(0.1) = 0.1094,\ \mathrm{R}(0.1) = 0.75 \times 10^{-6}$

$y'(x) = 1 + 2x - 2x^2 + \frac{7}{6}x^3,\ \mathrm{R}'(x) = -\frac{3}{8}x^4$

$y'(0.1) = 1.1812;\ \mathrm{R}'(x) = 0.37 \times 10^{-4};$

$y''(x) = 2 - 4x + \frac{7}{2}x^2,\ \mathrm{R}''(x) = -\frac{3}{2}x^3$

$y''(0.1) = 1.635,\ \mathrm{R}''(0.1) = 0.0015$

7.7 $f_x = 1,\ f_y = -1;\ y_1 = .9,\ \varepsilon_1 = \frac{h^2}{2}(1 - x + y) = 0.01$

$y_2 = 0.82;\ \varepsilon_2 = 0.018$

(anal. soln $y = 2e^{-x} + x - 1;\ y(.1) = 0.9097,\ y(0.2) = 0.8375$)

7.8 (*i*) 2.06667 (*ii*) $y(1.1) = 2.03333,\ y(1.2) = 2.06312$

(Implicit anal. soln $\tan^{-1}\frac{y}{x} + \frac{1}{2}\ln(x^2 + y^2) = 1.9122$)

7.9 $y_1^* = 1.2;\ y_1 = 1.2295$

7.10 $k_1 = 0,\ k_2 = 0.04,\ k_3 = 0.0404,\ k_4 = 0.081616;\ k = 0.0404;\ y_1 = 1.0404$

$k_1 = 0.0816,\ k_2 = 0.1249,\ k_3 = 0.1262,\ k_4 = 0.1733;\ k = 0.1262;\ y_2 = 1.16667.$

(anal. soln $y = 2e^{x^2/2} - 1.\ y(0.2) = 1.0404,\ y(0.4) = 1.16667$)

7.11 $k_1 = 0.4,\ l_1 = 0.2;\ k_2 = 0.42,\ l_2 = 0.22;\ k_3 = 0.422,\ l_3 = 0.222$

$k_4 = 0.4444,\ l_4 = 0.2444;\ k = 0.4214,\ l = 0.2214;\ y_1 = 1.4214,\ z_1 = 1.2214$

(anal. soln $y = e^x + x,\ z = e^x$)

7.12 $y_1^* = 1.4,\ y_1 = 1.5554;\ y_2^* = 1.21406,\ y_2 = 1.2830$

$y_3^p = 1.2641,\ y_3^c = 1.08405$

[anal. soln $y = 2x/(2x^2 - 1)$]

7.13 $y'(0) = 0,\ y''(0) = 1,\ y'''(0) = 0,\ y^{iv}(0) = -3$

Taylor's series $y = 1 + \frac{x^2}{2} - \frac{x^4}{8}$

$y_1 = 1.0050,\ y_2 = 1.0199,\ y_3 = 1.0440$

$y_4^p = 1.0770;\ y_4^c = 1.0771.$

(anal. soln $y = \sqrt{x^2+1}$)

7.14 $h = 0.25; (14x_i + 2)y_{i-1} - (32x_i + 1)y_i + (78x_i - 2)y_{i+1} = 0$

$-9y_1 + 2.5 = -5.5;\ 9y_1 - 17y_2 + 7y_3 = 0;\ 5.5y_2 - 5y_3 = -6.9$

$y_0 = 1,\ y_1 = 0.9939,\ y_2 = 1.3782,\ y_3 = 2.0691,\ y_4 = 3.$

7.15 $y_1 - y_{-1} = 0.4;\ (25 + 2.5x_i)y_{i-1} - 48y_i + (25 - 2.5x_i) = 4 + x_i$

$-24y_0 + 25y_1 = 7;\ 25.5y_0 - 48y_1 + 24.5y_2 = 4.2;\ 13y_1 - 24y_2 = -21.32$

$y_0 = 1,\ y_1 = 1.24,\ y_2 = 1.56,\ y_3 = 1.96.$

(anal. soln $y = x^2 + x + 1$)

when b.c. at $x = 0$ replaced by forward difference

$y_1 - y_0 = h = 0.2;\ -22.5y_1 + 24.5y_2 = 9.3,\ 13y_1 - 24y_2 = -21.32$

$y_0 = 1,\ y_1 = 1.3506,\ y_2 = 1.6199,\ y_3 = 1.96.$

Exercise 8

8.2 $1.2M_1 + 0.4M_2 = -1.1535;\ .4M_1 + 1.6M_2 = -0.981;\ M_1 = -0.8257,\ M_2 = -0.4067$
$s(.8) = 0.5852$; [data taken from $f(x) = \ln(1+x)$]

8.3 $m_0 + 4m_1 + m_2 = -0.9$; $m_1 = -0.175$; $s(0.7) = 0.3123$

8.4 $-1.9272y_1 + 1.0104y_2 = 1.9375$; $1.0182y_1 - 1.9584y_2 = -1.5085$

$y_1 = 1.9374$, $y_2 = 1.7776$.

Exercise 9

9.1 $55a + 5b = 134$, $5a + 1.4636b = 17.1167$; $y = 1.9918x + 4.8903/x$.

$\boldsymbol{y}^{*\mathrm{T}} = (6.8821 \;\; 6.4288 \;\; 7.6055 \;\; 9.1898 \;\; 10.9371)$

$\boldsymbol{e}^{\mathrm{T}} = (0.1179 \;\; -0.4288 \;\; 0.3945 \;\; -0.1898 \;\; 0.0623)$

9.2 $5a + 10b = 70$, $10a + 30b = 224$; $y = 8.4x - 2.8$; $\Sigma e_i = 0$; rms $= 5.49$.

9.3 $5a + 10b + 30c = 70$, $10a + 30b + 100c = 224$, $30a + 100b + 354c = 798$;

$y = 3.2 - 3.6x + 3x^2$; $\boldsymbol{e}^{\mathrm{T}} = (-1.2 \;\; 2.4 \;\; 0 \;\; -2.4 \;\; 1.2)$; rms $= 1.70$.

9.4 $4a + 2.5c = 11.4$, $2.5b = 0$; $2.5a + 2.125c = 5.55$; $y = 4.6 - 2.8x^2$; $\boldsymbol{e}^{\mathrm{T}} = (0 \;\; 0 \;\; 0 \;\; 0)$.

9.5 $5a + 10b + 22.5c = 3.1136$, $10a + 22.5b + 55c = 7.5812$,
$22.5a + 55b + 142.125c = 19.2995$

$y = -1.0214 + 1.1826x - 0.1602x^2$; $\boldsymbol{e}^{\mathrm{T}} = (-0.0020 \;\; 0.0135 \;\; -0.0098 \;\; -0.0175 \;\; 0.0140)$.

9.6 $a = \mathrm{A}$, $b = \mathrm{B}$; $\gamma = \mathrm{A} + \mathrm{B}x$; $5\mathrm{A} + 5\mathrm{B} = 5.9910$, $5\mathrm{A} + 7.5\mathrm{B} = 7.2408$;

$y = 2.0117\exp(0.4992x)$; (Data taken from $y = 2e^{0.5x}$)

9.7 $7x_1 + 4x_2 = 17$, $4x_1 + 15x_2 - 2x_3 = 23$, $-2x_2 + 7x_3 = -8$;

$x_1 = 1.9109$, $x_2 = 0.9059$, $x_3 = -0.8840$;

$\boldsymbol{e}^{\mathrm{T}} = (0.6067 \;\; 0.0319 \;\; -0.4152 \;\; -0.2554)$

9.8 $y = ax + bx^2 + cx^3$; $y(1) = 1$ renders $y = a(x - x^3) + b(x^2 - x^3) + x^3$;

$e(x) = y'' - y - x = a(x^3 - 7x) + b(2 - 6x - x^2 + x^3) - (x^3 - 5x)$

$$\frac{\partial s}{\partial a} = 0 \Rightarrow 13.6762a + 6.6262b = 9.4095;$$

$$\frac{\partial s}{\partial b} = 0 \Rightarrow 6.6262a + 3.6095b = 4.5262;$$

$$y = 0.7254x - 0.0814x^2 + 0.3540x^2$$

(anal. soln $y = 0.8509(e^x - e^{-x}) - x \simeq 0.7018x + 0.2836x^3$)

Exercise 10

10.1 $A_1 = \begin{bmatrix} 2.7636 & 0.8506 & 0 \\ 0.8506 & 2 & 0.5257 \\ 0 & 0.5257 & 7.2352 \end{bmatrix}$, $A_2 = \begin{bmatrix} 3.3144 & 0 & 0.2875 \\ 0 & 1.4496 & 0.4413 \\ 0.2857 & 0.4413 & 7.2352 \end{bmatrix}$

$$A_3 = \begin{bmatrix} 3.3144 & 0.0216 & 0.2849 \\ 0.0216 & 1.4160 & 0 \\ 0.2849 & 0 & 7.2682 \end{bmatrix}$$

$\lambda_1 = 7.3$; $|A - \lambda_1 I| = -0.237$.

10.2 $A_1 = \begin{bmatrix} 5.0003 & 0 \\ 0 & 2.0001 \end{bmatrix}$, $S = \begin{bmatrix} 0.8165 & -0.5774 \\ 0.5774 & 0.8165 \end{bmatrix}$

$\lambda_1 = 5.0003$, $\lambda_2 = 2.0001$.

$\boldsymbol{x}_1 = (0.8165 \;\; 0.5774)$, $\boldsymbol{x}_2 = (-0.5774 \;\; 0.8165)$

10.3 $T = \begin{bmatrix} 5 & 3 & 0 \\ 3 & 4 & -1 \\ 0 & -1 & 2 \end{bmatrix}$

10.4. $T = \begin{bmatrix} 5 & 3.0000 & 0 \\ 3.0000 & 4.0000 & 1.0000 \\ 0 & 1.0000 & 2.0000 \end{bmatrix}$

10.5 $f_0(\lambda) = 1$; $f_1(\lambda) = \lambda - 5$; $f_2(\lambda) = (\lambda - 4)f_1(\lambda) - 9$; $f_3(\lambda) = (\lambda - 2)f_2(\lambda) - f_1(\lambda)$

Intervals are (0, 1), (2, 3), (7, 8)

10.6 $A_1 = A$

$$L_1 = \begin{bmatrix} 1 & 0 & 0 \\ 0 & 1 & 0 \\ 0.25 & 0.25 & 1 \end{bmatrix}, \quad R_1 = \begin{bmatrix} 8 & 4 & 4 \\ 0 & 4 & 2 \\ 0 & 0 & 3.5 \end{bmatrix}$$

$$A_2 = R_1L_1 = \begin{bmatrix} 9 & 5 & 4 \\ 0.5 & 4.5 & 2 \\ 0.875 & 0.875 & 3.5 \end{bmatrix}$$

$$L_2 = \begin{bmatrix} 1 & 0 & 0 \\ 0.0556 & 1 & 0 \\ 0.0926 & 0.0921 & 1 \end{bmatrix}, \quad R_2 = \begin{bmatrix} 9 & 5 & 4 \\ 0 & 4.2220 & 1.7776 \\ 0 & 0 & 2.9474 \end{bmatrix}$$

$$A_3 = R_2L_2 = \begin{bmatrix} 9.6668 & 5.3684 & 4 \\ 0.4075 & 4.3857 & 1.7776 \\ 0.2865 & 0.2714 & 2.9474 \end{bmatrix}$$

10.7 $A = A_1 = \begin{bmatrix} 4 & 3 \\ 1 & 2 \end{bmatrix}, \quad A_2 = \begin{bmatrix} 4.75 & 3 \\ 0.3125 & 1.25 \end{bmatrix}$

$$A_3 = \begin{bmatrix} 4.9474 & 3 \\ 0.0693 & 1.0526 \end{bmatrix}, \quad A_4 = \begin{bmatrix} 4.9894 & 3 \\ 0.0141 & 1.0106 \end{bmatrix}$$

$$A_5 = \begin{bmatrix} 4.9978 & 3 \\ 0.0028 & 1.0022 \end{bmatrix}$$

$4.9978 - 4.9894 = 0.0084 \le 0.05$

$1.0106 - 1.0022 = 0.0084 \le 0.05$

$\lambda_1 = 5.0,\ \lambda_2 = 1.0$

Exercise 11

11.4

t \ x	0	0.1	0.2	0.3	0.4	0.5	0.6
0.0	0	0.3091	0.5877	0.8090	0.9510	1.0	0.9510
0.005	0	0.2938	0.5590	0.7640	0.9045	0.9510	0.9045
0.010	0	0.2795	0.5289	0.7318	0.8575	0.9045	0.8575
0.015	0	0.2644	0.5056	0.6932	0.8182	0.8575	0.8182
0.020	0	0.2528	0.4788	0.6619	0.7754	0.8182	0.7754

11.5 Same answer as in 11.4

Or

Same value in reverse order

11.6 Values of u at $t = 0.020\,(r = 0.5)$

Method \ x	0	0.2	0.4	0.6
Explicit method	0	0.4755	0.7694	0.7694
C–N method	0	0.4853	0.7852	0.7852
Analytical soln	0	0.4825	0.7807	0.7807

The equation for C–N scheme are :

$$-u_{i-1,1} + 6u_{i,1} - u_{j+1,1} = u_{i-1,0} + 2u_{i,0} + u_{i+1,0}$$

$6u_{11} - u_{21} = 2.1264$

$-u_{11} + 5u_{21} = 3.4407$ using $u_{21} = u_{31}$.

11.7

$t \backslash x$	0	0.25	0.50	0.75	1.0
0.0	1.0	0.9375	0.7500	0.4375	0
0.03125	0.9375	0.8750	0.6875	0.3750	0
0.0625	0.8750	0.8125	0.6250	0.3488	0
0.0625	0.8763	0.8151	0.6340	0.3460	0

C–N scheme for $r = 1 : -u_{i-1,1} + 4u_{i,1} - u_{i+1,1} = u_{i-1,0} + u_{i+1,0}$.
Equations are

$$\begin{aligned}
2u_{01} - u_{11} &= 0.9375 \\
-u_{01} + 4u_{11} - u_{21} &= 1.75 \quad \text{or} \quad 3.5u_{11} - u_{21} = 2.21875 \\
- u_{11} + 4u_{21} - u_{31} &= 1.375 \quad \text{or} \quad 3.71428u_{21} - u_{31} = 2.00893 \\
- u_{21} + 4u_{31} &= 0.75 \quad \text{or} \quad 3.73077u_{31} = 1.29087
\end{aligned}$$

$u_{31} = 0.3460,\ u_{21} = 0.6340,\ u_{11} = 0.8151,\ u_{01} = 0.8763$

11.8 $\Delta x = \Delta y = 0.25$; $\Delta t = 0.01$; $r = \Delta t/\Delta x^2 = \Delta t/\Delta y^2 = 0.16 (r_1 = r_2 = r)$
Explicit formula:

$$\begin{aligned}
u_{i,j}^{k+1} &= r(u_{i-1,j}^k + u_{i+1,j}^k + u_{i,j-1}^k + u_{i,j+1}^k) + (1 - 4r)u_{i,j}^k \\
&= 0.16(u_{i-1,j}^k + u_{i+1,j}^k + u_{i,j-1}^k,\ u_{i,j+1}^k) + 0.36u_{i,j}^k
\end{aligned}$$

At $t = 0$, the initial distribution and boundary conditions are shown as follows:

Boundary conditions: $\frac{\partial u}{\partial y} = -u$ (top), $\frac{\partial u}{\partial x} = u$ (left), $u = 1$ (right), $u = 1$ (bottom)

$j \uparrow$ / $i \rightarrow$	0	1	2	3	4
4	1.0	1.25	1.50	1.75	2.0
3	0.75	1.0	1.25	1.50	1.75
2	0.50	0.75	1.0	1.25	1.50
1	0.25	0.50	0.75	1.0	1.25
0	0	0.25	0.50	0.75	1.0

$t = 0.01$

$j \uparrow$ / $i \rightarrow$	0	1	2	3	4
4	0.64	0.80	1.0	1.2	*
3	0.80	1.0	1.25	1.50	1.0
2	0.60	0.75	1.0	1.25	1.0
1	0.40	0.50	0.75	1.0	1.0
0	*	1.0	1.0	1.0	1.0

$t = 0.02$

$j \uparrow$ / $i \rightarrow$	0	1	2	3	4
4	0.5990	0.7488	0.9360	1.0336	*
3	0.7488	0.936	1.17	1.292	1.0
2	0.6128	0.766	1.0	1.25	1.0
1	0.5152	0.644	0.830	1.0	1.0
0	*	1.0	1.0	1.0	1.0

Note : * denotes singularity at the corner.

11.9

$t = 0$

y	j	$i \to 0$	1	2	3
0.6	3	1.8	2.0	2.6	3.6
0.4	2	0.8	1.0	1.0	2.6
0.2	1	0.2	0.4	1.0	2.0
0	0	0	0.2	0.8	1.8
	$x \to$	0	.2	.4	.6

$t = .02$

j	$i \to 0$	1	2	3
3	1	1	1	*
2	1	1.295	1.78	2.225
1	1	0.845	1.18	1.475
0	1	1	1	*

$t = .04$

j	$i \to 0$	1	2	3
3	1	1	1	*
2	1	1.22	1.4267	1.7834
1	1	1,1	1.1867	1.4834
0	1	1	1	*

Note : * denotes singularity

11.10

j	$i \to 0$	1	2	3	4
3	*	5	5	5	*
2	2.875	4.125	4.500	4.375	3.625
1	2.875	4.125	4.500	4.375	3.625
0	*	5	5	5	*

Note : * denotes singularity of boundary conditions.

Equations are :

$$-2u_{11} + u_{21} = -3.75;\ u_{11} - 3u_{21} + u_{31} = -5;\ u_{21} - 2u_{31} = -4.25$$

11.11 $u_4 = 3.6770,\ u_5 = 3.4153,\ u_7 = 3.9317$

$u_1 = 2.6770,\ u_2 = 2.4153,\ u_3 = 2.9317$

Equations are :

$$-5u_4 + u_5 = -14.97$$

$$2u_4 - 3u_5 + u_7 = 1.04$$

$$u_5 - 9u_7 = -31.97$$

11.12 $u_1 = u_5 = 2.4162,\ u_2 = u_4 = 2.5821,\ u_3 = 3.7750$

$$-9u_1 + 3u_2 = -14,\ 3u_1 - 18u_2 + 3u_3 = -27.904,\ u_2 - 2u_3 = -4.968.$$

11.13 $-2u_{11} + u_{21} + u_{12} = 0.4;\ u_{11} - 3u_{21} = -6.4;\ u_{11} - 3u_{12} = -7.6.$

$u_{11} = 2.6,\ u_{21} = 3.0,\ u_{12} = 3.4$

11.14

$i \rightarrow$	1	2	1	2
j	0.528	1.1013	3.8481	4.2035
$\downarrow$	1.7955	2.2367	3.8637	4.2140
1	2.7520	3.4728	3.8707	4.2186
	3.1543	3.7411	3.8738	4.2207
	3.5554	4.0085	3.8751	4.2216
	3.7337	4.1273	3.8757	4.2220
	3.8129	4.1801	3.8760	4.2222
	2.7013	3.9342	5.8036	6.0024
2	4.4352	4.8906	5.8140	6.0093
	4.4062	5.2930	5.8187	6.0124
	5.3411	5.6941	5.8207	6.0138
	5.6085	5.8743	5.8216	6.0144
	5.7273	5.9515	5.8220	6.0147
	5.7801	5.9867	5.8222	6.0147

11.15.

$i \rightarrow$	1	2	1	2
j	0.6336	1.3638	3.8732	4.2256
$\downarrow$	2.3659	3.8076	3.8776	4.2228
1	3.8209	3.6347	3.8767	4.2227
	3.6551	4.1719	3.8763	4.2224
	3.8325	4.1957	3.8763	4.2221
	3.2838	5.0590	5.8209	6.0161
2	5.8436	5.8787	5.8237	6.0154
	5.8295	5.8159	5.8225	6.0150
	5.6529	5.9667	5.8224	6.0149
	5.8195	6.0127	5.8224	6.0148

11.16 $\dfrac{dy}{dx} = 2,\ -4x,\ x_R = 0.281,\ y_R = 0.162.$

$p_R = 0.3949,\ q_R = 0.2026;\ u_R = 0.088$

Exercise 12

12.1. $(a)\ \bar{u} = 2x - \dfrac{5}{18}x(1-x) \quad (b)\ \bar{u} = 2x - \dfrac{55}{202}x(1-x)$

(c) Galerkin: $\bar{u}(0.5) = 0.9306$; Least squares: $\bar{u}(0.5) = 0.9319$

Analytical soln : $1 + x - \dfrac{\sin(1-x)}{\sin 1},\ u(0.5) = 0.9303$

12.3 $\bar{u} = a_1 x(1-x) + a_2 x^2(1-x);$

$\dfrac{3}{5}a_1 + \dfrac{3}{10}a_2 = \dfrac{1}{2};\ \dfrac{3}{10}a_1 + \dfrac{26}{105}a_2 = \dfrac{4}{15};\ a_1 = \dfrac{67}{93},\ a_2 = \dfrac{7}{31}$

Analytical soln: $u = \dfrac{2\sin x - \sin(1-x)}{\sin 1} - (1+x)$

12.5 $\bar{u} = 2x + a_1 x(1-x)$

$$\mathrm{I}\,[u] = \left(4 + \frac{1}{3}a_1^2\right) - \left(\frac{4}{3} + \frac{1}{30}a_1^2 + \frac{1}{3}a_1\right) + \left(\frac{10}{3} + \frac{a_1}{2}\right)$$

$$\frac{d\mathrm{I}}{da_1} = 0 \text{ gives } a_1 = -\frac{5}{18};\ \bar{u} = 2x - \frac{5}{18}x(1-x).$$

12.6 Galerkin method : $\mathrm{Re} = 2a(-x - y + x^2 + y^2),\ \phi_1 = (xy - x^2y - xy^2 + x^2y^2)$

$$-\frac{a}{45} - \frac{10}{144} = 0 \text{ or } a = -\frac{25}{8} = -3.125$$

12.7 Rayleigh–Ritz :

$$\frac{\partial u}{\partial x} = a\,(y - 2xy - y^2 + 2y^2x);\ \frac{\partial u}{\partial y} = a(x - x^2 - 2xy + 2x^2y)$$

$$\iint \left\{\left(\frac{\partial u}{\partial x}\right)^2 + \left(\frac{\partial u}{\partial y}\right)^2\right\} dxdy = \frac{a^2}{45};\ \iint 2\ gudxdy = -10 - a\frac{20}{16 \times 3 \times 3}$$

$$\mathrm{I}\,[u] = \frac{a^2}{45} + a\frac{5}{36} + 10;\ \frac{d\mathrm{I}}{da} = 0 \text{ gives } a = -\frac{25}{8} = -3.125$$

12.8 Least Square Method

$$\frac{d\mathrm{Re}^2}{da} = 0 \text{ gives } \frac{11}{90}a = -\frac{5}{12} \text{ or } a = -\frac{75}{22} = -3.4091$$

12.9 Assembley Table of PU = R + T

u_1	u_2	u_3	u_4	R	T
−3.2333	3.3833	—	—	0.165	$-Q_1$
3.3833	−5.6000	2.5667	—	0.4667	0
—	2.5667	−5.6000	3.3833	0.5833	0
—	—	3.3833	−3.2393	0.2850	$+Q_4$

$$-5.6000u_2 + 2.5667u_3 = 0.4667$$

$$2.5667u_2 - 5.6000u_3 = -6.1833$$

$u_2 = 0.5351,\ u_3 = 1.3494;\ -Q_1 = 1.6454,\ -Q_4 = 2.1862$

Analytical : $u_2 = 0.5345,\ u_3 = 1.3488;\ -Q_1 = 1.6421,\ -Q_4 = 2.1884.$

12.10

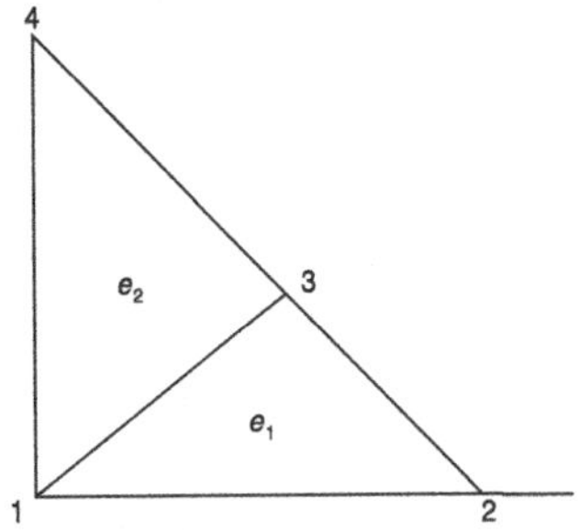

u_1	u_2	u_3	u_4	R	T
$2+2$	0	$-2-2$	0	0.0625 +0.0625	$\frac{1}{6}(2Q_1+Q_2)+\frac{1}{6}(Q_4+2Q_1)$
0	2	-2		0.0625	$\frac{1}{6}(Q_1+2Q_2)+\frac{1}{6\sqrt{2}}(2Q_2+Q_3)$
$-2-2$	-2	$4+4$	-2	0.0625 +0.0625	$\frac{1}{6\sqrt{2}}(Q_2+2Q_3)+\frac{1}{6\sqrt{2}}(2Q_3+Q_4)$
0		-2	2	0.0625	$\frac{1}{6}(2Q_4+Q_1)$

Given $Q_2 = u_2,\ Q_3 = u_3,\ Q_4 = u_4;\ u_1 = u_2 = u_4 = 1$

$$-4u_1 - 2u_2 + 8u_3 - 2u_4 = 0.125 + + \frac{1}{6\sqrt{2}}(u_2 + u_4) + \frac{1}{6\sqrt{2}}(4u_3)$$

$7.5286u_3 = 8.3607$ or $u_3 = 1.1105$

12.11

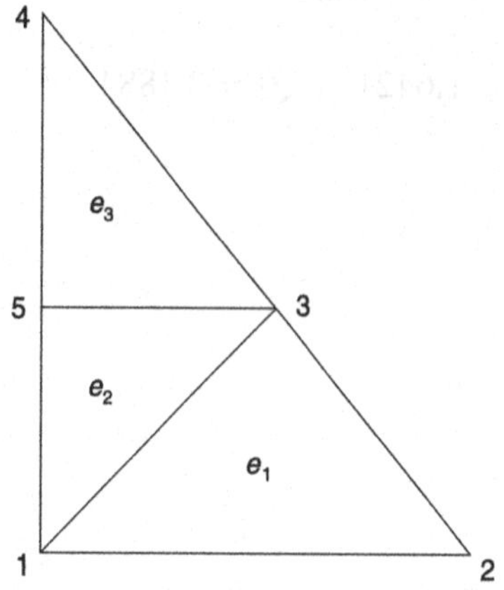

Given $u_2 = 0.5,\ u_3 = 0.75,\ u_4 = 1;\ Q_1 = 0,\ Q_2 = 0.5,\ Q_4 = 1,\ Q_5 = 0.5$

u_1	u_2	u_3	u_4	u_5	R	T
$5+4$	-3	$-2+0$	—	-4	0.0521 0.0260	$\frac{1}{12}(2Q_1+Q_2)+\frac{1}{12}(Q_5+2Q_1)$
-3	5	-2	—	—	0.0521 0.0521	$\frac{1}{12}(Q_1+2Q_2)+0.0932(2Q_2+Q_3)$
$-2+0$	-2	$4+16+16$	0	$-16-16$	0.0260 0.0469	$0.0932\ (Q_2+2Q_3)+0.0932(2Q_3+Q_4)$
—	—	0	4	-4	0.0469	$0.0932\ (Q_3+2Q_4)+\frac{1}{12}(2Q_4+Q_5)$
-4	—	$-16-16$	-4	$20+20$	0.0260 0.0469	$\frac{1}{12}(2Q_5+Q_1)+\frac{1}{12}(Q_4+2Q_5)$

From eqns (1) and (3)

$$9u_1 - 4u_5 = 3u_2 + 2u_3 - 0.0781 + \frac{1}{12}(Q_2 + Q_5 + 4Q_1) = 3.0052 \quad \text{.... (1)}$$

$$-4u_1 + 40u_5 = 32u_3 + 4u_4 - .0729 + \frac{1}{12}(Q_1 + Q_4 + 4Q_5) = 28.1771 \quad \text{.... (2)}$$

$86u_5 = 66.4037$ or $u_5 = 0.7721,\ u_1 = 0.6771.$

Exercise 13

13.1 $$\begin{bmatrix} 1-\frac{\lambda}{2} & -\lambda \\ -\frac{\lambda}{3} & 1-\frac{\lambda}{2} \end{bmatrix} \begin{bmatrix} c_1 \\ c_2 \end{bmatrix} = \begin{bmatrix} \frac{1}{2} \\ \frac{1}{3} \end{bmatrix}$$

$$y(x) = x - \frac{\lambda(\lambda+6)}{\lambda^2+12\lambda-12}x - \frac{4\lambda}{\lambda^2+12\lambda-12}, \ \lambda^2+12\lambda-12 \neq 0.$$

For $\lambda = \frac{1}{2}$, $y(x) = \frac{36}{23}x + \frac{8}{23}$

when right side is zero

$$\lambda_1 = -6+4\sqrt{3}, \ y(x) = \left(x + \frac{1}{\sqrt{3}}\right)$$

$$\lambda_2 = -6-4\sqrt{3}, \ y(x) = c\left(x - \frac{1}{\sqrt{3}}\right)$$

13.2 $$y_1 - \frac{\lambda}{3}y_2 - \frac{\lambda}{6}y_3 = 0$$

$$-\frac{\lambda}{12}y_1 + \left(1 - \frac{2\lambda}{3}\right)y_2 - \frac{\lambda}{4}y_3 = 0.5$$

$$-\frac{\lambda}{6}y_1 - \lambda y_2 + \left(1 - \frac{\lambda}{3}\right)y_3 = 1.0$$

For $\lambda = \frac{1}{2}$, $y_1 = \frac{8}{23}$, $y_2 = \frac{26}{23}$, $y_3 = \frac{44}{23}$

Method \ x	0	.5	1.0
Analytical	0.3478	1.1304	1.9130
Simpson's	0.3478	1.1304	1.9130
Trapezoidal	0.4000	1.2000	2.000

13.3 $y(x) = x + \frac{x^2}{2} + \frac{7}{6}x^3 + \frac{3}{8}x^4 + \frac{23}{120}x^5$

Analytical solution is

$$y(x) = \frac{2}{3}e^{2x} - \frac{5}{3}e^{-x} - 2x + 1$$

$$\simeq x + \frac{x^2}{2} + \frac{7}{6}x^3 + \frac{3}{8}x^4 + \frac{23}{120}x^5 + \cdots$$

13.4 $y(x) = x + \frac{x^2}{2} + \frac{7}{6}x^3 + \frac{3}{8}x^4$

13.5 $3xy(x) + \int_0^x y(\xi)d\xi = 4x - \frac{5}{3}x^3 + \frac{2}{15}x^5$

$4y(x) + 3xy'(x) = 4 - 5x^2 + \frac{2}{3}x^4$ giving $y(0) = 1$.

$y_1 = y(0.2) = 0.98101$; $y_2 = y(0.4) = 0.92192$; $y_3 = y(0.6) = 0.82074$

Analytical solution is the first four terms in the expansion of

$$y(x) = \cos(x) \simeq 1 - \frac{x^2}{2} + \frac{x^4}{6} - \frac{x^6}{24}$$

$\cos(0.2) = 0.9801$, $\cos(0.4) = 0.9211$, $\cos(0.6) = 0.8253$

Exercise 14

14.1 $na + \frac{n(n-1)}{2}d$

14.2 $\frac{n(n-1)}{2}$

14.3 $2n^2 - n$

14.4 $n\log_2 n - \frac{n}{2}$

14.5 $3n - 2 - \log_2 n$

14.6 $A = \begin{bmatrix} 0 & 3 & 0 \\ 0 & 4 & -1 \\ -2 & 3 & 2 \end{bmatrix}$

$\lambda_1 = 1,\ \lambda_2 = 2,\ \lambda_3 = 3,\ \lambda^4 = a_0 + a_1\lambda + a_2\lambda^2$

$a_0 = 36,\ a_1 = -60,\ a_2 = 25$

$A^4 = 36I - 60A + 25A^2$

$$A^2 = \begin{bmatrix} 0 & 12 & -3 \\ 2 & 13 & -6 \\ -4 & 12 & 1 \end{bmatrix};$$

$$A^4 = \begin{bmatrix} 36 & 120 & -75 \\ 30 & 121 & -90 \\ 20 & 120 & -59 \end{bmatrix} x(4) = \begin{bmatrix} 81 \\ 81 \\ 81 \end{bmatrix}$$

Exercise 15

15.1 $f(x) = a_0 + \sum_{n=1}^{\infty} (a_n \cos nx + b_n \sin nx)$

$$a_0 = \frac{1}{2\pi}\int_0^{2\pi} f(x)\,dx;\ a_n = \frac{1}{\pi}\int_0^{2\pi} f(x)\cos nx\,dx;\ b_n = \frac{1}{\pi}\int_0^{2\pi} f(x)\sin nx\,dx.$$

$a_0 = 0,\ a_n = 0$

$$b_n = \frac{2k}{n\pi}[1 - (-1)^n]$$

$= 2$ when n is odd and 0 when n is even

$$f(x) = 4k\left[\sin x + \frac{1}{3}\sin 3x + \frac{1}{5}\sin 5x + \ldots\right]$$

or, $f(x)$ is an odd function; hence only sine series.

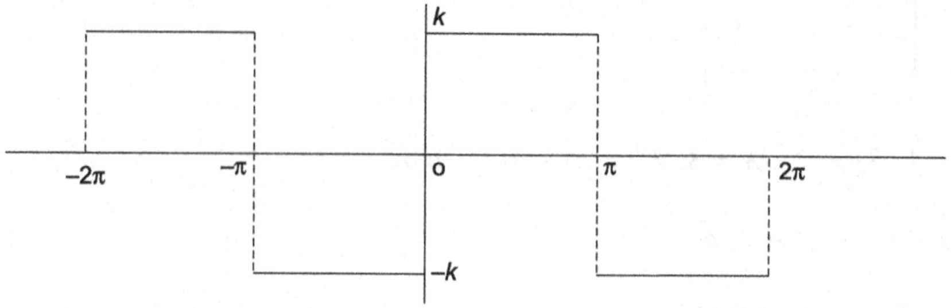

15.2 $f(x) = \sum_{n=1}^{\infty} b_n \sin\frac{n\pi x}{c}$ where $b_n = \frac{2}{c}\int_0^c f(x)\sin\frac{n\pi x}{c}dx$

$$b_n = (-1)^{n+1}.\frac{2c}{n\pi}$$

$$f(x) = \frac{2c}{\pi}\left[\sum_{n=1}^{\infty}(-1)^{n+1}\frac{1}{n}.\sin\frac{n\pi x}{c}\right]$$

$$= \frac{2c}{\pi}\left[\sin\frac{\pi x}{c} - \frac{1}{2}\sin\frac{2\pi x}{c} + \frac{1}{3}\sin\frac{3\pi x}{c} - \frac{1}{5}\sin\frac{5\pi x}{c} + \ldots\right]$$

when $c = \pi$

$$f(x) = 2\left[\sin x + \frac{1}{3}\sin 3x + \frac{1}{5}\sin 5x + \ldots\right]$$

Put $x = \pi/2$ and $f(x) = x$, getting

$$\frac{\pi}{4} = 1 - \frac{1}{3} + \frac{1}{5} - \ldots$$

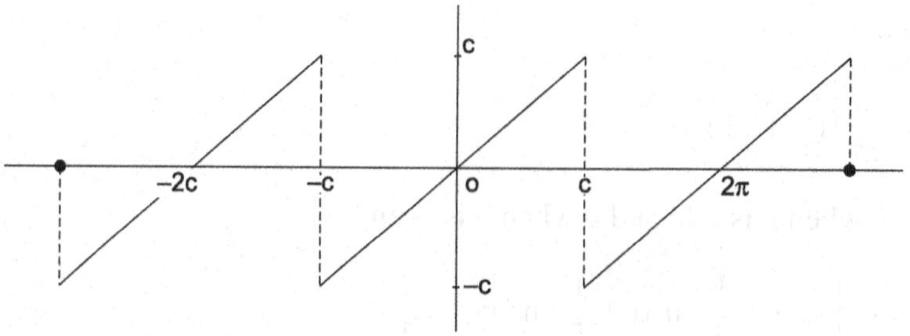

15.3 $f(x) = a_0 + \sum_{n=1}^{\infty} a_n \cos nx$

where $a_0 = \frac{1}{\pi}\int_0^{\pi} f(x)\,dx$; $a_n = \frac{2}{\pi}\int_0^{\pi} f(x).\cos nx dx.$

$$a_0 = \frac{1}{\pi}\int_0^{\pi} e^{-x} dx = \frac{1}{\pi}\left(1 - e^{-\pi}\right)$$

$$a_n = \frac{2}{\pi}\int_0^{\pi} e^{-x}.\cos nx dx = \frac{2}{\pi}.\frac{1}{1+n^2}\left\{1 - e^{-\pi}.(-1)^n\right\}$$

Index

Printed in the United States
By Bookmasters